ROB

THE KLUWER INTERNATIONAL SERIES IN ENGINEERING AND COMPUTER SCIENCE

ROBOTICS: VISION, MANIPULATION AND SENSORS

Consulting Editor: **Takeo Kanade**

ROBOTIC GRASPING AND FINE MANIPULATION, M. Cutkosky
ISBN: 0-89838-200-9

SHADOWS AND SILHOUETTES IN COMPUTER VISION, S. Shafer
ISBN: 0-89838-167-3

PERCEPTUAL ORGANIZATION AND VISUAL RECOGNITION, D. Lowe
ISBN: 0-89838-172-X

ROBOT DYNAMICS ALGORITHMS, F. Featherstone
ISBN: 0-89838-230-0

THREE- DIMENSIONAL MACHINE VISION, T. Kanade (editor)
ISBN: 0-89838-188-6

KINEMATIC* MODELING, IDENTIFICATION AND CONTROL OF *ROBOT MANIPULATORS, H.W. Stone
ISBN: 0-89838-237-8

OBJECT RECOGNITION USING VISION AND TOUCH, P. Allen
ISBN: 0-89838-245-9

INTEGRATION, COORDINATION AND CONTROL OF MULTI-SENSOR ROBOT SYSTEMS, H.F. Durrant-Whyte
ISBN: 0-89838-247-5

MOTION UNDERSTANDING: Robot and Human Vision, W.N. Martin and J. K. Aggrawal (editors)
ISBN: 0-89838-258-0

BAYESIAN MODELING OF UNCERTAINTY IN LOW-LEVEL VISION, R. Szeliski
ISBN 0-7923-9039-3

VISION AND NAVIGATION: THE CMU NAVLAB, C. Thorpe (editor)
ISBN 0-7923-9068-7

***TASK-DIRECTED SENSOR FUSION AND PLANNING: A Computational* Approach,** G. D. Hager
ISBN: 0-7923-9108-X

COMPUTER ANALYSIS OF VISUAL TEXTURES, F. Tomita and S. Tsuji
ISBN: 0-7923-9114-4

DATA FUSION FOR SENSORY INFORMATION PROCESSING *SYSTEMS,* J. Clark and A. Yuille
ISBN: 0-7923-9120-9

PARALLEL ARCHITECTURES AND PARALLEL ALGORITHMS FOR INTEGRATED VISION SYSTEMS, A.N. Choudhary, J. H. Patel
ISBN: 0-7923-9078-4

Robot Motion Planning

Jean-Claude Latombe
Stanford University

SPRINGER SCIENCE+BUSINESS MEDIA, LLC

Library of Congress Cataloging-in-Publication Data

Latombe, Jean Claude.
Robot motion planning / Jean-Claude Latombe.
p. cm. - (The Kluwer international series in engineering and computer science ; SECS 0124)
Includes bibliographical references and index.
ISBN 978-0-7923-9206-4 ISBN 978-1-4615-4022-9 (eBook)
DOI 10.1007/978-1-4615-4022-9
1. Robots - Motion. I. Title. II. Series.
TJ211.4.L38 1991
629.8'92 - dc20
90-49962
CIP

Originally published by Kluwer Academic Publishers, New York in 1991

Second printing, 1991.

Printed on acid-free paper.

To Claudine, Laurence, and Emmanuel

Contents

Preface

One of the ultimate goals in Robotics is to create *autonomous robots.* Such robots will accept high-level descriptions of tasks and will execute them without further human intervention. The input descriptions will specify *what* the user wants done rather than *how* to do it. The robots will be any kind of versatile mechanical device equipped with actuators and sensors under the control of a computing system.

Making progress toward autonomous robots is of major practical interest in a wide variety of application domains including manufacturing, construction, waste management, space exploration, undersea work, assistance for the disabled, and medical surgery. It is also of great technical interest, especially for Computer Science, because it raises challenging and rich computational issues from which new concepts of broad usefulness are likely to emerge.

Developing the technologies necessary for autonomous robots is a formidable undertaking with deep interweaved ramifications in automated reasoning, perception and control. It raises many important problems. One of them — *motion planning* — is the central theme of this book. It can be loosely stated as follows: How can a robot decide what motions to perform in order to achieve goal arrangements of physical objects? This capability is eminently necessary since, by definition, a robot accomplishes tasks by moving in the real world. The minimum one would expect from an autonomous robot is the ability to plan its own motions.

At first glance motion planning looks relatively simple, since humans deal with it with no apparent difficulty in their everyday lives. In fact, as is also the case with perception, the elementary operative intelligence that people use unconsciously to interact with their environment — e.g. preparing and serving coffee, assembling a device, moving in a building — turns out to be extremely difficult to duplicate using a computer-controlled robot. It is true that some naive methods can produce apparently impressive results, but the limitations of these methods quickly become obvious. The unaware reader will be surprised by the amount of nontrivial mathematical and algorithmic techniques that are necessary to build a reasonably general and reliable motion planner.

The research in robot motion planning can be traced back to the late 60's, during the early stages of the development of computer-controlled robots. Nevertheless, most of the effort is more recent and has been conducted during the 80's. Over the last few years, the theoretical and practical understanding of some of the issues has increased rapidly thanks to the combined work of researchers in Artificial Intelligence, theoretical Computer Science, Mathematics, and Mechanical Engineering. In addition to producing effective planning methods, this work has contributed to advancing our knowledge of the mathematical structure of the problems and to pinpointing their inherent computational complexity. This book is an attempt to give a comprehensive account of recent results in motion planning and to organize them in a unified framework.

An incorrect, but still widespread view is that motion planning essentially consists of doing some sort of collision checking or collision avoidance. In fact, motion planning is much more than that. It involves such diverse aspects as computing collision-free paths among possibly moving obstacles, coordinating the motions of several robots, planning sliding and pushing motions to achieve precise relations among objects, reasoning about uncertainty to build reliable sensory-based motion strategies, dealing with models of physical properties such as mass, gravity and friction, and planning stable grasps of objects. Therefore, motion planning requires the robot to consider geometrical constraints, as well as physical and temporal constraints. In addition, uncertainty may require that it plan not only motion commands, but also their interaction with sensing. When knowledge at planning time is too incomplete, it may become necessary to interweave planning and execution in order to collect appropriate information through sensing.

The concept of *configuration space* is used throughout this book in order to organize the various facets of motion planning in a coherent framework. This concept, which has been popularized in Robotics by Lozano-Pérez, is essentially a representational tool. It consists of treating the robot as a point in an appropriate space, the configuration space of the robot. The geometry of the task can be mapped in this configuration space. Physical concepts, such as force and friction, can also be represented in this space as additional geometrical constructs. Various tools from Geometry, Topology and Algebra apply nicely to this representation and provide the theoretical basis of motion planning.

Configuration space is sometimes regarded in the literature as a motion planning approach. This is an incorrect view. Configuration space is only a tool for formulating motion planning problems precisely. Planning methods that have been viewed as alternatives to the "configuration space approach", such as the "potential field" methods, can be consistently and usefully described using this tool. Configuration space has also been criticized for its inefficiency. As a representational tool, it cannot be efficient or inefficient. This unfounded criticism is probably a reflection of the complexity of motion planning in general. Hopefully, these facts will rapidly become obvious to the reader.

The concept of configuration space will make it possible to study the various aspects of motion planning in a unified fashion, hence facilitating the presentation and comparison of various planning approaches which are often considered separately in the literature. In addition, the description of the methods will proceed incrementally by considering a *basic motion planning problem* and extensions of this problem. Taking the basic problem as the first target, the major concepts and computational approaches to motion planning will be exposed in detail (the first half of this book). Then, several extensions of the basic problems will be studied; the methods presented previously will be extended and new ones will be described (the second half of the book).

The book consists of 11 chapters:

- Chapter 1 is an introduction to motion planning. It presents the basic motion planning problem, which is to plan a collision-free path for a rigid object translating and rotating among static obstacles. It introduces the concept of configuration space and the main computational approaches to motion planning. It discusses various extensions of the basic problem and gives an account of the important results relative to the computational complexity of motion planning. It reviews some key publications which have contributed to the development of the motion planning field.

- Chapters 2 and 3 develop the notion of configuration space for a rigid object that can translate and rotate freely among obstacles in a two- or three-dimensional workspace, i.e. the type of robot considered in the basic motion planning problem. Chapter 2 focuses on the differential and topological structure of this space. It introduces most of the mathematical concepts that will be used in the rest of the book. Chapter 3 describes how the objects contained in the robot's workspace (the "obstacles") map into configuration space when all the objects (including the robot) are described as semi-algebraic sets. The particular case where the objects are modeled as polygons and polyhedra are analyzed in detail.

- Chapters 4 through 7 describe four computational approaches for solving the basic motion planning problem. Chapter 4 presents the roadmap approach, which consists of capturing the global topology of the set of collision-free configurations of the robot in the form of a network of one-dimensional curves. Chapters 5 and 6 describe the exact and approximate cell decomposition approaches, which represent the set of collision-free configurations as a collection of cells and search the graph representing the adjacency relation among these cells. Chapter 7 describes the potential field approach, which regards the robot as a particle moving under the action of forces generated by an artificial potential field attracting the robot toward the goal while pushing it away from the obstacles.

- Chapters 8 through 11 present various extensions of the basic motion planning problem, and describe both modifications of the previous methods and new planning methods applying to these extensions. Chapter 8 considers the case where there are moving obstacles, multiple robots, and/or articulated robots. Chapter 9 analyzes the impact of nonholonomic kinematic constraints (the kind of constraints which apply to an automobile car) on motion planning. Chapter 10 investigates motion planning in the presence of uncertainty, in particular in robot control and in sensing. Chapter 11 considers the case where the robot can grasp/ungrasp movable objects in order to accomplish sophisticated manipulation tasks.

All the chapters are deliberately oriented toward presenting the fundamentals of motion planning. System engineering and implementation choices, which often depend closely on the domain of application and the features of the hardware systems, are considered only in a few places, and to the extent that they help understand the basic themes. The material presented, once understood, should be highly useful for designing practical systems.

Taken together, the methods presented in this book go beyond Robotics, and can be regarded as a contribution to the mechanization of *spatial reasoning*, i.e. the capacity to represent and reason about the geometry, spatial arrangement, and physics of the objects in the real world. Although the mainstream of Artificial Intelligence has been quite successful in developing various reasoning paradigms leading notably to efficient expert systems, it has done relatively little to automate spatial reasoning in a realistic fashion. Nonetheless, automatic spatial reasoning capabilities will be critical for building new, more advanced computer systems in a variety of domains related to design, manufacturing and engineering. A comprehensive computational theory of spatial reasoning will obviously increase our understanding of the interaction between design and manufacturing, eventually opening a new era of better designed and better manufactured products. The ideas and methods developed for robot motion planning can be a source of knowledge and inspiration for future developments in these domains.

This book is aimed at graduate students intending to study further in Robotics and Spatial Reasoning, and at engineers having to develop advanced computer-based systems in design, manufacturing, construction and automation. It assumes good "undergraduate-level" knowledge in Mathematics (Geometry, Topology, Algebra) and in Algorithms. A glossary of mathematical definitions is provided in Appendix A for the reader's convenience. Some background in Algorithms is also given in Appendices B (computational complexity), C (graph searching), and D (line sweeping). The book assumes elementary knowledge in Robotics and Mechanics.

A collection of exercises is included at the end of each chapter. Three types of exercises are typically proposed. Some exercises are quite formal and call for precise answers. Other exercises consist of writing programs and experimenting with them. They are intended to familiarize the reader with some implementation issues, and may be a good preparation before starting to work on a larger implementation project. They can be performed using almost any kind of personal computer or workstation with such programming languages as C, Pascal, and Lisp. Graphical inputs/outputs are suitable for several programming exercises. Finally, other exercises suggest topics for informal reflection, and are often used to introduce ideas not presented in the text.

I would like to acknowledge the help of many people who contributed their time reading and commenting on early drafts of this book. Many thanks to José Bañon, Jérôme Barraquand, Philippe Caloud, Wonyun Choi, Bill Dixon, Joseph Friedman, Maria Gini, François Ingrand, Ramesh Jain, Krasimir Kolarov, Oussama Khatib, Jean-Paul Laumond, Anthony Lazanas, Claude Le Pape, Jocelyne Pertin-Troccaz, Sean Quinlan, Jean-François Rit, Bernie Roth, Shashank Shekhar, Yoav Shoham, Dominique Snyers, Tom Strat, Yong Se Kim, Mark Yim, Randall Wilson, and David Zhu.

Chapter 6 (Approximate Cell Decomposition) greatly benefited from the work of David Zhu. Chapter 7 (Potential Field Methods) was considerably enhanced by discussions with Oussama Khatib and by the work of Jérôme Barraquand. Chapter 9 (Kinematic Constraints) would not have been possible without the contribution of Jérôme Barraquand. Chapter 10 (Dealing with Uncertainty) was improved by many suggestions from Bruce Donald, Mike Erdmann, Joseph Friedman, Anthony Lazanas, and Shashank Shekhar. Chapters 9 and 11 especially benefited from Jean-Paul Laumond's comments. Jean-François Rit pointed out many errors.

Thanks also to the students of Stanford's CS 327C in the Spring of 1987, 1989 and 1990, and the Winter of 1988. They suffered through early drafts of this book. I am grateful for their patience, enthusiasm, and comments. Leo Guibas attended the class in the Spring 1989 and made pertinent suggestions related to the organization of the book.

Cheng-Hsiu Wu carefully drew most of the figures. He did outstanding work and deserves special thanks. Wonyun Choi wrote the programs that produced some of the figures shown in Chapter 7. Jonas Karlsson and Irwin Welker helped eliminate the most blatant linguistic mistakes. However, since I rewrote some of the material after they read it, I have certainly reintroduced many gallicisms.

Many thanks to Jutta McCormick who have helped so efficiently with the administration of the Robotics Laboratory while I was impatiently busy writing this book.

Many parts of this book result from research supported by the Defense Advanced Research Projects Agency (DARPA), the Office for Naval Research (ONR) and the Army. Thanks also to Digital Equipment Corporation (DEC), the Stanford Institute of Manufacturing and Automation (SIMA), the Center for Integrated Systems (CIS), and the Center for Integrated Facility Engineering (CIFE) for their support.

The book was printed with the assistance of Mell Hall at the Stanford Publication Services.

Jean-Claude Latombe

Notations

The following conventions and notations are used throughout the book. Most of them are introduced in Chapter 1.

1. The robot is called $\mathcal{A}$. If there are several robots, they are called $\mathcal{A}_i$ $(i = 1, 2, ...)$.

2. The robots' workspace is denoted by $\mathcal{W}$ and is modeled as the Euclidean space $\mathbf{R}^N$, with $N = 2$ or 3. ($\mathbf{R}$ is the set of the real numbers.)

3. A Cartesian frame $\mathcal{F}_\mathcal{A}$ is attached to $\mathcal{A}$. Another Cartesian frame $\mathcal{F}_\mathcal{W}$ is attached to $\mathcal{W}$. The origin of $\mathcal{F}_\mathcal{A}$ (resp. $\mathcal{F}_\mathcal{W}$) is denoted by $O_\mathcal{A}$ (resp. $O_\mathcal{W}$).

4. The obstacles in $\mathcal{W}$ are denoted by $\mathcal{B}_i$ $(i = 1, 2, ...)$. The symbol $\mathcal{B}$ sometimes denotes a particular obstacle. More often, it denotes the union of all the obstacles (called the obstacle region).

5. The configuration space of a robot $\mathcal{A}$ is denoted by $\mathcal{C}$. It is a manifold of dimension m. An element of $\mathcal{C}$ (i.e. a configuration) is denoted by $\mathbf{q}$. The region of the workspace occupied by the robot $\mathcal{A}$ at configuration $\mathbf{q}$ is denoted by $\mathcal{A}(\mathbf{q})$.

6. An obstacle $\mathcal{B}$ in the workspace maps in configuration space to a region called C-obstacle and denoted by $\mathcal{CB}$.

7. The free space in configuration space is denoted by $\mathcal{C}_{free}$; the contact space by $\mathcal{C}_{contact}$. $\mathcal{C}_{valid} = \mathcal{C}_{free} \cup \mathcal{C}_{contact}$ is the valid space.

8. A path is denoted by τ. It is usually a function of a parameter denoted by s that takes its values in the interval $[0, 1]$. When τ is expressed as a function of time, it is called a trajectory.

9. A vector is denoted by a symbol with an arrow on top of it, e.g. $\vec{\nu}$. Its modulus is written $\|\vec{\nu}\|$. The inner product of two vectors $\vec{\nu}_1$ and $\vec{\nu}_2$ is denoted by $< \vec{\nu}_1, \vec{\nu}_2 >$ or $\vec{\nu}_1 \cdot \vec{\nu}_2$. The outer product of $\vec{\nu}_1$ and $\vec{\nu}_2$ is denoted by $\vec{\nu}_1 \wedge \vec{\nu}_2$. The angle between $\vec{\nu}_1$ and $\vec{\nu}_2$ is written $angle(\vec{\nu}_1, \vec{\nu}_2)$.

10. The distance between two points x and y of a Euclidean space is denoted by $\|x - y\|$.

11. An interval in $\mathbf{R}$ bounded by a and b is denoted by $[a, b]$ if it is closed at both ends and by (a, b) if it is open at both ends. It is denoted by $[a, b)$ or $(a, b]$ if it is closed at one end and open at the other.

12. Let a be a real number. $sign(a)$ denotes the "sign" of a; it is $+1$ if $a > 0$, -1 if $a < 0$, and 0 otherwise.

13. The symbols $\oplus$ and $\ominus$ are the Minkowski operators for affine set addition and subtraction, respectively.

14. Let E be a topological space and F be a subset of it. $cl(F)$, $int(F)$, and $\partial(F)$ denote the closure, the interior, and the boundary of F, respectively. $E \backslash F$ denotes the complement of F in E.

15. Most of the time, calligraphic letters, e.g. $\mathcal{S}$, denote sets.

16. Symbols in typewriter type style are used to denote boolean expressions or predicates, e.g. `CB`, `Achieve`. The symbols `true` and `false` are the logical values "true" and "false".

ROBOT MOTION PLANNING

Chapter 1

Introduction and Overview

A robot is a versatile mechanical device — for example, a manipulator arm, a multi-joint multi-fingered hand, a wheeled or legged vehicle, a free-flying platform, or a combination of these — equipped with actuators and sensors under the control of a computing system. It operates in a workspace within the real world. This workspace is populated by physical objects and is subject to the laws of nature. The robot performs tasks by executing motions in the workspace.

In this book we are interested in giving the robot the capability of planning its own motions, i.e. deciding automatically what motions to execute in order to achieve a task specified by initial and goal spatial arrangements of physical objects. Creating autonomous robots is a major undertaking in Robotics. It definitively requires that the ability to plan motions automatically be developed. Indeed, except in limited and carefully engineered environments, it is not realistic to anticipate and explicitly describe to the robot all the possible motions that it may have to execute in order to accomplish requested tasks. Even in those cases where such a description is feasible, it would certainly be useful to incorporate automatic motion planning tools in off-line robot programming systems. This would allow the user to specify tasks more declaratively, by stating *what* he/she wants done rather than *how* to do it. As robots become more dexterous, the need for motion planning tools will become more critical.

A still widespread view is that motion planning is essentially some sort of simple collision checking. Motion planning is much more than that. As we will see in this chapter, it presents an unexpected variety of facets, the simplest of which raise provably difficult computational issues. In fact, the kind of operative intelligence that people use unconsciously to interact with their environment, which is needed for perception and motion planning, turns out to be extremely difficult to duplicate in a computer program.

This chapter is both an introduction to robot motion planning and an overview of the rest of the book.

We first illustrate the various aspects of motion planning (Section 1). This illustration motivates our next step, the definition of a "basic problem" which is in some way the simplest useful motion planning problem that we may investigate (Section 2). Using this problem, we present the concept of configuration space, which is a unifying concept throughout the book (Section 3) and we outline the principles of the main computational approaches to motion planning (Section 4). We then introduce the extensions of the basic problem which will be studied in greater detail later (Section 5).

An important issue in motion planning is the computational complexity of the algorithms. In the following chapters of this book, we will analyze the complexity of particular planning methods. However, it is useful to have some preliminary idea of the intrinsic complexity of the problems which we are addressing. During the last few years, significant results have been obtained on this topic. We review them in Section 6. We also analyze the impact of these results on the design of operational systems, by proposing various guidelines for simplifying problems and implementing planners of increased practical efficiency (Section 7).

Motion planning is one major problem in the creation of autonomous robots. It is not the only one. It interacts with other important problems, including real-time motion control, sensing and task planning. We briefly discuss these interactions in Section 8.

The final section of this chapter reviews some key publications (Section 9). It is an historical account of the development of the motion planning field from the late 60's to the present time. Of course, other publications not mentioned in that section have also brought significant contributions to the field. We will reference many of them later in the book.

1 Aspects of Motion Planning

Consider a robot arm whose task is to build an assembly from a set of individual mechanical parts. Assume that the task is described as both a set of models of the parts, including their geometry and perhaps some physical properties, and a sequence of spatial relations [Latombe, 1984]. Figure 1 illustrates this description with a simple construction task. The sequence determines the order in which the parts are to be assembled. Each relation describes the position of a new part with respect to the current subassembly. The automatic transformation of this sequence into a description of the robot motions to be executed is one possible application of the methods presented in the book.

Let us analyze the example in more detail. In order to assemble the parts according to the input relations, the robot has to successively grasp each part, transfer it to the current subassembly, and position it on the subassembly:

- **Grasping** the part requires the position of the robot's gripper on the object to be selected and a path to this position to be generated. The grasp position must be accessible; it must also be stable and robust enough to resist some external forces. Sometimes, a satisfactory grasping position requires the part to be grasped, put down, and re-grasped.

- **Transferring** the part requires the geometry of a path for the arm to be computed. Since the motion will typically be executed at high speed, the path should avoid any contact with obstacles. Due to imprecision in motion control, it may even have to guarantee some minimal clearance.

- **Positioning** the part on the subassembly (part mating) requires motion commands to be selected for achieving a goal with high precision in a very constrained space. In general, due to uncertainty, a single

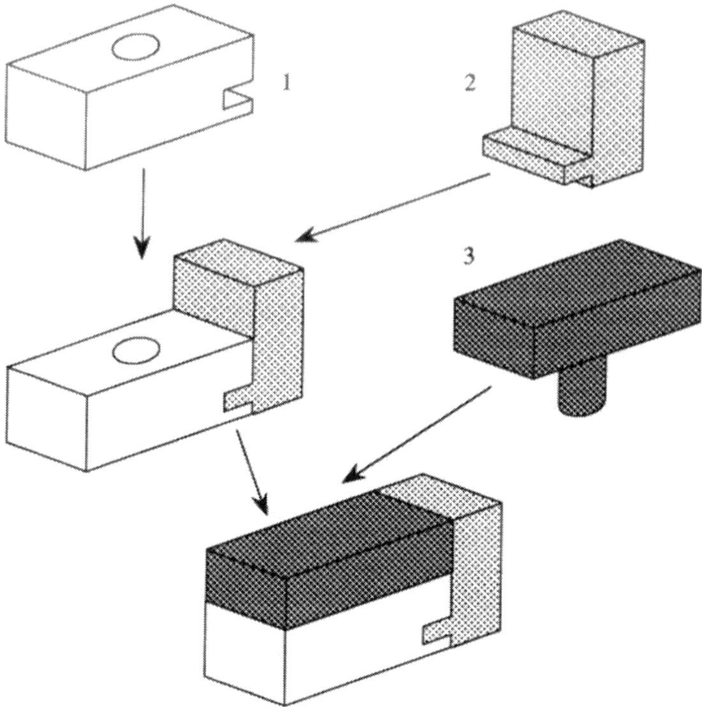

Figure 1. This simple construction task is defined by both the models of three constituent parts and a sequence of two spatial relations. Each relation may be input as a set of primitive relations among geometrical features, e.g. "Face A of Object 1 is against Face B of Object 2", "Shaft C of Object 3 fits Hole D of Object 1" [Ambler and Popplestone, 1975] [Mazer, 1983]. It may then be converted into a quantitative internal representation, e.g. the "homogeneous coordinate matrix" describing the transform between two Cartesian coordinate frames attached to the objects [Craig, 1986].

motion command is not enough; instead, a motion strategy involving several commands and using sensory inputs is necessary.

Prior knowledge of the spatial arrangement of the robot workspace is needed to plan these operations. For example, planning a collision-free path requires knowing where the obstacles are located. If complete knowledge is available (it may have been provided by the user or by a computer-aided design system, or obtained through sensing), motion plans can be fully generated and then executed. If this is not the case, some interweaving of planning and execution is necessary. When uncertainty is well-bounded (this is usually the case during part mating), robust plans anticipating various contingencies may be suitable.

The above example can easily be made more complicated. For instance, there might be more than one robot to achieve the task, or there might be mobile obstacles in the workspace. Then, some motion coordination is necessary, implying that motions are planned not just as geometrical paths, but as functions of time. A robot might also be equipped with a dexterous multi-joint multi-fingered hand allowing complex types of motions within the hand. The task might include some performance criterion to optimize, e.g. minimize the total amount of time needed to perform the task. Rather than specifying a sequence of spatial relations, the input description might only describe the spatial arrangement of the parts in the goal assembly, leaving to the planner the generation of the appropriate sequence.

The example stresses the fact that motion planning does not consist of a single, well-delimited problem. Instead, it looks more like a collection of many problems, which are more or less variants of each other. Although one may envision that a general formulation covering all possible cases could be ultimately established, such a formulation is certainly beyond our current understanding of the conceptual and computational issues underlying motion planning.

In order to organize the book, we will first consider a "basic motion planning problem" which deals with geometrical issues only, and then several extensions of this problem which involve more sophisticated geometrical and non-geometrical issues. The basic problem is defined in Section 2 and several extensions are presented in Section 5.

2 Basic Problem

The goal of defining a basic motion planning problem is to isolate some central issues and investigate them in depth before considering additional difficulties.

In the basic problem, we assume that the robot is the only moving object in the workspace and we ignore the dynamic properties of the robot, thus avoiding temporal issues. We also restrict motions to non-contact motions, so that the issues related to the mechanical interaction between two physical objects in contact can be ignored. These assumptions essentially transform the "physical" motion planning problem into a purely geometrical path planning problem. We simplify even further the geometrical issues by assuming that the robot is a single rigid object, i.e. an object whose points are fixed with respect to each other. The motions of this object are only constrained by the obstacles.

The **basic motion planning problem** resulting from these simplifications is the following:

> Let $\mathcal{A}$ be a single rigid object — the *robot* — moving in a Euclidean space $\mathcal{W}$, called *workspace*, represented as $\mathbf{R}^N$, with $N = 2$ or 3.
>
> Let $\mathcal{B}_1, ..., \mathcal{B}_q$ be fixed rigid objects distributed in $\mathcal{W}$. The $\mathcal{B}_i$'s are called *obstacles*.
>
> Assume that both the geometry of $\mathcal{A}, \mathcal{B}_1, ..., \mathcal{B}_q$ and the locations of the $\mathcal{B}_i$'s in $\mathcal{W}$ are accurately known. Assume further that no kinematic constraints limit the motions of $\mathcal{A}$ (we say that $\mathcal{A}$ is a *free-flying object*).
>
> The problem is: Given an initial position and orientation and a goal position and orientation of $\mathcal{A}$ in $\mathcal{W}$, generate a *path* τ specifying a continuous sequence of positions and orientations of $\mathcal{A}$ avoiding contact with the $\mathcal{B}_i$'s, starting at the initial position and orientation, and terminating at the goal position and orientation. Report failure if no such path exists.

Figure 2 illustrates two instances of the basic motion planning problem. It shows the paths of a rectangle (Figure 2.a) and an L-shaped polygon (Figure 2.b) among polygonal obstacles in a two-dimensional workspace.

Of course, the basic problem is somewhat oversimplified. We will see that it is nonetheless a difficult problem and that several of its solutions have straightforward extensions to more involved problems. Furthermore, solving it is also of practical interest. For instance, some mobile robot navigation problems can be realistically expressed as instances of the basic problem.

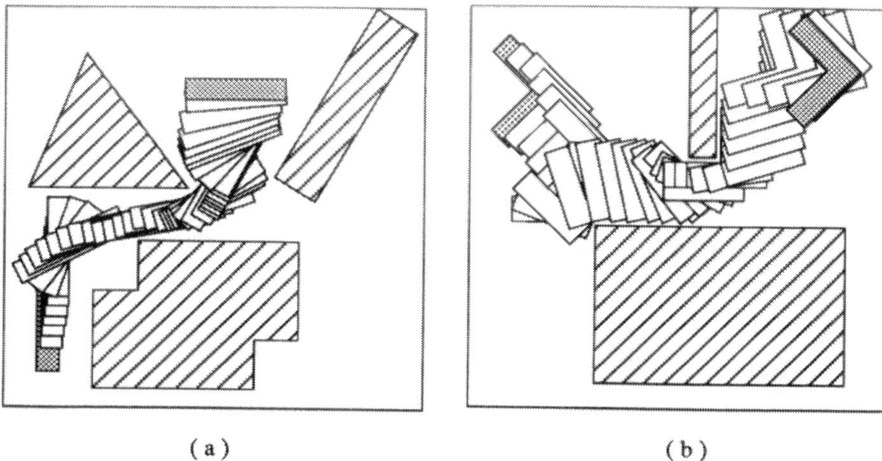

Figure 2. This figure illustrates the basic motion planning problem with two robots — a rectangle in Figure a and an L-shaped polygon in Figure b. The contact-free paths are displayed as discrete sequences of positions and orientations of the robot.

Although intuitively simple, some aspects of the above definition are still too informal and must be made more precise. For instance, the concept of a "free-flying" object is probably not very clear at this point. One possible formal statement of the problem is known as the *configuration space formulation.*

3 Configuration Space Formulation

The underlying idea of configuration space is to represent the robot as a point in an appropriate space — the robot's configuration space — and to map the obstacles in this space. This mapping transforms the problem of planning the motion of a dimensioned object into the problem of planning the motion of a point. It makes the constraints on the motions of the robot more explicit.

3.1 Notion of Configuration Space

Consider the basic problem. Let the robot $\mathcal{A}$ (at a certain position and orientation) be described as a compact (i.e. closed and bounded) subset of $\mathcal{W} = \mathbf{R}^N$, $N = 2$ or 3, and the obstacles $\mathcal{B}_1, ..., \mathcal{B}_q$ be closed subsets of $\mathcal{W}$. In addition, let $\mathcal{F_A}$ and $\mathcal{F_W}$ be Cartesian frames embedded in $\mathcal{A}$ and $\mathcal{W}$, respectively. $\mathcal{F_A}$ is a moving frame, while $\mathcal{F_W}$ is a fixed one. By definition, since $\mathcal{A}$ is rigid, every point a of $\mathcal{A}$ has a fixed position with respect to $\mathcal{F_A}$. But a's position in $\mathcal{W}$ depends on the position and orientation of $\mathcal{F_A}$ relative to $\mathcal{F_W}$. Since the $\mathcal{B}_i$'s are both rigid and fixed in $\mathcal{W}$, every point of $\mathcal{B}_i$, for all $i \in [1, q]$, has a fixed position with respect to $\mathcal{F_W}$.

A configuration of an arbitrary object is a specification of the position of every point in this object relative to a fixed reference frame [Arnold, 1978]. Therefore, a **configuration q** of $\mathcal{A}$ is a specification of the position $\mathcal{T}$ and orientation Θ of $\mathcal{F_A}$ with respect to $\mathcal{F_W}$. The **configuration space** of $\mathcal{A}$ is the space $\mathcal{C}$ of all the configurations of $\mathcal{A}$. The subset of $\mathcal{W}$ occupied by $\mathcal{A}$ at configuration $\mathbf{q}$ is denoted by $\mathcal{A}(\mathbf{q})$. In the same fashion, the point a on $\mathcal{A}$ at configuration $\mathbf{q}$ is denoted by $a(\mathbf{q})$ in $\mathcal{W}$.

One may describe a configuration by a list of real parameters. For example, the position $\mathcal{T}$ can be simply described by the vector of the N coordinates of the origin of $\mathcal{F_A}$ in $\mathcal{F_W}$. The orientation Θ can be described as the $N \times N$ matrix whose columns are the components, in $\mathcal{F_W}$, of the unit vectors along the axes of $\mathcal{F_A}$. Then, $\mathbf{q} = (\mathcal{T}, \Theta)$ is uniquely represented as a list of $N(N+1)$ parameters. Clearly, however, this representation is redundant, since the matrix describing Θ must have orthonormal columns and determinant +1. Therefore, $\mathcal{C}$ is only a subset of $\mathbf{R}^{N(N+1)}$.

One may be tempted to describe Θ by a list of parameters of minimal cardinality, i.e. one angle θ, if $N = 2$ (for example, the angle between the x-axes of $\mathcal{F_W}$ and $\mathcal{F_A}$), and three angles ϕ, θ and ψ, if $N = 3$ (for example, the Euler angles). However, such a minimal description is not injective, i.e. the same configuration may be described by different values of the angles. For example, in the two-dimensional case ($N = 2$), two angles whose difference is a multiple of 2π represent the same Θ. In this case, the difficulty can easily be fixed by restricting the values of the angle θ to the $[0, 2\pi)$ interval with modulo 2π arithmetic. In the three-dimensional case ($N = 3$), the difficulty is slightly harder to remedy.

Describing a configuration as a list of m independent parameters, with

$m = 3$ (if $N = 2$) and $m = 6$ (if $N = 3$), corresponds to representing $\mathcal{C}$ as $\mathbf{R}^m$. However, as we will see at greater length in Chapter 2, while $\mathcal{C}$ is locally "like" $\mathbf{R}^m$, it is globally different. This means that $\mathcal{C}$ can be decomposed into a finite union of p ($p > 1$) slightly overlapping "patches", each represented as a copy of $\mathbf{R}^m$ (called a *chart*). Hence, a configuration can locally be represented as a non-redundant list of m real parameters, but some administration is required to switch between representations when the robot's configuration moves from one patch to another. The number m is called the **dimension** of $\mathcal{C}$.

3.2 Notion of a Path

The notion of *continuity* is fundamental in the definition of a path. Its formalization requires us to define a topology in $\mathcal{C}$. One classical way to do so is to specify a distance function $d : \mathcal{C} \times \mathcal{C} \to \mathbf{R}$, and to let the topology in $\mathcal{C}$ be the metric topology induced by d. In order to match our intuition of the real-world physics, the distance between two configurations $\mathbf{q}$ and $\mathbf{q}'$ should decrease and tend toward zero when the regions $\mathcal{A}(\mathbf{q})$ and $\mathcal{A}(\mathbf{q}')$ get closer to each other and tend to coincide. A simple distance function that satisfies this condition is:

$$d(\mathbf{q}, \mathbf{q}') = \max_{a \in \mathcal{A}} \|a(\mathbf{q}) - a(\mathbf{q}')\|$$

where $\|x - x'\|$ denotes the Euclidean distance between any two points x and x' in $\mathbf{R}^N$. Throughout this book, the topology in $\mathcal{C}$ will be the topology induced by this distance.[1]

A **path** of $\mathcal{A}$ from the configuration $\mathbf{q}_{init}$ to the configuration $\mathbf{q}_{goal}$ is a continuous map:

$$\tau : [0,1] \to \mathcal{C}$$

with:

$$\tau(0) = \mathbf{q}_{init} \quad \text{and} \quad \tau(1) = \mathbf{q}_{goal}.$$

$\mathbf{q}_{init}$ and $\mathbf{q}_{goal}$ are the **initial** and **goal** configurations of the path, respectively. The continuity of τ entails that for all $s_0 \in [0,1]$, $\lim_{s \to s_0} \max_{a \in \mathcal{A}} \|a(\tau(s)) - a(\tau(s_0))\| = 0$, with s taking its values in $[0,1]$. Saying that $\mathcal{A}$ is a "free-flying object" means that, in the absence of obstacles, any path defined as above is feasible.

[1] As is well-known, the notion of a topology is more primitive than the notion of a metric. In Chapter 2 we will give other metrics which all induce the same topology as d.

3.3 Obstacles in Configuration Space

The above definition of a path does not take obstacles into consideration. We now want to characterize the set of paths which are solutions to the basic problem when there are obstacles in the workspace. To that purpose, we map the obstacles in configuration space and we consider the subset of $\mathcal{C}$ that is made of contact-free configurations.

Every obstacle $\mathcal{B}_i$, $i = 1$ to q, in the workspace $\mathcal{W}$ maps in $\mathcal{C}$ to a region:

$$\mathcal{CB}_i = \{\mathbf{q} \in \mathcal{C} \ / \ \mathcal{A}(\mathbf{q}) \cap \mathcal{B}_i \neq \emptyset\}$$

which is called a **C-obstacle.** The union of all the C-obstacles:

$$\bigcup_{i=1}^{q} \mathcal{CB}_i$$

is called the **C-obstacle region**, and the set:

$$\mathcal{C}_{free} = \mathcal{C} \setminus \bigcup_{i=1}^{q} \mathcal{CB}_i = \{\mathbf{q} \in \mathcal{C} \ / \ \mathcal{A}(\mathbf{q}) \cap (\bigcup_{i=1}^{q} \mathcal{B}_i) = \emptyset\}$$

is called the **free space**. Any configuration in $\mathcal{C}_{free}$ is called a **free configuration.**

A **free path** between two free configurations $\mathbf{q}_{init}$ and $\mathbf{q}_{goal}$ is a continuous map $\tau : [0,1] \rightarrow \mathcal{C}_{free}$, with $\tau(0) = \mathbf{q}_{init}$ and $\tau(1) = \mathbf{q}_{goal}$. Two configurations belong to the same connected component of $\mathcal{C}_{free}$ if and only if they are connected by a free path.

Given an initial and a goal configuration, the basic motion planning problem is to generate a free path between the two configurations, if they belong to the same connected component of $\mathcal{C}_{free}$, and to report failure otherwise.

Several planning methods described in Chapters 4 through 7 generate semi-free paths, rather than free paths. A **semi-free path** is a continuous map $\tau : [0,1] \rightarrow cl(\mathcal{C}_{free})$, where $cl(\mathcal{C}_{free})$ denotes the closure of $\mathcal{C}_{free}$. Hence, as it moves along such a path, the robot may touch obstacles. Under simple assumptions stated in Chapter 2, any semi-free path τ can be transformed into a free path that can be made arbitrarily close to τ.

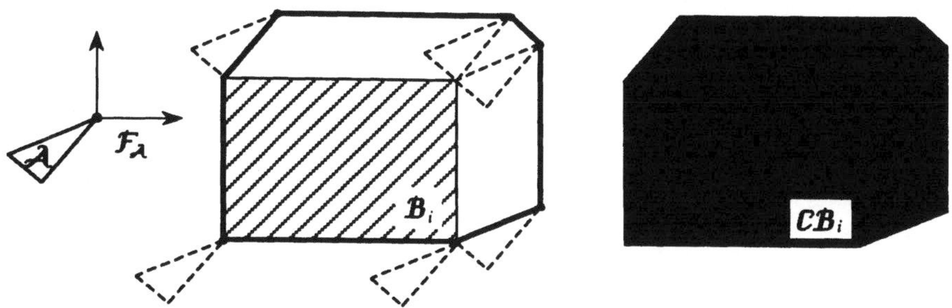

Figure 3. The robot $\mathcal{A}$ (a triangle) can translate freely in the plane at fixed orientation. Its configuration is represented as $\mathbf{q} = (x, y)$, the coordinates in $\mathcal{F}_{\mathcal{W}}$ of the vertex of $\mathcal{A}$ marked as a small circle (the origin of $\mathcal{F}_{\mathcal{A}}$). Hence, $\mathcal{A}$'s configuration space is $\mathcal{C} = \mathbf{R}^2$. The C-obstacle $\mathcal{CB}_i$ (shown dark) is obtained by "growing" the corresponding workspace obstacle $\mathcal{B}_i$ (a rectangle) by the shape of $\mathcal{A}$. Planning a motion of $\mathcal{A}$ relative to $\mathcal{B}_i$ is equivalent to planning a motion of the marked vertex of $\mathcal{A}$ relative to $\mathcal{CB}_i$.

3.4 Translational Case

When $\mathcal{A}$ consists of a single point (*point robot*), it is meaningless to talk about its orientation. Then, like $\mathcal{W}$, the configuration space of $\mathcal{A}$ is a copy of $\mathbf{R}^N$, and hence is a Euclidean space. The C-obstacles are identical to the obstacles in the workspace.

When $\mathcal{A}$ is a disc or when $\mathcal{A}$ is a dimensioned object allowed to translate freely (i.e. the origin of $\mathcal{F}_{\mathcal{A}}$ can follow any path in $\mathcal{W}$) without rotation (i.e. $\mathcal{F}_{\mathcal{A}}$ has a fixed orientation with respect to $\mathcal{F}_{\mathcal{W}}$), the configuration space $\mathcal{C}$ is also $\mathbf{R}^N$. But, in these cases, the C-obstacles are obtained by "growing" the obstacles by the shape of $\mathcal{A}$ as illustrated in Figure 3.

In the following, all these cases, where $\mathcal{C} = \mathbf{R}^N$, will be considered as particular cases of the basic motion planning problem. They will be convenient for introducing motion planning methods with simple examples.

4 Planning Approaches

There exists a large number of methods for solving the basic motion planning problem. Not all of them solve the problem in its full generality, however. For instance, some methods require the workspace to be two-dimensional and the objects to be polygonal. Despite many external differences, the methods are based on few different general approaches: *roadmap*, *cell decomposition*, and *potential field*. We briefly introduce these approaches below and we illustrate them with simple examples in two-dimensional configuration spaces populated by polygonal C-obstacles. As shown in Figure 3, such configuration spaces correspond to the case where the robot $\mathcal{A}$ is a polygon (or a point) translating in a workspace containing polygonal obstacles. These approaches will be described in detail in Chapters 4 through 7.

4.1 Roadmap

The roadmap approach to path planning consists of capturing the connectivity of the robot's free space in a network of one-dimensional curves, called the *roadmap*, lying in the free space $\mathcal{C}_{free}$ or its closure $cl(\mathcal{C}_{free})$. Once a roadmap $\mathcal{R}$ has been constructed, it is used as a set of standardized paths. Path planning is thus reduced to connecting the initial and goal configurations to points in $\mathcal{R}$ and searching $\mathcal{R}$ for a path between these points. The constructed path, if any, is the concatenation of three subpaths: a subpath connecting the initial configuration to the roadmap, a subpath contained in the roadmap, and a subpath connecting the roadmap to the goal configuration.

Various methods based on this general idea have been proposed. They compute different types of roadmaps called *visibility graph*, *Voronoi diagram*, *freeway net* and *silhouette*. We illustrate two of them below.

The *visibility graph method* is one of the earliest path planning methods. It mainly applies to two-dimensional configuration spaces with a polygonal C-obstacle region. The visibility graph is the non-directed graph G whose nodes are the initial and goal configurations $\mathbf{q}_{init}$ and $\mathbf{q}_{goal}$, and all the C-obstacle vertices. The links of G are all the straight line segments connecting two nodes that do not intersect the interior of the C-obstacle region. Figure 4 shows an example of a visibility graph. The links of the subgraph of G that is restricted to the C-obstacle vertices determine the roadmap $\mathcal{R}$. The other links of G connect the initial and goal configurations to the roadmap. G can be searched for the shortest semi-free path between $\mathbf{q}_{init}$ and $\mathbf{q}_{goal}$ (according to the Euclidean metric in $\mathbf{R}^2$). This path, if it exists, is a polygonal line connecting $\mathbf{q}_{init}$ to $\mathbf{q}_{goal}$ through C-obstacle vertices.

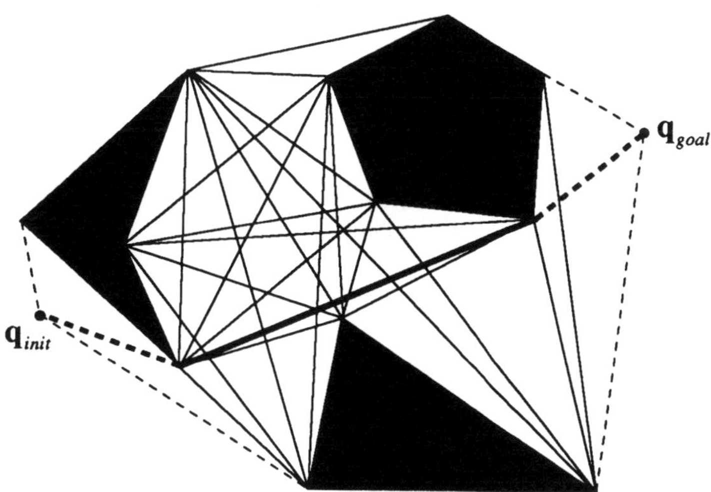

Figure 4. This figure shows the visibility graph in a two-dimensional configuration space with polygonal C-obstacles. The nodes of the graph are the initial and goal configurations and the C-obstacles' vertices. Two nodes are connected by a link if the straight line segment joining them does not intersect the C-obstacles' interior. (Hence, the links of the graph also include the edges of the C-obstacles.) The plain links connect C-obstacles' vertices. They are independent of the initial and goal configurations, and form the roadmap. The dashed links connect the initial and goal configurations to the roadmap. Any path in the visibility graph determines a semi-free path in configuration space. The graph contains the shortest semi-free path (according to the Euclidean metric in $\mathbf{R}^2$). This path is shown in both plain and dashed bold lines.

Another roadmap method, called *retraction*, consists of defining a continuous function of $\mathcal{C}_{free}$ onto a one-dimensional subset of itself (the roadmap) such that the restriction of this function to this subset is the identity map (in Topology such a function is called a "retraction"). In a two-dimensional configuration space, $\mathcal{C}_{free}$ is typically "retracted" onto its *Voronoi diagram*. This diagram is the set of all the free configurations whose minimal distance to the C-obstacle region $\mathcal{CB}$ is achieved with at least two points in the boundary of $\mathcal{CB}$ (Figure 5). The advantage of this diagram is that it yields free paths which tend to maximize the clearance between the robot and the obstacles. When the C-obstacles are polygons, the Voronoi diagram consists of straight and parabolic segments.

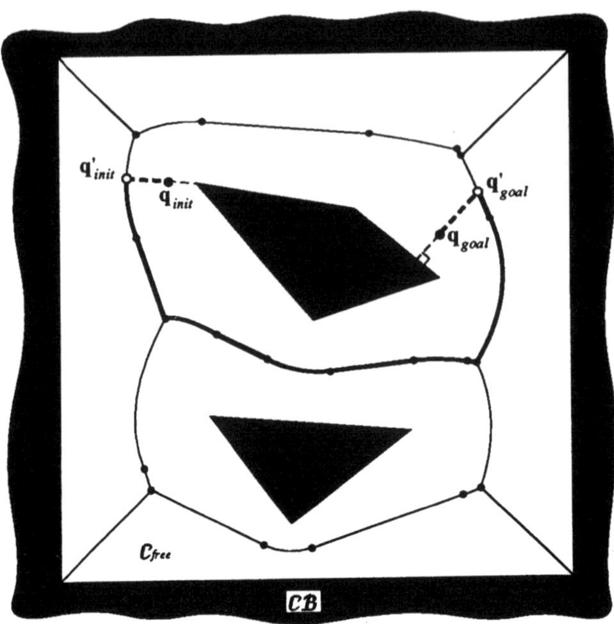

Figure 5. This figure shows the Voronoi diagram in a configuration space with a polygonal C-obstacle region. The free space in externally bounded by a rectangle (the boundary of a non-compact C-obstacle). The initial and goal configurations, $\mathbf{q}_{init}$ and $\mathbf{q}_{goal}$, are mapped in the Voronoi diagram to $\mathbf{q}'_{init}$ and $\mathbf{q}'_{goal}$, each by drawing the line along which its distance to the boundary of the C-obstacle region increases the fastest.

The initial and goal configurations are retracted in the diagram to $\mathbf{q}'_{init}$ and $\mathbf{q}'_{goal}$ as shown in Figure 5. A path is searched in the diagram between $\mathbf{q}'_{init}$ and $\mathbf{q}'_{goal}$. The free path between $\mathbf{q}_{init}$ and $\mathbf{q}_{goal}$ produced by the method, if any, consists of three subpaths: the straight path from $\mathbf{q}_{init}$ to $\mathbf{q}'_{init}$, a path in the diagram from $\mathbf{q}'_{init}$ to $\mathbf{q}'_{goal}$, and the straight path from $\mathbf{q}'_{goal}$ to $\mathbf{q}_{goal}$.

4.2 Cell Decomposition

Cell decomposition methods are perhaps the motion planning methods which have been the most extensively studied so far. They consist of decomposing the robot's free space into simple regions, called *cells*, such that a path between any two configurations in a cell can be easily generated. A non-directed graph representing the adjacency relation between the cells is then constructed and searched. This graph is called the *connectivity graph.* Its nodes are the cells extracted from the free space and two nodes are connected by a link if and only if the two corresponding cells are adjacent. The outcome of the search is a sequence of cells called a *channel.* A continuous free path can be computed from this sequence.

Cell decomposition methods can be broken down further into *exact* and *approximate* methods:

- Exact cell decomposition methods decompose the free space into cells whose union is exactly[2] the free space. The boundary of a cell corresponds to a criticality of some sort, i.e. a sudden change in the constraints applying to the motion of the robot.

- Approximate cell decomposition methods produce cells of predefined shape (e.g. rectangloids) whose union is strictly included in the free space. The boundary of a cell does not characterize a discontinuity of some sort and has no physical meaning.

Figures 6.1 and 6.2 illustrate an exact cell decomposition method in a two-dimensional configuration space. The free space is externally bounded by a polygon and internally bounded by three polygons. The free space is exactly decomposed into trapezoidal and triangular cells. These cells are built by drawing vertical rays from the C-obstacles' vertices. Two cells are adjacent if they share a portion of an edge of non-zero length. Once the connectivity graph has been constructed and a channel found, a free path is computed by connecting the initial configuration to the goal configuration through the midpoints of the intersections of every two successive cells.

Figure 7 illustrates an approximate cell decomposition method. The free space is externally bounded by a rectangle R and internally bounded by three polygons. The rectangle R is recursively decomposed into smaller rectangles. Each decomposition generates four identical new rectangles. (This type of decomposition is called a "quadtree" because it can be represented as a tree of degree 4.) At some level of resolution, only the

[2]The term "exact" refers to the input description of the motion planning problem, not to the actual physical problem, which can only be approximated in the input problem formulation.

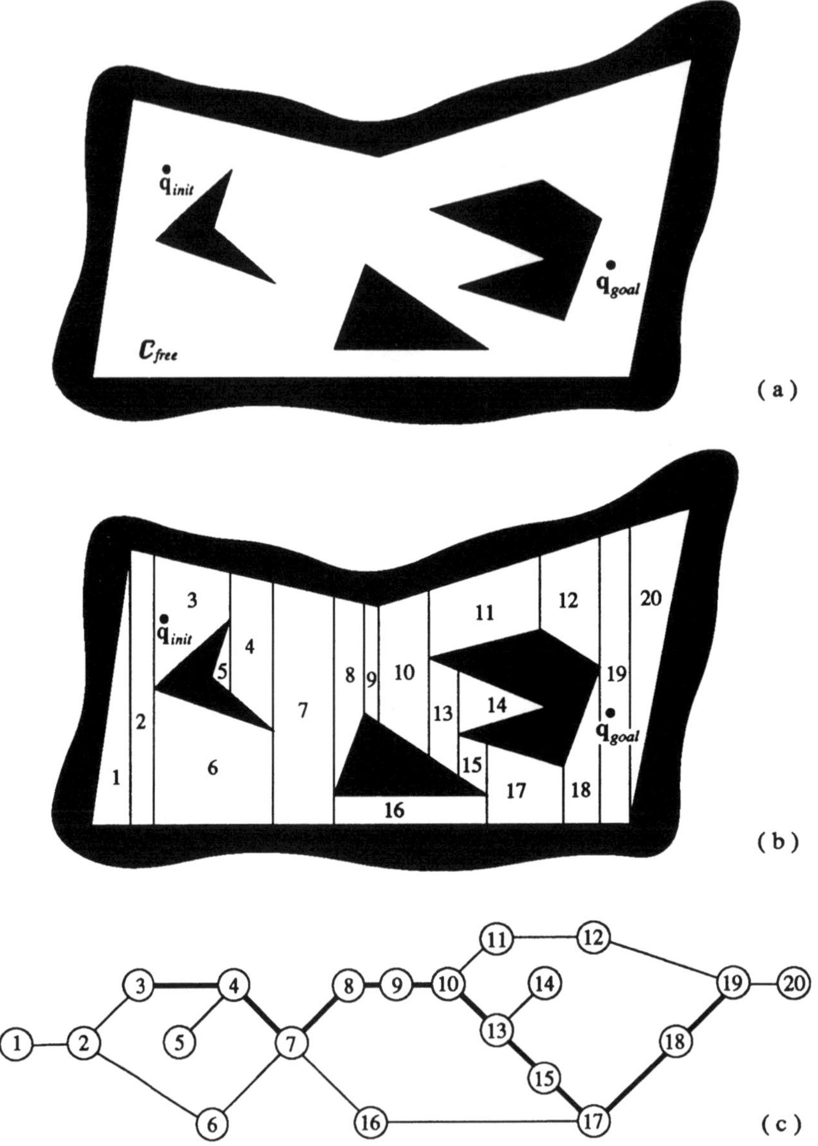

Figure 6.1. This figure and the next one illustrate an exact cell decomposition method. The free space is externally bounded by a polygon and internally bounded by three polygons (Figure a). The first step of the method consists of decomposing the free space into trapezoidal and triangular cells (Figure b). The second step is to construct the connectivity graph representing the adjacency relation between the cells (Figure c) and search this graph for a path (thick lines).

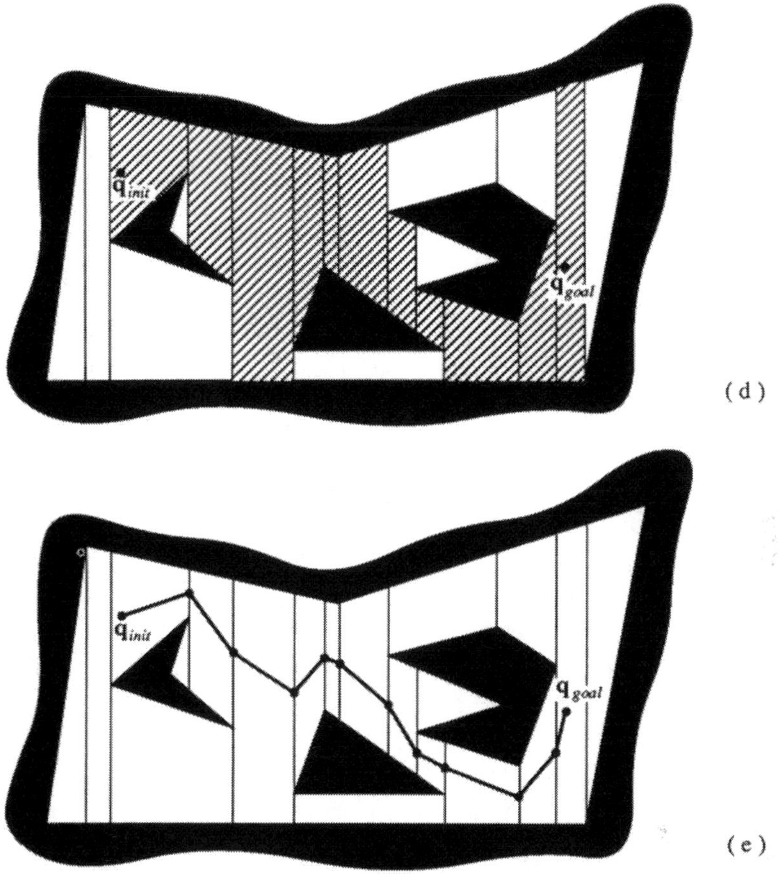

Figure 6.2. A path in the connectivity graph determines a channel (the sequence of striped cells in Figure d) in free space. The third step of the method consists of transforming this channel into a free path, for example by connecting the initial configuration to the goal configuration through the midpoints of the intersections of every two successive cells (Figure e).

cells whose interiors lie entirely in the free space are used to construct the connectivity graph. If the search of this graph terminates successfully, a path in the free space is easily generated. Otherwise, it may mean that either the resolution of the decomposition is insufficient, or no free path exists between the initial and goal configurations. Often, an approximate cell decomposition method operates in a hierarchical fashion by using a

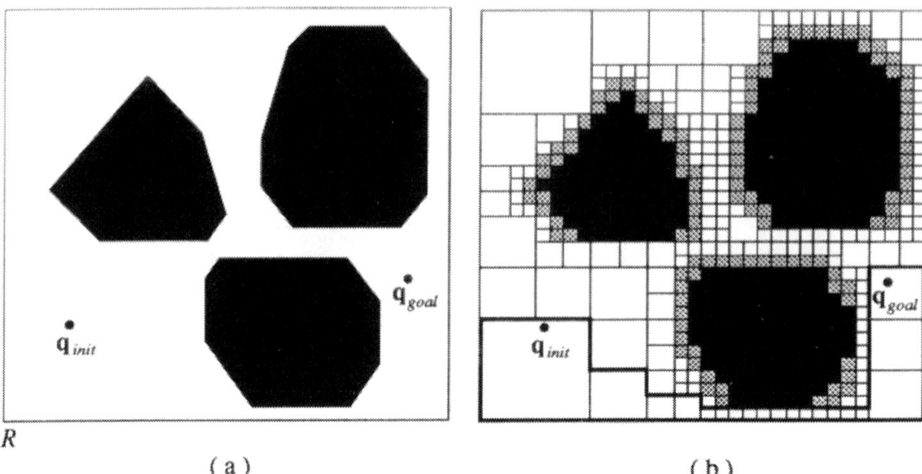

Figure 7. This figure illustrates an approximate cell decomposition method. The free space is externally bounded by a rectangle R and internally bounded by three polygons (Figure a). The rectangle R is decomposed into four identical rectangles. If the interior of a rectangle lies completely in free space or in the C-obstacle region, it is not decomposed further. Otherwise, it is recursively decomposed into four rectangles until some predefined resolution is attained. The result is shown in Figure b. White cells lie outside the C-obstacle region; black cells lie inside the C-obstacle region; grey cells are cut by the C-obstacle region. This kind of decomposition is called a "quadtree" decomposition. A channel extracted from this decomposition is shown in bold contour.

rather coarse resolution at the beginning, and refining it until either a free path is found or a limit resolution is attained.

Provided that they are equipped with both appropriate search techniques and exact numerical computation techniques, exact cell decomposition methods are complete — i.e. they are guaranteed to find a free path whenever one exists and to return failure otherwise. Approximate methods may not be complete; but, for most of them, the precision of the approximation can be tuned and made arbitrarily small (at the expense of the running time), so that the methods are said to be "resolution-complete". On the other hand, except in very simple cases, exact methods are mathematically more involved than approximate ones. Hence, the latter are usually easier to implement.

4.3 Potential Field

A straightforward approach to motion planning is to discretize the configuration space into a fine regular grid of configurations and to search this grid for a free path. This approach requires powerful heuristics to guide the search, since the grid is in general enormous. Several types of heuristics have been proposed. The most successful ones take the form of functions that are interpreted as *potential fields.*

The metaphor suggested by this terminology is that the robot represented as a point in configuration space is a particle moving under the influence of an artificial potential produced by the goal configuration and the C-obstacles. Typically the goal configuration generates an "attractive potential" which pulls the robot toward the goal, and the C-obstacles produce a "repulsive potential" which pushes the robot away from them. The negated gradient of the total potential is treated as an artificial force applied to the robot. At every configuration, the direction of this force is considered the most promising direction of motion.

Figure 8 illustrates the notion of attractive and repulsive potentials, and their combination. The attractive potential (Figure 8.b) is a quadratic well with its minimum at the goal configuration. The repulsive potential (Figure 8.c) is non-zero only within some distance from the C-obstacles and tends toward infinity when the distance to the C-obstacles tends toward zero (hence, is truncated in the figure). A path between the initial and goal configurations is constructed by tracking the negated gradient of the total potential (Figure 8.e). A matrix of orientations of the negated gradient vector field (orientations of the artifical forces induced by the potential field) is displayed in Figure 8.f.

In comparison to other methods, potential field methods can be very efficient. However, they have a major drawback. Since they are essentially fastest descent optimization methods, they can get trapped into local minima of the potential function other than the goal configuration. One attack of this problem is to design potential functions that have no local minima other than the goal configuration in the connected subset of free space which contains the goal configuration. Another approach is to complement the basic potential field approach with powerful mechanisms to escape from local minima.

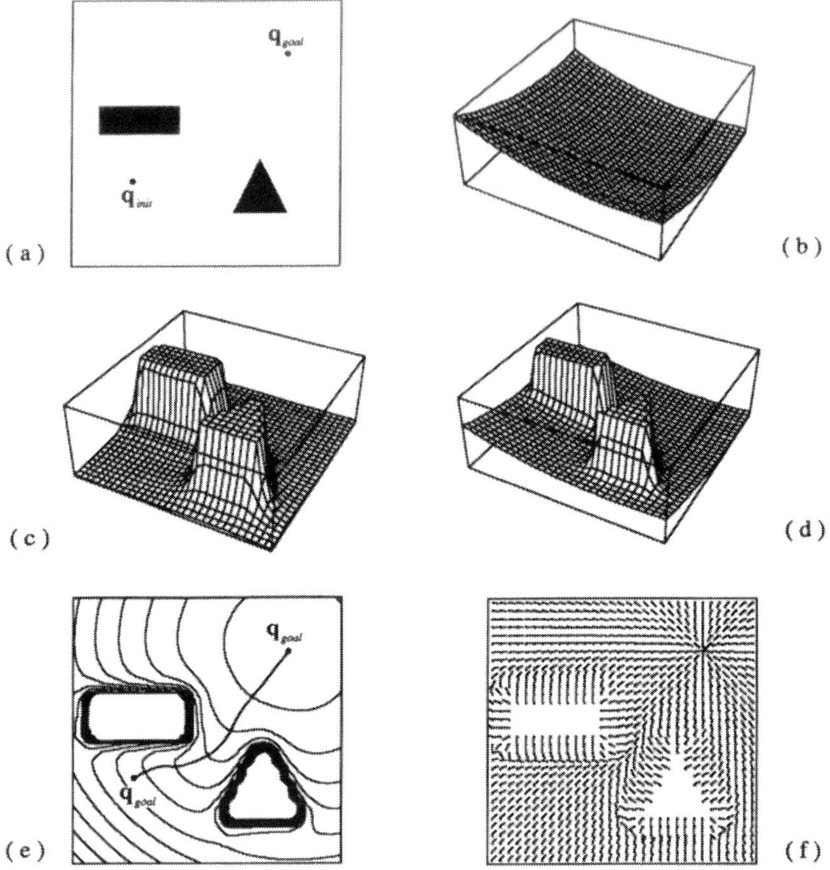

Figure 8. This series of diagrams illustrates the potential field approach. A simple two-dimensional configuration space with two polygonal C-obstacles is depicted in Figure a. Figure b shows an attractive potential generated by the goal configuration $\mathbf{q}_{goal}$. Figure c shows a repulsive potential generated by the C-obstacles. Figure d shows the sum of the two potentials. Figure e displays equipotential contours of the total potential and a path obtained by following its negated gradient. Figure f shows orientations of the negated gradient of the total potential over free space.

4.4 Global versus Local Methods

The roadmap and cell decomposition methods reduce the problem of finding a continuous free path to that of searching a graph (e.g. the visibility graph, the Voronoi diagram, the connectivity graph) by first analyzing the connectivity of the free space. Unlike these methods, potential field methods do not include an initial processing step aimed at capturing the connectivity of the free space in a concise representation. Instead, they search a much larger graph representing the adjacency among the configurations contained in the grid thrown over configuration space. At each step, they move from one configuration in the grid to another one, basing their choice on the computation of the potential gradient. It is typical for this gradient to depend only on the contents of the configuration space in the neighborhood of the current configuration of the robot. (As illustrated in Figure 8.c, the repulsive potential is usually defined with a limited range of influence around the C-obstacles.) Therefore, potential field methods are often called *local* methods, while the roadmap and cell decomposition methods are called *global* methods.

In practice, this distinction is certainly valid. For example, a potential field method can be run while the robot is moving, by tracking the negated gradient of a potential field defined as a function of both the distance to the goal configuration and the distances to the obstacles as measured by range sensors. The distinction is nevertheless intuitive and has no solid theoretical basis. A potential field method may use a potential function that is free of local minima (in the connected subset containing the goal configuration). Computing such a function certainly requires the geometry of the whole free space to be taken into account. Then the method is as global as any roadmap or cell decomposition method. On the other hand, a roadmap or cell decomposition method can be made local by restricting its application to a subset of the configuration space around the current configuration of the robot. A path is then generated in an iterative fashion by concatenating subpaths, each achieving an intermediate configuration in such a restricted subset.

5 Extensions of the Basic Problem

The basic motion planning problem makes assumptions that significantly limit the practicality of its solutions. In other words, it may be quite difficult to reduce an actual robotic problem to an instance of the basic problem, to solve this instance, and to adapt the produced solution in order to match the conditions of the original problem. In this section we introduce and discuss informally several important extensions that we will investigate in detail in Chapters 8 through 11. Other extensions can certainly be envisioned, but they have received too little attention so far to be treated in this book.

5.1 Multiple Moving Objects

In the basic problem we assumed that the obstacles were fixed, that there was a single robot, and that this robot was made of a single rigid object. Under the title "multiple moving objects", we consider a series of extensions that removes these assumptions. One extension consists of including *moving obstacles* in the workspace. The second one allows *multiple robots* to operate in the same workspace, each being a potential obstacle to the others. The third extension considers *articulated robots*, which are made of several rigid objects connected by joints.

The first extension (and possibly the second) requires time to be explicitly considered in the motion plans. The second and third extensions yield configuration spaces of arbitrarily large dimensions.[3]

Moving Obstacle. In the presence of moving obstacles, the motion planning problem can no longer be solved by merely constructing a geometric path. A continuous function of time specifying the robot's configuration at each instant must be generated instead. This can be done by adding one dimension representing time to the robot's configuration space. The new space, denoted by $\mathcal{CT}$, is called *configuration-time space*[4]. The workspace obstacles map in $\mathcal{CT}$ to static regions called *$\mathcal{CT}$-obstacles.* Every cross-section through $\mathcal{CT}$ at time t is the configuration space of the robot at instant t. It cuts the CT-obstacles according to the C-obstacles corresponding to the workspace obstacles at their locations at instant t. Motion planning consists of finding a path among the CT-obstacles in $\mathcal{CT}$. Since time is irreversible, this path must have the property of never going backward along the time axis. Hence, path planning methods have to be modified in order to take this specificity of the time dimension into account.

[3]In Section 6 we will see that the inherent computational complexity of path planning is exponential in the dimension of the robot's configuration space.

[4]In the literature it is also called configuration space-time.

If there is no constraint on the robot's velocity and acceleration, and if the motion of every obstacle is known beforehand, it is rather straightforward to extend some basic planning methods to handle this new problem. If constraints apply to the robot's velocity and/or its acceleration, they translate to geometric constraints on the slope and the curvature of a path along the time dimension, and the problem turns out to be significantly harder.

Multiple Robots. Motion planning with multiple robots differs from planning in the presence of moving obstacles in that the motions of the robots have to be planned, while the motions of the moving obstacles are not under control.

One way to deal with multiple robots $\mathcal{A}_1, ..., \mathcal{A}_p$ operating in the same workspace $\mathcal{W}$ is to treat them as a single multi-bodied robot $\mathcal{A} = \{\mathcal{A}_1, ..., \mathcal{A}_p\}$. The *composite configuration space* of $\mathcal{A}$ is $\mathcal{C} = \mathcal{C}_1 \times ... \times \mathcal{C}_p$, i.e. the set product of the configuration spaces $\mathcal{C}_i$ of the individual robots $\mathcal{A}_1, ..., \mathcal{A}_p$. Every configuration in $\mathcal{C}$ determines a unique position and orientation for each robot $\mathcal{A}_i$. C-obstacles in this space result from the interaction of either a robot $\mathcal{A}_i$ and an obstacle $\mathcal{B}_j$, or two robots $\mathcal{A}_i$ and $\mathcal{A}_j$. Basic path planning methods can be used to plan a path of $\mathcal{A}$ in $\mathcal{C}$. This approach to multi-robot path planning is called *centralized planning.* A difficulty is that it may yield high-dimensional configuration spaces. Indeed, the dimension of the composite configuration space $\mathcal{C}$ is equal to the sum of the dimensions of the individual configuration spaces $\mathcal{C}_1$ through $\mathcal{C}_p$.

Another approach to motion planning with multiple robots, *decoupled planning*, is to plan the motion of each robot more or less independently of the other robots and to consider the interactions among the paths in a second phase of planning. By doing so, the amount of computation may be significantly reduced, but at the expense of completeness. Figure 9 illustrates a planning problem where the decoupled approach is likely to fail. The problem consists of interchanging the positions of two robots (discs) in a narrow corridor where they cannot pass each other. There is enough room at one end of the corridor for a permutation of the two robots. However, by considering each robot separately, the decoupled planning approach has no mechanism to infer that both robots must first move to this end of the corridor.

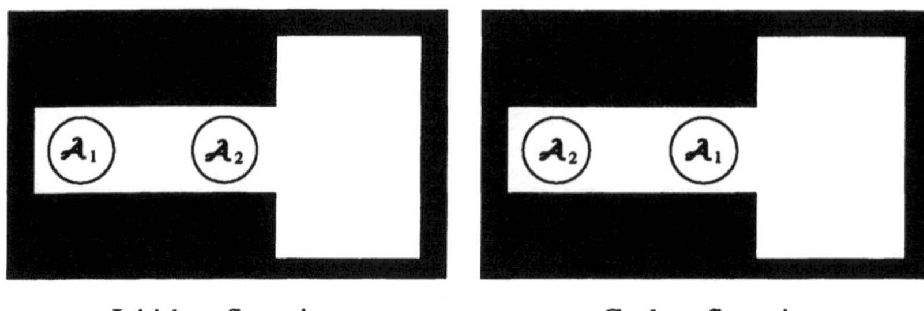

Figure 9. Two discs $\mathcal{A}_1$ and $\mathcal{A}_2$ have to interchange their positions in a narrow corridor where they cannot pass each other. However, there is enough room for permutation at one end of the corridor. Decoupled planning, which considers the two robots separately, would very likely fail to solve this problem.

Articulated Robots. An articulated robot $\mathcal{A}$ is made of several moving rigid objects $\mathcal{A}_1, ..., \mathcal{A}_p$ (called *links*) connected by *joints*, e.g. revolute joints (hinges) and prismatic joints (sliding joints). Each joint constrains the relative movements of the two objects it connects. A typical example of an articulated robot is a robot arm (see Figure 10).

$\mathcal{A}$ can be seen as a set of p moving rigid objects. The constraints imposed by the joints on the relative movements of the $\mathcal{A}_i$'s determine a subset of the composite configuration space of these objects, which is the actual configuration space of $\mathcal{A}$. Every configuration in this subset indeed determines a unique position and orientation for each of the p links. This subset is in general relatively easy to parameterize, for example by associating a distance or an angle with each joint. Several basic motion planning methods extend to this case in a straightforward fashion, at least in theory. A practical problem that has to be faced is that the dimension of the configuration space grows with the number of joints.

Figure 11 illustrates several configurations of a planar articulated robot along a path generated by a potential field method. In this example, the robot is a sequence of eight links. All the joints are revolute joints. Notice that for an articulated robot with many links and joints, it may be problematic to specify a free goal configuration. In the example of Figure 11, only the goal position of the endpoint of the arm was specified to the planner. This corresponds to defining a goal region in the robot's configuration space, rather than a unique goal configuration.

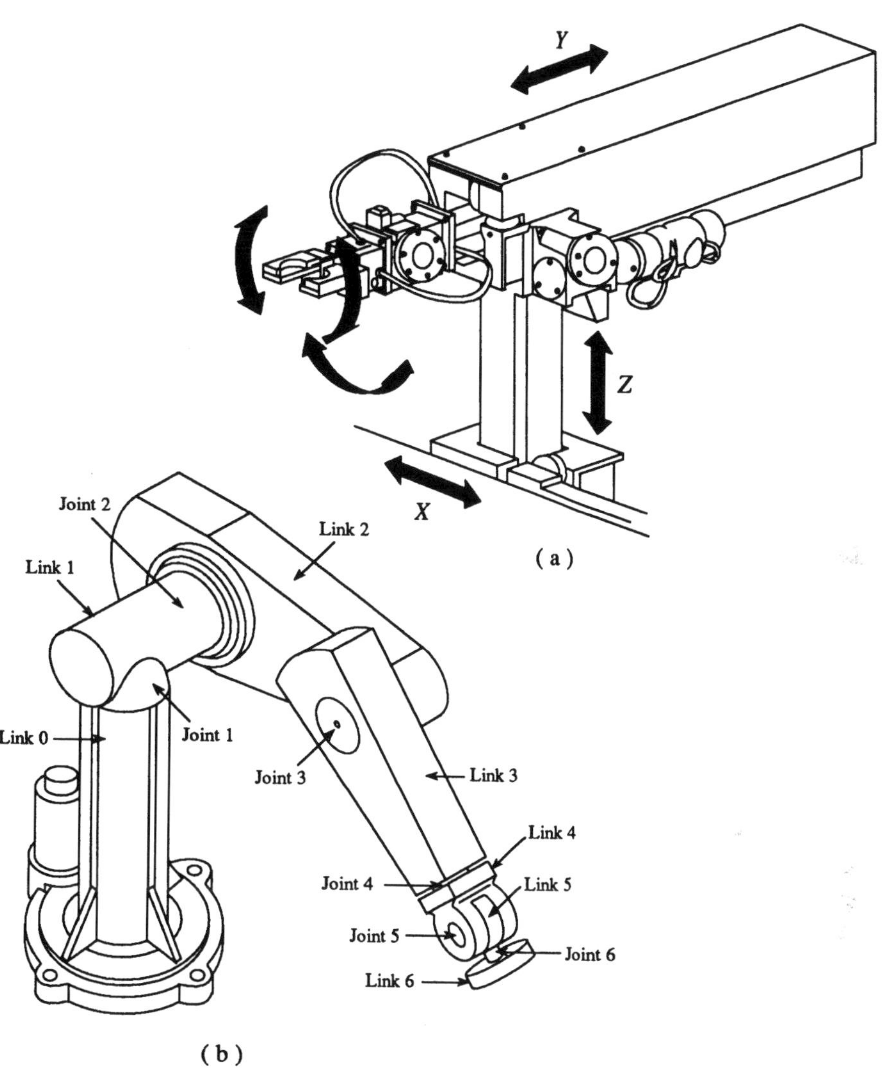

Figure 10. A robot arm is a typical example of an articulated robot. It is made of several rigid objects, called links, serially connected by joints. Most robot arms have revolute joints and/or prismatic joints. A revolute joint constrains the relative motion of two objects to be a rotation around an axis fixed with respect to both objects. A prismatic joint constrains the relative motion of two objects to be a translation along an axis fixed with respect to both objects. Figure a shows a Cartesian robot, with three prismatic joints and three revolute ones. Figure b shows a robot with six revolute joints.

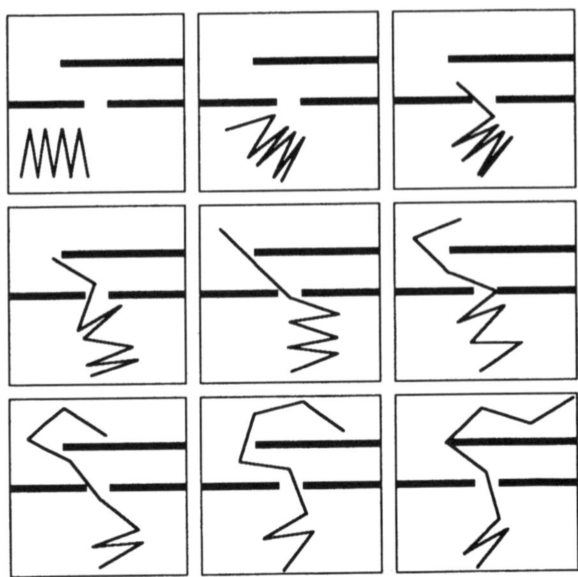

Figure 11. This figure shows a sequence of configurations of a planar eight-revolute-joint arm along a path generated using a potential field method.

5.2 Kinematic Constraints

In the basic problem we assumed that the robot was a free-flying object, i.e. the only constraints on its motions were due to the obstacles. In some problems we may want to impose additional kinematic constraints to the robot's motions. Two types of constraints, called *holonomic constraints* and *nonholonomic constraints*, can be considered.

Holonomic Constraints. Let us assume that a configuration is represented by a list of parameters of minimal cardinality (see Section 3). A holonomic equality constraint is an equality relation among these parameters, which can be solved for one of the parameters. Such a relation reduces the dimension of the actual configuration space of the robot by one. A set of k independent holonomic constraints reduces it by k.

Consider for example the case of a three-dimensional object $\mathcal{A}$ allowed to translate freely, but constrained to rotate around a fixed axis relative to $\mathcal{A}$. Let us represent the orientation of $\mathcal{A}$ by three angles. The constraints restricting the possible orientations of $\mathcal{A}$ can be expressed as two independent equations among these angles. While the dimension of the configuration space of a free-flying object in a three-dimensional workspace is 6, $\mathcal{A}$'s configuration space has dimension 4. The particular case where $\mathcal{A}$ can translate freely at fixed orientation could also be considered as a problem with holonomic constraints. However, since this problem is equivalent to the problem of planning the motion of a point robot in $\mathbf{R}^N$, we regard it as a particular case of the basic motion planning problem.

Articulated robots provide another example of holonomic constraints. For instance, a revolute joint determines two holonomic constraints. Indeed, while six parameters are necessary to define the configuration of two free-flying planar objects, four parameters suffice to determine the positions and orientations of two planar objects connected by a revolute joint. As another example, consider a six-joint arm carrying a glass of water. Requiring the glass to stay vertical during the motion corresponds to imposing two holonomic constraints to the arm.

Holonomic constraints certainly affect the definition of the robot's configuration space and may even change its global connectedness. Nonetheless, holonomic constraints do not raise new fundamental issues. Many basic planning methods remain applicable.

Nonholonomic Constraints. A nonholonomic equality constraint is a non-integrable equation involving the configuration parameters and their derivatives (velocity parameters). Such a constraint does not reduce the dimension of the space of configurations attainable by the robot, but reduces the dimension of the space of possible differential motions (i.e. the space of the velocity directions) at any given configuration.

Consider for example a car-like robot $\mathcal{A}$ rolling on a flat ground. We model it as a rectangular object moving in $\mathcal{W} = \mathbf{R}^2$ (see Figure 12). We know by experience that in an empty space we can drive the robot to any position with any orientation. Hence the robot's configuration space has three dimensions, two of translation and one of rotation. Let us represent a configuration of $\mathcal{A}$ by (x, y, θ), where x and y are the coordinates, in the frame $\mathcal{F}_\mathcal{W}$, of the midpoint R between the two rear wheels and $\theta \in [0, 2\pi)$ is the angle between the x-axis of $\mathcal{F}_\mathcal{W}$ and the main axis of $\mathcal{A}$. At any instant during a motion, assuming no slipping,

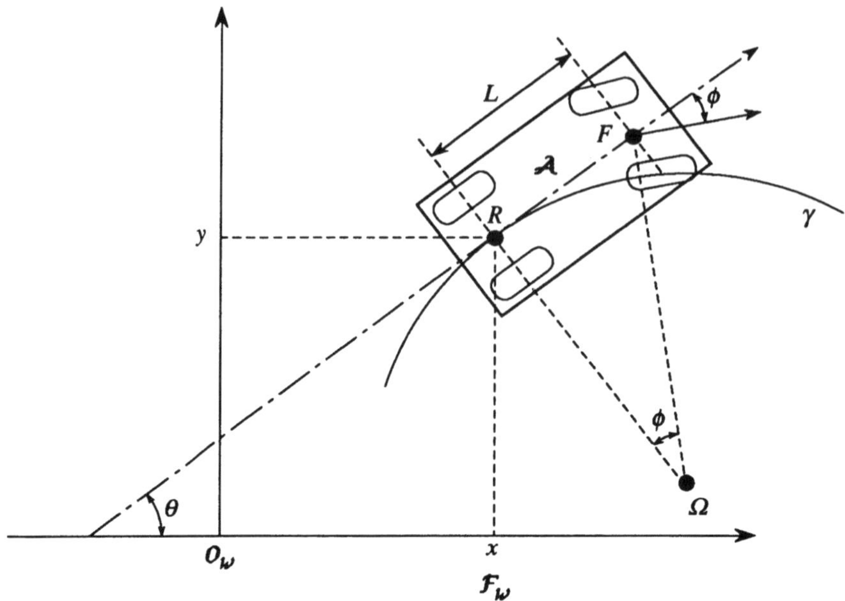

Figure 12. This car-like robot $\mathcal{A}$ is modeled as a rectangle. Any configuration of $\mathcal{A}$ is represented by three parameters $(x, y, \theta) \in \mathbf{R}^2 \times [0, 2\pi)$, with modulo 2π arithmetic on θ. Although in the absence of obstacles the robot can attain any configuration in $\mathbf{R}^2 \times [0, 2\pi)$, the projection of any path of $\mathcal{A}$ on the xy-plane is a curve γ tangent to the main axis of $\mathcal{A}$, thus implying that the relation $-\sin\theta\ dx\ +\ \cos\theta\ dy = 0$ be verified. This relation is a nonholonomic equality constraint. The curvature of γ is $L/\tan\phi$, where L is the distance between R and F, the midpoint between the two front wheels, and ϕ is the steering angle, i.e. the angle between $\mathcal{A}$'s main axis and the velocity vector of point F. In general, $|\phi| \leq \phi_{max} < \pi/2$.

the velocity of R has to point along the main axis of $\mathcal{A}$. Therefore, the motion is constrained by the relation:

$$-\sin\theta\ dx\ +\ \cos\theta\ dy = 0.$$

It can be shown that this equation is non-integrable, hence is a nonholonomic equality constraint. Due to this constraint, the space of differential motions $(dx, dy, d\theta)$ of the robot at any configuration (x, y, θ) is a two-dimensional space. If the robot was a free-flying object, this space would be three-dimensional. The instantaneous motion of the car-like robot is

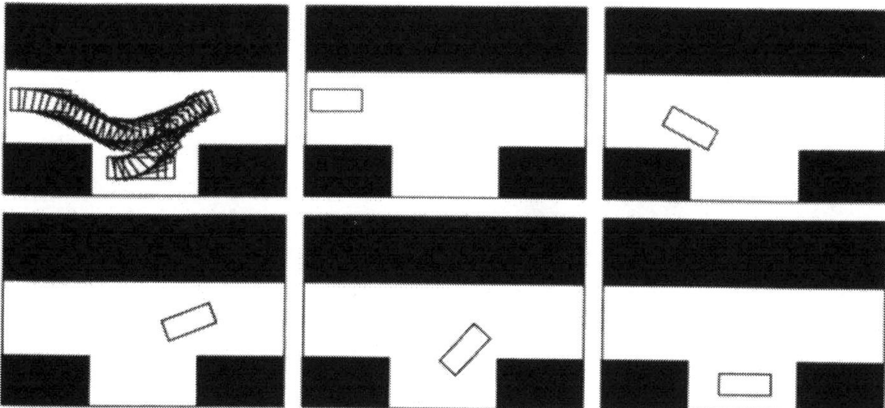

Figure 13. This figure illustrates a path generated by a planner described in Chapter 9 for a car-like robot in order to perform a parallel parking operation.

determined by two parameters: the linear velocity along its main axis and the steering angle. However, when the steering angle is non-zero, the robot changes orientation, and its linear velocity with it, allowing the robot's configurations to span a three-dimensional space.

Furthermore, the steering angle denoted by ϕ in Figure 12 is in general constrained to take values in an interval $[-\phi_{max}, +\phi_{max}]$, with $\phi_{max} < \pi/2$. This constraint can be rewritten as an inequality involving dx, dy and $d\theta$. It restricts the set of possible differential motions without changing its dimension. We call it a nonholonomic inequality constraint.

Nonholonomic constraints restrict the geometry of the feasible free paths between two configurations. They are much harder to deal with in a planner than holonomic constraints. Figure 13 illustrates a path generated for a car-like robot in order to perform a parallel parking operation.

5.3 Uncertainty

The basic problem assumes that the robot can follow the paths generated by the planner exactly. It also assumes that the geometry of the robot, the geometry of the obstacles, and the location of the obstacles are accurately known. No robot setting rigorously satisfies these assumptions, and both robot control and geometric models are imperfect. Fortunately, we often do not have to worry about these imperfections because they are small relative to the tolerance of the task being performed. But this is not always the case.

At the other extreme, the robot could have little or no prior knowledge about its workspace. Then, it would have to rely on its sensors at execution time to obtain the information needed to accomplish the task. This extreme case requires the robot to explore its workspace, and it tends to be outside the realm of motion planning. The robot may interweave planning and execution monitoring activities, but the more incomplete the prior knowledge, the less important the role of planning.

An intermediate situation is when there are errors in robot control and in the initial geometric models, but these errors are contained within bounded regions. For example, the actual obstacles' locations are slightly different from those in the robot's model, but the errors on the parameters defining these locations are bounded. Similarly, the robot moves along a direction that is different from the commanded one, but the actual direction of motion is contained in a narrow cone centered along the commanded direction. In order to deal with such bounded errors, we assume that the robot is equipped with sensors that can be used at execution time to acquire additional information. But sensors are not perfect, either. For example, a position sensor does not return the exact configuration of the robot. Again, it is realistic to assume that the sensing errors are contained in bounded uncertainty regions. When errors in control, sensing and model are reasonably small, it is interesting to generate motion plans that are tolerant to these errors, i.e. that achieve goals reliably by anticipating the various possible contingencies.

The motion planning problem with bounded uncertainty can be stated as follows: Given an *initial region* $\mathcal{I}$ and a *goal region* $\mathcal{G}$ in the robot's configuration space, generate a motion plan whose execution guarantees the robot to reach a configuration in $\mathcal{G}$ if it starts from any (unknown) configuration in $\mathcal{I}$, despite bounded uncertainty in control, sensing and model. A solution to this problem is a plan that combines motion commands and sensor readings that interact at execution time in order to reduce uncertainty and guide the robot toward the goal.

Planning in the presence of bounded uncertainty raises new issues explored neither in the basic problem, nor in the previous extensions. Due

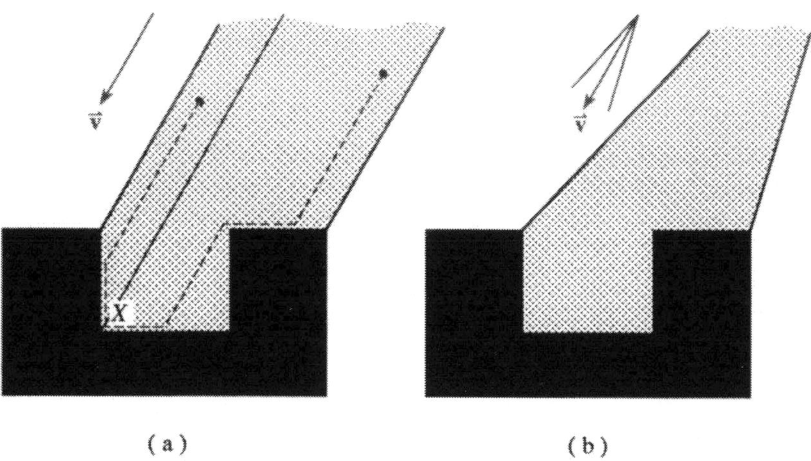

Figure 14. This figure illustrates a force-compliant motion in a two-dimensional configuration space with a polygonal C-obstacle. The robot is commanded to move along the direction $\vec{\mathbf{v}}$. Assuming perfect control (Figure a), the robot moves along a straight line parallel to $\vec{\mathbf{v}}$ as long as it is in free space. When it touches an obstacle, the robot switches to a path following the projection of $\vec{\mathbf{v}}$ in the C-obstacle's boundary. The grey region in Figure a shows the range of initial configurations from which any compliant motion commanded along $\vec{\mathbf{v}}$ reaches the vertex X. The two dashed lines illustrate two such motions. (The range of initial configurations from which a pure position-controlled motion commanded along $\vec{\mathbf{v}}$ reaches X is the half-line parallel to $\vec{\mathbf{v}}$ ending at X). If control is imperfect, e.g. the robot moves along a direction contained in a cone centered along the commanded direction $\vec{\mathbf{v}}$ (see Figure b), the grey region is slightly reduced. (In contrast, the range of initial configurations from which a pure position-controlled motion commanded along $\vec{\mathbf{v}}$ is guaranteed to reach X vanishes to the point X itself.)

to uncertainty in control, a motion command may produce any path among the infinitely many ones which are consistent with both the command and the uncertainty. In order for the planner to guarantee that execution will succeed, all of these paths must reach the goal (goal reachability). But this is only a sufficient condition. The plan must also terminate *in* the goal. Due to uncertainty in sensing, it may not be a trivial issue to recognize goal achievement reliably. The planner must make sure that enough information will be made available to the controller during execution (goal recognizability).

Uncertainty often leads to using sensor-based motion commands whose behavior is less sensitive to errors than purely position-controlled motion commands. *Force-compliant* motion commands are one example of such commands. When they are used, the robot may touch obstacle surfaces and slide on them (along the projection of the commanded direction of motion), rather than just stop. Figure 14 illustrates a compliant motion of a robot commanded along the direction $\vec{\mathbf{v}}$ in a two-dimensional configuration space. In Figure 14.a we assume perfect control; in Figure 14.b we assume imperfect control (the actual direction of motion is contained in a cone centered along $\vec{\mathbf{v}}$). In both cases, the grey region is the set of initial configurations from where the robot commanded along $\vec{\mathbf{v}}$ is guaranteed to reach the C-obstacle vertex X (assuming frictionless edges). The grey region in Figure 14.b is only slightly smaller than in Figure 14.a. In contrast, with uncertainty in control, no position-controlled motion commanded along $\vec{\mathbf{v}}$ is guaranteed to reach X. Planning such sensory-based motion commands may require the physics of the workspace to be taken into consideration. For example, when force sensing is used to detect contact with the obstacles and slide in their boundary, frictional effects ignored in Figure 14 may be critical.

5.4 Movable Objects

In the basic problem the objects in the workspace other than the robot are all stationary, non-movable obstacles. In one extension introduced in Subsection 5.1 the obstacles can move, but their motions are not under the robot's control. Another important extension is when the workspace contains *movable objects*, i.e. objects that the robot can displace, for example by grasping them.

The presence of movable objects requires the generation of sequences of motion commands, each intended to achieve some intermediate goal which may not have been specified in the input problem. For example, even if there exists no feasible path to a goal configuration in some given arrangement of the workspace, the robot may create one by moving movable objects to other locations. In order to do so, the planner must select these other locations (intermediate goals). Hence, a motion plan appears as an alternate sequence of *transit* and *transfer* motions. Along a transit motion the robot moves alone. Along a transfer motion it moves together with a movable object. The transit and transfer motions lie in different spaces. The transit motions occur in the configuration space of the robot, but different arrangements of the movable objects yield different geometries in this space. A transfer motion occurs in the configuration space of the union of the robot and the movable object moving with it. The main problem for the planner is to determine which spaces to use and when to change from one space to another.

The most common way for a robot to displace a movable object is to rigidly grasp it and to move with it. For example, if the robot is equipped with a multi-fingered gripper, grasping an object consists of placing the fingers of the gripper on the surface bounding the object and exerting some gripping force. Then, if the robot moves, the object moves with it as a single object. Grasping requires that the robot be capable of selecting the positions of the fingers on the object (another intermediate goal). Several constraints — force-closure, stability, tolerance to positioning errors, minimization of gripping forces, geometric feasibility — have to be satisfied by these positions. The grasp planning problem (a difficult problem in its own) is part of the motion planning problem in the presence of movable objects.

Grasping is not the only way of moving a movable object. The robot may also push an object on a table. It may also cause an object to move by tilting the surface supporting the object with respect to gravity. If it is equipped with a dexterous hand (i.e. a gripper with independent multi-joint fingers), it may also manipulate the object within its hand. However, these operations are very difficult to model and existing models still lack the required precision for constructing effective planners.

Movable objects allow new types of planning problems to be considered, where the important goals are not just to achieve given configurations of the robot. Arrangements of movable objects, such as assembling parts into a composite object or arranging pieces of furniture according to a given pattern, can now be the real goals.

6 Computational Complexity

Instances of the same type of motion planning problem may considerably differ in their "size", e.g. the dimension of configuration space and the number of obstacles in the workspace. It is important to quantify the performance of the planning methods — typically, the time they require to solve a problem — and the inherent computational complexity of the planning problems as a function of these parameters. Analysis of the methods is useful to assess their practicality and to detect opportunities for increased efficiency. Analysis of the problems is useful to suggest new ways of formulating them when there is substantial evidence that the original formulation only admits too costly solutions.

In this section we review important theoretical results relative to the computational complexity of robot motion planning. We mainly consider lower bounds on the time complexity of problems. In the following chapters we will analyze the complexity of specific planning methods (upper bounds). Basic notions in computational complexity analysis are surveyed in Appendix B.

The first lower bound was established by Reif for planning free paths in a configuration space of arbitrary dimension ("generalized mover's problem") [Reif, 1979]:

Planning a free path for a robot made of an arbitrary number of polyhedral bodies connected together at some joint vertices, among a finite set of polyhedral obstacles, between any two given configurations, is a PSPACE-hard problem.

PSPACE-hardness has also been established for a variety of path planning problems with simpler or more specific robotic systems, for example:

- *Planar linkages.* A planar linkage consists of an arbitrary number of one-dimensional rigid links connected by revolute joints, each joint connecting two or more links. The locations of some joints are fixed. Although all the links move in the same plane, they are allowed to cross each other. In the absence of obstacles, deciding whether a planar linkage in some initial configuration can be moved so that a certain joint reaches a given point in the plane is PSPACE-hard [Hopcroft, Joseph, and Whitesides, 1984]. This result reflects the complexity of the connectivity of the configuration space of a planar linkage.

- *Multiple rectangles.* The robotic system consists of arbitrarily many rectangles in an empty rectangular workspace. Each rectangle can only move in translation with its sides parallel to the sides of the workspace boundary. The problem is to plan the coordinated motion of the rectangles between two given configurations, so that they do not intersect. This problem, which is equivalent to planning a path in the configuration space of the multi-bodied robot consisting of all the rectangles, is both PSPACE-hard [Hopcroft, Schwartz and Sharir, 1984] and in PSPACE [Hopcroft and Wilfong, 1986b]. Hence, it is PSPACE-complete.

- *Planar arm.* Consider a planar arm consisting of arbitrarily many links serially connected by revolute joints such that all the links are constrained to move in the plane. Planning a free path for such an arm among a finite number of polygonal obstacles, between any two given configurations, is PSPACE-hard [Joseph and Plantiga, 1985].

The following result provides an upper bound on path planning complexity:

A free path in a configuration space of any fixed dimension m, when the free space is a set defined by n polynomial constraints of maximal degree d, can be computed by an algorithm whose time complexity is exponential in m and polynomial in both n ("geometrical complexity") and d ("algebraic complexity").

This bound was first established by Schwartz and Sharir [Schwartz and Sharir, 1983b], who gave an exact cell decomposition method based on the Collins decomposition algorithm. This method, which we will present in Chapter 5, is twice exponential in m. The most efficient algorithm to date is due to Canny [Canny, 1988] who proposed a roadmap method that is singly-exponential in m. Both methods reduce the planning problem to the problem of deciding the satisfiability of sentences in the first-order theory of the reals. As we will see in several chapters, the theory of the reals has been a powerful tool in establishing upper bounds for motion planning problems. Nonetheless, there exist no implementations of the decision algorithms that are fast enough to constitute practical solutions to these problems.

The above results strongly suggest that the complexity of path planning increases exponentially with the dimension of the configuration space. Although this statement is likely to be true for most robotic systems, there nevertheless exist simple settings for which it does not hold. For example, a polynomial time algorithm has been proposed to plan the paths of a planar arm with arbitrarily many joints within a circle and without obstacles other than the circle [Hopcroft, Joseph and Whitesides, 1985] [Kantabutra and Kosaraju, 1986].

On the other hand, if the dimension of the configuration space is relatively small and/or we accept incomplete planning algorithms (i.e. algorithms that may fail to generate a path, even if one exists), specific algorithms of reasonable complexity can be designed. For example, if the dimension of configuration space is limited to two, it has been shown that, under reasonable assumptions, the basic motion planning problem can be solved in time $O(\lambda_s(n)\log^2 n)$, where n is the number of polynomial constraints defining the free space, s is a parameter depending in particular on the maximum degree of these constraints, and $\lambda_s(n)$ is an almost linear function of n (the length of (n, s) Davenport-Schinzel sequences) [Guibas, Sharir and Sifrony, 1988].

Interesting results have also been obtained with moving obstacles:

Planning the motion of a rigid object translating without rotation in three dimensions among arbitrarily many moving obstacles that may both translate and rotate is a PSPACE-hard problem if the velocity modulus of the object is bounded, and an NP-hard problem otherwise.

This result was established in [Reif and Sharir, 1985]. A simpler problem — planning the motion of a point in the plane, with bounded velocity modulus, among arbitrarily many convex polygonal obstacles moving at constant linear velocity without rotation — has also been shown to be NP-hard [Canny, 1988]. However, if the velocity of the robot is not constrained and if the motions of the obstacles can be described algebraically, the problem admits a polynomial time algorithm.

Concerning motion planning with uncertainty, the following result has been established by Canny and Reif [Canny and Reif, 1987]:

Planning compliant motions for a point in the presence of uncertainty, in a three-dimensional polyhedral configuration space with an arbitrarily large number of faces, is a nondeterministic exponential time hard (NEXPTIME-hard) problem.

This result means that the problem is expected to require at least exponential time on a nondeterministic machine, hence doubly-exponential time on a deterministic machine. Previously, Natarajan [Natarajan,

1986] had shown that this problem is PSPACE-hard. With a few assumptions allowing the reduction of the planning problem in an m-dimensional configuration space to a decision problem in the theory of the reals, an r-step plan (if any) can be generated by an algorithm of complexity $2^{2^{O(rnm)}}$, where n is the number of polynomial constraints defining the boundary of the C-obstacle region [Canny, 1989]. More tractable subclasses of motion planning in the plane in the presence of uncertainty have been identified.

7 Reduction of Complexity

The complexity results surveyed in the previous section give strong evidence that the time required to solve a motion planning problem increases quickly with the dimension of the configuration space, the number of polynomial constraints on the robot's motion, and the degree of these constraints. Therefore, it is important that these parameters be as small as possible.

To that purpose, one may approximate the real problem by a simplified, but still realistic problem, before submitting it to the motion planner. If the robot task is reasonably well understood and if it is acceptable to trade some generality against better time performance, such a simplification is almost always feasible. The simplified problems can be provided by the user, or automatically generated by the planner from the original input description.

The following subsections present techniques for approximating motion planning problems into simpler ones. Most of the discussion focuses on planning free paths for rigid objects and robot arms. However, some of the techniques can also be applied to other motion planning problems.

7.1 Projection in Configuration Space

One can reduce the dimension of the configuration space $\mathcal{C}$ by replacing the robot $\mathcal{A}$ by the surface or volume it sweeps out when it moves along r independent axes. This corresponds to projecting the m-dimensional configuration space $\mathcal{C}$ along r of its dimensions. Let $\mathcal{C}'$ be the projected space. Its dimension is $m - r$. The C-obstacles in $\mathcal{C}$ are projected into $\mathcal{C}'$, and path planning is conducted in $\mathcal{C}'_{free}$, the free space in $\mathcal{C}'$.

Examples: - Consider a two-dimensional object $\mathcal{A}$ that can both translate and rotate freely in the plane. $\mathcal{A}$ can be replaced by a disc $\mathcal{A}'$ containing $\mathcal{A}$ (ideally, this disc is bounded by the minimum spanning circle of $\mathcal{A}$, i.e. the smallest circle that encloses all the points of $\mathcal{A}$). This approximation reduces the configuration space dimension from 3 to 2. It is a realistic one only when $\mathcal{A}$ is not an elongated object.

- Consider a robot arm with six revolute joints, the axes of the last three joints intersecting at a single point. It is rather typical to replace the last three links, the end-effector and the payload by the volume they sweep out when the last three revolute joints span their angular ranges. This reduces the configuration space dimension from 6 to 3. This approximation is realistic only when the payload is not a long object. ∎

If a free path τ is found in $\mathcal{C}'$, it determines a family of free paths in $\mathcal{C}$. The configuration parameters not specified in τ can be selected arbitrarily. But, if no free path is found in $\mathcal{C}'$, this does not guarantee that there is no free path in $\mathcal{C}$.

If most of the motions of $\mathcal{A}$ occur in a rather uncluttered workspace and if the ratio of the area/volume of $\mathcal{A}$ by the area/volume of $\mathcal{A}'$ is close to 1, it is probably more efficient to plan a motion for $\mathcal{A}'$, rather than for $\mathcal{A}$. It is also possible to build a two-stage planner that plans the path of $\mathcal{A}$ only when it previously failed to plan the path of $\mathcal{A}'$.

7.2 Slicing in Configuration Space

Another way to reduce the dimension of configuration space is to consider a cross-section of dimension $m - r$ through $\mathcal{C}$ and to plan a free path in the space defined by this cross-section. This corresponds to forbidding the motion of $\mathcal{A}$ along r axes.

Examples: - Let $\mathcal{A}$ be a free-flying three-dimensional object. In order to plan a free path from the initial configuration $\mathbf{q}_{init} = (\mathcal{T}_{init}, \Theta_{init})$ to the goal configuration $\mathbf{q}_{goal} = (\mathcal{T}_{goal}, \Theta_{goal})$, one may consider the intermediate configuration $\mathbf{q}_{inter} = (\mathcal{T}_{goal}, \Theta_{init})$ at which $\mathcal{A}$ has the same orientation Θ_{init} as at the initial configuration and the same position $\mathcal{T}_{goal}$ as at the goal configuration. Then, the planner may attempt to generate a first path from $\mathbf{q}_{init}$ to $\mathbf{q}_{inter}$ with a fixed orientation of $\mathcal{A}$ and a second path from $\mathbf{q}_{inter}$ to $\mathbf{q}_{goal}$ with a fixed position of $\mathcal{A}$. Each of the two paths lies in a three-dimensional cross-section of the configuration

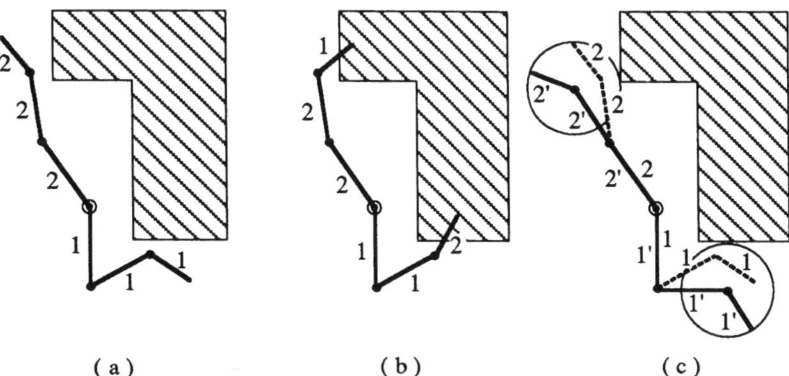

Figure 15. A planar three-joint arm has to move from the initial configuration marked **1** to the goal configuration marked **2**. Both configurations are shown in Figure a. The double-circled joint is fixed and there is a single obstacle. When the first two links are in the initial (resp. goal) configuration, and the third link in the goal (resp. initial) configuration, the arm collides with the obstacle as shown in Figure b. Therefore, the slicing technique alone does not solve this example. However, one can define two intermediate configurations (marked **1′** and **2′** in Figure c) which are close to the initial and goal configurations, respectively, such that the projection technique of Subsection 7.1 applies from **1′** to **2′**, and the slicing technique applies from **1** to **1′** and from **2′** to **2**.

space.

- Let us consider a slightly more complicated example, requiring the projection and slicing techniques to be combined. The robot is a planar arm with three revolute joints (see Figure 15). The initial and goal configurations of the arm, $\mathbf{q}_{init}$ and $\mathbf{q}_{goal}$, are marked **1** and **2** in the figure. Neither the projection technique of Subsection 7.1 applied to the third joint, nor the slicing technique applied by rigidifying the third joint allows us to plan a free path. Indeed, the initial (resp. goal) orientation of the third link cannot be achieved without collision at the goal (resp. initial) orientation of the first two links. However, the projection technique applied to the third joint makes it possible to generate a path from a free configuration $\mathbf{q}'_{init}$ (marked **1′** in the figure) close to $\mathbf{q}_{init}$, to a free configuration $\mathbf{q}'_{goal}$ (marked **2′**) close to $\mathbf{q}_{goal}$. Let $\mathcal{C}'$ be the configuration space of the first two links. $\mathbf{q}'_{init}$ (resp. $\mathbf{q}'_{goal}$) can be selected as a configuration that (1) projects in the free subspace of $\mathcal{C}'$ at a configuration close to the projection of $\mathbf{q}_{init}$ (resp. $\mathbf{q}_{goal}$), and (2) produces the same orientation of the third link with respect to the second link as $\mathbf{q}_{init}$ (resp. $\mathbf{q}_{goal}$). Then, it remains to plan a free path from $\mathbf{q}_{init}$ to $\mathbf{q}'_{init}$, and another one from $\mathbf{q}'_{goal}$ to $\mathbf{q}_{goal}$. This can be done by rigidifying the third joint and using the slicing technique. In the particular case of Figure 15, these two paths are rotations of the second joint. The orientation of the third link is changed along the path from $\mathbf{q}'_{init}$ to $\mathbf{q}'_{goal}$. ■

If planning in a cross-section of configuration space produces a free path, then this path is a free path in $\mathcal{C}$. But, if planning fails in a cross-section, this does not imply that there is no free path in $\mathcal{C}$.

7.3 Simplification of the Shape of Objects

The complexity of motion planning increases with the number of polynomial constraints defining the free space and the maximal degree of these constraints. One way to simplify these constraints is to approximate the shapes of both the robot and the obstacles by reducing the number of surfaces defining their boundary and/or the degree of the equations of these surfaces.

Simple and often efficient techniques consist of approximating the objects by rectangloids or balls. One can also approximate objects by more sophisticated polygons or polyhedra. This reduces the maximal degree of the constraints, but if one is tempted to increase the accuracy of the approximation, it ultimately results in an increase in the number of curves/surfaces, i.e. the number of constraints.

One may usefully distinguish between two kinds of approximation: *bounding* and *bounded* ones. A bounding approximation consists of replacing objects by new ones which completely contain the original ones. If a free path is generated with such an approximation, it is also a free path with the original objects (but the converse is not true). A bounded approximation consists of replacing objects by new objects completely contained in the original ones. If no free path is generated with such an approximation, it implies that there is no free path with the original objects (again, the converse is not true). Thus, bounded approximation can be used to detect that a particular problem has no solution.

One can generalize the above ideas and consider several bounding and bounded approximations of increasing precision. Then the planner it-

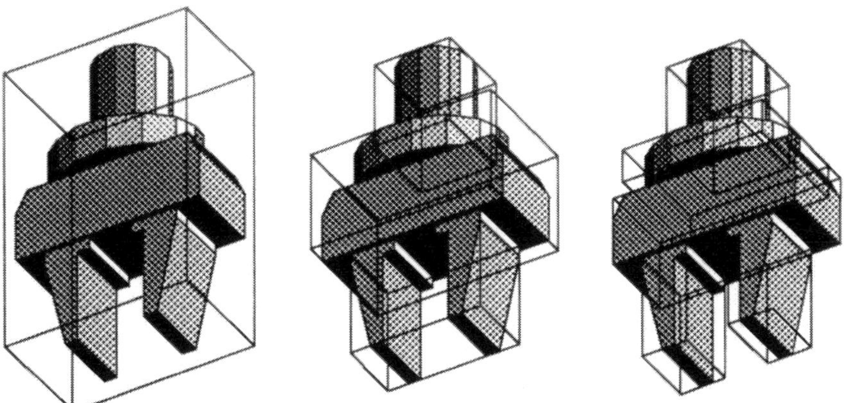

Figure 16. This figure shows three bounding approximations (from coarse to fine) of a robot gripper using rectangloids [Pasquier, 1989]. The motion planner may use such models in sequence until it finds a path.

eratively uses finer and finer approximations of the objects, until a free path is generated or there is evidence that no free path exists. Figure 16 illustrates a sequence of bounding approximations for a robot gripper.

7.4 Focusing Attention on a Subset of the Workspace

The techniques introduced in the three previous subsections are "simplification" heuristics. They exploit the fact that in most cases, motions take place in a relatively uncluttered space, so that there still remain solution paths after the various approximations have eliminated some. In many tasks, however, there are isolated localities that are quite cluttered and/or where contacts with obstacles have to be made (e.g. grasping, part mating). Simplification heuristics cannot be applied to these cluttered localities.

If cluttered localities can be identified in advance, one can break the original planning problem into subproblems. The subproblems involving cluttered localities only require a restricted subset of the workspace to be considered. Hence, the number of constraints on the motions is considerably reduced with respect to the original problem.

Example: Consider the problem of adding a part to an assembly with a robot arm. It can be broken down into five subproblems: (1) move the arm to a configuration near the part, (2) grasp the part, (3) transfer the part to the current subassembly, (4) mate the part with the subassembly, and (5) move away to some rest configuration. In general, only subproblems (2) and (4) involve cluttered localities. In both subproblems, the motions usually have small amplitude, so that motion planning can be restricted to a small subset of the workspace. For instance, in (4), one can consider the part to be positioned (and perhaps also the gripper) as the robot, rather than the whole arm, and restrict the workspace of this object to a small volume around the assembly. If one wants to be sure that the generated plan is safe, the motion of the whole arm can then be checked for possible collision. In case a collision is possible, the planner would have to find another solution for (4). ■

It can be further observed that in many cases cluttered localities are stereotyped situations (e.g. a mobile robot moving through a door or turning in a corridor). It may be worthwhile to attempt to recognize these situations and to apply specific (and more efficient) planning techniques to them. Along this line, Maddila and Yap explored the problems of moving a polygon around the corner in a corridor [Maddila and Yap, 1986] and through a door [Yap, 1987b]. Specific planning techniques applicable to stereotyped situations are called *local experts*.

8 Relation to Other Problems

An automatic motion planner can be used in a variety of ways. For instance, it can be included in an off-line robot programming system [Latombe et al., 1985] in order to simplify the task of describing robot trajectories. It can also be part of an interactive graphic process planning system (e.g. for planning the machining of a mechanical part, the assembly of an electromechanical device, or the construction of a power plant), and be used to check the geometrical feasibility of the planned operations (e.g. drilling a hole, mating two parts, erecting a pipe). More generally, it can be used to automatically generate graphic animated scenes. Nonetheless, beyond these "short-term" applications, the ultimate goal of automating motion planning is to create autonomous robots.

As mentioned earlier in this chapter, automatic path planning is a critical problem in the creation of autonomous robots, but it is not the only one. Other major problems include real-time motion control, sensing,

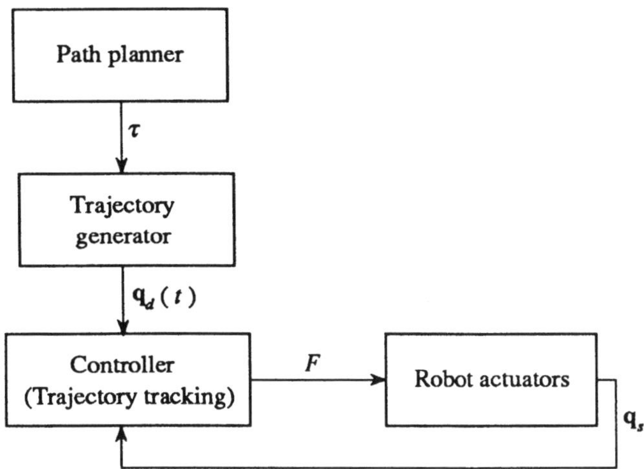

Figure 17. The path τ produced by the path planner is input into a trajectory generator that determines the time dependence of the configuration parameters. The output of the trajectory generator — the desired configurations $\mathbf{q}_d$ as a function of time — is fed into the controller. The controller computes the force F to be exerted by each actuator from the deviation of the current configuration $\mathbf{q}_s$ as measured by the sensors, relative to $\mathbf{q}_d$.

and task-level planning. Because of the potentially complex interactions among them, the design of a truly autonomous robot will certainly require all these problems to be solved concurrently. But, today, our current understanding of the issues is insufficient to present a comprehensive picture of such a design. In the following subsections we therefore limit our presentation to a brief analysis of some possible interactions between motion planning and the three other problems mentioned above.

8.1 Interaction with Real-Time Motion Control

A motion plan specifies a motion to be executed by the robot. For example, a path specifies a continuous sequence of configurations that the robot should traverse toward the goal configuration. The basic task of the real-time controller is to make the robot execute the motion plan, e.g. follow the generated path.

Let us first assume that the motion plan is a free path τ. The task of the controller is to transform τ into forces[5] to be exerted by the robot actuators. Typically, this transformation is broken down into two steps. The first step, called *trajectory generation*, consists of deciding on the velocity profile along the path. It can be done prior to motion execution. The second step, called *trajectory tracking*, consists of computing the forces to be exerted by the actuators at each time in order to perform the desired motion. The trajectory tracking step may use the dynamic equation of the robot — the equation expressing that the forces applied by the actuators equal the resultant of the various forces acting on the robot during motion, i.e. the gravitational, frictional, inertial, centrifugal and Coriolis forces — to compute the force that has to be delivered by each actuator. If the dynamic equation used by the controller was a perfect model, no feedback would be necessary. However, due to various disturbances, sensing is necessary to determine the deviation between the desired state and the actual state of the robot. While the motion is being executed, the controller computes the actuator forces which tend to eliminate this deviation at an update rate which typically ranges between 10 and 1000 Hz. Figure 17 shows the relationship between the path planner, the trajectory generator, the controller and the robot.

The simple architecture of Figure 17 is a classical one. Its advantage is to break the problem of deciding on a motion into clearly defined subproblems. But it may lead to rather inefficient motions. Even if the controller attempts to minimize execution time, the geometry of the paths may not allow the controller to fully exploit the capabilities of the actuators. A better approach is probably to take the dynamic equation of the robot, together with the actuators' saturation limits, into account during motion planning in order to generate paths allowing (quasi-)time-optimal motions to be executed. Several techniques have been recently proposed for implementing this approach [Donald and Xavier, 1989] [Barraquand, Langlois and Latombe, 1989a] [Shiller and Dubowsky, 1989] [Jacobs et al., 1989]. Such techniques may become particularly important when the robot operates in a dynamic workspace among mobile obstacles, when the task has to be accomplished within tight time bounds (e.g. because it interacts with an external process), and/or when the robot's productivity has to be optimized (e.g. in manufacturing).

---

[5] Here we use the word "force" in the generalized sense. It is either a force or a torque depending on the nature of the actuator.

Paths are the simplest sort of motion plans. We saw in Subsection 5.3 that in the presence of uncertainty, motion plans may include sensory-based motion commands, e.g. force-compliant motion commands, which are more tolerant to errors than classical position-controlled motion commands. The availability of such commands may make the task of the motion planner considerably simpler [Shekhar and Khatib, 1987]. The concurrent design of sensory-based motion primitives and motion planning methods using them would be a major step toward a better integration of motion planning and control. It has received little attention so far.

8.2 Interaction with Sensing

In the basic motion planning problem, we assumed that the robot has complete and exact knowledge of its environment. As mentioned in Subsection 5.3, this assumption is not always realistic. There are many real-life problems where the geometry of the workspace can only be partially known at planning time. Since planning requires the robot to have a model of its environment, the more incomplete this model, the more limited the role of planning.

If uncertainty in the prior knowledge is small, it is reasonable to anticipate all possible contingencies and to generate sensory-based motion plans that can deal with them (see Subsection 5.3). Sensing is used to guide the motions and monitor their execution. Planning makes sure that the motion commands will make enough information available at execution time so that the controller will be able to both monitor progress toward the goal and recognize goal achievement reliably.

On the other hand, if the robot has no prior knowledge about its environment, planning is useless. Then, the robot must rely heavily on sensing to acquire information while it is moving and it must react to the sensory data. For example, if the robot is equipped with proximity range sensors, a potential field method (see Subsection 4.3) can be applied on-line to guide the motion. The input goal configuration is used to create the attractive potential, while the data provided by the proximity sensors are used to create the repulsive potential. The negated gradient is treated as an external force, and the robot is controlled to comply to that force (i.e. to move along the flow of the negated gradient field) [Khatib, 1986]. There is no prior planning, and the robot can easily get

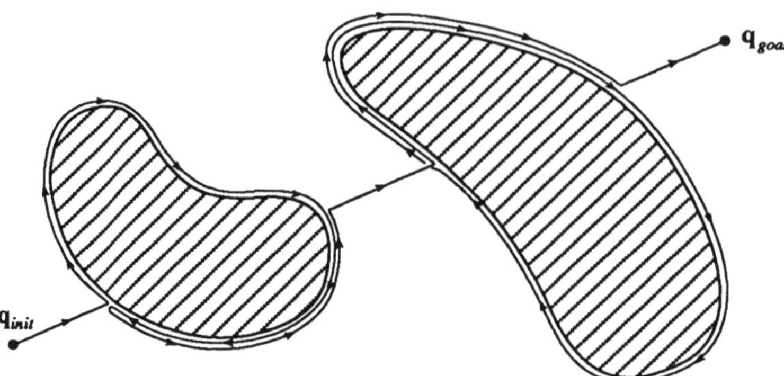

Figure 18. The robot has no prior knowledge of the obstacles. It follows the straight line segment joining $\mathbf{q}_{init}$ to $\mathbf{q}_{goal}$ until it hits an obstacle. Then it follows the C-obstacle's boundary, while keeping track of the intersections between this boundary and the line segment joining $\mathbf{q}_{init}$ and $\mathbf{q}_{goal}$. When it comes back to the hitting point, the robot returns to the intersection point that is closest to $\mathbf{q}_{goal}$ by following the shortest path along the C-obstacle's boundary. From this point, it resumes moving along the straight line segment toward $\mathbf{q}_{goal}$.

trapped into a local minimum of the potential field, e.g. a minimum created by an obstacle concavity. Escaping from such a local minimum requires some trial-and-error exploration by the robot.

Another reactive scheme has been proposed by Lumelsky when no prior knowledge is available [Lumelsky, 1987a]. The configuration space is assumed to be two-dimensional and the robot is equipped with a position sensor and a peripheral touch sensor that allows it to track the boundaries of the obstacles. Lumelsky described several specific algorithms which are variants of each other. One of them, illustrated in Figure 18, consists of having the robot move along a straight line toward the goal configuration $\mathbf{q}_{goal}$. If the robot hits an obstacle, it shifts to moving along the boundary of the corresponding C-obstacle in a predetermined direction (e.g. clockwise) until it eventually comes back to the collision point. During this motion, it identifies the intersection point of the C-obstacle's boundary and the line segment joining the initial and goal configurations that is closest to the goal. After having made the closed-loop path around the C-obstacle it returns to this point by following the

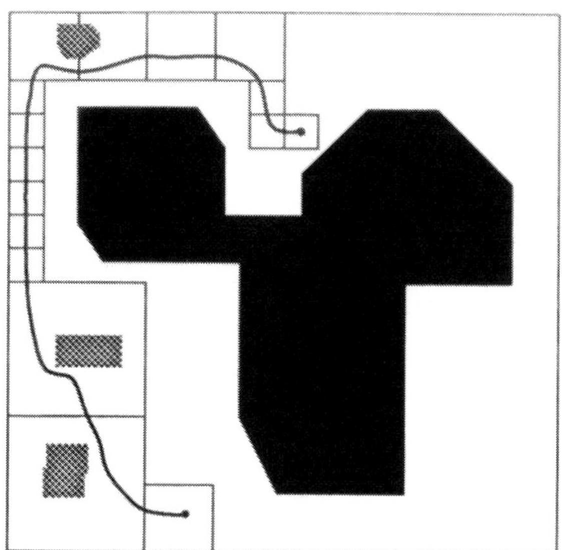

Figure 19. The region shown black is the C-obstacle region known at planning time. The regions shown grey depict unknown C-obstacles that are actually present. Knowledge available at planning time is used to compute a quadtree decomposition of the configuration space and extract a channel composed of adjacent cells lying outside the known C-obstacle region. At execution time, a sensory-based potential field is used to guide the robot through the channel away from the unexpected obstacles.

---

shortest path along the C-obstacle's boundary. From there it resumes moving along a straight line toward the goal. This simple algorithm is guaranteed to attain $\mathbf{q}_{goal}$ whenever every C-obstacle is bounded by a simple closed curve of finite length and any straight line segment intersects a finite number of C-obstacles. The length of the generated path is upper bounded by $D + 1.5\Sigma_i p_i$, where D is the distance between $\mathbf{q}_{init}$ and $\mathbf{q}_{goal}$ and the p_i's are the perimeters of the C-obstacles intersecting the disc of radius D centered at $\mathbf{q}_{init}$. Some steps toward generalizing this kind of algorithm to configuration spaces of dimensions greater than two have been made in [Sun and Lumelsky, 1989] and [Cheung and Lumelsky, 1990].

The extreme case where the robot has no prior knowledge of its environment is rare. A more realistic situation is when the robot knows the shape and location of the largest objects in its workspace, but has incomplete knowledge about smaller objects. An approach proposed in [Choi, Zhu and Latombe, 1989] to deal with such a situation is the following:

- first, a cell decomposition method is used to plan a channel among the known C-obstacles between the initial and goal configurations.
- Second, a potential field collision-avoidance method is used to guide the robot in the channel at execution time.

The potential field combines a dynamic attractive potential, generated by a target configuration that moves toward the goal configuration while the robot progresses in the channel, a repulsive potential, generated by the boundary of the channel (so that the robot's configuration does not escape the channel), and a repulsive potential, computed from the data provided by proximity range sensors. Figure 19 illustrates this approach in a two-dimensional configuration space. In this example, the channel was generated from a quadtree decomposition (the known C-obstacle region is shown black). The C-obstacles shown grey were not known at planning time and were detected during the motion. As long as the unexpected obstacles are small and sparsely distributed, the robot is unlikely to get trapped in a local minimum. However, there is no guarantee that this will never happen. When a local minimum is encountered, some replanning may be necessary using an updated model of the workspace.

There is a wide variety of possible software architecture interweaving motion planning and execution when prior knowledge is incomplete. The "best" architecture depends on both the nature and the extent of the lack of knowledge.

8.3 Interaction with Task-Level Planning

A motion planning goal is specified as a spatial arrangement of physical objects. Achieving such a goal contributes to the achievement of a more global task. For instance, the robot may be a troubleshooter whose task is to diagnose faults in pieces of electromechanical equipment and to repair them. Finding a fault typically requires the equipment to be inspected and operated, measuring devices to be connected, and probes to be inserted and removed. Each such operation can be specified by a spatial arrangement of objects, but deciding which operations must be

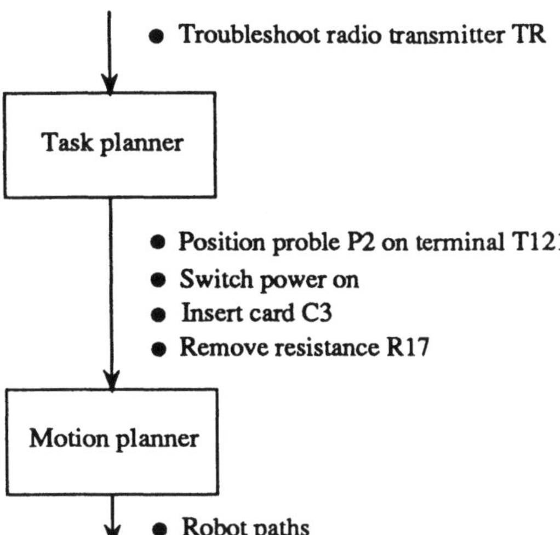

Figure 20. This figure illustrates the relationship between task planning and motion planning. The input to the task planner is a task description, e.g. "troubleshoot radio transmitter TR". Its output is a set of operations, e.g. "position proble P2 on terminal T121", "switch power on", "insert card C3", "remove resistance R17". Each operation can be specified as a goal spatial arrangement of objects (e.g. the goal position of the probe). The motion planner generates the motion plan for achieving this goal.

executed and their execution sequence — i.e. *task planning* — requires "higher-level" reasoning and planning capabilities than those possessed by a motion planner.

Although much research still remains to be done, Artificial Intelligence (AI) has produced a variety of methods for reasoning about tasks and for planning the high-level operations needed to accomplish them [Nilsson, 1980] [Genesereth and Nilsson, 1987] [Georgeff, 1987]. These methods are readily available to build planners whose outputs are sequences of operations, each specifying a new goal spatial relationship to the motion planner. Figure 20 illustrates the relationship between the task planner and the motion planner.

However, an AI task planner typically represents a task using logic-oriented constructs (e.g. predicate calculus language) which contain little information about the actual geometry of the objects and the spatial relations among them. Therefore, a plan generated by the task planner may turn out to be infeasible when developed at the motion level. For instance, there may exist no path for a probe to the terminal at which it should be connected. In such a situation, the task planner would have to generate another plan. Important questions are: What kind of pertinent information should the motion planner feed back to the task planner? How can the task planner use this information to generate another plan?

In addition, the behavior of the equipment to be repaired might not be stationary, hence requiring the troubleshooting operations to be executed within strict time constraints. For instance, a probe may have to be moved from one position to another so that two measurements can be done within a short interval of time. However, the durations of the motions are typically not known with good precision until the motion plans have been generated.

This discussion suggests that integrating task planning and motion planning is a much more ambitious enterprise than Figure 20 suggests. Little has been done so far in this domain.

9 Literature Landmarks

We close this introductory chapter with a brief discussion of key publications that can be considered to be landmarks in the development of robot motion planning. The citations are given approximately in chronological order, so that this section may also be regarded as a short historical account of the development of the motion planning field over the past 20 years. Many other references will be given throughout the book.

Nilsson [Nilsson, 1969] described a mobile robot system (Shakey) with planning capabilities. Although most of the planning done by Shakey used a model of the workspace and the robot actions expressed in the predicate calculus language (Strips planner), Nilsson introduced the visibility graph method for planning Shakey's motions.

Ambler and Popplestone [Ambler and Popplestone, 1975] described a method for inferring the quantitative location of objects in terms of homogeneous coordinate matrices, from input spatial symbolic relations among features of objects. This method was extended later to a larger set of relations and implemented as part of a high-level robot programming language (Rapt) [Popplestone, Ambler and Bellos, 1980].

Liebermann and Wesley [Liebermann and Wesley, 1977] and Lozano-Pérez [Lozano-Pérez, 1976] presented the first attempts to build integrated systems (Autopass and Lama, respectively) for automatically programming robot arms. The input to these systems was the description of a mechanical assembly, in the form of a set of geometrical models of the individual parts and goal assembly relations among the parts. The task of the systems was to automatically generate the robot programs for assembling the parts. Although these systems were never fully implemented, they have contributed in emphasizing the importance of geometrical reasoning in robot planning and in pointing out key motion planning problems in the context of mechanical assembly.

Taylor [Taylor, 1976] investigated the problem of planning motions in the presence of uncertainty. He proposed a method, known as "skeleton refinement", based on numerical uncertainty propagation techniques. A similar approach was also proposed in [Lozano-Pérez, 1976]. An improvement of the method, using symbolic propagation techniques rather than numerical ones, was later described by Brooks [Brooks, 1982].

Udupa [Udupa, 1977] introduced the idea of shrinking the robot to a point in an appropriate space (the term "configuration space" was not used yet). Lozano-Pérez and Wesley [Lozano-Pérez and Wesley, 1979] exploited this idea in a more systematic fashion, and proposed the first path planning algorithm for polygonal and polyhedral robots and obstacles without rotation. Their work is usually considered as the first contribution to "exact" motion planning. Later, Lozano-Pérez borrowed the notion of "configuration space" from Mechanics, and popularized it in motion planning. He also extended the ideas contained in [Lozano-Pérez and Wesley, 1979] and introduced the principle of the approximate cell decomposition approach [Lozano-Pérez, 1981 and 1983].

Reif [Reif, 1979] presented the first theoretical investigation of the inherent computational complexity of the path planning problem, showing that planning a free path in a configuration space of arbitrarily large dimension among fixed obstacles is PSPACE-hard. This work has contributed in attracting the interest of theoretical computer scientists toward motion planning. It has been followed by a flurry of papers on lower-bound complexity for a variety of motion planning problems (see

Section 6). Reif termed this area *Computational Robotics* [Reif, 1987].

Chatila [Chatila, 1982] investigated motion planning with incomplete knowledge for a mobile robot represented as a point in a two-dimensional workspace. He was the first to base his planner on an exact decomposition of the empty subset of the workspace into convex cells (an idea which was extended later to free space in configuration space). The planner was operating on-line, and the decomposition was periodically updated in order to take into account new information provided by the sensors.

Mason [Mason, 1982] stressed the importance of including pushing motions in a variety of manipulation tasks. Although his initial work was mainly aimed at modeling the mechanics of pushing under quasi-static assumptions, it has been followed by other publications by Mason and others directly related to motion planning.

Ó'Dúnlaing and Yap [Ó'Dúnlaing and Yap, 1982] introduced retraction as a new theoretical approach for path planning. Independently, and almost simultaneously, Brooks [Brooks, 1983a] presented a more empirical, "retraction-like" planning method, known as the freeway method.

In 1983 and 1984, Schwartz, Sharir, and Ariel-Sheffi published a series of five papers called the Piano Movers' Problem series [Schwartz and Sharir, 1983a, 1983b and 1983c] [Sharir and Ariel-Sheffi, 1983] [Schwartz and Sharir, 1983d]. The first paper of the series described the first exact method for planning free paths of a polygonal object allowed to both translate and rotate in a two-dimensional polygonal workspace. The second paper established the first upper bound on the time complexity of path planning in a semi-algebraic free space of any fixed dimension.

Laugier and Pertin [Laugier and Pertin, 1983] described an implemented method for planning the grasp position of a mechanical hand with two parallel fingers on an object bounded by planar, cylindrical, and spherical surfaces. In a first processing phase, the planner generates various grasps and assesses their local feasibility using a collection of empirical tests. In a second phase, it checks the global accessibility of each grasp by planning a path in the configuration space of the hand. These ideas were refined later by other authors.

Lozano-Pérez, Mason and Taylor [Lozano-Pérez, Mason and Taylor,

1984] described the preimage backchaining approach to motion planning with uncertainty in control and sensing. For the first time, this approach considered compliant motions in planning. In the original paper, the approach was essentially a theoretical framework, and was not implemented. Since then, the theory and the implementation of preimage backchaining have been investigated further.

Dufay and Latombe [Dufay and Latombe, 1984] described an implemented approach to motion planning with uncertainty based on inductive learning. The approach consists of executing the same task several times and combining the execution traces into a more general strategy. More recently, other researchers have investigated different learning techniques for motion planning and motion skill acquisition.

Gouzènes [Gouzènes, 1984a and 1984b] described the first implemented approximate cell decomposition method to plan the motion of robot arms with revolute joints. His method consists of approximating the free space as a collection of rectangloid cells obtained by dividing the range of possible angular positions of each link into small sub-intervals. The same type of method was later used and refined by several other researchers.

Khatib [Khatib, 1986] pioneered the potential field approach[6]. Although he implemented it as a real-time collision avoidance module in a robot controller, the approach is extendable to motion planning, which has been done since then. Koditschek [Koditschek, 1987] introduced the notion of a "navigation function", a local-minimum-free potential function. Using a variant of the potential field approach, Faverjon and Tournassoud [Faverjon and Tournassoud, 1987] implemented a practical system able to plan the motions of a manipulator arm with eight joints moving among vertical pipes. Barraquand and Latombe combined a potential field method with random techniques to escape from local minima and implemented a planner for generating free paths for robots with many degrees of freedom [Barraquand and Latombe, 1989a].

Laumond [Laumond, 1986 and 1987b] considered the problem of planning free paths for nonholonomic car-like mobile robots. He produced the interesting result that a free path for a free-flying robot in a two-

[6]Actually, Khatib introduced the principle of this approach in earlier publications (see [Khatib, 1980]). However, the paper cited here is the first comprehensive description of the approach.

dimensional workspace can always be transformed into a free path for a nonholonomic car-like robot having the same geometry and moving in the same workspace, by introducing simple backup maneuvers. Li and Canny [Li and Canny, 1989] were the first to apply tools developed in controllability theory for nonlinear systems to nonholonomic robots. These tools make it possible to generalize Laumond's results.

Lozano-Pérez et al. [Lozano-Pérez et al., 1987] described an implemented robot system, Handey, integrating a grasp planner and a path planner. This system is able to plan the motions of a manipulator robot for constructing simple assemblies made of polyhedral objects, and to execute the plans, assuming no uncertainty. Part of the initial knowledge about the workspace is obtained by using a vision system. In addition, the system is capable of planning the re-grasping of an object, if no grasp at the initial location of the object is compatible with a grasp at the goal location.

Exercises

1: Discuss in which respects motion planning differs from: (1) collision checking; (2) collision avoidance.

2: Design an algorithm for checking whether two polygons intersect.

3: Design an algorithm for checking whether two polyhedra intersect.

4: Consider the 3×3 matrix Θ defined in Subsection 3.1 for representing the orientation of a rigid object $\mathcal{A}$ in $\mathcal{W} = \mathbf{R}^3$. What equations relate the 9 parameters of the matrix? How many parameters are needed to represent a rotation? Comment on your results.

5: It is often said that representing the robot as a point in its configuration space transforms the problem of planning the motion of a dimensioned object into the "simpler" problem of planning the motion of a point. Do you agree or disagree with this assertion? Elaborate on your answer.

6: Let $\mathcal{C}$ be the configuration space of a compact rigid object $\mathcal{A}$ moving

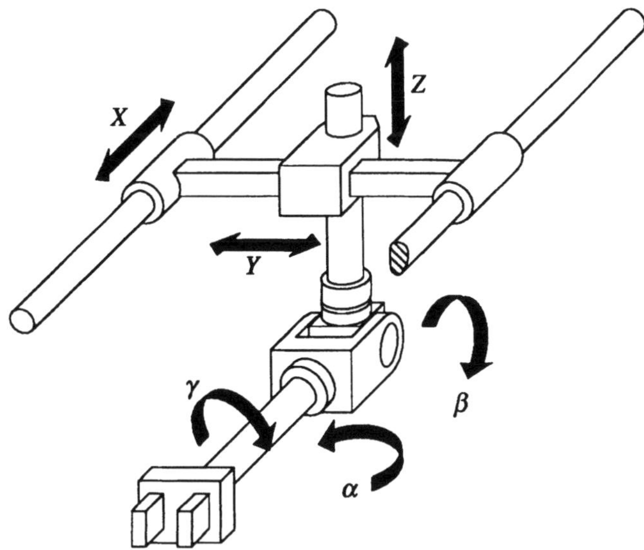

Figure 21. The first three joints of this robot are three perpendicular prismatic joints. The last three joints are revolute joints whose axes intersect at a single point.

freely in $\mathbf{R}^N$. Verify that the map $d : \mathcal{C} \times \mathcal{C} \rightarrow \mathbf{R}$ defined by:

$$\forall \mathbf{q}, \mathbf{q}' \in \mathcal{C} \; : \; d(\mathbf{q}, \mathbf{q}') = \max_{a \in \mathcal{A}} \|a(\mathbf{q}) - a(\mathbf{q}')\|,$$

where $\|x - x'\|$ denotes the Euclidean distance between two points x and x' in $\mathbf{R}^N$, determines a distance in $\mathcal{C}$.

7: Discuss the assumptions under which it is realistic to consider a wheeled mobile robot as a two-dimensional object moving freely in the plane.

8: Consider the simple six-joint gantry-type robot illustrated in Figure 21. The first three joints (denoted by X, Y and Z) are prismatic joints perpendicular to each other; the last three joints (denoted by α, β and γ) are revolute joints whose rotation axes intersect at a single point. Discuss the assumptions under which it is realistic to consider the end-effector of this robot as a free-flying object moving among obstacles in a three-dimensional workspace.

9: Consider the case where p free-flying objects $\mathcal{A}_1, ..., \mathcal{A}_p$ operate in the same workspace $\mathcal{W}$ among q fixed obstacles $\mathcal{B}_1, ..., \mathcal{B}_q$. Treat these moving objects as a multi-bodied robot $\mathcal{A} = \{\mathcal{A}_1, ..., \mathcal{A}_p\}$ whose configuration space is denoted by $\mathcal{C}$ (see Subsection 5.1). Specify the C-obstacles and the free space in $\mathcal{C}$ as we did for a single rigid object in Section 3.

10: Consider a point moving in $\mathbf{R}^2$. Assume that its motions are imperfectly controlled, but that its actual position is always within a distance less than a known constant ε from the position where it is desired to be. How must the obstacles be grown so that path planning is guaranteed to generate a path whose execution will be collision-free? Illustrate with an example the need for more sophisticated motion planning methods capable of generating sensory-based motion plans, rather than just paths.

11: Consider a part mating operation. What are the sources of errors that may prevent a position-controlled motion to execute the operation successfully? Is it reasonable to bound these errors by uncertainty intervals and to plan a robust motion plan as suggested in Subsection 5.3?

12: Propose and discuss one or two extensions of the basic motion planning problem not presented in Section 5.

13: Define (in a sentence) the words "kinematics" and "dynamics". [Hint: Look into a good dictionary.]

14: Identify and discuss briefly some typical motion planning problems for: (1) a mobile robot carrying objects in an office environment; (2) a space platform robot actuated by thrusters and equipped with two arms for assembling an orbital platform.

15: What does it mean for an algorithm to have time complexity $O(1)$? $n^{O(1)}$? $O(2^n)$? $2^{O(n)}$?

16: Sketch a few techniques for approximating objects by simple geometrical models and possibly reducing the complexity of motion planning.

17: Compare the projection and slicing techniques for reducing the dimension of configuration space (see Subsections 7.1 and 7.2). Are they *both* useful? Why?

18: Propose a software architecture for interweaving motion planning and motion execution when prior knowledge is incomplete.

19: Compare motion planning to the type of planning performed by an Artificial Intelligence planner such as Strips [Nilsson, 1980]. Describe, in general terms, a system integrating the two types of planning.

Chapter 2

Configuration Space of a Rigid Object

In Chapter 1 we introduced configuration space as a space in which the robot maps to a point. The mathematical structure of this space, however, is not completely straightforward, and deserves some specific consideration. The purpose of this chapter and the next one is to provide the reader with a general understanding of this structure when the robot is a rigid object not constrained by any kinematic or dynamic constraint. This chapter mainly focuses on topological and differential properties of the configuration space. More detailed algebraic and geometric properties related to the mapping of the obstacles into configuration space will be investigated in Chapter 3.

Although our primary goal at this stage is to provide the necessary background for investigating the basic motion planning problem, our presentation is substantially broader. Indeed, some concepts which are not directly related to this problem contribute to give an articulate vision of configuration space. These concepts may also help the reader to realize the depth of motion planning in general. Some other notions, which will be applied when we investigate extensions of the basic problem, are easily introduced here.

In Section 1 we define the configuration space $\mathcal{C}$ of a rigid object. In Section 2 we represent this space by "embedding" it in a Euclidean space $\mathbf{R}^M$. In Section 3 we discuss the local topological and differential structure of $\mathcal{C}$ and we show that $\mathcal{C}$ is a "smooth manifold", i.e. a space that is locally like $\mathbf{R}^m$, where m is the dimension of $\mathcal{C}$ $(m < M)$. In Section 4 we discuss several ways of representing a configuration as a list of independent real parameters (e.g. stereographic projection, Euler angles, quaternion). In Section 5 we develop the notion of a path and we show how the global topology of $\mathcal{C}$ differs from that of $\mathbf{R}^m$. In Section 6 we present the notion of a tangent space, which makes it possible to talk about directions in configuration space. The topology of $\mathcal{C}$ is defined in Section 3 as the topology induced by the Euclidean metric in $\mathbf{R}^M$ (the space in which we previously embedded $\mathcal{C}$). In Section 7 we propose other metrics that induce the same topology. In Section 8 we present several particular cases that have been overlooked in the previous sections; some of them yield particularly simple configuration spaces. In Section 9 we discuss set-theoretic and topological properties of the image of the workspace obstacles in configuration space. Finally, in Section 10 we describe how forces map into configuration space.

A large part of this chapter is built upon elementary notions in Differential Geometry and Topology. These notions are developed in more depth in textbooks such as [Guillemin and Pollack, 1974] and [Spivak, 1979].

1 Definition

We consider a rigid object $\mathcal{A}$ — the robot — moving in a physical workspace $\mathcal{W}$ (see Figure 1). We represent $\mathcal{W}$ as the N-dimensional Euclidean space $\mathbf{R}^N$, where $N = 2$ or 3, equipped with a fixed Cartesian coordinate system, or frame, denoted by $\mathcal{F}_{\mathcal{W}}$. We represent $\mathcal{A}$ at a reference position and orientation as a *compact* subset of $\mathbf{R}^N$. A moving frame, $\mathcal{F}_{\mathcal{A}}$, is attached to $\mathcal{A}$ so that each point in the robot has fixed coordinates in $\mathcal{F}_{\mathcal{A}}$. The origins of $\mathcal{F}_{\mathcal{W}}$ and $\mathcal{F}_{\mathcal{A}}$ are denoted by $O_{\mathcal{W}}$ and $O_{\mathcal{A}}$, respectively. $O_{\mathcal{A}}$ is called the **reference point** of $\mathcal{A}$. We assume that $\mathcal{A}$ has no symmetry[1] — hence, in particular, it is not a single point.

Strictly speaking, $\mathcal{W}$ is a "physical space", while $\mathbf{R}^N$ is a "mathematical

[1] Particular cases induced by symmetry will be examined in Section 8.

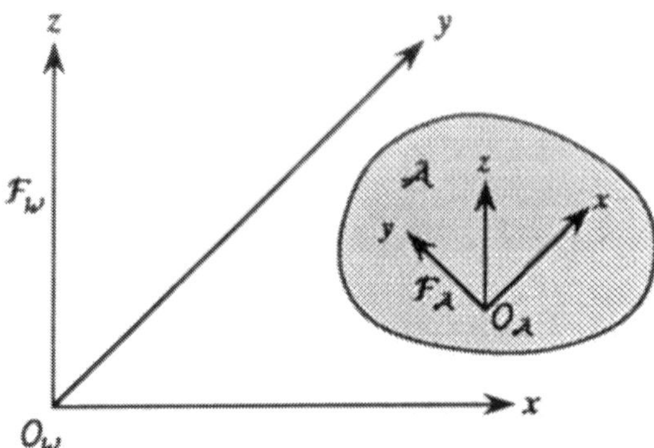

Figure 1. The robot $\mathcal{A}$ moves in a workspace $\mathcal{W} = \mathbf{R}^N$, with $N = 2$ or 3. $\mathcal{A}$ is modeled as a compact subset of $\mathbf{R}^N$. A fixed Cartesian frame $\mathcal{F}_\mathcal{W}$ is embedded in $\mathcal{W}$. A moving Cartesian frame $\mathcal{F}_\mathcal{A}$ is embedded in $\mathcal{A}$. The configuration of $\mathcal{A}$ specifies the position and orientation of $\mathcal{F}_\mathcal{A}$ with respect to $\mathcal{F}_\mathcal{W}$.

space" representing the former. However, we will not make this distinction any longer and we will use the same notation, $\mathcal{W}$, to denote both spaces.

The configuration space of $\mathcal{A}$ is defined as follows:

DEFINITION 1: *A* **configuration**[2] $\mathbf{q}$ *of $\mathcal{A}$ is a specification of the position and orientation of $\mathcal{F}_\mathcal{A}$ with respect to $\mathcal{F}_\mathcal{W}$. The* **configuration space** *of $\mathcal{A}$ is the space $\mathcal{C}$ of all the possible configurations of $\mathcal{A}$. A unique configuration of $\mathcal{C}$ is arbitrarily selected and is called the* **reference configuration** *of $\mathcal{A}$. It is denoted by* $\mathbf{0}$.

The configuration space $\mathcal{C}$ is intrinsically independent of the choice of the frames $\mathcal{F}_\mathcal{A}$ and $\mathcal{F}_\mathcal{W}$. Only the representation of $\mathcal{C}$ depends on these frames (see Section 2).

The subset of $\mathcal{W}$ occupied by $\mathcal{A}$ at configuration $\mathbf{q}$ is denoted by $\mathcal{A}(\mathbf{q})$. A point a of $\mathcal{A}$ is denoted by $a(\mathbf{q})$ in $\mathcal{W}$ when $\mathcal{A}$ is at configuration $\mathbf{q}$. Thus, for any two configurations $\mathbf{q}$ and $\mathbf{q}'$, $a(\mathbf{q})$ and $a(\mathbf{q}')$ are the same

[2]In some texts, the word "placement" is used instead of "configuration". The word "configuration" has been widely used in Mechanics for a long time. The use of the word "placement" seems more recent, with origins in Computational Geometry.

point in $\mathcal{A}$, but in general do not coincide in $\mathcal{W}$. More generally, if x is a geometric feature (e.g. a point, a vector, a set of points) in $\mathcal{A}$, $x(\mathbf{q})$ denotes the same feature in $\mathcal{W}$ when $\mathcal{A}$ is at configuration $\mathbf{q}$.

It is sometimes convenient to interpret a configuration as a rigid body transformation (also called a positive isometry), i.e. a transformation that preserves distance and orientation. The transformation $TR_{\mathbf{q}}$ corresponding to the configuration $\mathbf{q}$ in $\mathcal{C}$ rotates and translates $\mathcal{A}(0)$, i.e. the robot at its reference configuration, to $\mathcal{A}(\mathbf{q})$; we can write $TR_{\mathbf{q}}(\mathcal{A}(0)) = \mathcal{A}(\mathbf{q})$, where $TR_{\mathbf{q}}(x)$ denotes the result of applying $TR_{\mathbf{q}}$ to a geometrical feature x. For every $\mathbf{q} \in \mathcal{C}$, there exists a unique rotation r and translation t, such that $TR_{\mathbf{q}} = t \circ r$, where $t \circ r$ denotes the composition of t and r, with r applied first. $TR_{\mathbf{q}}$ is a continuous map of $\mathcal{W}$ onto itself. The space of rigid body transformations is known to have the structure of a non-commutative group, with the composition of two transformations as the internal binary operation.

2 Embedding in Euclidean Space

In this section we represent $\mathcal{C}$ as a subset of a Euclidean space $\mathbf{R}^M$, where M depends on the dimension N of $\mathcal{W}$. This space will be used in the next section to introduce the differential properties of $\mathcal{C}$. It will be called the *ambient space* of $\mathcal{C}$, and the representation of $\mathcal{C}$ will be referred to as the *embedding* of $\mathcal{C}$ in $\mathbf{R}^M$.

A configuration $\mathbf{q}$ of $\mathcal{A}$ can be bijectively represented, in a way that depends on the choice of the frames $\mathcal{F}_{\mathcal{A}}$ and $\mathcal{F}_{\mathcal{W}}$, as a pair $(\mathcal{T}, \Theta)$, where $\mathcal{T}$ determines the position of the origin $O_{\mathcal{A}}$ of $\mathcal{F}_{\mathcal{A}}$ in $\mathcal{F}_{\mathcal{W}}$ and Θ determines the orientation of $\mathcal{F}_{\mathcal{A}}$'s axes with respect to $\mathcal{F}_{\mathcal{W}}$.

Let us adopt the following convention: $\mathcal{T}$ is equal to the N-vector of the coordinates of $O_{\mathcal{A}}$ in $\mathcal{F}_{\mathcal{W}}$ and Θ is equal to the $N \times N$ matrix whose columns are the components of the unit vectors along $\mathcal{F}_{\mathcal{A}}$'s axes in $\mathcal{F}_{\mathcal{W}}$. Θ belongs to the so-called Special Orthogonal Group of the $N \times N$ matrices in $\mathbf{R}^{N^2}$ with orthonormal columns (and rows) and determinant $+1$. This group is denoted by $SO(N)$.

With the above convention, the configuration where $\mathcal{F}_{\mathcal{A}}$ and $\mathcal{F}_{\mathcal{W}}$ coincide corresponds to $\mathcal{T} = \vec{0}$ (null N-vector) and $\Theta = I$ (identity $N \times N$ matrix).

Choosing this configuration as the reference configuration 0, we have:

$$\forall a \in \mathcal{A} : \begin{cases} a & = & a(0) \\ a(\mathbf{q}) & = & \Theta\, a(0) + \mathcal{T} \end{cases}$$

where a denotes a point in $\mathcal{A}$ described by its N-dimensional coordinate vector in $\mathcal{F}_{\mathcal{A}}$ and $a(\mathbf{q})$ denotes the same point described by its coordinate vector in $\mathcal{F}_{\mathcal{W}}$ when $\mathcal{A}$ is at configuration $\mathbf{q} = (\mathcal{T}, \Theta)$.

The matrix Θ can be constructed as follows [Paul, 1981]:

- If $N = 2$:

$$\Theta = \begin{pmatrix} r_{11} & r_{12} \\ r_{21} & r_{22} \end{pmatrix} = \begin{pmatrix} \cos\theta & -\sin\theta \\ \sin\theta & \cos\theta \end{pmatrix}.$$

  This matrix represents the orientation of $\mathcal{F}_{\mathcal{A}}$ after a rotation of angle θ around $O_{\mathcal{W}}$, starting at the reference configuration.

- If $N = 3$:

$$\Theta = \begin{pmatrix} r_{11} & r_{12} & r_{13} \\ r_{21} & r_{22} & r_{23} \\ r_{31} & r_{32} & r_{33} \end{pmatrix} =$$

$$\begin{pmatrix} n_x^2(1-\mathrm{c}\theta)+\mathrm{c}\theta & n_x n_y(1-\mathrm{c}\theta)-n_z\mathrm{s}\theta & n_x n_z(1-\mathrm{c}\theta)+n_y\mathrm{s}\theta \\ n_x n_y(1-\mathrm{c}\theta)+n_z\mathrm{s}\theta & n_y^2(1-\mathrm{c}\theta)+\mathrm{c}\theta & n_y n_z(1-\mathrm{c}\theta)-n_x\mathrm{s}\theta \\ n_x n_z(1-\mathrm{c}\theta)-n_y\mathrm{s}\theta & n_y n_z(1-\mathrm{c}\theta)+n_x\sin\theta & n_z^2(1-\mathrm{c}\theta)+\mathrm{c}\theta \end{pmatrix}$$

  where $\mathrm{c}\theta = \cos\theta$ and $\mathrm{s}\theta = \sin\theta$. This matrix represents the orientation of $\mathcal{F}_{\mathcal{A}}$ after a rotation of angle θ starting at the reference configuration and performed around an axis passing through $O_{\mathcal{W}}$ and oriented along a unit vector $\vec{n}$ whose components in $\mathcal{F}_{\mathcal{W}}$ are n_x, n_y, and n_z.

In the following, $\mathcal{M}_{\mathcal{C}}$ will denote the subset $\mathbf{R}^N \times SO(N) \subset \mathbf{R}^M$, with $M = N + N^2$. $\mathcal{M}_{\mathcal{C}}$ represents $\mathcal{C}$ in $\mathbf{R}^M$.

In the case where $N = 2$, $\mathcal{M}_{\mathcal{C}}$ is a subset of $\mathbf{R}^6$, with each configuration $\mathbf{q} \in \mathcal{C}$ identified by 6 real numbers $(x, y, r_{11}, r_{12}, r_{21}, r_{22}) \in \mathcal{M}_{\mathcal{C}}$. The constraints defining $\mathcal{M}_{\mathcal{C}}$ are the three polynomial equations:

$$\begin{aligned} h_1(x, y, r_{11}, \ldots, r_{22}) &= r_{11}^2 + r_{21}^2 - 1 = 0 \\ h_2(x, y, r_{11}, \ldots, r_{22}) &= r_{12}^2 + r_{22}^2 - 1 = 0 \\ h_3(x, y, r_{11}, \ldots, r_{22}) &= r_{11}r_{12} + r_{21}r_{22} = 0 \end{aligned}$$

expressing that the matrix:

$$\begin{pmatrix} r_{11} & r_{12} \\ r_{21} & r_{22} \end{pmatrix}$$

is orthonormal, and the polynomial equation:

$$r_{11}r_{22} - r_{21}r_{12} - 1 = 0$$

expressing that its determinant is $+1$. The first three equations define two disjoint subsets in $\mathbf{R}^6$ corresponding to matrices with determinants $+1$ and -1, respectively. The fourth equation merely selects the first of these two subsets.

In the case where $N = 3$, $\mathcal{M}_C$ is a subset of $\mathbf{R}^{12}$, with each configuration identified by 12 numbers $(x, y, z, r_{11}, ..., r_{33}) \in \mathcal{M}_C$. In the same fashion as above, expressing that the r_{ij}'s are elements of an orthonormal matrix 3×3 leads to six polynomial equations, $h_k(x, y, z, r_{11}, ..., r_{33}) = 0$, $k = 1, ..., 6$, which determine two disjoint subsets of $\mathbf{R}^{12}$. These two subsets again correspond to matrices with determinants $+1$ and -1, respectively. The additional constraint that the determinant be $+1$ selects one subset.

Notice that $SO(N)$ is a compact subset of $\mathbf{R}^{N^2}$. Indeed, it is closed, since it is the intersection of a finite number of closed sets. It is also bounded, since each element r_{ij} is contained in the interval $[-1, +1]$.

3 Manifold Structure

We now know how to represent the configuration space $\mathcal{C}$ of a rigid object $\mathcal{A}$ as the subset $\mathcal{M}_C = \mathbf{R}^N \times SO(N)$ of $\mathbf{R}^M$, with $M = N + N^2$. In this section we show that $\mathcal{C}$ is locally "like" $\mathbf{R}^m$, for some fixed m depending on M. More precisely, every element of $\mathcal{C}$ has an open neighborhood that is diffeomorphic to an open subset of $\mathbf{R}^m$. This gives $\mathcal{C}$ the structure of a smooth m-dimensional manifold.

3.1 $\mathcal{C}$ is a Manifold

Let the topology of $\mathcal{C}$ be the subset topology induced by the Euclidean metric topology in $\mathbf{R}^M$, i.e. $X \subseteq \mathcal{M}_C$ represents an open subset of $\mathcal{C}$ if and only if there exists an open subset Y of $\mathbf{R}^M$ such that $X = Y \cap \mathcal{M}_C$.

Although this topology derives from the Euclidean metric in $\mathbf{R}^M$, it is considerably more general and can be induced by other metrics (see Section 7), including metrics which are independent of the representation of $\mathcal{C}$, such as the following distance function:

$$d(\mathbf{q},\mathbf{q}') = \max_{a\in\mathcal{A}} \|a(\mathbf{q}) - a(\mathbf{q}')\|$$

where $\|x - x'\|$ denotes the Euclidean distance between any two points x and x' in a Euclidean space. All these metrics share the following "natural" property: The distance between two configurations $\mathbf{q}$ and $\mathbf{q}'$ tends toward zero when the regions $\mathcal{A}(\mathbf{q})$ and $\mathcal{A}(\mathbf{q}')$ tend to coincide, and conversely. In the following, we will consider the topology of $\mathcal{C}$ defined above independently of any particular metric.

The subset $\mathcal{M}_{\mathcal{C}}$ of $\mathbf{R}^M$ is defined by $\frac{1}{2}N(N+1)$ polynomial equations $h_k(u_1, u_2, ..., u_M) = 0$, with $k = 1$ to $\frac{1}{2}N(N+1)$, where $u_1, u_2, ..., u_M$ are merely new symbols for $x, ..., r_{11}, ..., r_{NN}$. (Actually, there is an additional equation: the equation that selects the matrices with determinant $+1$; this equation, however, has no effect on the local properties of $\mathcal{M}_{\mathcal{C}}$.) One can verify that, whether $N = 2$ or 3, the smooth functions h_k admit gradient vectors

$$\vec{\nabla} h_k = \left(\frac{\partial h_k}{\partial u_i}\right)_{i=1,...,M}$$

which are linearly independent at every $(u_1, ..., u_M) \in \mathcal{M}_{\mathcal{C}}$. In other words, $\mathcal{M}_{\mathcal{C}}$ is at the intersection of $\frac{1}{2}N(N+1)$ hypersurfaces of dimension $M-1$, such that at every intersection point, the tangent planes of these surfaces are distinct.

Let $m = M - \frac{1}{2}N(N+1) = \frac{1}{2}N(N+1)$. Every point $\mathbf{u} = (u_1, ..., u_M)$ in $\mathcal{M}_{\mathcal{C}}$ admits an open neighborhood U diffeomorphic to an open subset X of $\mathbf{R}^m$. Indeed, according to the Implicit Function Theorem (see Appendix A), there exists an open neighborhood U of $\mathbf{u}$ in $\mathcal{M}_{\mathcal{C}}$ with the following property: $\frac{1}{2}N(N+1) = M - m$ variables u_i, say $u_{m+1}, ..., u_M$, can be expressed as smooth functions g_i, with $i = m+1, ..., M$, of the m other variables $u_1, ..., u_m$; these functions are unique. The map $\phi : U \subset \mathcal{M}_{\mathcal{C}} \rightarrow \mathbf{R}^m$ defined by:

$$\phi(u_1, ..., u_M) = (u_1, ..., u_m)$$

is a smooth one-to-one map of U onto an open subset X of $\mathbf{R}^m$. In X, it admits a smooth inverse $\phi^{-1} : X \subseteq \mathbf{R}^m \rightarrow U \subset \mathcal{M}_C$, defined by:

$$\phi^{-1}(u_1, ..., u_m) = (u_1, ..., u_m, g_{m+1}(u_1, ..., u_m), ..., g_M(u_1, ..., u_m)).$$

Hence, ϕ is a diffeomorphism (see Appendix A).

A subset of $\mathbf{R}^M$ is a smooth m-dimensional manifold if and only if it is locally diffeomorphic to $\mathbf{R}^m$. By extension, any set that is related to this subset by a bijective map is also a smooth m-dimensional manifold. Thus:

PROPOSITION 1: *The configuration space $\mathcal{C}$ of a rigid object $\mathcal{A}$ moving in an N-dimensional workspace is a smooth manifold of dimension $m = \frac{1}{2}N(N+1)$. The bijective application that maps any configuration of $\mathcal{C}$ to its representation in $\mathcal{M}_C \subset \mathbf{R}^M$, with $M = N(N+1)$, is called an* **embedding** *of $\mathcal{C}$ in $\mathbf{R}^M$. $\mathbf{R}^M$ is called the* **ambient** *Euclidean space.*

That $\mathcal{C}$ is a manifold[3] means that its local topological and differential structures are identical to those of $\mathbf{R}^m$. It is directly related to the well-known fact that infinitesimal rotations (equivalently, angular velocities) form a vector space [Paul, 1981].

The local likeness of $\mathcal{C}$ and $\mathbf{R}^m$ is emphasized by the standard definitions given in the following subsection.

3.2 Charts and Atlas

Let $u : \mathcal{C} \rightarrow \mathcal{M}_C$ be the bijective application that maps any configuration $\mathbf{q}$ in $\mathcal{C}$ to its representation $\mathbf{u}$ in $\mathcal{M}_C = \mathbf{R}^N \times SO(N)$.

A pair (U, ϕ) consisting of an open subset U of $\mathcal{C}$ and a diffeomorphism ϕ of the image of U in $\mathcal{M}_C$ onto X, an open subset of $\mathbf{R}^m$, is called a **chart** or a **coordinate system** on U. For any $\mathbf{q}$ in U, let us write $\phi(u(\mathbf{q})) = (x_1(u(\mathbf{q})), ..., x_m(u(\mathbf{q})))$. The m smooth functions $x_1, ..., x_m$ are called the **coordinate functions** or more simply the **coordinates** of $\mathbf{q}$ in the chart. The inverse diffeomorphism ϕ^{-1} is called a **parameterization** of U. In the following, we will simply write $\phi(\mathbf{q}) = (x_1(\mathbf{q}), ..., x_m(\mathbf{q}))$.

Consider two charts (U, ϕ) and (V, ψ) such that $U \cap V \neq \emptyset$ (see Figure 2). Let X and Y be the images of U and V by ϕ and ψ, respectively. The

[3] From now on, we will omit the word "smooth" and simply say that $\mathcal{C}$ is a manifold.

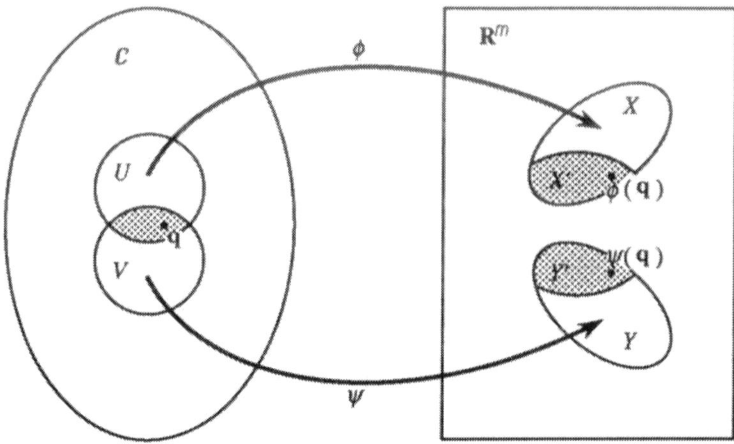

Figure 2. Two intersecting open subsets of $\mathcal{C}$, U and V, are mapped diffeomorphically onto two open subsets X and Y of $\mathbf{R}^m$ by ϕ and ψ, respectively. X' and Y' denote the images of $U \cap V$ by ϕ and ψ. The composed applications $\psi \circ \phi^{-1} : X' \rightarrow Y'$ and $\phi \circ \psi^{-1} : Y' \rightarrow X'$ are smooth, and the charts (U, ϕ) and (V, ψ) are said to be C^∞-related.

intersection $U \cap V$ is open and thus has two different charts $(U \cap V, \phi)$ and $(U \cap V, \psi)$. Let $X' \subseteq X$ and $Y' \subseteq Y$ be the images of $U \cap V$ by ϕ and ψ, respectively. For every $\mathbf{q} \in U \cap V$, we write:

$$\phi(\mathbf{q}) = (x_1(\mathbf{q}), ..., x_m(\mathbf{q})),$$

$$\psi(\mathbf{q}) = (y_1(\mathbf{q}), ..., y_m(\mathbf{q})).$$

ϕ has a smooth inverse ϕ^{-1}, so that the composed application $\psi \circ \phi^{-1} : X' \rightarrow Y'$, with:

$$\psi \circ \phi^{-1}(x_1(\mathbf{q}), ..., x_m(\mathbf{q})) = (y_1(\mathbf{q}), ... y_m(\mathbf{q})),$$

is smooth. In the same way, the inverse application $\phi \circ \psi^{-1} : Y' \rightarrow X'$ is also smooth. The charts (U, ϕ) and (V, ψ) are said to be C^∞-**related.**

A set of charts $\{(U_\alpha, \phi_\alpha)\}$ whose domains U_α cover $\mathcal{C}$ is called an **atlas** of $\mathcal{C}$. Once an atlas has been defined for $\mathcal{C}$, we can represent $\mathcal{C}$ as a collection of copies of $\mathbf{R}^m$ without referring to its embedding in $\mathbf{R}^M$. Since $\mathbf{R}^N$ can be covered by a single chart, and $SO(N)$ is compact, $\mathcal{M}_\mathcal{C}$, hence $\mathcal{C}$, can be covered by a finite number of charts. (We recall that in a compact set $\mathcal{E}$, any infinite collection of open subsets covering $\mathcal{E}$ contains a finite subcollection covering $\mathcal{E}$.)

The fact that the charts in an atlas are C^∞-related will allow us to extend the differential properties established in a chart of the atlas to all the other charts.

3.3 Other Embeddings

As we mentioned above, we can consider $\mathcal{C}$ as an abstract space independently of the ambient space $\mathbf{R}^M$. The latter then looks somewhat arbitrary. Indeed, we could now pick any space $\mathbf{R}^{M'}$, with $M' > M$, and easily represent $\mathcal{C}$ as a subset $\mathcal{M}'_{\mathcal{C}}$ of $\mathbf{R}^{M'}$ equivalent (i.e. globally diffeomorphic) to $\mathcal{M}_{\mathcal{C}}$. $\mathbf{R}^{M'}$ would then be another ambient space of $\mathcal{C}$.

Since M is significantly greater than m, a legitimate question is the following: Does there exist $M'' < M$ such that $\mathbf{R}^{M''}$ contains a globally diffeomorphic copy $\mathcal{M}''_{\mathcal{C}}$ of $\mathcal{M}_{\mathcal{C}}$? A partial answer to this question is given by the Whitney Embedding Theorem: Every r-dimensional manifold can be embedded in $\mathbf{R}^{2r}$ [Guillemin and Pollack, 1974]. In other words, there is enough "twisting room" in $\mathbf{R}^{2r}$ for embedding any r-dimensional manifold. (Although the Whitney Theorem is optimal, it does not mean that a particular r-dimensional manifold cannot be embedded in a Euclidean space of dimension lower than $2r$; for instance, $\mathbf{R}^r$ is embedded in itself.)

Therefore, when we embed the one-dimensional manifold $SO(2)$ in $\mathbf{R}^4$ and the three-dimensional manifold $SO(3)$ in $\mathbf{R}^9$, there is some extra twisting room in the ambient spaces, which we do not use. We should at least be able to embed these manifolds in two- and six-dimensional Euclidean spaces, respectively.

Matrices in $SO(2)$ are of the form:

$$\begin{pmatrix} r_{11} & r_{12} \\ r_{21} & r_{22} \end{pmatrix} = \begin{pmatrix} \cos\theta & -\sin\theta \\ \sin\theta & \cos\theta \end{pmatrix} = \begin{pmatrix} r_{11} & -r_{21} \\ r_{21} & r_{11} \end{pmatrix}.$$

$SO(2)$ can be mapped to the unit circle in $\mathbf{R}^2$, denoted by S^1, with the following mapping:

$$(r_{ij})_{i,j\in[1,2]} \in \mathbf{R}^4 \mapsto (r_{11}, r_{21}) \in \mathbf{R}^2.$$

This mapping is a smooth bijective mapping whose inverse is also smooth. Hence, $SO(2)$ and S^1 are diffeomorphic, and S^1 can be regarded as an embedding of $SO(2)$ in $\mathbf{R}^2$. S^1 is so easy to visualize as a unit circle that we will often use S^1 in place of $SO(2)$.

Matrices in $SO(3)$ can be mapped to vectors in $\mathbf{R}^6$ by selecting the first two columns, i.e. :

$$(r_{ij})_{i,j\in[1,3]} \in \mathbf{R}^9 \mapsto (r_{11}, r_{21}, r_{31}, r_{12}, r_{22}, r_{32}) \in \mathbf{R}^6.$$

This mapping is smooth. In addition, since the third column can be obtained as the outer product of the first two, it is bijective, and its inverse is also smooth. Thus, it allows us to embed $SO(3)$ in $\mathbf{R}^6$. However, in the rest of the book, we will not use this embedding.

4 Parameterizations

In this section we describe various ways to chose charts and to construct an atlas of $\mathcal{C}$. Once an atlas has been constructed, every configuration can be represented by a vector of m parameters.

Since the parameterization of $\mathbf{R}^N$ is trivial, we only consider $SO(N)$. We first treat the case where $N = 2$, then the case where $N = 3$. Several parameterizations other than those presented below can be found in textbooks on Kinematics (e.g. [Bottema and Roth, 1979]) and Robotics (e.g. [Paul, 1981] [Craig, 1986]).

4.1 Parameterizations of $SO(2)$

A straightforward parameterization of $SO(2)$ consists of representing an orientation of the robot $\mathcal{A}$ by a single angle θ, say the angle between the x-axes of $\mathcal{F}_\mathcal{W}$ and $\mathcal{F}_\mathcal{A}$. The map is obviously not a one-to-one map, since all the angles $\theta + 2k\pi$, with $k \in \mathbf{Z}$, represent the same orientation of $\mathcal{A}$. One way to get a chart is to restrict the values of θ to an open interval of length 2π. Two such charts, say $0 < \theta < 2\pi$ and $-\pi < \theta' < +\pi$, are needed to have an atlas. However, in most path planning algorithms, it is sufficient to consider a single interval closed at one end, say $\theta \in [0, 2\pi)$, together with modulo 2π arithmetic on θ; but this parameterization is not a chart.

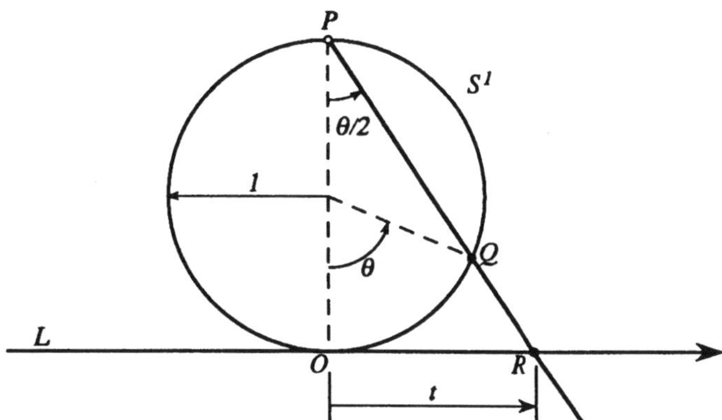

Figure 3. The stereographic projection allows us to map algebraically the circle S^1 onto the line of the reals (see text). It defines a chart valid over all S^1, except the point P. Two stereographic charts with two different points P are necessary and sufficient to build an atlas of S^1, or equivalently, $SO(2)$.

Some path planning methods require motion constraints to be expressed in algebraic form. Using the above parameterization yields expressions depending on θ through cosine and sine functions, i.e. non-algebraic functions. Another parameterization of $SO(2)$, which allows to express constraints in algebraic form, is known as the **stereographic projection.** It consists of projecting the whole unit circle S^1 (with the exception of one point) onto a line L tangent to the circle. L represents the set $\mathbf{R}$, with the origin at the point of tangency O (see Figure 3). Another line is drawn from the point P on the circle diametrically opposite the point of tangency. It intersects the circle at Q and L at R. The stereographic projection maps every point Q on the circle, except P, to the point R on L. In every open subset of S^1 not including P, the map is diffeomorphic, and hence defines a chart. Two stereographic charts, built with two different points P, are needed to form an atlas of S^1.

Let t be the abscissa of R along L and O correspond to $\theta = 0 \pmod{2\pi}$. We have:

$$t = 2\tan\frac{\theta}{2}.$$

Writing $w = \frac{t}{2}$, we can express $\cos\theta$ and $\sin\theta$ as the following rational

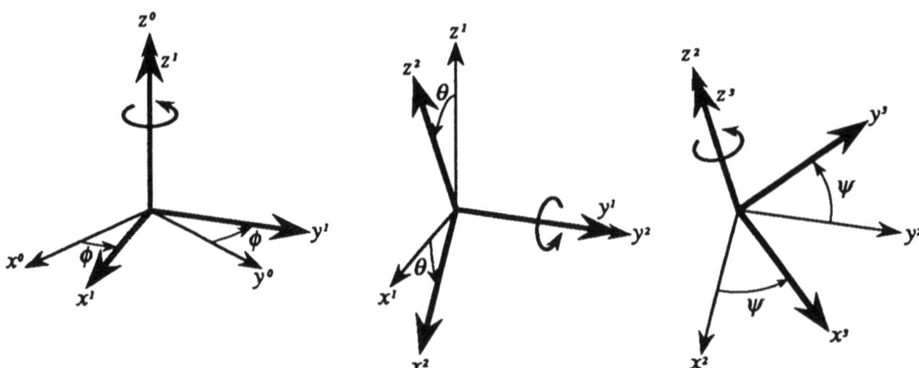

Figure 4. $\mathcal{F}_{\mathcal{A}}^i = (O_{\mathcal{A}}, x^i, y^i, z^i)$, $i = 0,1,2,3$, denotes the frame $\mathcal{F}_{\mathcal{A}}$ at four successive orientations of $\mathcal{A}$. (ϕ, θ, ψ) are the Euler angles defining the orientation of $\mathcal{F}_{\mathcal{A}}^3$ relative to $\mathcal{F}_{\mathcal{A}}^0$. First, $\mathcal{A}$ is rotated by ϕ around the z^0-axis. Second, $\mathcal{A}$ is rotated by θ around the y^1-axis. Third, $\mathcal{A}$ is rotated by ψ around the z^2-axis.

fractions:

$$\cos\theta = \frac{1-w^2}{1+w^2} \quad \text{and} \quad \sin\theta = \frac{2w}{1+w^2}$$

leading to algebraic expressions in t of the constraints applying to $\mathcal{A}$'s motion.

There is no way to build an atlas of $SO(2)$ containing a single chart.

4.2 Euler Angles

The **Euler angles** are a widely used parameterization of $SO(3)$. They consist of three angles (ϕ, θ, ψ). The orientation of $\mathcal{A}$ they specify is obtained by applying three successive rotations to $\mathcal{A}$, starting at its reference orientation. Let $\mathcal{F}_{\mathcal{A}}^0 = (O_{\mathcal{A}}, x^0, y^0, z^0)$ denote the frame $\mathcal{F}_{\mathcal{A}}$ at its reference orientation (see Figure 4). The rotations defined by ϕ, θ, and ψ are the following:[4]

- First, $\mathcal{A}$ is rotated by ϕ around the z^0-axis. Let $\mathcal{F}_{\mathcal{A}}^1 = (O_{\mathcal{A}}, x^1, y^1, z^1)$ denote the frame $\mathcal{F}_{\mathcal{A}}$ at its new orientation.

[4] Slightly different specifications of these rotations are proposed in some textbooks. They simply determine different charts on $SO(3)$.

- Second, $\mathcal{A}$ is rotated by θ around the y^1-axis. Let $\mathcal{F}_{\mathcal{A}}^2 = (O_{\mathcal{A}}, x^2, y^2, z^2)$ denote the frame $\mathcal{F}_{\mathcal{A}}$ at its new orientation.

- Third, $\mathcal{A}$ is rotated by ψ around the z^2-axis. Let $\mathcal{F}_{\mathcal{A}}^3 = (O_{\mathcal{A}}, x^3, y^3, z^3)$ denote the frame $\mathcal{F}_{\mathcal{A}}$ at its new orientation.

The orientation of $\mathcal{F}_{\mathcal{A}}^3$ is the orientation of $\mathcal{A}$ defined by the Euler angles (ϕ, θ, ψ).

The parameterization of $SO(3)$ using (ϕ, θ, ψ) is not a one-to-one mapping. The same orientation is represented by $(\phi + 2k_1\pi, \theta + 2k_2\pi, \psi + 2k_3\pi)$, for any $k_1, k_2, k_3 \in \mathbf{Z}$. As for $SO(2)$, we may be tempted to apply modulo 2π arithmetic to every Euler angle. However, it is not so simple. Indeed, (ϕ, θ, ψ) represents the same orientation as $(\phi, \theta + \pi, 2\pi - \psi)$, so that a unique representation requires us to restrict the values of the three Euler angles to, say, $\phi \in [0, 2\pi)$, $\theta \in [0, \pi)$, and $\psi \in [0, 2\pi)$. In addition, even with this restriction, there still exists a singularity at $\theta = 0$. Indeed, if $\theta = 0$, the first and third rotations are made around the same axis. Thus, any two pairs (ϕ_1, ψ_1) and (ϕ_2, ψ_2), such that $\phi_1 + \psi_1 = \phi_2 + \psi_2$, determine the same orientation of $\mathcal{A}$. This means that in the $\phi\theta\psi$-space, a motion along a line $\phi = -\psi + c \pmod{2\pi}$ at $\theta = 0$, where c is a constant, corresponds to no motion of $\mathcal{A}$.

A single set of Euler angles (ϕ, θ, ψ), with appropriate restrictions on the values of ϕ, θ and ψ, is called an Eulerian chart. Several Eulerian charts can be defined by applying three successive rotations to $\mathcal{A}$ around different axes. However, no atlas on $\mathcal{C}$ can be constructed with Eulerian charts only, since the null rotation corresponds to a singularity in all these charts. When Euler angles are used to represent the robot's orientation in a motion planner, it is usually appropriate to use a single set of Euler angles $0 \leq \phi < 2\pi$, $0 \leq \theta < \pi$, $0 \leq \psi < 2\pi$, taking precautions at the interval extremities. If the singularities cannot be tolerated, it is preferable to use the quaternion parameterization presented in the next subsection.

Another drawback of Euler angles is that they do not allow us to express motion constraints in algebraic form. Applying a rotation specified by (ϕ, θ, ψ) to a geometric feature, say a point x, is equivalent to multiplying the coordinate vector of x by the corresponding matrix of $SO(3)$. The new coordinates of x are non-algebraic functions of ϕ, θ, and ψ. Hence, some motion planning methods which require the motion constraints to be algebraic cannot make use of this parameterization.

The matrix $\mathbf{E}_{\phi\theta\psi}$ of $SO(3)$ corresponding to the three Euler angles defined above, with the conventions taken in Section 2, is:

$$\begin{pmatrix} \cos\phi & -\sin\phi & 0 \\ \sin\phi & \cos\phi & 0 \\ 0 & 0 & 1 \end{pmatrix} \begin{pmatrix} \cos\theta & 0 & \sin\theta \\ 0 & 1 & 0 \\ -\sin\theta & 0 & \cos\theta \end{pmatrix} \begin{pmatrix} \cos\psi & -\sin\psi & 0 \\ \sin\psi & \cos\psi & 0 \\ 0 & 0 & 1 \end{pmatrix} =$$

$$\begin{pmatrix} \cos\phi\cos\theta\cos\psi - \sin\phi\sin\psi & -\cos\phi\cos\theta\sin\psi - \sin\phi\cos\psi & \cos\phi\sin\theta \\ \sin\phi\cos\theta\cos\psi + \cos\phi\sin\psi & -\sin\phi\cos\theta\sin\psi + \cos\phi\cos\psi & \sin\phi\sin\theta \\ -\sin\theta\cos\psi & \sin\theta\sin\psi & \cos\theta \end{pmatrix}.$$

By restricting ϕ and ψ to $[0, 2\pi)$ and θ to $(0, \pi)$, and given a matrix of $SO(3)$:

$$\begin{pmatrix} r_{11} & r_{12} & r_{13} \\ r_{21} & r_{22} & r_{23} \\ r_{31} & r_{32} & r_{33} \end{pmatrix}$$

the three Euler angles (ϕ, θ, ψ) are determined by the following relations:

$$\begin{aligned} \phi &= atan2(r_{23}, r_{13}) \\ \theta &= atan2(\sqrt{r_{31}^2 + r_{32}^2}, r_{33}) \\ \psi &= atan2(r_{32}, -r_{31}) \end{aligned}$$

where $atan2(y, x)$ is defined as the argument of the complex number $x + \vec{i}y$.

4.3 Quaternions

Quaternions encode orientations in a three-dimensional workspace with four parameters.[5]

A quaternion $\mathbf{X}$ is an element of $\mathbf{R}^4$ which can be split into a scalar part x_0 and a 3-vector $\vec{x}$. To simplify notation, the scalar part and the vector part are combined using the addition symbol, i.e. $\mathbf{X} = x_0 + \vec{x}$. It

[5]The theory of quaternions is described in many textbooks. The most complete reference is [Hamilton, 1969].

is convenient to treat 3-vectors as quaternions having a zero scalar part, and scalars as quaternions having a zero vector part.

The product of two quaternions **P** and **Q** is the quaternion $\mathbf{R} = \mathbf{PQ}$ specified by:

$$\mathbf{R} = r_0 + \vec{r}$$

with:

$$r_0 = p_0 q_0 - \vec{p} \cdot \vec{q}$$

$$\vec{r} = p_0 \vec{q} + q_0 \vec{p} + \vec{p} \wedge \vec{q}$$

where $\vec{p} \cdot \vec{q}$ and $\vec{p} \wedge \vec{q}$ denote the inner and outer products of $\vec{p}$ and $\vec{q}$. The product **PQ** is linear in **P** and **Q**.

The conjugate $\mathbf{X}^*$ of a quaternion $\mathbf{X} = x_0 + \vec{x}$ is defined by:

$$\mathbf{X}^* = x_0 - \vec{x}.$$

Notice that:

$$\mathbf{XX}^* = x_0^2 + \|\vec{x}\|^2$$

is a positive or null scalar. $\|\mathbf{X}\| = \sqrt{\mathbf{XX}^*}$ is the Euclidean length of **X** in $\mathbf{R}^4$. **X** is called a **unit quaternion** if it satisfies $\mathbf{XX}^* = 1$.

A rotation around a unit vector $\vec{n}$ by an angle θ can be represented by the following unit quaternion:

$$\mathbf{X}(\vec{n}, \theta) = \cos\frac{\theta}{2} + \vec{n}\sin\frac{\theta}{2}.$$

With this representation, one can verify that:

$$\forall a \in \mathcal{A} : a(\mathbf{q}) = \mathbf{X}_\Theta a(0) \mathbf{X}_\Theta^* + \mathcal{T}$$

where $a(\mathbf{q})$ denotes the coordinate vector of a in $\mathcal{F}_\mathcal{W}$ when $\mathcal{A}$ is at configuration $\mathbf{q} = (\mathcal{T}, \Theta) \in \mathbf{R}^3 \times SO(3)$, and $\mathbf{X}_\Theta$ denotes the unit quaternion representing the orientation Θ in $\mathbf{R}^4$. The expression $\mathbf{X}_\Theta a(0) \mathbf{X}_\Theta^* + \mathcal{T}$ is worth some explanation. In the product $\mathbf{X}_\Theta a(0) \mathbf{X}_\Theta^*$, $a(0)$ is treated as a quaternion having a zero scalar part, i.e. $0 + a(0)$. It is straightforward to verify that the product is also a quaternion having a zero scalar part. The three components of the vector part of $\mathbf{X}_\Theta a(0) \mathbf{X}_\Theta^*$ are added to the components of $\mathcal{T}$, yielding the coordinate vector $a(\mathbf{q})$.

The space of unit quaternions is S^3, the unit 3-sphere in $\mathbf{R}^4$. The same orientation can be represented by two unit quaternions $\mathbf{X}$ and $-\mathbf{X}$. Thus, the application $S^3 \rightarrow SO(3)$ that maps a quaternion to the corresponding orientation in $SO(3)$ is a two-to-one smooth map. Although S^3 is locally diffeomorphic to $SO(3)$, the application $S^3 \rightarrow SO(3)$ is not a diffeomorphism, and S^3 is not an embedding of $SO(3)$ in $\mathbf{R}^4$.

Let $\sim$ be the equivalence relation such that for all $X \in S^3$, $X \sim -X$. The quotient space $S^3/\sim$ with the quotient topology (see Appendix A) is a topological space denoted by P^3 and called **projective 3-space**. It can be regarded as the 3-sphere S^3 with antipodal points identified, or as the space of all lines in $\mathbf{R}^4$ passing through the origin. $SO(3)$ and P^3 are homeomorphic, i.e. related by a continuous bijective map. Hence, they are topologically equivalent.

One can use unit quaternions as follows to construct an atlas of $SO(3)$ made of four charts (U_i, ϕ_i), $i = 0, ..., 3$ [Khatib, 1980] [Canny, 1988]. Let $\mathbf{u}$ be an element of $SO(3)$ represented as a unit quaternion $\mathbf{U} = u_0 + \vec{u} = (u_0, u_1, u_2, u_3)$. Assume that the ith component of $\mathbf{U}$ has the largest magnitude. Then $\mathbf{u} \in U_i$ and $\phi_i(\mathbf{u}) = (\frac{u_j}{|u_i|})_{j=0,...,3\,;\,j \neq i}$. For instance, if $i = 0$, $\phi_0(\mathbf{u}) = (\frac{u_1}{|u_0|}, \frac{u_2}{|u_0|}, \frac{u_3}{|u_0|})$. The four U_i's cover $SO(3)$, but they are closed sets which intersect at their boundaries. One way to obtain open sets covering $SO(3)$ is to define U_i as the subset of all the elements $\mathbf{u} = (u_0, ..., u_3)$ of $SO(3)$ such that for all $j \in [0,3]$, $j \neq i$, $|u_j| - |u_i| < \varepsilon$ for a given small $\varepsilon > 0$.

The matrix of $SO(3)$ corresponding to the unit quaternion $\mathbf{U} = (u_0, u_1, u_2, u_3)$ is:

$$\begin{pmatrix} 2(u_0^2 + u_1^2) - 1 & 2(u_1 u_2 - u_0 u_4) & 2(u_1 u_3 + u_0 u_2) \\ 2(u_1 u_2 + u_0 u_3) & 2(u_0^2 + u_2^2) - 1 & 2(u_2 u_3 - u_0 u_1) \\ 2(u_1 u_3 - u_0 u_2) & 2(u_2 u_3 + u_0 u_1) & 2(u_0^2 + u_3^2) - 1 \end{pmatrix}.$$

5 Paths and Multiple Connectedness

In this section we define the notion of a path in configuration space. In a Euclidean space, such as $\mathbf{R}^m$, if two paths connect the same initial point to the same final point, either one of them can be "continuously" deformed to the other. In $\mathcal{C}$, this is not true, and $\mathcal{C}$ is said to be multiply connected. This provides evidence that the global topology of $\mathcal{C}$ is different from that of $\mathbf{R}^m$.

5.1 Path in $\mathcal{C}$

The definition of a path in $\mathcal{C}$ is the standard definition of a path in a topological space [Jänich, 1984], i.e. :

DEFINITION 2: *A* **path** *of the robot $\mathcal{A}$ from $\mathbf{q}_{init}$ to $\mathbf{q}_{goal}$ is a continuous map $\tau : [0,1] \to \mathcal{C}$, with $\tau(0) = \mathbf{q}_{init}$ and $\tau(1) = \mathbf{q}_{goal}$. $\mathbf{q}_{init}$ and $\mathbf{q}_{goal}$ are called the* **initial** *and* **goal** *configurations of the path, respectively.*

This definition directly entails that the set $\{\tau(s) \,/\, s \in [0,1]\}$ is compact. (Indeed, the image of a compact set by a continuous map is a compact set.) This implies that $\tau(s)$ is contained in a bounded domain of $\mathbf{R}^M$.

Now we can formally state the **free-flying object assumption** of the basic motion planning problem as follows: In the absence of obstacles, any path defined as above is feasible. It corresponds to assuming that the motions of $\mathcal{A}$ are not limited by any kinematic or dynamic constraints.

Let us consider any two configurations $\mathbf{q}_{init}$ and $\mathbf{q}_{goal}$ in $\mathcal{C}$. The rigid body transformation that moves $\mathcal{A}$ from $\mathbf{q}_{init}$ to $\mathbf{q}_{goal}$ can be uniquely represented as the composition of a rotation $Rot(\theta)$ by some angle θ around a fixed axis and a translation $Trsl(\delta)$ by some distance δ along a fixed vector. We can write:

$$TR_{\mathbf{q}_{goal}} = Trsl(\delta) \circ Rot(\theta) \circ TR_{\mathbf{q}_{init}},$$

where $TR_{\mathbf{q}_{init}}$ and $TR_{\mathbf{q}_{goal}}$ are the transformations corresponding to $\mathbf{q}_{init}$ and $\mathbf{q}_{goal}$, respectively (see Section 1). The continuous map $\tau : s \in [0,1] \mapsto \tau(s) \in \mathcal{C}$, with

$$TR_{\tau(s)} = Trsl(s\delta) \circ Rot(s\theta) \circ TR_{\mathbf{q}_{init}},$$

is a path from $\mathbf{q}_{init}$ to $\mathbf{q}_{goal}$. Therefore:

PROPOSITION 2: *$\mathcal{C}$ is* **connected**, *i.e. any two configurations are connected by a path.*[6]

[6] In any manifold, connectedness and path-connectedness are two equivalent notions [Bishop and Goldberg, 1980].

This proposition simply states "physical evidence". If $\mathcal{A}$ is a rigid object moving freely in a workspace without obstacles, and $\mathbf{q}_{init}$ and $\mathbf{q}_{goal}$ are configurations of $\mathcal{A}$, then there is a way to move $\mathcal{A}$ from $\mathbf{q}_{init}$ to $\mathbf{q}_{goal}$.

Let τ and τ' be two paths in $\mathcal{C}$ such that $\tau(1) = \tau'(0)$. The path $\tau \bullet \tau'$ defined by:

$$\tau \bullet \tau'(s) = \begin{cases} \tau(2s) & \text{if } 0 \leq s \leq \frac{1}{2}, \\ \tau'(2s-1) & \text{if } \frac{1}{2} \leq s \leq 1, \end{cases}$$

is called the **product** (or composition) of τ by τ'.

The path $\varepsilon_{\mathbf{q}_0}$ such that $\varepsilon_{\mathbf{q}_0}(s) = \mathbf{q}_0$ for all $s \in [0,1]$ is called the **null path** at $\mathbf{q}_0$.

Let τ be a path. The path $\overline{\tau}$, with $\overline{\tau}(s) = \tau(1-s)$ for all $s \in [0,1]$, is called the **inverse path** of τ.

5.2 Homotopy and Fundamental Group

In this subsection we give standard definitions and results from Topology (e.g. see [Massey, 1967] [Jänich, 1984]).

Two paths τ and τ' in a topological space $\mathcal{X}$ are **homotopic** (with endpoints fixed) if there exists a continuous map $h : [0,1] \times [0,1] \rightarrow \mathcal{X}$, with $h(s,0) = \tau(s)$ and $h(s,1) = \tau'(s)$ for all $s \in [0,1]$, and $h(0,t) = h(0,0)$ and $h(1,t) = h(1,0)$ for all $t \in [0,1]$. Homotopy determines a relation of equivalence denoted by $\simeq$ among the paths having the same origin and the same final extremity. Obviously, if $\tau_1 \simeq \tau_1'$ and $\tau_2 \simeq \tau_2'$, and $\tau_1(1) = \tau_2(0)$, then $\tau_1 \bullet \tau_2 \simeq \tau_1' \bullet \tau_2'$.

A path-connected space $\mathcal{X}$ is **simply connected** if every two paths τ and τ' such that $\tau(0) = \tau'(0)$ and $\tau(1) = \tau'(1)$ are homotopic; otherwise, it is **multiply connected**. The tool used for characterizing the connectedness of a topological space is the *fundamental group*, which we introduce below.

Let x_0 be an arbitrary point in $\mathcal{X}$. A **loop** based at x_0 is a path λ in $\mathcal{X}$ such that $\lambda(0) = \lambda(1) = x_0$. Let $\Omega(\mathcal{X}, x_0)$ be the set of all the loops in $\mathcal{X}$ based at x_0, and $\pi_1(\mathcal{X}, x_0) = \Omega(\mathcal{X}, x_0)/\simeq$ the set of homotopic classes of loops based at x_0. If λ is a loop of $\Omega(\mathcal{X}, x_0)$, $[\lambda]$ denotes the loop class in $\pi_1(\mathcal{X}, x_0)$ to which it belongs.

The operation $[\lambda] \bullet [\lambda'] \stackrel{\text{def}}{=} [\lambda \bullet \lambda']$ provides $\pi_1(\mathcal{X}, x_0)$ with the structure of a group. The identity element, called the null-homotopic element, is $[\varepsilon_{x_0}]$, where ε_{x_0} is the null path at x_0. The inverse of any $[\lambda] \in \pi_1(\mathcal{X}, x_0)$ is $[\overline{\lambda}]$. If $\mathcal{X}$ is path-connected, the groups $\pi_1(\mathcal{X}, x_0)$ and $\pi_1(\mathcal{X}, x_1)$ are isomorphic for any two points $x_0, x_1 \in \mathcal{X}$. Hence, we will use the notation $\pi_1(\mathcal{X})$ for denoting any of these groups. $\pi_1(\mathcal{X})$ is called the **fundamental group** of $\mathcal{X}$. One can verify that the number of classes of homotopic paths joining any two given points is equal to the cardinality of the fundamental group of the topological space. Thus, a path-connected topological space is simply connected if and only if its fundamental group contains the null-homotopic element as its sole element. It is multiply connected otherwise.

If $\mathcal{X}$ and $\mathcal{Y}$ are two topological spaces, the fundamental group of the product space $\mathcal{X} \times \mathcal{Y}$ is the product of the fundamental groups of the two spaces, i.e. $\pi_1(\mathcal{X} \times \mathcal{Y}) = \pi_1(\mathcal{X}) \times \pi_1(\mathcal{Y})$. In the particular case of $\mathcal{C}$, $\pi_1(\mathcal{C}) = \pi_1(\mathbf{R}^N \times SO(N)) = \pi_1(\mathbf{R}^N) \times \pi_1(SO(N))$.

The fundamental group of $\mathbf{R}^n$ ($n > 0$) is $\{0\}$, where 0 denotes the null-homotopic element. Hence, every two paths in $\mathbf{R}^n$ with the same origin and the same extremity are homotopic.

In the next two subsections we discuss informally the connectedness of $\mathbf{R}^N \times SO(N)$. A rigorous treatment of this topic can be found in [Massey, 1967].

5.3 Connectedness of $\mathbf{R}^2 \times SO(2)$

We noticed in Subsection 3.3 that $SO(2)$ is diffeomorphic (hence, homeomorphic) to S^1, the unit circle in $\mathbf{R}^2$.

The fundamental group of S^1 is known to be isomorphic to $\mathbf{Z}$, the set of integers with addition as the group operation. This means that there is a countable infinity of homotopic classes of paths between any two configurations $\mathbf{q}_{init}$ and $\mathbf{q}_{goal}$ in $\mathbf{R}^2 \times SO(2)$ (or equivalently, $\mathbf{R}^2 \times S^1$). Each class is uniquely identified by the sign of the net rotation around the circle S^1 and the number of net complete rotations around S^1.

We can graphically illustrate the multiple connectedness of $\mathbf{R}^2 \times S^1$ by considering the space $\mathbf{R} \times S^1$, a cylinder in $\mathbf{R}^3$ shown in Figure 5. No two of the three paths τ_1, τ_2, and τ_3 drawn in the cylinder are homotopic.

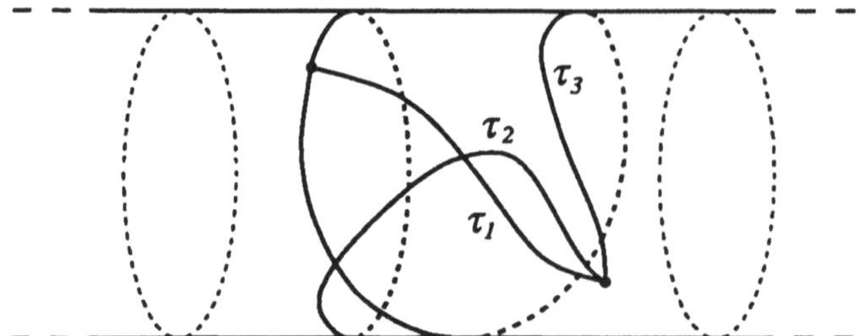

Figure 5. $SO(2)$ (or S^1) is a multiply connected topological space, with $\mathbf{Z}$ as the fundamental group. The three paths τ_1, τ_2, and τ_3 drawn in the infinite cylinder $\mathbf{R} \times S^1$ are not homotopic. Indeed, no two of them wind around the cylinder with the same number of net rotations (with the same sign).

5.4 Connectedness of $\mathbf{R}^3 \times SO(3)$

We noticed in Subsection 4.3 that $SO(3)$ is homeomorphic to the quotient space of the 3-sphere S^3 in $\mathbf{R}^4$ obtained by identifying its antipodal points, i.e. $P^3 = S^3/\sim$.

The fundamental group of $P^n = S^n/\sim$ $(n \geq 2)$ is known to be isomorphic to the quotient set $\mathbf{Z}$ mod 2, i.e. $\mathbf{Z}_2 = \{0, 1\}$. This means that there are two homotopic classes of paths[7] between any two configurations $\mathbf{q}_{init}$ and $\mathbf{q}_{goal}$ in $\mathbf{R}^3 \times P^3$, hence $\mathbf{R}^3 \times SO(3)$.

The null loop class, 0, contains all the loops along which $\mathcal{A}$ undergoes a net rotation of $4k\pi$, with $k \in \mathbf{Z}$, i.e. a net rotation of 0 or an even multiple of 2π. The other loop class, 1, contains all the loops along which $\mathcal{A}$ undergoes a net rotation of $2(2k+1)\pi$, with $k \in \mathbf{Z}$, i.e. a net rotation of an odd multiple of 2π. The fact that any loop along which $\mathcal{A}$ undergoes a net rotation of $4k\pi$ is null-homotopic can be illustrated as follows. Assume that $k = 1$. The rotation during the loop can be expressed as a rotation of 4π around a single axis. Let us decompose this rotation into two successive rotations of 2π each, around this same axis. The second rotation can be homotopically deformed by a continuous change of the orientation of its axis. (This change leaves the initial and final orientations of $\mathcal{A}$ unchanged because the rotation is by an angle of 2π.) Let the change in orientation proceed until the axis of the second rotation

[7] See [Burdick, 1988] for a "robotic-oriented" description of the topology of $SO(3)$.

gets colinear with the axis of the first rotation, with opposite direction. Then the composition of the two rotations produces a net rotation of 0, showing that the original loop is null-homotopic. The same deformation as above cannot be applied to a loop of $2(2k+1)\pi$ net rotation.

Assume that the base configuration of the loop is $\mathbf{q}_0$ and let us select a point p_0 representing this configuration in S^3. A loop in the null loop class corresponds to a continuous curve in S^3 that terminates at p_0, while a loop in the non-null loop class corresponds to a continuous curve in S^3 that terminates at $-p_0$, the antipodal point of p_0. In Topology, S^3 is said to be a twofold covering space of $SO(3)$, and the antipodal identification is the covering map.

Let us consider two paths τ_1 and τ_2, such that $\tau_1(0) = \tau_2(0)$ and $\tau_1(1) = \tau_2(1)$. Let ω_1 be the net rotation that $\mathcal{A}$ undergoes along τ_1 and ω_2 the net rotation along τ_2. If ω_1 and ω_2 differ by 0 or an even multiple of 2π, τ_1 and τ_2 are homotopic; otherwise they are not. Viewed in S^3, if τ_1 and τ_2 are non homotopic, they start at the same point and end at two antipodal points.

6 Directions of Motion

In this section we introduce the necessary notions that make it possible to talk about directions in configuration space. We define the tangent vector to a differentiable curve contained in $\mathcal{C}$. The space spanned by all the possible tangent vectors at a given configuration is a vector space, called the *tangent space*. One such space can be attached with each configuration $\mathbf{q}$ in $\mathcal{C}$.

6.1 Tangent Space

Consider a curve γ:

$$s \in \mathbf{R} \mapsto \gamma(s) \in \mathcal{C}.$$

Let (U, ϕ) be a chart in $\mathcal{C}$ such that $\gamma(s_0) \in U$. We denote by $x_i(\mathbf{q})$, $i = 1, ..., m$, the coordinates of any configuration $\mathbf{q} \in U$ given by ϕ. γ is **differentiable** at $s = s_0$ if and only if for every $i \in [1, m]$, $x_i(\gamma(s))$ is a differentiable function of s at s_0. Notice that if this property is verified for a chart (U, ϕ), then it also holds for any other chart (V, ψ) in the atlas of $\mathcal{C}$ such that $\gamma(s_0) \in V$, since (U, ϕ) and (V, ψ) are C^∞-related. A curve is differentiable[8] if it is differentiable at all the values of s in its definition domain ($\mathbf{R}$ or an open interval of $\mathbf{R}$). The notion of differentiability is extended in the usual fashion to curves γ defined on a closed interval by using the right and left derivatives at the extremities of the interval.

Let us consider the embedding $\mathcal{M}_{\mathcal{C}}$ of $\mathcal{C}$ in $\mathbf{R}^M$ that we defined in Section 2. At every point $\mathbf{u}$ in $\mathcal{M}_{\mathcal{C}}$, the $M-m$ independent gradient vectors $\vec{\nabla} h_1, \ldots, \vec{\nabla} h_{M-m}$ (see Subsection 3.1) span a vector subspace of $\mathbf{R}^M$ of dimension $M-m$. The orthogonal complement to this subspace in $\mathbf{R}^M$ is a subspace of dimension m, called the tangent space of $\mathcal{M}_{\mathcal{C}}$ at $\mathbf{u}$. For a moment, let us denote this space by $T_{\mathbf{u}}$.

Let u be the map embedding $\mathcal{C}$ in $\mathbf{R}^M$, with:

$$u : \mathbf{q} \in \mathcal{C} \mapsto u(\mathbf{q}) = (u_1(\mathbf{q}), \ldots, u_M(\mathbf{q})) \in \mathcal{M}_{\mathcal{C}}.$$

Let γ be a differentiable curve in $\mathcal{C}$. The representation of γ in $\mathcal{M}_{\mathcal{C}}$, γ', is defined by:

$$s \in \mathbf{R} \mapsto \gamma'(s) = (u_1(\gamma(s)), \ldots, u_M(\gamma(s))) \in \mathcal{M}_{\mathcal{C}}.$$

It is a differentiable function of s and the M-vector

$$\frac{d\vec{\gamma}'(s)}{ds}_{|s=s_0} = \left(\frac{du_i(\gamma(s))}{ds}_{|s=s_0}\right)_{i=1,\ldots,M}$$

is the tangent vector to γ' at $s = s_0$. This vector belongs to $T_{\mathbf{u}_0}$, where $\mathbf{u}_0 = \gamma'(s_0) = u(\gamma(s_0))$. Thus, the tangent space $T_{\mathbf{u}_0}$ can be regarded as the space spanned by the tangent vectors to all the differentiable curves at $\mathbf{u}_0 \in \mathcal{M}_{\mathcal{C}}$.

Tangent space and tangent vectors can be defined independently of the embedding of $\mathcal{C}$ into $\mathbf{R}^M$. Two differentiable curves γ_1 and γ_2 in $\mathcal{C}$ are tangent at $s = s_0$ if $\gamma_1(s_0) = \gamma_2(s_0) = \mathbf{q}_0$ and, if in some chart (U, ϕ) in the atlas of $\mathcal{C}$ such that $\mathbf{q}_0 \in U$, we have:

$$\forall i \in [1, m] : \lim_{s \to s_0} \frac{x_i(\gamma_1(s)) - x_i(\gamma_2(s))}{s - s_0} = 0,$$

i.e. $(dx_i(\gamma_1(s))/ds)_{|s=s_0} = (dx_i(\gamma_2(s))/ds)_{|s=s_0}$, for $i = 1, \ldots, m$. Again, since overlapping charts are C^∞-related, this relationship is independent

[8] We say that the curve is smooth when the functions $x_i(\gamma(s))$ are smooth, i.e. have derivatives of all orders.

of the chart at $\mathbf{q}_0$ in the atlas. It defines an equivalence class of paths passing through $\mathbf{q}_0$ in $\mathcal{C}$, which is characterized by the m numbers:

$$\xi_i = \frac{dx_i(\gamma_1(s))}{ds}_{|s=s_0} = \frac{dx_i(\gamma_2(s))}{ds}_{|s=s_0}.$$

Such an equivalent class is called a **tangent vector** of $\mathcal{C}$ at $\mathbf{q}_0$.

Given two tangent vectors $\vec{\xi}$ and $\vec{\zeta}$ at $\mathbf{q}$, it is easy to construct curves γ and γ' passing through $\mathbf{q}$, such that γ and γ' admit $\lambda\vec{\xi}$ and $\vec{\xi}+\vec{\zeta}$ as their respective tangent vectors at $\mathbf{q}$. Hence, the set of tangent vectors of $\mathcal{C}$ at any $\mathbf{q} \in \mathcal{C}$ forms an m-dimensional vector space. This space is called the **tangent space** of $\mathcal{C}$ at $\mathbf{q}$. From now on, we will denote it by $T_{\mathbf{q}}(\mathcal{C})$.

We can define an inner product in a tangent space. Such an inner product induces a norm and makes it possible to compare the orientations of two differentiable curves at an intersection point. A manifold with an inner product on every tangent space is called a **Riemannian manifold**. The norm associated with this inner product is called the **Riemannian metric** [Arnold, 1978].

Let $\gamma : t \in [0,T] \mapsto \gamma(t) \in \mathcal{C}$, where t denotes time, be a differentiable curve representing a motion of $\mathcal{A}$. At any time $t \geq 0$, the tangent vector to γ at the current configuration represents the velocity of $\mathcal{A}$. Thus, the tangent space is also called *velocity space*.

It is sometimes convenient to blend the tangent spaces at the various points of $\mathcal{C}$ into a single structure. The union of the tangent spaces at the various points of $\mathcal{C}$, $T(\mathcal{C}) = \bigcup_{\mathbf{q}\in\mathcal{C}} T_{\mathbf{q}}(\mathcal{C})$, is called the **tangent bundle** of $\mathcal{C}$. It is a smooth manifold of dimension $2m$ [Spivak, 1979]. Since a pair (configuration, velocity) is called a phase in Physics and a state in Control Theory, the tangent bundle of $\mathcal{C}$ is also named *phase space* and *state space*.

6.2 Basis Induced by a Chart

Let (U,ϕ) be a chart and X be the image of U in $\mathbf{R}^m$ by ϕ. A basis of $\mathbf{R}^m$ and the chart (U,ϕ) **induce** an ordered basis[9] $(\vec{t}_1, ..., \vec{t}_m)$ in $T_{\mathbf{q}}(\mathcal{C})$, which is defined as follows. Each vector $\vec{t}_i$ is the tangent vector at $\mathbf{q}$ to

[9]Ordering the vectors $\vec{t}_i$, $i = 1, ..., m$, makes sense because $\mathbf{R}^N \times SO(N)$ is orientable [Guillemin and Pollack, 1974] [Spivak, 1979].

any curve γ that maps in $\mathbf{R}^m$ to a curve $\phi \circ \gamma$ tangent at $\phi(\mathbf{q})$ to the ith vector of the basis of $\mathbf{R}^m$. The m numbers $\xi_i = \frac{dx_i(\gamma(s))}{ds}_{|s=s_0}$ are the **components** of the tangent vector of γ at $\mathbf{q}_0 = \gamma(s_0)$ in this basis.

Consider two charts (U, ϕ) and (U', ϕ'), with $\mathbf{q} \in U \cap U'$. Let β and β' be the bases induced by these charts in $T_\mathbf{q}(\mathcal{C})$ and $f = \phi' \circ \phi^{-1}$, with:

$$f : (x_1, ..., x_m) \in \mathbf{R}^m \mapsto (f_1(x_1, ..., x_m), ..., f_m(x_1, ..., x_m)) \in \mathbf{R}^m.$$

The **Jacobian matrix** of the map f, denoted by $J_f(x_1, ..., x_m)$, is the $m \times m$ matrix of the first derivatives of f, i.e. :

$$J_f(x_1, ..., x_m) = \begin{pmatrix} \frac{\partial f_1}{\partial x_1}(x_1, ..., x_m) & \cdots & \frac{\partial f_1}{\partial x_m}(x_1, ..., x_m) \\ \cdot & & \cdot \\ \cdot & & \cdot \\ \cdot & & \cdot \\ \frac{\partial f_m}{\partial x_1}(x_1, ..., x_m) & \cdots & \frac{\partial f_m}{\partial x_m}(x_1, ..., x_m) \end{pmatrix}.$$

Let $\vec{t}$ be a vector of $T_\mathbf{q}(\mathcal{C})$. We denote by $[\,\vec{t}\,]_\beta$ and $[\,\vec{t}\,]_{\beta'}$ the m-vectors of $\vec{t}$'s components in β and β'. We have:

$$[\,\vec{t}\,]_{\beta'} = J_f(x_1(\mathbf{q}), ..., x_m(\mathbf{q}))\,[\,\vec{t}\,]_\beta.$$

The basis induced in $T_\mathbf{q}(\mathcal{C})$ by the chart (U, ϕ) makes it possible to use the same copy of $\mathbf{R}^m$ for representing both the set $\phi(U)$ and the tangent spaces of $\mathcal{C}$ at all the configurations in U. However, a single vector of $\mathbf{R}^m$ does not determine a unique direction of motion relative to $\mathcal{F}_\mathcal{A}$. This direction depends on the current configuration of $\mathcal{A}$.

6.3 Principal Axis Basis

Consider the case where $N = 3$. It is often convenient to express the velocity of $\mathcal{A}$ as a six-dimensional vector $(\dot{x}_\mathcal{A}, \dot{y}_\mathcal{A}, \dot{z}_\mathcal{A}, \dot{\alpha}, \dot{\beta}, \dot{\gamma})$, where $\dot{x}_\mathcal{A}$, $\dot{y}_\mathcal{A}$, and $\dot{z}_\mathcal{A}$ are the linear velocities along the axes of $\mathcal{F}_\mathcal{A}$, while $\dot{\alpha}$, $\dot{\beta}$, and $\dot{\gamma}$ are the angular velocities around these axes. Pure instantaneous translations along each of the axes of $\mathcal{F}_\mathcal{A}$ and pure instantaneous rotations around each of these axes determine a vector basis in every tangent space, which we call the **principal axis basis.**[10]

[10]This name derives from the fact that $\mathcal{A}$'s principal axes of inertia are a natural choice for $\mathcal{F}_\mathcal{A}$'s axes.

Similarly, in the two-dimensional case, the principal axis basis consists of three vectors corresponding to two instantaneous translations along the axes of $\mathcal{F}_{\mathcal{A}}$ and an instantaneous rotation around the origin $O_{\mathcal{A}}$.

As we will verify below, when $N = 3$, the principal axis basis is not induced by any chart in $\mathcal{C}$. However, its relationship with a basis induced by a chart can be established. For instance, let $(\dot{x}_{\mathcal{A}}, \dot{y}_{\mathcal{A}}, \dot{z}_{\mathcal{A}}, \dot{\alpha}, \dot{\beta}, \dot{\gamma})$ be a tangent vector represented in the principal axis basis and $(\dot{x}_{\mathcal{W}}, \dot{y}_{\mathcal{W}}, \dot{z}_{\mathcal{W}}, \dot{\phi}, \dot{\theta}, \dot{\psi})$ be the same vector represented in the chart induced by the Eulerian chart defined in Subsection 4.2. We have:

$$\begin{pmatrix} \dot{x}_{\mathcal{A}} \\ \dot{y}_{\mathcal{A}} \\ \dot{z}_{\mathcal{A}} \end{pmatrix} = \mathbf{E}^T_{\phi\theta\psi} \begin{pmatrix} \dot{x}_{\mathcal{W}} \\ \dot{y}_{\mathcal{W}} \\ \dot{z}_{\mathcal{W}} \end{pmatrix}$$

where $\mathbf{E}^T_{\phi\theta\psi}$ denotes the transpose of the $SO(3)$ matrix[11] corresponding to the Euler angles (ϕ, θ, ψ), and:

$$\begin{pmatrix} \dot{\alpha} \\ \dot{\beta} \\ \dot{\gamma} \end{pmatrix} = \begin{pmatrix} -\sin\theta\cos\psi & \sin\psi & 0 \\ \sin\theta\sin\psi & \cos\psi & 0 \\ \cos\theta & 0 & 1 \end{pmatrix} \begin{pmatrix} \dot{\phi} \\ \dot{\theta} \\ \dot{\psi} \end{pmatrix}.$$

The determinant of the second system is $-\sin\theta$, which is non-zero for $0 < \theta < \pi$. One can also verify that the second system cannot be integrated, i.e. $\dot{\alpha}$, $\dot{\beta}$ and $\dot{\gamma}$ are not the differentials of functions of the Eulerian angles. Indeed, from the third equation, it follows that $\frac{\partial\gamma}{\partial\phi} = \cos\theta$ and $\frac{\partial\gamma}{\partial\theta} = 0$. Hence $\frac{\partial}{\partial\theta}\frac{\partial\gamma}{\partial\phi} \neq \frac{\partial}{\partial\phi}\frac{\partial\gamma}{\partial\theta}$. If the principal axis basis was induced by a chart in $\mathcal{C}$, $\dot{\alpha}$, $\dot{\beta}$ and $\dot{\gamma}$ would be the differentials of coordinates α, β and γ, which could be expressed in terms of ϕ, θ and ψ.

When the principal axis basis is used, a vector of $\mathbf{R}^m$, with $\mathbf{R}^m$ being considered as a representation of the tangent spaces of $\mathcal{C}$, determines a unique direction of motion relative to $\mathcal{F}_{\mathcal{A}}$, independent of the configuration of $\mathcal{A}$.

6.4 Remark on the Differentiability of Paths

Let us consider a path $\tau : s \in [0,1] \mapsto \tau(s) \in \mathcal{C}$. A bijective function $f : s \in [0,1] \mapsto f(s) \in [0,1]$, with $f(0) = 0$ and $f(1) = 1$, is called a

[11]The inverse of a matrix in $SO(N)$ is equal to the transpose of this matrix.

reparameterization of τ. The path $\tau' : s \in [0,1] \mapsto \tau'(s) = \tau(f(s)) \in \mathcal{C}$ produces the same geometric curve segment as τ, but the rate at which this segment is traced out when s varies from 0 to 1 is different.

Assume that τ is differentiable at every $s \in [0,1]$, except possibly at a finite number of values of s. The non-differentiability of τ at some points may simply be an effect of the parameterization of the path. For example, if τ is not differentiable at s, but if its left and right tangent vectors at s:

$$\lim_{\varepsilon \to 0} \frac{d\vec{\tau}}{ds}_{|s-\varepsilon} \quad \text{and} \quad \lim_{\varepsilon \to 0} \frac{d\vec{\tau}}{ds}_{|s+\varepsilon}$$

exist, are non-zero, and have the same orientation, i.e. :

$$\lim_{\varepsilon \to 0} \frac{d\vec{\tau}}{ds}_{|s-\varepsilon} = \alpha \lim_{\varepsilon \to 0} \frac{d\vec{\tau}}{ds}_{|s+\varepsilon} \quad \text{with} \quad \alpha > 0,$$

then we can reparameterize τ into a path τ' that produces the same curve segment as τ and is differentiable at s.

If τ is differentiable everywhere except possibly at a finite number of points, it has a finite length (using, for instance, the Euclidean metric in $\mathbf{R}^M$). Let $l : [0,1] \rightarrow \mathbf{R}$ be the function that measures the length between $\tau(0)$ and $\tau(s)$ along τ. $L = \tau(1)$ is the length of the curve segment representing τ. The function $l(s)/L$ (normalized length) is a reparameterization of τ.

Let τ be a path whose parameter denotes the normalized length. If τ is differentiable at s, its tangent vector at s is non-zero. If τ is non-differentiable at s, neither the left nor the right tangent vector at s are zero and they do not have the same orientation. The direction of the tangent vector at $\tau(s)$, if τ is differentiable at s, determines the direction of the velocity of $\mathcal{A}$ for any execution of the path. The velocity of a physical object $\mathcal{A}$ cannot change instantaneously without consuming infinite energy. Therefore, the points where τ is not differentiable determine configurations where $\mathcal{A}$ has to stop.

A path τ is said to be of **class** C^1 if the reparameterized path using the normalized length is differentiable everywhere. It is said to be **piecewise of class** C^1 if the reparameterized path is differentiable everywhere except possibly at a finite number of values of s.

A path planner should only generate paths of class C^1 or piecewise of class C^1. Indeed, although we assume no dynamic constraint in the basic motion planning problem, there is little interest in generating paths which we know in advance to be infeasible.

All the path planning methods presented later directly generate piecewise differentiable paths. There is nothing surprising in that. On the one hand, non-piecewise differentiable paths are not the easiest to build. On the other hand, they are not needed. Indeed, every path τ in $\mathcal{C}$ is homotopic to a piecewise smooth path, which can be made arbitrarily close to it — i.e. for every $\varepsilon > 0$, there exists a piecewise smooth path τ' such that:

$$\forall s \in [0,1]: \; d(\tau(s), \tau'(s)) < \varepsilon,$$

where d is the metric in use in $\mathcal{C}$ (see Section 7). This may be seen by covering τ with a finite number of open subsets U_i, $i = 1, 2, \ldots$, of $\mathcal{C}$ (remember that τ is compact). Using charts of $\mathcal{C}$, each U_i can be mapped to an open subset X_i of $\mathbf{R}^m$. Thus, τ is homotopic to a piecewise smooth path τ', defined as the inverse image of a finite sequence of straight segments contained in the X_i's. The path τ' can be made as close as one wishes to τ by taking each of the U_i's small enough. More generally, using Weierstrass' Approximation Theorem, one can show that any path τ in $\mathcal{C}$ is homotopic to a smooth path, which can be made as close as one wishes to τ. Most path planning methods, however, do not produce smooth paths directly.

7 Metrics

We mentioned earlier that several metrics yield the same topology of $\mathcal{C}$ as the one previously defined. In this section we give several such metrics. Metrics play no role in the definition of the basic motion planning problem. However, some planning methods use distances to the obstacles and to the goal configuration as heuristic functions in order to guide the search process.

One straightforward way to define a distance function in $\mathcal{C}$ is to let $\mathcal{C}$ inherit the Euclidean distance of the ambient space $\mathbf{R}^M$, i.e. if $\mathbf{q}_1$ and $\mathbf{q}_2$ are two configurations mapping to $\mathbf{u}_1$ and $\mathbf{u}_2$ in $\mathcal{M}_\mathcal{C} \subset \mathbf{R}^M$, then:

$$d(\mathbf{q}_1, \mathbf{q}_2) = \|\mathbf{u}_1 - \mathbf{u}_2\|.$$

A second way to define a distance in $\mathcal{C}$ is by making use of the Euclidean distance in the workspace $\mathcal{W}$. For example, the two functions $d : \mathcal{C} \times \mathcal{C} \rightarrow \mathbf{R}$ specified by:

$$d(\mathbf{q}_1, \mathbf{q}_2) = \max_{a \in \mathcal{A}} \|a(\mathbf{q}_1) - a(\mathbf{q}_2)\|$$

$$d(\mathbf{q}_1, \mathbf{q}_2) = \sum_{k \in [1,N]} \|a_k(\mathbf{q}_1) - a_k(\mathbf{q}_2)\|$$

where a_k, $k = 1, ..., N$, are distinct points in $\mathcal{A}$ (if $N = 3$, the points should not be aligned), are distances in $\mathcal{C}$. (The first of these two functions is a distance because $\mathcal{A}$ is compact.)

Another example of a metric in $\mathcal{C}$ whose definition makes use of the Euclidean distance in the workspace is known as the Hausdorff distance [Grünbaum, 1967]. It requires $\mathcal{A}$ to be compact and is defined as follows:

$$d(\mathbf{q}_1, \mathbf{q}_2) = \max\{d_h(\mathcal{A}(\mathbf{q}_1), \mathcal{A}(\mathbf{q}_2)), d_h(\mathcal{A}(\mathbf{q}_2), \mathcal{A}(\mathbf{q}_1))\}$$

with:

$$d_h(U, V) = \max_{u \in U} \min_{v \in V} \|u - v\|.$$

(Notice that in general: $d_h(U, V) \neq d_h(V, U)$.) For any compact subset S of $\mathbf{R}^N$ and $\varepsilon > 0$, let $S_\varepsilon = \{x \in \mathbf{R}^N \;/\; \exists y \in S : \|x - y\| \leq \varepsilon\}$. Then d_h can also be defined by: $d_h(U, V) = \min\{\varepsilon > 0 \;/\; U \subseteq V_\varepsilon \text{ and } V \subseteq U_\varepsilon\}$.

One can easily verify that all the metrics proposed above induce the same topology of $\mathcal{C}$.

A third way to define a metric in $\mathcal{C}$ is to make $\mathcal{C}$ into a Riemannian manifold by defining an inner product on every tangent space $T_{\mathbf{q}}(\mathcal{C})$. (We denote by $\|\vec{\xi}\|$ the norm of a vector $\vec{\xi} \in T_{\mathbf{q}}(\mathcal{C})$ associated with this inner product.) Let $\gamma : s \in [0, 1] \mapsto \gamma(s) \in \mathcal{C}$ be a smooth curve in $\mathcal{C}$ and $\frac{d\vec{\gamma}(s)}{ds}$ be the tangent vector of γ at s. The integral:

$$L(\gamma) = \int_0^1 \|\frac{d\vec{\gamma}(s)}{ds}\| ds$$

is well-defined and is called the *length* of γ induced by the Riemannian metric. Since $\mathcal{C}$ is connected:

$$d(\mathbf{q}_1, \mathbf{q}_2) = \inf_{\gamma}\{L(\gamma)\}$$

is a distance in $\mathcal{C}$ [Spivak, 1979].

One can show that the above distance induces the same topology in $\mathcal{C}$ as the previous metrics (see [Spivak, 1979], pp. 428–429). $\mathcal{C}$ is also "geodesically complete" for this metric (see [Spivak, 1979], pp. 462–464). This means that between any two configurations $\mathbf{q}_1$ and $\mathbf{q}_2$, there exists at least one curve, called the *geodesic of minimal length*, whose length is equal to $d(\mathbf{q}_1, \mathbf{q}_2)$. This curve may not be unique, e.g. there are two curves of minimal length between any two antipodal points of S^1.

8 Particular Cases

So far, we have assumed that the robot $\mathcal{A}$ has no symmetry. Removing this restriction yields particular cases considered below. Some of these cases are of a particular interest because they result in configuration spaces of lower dimensions than $\mathbf{R}^N \times SO(N)$, or even in Euclidean configuration spaces, in which path planning concepts and methods are easier to introduce.

If $\mathcal{A}$ is an N-ball[12] with no mark on it, we cannot distinguish the configurations that only differ by a rotation around the center of the ball. Then, by choosing $O_{\mathcal{A}}$ at the center of the ball, the configuration space can be represented as the N-dimensional Euclidean space $\mathbf{R}^N$.

If $\mathcal{A}$ is a cylinder of finite length[13] in $\mathbf{R}^3$, we cannot distinguish between two configurations that only differ by a rotation around the axis of the cylinder. Then, choosing $O_{\mathcal{A}}$ on $\mathcal{A}$'s axis and one of $\mathcal{F}_{\mathcal{A}}$'s axes along $\mathcal{A}$'s axis leads to representing $\mathcal{C}$ as $\mathbf{R}^3 \times S^2$, where S^2 denotes the unit sphere in $\mathbf{R}^3$. Thus, in this case, $\mathcal{C}$ is a five-dimensional manifold. S^2 can be parameterized using the stereographic projection or the spherical angles.

Non-revolute axial symmetries in $\mathcal{A}$ do not reduce the dimension of $\mathcal{C}$. However, they transform the configuration space into a quotient space of $\mathbf{R}^N \times SO(N)$. For instance, a square $\mathcal{A}$ in $\mathbf{R}^2$ is invariant under a rotation by $\pi/2$ around its center. By choosing $O_{\mathcal{A}}$ at the center of the square, we can represent $\mathcal{C}$ as $\mathbf{R}^2 \times [SO(2)/\{\theta \sim \theta + k\pi/2\}]$, where

[12]This includes the case where $\mathcal{A}$ is a single point in $\mathbf{R}^N$.

[13]This includes the case where $\mathcal{A}$ is a line segment in $\mathbf{R}^3$.

$SO(2)/\{\theta \sim \theta + k\pi/2\}$ is the group of planar rotations modulo $\pi/2$. We will not consider this particular case further in the book.

An interesting case occurs when $\mathcal{A}$ is a dimensioned object which can translate but not rotate. Then $\mathcal{C}$ can be represented as $\mathbf{R}^N$, with a configuration $\mathbf{q}$ being defined as $O_{\mathcal{A}}$'s coordinates in $\mathcal{F}_{\mathcal{W}}$. If any path in $\mathcal{C} = \mathbf{R}^N$ is possible, we say that $\mathcal{A}$ is a **free-translating object**. This situation is similar to that obtained when $\mathcal{A}$ is a free-flying N-ball.

Whenever $\mathcal{C} = \mathbf{R}^N$, we can use a single copy of $\mathbf{R}^N$ to represent both $\mathcal{C}$ and all the tangent spaces.

9 Obstacles

Let $\mathcal{W}$ now contain fixed obstacles $\mathcal{B}_i$, $i = 1, 2, ..., q$, which we represent as closed, but not necessarily bounded, regions of $\mathbf{R}^N$. The $\mathcal{B}_i$'s denote both the "physical" obstacles and the subsets of $\mathbf{R}^N$ that represent them.

Given a region $\mathcal{S}$ in a topological space $\mathcal{X}$, we denote the interior, boundary, and closure of $\mathcal{S}$, by $int(\mathcal{S})$, $\partial\mathcal{S}$, and $cl(\mathcal{S})$, respectively, and the complement of $\mathcal{S}$ in $\mathcal{X}$ by $\mathcal{X}\backslash\mathcal{S}$. By definition, $\partial\mathcal{S} = cl(\mathcal{S})\backslash int(\mathcal{S})$.

9.1 C-Obstacles

DEFINITION 3: *The obstacle $\mathcal{B}_i$ in $\mathcal{W}$ maps in $\mathcal{C}$ to the region $\mathcal{CB}_i = \{\mathbf{q} \in \mathcal{C} \;/\; \mathcal{A}(\mathbf{q}) \cap \mathcal{B}_i \neq \emptyset\}$. $\mathcal{CB}_i$ is called a* **C-obstacle.**

The union of all the C-obstacles in $\mathcal{C}$ is called the **C-obstacle region.** We will investigate the representation of C-obstacles in detail in Chapter 3, when both $\mathcal{A}$ and $\mathcal{B}$ are represented as semi-algebraic subsets of $\mathbf{R}^N$. Below, we illustrate the notion of a C-obstacle with very simple examples.

Examples:

- Let $\mathcal{A}$ be a disc, with $O_{\mathcal{A}}$ at the center of the disc, and $\mathcal{B}$ be a polygon in $\mathcal{W} = \mathbf{R}^2$. Then $\mathcal{C} = \mathbf{R}^2$. $\mathcal{CB}$ is obtained by "growing" $\mathcal{B}$ isotropically by the radius of $\mathcal{A}$ (see Figure 6). The boundary of $\mathcal{CB}$ is the curve followed by $O_{\mathcal{A}}$ when $\mathcal{A}$ "rolls" over the boundary of $\mathcal{B}$.

- Figure 7 illustrates the case where both $\mathcal{A}$ and $\mathcal{B}$ are convex polygons in $\mathbf{R}^2$, $\mathcal{A}$ being able to translate freely at fixed orientation. It shows

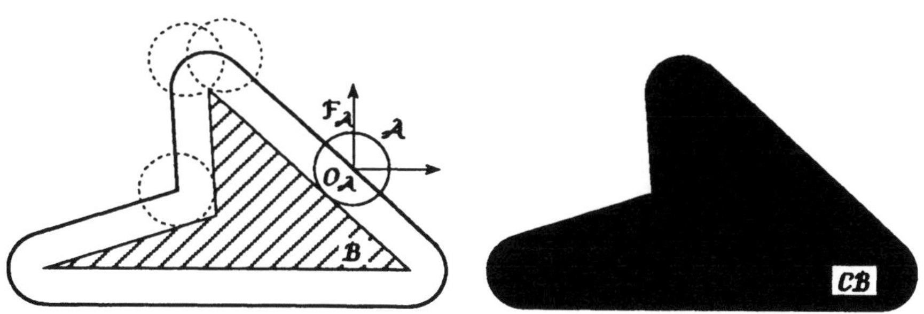

Figure 6. $\mathcal{A}$ is a disc in $\mathbf{R}^2$ and the origin of $\mathcal{F}_{\mathcal{A}}$ is chosen at the center of the disc. $\mathcal{B}$ is a polygon. The C-obstacle $\mathcal{CB}$ is obtained by growing $\mathcal{B}$ isotropically by the radius of $\mathcal{A}$.

the C-obstacles in $\mathbf{R}^2$ obtained for two different fixed orientations of $\mathcal{A}$. The boundary of each C-obstacle is the curve followed by $O_{\mathcal{A}}$ when $\mathcal{A}$ slides in contact with $\mathcal{B}$'s boundary, without overlapping of $\mathcal{A}$'s and $\mathcal{B}$'s interiors. Each C-obstacle is a convex polygon.

- Figure 8 displays a representation of the C-obstacle $\mathcal{CB}$ when $\mathcal{A}$ and $\mathcal{B}$ are the same convex polygons as in the previous example, but $\mathcal{A}$ can both translate and rotate. The representation is given by parameterizing each configuration $\mathbf{q}$ by $(x, y, \theta) \in \mathbf{R}^2 \times [0, 2\pi]$. Each cross-section through the C-obstacle, perpendicular to the θ-axis, is the two-dimensional C-obstacle at the corresponding fixed orientation of $\mathcal{A}$. ∎

Let us denote by $\mathcal{CB}_\Theta$ the cross-section of the representation of $\mathcal{CB}$ in $\mathbf{R}^N \times SO(N)$ at a fixed orientation $\Theta \in SO(N)$ of $\mathcal{A}$, by $\mathcal{A}_\Theta$ the object $\mathcal{A}$ at that orientation, and by $\mathcal{A}_\Theta(\mathcal{T})$ the region of $\mathbf{R}^N$ occupied by $\mathcal{A}_\Theta$ at position $\mathcal{T} \in \mathbf{R}^N$. The following proposition relates the shape $\mathcal{CB}_\Theta$ to the shapes of $\mathcal{A}$ and $\mathcal{B}$ [Lozano-Pérez, 1981]. It is illustrated by the above three examples.

PROPOSITION 3: *Posing:*

$$\mathcal{B} \ominus \mathcal{A}_\Theta(0) = \{x \in \mathcal{W} \; / \; \exists b \in \mathcal{B}, \exists a_0 \in \mathcal{A}_\Theta(0) : x = b - a_0\},$$

we have: $\mathcal{CB}_\Theta = \mathcal{B} \ominus \mathcal{A}_\Theta(0)$.

($\ominus$ denotes the Minkowski operator for affine set difference.)

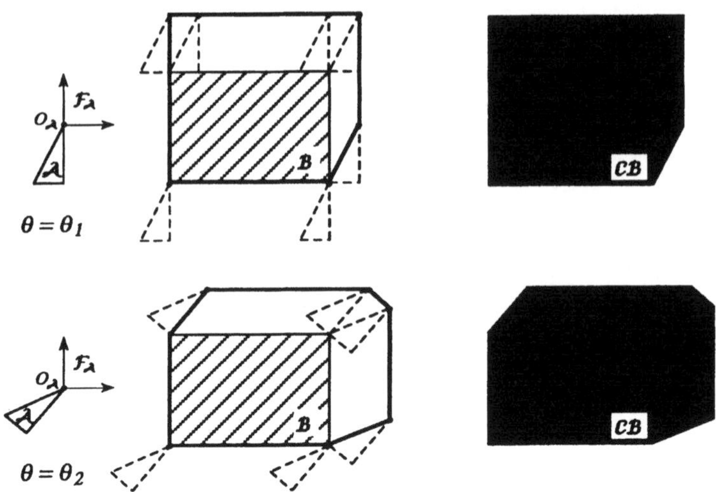

Figure 7. $\mathcal{A}$ and $\mathcal{B}$ are convex polygons. $\mathcal{A}$ can translate freely, but cannot rotate. Therefore, $\mathcal{C} = \mathbf{R}^2$. The figure shows two C-obstacles corresponding to two different fixed orientations of $\mathcal{A}$.

Proof:

1. $\forall a \in \mathcal{A}_\Theta(\mathcal{T})$: $a = a_0 + \mathcal{T}$, with $a_0 \in \mathcal{A}_\Theta(0)$.

   $\forall b \in \mathcal{A}_\Theta(\mathcal{T}) \cap \mathcal{B}$: $\exists a_0 \in \mathcal{A}_\Theta(0)$ such that $b = a_0 + \mathcal{T}$.

   $\forall \mathcal{T} \in \mathcal{CB}_\Theta$: $\mathcal{A}_\Theta(\mathcal{T}) \cap \mathcal{B} \neq \emptyset$.

   Hence, $\mathcal{T} \in \mathcal{B} \ominus \mathcal{A}_\Theta(0)$ and $\mathcal{CB}_\Theta \subseteq \mathcal{B} \ominus \mathcal{A}_\Theta(0)$.

2. $\forall a_0 \in \mathcal{A}_\Theta(0)$: $a_0 + \mathcal{T} \in \mathcal{A}_\Theta(\mathcal{T})$.

   $\forall \mathcal{T} \in \mathcal{B} \ominus \mathcal{A}_\Theta(0)$: $\exists b \in \mathcal{B}$ and $\exists a_0 \in \mathcal{A}_\Theta(0)$ such that $\mathcal{T} = b - a_0$.

   Hence, this same b is an element of $\mathcal{A}_\Theta(\mathcal{T}) \cap \mathcal{B}$, and $\mathcal{B} \ominus \mathcal{A}_\Theta(0) \subseteq \mathcal{CB}_\Theta$.

Therefore: $\mathcal{CB}_\Theta = \mathcal{B} \ominus \mathcal{A}_\Theta(0)$. ∎

Since the Minkowski sum of two convex sets is convex, $\mathcal{CB}_\Theta$ is convex whenever both $\mathcal{A}$ and $\mathcal{B}$ are convex.

9.2 Compactness and Connectedness

We show below that:

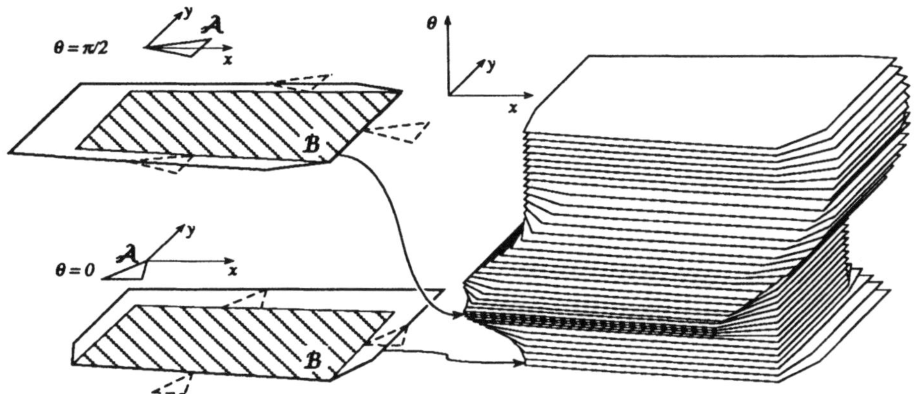

Figure 8. $\mathcal{A}$ and $\mathcal{B}$ are convex polygons. $\mathcal{A}$ is a free-flying object. $\mathcal{C}$ is represented by parameterizing each configuration $\mathbf{q}$ by $(x, y, \theta) \in \mathbf{R}^2 \times [0, 2\pi]$. The represented C-obstacle is a volume bounded by patches of ruled surfaces (see Chapter 3).

- $\mathcal{CB}$ is compact if $\mathcal{B}$ is compact (we already required that $\mathcal{A}$ be compact);
- $\mathcal{CB}$ is closed;
- $\mathcal{CB}$ is connected if both $\mathcal{A}$ and $\mathcal{B}$ are connected.

PROPOSITION 4: *If $\mathcal{B}$ is compact, so is $\mathcal{CB}$.*

Proof: We first show that $\mathcal{CB}$ is closed. Then, embedding $\mathcal{C}$ into $\mathbf{R}^M$, we show that $\mathcal{CB}$ is a bounded subset of $\mathbf{R}^M$. As a closed bounded subset of $\mathbf{R}^M$, $\mathcal{CB}$ is compact.

(1) If for any sequence $\{\mathbf{q}_i\}$, $i = 1, 2, ...,$ of configurations in $\mathcal{CB}$ such that $\lim_{i\to\infty} \mathbf{q}_i = \mathbf{q}$, we can prove that $\mathbf{q}$ is in $\mathcal{CB}$, then $\mathcal{CB}$ is closed. Indeed, the configuration $\mathbf{q}$ defined as the limit of a sequence of configurations in $\mathcal{CB}$ is called a *limit point* of $\mathcal{CB}$, and a set that contains all its limit points is a closed set.

Given $\{\mathbf{q}_i\}$, we can define two sequences of points, $\{a_i\}$ in $\mathcal{A}$ and $\{b_i\}$ in $\mathcal{B}$, such that for every i, $a_i(\mathbf{q}_i)$ and b_i coincide in $\mathcal{W}$. These two sequences exist since $\mathbf{q}_i \in \mathcal{CB}$ for all i. Since $\mathcal{A}$ and $\mathcal{B}$ are both compact subsets of the Euclidean space $\mathbf{R}^N$, they are both compact metric sets, and their cross-product is also a compact metric set. A compact metric set is also sequentially compact, which means that every sequence in the set has a convergent subsequence [Jänich, 1984]. Thus, there exists a convergent subsequence $\{(a_j, b_j)\} \subseteq \{(a_i, b_i)\}$. Let $lim_{j\to\infty}(a_j, b_j) = (a, b)$. Since both $\mathcal{A}$ and $\mathcal{B}$ are closed sets, $a \in \mathcal{A}$ and $b \in \mathcal{B}$.

By definition of $\{a_i\}$ and $\{b_i\}$, we have $a_j(\mathbf{q}_j) = b_j \in \mathcal{A}(\mathbf{q}_j) \cap \mathcal{B}$. Hence $\|a_j(\mathbf{q}_j) - b_j\| = 0$, and $\lim_{j\to\infty} \|a_j(\mathbf{q}_j) - b_j\| = 0$. Since the Euclidean distance $\|x - y\|$ is a continuous function of both x and y and $f(a_j, \mathbf{q}_j) = a_j(\mathbf{q}_j)$ is a continuous function of both $\mathbf{q}_j$ and a_j, we can write:

$$\lim_{j\to\infty} \|a_j(\mathbf{q}_j) - b_j\| = \|a(\mathbf{q}) - b\|.$$

Therefore: $\|a(\mathbf{q}) - b\| = 0$, $\mathbf{q} \in \mathcal{CB}$, and $\mathcal{CB}$ is closed.

(2) Let $\mathcal{C}$ be represented as $\mathbf{R}^N \times SO(N)$. Let $\rho = \max_{a\in\mathcal{A}} \|O_{\mathcal{A}} - a\|$ and $ball(\rho, 0)$ be the N-ball of radius ρ centered at the origin of $\mathcal{F_W}$. For all $\Theta \in SO(N)$, we have $\mathcal{CB}_\Theta \subseteq \mathcal{X}$, with $\mathcal{X} = \mathcal{B} \ominus ball(\rho, 0)$. Hence, $\mathcal{CB} \subseteq \mathcal{X} \times SO(N)$. Since $SO(N)$ is compact, $\mathcal{CB}$ is bounded. ∎

PROPOSITION 5: *$\mathcal{CB}$ is closed.*

Proof: This proposition can be established by slightly modifying the first part of the above proof to handle the case where $\mathcal{B}$ is not bounded. Since the sequence $\{\mathbf{q}_i\}$ converges, there exists Q such that the set of all $\mathbf{q}_i$, with $i > Q$, maps in $\mathbf{R}^N \times SO(N)$ to a bounded subset. The area $\bigcup_{i>Q} \mathcal{A}(\mathbf{q}_i) \subset \mathcal{W}$ is bounded, and hence is contained in a compact subset S of $\mathcal{W}$. The sequence $\{b_i\}$ is contained in $\mathcal{B} \cap S$, which is a compact metric set, hence a sequentially compact set. ∎

PROPOSITION 6: *If both $\mathcal{A}$ and $\mathcal{B}$ are connected, so is $\mathcal{CB}$.*

Proof: Let us consider any two configurations $\mathbf{q}_1 = (X_1, \Theta_1)$ and $\mathbf{q}_2 = (X_2, \Theta_2)$ in $\mathcal{CB}$. By definition, $\mathcal{A}(\mathbf{q}_1) \cap \mathcal{B} \neq \emptyset$ and $\mathcal{A}(\mathbf{q}_2) \cap \mathcal{B} \neq \emptyset$. Assume that $\mathcal{A}$ is at configuration $\mathbf{q}_1$. We can construct a path τ such that $\tau(0) = \mathbf{q}_1$, $\tau(1) = \mathbf{q}_2$, and, for all $s \in [0, 1]$, $\tau(s) \in \mathcal{CB}$. This path is the product $\tau = \tau_1 \bullet \tau_2 \bullet \tau_3$ of three paths τ_1, τ_2, and τ_3 defined below.

Let a_1 be a point of $\mathcal{A}$ and b_1 a point of $\mathcal{B}$ such that $a_1(\mathbf{q}_1)$ and b_1 coincide. Let a_2 be a point of $\mathcal{A}$ and b_2 a point of $\mathcal{B}$ such that $a_2(\mathbf{q}_2)$

and b_2 coincide.

- The path τ_1 consists of rotating $\mathcal{A}$ around a fixed axis passing through a_1 from the orientation Θ_1 to the orientation Θ_2.
- Let σ be a continuous curve segment included in $\mathcal{B}$ and connecting b_1 to b_2. τ_2 is obtained when $\mathcal{A}$ moves at the fixed orientation Θ_2 in such a way that a_1 follows σ.
- Let σ' be a continuous curve segment included in $\mathcal{A}$ and connecting a_1 to a_2. τ_3 is obtained when $\mathcal{A}$ moves at the fixed orientation Θ_2 in such a way that b_2 follows σ'.

Thus, $\mathcal{CB}$ is connected. ■

Notice that the last proof requires that $\mathcal{A}$ can translate freely at any fixed orientation.

9.3 Regularity

In this subsection we assume that both $\mathcal{A}$ and $\mathcal{B}$ are **regular** sets, i.e. $\mathcal{A} = cl(int(\mathcal{A}))$ and $\mathcal{B} = cl(int(\mathcal{B}))$ [Requicha, 1977]. We show that under this assumption, $\mathcal{CB}$ is also regular.

Let $\mathcal{CB}_{overlap}$ denote the subset of configurations where the interior of $\mathcal{A}$ intersects the interior of $\mathcal{B}$, i.e. :

$$\mathcal{CB}_{overlap} = \{\mathbf{q} \in \mathcal{C} \ / \ int(\mathcal{A}(\mathbf{q})) \cap int(\mathcal{B}) \neq \emptyset\}.$$

Let $\mathcal{CB}_{contact}$ denote the subset of configurations where $\mathcal{A}$ and $\mathcal{B}$, but not their interiors, intersect, i.e. :

$$\mathcal{CB}_{contact} = \{\mathbf{q} \in \mathcal{C} \ / \ \mathcal{A}(\mathbf{q}) \cap \mathcal{B} \neq \emptyset \text{ and } int(\mathcal{A}(\mathbf{q})) \cap int(\mathcal{B}) = \emptyset\}.$$

We have: $\mathcal{CB}_{overlap} = \mathcal{CB} \backslash \mathcal{CB}_{contact}$.

The boundary of $\mathcal{CB}$ is part of $\mathcal{CB}_{contact}$. Indeed, since $\mathcal{A}$ and $\mathcal{B}$ are regular, if $\mathbf{q}$ is in $\partial\mathcal{CB}$, then an arbitrarily small displacement of $\mathcal{A}$ can make $\mathcal{A}$ and $\mathcal{B}$ disjoint, which would not be possible if the interiors of $\mathcal{A}(\mathbf{q})$ and $\mathcal{B}$ were intersecting. However, $\partial\mathcal{CB}$ may be different from $\mathcal{CB}_{contact}$, and the only general relation between them is:

$$\partial\mathcal{CB} \subseteq \mathcal{CB}_{contact}.$$

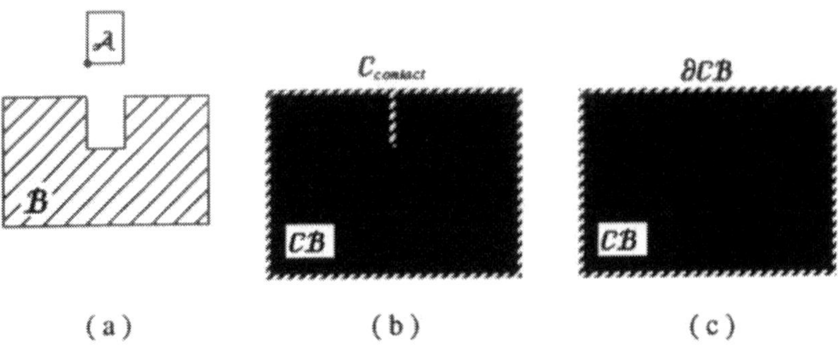

Figure 9. This figure illustrates an example where $\partial\mathcal{CB} \neq \mathcal{C}_{contact}$. Figure a depicts the objects $\mathcal{A}$ and $\mathcal{B}$ in the workspace. $\mathcal{C}_{contact}$ and $\partial\mathcal{CB}$ are shown in striped lines in Figures b and c, respectively.

Figure 9 gives a simple example where $\partial\mathcal{CB}$ is strictly included in $\mathcal{CB}_{contact}$.

PROPOSITION 7: $\mathcal{CB} = cl(\mathcal{CB}_{overlap})$.

Proof: Let $\mathbf{q} = (\mathcal{T}, \Theta)$ be a configuration in $\mathcal{CB}$. Let a be a point in $\mathcal{A}$ and b a point in $\mathcal{B}$ such that $a(\mathbf{q})$ and b coincide in $\mathcal{W}$. Since both $\mathcal{A}$ and $\mathcal{B}$ are regular sets, there are sequences of points $\{a_k\}$ in $int(\mathcal{A})$ and $\{b_k\}$ in $int(\mathcal{B})$ such that $\lim_{k\to\infty} a_k = a$ and $\lim_{k\to\infty} b_k = b$. Let $\{\mathbf{q}_k\} = \{(\mathcal{T}_k, \Theta)\}$ be a sequence of configurations in $\mathcal{CB}_{overlap}$ such that $\mathcal{A}(\mathbf{q}_k)$ and $\mathcal{A}(\mathbf{q})$ have the same orientation Θ and $a_k(\mathbf{q}_k)$ and b_k coincide. We have: $\lim_{k\to\infty} \mathbf{q}_k = \mathbf{q}$. Therefore: $\mathbf{q} \in cl(\mathcal{CB}_{overlap})$ and $\mathcal{CB} \subseteq cl(\mathcal{CB}_{overlap})$.

Since $\mathcal{CB}_{overlap} \subset \mathcal{CB}$ and $\mathcal{CB}$ is closed, we also have: $cl(\mathcal{CB}_{overlap}) \subseteq \mathcal{CB}$.

Hence: $\mathcal{CB} = cl(\mathcal{CB}_{overlap})$. ∎

Notice that this proof requires that $\mathcal{A}$ be able to translate freely at fixed orientation.

PROPOSITION 8: *$\mathcal{CB}$ is regular.*

Proof: The two relations:

$$\mathcal{CB}_{overlap} = \mathcal{CB} \backslash \mathcal{CB}_{contact}$$

and

$$\partial\mathcal{CB} \subseteq \mathcal{CB}_{contact}$$

entail:

$$\mathcal{CB}_{overlap} \subseteq int(\mathcal{CB}) \subset \mathcal{CB}.$$

Taking the closure of each set, we obtain:

$$cl(\mathcal{CB}_{overlap}) \subseteq cl(int(\mathcal{CB})) \subseteq \mathcal{CB}.$$

This relation combined with Proposition 7 implies that $cl(int(\mathcal{CB})) = \mathcal{CB}$. Hence, $\mathcal{CB}$ is regular. ■

9.4 Free and Semi-Free Paths

A free path is a path along which $\mathcal{A}$ touches or overlaps no obstacle. This is formalized as follows:

DEFINITION 4: Free space *is the subset of* $\mathcal{C}$ *defined by:*

$$\mathcal{C}_{free} = \mathcal{C} \setminus \bigcup_{i\in[1,q]} \mathcal{CB}_i.$$

Any configuration in $\mathcal{C}_{free}$ *is called a* **free configuration**. *A* **free path** *between two free configurations* $\mathbf{q}_{init}$ *and* $\mathbf{q}_{goal}$ *is a continuous map* $\tau : [0,1] \rightarrow \mathcal{C}_{free}$, *with* $\tau(0) = \mathbf{q}_{init}$ *and* $\tau(1) = \mathbf{q}_{goal}$.

Since the C-obstacles are closed subsets of $\mathcal{C}$, $\mathcal{C}_{free}$ is an open subset of $\mathcal{C}$, hence is a smooth manifold.

Two configurations belong to the same connected components of $\mathcal{C}_{free}$ if and only if they are connected by a free path. Thus, the basic motion planning problem — finding a free path between two free configurations — includes the pure topological problem of deciding whether the two configurations belong to the same connected component of $\mathcal{C}_{free}$.

Two free paths τ and τ' are **topologically equivalent** if and only if they are homotopic in $\mathcal{C}_{free}$, i.e. if there exists a continuous map $h : [0,1] \times [0,1] \rightarrow \mathcal{C}_{free}$, with $h(s,0) = \tau(s)$ and $h(s,1) = \tau'(s)$ for all $s \in [0,1]$.

Example: Consider the simple two-dimensional configuration space $\mathcal{C} = \mathbf{R}^2$ shown in Figure 10 with three disjoint compact C-obstacles.

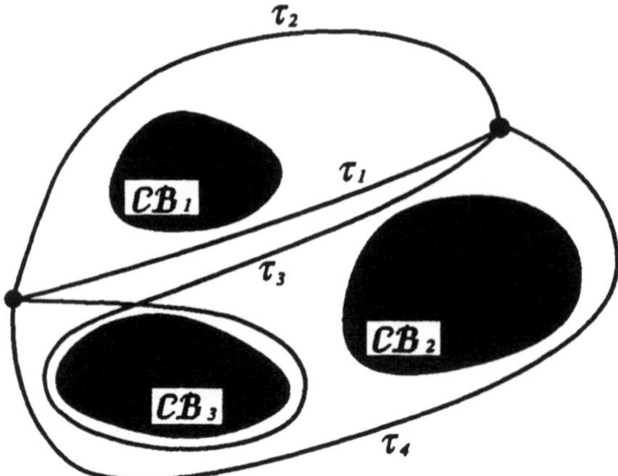

Figure 10. In $\mathcal{C} = \mathbf{R}^2$, every compact, connected component of the C-obstacle region induces a countable infinity of classes of topologically equivalent paths. The above figure shows four paths belonging to four distinct classes.

Although $\mathbf{R}^2$ is simply connected, $\mathcal{C}_{free}$ is multiply connected in this example. Between any two configurations in $\mathcal{C}_{free}$ there is a countable infinity of different classes of topologically equivalent paths. ■

Any free path is topologically equivalent to a (piecewise) smooth free path that is arbitrarily close to it (see Subsection 6.4).

DEFINITION 5: *A* **semi-free path** *between two configurations* $\mathbf{q}_{init}$ *and* $\mathbf{q}_{goal}$ *is a continuous map* $\tau : [0,1] \rightarrow cl(\mathcal{C}_{free})$, *with* $\tau(0) = \mathbf{q}_{init}$ *and* $\tau(1) = \mathbf{q}_{goal}$.

It can be verified that if there exists a semi-free path between two configurations, then there also exists a path between the same two configurations that lies in free space, except perhaps at isolated configurations [Laumond, 1986].

If $cl(\mathcal{C}_{free})$ is a *manifold with boundary* [Guillemin and Pollack, 1974] — i.e. if every configuration in $cl(\mathcal{C}_{free})$ has a neighborhood homeomorphic to either an open ball or a closed half-space of $\mathbf{R}^m$ — then the existence of a semi-free path τ' between two free configurations entails the existence of a free path τ. One such path can be obtained by an arbitrarily small deformation of τ'.

9.5 Contact Space and Valid Space

In this subsection we assume that both $\mathcal{A}$ and the $\mathcal{B}_i$'s are regular sets.

Contact space is the subset of $\mathcal{C}$ made of configurations at which $\mathcal{A}$ touches one or several obstacles without overlapping any. The union of free space and contact space is *valid space*, the set of configurations achievable by $\mathcal{A}$ when we accept contacts between $\mathcal{A}$ and the obstacles.

DEFINITION 6: Contact space *is the subset of* $\mathcal{C}$ *defined as:*

$$\mathcal{C}_{contact} = \{\mathbf{q} \in \mathcal{C} \; / \; \mathcal{A}(\mathbf{q}) \cap \bigcup_{i=1}^{q} \mathcal{B}_i \neq \emptyset \text{ and } int(\mathcal{A}(\mathbf{q})) \cap \bigcup_{i=1}^{q} int(\mathcal{B}_i) = \emptyset\}.$$

Valid space *is:* $\mathcal{C}_{valid} = \mathcal{C}_{free} \cup \mathcal{C}_{contact}$.

One can easily verify that $\mathcal{C}_{valid}$ is a closed subset of $\mathcal{C}$ and that $\partial(\bigcup_i \mathcal{CB}_i) \subseteq \mathcal{C}_{contact}$. Related properties include: $\partial\mathcal{C}_{free} \subseteq \mathcal{C}_{contact}$, $cl(\mathcal{C}_{free}) \subseteq \mathcal{C}_{valid}$, and $\partial\mathcal{C}_{free} \subseteq \partial\mathcal{C}_{valid}$.

DEFINITION 7: *A* **valid path** *(resp.* **contact path***) between two configurations* $\mathbf{q}_{init}$ *and* $\mathbf{q}_{goal}$ *in* $\mathcal{C}_{valid}$ *(resp.* $\mathcal{C}_{contact}$*) is a continuous map* $\tau : [0,1] \rightarrow \mathcal{C}_{valid}$ *(resp.* $\mathcal{C}_{contact}$*), with* $\tau(0) = \mathbf{q}_{init}$ *and* $\tau(1) = \mathbf{q}_{goal}$.

Let $\mathcal{A}$ and $\mathcal{B}$ be two connected objects. It can be shown, under the condition that $\mathcal{A}$ is free-flying, that if there exists a valid path of $\mathcal{A}$ between two contact configurations where $\mathcal{A}$ touches $\mathcal{B}$, then there also exists a contact path between the same two configurations [Hopcroft and Wilfong, 1986a].

10 Representation of Forces

Forces play no role in the definition of the basic motion planning problem. However, an important planning approach, potential field, uses artificial forces to guide planing. Forces will also be considered in Chapter 10 when we discuss motion planning with uncertainty in control.

Forces exerted on $\mathcal{A}$ at configuration $\mathbf{q}$ can be represented as vectors in the tangent space $T_{\mathbf{q}}(\mathcal{C})$. This representation requires that an inner product be defined in the tangent spaces, and it depends on the choice of this inner product. This results from the fact that a force is defined through the *virtual work* that it produces along virtual displacements [Greenwood, 1965]. This work, which is independent of the representation of the force, is expressed as an inner product. Changing the inner product requires us to change the representation of the force.

Let $\vec{\mathbf{F}}(\mathbf{x})$ be a field of N-dimensional vectors representing a field of forces in $\mathcal{W}$. The work of these forces along a curve $\gamma : s \in [0,1] \mapsto \gamma(s) \in \mathcal{W}$ is:

$$\int_0^1 \vec{\mathbf{F}}(\gamma(s)) \cdot \frac{d\vec{\gamma}(s)}{ds} ds$$

where "$\cdot$" denotes the inner product in $\mathcal{W}$ and $\frac{d\vec{\gamma}(s)}{ds}$ is the tangent vector to γ at s.

Assume that the forces $\vec{\mathbf{F}}(\mathbf{x})$ are applied only at point a of $\mathcal{A}$. Thus, when $\mathcal{A}$ is at configuration $\mathbf{q}$, the force $\vec{\mathbf{F}}(a(\mathbf{q}))$ is applied to $\mathcal{A}$ at a. Let "$< , >$" denote the inner product placed in $T_{\mathbf{q}}(\mathcal{C})$. The mapping of $\vec{\mathbf{F}}(\mathbf{x})$ in $T_{\mathbf{q}}(\mathcal{C})$ is defined as the vector field $\vec{F}(\mathbf{q}) \in T_{\mathbf{q}}(\mathcal{C})$ such that:

$$< \vec{F}(\mathbf{q}), \vec{\delta t} > \; = \; \vec{\mathbf{F}}(a(\mathbf{q})) \cdot \vec{\delta \mathbf{x}}$$

where $\vec{\delta t}$ denotes any virtual displacement of $\mathcal{A}$ along any vector $\vec{t} \in T_{\mathbf{q}}(\mathcal{C})$ and $\vec{\delta \mathbf{x}}$ denotes the corresponding virtual displacement of a in $\mathcal{W}$. This relation expresses that the work of $\vec{\mathbf{F}}$ along $\vec{\delta \mathbf{x}}$ is equal to the work of $\vec{F}$ along $\vec{\delta t}$. Hence, for any path $\tau : s \in [0,1] \mapsto \tau(s) \in \mathcal{C}$, the work of $\vec{F}$ along τ is equal to the work of $\vec{\mathbf{F}}$ along the path $\tau_a : s \in [0,1] \mapsto a(\tau(s)) \in \mathcal{W}$ followed by a when $\mathcal{A}$ moves along τ, i.e. :

$$\int_0^1 < \vec{F}(\tau(s)), \frac{d\vec{\tau}(s)}{ds} > ds = \int_0^1 \vec{\mathbf{F}}(\tau_a(s)) \cdot \frac{d\vec{\tau}_a(s)}{ds} ds.$$

Below, we consider that $\mathcal{A}$ is at configuration $\mathbf{q}$; we abbreviate $\vec{F}(\mathbf{q})$ to $\vec{F}$ and $\vec{\mathbf{F}}(a(\mathbf{q}))$ to $\vec{\mathbf{F}}$.

Consider the case where $N = 3$. Assume that the inner product $< , >$ in $T_{\mathbf{q}}(\mathcal{C})$ makes the principal axis basis into an orthonormal basis, i.e. if $(\xi_1 ... \xi_m)^T$ and $(\zeta_1 ... \zeta_m)^T$ are vectors $\vec{\xi}$ and $\vec{\zeta}$ of $T_{\mathbf{q}}(\mathcal{C})$ described by their components in the principal axis basis, the inner product of $\vec{\xi}$ and $\vec{\zeta}$ is:

$$<\vec{\xi},\vec{\zeta}> = \begin{pmatrix} \xi_1 & \dots & \xi_m \end{pmatrix} \begin{pmatrix} \zeta_1 \\ \cdot \\ \cdot \\ \cdot \\ \zeta_m \end{pmatrix}.$$

Let $\vec{F}$ be the vector of $T_{\mathbf{q}}(\mathcal{C})$ whose components in the principal axis basis are $f_1, f_2, f_3, m_1, m_2, m_3$, with f_i, $i = 1,2,3$, being the components of $\vec{\mathbf{F}}$ along the x-, y-, and z-axes of $\mathcal{F}_\mathcal{A}$, and m_j, $j = 1,2,3$, being the components of the vector $\vec{O_\mathcal{A}a} \wedge \vec{\mathbf{F}}$ along these axes. Let $(\delta x_\mathcal{A}\ \delta y_\mathcal{A}\ \delta z_\mathcal{A}\ \delta\alpha\ \delta\beta\ \delta\gamma)^T$ be any virtual displacement $\vec{\delta t}$ of $\mathcal{A}$ expressed in the principal axis basis. The corresponding displacement $\vec{\delta \mathbf{x}}$ of a is the 3-vector:

$$\begin{pmatrix} \delta x_\mathcal{A} \\ \delta y_\mathcal{A} \\ \delta z_\mathcal{A} \end{pmatrix} + \begin{pmatrix} \delta\alpha \\ \delta\beta \\ \delta\gamma \end{pmatrix} \wedge \vec{O_\mathcal{A}a}.$$

We have:

$$<\vec{F},\vec{\delta t}> = \vec{\mathbf{F}} \cdot \vec{\delta \mathbf{x}}.$$

Therefore, the six-dimensional vector $\vec{F}$ defined as above represents the force $\vec{\mathbf{F}}$ in the principal axis basis of $T_{\mathbf{q}}(\mathcal{C})$.

In a similar fashion, if $N = 2$, the force $\vec{\mathbf{F}}$ maps in $T_{\mathbf{q}}(\mathcal{C})$ to a 3-vector $\vec{F}$, whose components in the principal axis basis of the tangent space are f_1, f_2 and m, with f_1 and f_2 being the components of $\vec{\mathbf{F}}$ along the x- and y-axes of $\mathcal{F}_\mathcal{A}$ and $m = \vec{O_\mathcal{A}a} \wedge \vec{\mathbf{F}}$.

Consider now two distinct inner products in $T_{\mathbf{q}}(\mathcal{C})$, denoted by $< , >_1$ and $< , >_2$. Let β_1 and β_2 be two bases of $T_{\mathbf{q}}(\mathcal{C})$ such that β_i, $i = 1,2$, is an orthonormal basis when the inner product $< , >_i$ is used. Let $\vec{t}$ be any vector in $T_{\mathbf{q}}(\mathcal{C})$. We denote by $[\,\vec{t}\,]_i$ the m-vector of the components of $\vec{t}$ in β_i. J denotes the $m \times m$ matrix such that:

$$[\,\vec{t}\,]_2 = J\,[\,\vec{t}\,]_1.$$

(In the case where the bases β_1 and β_2 are induced by charts in $\mathcal{C}$, say (U_1, ϕ_1) and (U_2, ϕ_2), J is the Jacobian matrix of the map $\phi_2 \circ \phi_1^{-1}$.)

Let the same force be represented by two vectors: $\vec{F_1}$ when $<\ ,\ >_1$ is used, and $\vec{F_2}$ when $<\ ,\ >_2$ is used. For all virtual vectors $\vec{\delta t} \in T_{\mathbf{q}}(\mathcal{C})$, we have:

$$< \vec{F_1}, \vec{\delta t} >_1 = < \vec{F_2}, \vec{\delta t} >_2$$

since both expressions represent the same virtual work. Hence:

$$[\ \vec{F_1}\ ]_1^T\ [\ \vec{\delta t}\ ]_1 = [\ \vec{F_2}\ ]_2^T\ [\ \vec{\delta t}\ ]_2.$$

Using the relation $[\ \vec{\delta t}\ ]_2 = J\ [\ \vec{\delta t}\ ]_1$, we obtain:

$$[\ \vec{F_1}\ ]_1^T\ [\ \vec{\delta t}\ ]_1 = [\ \vec{F_2}\ ]_2^T\ J\ [\ \vec{\delta t}\ ]_1$$

which leads to:

$$[\ \vec{F_1}\ ]_1 = J^T\ [\ \vec{F_2}\ ]_2.$$

Let (U, ϕ) be a chart in $\mathcal{C}$. Let $g : \mathbf{R}^m \rightarrow \mathbf{R}^N$ be the application that maps the coordinates x_i, $i = 1, ..., m$, of $\mathbf{q}$ in the chart to the N coordinates of $a(\mathbf{q})$ in $\mathcal{W}$. Assume that the inner product in $T_{\mathbf{q}}(\mathcal{C})$ makes the basis β induced by the chart (U, ϕ) into an orthonormal basis. Let J_g be the Jacobian matrix of g, i.e. the $N \times m$ matrix $\left(\frac{dg_i}{dx_j}\right)$, $i = 1, ..., N$, $j = 1, ..., m$, of first derivatives. From:

$$< \vec{F}, \vec{\delta t} > = \ \vec{\mathbf{F}} \cdot \vec{\delta \mathbf{x}}$$

we obtain:

$$[\ \vec{F}\ ]_\beta^T \ = \ (f_1 \ ...\ f_N)\ J_g$$

and:

$$[\ \vec{F}\ ]_\beta \ = \ J_g^T \begin{pmatrix} f_1 \\ \cdot \\ \cdot \\ \cdot \\ f_N \end{pmatrix}$$

where f_1,...,f_N denote the components of $\vec{\mathbf{F}}$ in $\mathcal{F}_{\mathcal{W}}$.

An interesting case occurs when the force $\vec{\mathbf{F}}$ is the reaction force exerted by an obstacle $\mathcal{B}$ on $\mathcal{A}$. For simplification, we assume that $a \in \mathcal{A}$ is the only contact point of $\mathcal{A}$ with the obstacles and that the contact is between a vertex X of $\mathcal{A}$ (or $\mathcal{B}$) and a smooth $(N-1)$-dimensional subset S of $\mathcal{B}$'s (or $\mathcal{A}$'s) boundary. We denote the configuration of $\mathcal{A}$

by $\mathbf{q}_c \in \mathcal{C}_{contact}$. Let (U, ϕ) be any chart of $\mathcal{C}$ such that $\mathbf{q}_c \in U$. U is taken sufficiently small so that ϕ maps $\mathcal{C}_{contact} \cap U$ to a region of $\mathbf{R}^m$ determined by a single equation $f(x_1, ..., x_m) = 0$, where f is a smooth function of the variables $x_1, ..., x_m$, expressing that X is in S. Thus, $\mathcal{C}_{contact} \cap U$ is diffeomorphic to an open subset of $\mathbf{R}^{m-1}$. Let $\mathcal{A}$ move slightly in such a way that X stays in contact with S. This motion can be described as a differentiable curve γ_c in $\mathcal{C}_{contact} \cap U$. The tangent vectors of all the possible curves γ_c at $\mathbf{q}_c$ span an $(m-1)$-subspace of $T_{\mathbf{q}_c}(\mathcal{C})$, which is nothing other than the tangent space $T_{\mathbf{q}_c}(\mathcal{C}_{contact})$ of $\mathcal{C}_{contact}$ at $\mathbf{q}_c$. Assuming that the contact between X and S is frictionless, the work of the reaction force when $\mathcal{A}$ moves along γ_c is zero. Therefore, whatever inner product is used in $T_{\mathbf{q}}(\mathcal{C})$, the reaction force exerted by $\mathcal{B}$ on $\mathcal{A}$ maps in $T_{\mathbf{q}_c}(\mathcal{C})$ to a vector perpendicular to $T_{\mathbf{q}_c}(\mathcal{C}_{contact})$.

Exercises

1: Let $\mathcal{F}_\mathcal{W}$ and $\mathcal{F}_\mathcal{A}$ be two Cartesian frames in $\mathbf{R}^N$, $N = 2$ or 3. Assume that the origins of the two frames coincide, but not their axes. Let Θ be the $N \times N$ matrix whose columns are the components, in $\mathcal{F}_\mathcal{W}$, of the unit vectors along the axes of $\mathcal{F}_\mathcal{A}$. Show that Θ can be interpreted as a rotational operator that moves the axes of $\mathcal{F}_\mathcal{W}$ onto those of $\mathcal{F}_\mathcal{A}$. Let Θ_1 and Θ_2 be two matrices specified as Θ. Which rotational operators do the matrix products $\Theta_1 \times \Theta_2$ and $\Theta_2 \times \Theta_1$ represent?

2: Let $\mathcal{F}_\mathcal{W}$ and $\mathcal{F}_\mathcal{A}$ be two Cartesian frames in $\mathbf{R}^N$, $N = 2$ or 3. Let $O_\mathcal{W}$ and $O_\mathcal{A}$ denote the origins of $\mathcal{F}_\mathcal{W}$ and $\mathcal{F}_\mathcal{A}$, respectively. Let $\mathcal{T}$ be the N-vector of the coordinates of $O_\mathcal{A}$ in $\mathcal{F}_\mathcal{W}$. Let Θ be the $N \times N$ matrix whose columns are the components, in $\mathcal{F}_\mathcal{W}$, of the unit vectors along the axes of $\mathcal{F}_\mathcal{A}$. Using $\mathcal{T}$ and Θ, express the relation between the coordinates of any point of $\mathbf{R}^N$ in $\mathcal{F}_\mathcal{W}$ and its coordinates in $\mathcal{F}_\mathcal{A}$.

3: Prove that any matrix in $SO(3)$ determines a single rotation through an angle θ about some axis $\vec{n}$.

4: Show that $SO(2)$ is a commutative (abelian) group. Show that $SO(3)$ is a non-commutative group.

5: The configuration space of a rigid object in $\mathcal{W} = \mathbf{R}^N$, $N = 2$ or 3, is $\mathcal{C} = \mathbf{R}^N \times SO(N)$. Its dimension is $m = \frac{1}{2}N(N+1)$. Explain why representing $\mathcal{C}$ as a single copy of $\mathbf{R}^m$ (or a subset of $\mathbf{R}^m$) does not allow us to model $\mathcal{A}$'s motions adequately?

6: Consider the distance function $d : \mathcal{C} \times \mathcal{C} \rightarrow \mathbf{R}$ defined as:

$$d(\mathbf{q}_1, \mathbf{q}_2) = \max_{a \in \mathcal{A}} \|a(\mathbf{q}_1) - a(\mathbf{q}_2)\|.$$

Verify that it induces the same topology of $\mathcal{C}$ as the Euclidean metric in the ambient space $\mathbf{R}^M$.

7: Consider the subset S of $\mathbf{R}^2$ given by:

$$S = \{(x, \sin\frac{1}{x}) \;/\; 0 < x \leq 1\} \cup \{(0, y) \;/\; y \in [-1, +1]\}.$$

Are any two points in S connected by a path in S?

8: Show that infinitesimal rotations form a vector space. [Hint: Consider two matrices Θ_1 and Θ_2 of $SO(3)$ representing rotations about two vectors $\vec{n}_1$ and $\vec{n}_2$ through infinitesimal angles $d\theta_1$ and $d\theta_2$, respectively. Compute $\Theta_1 \times \Theta_2$ and $\Theta_2 \times \Theta_1$.] Relate this result to the fact that $\mathcal{C}$ is a manifold.

9: Verify that the following relation (see Subsection 4.3) holds:

$$\forall a \in \mathcal{A} : a(\mathbf{q}) = \mathbf{X}_\Theta a(0) \mathbf{X}^*_\Theta + \mathcal{T}$$

10: Consider S^3, the unit 3-sphere in $\mathbf{R}^4$ that represents the space of unit quaternions. Let $\mathbf{X}$ be a unit quaternion. Show that the curve in S^3 representing a rotation of 4π about a fixed axis begins and ends at the same point $\mathbf{X}$ of S^3, and that the curve representing a rotation of 2π begins and ends at two antipodal points $\mathbf{X}$ and $-\mathbf{X}$.

11: Define the stereographic projection mapping the sphere S^3 onto $\mathbf{R}^3$ as a way to algebraically represent elements of $SO(3)$.

12: Verify that homotopy determines a relation of equivalence on the set of all paths having the same initial configuration and the same final configuration.

13: Let $\tau_1 \bullet \tau_2$ denote the product of two paths τ_1 and τ_2, and $\overline{\tau}$ the inverse of the path τ (see Subsection 5.1).

a. Show that if τ_1 and τ_1' are homotopic and if τ_2 and τ_2' are homotopic, then $\tau_1 \bullet \tau_2$ and $\tau_1' \bullet \tau_2'$ are also homotopic.

b. Show that if two paths τ_1 and τ_2 are homotopic, the loop $\tau_1 \bullet \overline{\tau_2}$ is null-homotopic.

c. Show that if the loop $\tau_1 \bullet \overline{\tau_2}$ is null-homotopic, then the two paths τ_1 and τ_2 are homotopic. [Hint: Consider the two paths τ_1 and $\tau_1 \bullet \overline{\tau_2} \bullet \tau_2$.]

14: Let $s \in \mathbf{R} \mapsto \gamma(s) \in \mathcal{C}$ be a curve in the smooth manifold $\mathcal{C}$. Show that the property of γ of being differentiable at $s = s_0$ is independent of the selected chart in which the property is verified.

15: Show that if a path τ is piecewise of class C^1, it has a finite length in any metric inducing the topology of $\mathcal{C}$.

16: Consider a two-dimensional polygonal object $\mathcal{A}$ translating in the plane at fixed orientation among polygonal obstacles. With an example, show that different polygonal shapes for both $\mathcal{A}$ and the obstacles can yield C-obstacles with the same geometry in $\mathcal{C} = \mathbf{R}^2$.

17: Show that the Minkowski sum $E \oplus F$ of two convex subsets of $\mathbf{R}^N$ is a convex subset of $\mathbf{R}^N$.

18: Let $\mathcal{A}$ be a free-flying object in $\mathcal{W} = \mathbf{R}^2$. Its configuration space is $\mathcal{C} = \mathbf{R}^2 \times S^1$. Configurations are parameterized by $\mathbf{q} = (x, y, \theta)$, with $x \in \mathbf{R}$, $y \in \mathbf{R}$, and $\theta \in [0, 2\pi)$. x and y denote the coordinates of $O_{\mathcal{A}}$ in $\mathcal{F}_{\mathcal{W}}$. θ denotes the angle (mod 2π) between the x-axes of $\mathcal{F}_{\mathcal{W}}$ and $\mathcal{F}_{\mathcal{A}}$.

Let $\mathcal{B}$ be an obstacle in $\mathcal{W}$. $\mathcal{CB}$ is the corresponding C-obstacle in $\mathcal{C}$.

Let $\mathcal{A}'$ be the region of $\mathbf{R}^2$ swept out by $\mathcal{A}$ when it rotates around $O_{\mathcal{A}}$ from the orientation θ_1 to the orientation θ_2. Let $O_{\mathcal{A}'}$ be the point of $\mathcal{A}'$ that coincides with $O_{\mathcal{A}}$ during the rotation. We treat $\mathcal{A}'$ as an object that can only translate in $\mathbf{R}^2$. We represent a configuration of $\mathcal{A}'$ as the coordinates of $O_{\mathcal{A}'}$ in $\mathcal{F}_{\mathcal{W}}$. The configuration space of $\mathcal{A}'$ is $\mathcal{C}' = \mathbf{R}^2$ and the C-obstacle corresponding to $\mathcal{B}$ in this space is denoted by $\mathcal{CB}'$.

$\pi_{[\theta_1,\theta_2]}(\mathcal{CB})$ denotes the projection of the slice of the volume $\mathcal{CB}$ comprised between $\theta = \theta_1$ and $\theta = \theta_2$ on the xy-plane.

Prove that: $\pi_{[\theta_1,\theta_2]}(\mathcal{CB}) = \mathcal{CB}'$.

19: Show that every path in the configuration space of a free-flying rigid object is homotopic to a path that is the product of a finite number of paths, each being either a pure translation or a pure rotation of the object. Show that if two configurations are connected by a free path, they are also connected by a free path made of a finite succession of pure translations and pure rotations.

20: Define the subset $\mathcal{C}^{\varepsilon}_{free}$ of $\mathcal{C}$ made of all configurations at which the robot $\mathcal{A}$ is distant from all the obstacles $\mathcal{B}_i$ by more than ε, with $\varepsilon > 0$. A path in $\mathcal{C}^{\varepsilon}_{free}$ is said to be *ε-free*. Discuss the usefulness of this notion in practical path planning problems.

21: Consider the closure $cl(\mathcal{C}_{free})$ of $\mathcal{C}_{free}$. Let $\mathbf{q}_1$ and $\mathbf{q}_2$ be two configurations such that there exists a semi-free path $\tau' : [0,1] \rightarrow cl(\mathcal{C}_{free})$, with $\tau'(0) = \mathbf{q}_1$ and $\tau'(1) = \mathbf{q}_2$. Show that there exists $\tau : [0,1] \rightarrow cl(\mathcal{C}_{free})$, with $\tau(0) = \mathbf{q}_1$ and $\tau(1) = \mathbf{q}_2$, such that $\forall s \in [0,1] : \tau(s) \in \mathcal{C}_{free}$, except maybe at isolated values of s [Laumond, 1986]. Illustrate this result with an example in $\mathcal{C} = \mathbf{R}^2$.

22: Show that if $cl(\mathcal{C}_{free})$ is a manifold with boundary, then the existence of a semi-free path between any two free configurations entails the existence of a free path between these configurations.

23: Let $\mathcal{A}$ and $\mathcal{B}$ be two convex polygons. Assume that $\mathcal{A}$ can translate freely at fixed orientation. How the reaction force exerted by $\mathcal{B}$ on $\mathcal{A}$, when they are in contact, maps in configuration space (assume no friction)? Justify your answer.

Chapter 3

Obstacles in Configuration Space

Obstacles in the workspace $\mathcal{W}$ map in the configuration space $\mathcal{C}$ to regions called C-obstacles. In Chapter 2 we defined the C-obstacle $\mathcal{CB}$ corresponding to a workspace obstacle $\mathcal{B}$ as the following region in $\mathcal{C}$:

$$\mathcal{CB} = \{\mathbf{q} \in \mathcal{C} \ / \ \mathcal{A}(\mathbf{q}) \cap \mathcal{B} \neq \emptyset\}.$$

We also established general set-theoretic and topological properties of C-obstacles. In this chapter we construct the representation of the C-obstacles given that of the workspace objects, i.e. the robot $\mathcal{A}$ and the obstacles.

In Section 1 we analyze in detail the case where the workspace is two--dimensional and the objects are modeled as polygonal regions. We show that when the robot can only translate, the C-obstacles are also polygonal regions of $\mathcal{C}$. Each C-obstacle edge corresponds to the contact between an edge of $\mathcal{A}$ and an obstacle vertex, or the contact between a vertex of $\mathcal{A}$ and an obstacle edge. When $\mathcal{A}$ is allowed to rotate, the C-obstacles become three-dimensional regions bounded by patches of curved ruled surfaces generated by translating and rotating lines.

In Section 2 we treat the case where the workspace is three-dimensional and the objects are modeled as polyhedral regions. The main difference with the polygonal case comes from the existence of three types of contact (face-vertex, vertex-face, and edge-edge), instead of two. When $\mathcal{A}$ is allowed to rotate freely, the C-obstacles are six-dimensional regions bounded by patches of curved surfaces generated by translating and rotating planes.

Finally, in Section 3 we present the more general case where the objects in the workspace are modeled as arbitrary semi-algebraic regions, i.e. are defined by boolean combinations of polynomial equations and inequations, with rational coefficients.

Thus, throughout the chapter, we only consider algebraic models of objects. As a matter of fact, these models are particularly attractive. They give flexibility for describing any physical objects at any desired precision with a reasonably small number of parameters, and they admit a very simple and useful particular case — the polygonal/polyhedral representation. Algebraic models also make it possible to use standard mathematical tools, exact calculation techniques (see [Akritas, 1980] [Schwartz and Sharir, 1983b]) and algebraic decision methods (e.g. the Collins decomposition [Collins, 1975]), which are useful for building motion planning methods. Furthermore, in several circumstances, algebraic models are directly available from Computer-Aided Design systems.

Not surprisingly, most of the current results in motion planning use semi-algebraic representations. It is possible, however, that other models, for example bitmap representation, will attract more interest in the future than they currently do. In Chapter 7 (Potential Field) we will describe an efficient path planner that works from a bitmap representation of the workspace objects.

1 Case of a Polygonal Workspace

In this section we examine the case where $\mathcal{A}$ is a polygon moving freely in the plane among obstacles modeled as polygonal regions. We construct two different representations of the C-obstacles. One representation is in the form of a predicate `CB` taking a configuration $\mathbf{q}$ as argument. `CB`($\mathbf{q}$) evaluates to `true` if and only if $\mathbf{q} \in \mathcal{CB}$, where $\mathcal{CB}$ is the C-obstacle region represented by CB. The other representation is an explicit description of the boundary of $\mathcal{CB}$ in terms of (curved) faces, edges and vertices, and their adjacency relation. For both representations, we also analyze the particular case where $\mathcal{A}$ can only translate.

1.1 Polygonal Regions and Polygons

An infinite straight line L in $\mathbf{R}^2$ decomposes the plane into two half-planes. Let $h(x, y) = 0$ be the equation of L. We denote by $\overline{h}^-$ and $\overline{h}^+$ the two closed half-planes determined by $h(x, y) \leq 0$ and $h(x, y) \geq 0$, respectively.

DEFINITION 1: *A* **convex polygonal region** *in* $\mathbf{R}^2$ *is the intersection of a finite number of closed half-planes. A* **polygonal region** *is any subset of* $\mathbf{R}^2$ *obtained by taking the union of a finite number of convex polygonal regions.*

A (convex) polygonal region may not be bounded, nor connected. Its connected components may not be simply-connected.

DEFINITION 2: *A* **polygon** *is any polygonal region that is homeomorphic to the closed unit disc.*

Hence, a polygon is a compact, regular, simply-connected region of $\mathbf{R}^2$. Its boundary is a closed-loop sequence of line segments forming a Jordan curve, i.e. a curve homeomorphic to the unit circle.

Let $\mathcal{P}$ be a regular polygonal region. Each maximal connected line segment in $\mathcal{P}$'s boundary that is not intersected by the rest of $\mathcal{P}$'s boundary, except at its endpoints, is called an **edge**. Every endpoint of an edge is called a **vertex**. An edge is said to be **closed** (resp. **open**) if we include (resp. do not include) its endpoints. If $\mathcal{P}$ is a manifold with boundary, each vertex is the endpoint of exactly two edges.

Every edge E_i of $\mathcal{P}$ is contained in an infinite line L_i called the **supporting line** of E_i. Let $h_i(x, y) = 0$ be the equation of L_i. Without loss of generality, we write this equation so that every point in the open edge E_i has a neighborhood whose intersection with $\mathcal{P}$ only contains points (x, y) such that $h_i(x, y) \leq 0$. The half-plane $\overline{h}_i^+$ (resp. $\overline{h}_i^-$) is called the **outer** (resp. **inner**) half-plane of E_i. The **outgoing** (resp. **ingoing**) **normal** of E_i is the unit vector normal to E_i pointing toward the interior of the outer (resp. inner) half-plane of E_i.

A polygon is conveniently represented by the list of its vertices in counterclockwise order. Each vertex is described by its coordinates. A convex polygonal region may also be represented by the set of inequalities defining the half-planes $\overline{h_i^-}$, i.e. $\bigcap_i \overline{h_i^-}$.

The **complexity** of a polygonal region is defined as the number of its edges and vertices.

1.2 Type A and Type B Contacts

In this subsection we consider an obstacle $\mathcal{B}$ and we construct the equations of the surfaces bounding the corresponding C-obstacle $\mathcal{CB}$. We assume that both $\mathcal{A}$ and $\mathcal{B}$ are convex polygons.

The vertices of $\mathcal{A}$ are denoted by a_i, $i = 1, ..., n_{\mathcal{A}}$, and those of $\mathcal{B}$ by b_j, $j = 1, ..., n_{\mathcal{B}}$. The edge of $\mathcal{A}$ connecting a_i to a_{i+1} and the edge of $\mathcal{B}$ connecting b_j to b_{j+1} are denoted by $E_i^{\mathcal{A}}$ and $E_j^{\mathcal{B}}$, respectively. (We apply modulo $n_{\mathcal{A}}$ arithmetic to index i and modulo $n_{\mathcal{B}}$ arithmetic to index j, so that $a_{n_{\mathcal{A}}+1} \equiv a_1$ and $b_{n_{\mathcal{B}}+1} \equiv b_1$.) All edges are treated as closed edges, i.e. they include their endpoints.

We denote the outgoing normal of $E_i^{\mathcal{A}}(\mathbf{q})$ (resp. $E_j^{\mathcal{B}}$) by $\vec{\nu}_i^{\mathcal{A}}(\mathbf{q})$ (resp. $\vec{\nu}_j^{\mathcal{B}}$). If $\mathbf{q}$ is expressed as $(\mathcal{T}, \Theta)$, where Θ specifies the orientation of $\mathcal{F}_{\mathcal{A}}$ with respect to $\mathcal{F}_{\mathcal{W}}$, the outgoing normal of $E_i^{\mathcal{A}}(\mathbf{q})$ depends only on the orientation Θ of $\mathcal{A}$ at $\mathbf{q}$, though we denote it by $\vec{\nu}_i^{\mathcal{A}}(\mathbf{q})$.

Let us now suppose that $\mathcal{A}$ and $\mathcal{B}$ are in contact, i.e. :

$$\mathcal{A}(\mathbf{q}) \cap \mathcal{B} \neq \emptyset \quad \text{and} \quad int(\mathcal{A}(\mathbf{q})) \cap int(\mathcal{B}) = \emptyset.$$

We distinguish two types of contact [Lozano-Pérez, 1983] [Donald, 1984]. When an edge $E_i^{\mathcal{A}}$ of $\mathcal{A}$ contains a vertex b_j of $\mathcal{B}$, the contact is called a **type A contact** (Figure 1). When a vertex a_i of $\mathcal{A}$ is contained in an edge $E_j^{\mathcal{B}}$ of $\mathcal{B}$, the contact is called a **type B contact** (Figure 2). When the contact zone $\mathcal{A}(\mathbf{q}) \cap \mathcal{B}$ contains both a vertex of $\mathcal{A}$ and a vertex of $\mathcal{B}$, the type of the contact is both A and B.

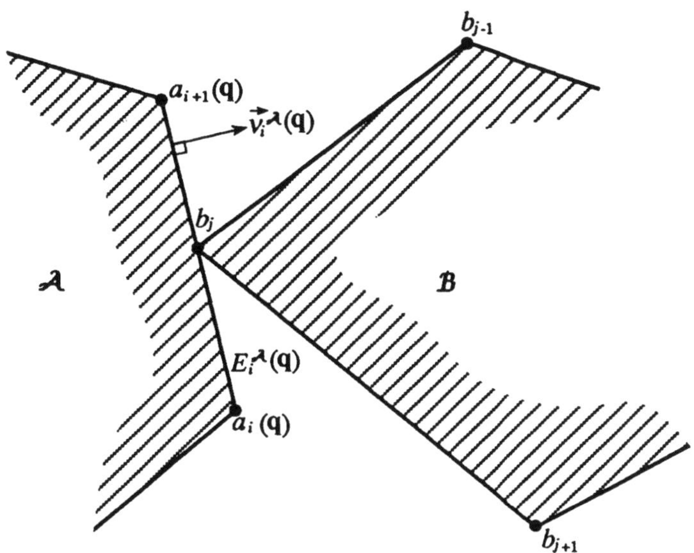

Figure 1. A type A contact between $\mathcal{A}$ and an obstacle $\mathcal{B}$ occurs when an edge $E_i^{\mathcal{A}}$ of $\mathcal{A}$ contains a vertex b_j of $\mathcal{B}$, without overlapping of the interiors of $\mathcal{A}$ and $\mathcal{B}$.

Type A Contact. The constraint that the interiors of $\mathcal{A}$ and $\mathcal{B}$ do not overlap makes a type A contact between $E_i^{\mathcal{A}}$ and b_j feasible only for a subrange of orientations of $\mathcal{A}$. This subrange is determined by the following two conditions (see Figure 1):

$$\vec{\nu}_i^{\mathcal{A}}(\mathbf{q}) \cdot (b_{j-1} - b_j) \geq 0$$

$$\vec{\nu}_i^{\mathcal{A}}(\mathbf{q}) \cdot (b_{j+1} - b_j) \geq 0$$

where $b_{j-1} - b_j$ is the vector connecting b_j to b_{j-1}, $b_{j+1} - b_j$ is the vector connecting b_j to b_{j+1}, and $\vec{u} \cdot \vec{v}$ is the inner product of two vectors in $\mathcal{W}$. We denote the conjunction of these two conditions by $\mathtt{APPL}_{i,j}^{A}(\mathbf{q})$, i.e. :

$$\mathtt{APPL}_{i,j}^{A}(\mathbf{q}) = [\vec{\nu}_i^{\mathcal{A}}(\mathbf{q}) \cdot (b_{j-1} - b_j) \geq 0] \bigwedge [\vec{\nu}_i^{\mathcal{A}}(\mathbf{q}) \cdot (b_{j+1} - b_j) \geq 0].$$

It is called the **applicability condition** of the type A contact between $E_i^{\mathcal{A}}$ and b_j. When $\mathbf{q}$ is expressed as $(\mathcal{T}, \Theta)$, this condition depends only on Θ.

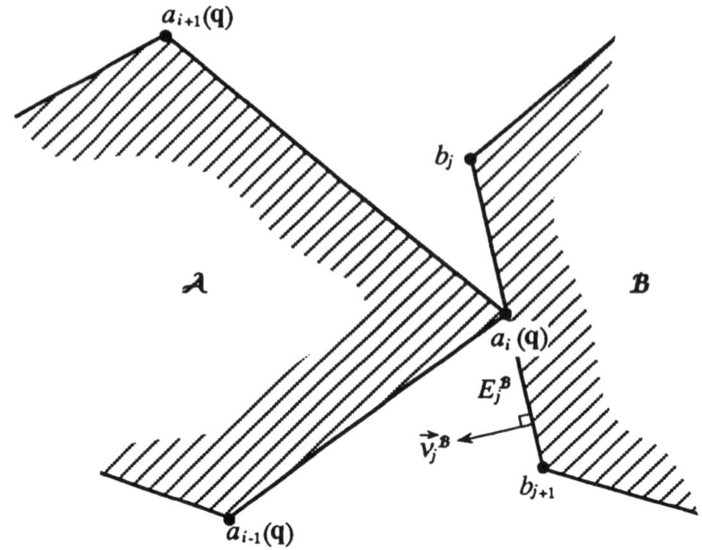

Figure 2. A type B contact between $\mathcal{A}$ and an obstacle $\mathcal{B}$ occurs when a vertex a_i of $\mathcal{A}$ is contained in an edge $E_j^{\mathcal{B}}$ of $\mathcal{B}$, without overlapping of the interiors of $\mathcal{A}$ and $\mathcal{B}$.

Let us displace $\mathcal{A}$ in such a way that the type A contact between $E_i^{\mathcal{A}}$ and b_j is maintained. During the displacement, the robot's configuration moves along a surface in $\mathcal{C}$ (i.e. a two-dimensional submanifold of $\mathcal{C}$) whose equation is:

$$f_{i,j}^{A}(\mathbf{q}) = 0$$

with:

$$f_{i,j}^{A}(\mathbf{q}) = \vec{\nu}_i^{\mathcal{A}}(\mathbf{q}) \cdot (b_j - a_i(\mathbf{q})).$$

This equation expresses the fact that when b_j is in $E_i^{\mathcal{A}}(\mathbf{q})$, the two vectors $\vec{\nu}_i^{\mathcal{A}}(\mathbf{q})$ and $(b_j - a_i(\mathbf{q}))$ are perpendicular or $b_j - a_i(\mathbf{q}) = 0$. The surface it represents is called a **C-surface** of type A. It separates $\mathcal{C}$ into two half-spaces. The C-obstacle $\mathcal{CB}$ lies completely within the closed half-space determined by $f_{i,j}^{A}(\mathbf{q}) \leq 0$.

DEFINITION 3: *The expression:*

$$\texttt{APPL}_{i,j}^{A}(\mathbf{q}) \Rightarrow [f_{i,j}^{A}(\mathbf{q}) \leq 0]$$

where:

$$\text{APPL}^{A}_{i,j}(\mathbf{q}) = [\vec{\nu}^{\mathcal{A}}_{i}(\mathbf{q}) \cdot (b_{j-1} - b_j) \geq 0] \bigwedge [\vec{\nu}^{\mathcal{A}}_{i}(\mathbf{q}) \cdot (b_{j+1} - b_j) \geq 0]$$

and:

$$f^{A}_{i,j}(\mathbf{q}) = \vec{\nu}^{\mathcal{A}}_{i}(\mathbf{q}) \cdot (b_j - a_i(\mathbf{q}))$$

is called a **C-constraint of type A**. *It is denoted by* $\text{CONST}^{A}_{i,j}(\mathbf{q})$.

Type B Contact. In the same manner, a type B contact between a_i and $E^{\mathcal{B}}_j$ is feasible for a subrange of orientations of $\mathcal{A}$ determined by two conditions (see Figure 2):

$$(a_{i-1}(\mathbf{q}) - a_i(\mathbf{q})) \cdot \vec{\nu}^{\mathcal{B}}_j \geq 0$$

$$(a_{i+1}(\mathbf{q}) - a_i(\mathbf{q})) \cdot \vec{\nu}^{\mathcal{B}}_j \geq 0.$$

The conjunction of these two conditions is denoted by $\text{APPL}^{B}_{i,j}(\mathbf{q})$ and is called the applicability condition of the type B contact between a_i and $E^{\mathcal{B}}_j$. When $\mathbf{q}$ is expressed as $(\mathcal{T}, \Theta)$, it depends only on Θ.

When $\mathcal{A}$ moves in such a way that the type B contact between a_i and $E^{\mathcal{B}}_j$ is maintained, the robot's configuration moves in a surface, called a **C-surface** of type B, whose equation is:

$$f^{B}_{i,j}(\mathbf{q}) = 0$$

where:

$$f^{B}_{i,j}(\mathbf{q}) = \vec{\nu}^{\mathcal{B}}_j \cdot (a_i(\mathbf{q}) - b_j).$$

$\mathcal{CB}$ lies completely within the closed half-space determined by $f^{B}_{i,j}(\mathbf{q}) \leq 0$.

DEFINITION 4: *The expression:*

$$\text{APPL}^{B}_{i,j}(\mathbf{q}) \Rightarrow [f^{B}_{i,j}(\mathbf{q}) \leq 0]$$

where:

$$\text{APPL}^{B}_{i,j}(\mathbf{q}) = [(a_{i-1}(\mathbf{q}) - a_i(\mathbf{q})) \cdot \vec{\nu}^{\mathcal{B}}_j \geq 0] \bigwedge [(a_{i+1}(\mathbf{q}) - a_i(\mathbf{q})) \cdot \vec{\nu}^{\mathcal{B}}_j \geq 0]$$

and:

$$f^{B}_{i,j}(\mathbf{q}) = \vec{\nu}^{\mathcal{B}}_j \cdot (a_i(\mathbf{q}) - b_j)$$

is called a **C-constraint of type B**. *It is denoted by* $\text{CONST}^{B}_{i,j}(\mathbf{q})$.

1.3 Representation of a C-obstacle with C-constraints

The main result presented below is Theorem 1 [Lozano-Pérez, 1983] [Donald, 1984], which describes the C-obstacle $\mathcal{CB}$ as a predicate $\texttt{CB}(\mathbf{q})$. $\texttt{CB}(\mathbf{q})$ is constructed as the conjunction of all the C-constraints of types A and B generated by the edges and the vertices of $\mathcal{A}$ and $\mathcal{B}$. We first state a simple lemma that will be used in the proof of Theorem 1.

LEMMA 1: *Two convex polygons do not intersect if and only if one of them contains an edge E such that the other polygon lies entirely in the open outer half-plane of E.*

Proof: If one of the polygons contains an edge E such that the other polygon lies entirely in the open outer half-plane of E, then it is clear that the two polygons do not intersect.

On the other hand, if the two polygons do not intersect, then the shortest distance d between the two polygons is non-zero. Three cases illustrated in Figure 3 are possible:

- The shortest distance d is (uniquely) achieved between a vertex p of a polygon and an open edge E of the other polygon (Figure 3.a). Then the polygon containing p lies entirely in the open outer half-plane of E.

- The shortest distance d is achieved (non-uniquely) between two parallel open edges E and E' (Figure 3.b). Then the polygon containing E' lies in the open outer half-plane of E.

- The shortest distance d is (uniquely) achieved between a vertex p of a polygon — call it $\mathcal{P}$ — and a vertex p' of the other polygon — call it $\mathcal{P}'$ (Figure 3.c). Let E_1 and E_2 (resp. E'_1 and E'_2) be the two edges of $\mathcal{P}$ (resp. $\mathcal{P}'$) abutting p (resp. p') in counterclockwise order. Assume that the supporting line of each of the two edges E_1 and E_2 intersects either one of the two edges E'_1 and E'_2. (If this assumption is not verified, then either E_1 or E_2, or both, satisfy the test of the lemma.) In order for the shortest distance d between $\mathcal{P}$ and $\mathcal{P}'$ to be uniquely achieved between p and p', it is necessary that p lie in a circular arc C of radius d centered at p' (arc in plain line in Figure 3.c); each endpoint of C is a point where the tangent to C is parallel to either E'_1 or E'_2.

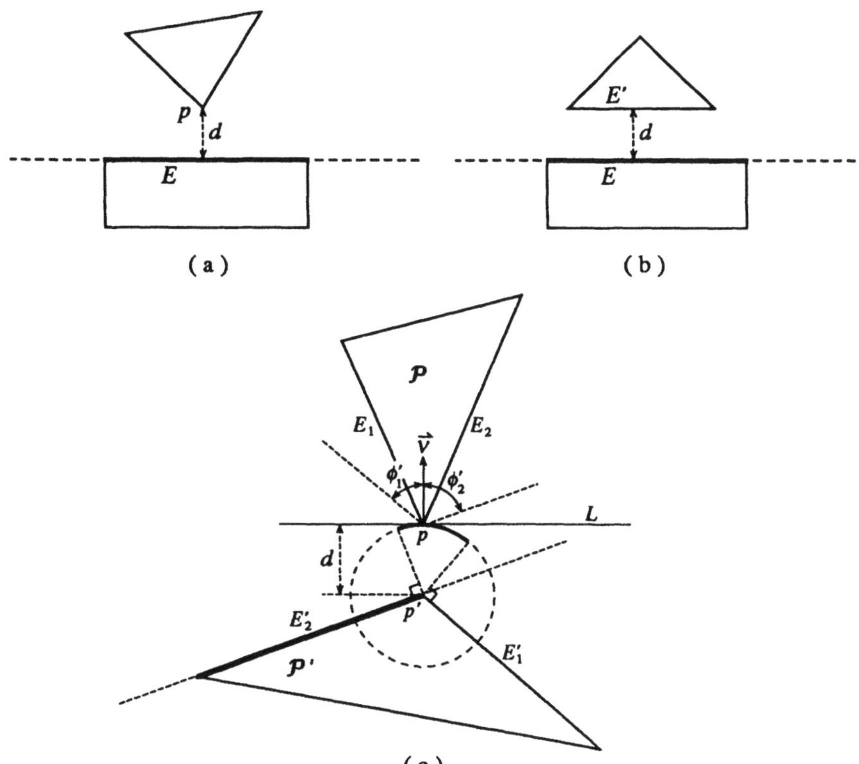

Figure 3. If two convex polygons do not intersect, the shortest distance between them is non-zero and is achieved either between a vertex and an open edge (Figure a), or between two open edges (Figure b), or between two vertices (Figure c). In every case, one polygon contains an edge (shown in bold line) such that the other polygon lies entirely in the open outer half-plane of that edge.

Further, let L be the line tangent to C at p; $\mathcal{P}$ must also lie in one of the two closed half-planes bounded by L, while $\mathcal{P}'$ must lie in the other open half-plane. Let $\overline{L}_{\mathcal{P}}$ denote the closed half-plane containing $\mathcal{P}$ and $\vec{\nu}$ the normal vector to L pointing into $\overline{L}_{\mathcal{P}}$. At p, we erect two half-lines contained in $\overline{L}_{\mathcal{P}}$ and parallel to E'_1 and E'_2, respectively. Let ϕ'_1 and ϕ'_2 be the angles made by these half-lines with $\vec{\nu}$. In order for the supporting lines of E_1 and E_2 to intersect either E'_1 or E'_2, $\mathcal{P}$ must be contained in a cone (not shown in the figure) defined as follows: its apex is p, its axis points along $\vec{\nu}$, and its sides make the angle $\max\{\phi_1', \phi_2'\}$ with $\vec{\nu}$. Now, if $\phi_1' \geq \phi_2'$ then $\mathcal{P}$ lies entirely in the open outer half-plane of E_1'. (This is obvious if p lies in the open outer half-plane of E_1'; on the other hand, if p does not lie in this half-plane, we cannot have $\phi_1' \geq \phi_2'$.) If $\phi_2' \geq \phi_1'$, $\mathcal{P}$ lies entirely in the open outer half-plane of E_2'. ∎

THEOREM 1: *Let $\mathcal{A}$ and $\mathcal{B}$ be two convex polygons. The C-obstacle:*

$$\mathcal{CB} = \{\mathbf{q} \in \mathcal{C} \;/\; \mathcal{A}(\mathbf{q}) \cap \mathcal{B} \neq \emptyset\}$$

is such that:

$$\mathbf{q} \in \mathcal{CB} \Leftrightarrow \mathtt{CB}(\mathbf{q})$$

with:

$$\mathtt{CB}(\mathbf{q}) \equiv (\bigwedge_{i,j} \mathtt{CONST}^{A}_{i,j}(\mathbf{q})) \bigwedge (\bigwedge_{i,j} \mathtt{CONST}^{B}_{i,j}(\mathbf{q})).$$

Proof: The expression:

$$\mathbf{q} \in \mathcal{CB} \Leftrightarrow \mathtt{CB}(\mathbf{q})$$

is equivalent to:

$$\mathbf{q} \notin \mathcal{CB} \Leftrightarrow \neg\mathtt{CB}(\mathbf{q})$$

with:

$$\neg\mathtt{CB}(\mathbf{q}) \equiv (\bigvee_{i,j} \neg\mathtt{CONST}^{A}_{i,j}(\mathbf{q})) \bigvee (\bigvee_{i,j} \neg\mathtt{CONST}^{B}_{i,j}(\mathbf{q})).$$

(1) Let $\mathbf{q} \notin \mathcal{CB}$. Hence, $\mathcal{A}(\mathbf{q})$ and $\mathcal{B}$ do not intersect. According to the previous lemma, there exists an edge $E_i^{\mathcal{A}}(\mathbf{q})$ or $E_j^{\mathcal{B}}$ contained in a line L such that $\mathcal{A}(\mathbf{q})$ lies on one side of L and $\mathcal{B}$ on the other side, with either $\mathcal{A}(\mathbf{q})$ or $\mathcal{B}$, but not both, having a non-empty intersection with L. Assume that L contains $E_i^{\mathcal{A}}(\mathbf{q})$ (a similar development could be made if L contained $E_j^{\mathcal{B}}$). Then there exists a vertex b_j of $\mathcal{B}$ such that $\mathtt{APPL}^{A}_{i,j}(\mathbf{q})$ holds. Furthermore, this vertex verifies: $f^{A}_{i,j}(\mathbf{q}) > 0$. Thus, there exists a pair (i, j) such that the C-constraint:

$$\mathtt{CONST}^{A}_{i,j}(\mathbf{q}) \equiv (\mathtt{APPL}^{A}_{i,j}(\mathbf{q}) \Rightarrow f^{A}_{i,j}(\mathbf{q}) \leq 0)$$

does not hold. Therefore:

$$\mathbf{q} \notin \mathcal{CB} \Rightarrow \neg\mathtt{CB}(\mathbf{q}).$$

(2) Let $\neg\mathtt{CB}(\mathbf{q})$ hold. Then there exists at least one pair (i,j) verifying:

$$\mathtt{APPL}^{A}_{i,j}(\mathbf{q}) \quad \text{and} \quad f^{A}_{i,j}(\mathbf{q}) > 0$$

or:

$$\mathtt{APPL}^{B}_{i,j}(\mathbf{q}) \quad \text{and} \quad f^{B}_{i,j}(\mathbf{q}) > 0.$$

In both cases, $\mathcal{A}(\mathbf{q})$ and $\mathcal{B}$ cannot intersect, since they are convex. Hence, $\mathbf{q} \notin \mathcal{CB}$, and:

$$\neg\mathtt{CB}(\mathbf{q}) \Rightarrow \mathbf{q} \notin \mathcal{CB}.$$

■

Using the predicate CB, we can compute whether a configuration $\mathbf{q}$ lies in $\mathcal{CB}$ or not by evaluating every individual C-constraint until one of them evaluates to **false** (then $\mathbf{q} \notin \mathcal{CB}$) or all of them evaluate to **true** (then $\mathbf{q} \in \mathcal{CB}$). The number of C-constraints in $\mathtt{CB}(\mathbf{q})$ (i.e. the worst-case time complexity of this computation) is $2n_{\mathcal{A}}n_{\mathcal{B}}$.

A configuration $\mathbf{q}$ is in $\mathcal{CB}$'s boundary if and only if:

- $\exists i,j$ such that $\mathtt{APPL}^{A}_{i,j}(\mathbf{q})$ (or $\mathtt{APPL}^{B}_{i,j}(\mathbf{q})$) holds, and $f^{A}_{i,j}(\mathbf{q}) = 0$ (or $f^{B}_{i,j}(\mathbf{q}) = 0$),
- $\forall i',j'$ such that $\mathtt{APPL}^{A}_{i',j'}(\mathbf{q})$ holds, $f^{A}_{i',j'}(\mathbf{q}) \leq 0$,
- $\forall i',j'$ such that $\mathtt{APPL}^{B}_{i',j'}(\mathbf{q})$ holds, $f^{B}_{i',j'}(\mathbf{q}) \leq 0$.

If $\mathcal{A}$ and $\mathcal{B}$ are non-convex polygons, they can be represented as finite unions of convex polygons $\mathcal{A}_k$ and $\mathcal{B}_l$:

$$\mathcal{A} = \bigcup_k \mathcal{A}_k \quad \text{and} \quad \mathcal{B} = \bigcup_l \mathcal{B}_l.$$

By defining:

$$\mathcal{CB}_{kl} = \{\mathbf{q} \in \mathcal{C} \; / \; \mathcal{A}_k(\mathbf{q}) \cap \mathcal{B}_l \neq \emptyset\}$$

we get:

$$\mathcal{CB} = \bigcup_{k,l} \mathcal{CB}_{kl}.$$

The predicate $\texttt{CB}_{kl}$ representing $\mathcal{CB}_{kl}$ can be constructed as in Theorem 1. We have:

$$\texttt{CB}(\mathbf{q}) \equiv \bigvee_{k,l} \texttt{CB}_{kl}(\mathbf{q}). \tag{1}$$

The problem of decomposing a non-convex polygon into a union of convex polygons is a classical one in Computational Geometry. Several algorithms have been proposed (e.g. see [Schachter, 1978] [Garey et al., 1978] [Chazelle and Dobkin, 1979 and 1985] [Keil and Sack, 1985] [Chazelle, 1987]). Two types of decomposition can be considered: covering, which does allow the interiors of the convex components to overlap, and partitioning, which disallows the interiors of the convex components from overlapping. Both types of decomposition, including the optimal decomposition into the smallest number of convex polygons, can be performed in polynomial time. In particular, a sweep-line algorithm partitions a non-convex polygon into (non-overlapping) triangles and trapezoids in $O(n \log n)$ time, where n is the number of vertices of the decomposed polygon (see Section 1 of Chapter 5).

When $\mathcal{A}$ is a convex polygon and $\mathcal{B}$ is a regular non-bounded convex polygonal region, we can still use the result of Theorem 1, if we take some precautions. Some vertices b_j are at infinity and do not produce type A contacts. The expressions $\texttt{CONST}^{A}_{i,j}(\mathbf{q})$ involving these vertices should be modified appropriately, for instance by introducing fictitious vertices at finite distances. If $\mathcal{A}$ is a non-convex polygon and $\mathcal{B}$ is a regular non-bounded non-convex polygonal region, both $\mathcal{A}$ and $\mathcal{B}$ can be decomposed into a possibly non-minimal number of convex parts in polynomial time[1], and a predicate CB can be constructed as in expression (1) to represent $\mathcal{CB}$. It is also not difficult to extend the construction of CB to non-regular polygonal regions.

Since the union of a finite number of polygonal regions is also a polygonal region, we can construct a predicate CB representing the C-obstacle region in $\mathcal{C}$. The free space is then defined by:

$$\mathbf{q} \in \mathcal{C}_{free} \Leftrightarrow \neg\texttt{CB}(\mathbf{q})$$

[1]The optimal decomposition of a polygon with holes is NP-hard [Lingas, 1982].

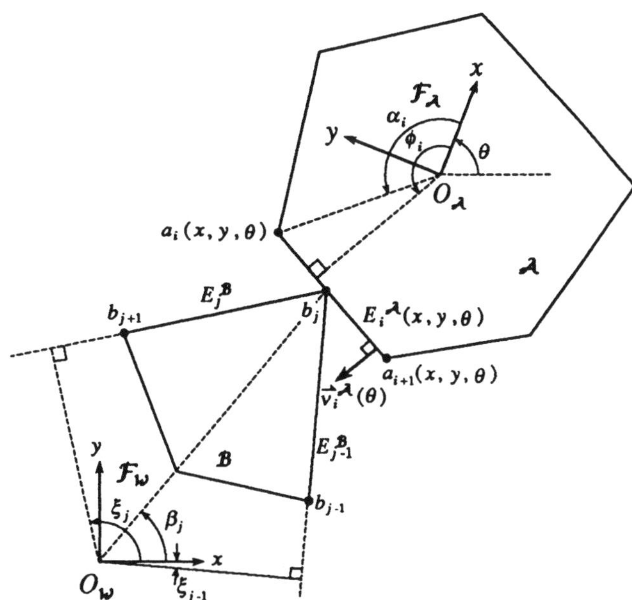

Figure 4. This figure illustrates the notations used in the text for establishing the parameterized form of a C-constraint of type A.

1.4 Parameterization of C-Constraints

In Subsection 1.2 we expressed the C-constraints without assuming any specific parameterization of $\mathcal{C}$. By selecting a representation of the configurations in the form of a list of real parameters, as discussed in Section 4 of Chapter 2, we can derive the corresponding parameterized form of the C-constraints [Brooks and Lozano-Pérez, 1982 and 1983].

For example, let us represent a configuration $\mathbf{q} \in \mathbf{R}^2 \times SO(2)$ by (x, y, θ), where x and $y \in \mathbf{R}$ are the coordinates of $O_\mathcal{A}$ in $\mathcal{F_W}$, and $\theta \in [0, 2\pi)$ is the angle between the x-axes of $\mathcal{F_W}$ and $\mathcal{F_A}$. Figures 4 and 5 illustrate the notations used below.

The outgoing normal vector to $E_i^\mathcal{A}(x, y, \theta)$ is (in $\mathcal{F_W}$):

$$\vec{\nu}_i^\mathcal{A}(\theta) = (\cos(\phi_i + \theta), \sin(\phi_i + \theta)).$$

The outgoing normal vector to $E_j^\mathcal{B}$ is:

$$\vec{\nu}_j^\mathcal{B} = (\cos \xi_j, \sin \xi_j).$$

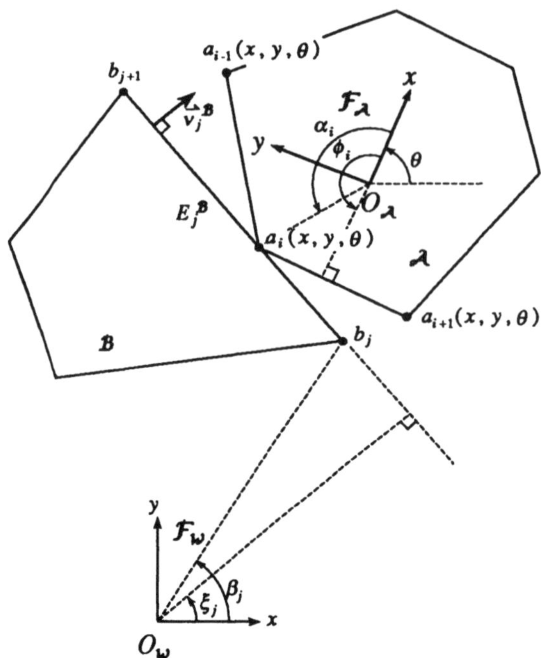

Figure 5. This figure illustrates the notations used in the text for establishing the parameterized form of a C-constraint of type B.

The coordinates of a_i in $\mathcal{F}_\mathcal{A}$ are $\|a_i\| \cos \alpha_i$ and $\|a_i\| \sin \alpha_i$, where $\|a_i\|$ is the distance from the origin of $\mathcal{F}_\mathcal{A}$ to the vertex a_i. Hence:

$$a_i(x, y, \theta) = (x + \|a_i\| \cos(\alpha_i + \theta), y + \|a_i\| \sin(\alpha_i + \theta)).$$

The vertex b_j is given by:

$$b_j = (\|b_j\| \cos \beta_j, \|b_j\| \sin \beta_j)$$

where $\|b_j\|$ is the distance from the origin of $\mathcal{F}_\mathcal{W}$ to b_j.

Consider the type A contact between the edge E_i^A and the vertex b_j shown in Figure 4. The corresponding parameterized C-constraint is:

$$\text{APPL}_{i,j}^A(\theta) \Rightarrow [f_{i,j}^A(x, y, \theta) \leq 0].$$

The applicability condition:

$$\text{APPL}_{i,j}^A(\theta) = [\vec{\nu}_i^A(\theta) \cdot (b_{j-1} - b_j) \geq 0] \bigwedge [\vec{\nu}_i^A(\theta) \cdot (b_{j+1} - b_j) \geq 0]$$

entails:

$$\theta \in [\xi_{j-1} - \phi_i - \pi, \xi_j - \phi_i - \pi] \quad (\text{mod } 2\pi).$$

The expression $f^A_{i,j}(x, y, \theta) = \vec{\nu}^A_i(\theta) \cdot (b_j - a_i(x, y, \theta))$ can be written:

$$-x\cos(\phi_i + \theta) - y\sin(\phi_i + \theta) + \|b_j\|\cos(\phi_i + \theta - \beta_j) - \|a_i\|\cos(\phi_i - \alpha_i).$$

Consider the type B contact between the vertex a_i and the edge $E^{\mathcal{B}}_j$ shown in Figure 5. The corresponding parameterized C-constraint is:

$$\texttt{APPL}^B_{i,j}(\theta) \Rightarrow [f^B_{i,j}(x, y, \theta) \leq 0].$$

The applicability condition:[2]

$$\begin{aligned} \texttt{APPL}^B_{i,j}(\theta) &= [(a_{i-1}(x, y, \theta) - a_i(x, y, \theta)) \cdot \vec{\nu}^{\mathcal{B}}_j \geq 0] \\ &\quad \bigwedge [(a_{i+1}(x, y, \theta) - a_i(x, y, \theta)) \cdot \vec{\nu}^{\mathcal{B}}_j \geq 0]. \end{aligned}$$

entails:

$$\theta \in [\xi_j - \phi_i - \pi, \xi_j - \phi_{i-1} - \pi] \quad (\text{mod } 2\pi).$$

The expression $f^B_{i,j}(x, y, \theta) = \vec{\nu}^{\mathcal{B}}_j \cdot (a_i(x, y, \theta) - b_j)$ can be written:

$$x\cos\xi_j + y\sin\xi_j + \|a_i\|\cos(\alpha_i + \theta - \xi_j) - \|b_j\|\cos(\beta_j - \xi_j).$$

Using a parameterization of $\mathcal{A}$'s orientation based on the stereographic projection, instead of the angle θ, yields a semi-algebraic expression of $\texttt{CB}(\mathbf{q})$.

1.5 Geometric Interpretation

Let $\mathcal{A}$ and $\mathcal{B}$ be convex polygons, and $\mathbf{q}$ be represented by $(x, y, \theta) \in \mathbf{R}^2 \times [0, 2\pi)$ as in the previous subsection. Each C-surface bounding $\mathcal{CB}$ is represented by $f^A_{i,j}(x, y, \theta) = 0$ or $f^B_{i,j}(x, y, \theta) = 0$ in $\mathbf{R}^2 \times [0, 2\pi)$. If we keep the orientation of $\mathcal{A}$ constant and equal to θ_0, then both $f^A_{i,j}(x, y, \theta_0) = 0$ and $f^B_{i,j}(x, y, \theta_0) = 0$ represent straight lines in $\mathbf{R}^2$:

- The line $f^A_{i,j}(x, y, \theta_0) = 0$, whose parameterized equation is:

$$-x\cos(\phi_i + \theta_0) - y\sin(\phi_i + \theta_0) + \|b_j\|\cos(\phi_i + \theta_0 - \beta_j) - \|a_i\|\cos(\phi_i - \alpha_i) = 0,$$

[2] Despite the appearance, $\texttt{APPL}^B_{i,j}(\theta)$ does not depend on x and y. Indeed, we could as well write:
$\texttt{APPL}^B_{i,j}(\theta) = [(a_{i-1}(0,0,\theta) - a_i(0,0,\theta)) \cdot \vec{\nu}^{\mathcal{B}}_j \geq 0] \bigwedge [(a_{i+1}(0,0,\theta) - a_i(0,0,\theta)) \cdot \vec{\nu}^{\mathcal{B}}_j \geq 0].$

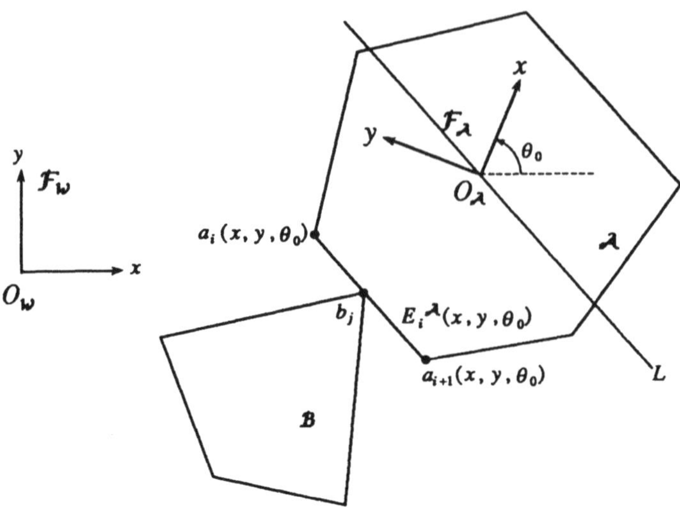

Figure 6. $\mathcal{A}$ is in type A contact with $\mathcal{B}$. If $\mathcal{A}$ translates at a fixed orientation θ_0 while maintaining this contact, the robot's configuration moves along a segment contained in a line L parallel to $E_i^{\mathcal{A}}(x, y, \theta_0)$. If $\mathcal{A}$ takes a slightly different orientation $\theta_0 + \delta\theta$, the line rotates by the same angle. The vertex b_j is the center of rotation.

is parallel to the edge $E_i^{\mathcal{A}}(x, y, \theta_0)$. It rotates when $\mathcal{A}$ rotates within the range of orientations where the contact between $E_i^{\mathcal{A}}$ and b_j is feasible (see Figure 6).

- The line $f_{i,j}^{B}(x, y, \theta_0) = 0$, whose equation is:

$$x \cos \xi_j + y \sin \xi_j + \|a_i\| \cos(\alpha_i + \theta_0 - \xi_j) - \|b_j\| \cos(\beta_j - \xi_j) = 0,$$

is parallel to the edge $E_j^{\mathcal{B}}$. It translates when $\mathcal{A}$ rotates within the range of orientations where the contact between a_i and $E_j^{\mathcal{B}}$ is feasible (see Figure 7).

Therefore, all the C-surfaces are ruled surfaces (the C-surfaces of type A are helicoids). Figure 8 shows a C-obstacle $\mathcal{CB}$ in $\mathbf{R}^2 \times [0, 2\pi]$. It is a three-dimensional volume bounded by patches of C-surfaces. These patches are called the **faces** of the C-obstacle.

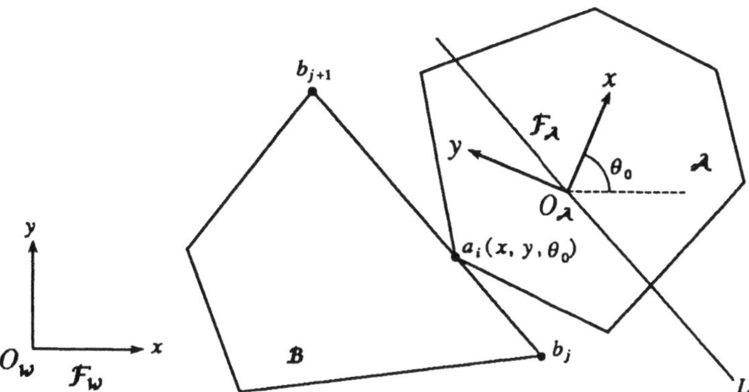

Figure 7. $\mathcal{A}$ is in type B contact with $\mathcal{B}$. If $\mathcal{A}$ translates at a fixed orientation, while maintaining this contact, the robot's configuration moves along a segment contained in a line L parallel to $E_j^{\mathcal{B}}$. If $\mathcal{A}$ takes a slightly different orientation, the line translates accordingly.

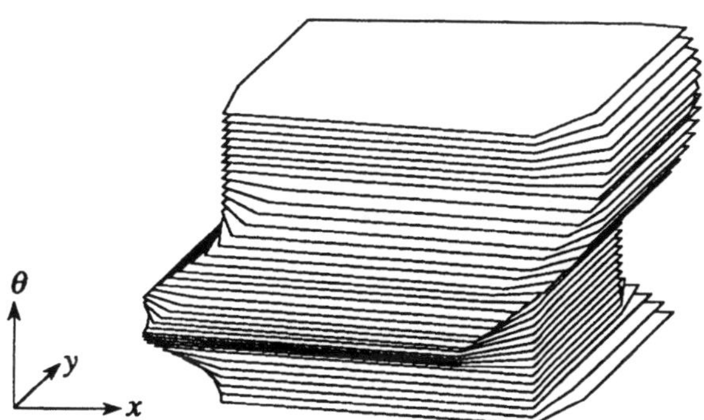

Figure 8. This figure represents a C-obstacle $\mathcal{CB}$ in $\mathbf{R} \times [0, 2\pi]$ when both $\mathcal{A}$ and $\mathcal{B}$ are convex polygons. Each cross-section at a fixed θ is a convex polygon, the C-obstacle in $\mathbf{R}^2$ when $\mathcal{A}$ translates at this fixed orientation. The boundary of $\mathcal{CB}$ is made of patches of ruled surfaces, called faces. Each face corresponds to a type A or type B contact, and is generated either by a rotating line (type A contact) or a translating line (type B contact).

The cross-section $\mathcal{CB}_{\theta_0}$ of $\mathcal{CB}$ at $\theta = \theta_0$ is a region of $\mathbf{R}^2$ defined by:

$$(x, y) \in \mathcal{CB}_{\theta_0} \Leftrightarrow [\bigwedge_{i,j} \mathtt{CONST}_{i,j}^{A}(x, y, \theta_0)] \bigwedge [\bigwedge_{i,j} \mathtt{CONST}_{i,j}^{B}(x, y, \theta_0)].$$

$\mathcal{CB}_{\theta_0}$ is the intersection of a finite number of closed half-planes, each bounded by a straight line $f_{i,j}^A(x,y,\theta_0) = 0$ or $f_{i,j}^B(x,y,\theta_0) = 0$. Since $\mathcal{CB}$ is compact and regular (see Propositions 4 and 8 in Chapter 2), $\mathcal{CB}_{\theta_0}$ is a convex polygon equal to the Minkowski difference $\mathcal{B} \ominus \mathcal{A}(0,0,\theta_0)$ (see Proposition 3 in Chapter 2).

We can also remark that:

$$\begin{aligned} |f_{i,j}^A(x,y,\theta_0)| &= |\vec{\nu}_i^A(\theta_0) \cdot (b_j - a_i(x,y,\theta_0))| \\ |f_{i,j}^B(x,y,\theta_0)| &= |\vec{\nu}_j^B \cdot (a_i(x,y,\theta_0) - b_j)| \end{aligned}$$

are the Euclidean distances from the point$(x,y) \in \mathbf{R}^2$ to the lines $f_{i,j}^A(x,y,\theta_0) = 0$ and $f_{i,j}^B(x,y,\theta_0) = 0$, respectively. Thus, when $(x,y,\theta_0) \notin \mathcal{CB}_{\theta_0}$, these values can be used to compute the distance from (x,y,θ_0) to $\mathcal{CB}$ in the subspace $\theta = \theta_0$.

1.6 Boundary Representation (Translational Case)

The predicate `CB` is conceptually simple. But it does not provide an explicit representation of the boundary of the C-obstacle $\mathcal{CB}$, such as a list of faces, edges and vertices, with both their equations and topological adjacency relation. In this subsection we present a method for constructing such a boundary representation in the case where $\mathcal{A}$ can only translate. In the next subsection, we will describe a more general method that is applicable when $\mathcal{A}$ can both translate and rotate.

Let us first assume that $\mathcal{A}$ and $\mathcal{B}$ are convex polygons and that $\mathcal{A}$ translates at a fixed orientation θ_0. We saw previously that in this case, $\mathcal{CB}_{\theta_0}$ is also a convex polygon. One can easily verify that, if θ_0 is a general orientation of $\mathcal{A}$ relative to $\mathcal{B}$ (i.e. no edge of $\mathcal{A}(x,y,\theta_0)$ is parallel to an edge of $\mathcal{B}$), then the vertices of $\mathcal{CB}_{\theta_0}$ are all the points obtained as follows:

- If $\mathtt{APPL}_{i,j}^A(\theta_0)$ holds, then the points

$$b_j - a_i(0,0,\theta_0) \text{ and } b_j - a_{i+1}(0,0,\theta_0)$$

  are vertices of $\mathcal{CB}_{\theta_0}$.

- If $\mathtt{APPL}_{i,j}^B(\theta_0)$ holds, then the points

$$b_j - a_i(0,0,\theta_0) \text{ and } b_{j+1} - a_i(0,0,\theta_0)$$

  are vertices of $\mathcal{CB}_{\theta_0}$.

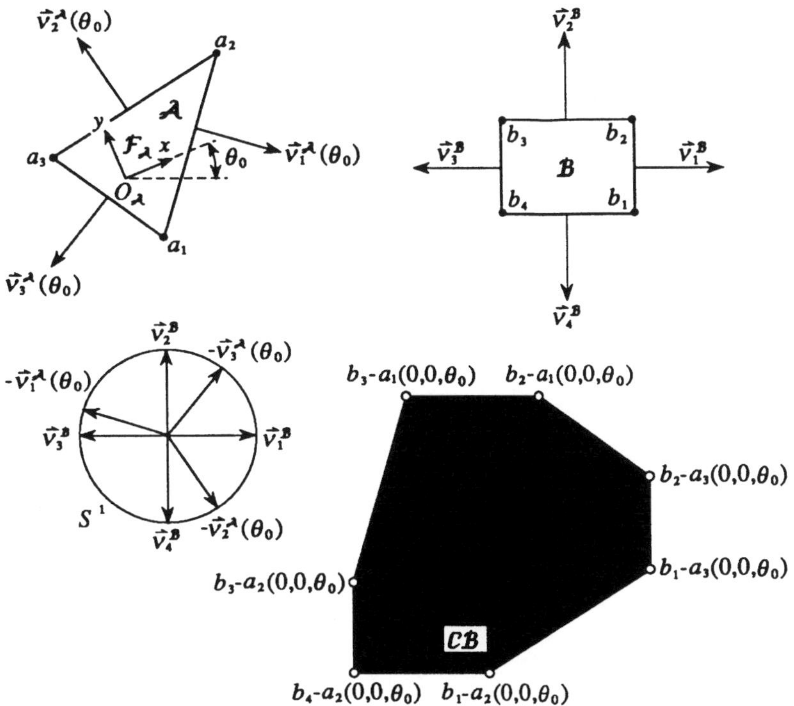

Figure 9. This figure illustrates an efficient algorithm for computing $\mathcal{CB}_{\theta_0}$. The unit vectors $-\vec{\nu}_i^{\mathcal{A}}(\theta_0)$, $i = 1, ..., n_{\mathcal{A}}$, and $\vec{\nu}_j^{\mathcal{B}}$, $j = 1, ..., n_{\mathcal{B}}$ are affixed at the center of a unit circle S^1. By scanning the vectors $-\vec{\nu}_i^{\mathcal{A}}(\theta_0)$ and $\vec{\nu}_j^{\mathcal{B}}$, say in counterclockwise order, one constructs the successive $n_{\mathcal{A}} + n_{\mathcal{B}}$ vertices of $\mathcal{CB}_{\theta_0}$.

This yields the following efficient algorithm for constructing $\mathcal{CB}_{\theta_0}$. This algorithm was originally proposed by Lozano-Pérez [Lozano-Pérez, 1983]. (See also [Guibas, Ramshaw and Stolfi, 1983].)

Let us affix the vectors $-\vec{\nu}_i^{\mathcal{A}}(\theta_0)$, $i = 1, ..., n_{\mathcal{A}}$, and $\vec{\nu}_j^{\mathcal{B}}$, $j = 1, ..., n_{\mathcal{B}}$ at the center of the unit circle S^1 (see Figure 9). Since both $\mathcal{A}$ and $\mathcal{B}$ are convex polygons, the counterclockwise angle between two successive vectors $-\vec{\nu}_i^{\mathcal{A}}(\theta_0)$ and $-\vec{\nu}_{i+1}^{\mathcal{A}}(\theta_0)$, or $\vec{\nu}_j^{\mathcal{B}}$ and $\vec{\nu}_{j+1}^{\mathcal{B}}$, is smaller than π. Therefore, θ_0 being a general orientation of $\mathcal{A}$ relative to $\mathcal{B}$, $\mathtt{APPL}_{i,j}^{A}(\theta_0)$ holds if and only if $-\vec{\nu}_i^{\mathcal{A}}(\theta_0)$ points between $\vec{\nu}_{j-1}^{\mathcal{B}}$ and $\vec{\nu}_j^{\mathcal{B}}$, i.e. $-\vec{\nu}_i^{\mathcal{A}}(\theta_0)$ is a positive linear combination of $\vec{\nu}_{j-1}^{\mathcal{B}}$ and $\vec{\nu}_j^{\mathcal{B}}$. In the same way, $\mathtt{APPL}_{i,j}^{B}(\theta_0)$ holds if and only if $\vec{\nu}_j^{\mathcal{B}}$ points between $-\vec{\nu}_{i-1}^{\mathcal{A}}(\theta_0)$ and $-\vec{\nu}_i^{\mathcal{A}}(\theta_0)$. By scanning the vectors $-\vec{\nu}_i^{\mathcal{A}}(\theta_0)$ and $\vec{\nu}_j^{\mathcal{B}}$, say in counterclockwise order, one constructs the successive $n_{\mathcal{A}}+n_{\mathcal{B}}$ vertices of $\mathcal{CB}_{\theta_0}$. For instance, if $-\vec{\nu}_i^{\mathcal{A}}(\theta_0)$ is being considered, let $\vec{\nu}_{j-1}^{\mathcal{B}}$ and $\vec{\nu}_j^{\mathcal{B}}$ be the two normals to edges of $\mathcal{B}$ between which it points in S^1. The vertex $b_j - a_{i+1}(0,0,\theta_0)$ is then generated. If $\mathcal{A}$ and $\mathcal{B}$ are given as sequences of vertices specified by their coordinates, the complete boundary of $\mathcal{CB}_{\theta_0}$ can thus be traced out in $O(n_{\mathcal{A}} + n_{\mathcal{B}})$ time, which is optimal.

When $\mathcal{A}$ is at a critical orientation θ_0 such that an edge $E_i^{\mathcal{A}}(x,y,\theta_0)$ of $\mathcal{A}(x,y,\theta_0)$ is parallel to an edge $E_j^{\mathcal{B}}$ of $\mathcal{B}$ and the contact between the two edges is feasible, the points $b_j - a_i(0,0,\theta_0)$, $b_{j+1} - a_i(0,0,\theta_0)$, $b_j - a_{i+1}(0,0,\theta_0)$, and $b_{j+1} - a_{i+1}(0,0,\theta_0)$ are colinear, and the two points $b_{j+1} - a_i(0,0,\theta_0)$ and $b_j - a_{i+1}(0,0,\theta_0)$ are not vertices of $\mathcal{CB}_{\theta_0}$. The situation can easily be recognized (since $-\vec{\nu}_i^{\mathcal{A}}(\theta_0) = \vec{\nu}_j^{\mathcal{B}}$), and the algorithm adapted. The modified algorithm runs in the same time $O(n_{\mathcal{A}} + n_{\mathcal{B}})$.

(Lozano-Pérez [Lozano-Pérez, 1983] proposed another method for computing $\mathcal{CB}_{\theta_0}$ which is slightly less efficient. He proved that:

$$\mathcal{CB}_{\theta_0} = conv(vert(\mathcal{B}) \ominus vert(\mathcal{A}(0,0,\theta_0)))$$

where $vert(\mathcal{P})$ denotes the set of vertices of a polygon $\mathcal{P}$, and $conv(S)$ denotes the convex hull of a set of points S. The computation of the convex hull of a set of points is a well-known problem in Computational Geometry [Preparata and Hong, 1977] [Preparata and Shamos, 1985]. Algorithms have been proposed which run in $O(n \log n)$ time, where n is the number of points in the input set. Here, $n = n_{\mathcal{A}}n_{\mathcal{B}}$.)

If $\mathcal{A}$ and $\mathcal{B}$ are non-convex polygons, $\mathcal{CB}_{\theta_0}$ is still a polygon. Indeed, one can decompose both $\mathcal{A}$ and $\mathcal{B}$ into convex polygons and compute the C-obstacle corresponding to every pair of polygons. The union of these C-obstacles, $\mathcal{CB}_{\theta_0}$, is a polygon. Its boundary is made of edges contained in $O(n_{\mathcal{A}}n_{\mathcal{B}})$ lines. These lines may intersect at $O(n_{\mathcal{A}}^2 n_{\mathcal{B}}^2)$ points. Hence, $\mathcal{CB}_{\theta_0}$ has at most $O(n_{\mathcal{A}}^2 n_{\mathcal{B}}^2)$ vertices. Figure 10 shows an example where the number of vertices of $\mathcal{CB}_{\theta_0}$ is actually proportional to $n_{\mathcal{A}}^2 n_{\mathcal{B}}^2$. The following method computes the representation of $\mathcal{CB}_{\theta_0}$'s boundary in $O(n_{\mathcal{A}}^2 n_{\mathcal{B}}^2 \log n_{\mathcal{A}}n_{\mathcal{B}})$ time [Sharir, 1987]. First, $\mathcal{A}$ (resp. $\mathcal{B}$) is partitioned into $O(n_{\mathcal{A}})$ (resp. $O(n_{\mathcal{B}})$) trapezoids and triangles in $O(n_{\mathcal{A}} \log n_{\mathcal{A}})$ (resp. $O(n_{\mathcal{B}} \log n_{\mathcal{B}})$) time, using a sweep-line technique. By considering the trapezoids and the triangles of the two decompositions pairwise, $\mathcal{CB}_{\theta_0}$ is the union of $O(n_{\mathcal{A}}n_{\mathcal{B}})$ convex polygons, each of

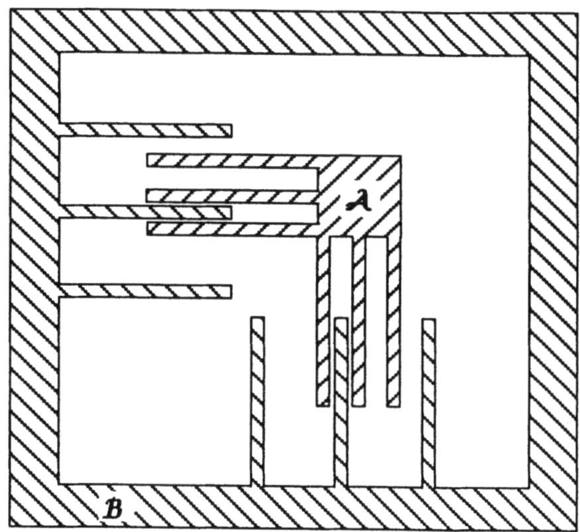

Figure 10. If $\mathcal{A}$ and $\mathcal{B}$ are non-convex polygons, $\mathcal{CB}_{\theta_0}$ is still a polygon. This figure illustrates the case where the number of vertices of $\mathcal{CB}_{\theta_0}$ is proportional to $n_\mathcal{A}^2 n_\mathcal{B}^2$, where $n_\mathcal{A}$ and $n_\mathcal{B}$ are the respective numbers of vertices of $\mathcal{A}$ and $\mathcal{B}$.

constant complexity. The boundary of each of these polygons can be computed in constant time (using the method presented above for convex polygons). The $O(n_\mathcal{A} n_\mathcal{B})$ boundaries can be intersected and $\mathcal{CB}_{\theta_0}$ traced out in $O((n_\mathcal{A} n_\mathcal{B} + c) \log n_\mathcal{A} n_\mathcal{B})$ time, with $c \in O(n_\mathcal{A}^2 n_\mathcal{B}^2)$, using a sweep-line algorithm.

The above method also applies when $\mathcal{B}$ is any regular polygonal region. Hence, when $\mathcal{A}$ can only translate in a polygonal workspace, the boundary of $\mathcal{C}_{free}$ can be computed in $O((n_\mathcal{A} n_\mathcal{B} + c) \log n_\mathcal{A} n_\mathcal{B})$ time, where $n_\mathcal{B}$ is the total complexity of the workspace obstacles and $c \in O(n_\mathcal{A}^2 n_\mathcal{B}^2)$. The complexity of this boundary (number of edges and vertices) is $O(n_\mathcal{A}^2 n_\mathcal{B}^2)$.

1.7 Boundary Representation (General Case)

Let $\mathcal{A}$ and $\mathcal{B}$ be convex polygons. $\mathcal{A}$ can both translate and rotate. The method presented above for computing the boundary of $\mathcal{CB}_{\theta_0}$ in $O(n_\mathcal{A} + n_\mathcal{B})$ time can be extended to compute the faces, edges and vertices in the boundary of $\mathcal{CB}$, and their adjacency relation. Each face corresponds to the contact of some edge of $\mathcal{A}$ (resp. $\mathcal{B}$) with some vertex of $\mathcal{B}$ (resp.

$\mathcal{A}$). The intersection of the face with a plane at some fixed value of θ is a line segment. The coordinates of its endpoints are functions of θ computable in constant time. These functions define the equations of two of the edges bounding the face. The cross-section through $\mathcal{CB}$ at the same orientation also allows us to compute the adjacency of the face with other faces at this orientation. Some edge equations, the topology of the cross-section, and the adjacency relationships change at $O(n_{\mathcal{A}}n_{\mathcal{B}})$ critical orientations where an edge of $\mathcal{A}$ is aligned with an edge of $\mathcal{B}$. By sweeping a plane at constant θ from $\theta = 0$ to $\theta = 2\pi$ and stopping at every critical orientation, we can construct an explicit representation of the boundary of $\mathcal{CB}$ in time $O(n_{\mathcal{A}}n_{\mathcal{B}}(n_{\mathcal{A}} + n_{\mathcal{B}}))$. This representation has complexity $O(n_{\mathcal{A}}n_{\mathcal{B}}(n_{\mathcal{A}}+n_{\mathcal{B}}))$ (number of faces, edges and vertices). Assuming that $n_{\mathcal{A}} < n_{\mathcal{B}}$, the time and size complexities are $O(n_{\mathcal{A}}n_{\mathcal{B}}^2)$.

If one or both of $\mathcal{A}$ and $\mathcal{B}$ are non-convex, one could think of decomposing them into convex parts, computing the boundaries of the corresponding C-obstacles and intersecting them. However, intersecting boundaries that consist of faces of type A and type B is not easy. Instead, one could try to compute the boundary of the cross-section of the C-obstacle at a fixed orientation of $\mathcal{A}$, and analyze how this boundary varies when $\mathcal{A}$'s orientation changes. But, unlike in the convex case, the cross-section may undergo significant topological changes at orientations other than those where an edge of $\mathcal{A}$ is parallel to an edge of $\mathcal{B}$. Below, we describe another method which directly constructs the boundary of $\mathcal{CB}$ from the description of $\mathcal{A}$ and $\mathcal{B}$. The method is due to Avnaim and Boissonnat [Avnaim and Boissonnat, 1988]. (See also [Koutsou, 1986] and [Brost, 1989] for related work and methods.)

We assume that $\mathcal{A}$ is a polygon and $\mathcal{B}$ is a regular (possibly non-bounded) polygonal region whose edges are all segments of bounded lengths. (If $\mathcal{B}$ is the union of all the obstacles in the workspace, the method described below computes the boundary of $\mathcal{C}_{free}$.) We denote the vertices of $\mathcal{A}$ by a_i, $i = 1, ..., n_{\mathcal{A}}$ and the vertices of $\mathcal{B}$ by b_j, $j = 1, ..., n_{\mathcal{B}}$. We denote the edge of $\mathcal{A}$ connecting a_i to a_{i+1} by $E_i^{\mathcal{A}}$ and the edge of $\mathcal{B}$ connecting b_j to b_{j+1} by $E_j^{\mathcal{B}}$. $\mathcal{B}$'s boundary may contain several closed-loop sequences of edges. In each loop modulo p_k arithmetic is applied to index j, where p_k is the number of vertices in the loop.

We represent a configuration $\mathbf{q}$ as $(x, y, \theta) \in \mathbf{R}^2 \times [0, 2\pi)$, with x and y being the coordinates of $O_{\mathcal{A}}$ in $\mathcal{F}_{\mathcal{W}}$ and θ being the angle between the

x-axes of $\mathcal{F}_{\mathcal{W}}$ and $\mathcal{F}_{\mathcal{A}}$. Through modulo 2π arithmetic, $[0, 2\pi)$ is treated as a cyclic interval. As discussed above, the C-obstacle $\mathcal{CB}$ is a regular three-dimensional volume bounded by patches of ruled surfaces (faces). Our goal is to determine these faces, their edges and vertices, and their adjacency relation.

Each face of $\mathcal{CB}$ is either of type A or of type B. In the following, we first examine in detail the faces of type B. After this study, we will consider the faces of type A.

Contact Regions. We define $\mathcal{R}^B_{i,j}$ as the set of all the configurations (x, y, θ) such that $a_i(x, y, \theta) \in E^{\mathcal{B}}_j$ and $int(\mathcal{A}(x, y, \theta)) \cap int(\mathcal{B}) = \emptyset$. $\mathcal{R}^B_{i,j}$ is called a **contact region of type B**. It consists of one or several faces of type B.

As established in Subsection 1.2, $\texttt{APPL}^B_{i,j}(x, y, \theta)$ is a necessary condition for (x, y, θ) to be in $\mathcal{R}^B_{i,j}$. This condition entails that:

$$\theta \in [\theta^{B,-}_{i,j}, \theta^{B,+}_{i,j}]$$

where $\theta^{B,-}_{i,j} = \xi_j - \phi_i - \pi$ and $\theta^{B,+}_{i,j} = \xi_j - \phi_{i-1} - \pi$ as computed in Subsection 1.4.

Let $\theta_0 \in [\theta^{B,-}_{i,j}, \theta^{B,+}_{i,j}]$. The cross-section of $\mathcal{R}^B_{i,j}$ by the plane $\theta = \theta_0$ is contained in a line segment whose endpoints are $b_j - a_i(0, 0, \theta_0)$ and $b_{j+1} - a_i(0, 0, \theta_0)$. We denote this line segment by $F^B_{i,j}(\theta_0)$.

In order for (x, y, θ_0) to be in $\mathcal{R}^B_{i,j}$, it must also be the case that no open edge of $\mathcal{A}(x, y, \theta_0)$ intersects an open edge of $\mathcal{B}$. Let $P_{k,l}(\theta_0)$ denote the set of configurations (x, y, θ_0) at which the open edge $E^{\mathcal{A}}_k(x, y, \theta_0)$ intersects the open edge $E^{\mathcal{B}}_l$. $P_{k,l}(\theta_0)$ is the interior of a parallelogram whose vertices are: $b_l - a_k(0, 0, \theta_0)$, $b_l - a_{k+1}(0, 0, \theta_0)$, $b_{l+1} - a_{k+1}(0, 0, \theta_0)$, $b_{l+1} - a_k(0, 0, \theta_0)$ (see Figure 11). $P_{k,l}(\theta_0)$ is the intersection of four open half-spaces bounded by four lines with the following respective equations:

$$\begin{aligned} f^A_{k,l}(x, y, \theta_0) &= 0, \\ f^A_{k+1,l}(x, y, \theta_0) &= 0, \\ f^B_{k,l}(x, y, \theta_0) &= 0, \\ f^B_{k,l+1}(x, y, \theta_0) &= 0. \end{aligned}$$

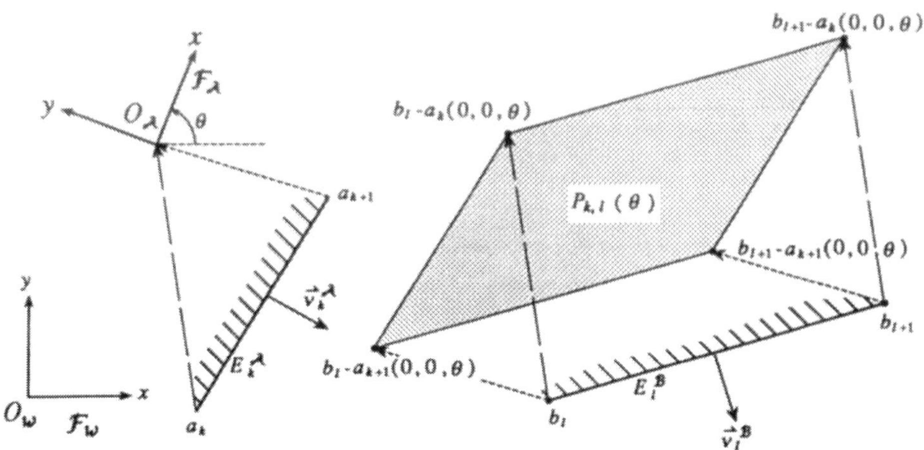

Figure 11. The set of configurations (x, y, θ_0) at which the open edge $E_k^{\mathcal{A}}(x, y, \theta_0)$ intersects the open edge $E_l^{\mathcal{B}}$ is the interior $P_{k,l}(\theta_0)$ of a parallelogram whose vertices are: $b_l - a_k(0,0,\theta_0)$, $b_l - a_{k+1}(0,0,\theta_0)$, $b_{l+1} - a_{k+1}(0,0,\theta_0)$, $b_{l+1} - a_k(0,0,\theta_0)$.

These equations have been developed in Subsection 1.4. Their forms are:

$$cs(\theta)x + cs(\theta)y + cs(\theta) = 0$$

for the first two, and:

$$\aleph x + \aleph y + cs(\theta) = 0$$

for the last two, where $cs(\theta)$ is an expression of the form:

$$\aleph \cos\theta + \aleph \sin\theta + \aleph$$

and $\aleph$ is a constant. (Each occurrence of $\aleph$ may be a different constant and each occurrence of $cs(\theta)$ may be a different expression.)

We have:

$$\mathcal{R}^{B}_{i,j} = \bigcup_{\theta \in [\theta^{B,-}_{i,j}, \theta^{B,+}_{i,j}]} \left(F^{B}_{i,j}(\theta) \setminus \bigcup_{k=1}^{n_{\mathcal{A}}} \bigcup_{l=1}^{n_{\mathcal{B}}} P_{k,l}(\theta) \right).$$

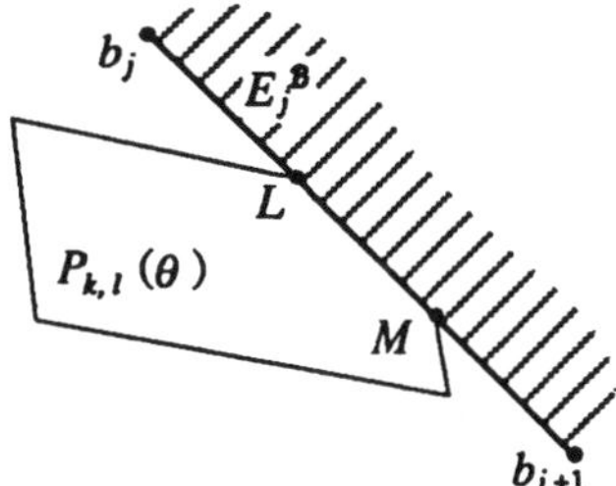

Figure 12. We denote the endpoints of $E_j^{\mathcal{B}} \cap P_{k,l}(\theta)$ by L and M. L is the endpoint which is the closest to b_j.

Intersection of $F_{i,j}^{B}(\theta)$ with $P_{k,l}(\theta)$. We now compute the intersection of $F_{i,j}^{B}(\theta)$ with $P_{k,l}(\theta)$ for any orientation $\theta \in [\theta_{i,j}^{B,-}, \theta_{i,j}^{B,+}]$. This intersection is a possibly empty, open segment.

Let us assume for a moment that the origin $O_{\mathcal{A}}$ of the frame $\mathcal{F}_{\mathcal{A}}$ coincides with a_i, so that $F_{i,j}^{B}(\theta)$ coincides with the edge $E_j^{\mathcal{B}}$, which does not depend on θ. We denote the endpoints of $E_j^{\mathcal{B}} \cap P_{k,l}(\theta)$ by L and M, as illustrated in Figure 12 (L is closer to b_j than M). The coordinates of L and M are of the form $cs(\theta)$.

Let $\lambda_{k,l}^{(i,j)}$ and $\mu_{k,l}^{(i,j)}$ (which we abbreviate in this paragraph by λ and μ) denote the two functions:

$$\lambda_{k,l}^{(i,j)}, \mu_{k,l}^{(i,j)} : [\theta_{i,j}^{B,-}, \theta_{i,j}^{B,+}] \rightarrow [0,1]$$

defined by:

$$\begin{aligned} \lambda_{k,l}^{(i,j)}(\theta) &= \lambda(\theta) &= (L - b_j)/(b_{j+1} - b_j), \\ \mu_{k,l}^{(i,j)}(\theta) &= \mu(\theta) &= (M - b_j)/(b_{j+1} - b_j). \end{aligned}$$

For every θ where the two functions are defined, we have: $\lambda(\theta) < \mu(\theta)$.

We also know that $\lambda(\theta) = 0$ if and only if $b_j \in cl(P_{k,l}(\theta))$. From the four equations of the lines supporting $P_{k,l}(\theta)$'s edges, we derive four inequalities expressing that $b_j \in cl(P_{k,l}(\theta))$. The inequalities are all of the form $cs(\theta) \leq 0$. The set of values of θ that satisfy a constraint of the form $cs(\theta) \leq 0$ is a single closed connected sub-interval of $[0, 2\pi)$, with 0 and 2π procedurally identified. Therefore, the set of values of θ such that $\lambda(\theta) = 0$ consists of a finite number of closed disjoint sub-intervals of $[\theta_{i,j}^{B,-}, \theta_{i,j}^{B,+}]$ which can be computed in constant time.

In a similar fashion, the set of values of θ such that $\mu(\theta) = 1$ consists of a finite number of closed disjoint sub-intervals of $[\theta_{i,j}^{B,-}, \theta_{i,j}^{B,+}]$ which can be computed in constant time.

Let us now assume that $\lambda(\theta) \neq 0$ (resp. $\mu(\theta) \neq 1$). Then L (resp. M) is the intersection of an edge E of $P_{k,l}(\theta)$ with $E_j^{\mathcal{B}}$. Two segments intersect if and only if the two endpoints of both segments are not on the same open side of the line supporting the other segment. Given two points (x_1, y_1) and (x_2, y_2) and a line with equation $f(x, y) = 0$, the two points are not on the same open side of the line if and only if $f(x_1, y_1)f(x_2, y_2) \leq 0$. Therefore, E and $E_j^{\mathcal{B}}$ intersect if and only if two constraints of the form $cs(\theta)cs(\theta) \leq 0$ are satisfied. In order to be satisfied, a constraint of the form $cs(\theta)cs(\theta) \leq 0$ requires either one of two pairs of constraints of the form $cs(\theta) \leq 0$ and $cs(\theta) \geq 0$ to be simultaneously satisfied. Hence, it is satisfied over a finite number of closed disjoint sub-intervals of $[0, 2\pi)$. If E and $E_j^{\mathcal{B}}$ intersect, the coordinates of the intersection point are of the form $cs(\theta)$ if the equation of the line supporting E is of the form $\aleph x + \aleph y + cs(\theta) = 0$, and of the form $cs(\theta)/cs(\theta)$ if the equation of the line supporting E is of the form $cs(\theta)x + cs(\theta)y + cs(\theta) = 0$. While λ is determined by the intersection of the edges of $P_{k,l}(\theta)$ and $E_j^{\mathcal{B}}$, which is the closest to b_j, μ is determined by the intersection that is the closest to b_{j+1}.

Therefore, λ and μ are defined for the same values of θ (those for which $P_{k,l}(\theta)$ intersects $E_j^{\mathcal{B}}$). We can divide the domain of definition of λ and μ, which is a closed subset of $[\theta_{i,j}^{B,-}, \theta_{i,j}^{B,+}]$, into a finite number of non-overlapping closed sub-intervals such that in each sub-interval both $\lambda(\theta)$ and $\mu(\theta)$ keep the same analytical expression. The analytical expression of $\lambda(\theta)$ (resp. $\mu(\theta)$) over a sub-interval is one of the following forms:

- 0 (resp. 1),

- $cs(\theta)$,

- $cs(\theta)/cs(\theta)$.

The sub-intervals and the expressions of λ and μ over each of them can be computed in constant time.

Computation of $\mathcal{R}^{\mathcal{B}}_{i,j}$. For each parallelogram[3] $P_{k,l}(\theta)$, $k = 1, ..., n_{\mathcal{A}}$, $l = 1, ..., n_{\mathcal{B}}$, we can compute the functions $\lambda^{(i,j)}_{k,l}$ and $\mu^{(i,j)}_{k,l}$ and consider the curves $\xi = \lambda^{(i,j)}_{k,l}(\theta)$ and $\xi = \mu^{(i,j)}_{k,l}(\theta)$ in the rectangle $[0, 2\pi) \times [0, 1]$ over the intervals in which the functions are defined. For each pair (k, l), let $\phi_{k,l}$ denote the region comprised between the curves $\xi = \lambda^{(i,j)}_{k,l}(\theta)$ and $\xi = \mu^{(i,j)}_{k,l}(\theta)$. Let $\Phi^{\mathcal{B}}_{i,j}$ be the region defined as:

$$\Phi^{\mathcal{B}}_{i,j} = [\theta^{\mathcal{B},-}_{i,j}, \theta^{\mathcal{B},+}_{i,j}] \times [0, 1] \setminus \bigcup_{k,l} \phi_{k,l}.$$

Figure 13.a illustrates a region $\Phi^{\mathcal{B}}_{i,j}$. The boundary of $\Phi^{\mathcal{B}}_{i,j}$ corresponds to a double contact between $\mathcal{A}$ and $\mathcal{B}$ (two contacts of type B, or a contact of type B and a contact of type A).

Let $h^{\mathcal{B}}_{i,j}$ be the bijective application defined as follows. For every (θ, ξ) in $[0, 2\pi) \times [0, 1]$, $h^{\mathcal{B}}_{i,j}(\theta, \xi) = (x, y, \theta)$, where x and y are the coordinates of the point $b_j + \xi(b_{j+1} - b_j)$ in $\mathcal{F}_{\mathcal{W}}$. Thus, the image of $\Phi^{\mathcal{B}}_{i,j}$ by $h^{\mathcal{B}}_{i,j}$ is exactly the contact region $\mathcal{R}^{\mathcal{B}}_{i,j}$. In order to compute $\mathcal{R}^{\mathcal{B}}_{i,j}$, we first compute the region $\Phi^{\mathcal{B}}_{i,j}$ and then apply the map $h^{\mathcal{B}}_{i,j}$ to $\Phi^{\mathcal{B}}_{i,j}$.

Let Ψ be a maximal connected subregion of $\Phi^{\mathcal{B}}_{i,j}$ comprised between two orientations θ_1 and θ_2, and between two curves $\xi = \mu^{(i,j)}_{k',l'}(\theta)$ and $\xi = \lambda^{(i,j)}_{k,l}(\theta)$ which keep the same analytical expression in $[\theta_1, \theta_2]$. Ψ is called a **face** of $\Phi^{\mathcal{B}}_{i,j}$ and its mapping in $\mathcal{C}$ by $h^{\mathcal{B}}_{i,j}$ is called a **face** of $\mathcal{CB}$. Such a region Ψ is shown in Figure 13.b.

The faces of the diagram of $\Phi^{\mathcal{B}}_{i,j}$ can be constructed by sweeping a line perpendicular to the θ-axis across the rectangle $[0, 2\pi) \times [0, 1]$. The critical orientations θ_c where the faces may change are all the orientations where:

- a function $\lambda^{(i,j)}_{k,l}$ (and the associated function $\mu^{(i,j)}_{k,l}$) starts/stops being defined,
- the analytical expression of a function $\lambda^{(i,j)}_{k,l}$ or $\mu^{(i,j)}_{k,l}$ changes,
- two functions $\lambda^{(i,j)}_{k,l}$ and $\lambda^{(i,j)}_{k',l'}$, or $\mu^{(i,j)}_{k,l}$ and $\mu^{(i,j)}_{k',l'}$, intersect.

[3]We may discard $P_{i,j}(\theta)$ and $P_{i+1,j}(\theta)$, which cannot intersect $E^{\mathcal{B}}_j$.

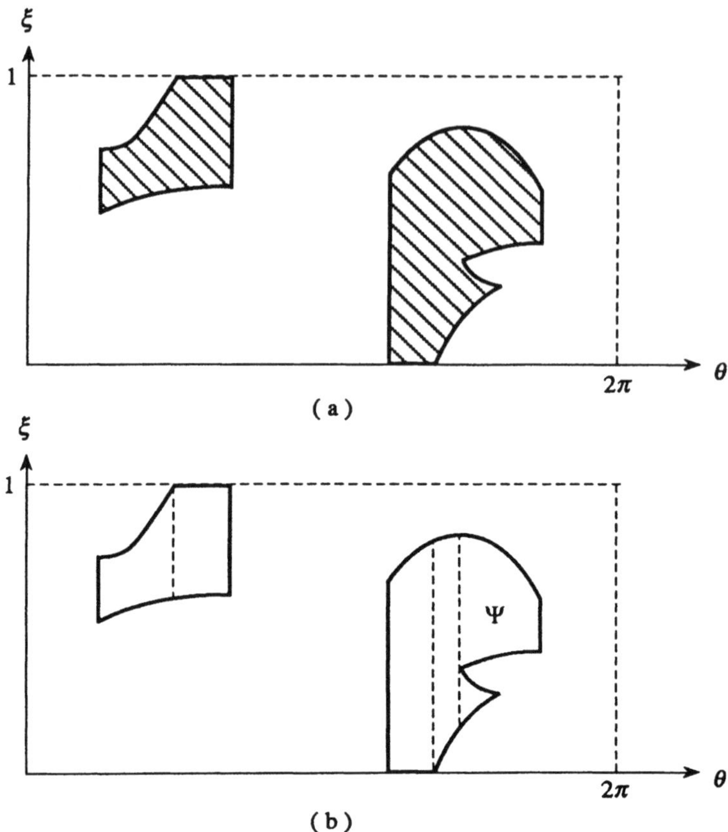

Figure 13. The striped areas in Figure a form a region $\Phi^B_{i,j}$ (see text). Figure b shows the decomposition of this region into six faces. One of them is denoted by Ψ. Each face has at most four edges. The edges contained in the boundary of $\Phi^B_{i,j}$ are called type 2 edges. The other edges (shown in dashed line) are called type 1 edges.

We already know how to compute the critical orientations of the first and second types. The critical orientations of the third types are the intersections of two curves, each having one of the following forms: $\xi = 0$ (or 1), $\xi = cs(\theta)$, and $\xi = cs(\theta)/cs(\theta)$. Every two curves may have up to two intersection points in their common sub-domain of definition. The computation of these points reduces to either the computation of the intersection between a line and a circle, or the computation of the intersection between a circle and a general conic.

There are $n_{\mathcal{A}} n_{\mathcal{B}}$ critical orientations of the first and second types. There are $O(n_{\mathcal{A}}^2 n_{\mathcal{B}}^2)$ orientations of the third type. Hence, the sweep-line algorithm computes the faces of $\mathcal{R}_{i,j}^B$ in $O((n_{\mathcal{A}} n_{\mathcal{B}} + c)\log n_{\mathcal{A}} n_{\mathcal{B}})$ time, with $c \in O(n_{\mathcal{A}}^2 n_{\mathcal{B}}^2)$.

Faces, Edges and Vertices of $\mathcal{CB}$. Given a face Ψ of $\Phi_{i,j}^B$, the corresponding face of $\mathcal{CB}$ is obtained by applying the map $h_{i,j}^B$. It is denoted by $h_{i,j}^B(\Psi)$.

The region Ψ is bounded by two curve segments contained in $\xi = \lambda_{k,l}^{(i,j)}(\theta)$ and $\xi = \mu_{k',l'}^{(i,j)}(\theta)$, and possibly by two line segments parallel to the ξ-axis. The application $h_{i,j}^B$ maps these curve segments to curve segments in $\mathbf{R}^2 \times [0, 2\pi)$ which bound the face $h_{i,j}^B(\Psi)$ and form its edges. The intersections of these edges are the vertices of the face.

In order to compute $\Phi_{i,j}^B$, we assumed that a_i and $O_{\mathcal{A}}$ were coincident. If this is not the case, we can compute $\Phi_{i,j}^B$ in the same manner by translating $O_{\mathcal{A}}$ onto a_i; $\mathcal{R}_{i,j}^B$ is then obtained by applying the map $T_{-a_i} \circ h_{i,j}^B$ to $\Phi_{i,j}^B$, where T_{-a_i} denotes the translation of vector $-a_i$. In this way, we can compute the faces of all the type B contact regions $\mathcal{R}_{i,j}^B$, $i = 1, \ldots, n_{\mathcal{A}}$, $j = 1, \ldots, n_{\mathcal{B}}$, their edges and their vertices.

Now, we still have to compute the faces of $\mathcal{CB}$ produced by type A contacts. Let:

$$\mathcal{R}_{i,j}^A = \{(x, y, \theta) \in \mathbf{R}^2 \times [0, 2\pi) \;/\; b_j \in E_i^A(x, y, \theta) \;;\; int(\mathcal{A}(x, y, \theta)) \cap int(\mathcal{B}) = \emptyset\}.$$

We can define and compute the region $\Phi_{i,j}^A$ corresponding to $\mathcal{R}_{i,j}^A$ in the same way as $\Phi_{i,j}^B$, by considering that $\mathcal{A}$ is the obstacle and $\mathcal{B}$ is the moving object. This corresponds to taking $\mathbf{q} = (x, y, \theta)$, with x and y being the coordinates of $O_{\mathcal{W}}$ in $\mathcal{F}_{\mathcal{A}}$ and θ being the angle between the x-axis of $\mathcal{F}_{\mathcal{A}}$ and the x-axis of $\mathcal{F}_{\mathcal{W}}$. In order to return to our original parameterization of $\mathcal{C}$, we apply the map INV defined by:

$$INV(x, y, \theta) = (-x\cos\theta - y\sin\theta,\; x\sin\theta - y\cos\theta,\; -\theta).$$

Once the region $\Phi_{i,j}^A$ and its faces have been computed, they are mapped in configuration space by applying $INV \circ T_{-b_j} \circ h_{i,j}^A$.

The computation of all the faces, with their edges and vertices, takes time $O(n_{\mathcal{A}}^3 n_{\mathcal{B}}^3 \log n_{\mathcal{A}} n_{\mathcal{B}})$. The vertices of $\mathcal{CB}$ correspond to triple contacts where the interiors of $\mathcal{A}$ and $\mathcal{B}$ do not intersect. Avnaim and Boissonnat [Avnaim and Boissonnat, 1988] gave an example where the number of such triple contacts is actually proportional to $n_{\mathcal{A}}^3 n_{\mathcal{B}}^3$. Hence, the algorithm is close to optimal.

Adjacency Graph At this point, we have computed all the faces, edges and vertices of $\mathcal{CB}$. It remains to determine their topological adjacency relation. Two faces are adjacent if and only if they share a subset of an edge of non-zero measure in $\mathbf{R}^2$.

We distinguish two types of edges in a face (see Figure 13.b). An edge of type 1 belongs to two different faces of the same contact region $\mathcal{R}$ ($\mathcal{R}_{i,j}^A$ or $\mathcal{R}_{i,j}^B$). An edge of type 2 lies in the boundary of a contact region and thus corresponds to a double contact between $\mathcal{A}$ and $\mathcal{B}$. An edge of type 1 lies in a plane at constant θ; hence, it is always a straight edge. An edge of type 2 may or may not lie in a plane at constant θ; it may or may not be a straight edge.

The adjacency of two faces by an edge of type 1 can be computed during the construction of the Φ regions ($\Phi_{i,j}^A$'s and $\Phi_{i,j}^B$'s) with the sweep-line algorithm, within the same time complexity.

An edge of type 2 corresponds to a double contact that can be identified in constant time during the construction of the Φ regions. During the decomposition of a region Φ into faces, we associate with each double contact the list of the faces of Φ which have an edge of type 2 corresponding to this double contact. By construction, the list is sorted by increasing θ or ξ. Once all the regions have been decomposed into faces, each double contact is labeled by two (possibly empty) lists of faces, which we denote by $\mathcal{L}$ and $\mathcal{L}'$. The double contact may or may not occur at constant θ. If it occurs at a constant θ, say θ_0, $\mathcal{L}$ and $\mathcal{L}'$ are sorted by increasing ξ; a face of $\mathcal{L}$ and a face of $\mathcal{L}'$ are adjacent if their straight edges at θ_0 intersect along a subsegment of non-zero length. If the double contact occurs at non-constant θ, $\mathcal{L}$ and $\mathcal{L}'$ are sorted by increasing θ; a face of $\mathcal{L}$ is adjacent to a face of $\mathcal{L}'$ if and only if the two faces are both defined in an interval of values of θ having non-zero length. In both cases, the adjacency relation among the faces of the two lists can be computed in worst-case time equal to the sum of the lengths of the two lists. Each face has at most four edges of type 2. Hence, the adjacency of faces by edges of type 2 can be computed in time linear in the total number of faces, which is in $O(n_{\mathcal{A}}^3 n_{\mathcal{B}}^3)$.

Thus, the computation of the boundary of $\mathcal{CB}$ takes time $O(n_{\mathcal{A}}^3 n_{\mathcal{B}}^3 \log n_{\mathcal{A}} n_{\mathcal{B}})$ and the resulting representation has size $O(n_{\mathcal{A}}^3 n_{\mathcal{B}}^3)$. Cases where the actual complexity of the method is much smaller are described in [Avnaim and Boissonnat, 1988]. If $\mathcal{A}$ is a line segment and $\mathcal{B}$ contains no three aligned vertices, then the time complexity of the method is $O(n_{\mathcal{B}}^2 \log n_{\mathcal{B}})$. If the number of vertices $\mathcal{A}$ is small, so that it can be considered constant, and if the edges of $\mathcal{B}$ are not concentrated near each other compared to the diameter of $\mathcal{A}$ (a situation of *bounded local complexity* [Sifrony and Sharir, 1986]), then the number of faces of $\mathcal{CB}$ is $O(n_{\mathcal{B}})$ and the method can be adapted to run in time $O(n_{\mathcal{B}} \log n_{\mathcal{B}})$.

2 Polyhedral Workspace

In this section we extend some of the results presented in the previous section to the case of a polyhedral workspace. The C-obstacle $\mathcal{CB}$ corresponding to an obstacle $\mathcal{B}$ is now a six-dimensional region of $\mathcal{C}$ bounded by five-dimensional C-surfaces generated by C-constraints. We will see that C-constraints arise from three types of contact between $\mathcal{A}$ and $\mathcal{B}$:

- **type A contacts**, each between a face of $\mathcal{A}$ and a vertex of $\mathcal{B}$,
- **type B contacts**, each between a vertex of $\mathcal{A}$ and a face of $\mathcal{B}$,
- **type C contacts**, each between an edge of $\mathcal{A}$ and an edge of $\mathcal{B}$.

Each type of contact produces a distinct C-constraint, and, as in the polygonal case, if $\mathcal{A}$ and $\mathcal{B}$ are convex, the predicate CB defined by $\mathbf{q} \in \mathcal{CB} \Leftrightarrow \text{CB}(\mathbf{q})$ can be constructed as the conjunction of all the C-constraints. However, there are usually many more C-constraints in the polyhedral case than in the polygonal one. For instance, if both $\mathcal{A}$ and $\mathcal{B}$ are parallelepipeds, there are 48 different contacts of type A, 48 different contacts of type B, and 144 different contacts of type C!

2.1 Polyhedral Regions and Polyhedra

A plane P in $\mathbf{R}^3$ decomposes the three-dimensional space into two half-spaces. Let $h(x, y, z) = 0$ be the equation of P. The two closed half-spaces determined by $h(x,y,z) \leq 0$ and $h(x,y,z) \geq 0$ are denoted by $\overline{h}^-$ and $\overline{h}^+$, respectively.

DEFINITION 5: *A* **convex polyhedral region** *$\mathcal{P}$ in $\mathbf{R}^3$ is the intersection of a finite number of closed half-spaces. A* **polyhedral region** *is any subset of $\mathbf{R}^3$ obtained by taking the union of a finite number of convex polyhedral regions.*

A polyhedral region may not be bounded, nor connected. Its connected components may not be simply-connected.

DEFINITION 6: *A* **polyhedron** *is any polyhedral region that is homeomorphic to the closed unit ball in $\mathbf{R}^3$.*

Hence, a polyhedron is compact, regular, and simply-connected. Its boundary is a surface homeomorphic to the unit sphere.

Let $\mathcal{P}$ be a regular polyhedral region. Each maximal connected planar region in $\mathcal{P}$'s boundary that is not intersected by the rest of $\mathcal{P}$'s boundary, except possibly along its own boundary, is called a **face**. Each face is a polygonal region. Its edges are called the **edges** of $\mathcal{P}$. Every endpoint of an edge is called a **vertex**. An edge is **closed** (resp. **open**) if we include (resp. do not include) its endpoints. A face is **closed** (resp. **open**) if we include (resp. do not include) its bounding edges. If $\mathcal{P}$ is a manifold with boundary, each edge is shared by exactly two faces.

Every face F_i of $\mathcal{P}$ is contained in a plane P_i called the **supporting plane** of F_i. Let $h_i(x,y,z) = 0$ be the equation of P_i. Without loss of generality, we write this equation so that every point in the open face F_i has a neighborhood whose intersection with $\mathcal{P}$ only contains points verifying $h_i(x,y,z) \leq 0$. The half-space $\overline{h}_i^+$ (resp. $\overline{h}_i^-$) is called the **outer** (resp. **inner**) half-space of F_i. The **outgoing** (resp. **ingoing**) normal of F_i is the unit vector normal to F_i and pointing toward the interior of the outer (resp. inner) half-space of F_i.

Let f, e, and v denote respectively the numbers of faces, edges and vertices of a polyhedron. The surface of a polyhedron is isomorphic to a planar subdivision, i.e. the graph whose nodes are the polyhedron's vertices and whose links are the polyhedron's edges is a graph which can

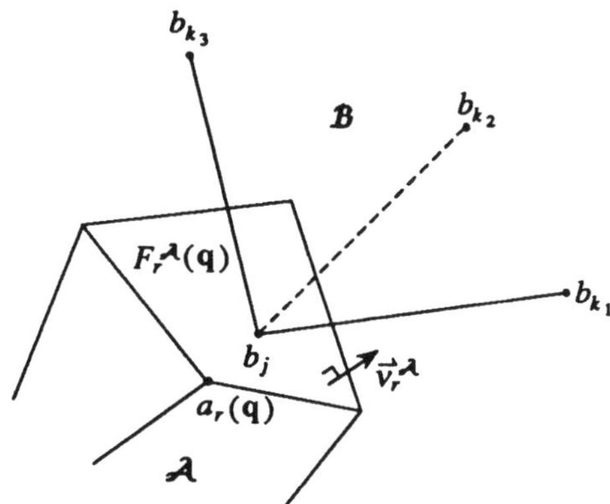

Figure 14. A type A contact between $\mathcal{A}$ and $\mathcal{B}$ occurs when a face $F_r^{\mathcal{A}}$ of $\mathcal{A}$ contains a vertex b_j of $\mathcal{B}$, without overlapping of the interiors of $\mathcal{A}$ and $\mathcal{B}$.

be embedded in the plane without crossings of the links (a planar graph). Hence, the three parameters, f, e and v, are related by Euler's formula:

$$f - e + v = 2.$$

Since each vertex is the extremity of at least three edges and each face is bounded by at least three edges, it can easily be proven that v, e and f are pairwise proportional.

A polyhedron is conveniently represented as a data structure that makes explicit the adjacency relation among its faces, edges, and vertices (e.g. the "doubly-connected-edge-list" (DCEL) [Preparata and Shamos, 1985]). A convex polyhedral region may also be represented by the set of the inequations that defines the half-spaces $\overline{h_i^-}$, i.e. $\bigcap_i \overline{h_i^-}$.

The **complexity** of a polyhedral region is defined as the number of its faces, edges and vertices.

2.2 C-constraints of Type A and Type B

C-constraints of types A and B are established in the same fashion as in the polygonal case. Figures 14 and 15 illustrate the notations used below. The subscript i (resp. j) is used for vertices of $\mathcal{A}$ (resp. $\mathcal{B}$). The subscript l (resp. m) is used for edges of $\mathcal{A}$ (resp. $\mathcal{B}$). The subscript r

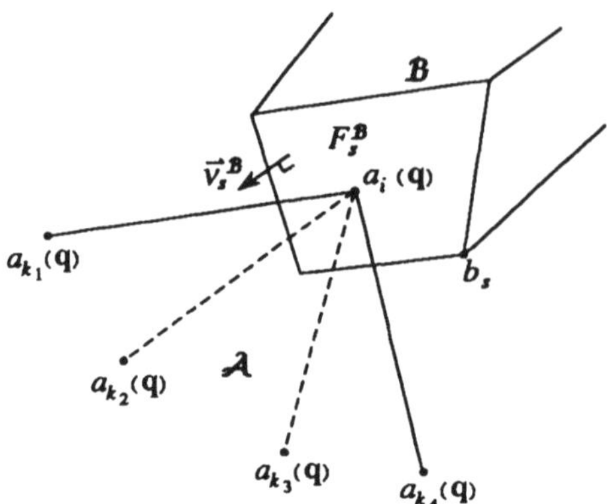

Figure 15. A type B contact between $\mathcal{A}$ and $\mathcal{B}$ occurs when a vertex a_i of $\mathcal{A}$ is contained in a face $F_s^{\mathcal{B}}$ of $\mathcal{B}$, without overlapping of the interiors of $\mathcal{A}$ and $\mathcal{B}$.

(resp. s) is used for faces of $\mathcal{A}$ (resp. $\mathcal{B}$). All edges are treated as closed edges, i.e. they include their endpoints. All faces are treated as closed faces, i.e. they include their bounding edges. In order to simplify the notations, we assume that one extremity of edge $E_l^{\mathcal{A}}$ (resp. $E_m^{\mathcal{B}}$) is a_l (resp. b_m), and that one vertex in the boundary of $\mathcal{A}$'s face $F_r^{\mathcal{A}}$ (resp. $\mathcal{B}$'s face $F_s^{\mathcal{B}}$) is a_r (resp. b_s). We denote the outgoing normal of $F_r^{\mathcal{A}}$ (resp. $F_s^{\mathcal{B}}$) by $\vec{\nu}_r^{\mathcal{A}}$ (resp. $\vec{\nu}_s^{\mathcal{B}}$).

Type A Contact. Consider the type A contact between the face $F_r^{\mathcal{A}}$ of $\mathcal{A}$ and the vertex b_j of $\mathcal{B}$ (Figure 14). The coboundary of b_j is defined as the set of all the edges of $\mathcal{B}$ having b_j as one of their endpoints. We let $\omega(b_j)$ denote the set of all the vertices (except b_j) at the extremities of the edges belonging to the coboundary of b_j. In Figure 14, $\omega(b_j) = \{b_{k_1}, b_{k_2}, b_{k_3}\}$.

DEFINITION 7: *The expression:*

$$\mathtt{APPL}_{r,j}^{A}(\mathbf{q}) \Rightarrow [f_{r,j}^{A}(\mathbf{q}) \leq 0]$$

where:

$$\mathtt{APPL}_{r,j}^{A}(\mathbf{q}) = \bigwedge_{b_k \in \omega(b_j)} [\vec{\nu}_r^{\mathcal{A}}(\mathbf{q}) \cdot (b_k - b_j) \geq 0]$$

and:

$$f_{r,j}^{A}(\mathbf{q}) = \vec{\nu}_r^{\mathcal{A}}(\mathbf{q}) \cdot (b_j - a_r(\mathbf{q}))$$

is called a **C-constraint of type A**. *It is denoted by* $\texttt{CONST}^{A}_{r,j}(\mathbf{q})$.

Type B Contact. Consider the type B contact between the vertex a_i of $\mathcal{A}$ and the face $F^{\mathcal{B}}_s$ of $\mathcal{B}$ (see Figure 15).

DEFINITION 8: *The expression:*

$$\texttt{APPL}^{B}_{i,s}(\mathbf{q}) \Rightarrow [f^{B}_{i,s}(\mathbf{q}) \leq 0]$$

where:

$$\texttt{APPL}^{B}_{i,s}(\mathbf{q}) = \bigwedge_{a_k \in \omega(a_i)} [(a_k(\mathbf{q}) - a_i(\mathbf{q})) \cdot \vec{\nu}^{\mathcal{B}}_s \geq 0]$$

and:

$$f^{B}_{i,s}(\mathbf{q}) = \vec{\nu}^{\mathcal{B}}_s \cdot (a_i(\mathbf{q}) - b_s)$$

is called a **C-constraint of type B**. *It is denoted by* $\texttt{CONST}^{B}_{i,s}(\mathbf{q})$.

The surface (i.e. the five-dimensional submanifold of $\mathcal{C}$) represented by $f^{A}_{r,j}(\mathbf{q}) = 0$ (resp. $f^{B}_{i,s}(\mathbf{q}) = 0$) is called a **C-surface** of type A (resp. type B).

When $\mathbf{q}$ is expressed as $(\mathcal{T}, \Theta)$, the applicability conditions $\texttt{APPL}^{A}_{r,j}(\mathbf{q})$ and $\texttt{APPL}^{B}_{i,s}(\mathbf{q})$ depend only on Θ.

2.3 C-constraints of Type C

A contact of type C occurs when an edge $E^{\mathcal{A}}_l$ of $\mathcal{A}$ is in contact with an edge $E^{\mathcal{B}}_m$ of $\mathcal{B}$, with no intersection of the interiors of the two objects. A type C contact is illustrated in Figure 16.

A C-contact between $E^{\mathcal{A}}_l$ and $E^{\mathcal{B}}_m$ is feasible only for a subrange of orientations of $\mathcal{A}$. This subrange can be determined as follows [Donald, 1984]. Let $P_{l,m}(\mathbf{q})$ be the plane containing both $E^{\mathcal{A}}_l(\mathbf{q})$ and $E^{\mathcal{B}}_m$ when the two edges are in contact (we assume here that $E^{\mathcal{A}}_l(\mathbf{q})$ and $E^{\mathcal{B}}_m$ are not colinear). There is no intersection of the interiors of $\mathcal{A}(\mathbf{q})$ and $\mathcal{B}$ if and only if $P_{l,m}(\mathbf{q})$ is a *separating plane* of $\mathcal{A}(\mathbf{q})$ and $\mathcal{B}$, i.e. if the interiors of $\mathcal{A}(\mathbf{q})$ and $\mathcal{B}$ lie on different sides of $P_{l,m}(\mathbf{q})$.

Let $F^{\mathcal{A}}_{l,1}$ and $F^{\mathcal{A}}_{l,2}$ be the two faces of $\mathcal{A}$ intersecting at $E^{\mathcal{A}}_l$ (see Figure 17.a). Let $F^{\mathcal{B}}_{m,1}$ and $F^{\mathcal{B}}_{m,2}$ be the two faces of $\mathcal{B}$ intersecting at $E^{\mathcal{B}}_m$ (see Figure 17.b). We denote by:

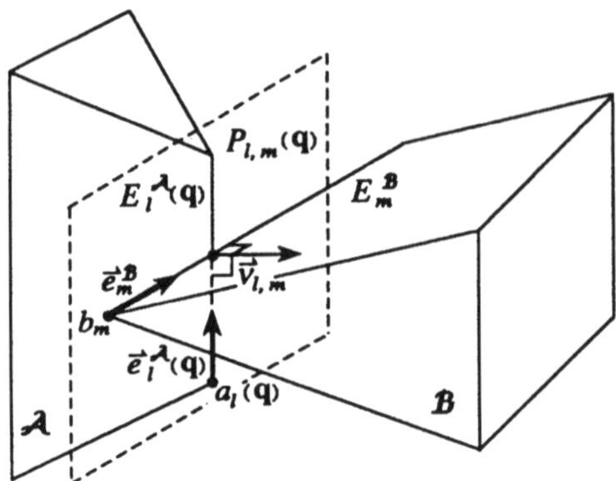

Figure 16. A type C contact between $\mathcal{A}$ and $\mathcal{B}$ occurs when an edge $E_l^{\mathcal{A}}$ of $\mathcal{A}$ intersects an edge $E_m^{\mathcal{B}}$ of $\mathcal{B}$, without overlapping of the interiors of $\mathcal{A}$ and $\mathcal{B}$.

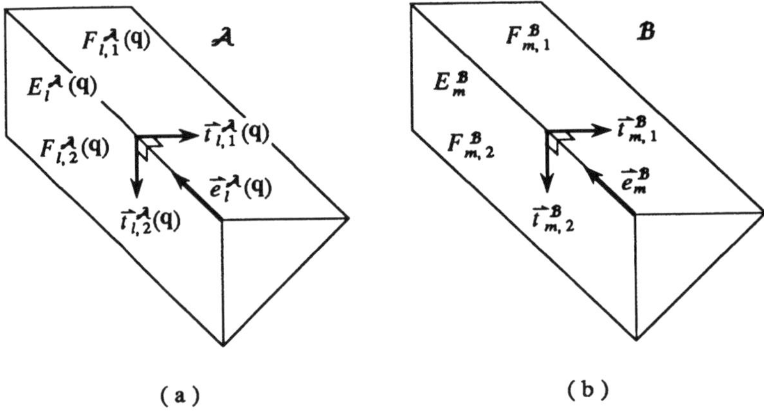

Figure 17. This figure illustrates some of the notations for establishing the general form of a C-constraint of type C (see text).

- $\vec{t}_{l,1}^{\mathcal{A}}(\mathbf{q})$ (resp. $\vec{t}_{l,2}^{\mathcal{A}}(\mathbf{q})$) the vector contained in $F_{l,1}^{\mathcal{A}}(\mathbf{q})$ (resp. $F_{l,2}^{\mathcal{A}}(\mathbf{q})$) and perpendicular to $E_l^{\mathcal{A}}(\mathbf{q})$,
- $\vec{t}_{m,1}^{\mathcal{B}}$ (resp. $\vec{t}_{m,2}^{\mathcal{B}}$) the vector contained in $F_{m,1}^{\mathcal{B}}$ (resp. $F_{m,2}^{\mathcal{B}}$) and perpendicular to $E_m^{\mathcal{B}}$.

Let:

$$\vec{\nu}_{l,m}(\mathbf{q}) = \epsilon_{l,m}\ \vec{e}_l^{\mathcal{A}}(\mathbf{q}) \wedge \vec{e}_m^{\mathcal{B}}$$

where:
- $\epsilon_{l,m} = \pm 1$ (the actual value of $\epsilon_{l,m}$ will be defined later),
- $\vec{e}_l^{\mathcal{A}}(\mathbf{q})$ (resp. $\vec{e}_m^{\mathcal{B}}$) denotes a unit vector parallel to $E_l^{\mathcal{A}}(\mathbf{q})$ (resp. $E_m^{\mathcal{B}}$),
- $\vec{u} \wedge \vec{v}$ denotes the outer product of two vectors.

Hence, $\vec{\nu}_{l,m}(\mathbf{q})$ is normal to the plane $P_{l,m}(\mathbf{q})$. Below, we will set $\epsilon_{l,m}$ so that it points toward $\mathcal{B}$.

$P_{l,m}(\mathbf{q})$ is a separating plane of $\mathcal{A}(\mathbf{q})$ and $\mathcal{B}$ if and only if $\vec{t}_{l,1}^{\mathcal{A}}(\mathbf{q})$ and $\vec{t}_{l,2}^{\mathcal{A}}(\mathbf{q})$ point away into one half-space bounded by $P_{l,m}(\mathbf{q})$, while $\vec{t}_{m,1}^{\mathcal{B}}$ and $\vec{t}_{m,2}^{\mathcal{B}}$ point away into the other half-space. This can be expressed by the following three conditions:

$$\begin{array}{lclcl} k_{\mathcal{A}} & = & sign(\vec{t}_{l,1}^{\mathcal{A}}(\mathbf{q}) \cdot \vec{\nu}_{l,m}(\mathbf{q})) & = & sign(\vec{t}_{l,2}^{\mathcal{A}}(\mathbf{q}) \cdot \vec{\nu}_{l,m}(\mathbf{q})), \\ k_{\mathcal{B}} & = & sign(\vec{t}_{m,1}^{\mathcal{B}} \cdot \vec{\nu}_{l,m}(\mathbf{q})) & = & sign(\vec{t}_{m,2}^{\mathcal{B}} \cdot \vec{\nu}_{l,m}(\mathbf{q})), \\ k_{\mathcal{A}} & \neq & k_{\mathcal{B}}. & & \end{array}$$

These conditions are called the **applicability conditions** of the type C contact between $E_l^{\mathcal{A}}$ and $E_m^{\mathcal{B}}$. We denote their conjunction by $\mathtt{APPL}_{l,m}^{C}(\mathbf{q})$.

Let $\mathcal{A}$ move in such a way that the type C contact between $E_l^{\mathcal{A}}$ and $E_m^{\mathcal{B}}$ is maintained. During the displacement, the robot's configuration moves in a five-dimensional submanifold of $\mathcal{C}$ whose equation is:

$$f_{l,m}^{C}(\mathbf{q}) = 0$$

where:

$$f_{l,m}^{C}(\mathbf{q}) = \vec{\nu}_{l,m}(\mathbf{q}) \cdot (b_m - a_l(\mathbf{q})).$$

This equation expresses the fact that when $E_l^{\mathcal{A}}(\mathbf{q})$ and $E_m^{\mathcal{B}}$ are in contact, the two vectors $\vec{\nu}_{l,m}(\mathbf{q})$ and $(b_m - a_l(\mathbf{q}))$ are perpendicular or $b_m - a_l(\mathbf{q}) = 0$. The surface it represents is called **C-surface** of type C. It separates $\mathcal{C}$ into two half-spaces. If we take:

$$\epsilon_{l,m} = sign((\vec{e}_l^{\mathcal{A}}(\mathbf{q}) \wedge \vec{e}_m^{\mathcal{B}}) \cdot \vec{t}_{m,1}^{\mathcal{B}})$$

then $\vec{\nu}_{l,m}$ points into the half-space bounded by $P_{l,m}(\mathbf{q})$ which contains $\mathcal{B}$, and the C-obstacle $\mathcal{CB}$ lies completely within the half-space determined by: $f_{l,m}^{C}(\mathbf{q}) \leq 0$.

DEFINITION 9: *Let:*

$$\vec{\nu}_{l,m}(\mathbf{q}) = \epsilon_{l,m}\ \vec{e}_l^{\mathcal{A}}(\mathbf{q}) \wedge \vec{e}_m^{\mathcal{B}}$$

with:

$$\epsilon_{l,m} = sign((\vec{e}_l^{\mathcal{A}}(\mathbf{q}) \wedge \vec{e}_m^{\mathcal{B}}) \cdot \vec{t}_{m,1}^{\mathcal{B}}).$$

The expression:

$$\text{APPL}_{l,m}^{C}(\mathbf{q}) \Rightarrow [f_{l,m}^{C}(\mathbf{q}) \leq 0]$$

where:

$$\begin{aligned}\text{APPL}_{l,m}^{C}(\mathbf{q}) = \; & [k_{\mathcal{A}} = sign(\vec{t}_{l,1}^{\mathcal{A}}(\mathbf{q}) \cdot \vec{\nu}_{l,m}(\mathbf{q})) = sign(\vec{t}_{l,2}^{\mathcal{A}}(\mathbf{q}) \cdot \vec{\nu}_{l,m}(\mathbf{q}))] \\ & \bigwedge [k_{\mathcal{B}} = sign(\vec{t}_{m,1}^{\mathcal{B}} \cdot \vec{\nu}_{l,m}(\mathbf{q})) = sign(\vec{t}_{m,2}^{\mathcal{B}} \cdot \vec{\nu}_{l,m}(\mathbf{q}))] \\ & \bigwedge k_{\mathcal{A}} \neq k_{\mathcal{B}}\end{aligned}$$

and:

$$f_{l,m}^{C}(\mathbf{q}) = \vec{\nu}_{l,m}(\mathbf{q}) \cdot (b_m - a_l(\mathbf{q}))$$

is called a **C-constraint of type C**. *It is denoted by* $\text{CONST}_{l,m}^{C}(\mathbf{q})$.

In this definition we assume that:

(1) The outer product of $\vec{e}_l^{\mathcal{A}}(\mathbf{q})$ and $\vec{e}_m^{\mathcal{B}}$ is non-zero.

(2) None of the inner products:

$$\vec{t}_{l,1}^{\mathcal{A}}(\mathbf{q}) \cdot \vec{\nu}_{l,m}(\mathbf{q})$$

$$\vec{t}_{l,2}^{\mathcal{A}}(\mathbf{q}) \cdot \vec{\nu}_{l,m}(\mathbf{q})$$

$$\vec{t}_{m,1}^{\mathcal{B}} \cdot \vec{\nu}_{l,m}(\mathbf{q})$$

$$\vec{t}_{m,2}^{\mathcal{B}} \cdot \vec{\nu}_{l,m}(\mathbf{q})$$

is zero.

If condition (1) is not satisfied, it means that $E_l^{\mathcal{A}}(\mathbf{q})$ and $E_m^{\mathcal{B}}$ are colinear. Then, the contact is also a type A and/or type B contact. Hence, one or several C-constraints of type A and/or B already take this situation into account. Similarly, if condition (2) is not satisfied, an edge of $\mathcal{A}$ (or $\mathcal{B}$) is contained in the plane of a face of $\mathcal{B}$ (or $\mathcal{A}$), and again the situation is already taken into account by one or several C-constraints of type A or B. Thus, the situations where the above conditions are not satisfied need not be included in the C-constraints of type C.

2.4 Representation of a C-obstacle with C-constraints

As in the polygonal case, we can express $\mathcal{CB}$ as the conjunction of all the C-constraints of types A, B, and C generated by the faces, vertices, and edges of $\mathcal{A}$ and $\mathcal{B}$ [Donald, 1984].

THEOREM 2: *Let $\mathcal{A}$ and $\mathcal{B}$ be two convex polyhedra. The C-obstacle:*

$$\mathcal{CB} = \{\mathbf{q} \in \mathcal{C} \;/\; \mathcal{A}(\mathbf{q}) \cap \mathcal{B} \neq \emptyset\}$$

is such that:

$$\mathbf{q} \in \mathcal{CB} \Leftrightarrow \mathtt{CB}(\mathbf{q})$$

with:

$$\mathtt{CB}(\mathbf{q}) \equiv (\bigwedge_{r,j} \mathtt{CONST}^{A}_{r,j}(\mathbf{q})) \bigwedge (\bigwedge_{i,s} \mathtt{CONST}^{B}_{i,s}(\mathbf{q})) \bigwedge (\bigwedge_{l,m} \mathtt{CONST}^{C}_{l,m}(\mathbf{q})).$$

Since the numbers of vertices, edges and faces in a polyhedron are piecewise proportional, the total number of C-constraints of Types A, B, and C in $\mathtt{CB}(\mathbf{q})$ (i.e. the time complexity of its evaluation) is proportional to $n_{\mathcal{A}} n_{\mathcal{B}}$, where $n_{\mathcal{A}}$ (resp. $n_{\mathcal{B}}$) is the number of vertices of $\mathcal{A}$ (resp. $\mathcal{B}$). Other forms of the predicate CB have been proposed in [Canny, 1988].

If $\mathcal{A}$ and $\mathcal{B}$ are non-convex polyhedra, then they can be decomposed into convex ones. As in the polygonal case, $\mathtt{CB}(\mathbf{q})$ becomes a disjunction. Every term of the disjunction describes the C-obstacle corresponding to one convex polyhedron resulting from the decomposition of $\mathcal{A}$ and one convex polyhedron resulting from the decomposition of $\mathcal{B}$. The optimal convex decomposition of a non-convex polyhedron is known to be *NP*-hard [Lingas, 1982]. However, a non-optimal decomposition can be produced in polynomial time [Chazelle, 1984] [Bajaj and Dey, 1988]. $\mathtt{CB}(\mathbf{q})$ can also be adapted to handle the case where $\mathcal{B}$ is a non-bounded and/or non-regular polyhedral region.

2.5 Parameterization of C-Constraints

Once a representation of $\mathbf{q}$ has been selected, it can be used to parameterize the C-constraints of types A, B, and C, as we did in the polygonal case (see Subsection 1.4). However, the resulting expressions are quite long and we do not give them here. In practice, they may be established using an automatic symbolic computation system such as Macsyma (see [Donald, 1984] [Canny, 1988]). If a quaternion-based representation is used, the parameterized expression of CB(**q**) can be established in semi-algebraic form.

2.6 Geometric Interpretation

Let us consider the expression of $\mathcal{CB}$ given in Theorem 2 ($\mathcal{A}$ and $\mathcal{B}$ are convex polyhedra), with **q** represented as $(\mathcal{T}, \Theta)$, where $\mathcal{T} \in \mathbf{R}^3$ consists of the three coordinates of $O_{\mathcal{A}}$ in $\mathcal{F}_{\mathcal{W}}$ and Θ is any parameterization of the orientation of $\mathcal{A}$. Each C-surface is represented by:

$$f^A_{r,j}(\mathcal{T}, \Theta) = 0 \quad \text{or} \quad f^B_{i,s}(\mathcal{T}, \Theta) = 0 \quad \text{or} \quad f^C_{l,m}(\mathcal{T}, \Theta) = 0.$$

If we fix the orientation of $\mathcal{A}$ equal to Θ_0, then:

$$f^A_{r,j}(\mathcal{T}, \Theta_0) = 0, \quad f^B_{i,s}(\mathcal{T}, \Theta_0) = 0 \quad \text{and} \quad f^C_{l,m}(\mathcal{T}, \Theta_0) = 0$$

represent planes in $\mathbf{R}^3$:

- The plane $f^A_{r,j}(\mathcal{T}, \Theta_0) = 0$ is parallel to $F^{\mathcal{A}}_r(\mathcal{T}, \Theta_0)$. It rotates if we slightly modify the orientation of $\mathcal{A}$.

- The plane $f^B_{i,s}(\mathcal{T}, \Theta_0) = 0$ is parallel to $F^{\mathcal{B}}_s$. It translates if we slightly modify the orientation of $\mathcal{A}$.

- The plane $f^C_{l,m}(\mathcal{T}, \Theta_0) = 0$ is parallel to both $E^{\mathcal{A}}_l(\mathcal{T}, \Theta_0)$ and $E^{\mathcal{B}}_m$. It rotates if we slightly modify the orientation of $E^{\mathcal{A}}_l(\mathcal{T}, \Theta_0)$.

The cross-section $\mathcal{CB}_{\Theta_0}$ at $\Theta = \Theta_0$ is the intersection of a finite number of closed half-spaces, each bounded by a plane $f^A_{r,j}(\mathcal{T}, \Theta_0) = 0$, $f^B_{i,s}(\mathcal{T}, \Theta_0) = 0$, or $f^C_{l,m}(\mathcal{T}, \Theta_0) = 0$. Since $\mathcal{CB}$ is compact and regular (see Propositions 4 and 8 in Chapter 2), $\mathcal{CB}_{\Theta_0}$ is a convex polyhedron equal to the Minkowski difference $\mathcal{B} \ominus \mathcal{A}(0, \Theta_0)$.

The values of $|f^A_{r,j}(\mathcal{T}, \Theta_0)|$, $|f^B_{i,s}(\mathcal{T}, \Theta_0)|$, and $|f^C_{l,m}(\mathcal{T}, \Theta_0)|$ are the Euclidean distances from $\mathcal{T} \in \mathbf{R}^3$ to the planes $f^A_{r,j}(\mathcal{T}, \Theta_0) = 0$, $f^B_{i,s}(\mathcal{T}, \Theta_0) = 0$, and $f^C_{l,m}(\mathcal{T}, \Theta_0) = 0$, respectively. When $(\mathcal{T}, \Theta_0) \notin \mathcal{CB}_{\Theta_0}$, these values can be used to compute the distance from $(\mathcal{T}, \Theta_0)$ to $\mathcal{CB}$ in the subspace $\Theta = \Theta_0$.

2.7 Boundary Representation

Various algorithms have been proposed to compute the explicit representation of the boundary of a C-obstacle $\mathcal{CB}$ when the robot $\mathcal{A}$ can only translate.

If both $\mathcal{A}$ and $\mathcal{B}$ are convex polyhedra, $\mathcal{CB}$ can be expressed as ([Lozano-Pérez, 1983]):

$$\mathcal{CB}_{\Theta_0} = conv(vert(\mathcal{B}) \ominus vert(\mathcal{A}(0, \Theta_0))).$$

There exist algorithms for computing the convex hull of a set of points in three-dimensional space which run in $O(n \log n)$ time, where n is the cardinality of the input set [Preparata and Shamos, 1985]. Here, $n = n_{\mathcal{A}} n_{\mathcal{B}}$.

When $\mathcal{A}$ and $\mathcal{B}$ are convex polyhedra, a more efficient algorithm for constructing $\mathcal{CB}_{\Theta_0}$'s boundary is given in [Guibas and Seidel, 1986]. It runs in $O(n_{\mathcal{A}} + n_{\mathcal{B}} + c)$ time, where c denotes the number of vertices of $\mathcal{CB}_{\Theta_0}$, which is $O(n_{\mathcal{A}} n_{\mathcal{B}})$ in the worst case.

Avnaim and Boissonnat [Avnaim and Boissonnat, 1988] proposed an adaptation of the method described in Subsection 1.7 to compute the boundary of $\mathcal{CB}_{\Theta_0}$ when $\mathcal{A}$ and $\mathcal{B}$ are (possibly non-convex) polyhedra bounded by triangular and/or trapezoidal faces. The adapted method runs in time $O(n_{\mathcal{A}}^3 n_{\mathcal{B}}^3 \log n_{\mathcal{A}} n_{\mathcal{B}})$. The method can be extended with the same asymptotic time complexity to any pair of arbitrary polyhedra by decomposing their faces into triangles and trapezoids. (Each face with n vertices can be decomposed into $O(n)$ triangles and trapezoids in time $O(n \log n)$. Hence, the faces of the two polyhedra can be decomposed into $O(n_{\mathcal{A}})$ and $O(n_{\mathcal{B}})$ components in times $O(n_{\mathcal{A}} \log n_{\mathcal{A}})$ and $O(n_{\mathcal{B}} \log n_{\mathcal{B}})$, respectively.)

3 Semi-Algebraic Workspace

In this section we consider the more general case where both $\mathcal{A}$ and $\mathcal{B}$ are described as arbitrary semi-algebraic subsets of $\mathcal{W} = \mathbf{R}^N$. We show that in this case the C-obstacle $\mathcal{CB}$ can also be represented as a semi-algebraic subset. The polygonal and polyhedral workspaces studied in the previous two sections are particular cases of semi-algebraic workspaces.

3.1 Semi-Algebraic Sets

Semi-algebraic sets are defined by means of first-order logic sentences whose predicates are $=$, $\neq$, $>$, $<$, $\geq$, and $\leq$, whose variables are real numbers and whose terms are multivariate polynomials with rational coefficients. $\mathbf{Q}[X_1; ...; X_n]$ denotes the ring of polynomials in n real variables with rational coefficients.

DEFINITION 10: *An* **atomic polynomial expression** *over* $\mathbf{R}^n$ *is one of the form:*

$$P(x) \bowtie 0$$

where $P \in \mathbf{Q}[X_1; ...; X_n]$, $x \in \mathbf{R}^n$, *and* $\bowtie$ *can be any comparator in* $\{=, \neq, >, <, \geq, \leq\}$. *A* **polynomial expression** *over* $\mathbf{R}^n$ *is any finite boolean combination of atomic polynomial expressions over* $\mathbf{R}^n$. *A subset of* $\mathbf{R}^n$ *whose points satisfy a polynomial expression is called a* **semi-algebraic set.**

DEFINITION 11: *A* **Tarski sentence** *is any polynomial expression prefixed by a finite number (possibly zero) of* $\exists$ *and* $\forall$ *quantifiers applying to some of the variables in the sentence over the set of the reals.*

For example:

$$\forall x,\ \exists y : (x^3 + y^3 > 0) \Rightarrow (x^2 + y^2 + z \leq 0)$$

is a Tarski sentence. A polynomial expression is a quantifier-free Tarski sentence.

The variables in a Tarski sentence which are not quantified (e.g. z in the above example) are called **free** variables. A Tarski sentence without free variables is either `true` or `false`. The **first-order theory of the real numbers**, denoted by Th($\mathbf{R}$), is the set of all the Tarski sentences without free variables which are `true`. For example:

$$\exists x : x^2 - 1 = 0$$

belongs to Th($\mathbf{R}$), but:

$$\exists x : x^2 + 1 \leq 0$$

does not.

Let $\phi(x_1, ..., x_n)$ denote a Tarski sentence in which all occurrences of the variables $x_1, ..., x_n$ are free, each x_i, $i \in [1, n]$, may or may not occur, and there are no free variables other than $x_1, ..., x_n$. $\phi(x_1, ..., x_n)$ is **satisfiable** if and only if the sentence $\exists x_1, ... \exists x_n : \phi(x_1, ..., x_n)$ is in $\mathrm{Th}(\mathbf{R})$. If it is satisfiable, $\phi(x_1, ..., x_n)$ defines a non-empty subset of $\mathbf{R}^n$. The points of this subset are all the lists $(a_1, ..., a_n)$ of constant real numbers such that $\phi(a_1, ..., a_n)$ is `true`.

Two Tarski sentences $\phi_1(x_1, ..., x_n)$ and $\phi_2(x_1, ..., x_n)$ are said to be **equivalent** if for all real numbers $a_1, ..., a_n$, $\phi_1(a_1, ..., a_n)$ is `true` if and only if $\phi_2(a_1, ..., a_n)$ is `true`, i.e. if the sentence:

$$\forall x_1, \ ... \ , \ \forall x_n : \phi_1(x_1, ..., x_n) \Leftrightarrow \phi_2(x_1, ..., x_n)$$

belongs to $\mathrm{Th}(\mathbf{R})$.

Tarski showed that any Tarski sentence is equivalent to a (quantifier-free) polynomial expression [Tarski, 1951]. For instance, if $P(x_1, ..., x_n, x_{n+1})$ is a polynomial expression over $\mathbf{R}^{n+1}$, there exist two polynomial expressions over $\mathbf{R}^n$, $Q(x_1, ..., x_n)$ and $R(x_1, ..., x_n)$, such that $Q(x_1, ..., x_n)$ is equivalent to $\exists x_{m+1} : P(x_1, ..., x_m, x_{m+1})$ and $R(x_1, ..., x_n)$ is equivalent to $\forall x_{m+1} : P(x_1, ..., x_m, x_{m+1})$. Several constructive proofs of this result have been given, including one by Tarski. In particular, Collins proposed a quantifier-elimination algorithm (known as the Collins decomposition; see Section 3 of Chapter 5) that takes double-exponential time in the number of variables [Collins, 1975]. Grigoryev [Grigoryev, 1988] described another method that takes simple-exponential time in the number of variables and double-exponential time only in the number of quantifier alternations. By using these algorithms, one can reduce any Tarski sentence containing no free variable to either $0 = 0$ (if the sentence is `true`) or $1 = 0$ (if the sentence is `false`), and thus compute whether the value of the sentence is `true` or `false` (hence, the first-order theory of the real numbers is decidable). One can also check whether a Tarski sentence containing free variables is satisfiable by existentially quantifying the free variables and eliminating them. In addition, if a Tarski sentence is `true` or satisfiable, the methods of Collins and Grigoryev return a sample list of values for the free and existentially quantified variables which satisfy the sentence (see Chapter 5). Various applications of quantifier elimination are presented in [Arnon, 1988].

Hence, any Tarski sentence $\phi(x_1, ..., x_n)$ containing n free variables defines a semi-algebraic subset X of $\mathbf{R}^n$. $\phi(x_1, ..., x_n)$ is called the **defining formula** of X.

Given two semi-algebraic subsets of $\mathbf{R}^n$, A and B, with defining formulae $\phi_A(x_1, ..., x_n)$ and $\phi_B(x_1, ..., x_n)$, then the sets $A \cup B$ and $A \cap B$ are semi-algebraic sets with defining formulae $\phi_A(x_1, ..., x_n) \vee \phi_B(x_1, ..., x_n)$ and $\phi_A(x_1, ..., x_n) \wedge \phi_B(x_1, ..., x_n)$. The expression $A \subseteq B$ is a logical expression that translates into the Tarski sentence $\forall x_1, ..., \forall x_n$: $\phi_A(x_1, ..., x_n) \Rightarrow \phi_B(x_1, ..., x_n)$. In a similar way, $(x_1, ..., x_n) \in A$ translates to the sentence $\phi_A(x_1, ..., x_n)$. These simple rewriting rules will allow us to construct semi-algebraic sets without explicitly giving all their defining formulae, and conversely, will allow us to write Tarski sentences using semi-algebraic sets.

3.2 C-Obstacles

Let every object in $\mathcal{W} = \mathbf{R}^N$ be represented as a semi-algebraic set. We denote the defining formulae of $\mathcal{A}$ and $\mathcal{B}$ by $\Lambda_{\mathcal{A}}(a)$ and $\Lambda_{\mathcal{B}}(b)$, respectively, with a and $b \in \mathbf{R}^N$. $\Lambda_{\mathcal{A}}(a)$ represents $\mathcal{A}$ at the reference configuration.

The C-obstacle $\mathcal{CB} = \{\mathbf{q} \in \mathcal{C} \ / \ \mathcal{A}(\mathbf{q}) \cap \mathcal{B} \neq \emptyset\}$ can be described as:

$$\mathcal{CB} = \{\mathbf{q} \in \mathcal{C} \ / \ \exists a,\ \exists b : \Lambda_{\mathcal{A}}(a) \wedge \Lambda_{\mathcal{B}}(b) \wedge (a(\mathbf{q}) = b)\}.$$

As shown in Section 2 of Chapter 2, any configuration $\mathbf{q}$ of $\mathcal{A}$ can be represented by a vector $\mathbf{u} = (\mathcal{T}, \Theta)$ of $N(N+1)$ real numbers related by $\frac{1}{2}N(N+1)+1$ polynomial equations with integer coefficients. These equations express the fact that the $N \times N$ matrix Θ representing the orientation of $\mathcal{A}$ has orthonormal columns and determinant $+1$. Let $\Gamma(\mathbf{u})$ denote the conjunction of these equations. Let $\Upsilon(\mathbf{u}, a, b)$, with $a \in \mathbf{R}^N$ and $b \in \mathbf{R}^N$, be the conjunction of the N polynomial equations expressing that the coordinates of a point a in $\mathcal{A}$ equal those of a point b in $\mathcal{B}$ when $\mathcal{A}$ is at a configuration represented by $\mathbf{u} = (\mathcal{T}, \Theta)$. These equations are derived from the relation $b = \Theta\ a\ +\ \mathcal{T}$.

Thus $\mathcal{CB}$ can be represented in $\mathbf{R}^{N(N+1)}$ as:

$$\mathcal{CB} = \{\mathbf{u} \in \mathbf{R}^{N(N+1)} \ / \ \mathsf{CB}(\mathbf{u})\}$$

with:

$$\mathtt{CB}(\mathbf{u}) \equiv \exists a,\ \exists b : \Lambda_{\mathcal{A}}(a) \wedge \Lambda_{\mathcal{B}}(b) \wedge \Upsilon(\mathbf{u}, a, b) \wedge \Gamma(\mathbf{u}).$$

The expression $\exists a,\ \exists b : \Lambda_{\mathcal{A}}(a) \wedge \Lambda_{\mathcal{B}}(b) \wedge \Upsilon(\mathbf{u}, a, b) \wedge \Gamma(\mathbf{u})$ is a Tarski sentence. Hence, $\mathcal{CB}$ is a semi-algebraic subset of $\mathbf{R}^{N(N+1)}$.

PROPOSITION 1: *If the robot and all the obstacles (in finite number) are represented as semi-algebraic subsets of* $\mathbf{R}^N$, *then every C-obstacle is a semi-algebraic subset of* $\mathbf{R}^{N(N+1)}$. $\mathcal{C}_{free}$, $\mathcal{C}_{contact}$, *and* $\mathcal{C}_{valid}$ *are also semi-algebraic subsets of* $\mathbf{R}^{N(N+1)}$.

The same result can be obtained in $\mathbf{R}^m$, with $m = \frac{1}{2}N(N+1)$, rather than in $\mathbf{R}^{N(N+1)}$, by using an appropriate non-redundant parameterization (chart) of $\mathcal{C}$, e.g. the stereographic projection in the two-dimensional case and unit quaternions in the three-dimensional case (see Section 4 of Chapter 2). Let $\mathbf{x}$ denote the representation of $\mathbf{q}$ in such a chart. The relation $a(\mathbf{q}) = b$ entails a polynomial expression $\Upsilon'(\mathbf{x}, a, b)$. We have:

$$\mathcal{CB} = \{\mathbf{x} \in \mathbf{R}^m \ / \ \mathtt{CB}(\mathbf{x})\}$$

with:

$$\mathtt{CB}(\mathbf{x}) \equiv \exists a,\ \exists b : \Lambda_{\mathcal{A}}(a) \wedge \Lambda_{\mathcal{B}}(b) \wedge \Upsilon'(\mathbf{x}, a, b).$$

Such a parameterization significantly reduces the number of variables in the semi-algebraic representation of $\mathcal{CB}$, and hence the cost of applying quantifier-elimination procedures. However, some extra-work may be needed in order to switch from one chart to another.

Exercises

1: $\mathcal{A}$ and $\mathcal{B}$ are two convex polygons. $\mathcal{A}$ can translate and rotate freely in the plane. Give a computable expression of the predicate $\partial\mathtt{CB}$ representing the boundary of $\mathcal{CB}$, i.e. :

$$\mathbf{q} \in \partial\mathcal{CB} \Leftrightarrow \partial\mathtt{CB}(\mathbf{q}).$$

Extend this result to the case where $\mathcal{A}$ and $\mathcal{B}$ are non-convex polygons.

2: Describe a sweep-line algorithm that partitions a non-convex polygon into triangles and trapezoids. [Hint: See Section 1 of Chapter 5.]

3: Using the representation of $\mathcal{CB}$ given in Theorem 1, propose a method for computing the intersection of a pure rotational path with the boundary of a C-obstacle $\mathcal{CB}$ [Donald, 1984]. (Assume that $\mathcal{A}$ and $\mathcal{B}$ are convex polygons.)

4: Let $\mathcal{A}$ and $\mathcal{B}$ be convex polygons. A configuration of $\mathcal{A}$ is parameterized by (x, y, θ), with x and y being the coordinates of $O_{\mathcal{A}}$ in $\mathcal{F_W}$ and θ being the angle between the x-axes of $\mathcal{F_W}$ and $\mathcal{F_A}$. The C-obstacle $\mathcal{CB}$ is a three-dimensional volume in $\mathbf{R}^2 \times [0, 2\pi)$ bounded by patches of C-surfaces of type A and type B (faces). Draw the projection of a face of type A (resp. type B) on the xy-plane. Specify the parameters of the figure (lengths of straight segments, radii of circular arcs). [Hint: See Subsection 3.2 of Chapter 6.]

5: Let $\mathcal{A}$ and $\mathcal{B}$ be two line segments of bounded lengths. Describe the C-obstacle $\mathcal{CB}$ in $\mathbf{R}^2 \times [0, 2\pi)$. How many faces does it have?

6: Let $\mathcal{A}$ and $\mathcal{B}$ be two convex polygons. $\mathcal{A}$ translates at a fixed orientation θ_0. Show that:

$$\mathcal{CB}_{\theta_0} = conv(vert(\mathcal{B}) \ominus vert(\mathcal{A}(0, 0, \theta_0)))$$

where $vert(\mathcal{P})$ denotes the set of vertices of a polygon$\mathcal{P}$ and $conv(S)$ denotes the convex hull of a planar set of points S. Show that the result still holds in the case where $\mathcal{A}$ and $\mathcal{B}$ are convex polyhedra, with $conv(S)$ denoting the convex hull of a set of points in $\mathbf{R}^3$ [Lozano-Pérez, 1983].

7: Let $\mathcal{A}$ and $\mathcal{B}$ be two convex polygons.

a. Let us first assume that $\mathcal{A}$ can only translate at a fixed orientation θ_0. Write a program that computes the boundary of the C-obstacle $\mathcal{CB}_{\theta_0}$ using the method described in Subsection 1.6.

b. Let us now consider the case where $\mathcal{A}$ can translate and rotate freely. Extend the previous program in order to compute the boundary of $\mathcal{CB}$ using the method sketched at the beginning of Subsection 1.7.

8: A *generalized polygon* is a region of $\mathbf{R}^2$ homeomorphic to the closed unit disc, whose boundary (a Jordan curve) is a closed-loop sequence of straight segments and circular arcs. Let $\mathcal{A}$ and $\mathcal{B}$ be two convex generalized polygons. Show that the C-obstacle $\mathcal{CB}$ is also a convex generalized polygon.

9: Explain how Figure 10 illustrates the fact that the complexity of $\mathcal{CB}_{\theta_0}$ can be proportional to $n_{\mathcal{A}}^2 n_{\mathcal{B}}^2$.

10: Show that the computation of the intersection of two curves $\xi = \aleph$ and $\xi = cs(\theta)$, where $cs(\theta)$ is an expression of the form $\aleph \cos\theta + \aleph \sin\theta + \aleph$ and $\aleph$ is a constant (each occurrence of $\aleph$ may be a different constant and each occurrence of $cs(\theta)$ may be a different expression), can be reduced to the computation of the intersection of a line and a circle. Show that the computation of the intersection of two curves of the form $\xi = cs(\theta)$ and/or $\xi = cs(\theta)/cs(\theta)$ can be reduced to the computation of a circle and a general conic (see Subsection 1.7).

11: Let $\mathcal{A}$ be a line segment of bounded length. $\mathcal{B}$ is a regular (possibly non-bounded) polygonal region whose edges are all segments of bounded lengths. A contact configuration between $\mathcal{A}$ and $\mathcal{B}$ is a configuration $\mathbf{q}$ for which there exists a pair (X, E) such that X is a vertex of $\mathcal{B}$ (resp. an endpoint of $\mathcal{A}(\mathbf{q})$), E is $\mathcal{A}(\mathbf{q})$ (resp. an edge of $\mathcal{B}$), and X is contained in E. A triple-contact configuration between $\mathcal{A}$ and $\mathcal{B}$ is a configuration $\mathbf{q}$ such that there exist three distinct such pairs (X_i, E_i), $i = 1, 2, 3$. A (triple-)contact configuration $\mathbf{q}$ is semi-free if and only if $\mathcal{A}(\mathbf{q})$ does not intersect the interior of $\mathcal{B}$.

Let $n_{\mathcal{B}}$ be the number of vertices and edges of $\mathcal{B}$. Assume that no three vertices of $\mathcal{B}$ are aligned. Analyze the various types of triple-contact configurations. Show that there are $O(n_{\mathcal{B}} \log n_{\mathcal{B}})$ semi-free triple-contact configurations of $\mathcal{A}$. Derive from these results the fact that the method of Avnaim and Boissonnat (Subsection 1.7) can compute the boundary of $\mathcal{CB}$ in $O(n_{\mathcal{B}}^2 \log n_{\mathcal{B}})$ time and that the size of the boundary is $O(n_{\mathcal{B}}^2)$ [Avnaim and Boissonnat, 1988].

12: Let f, e and v be the number of faces, edges and vertices in a polyhedron. Show that f, e and v are pairwise proportional.

13: Prove Theorem 2.

14: Let $\mathcal{A}$ and $\mathcal{B}$ be two convex polyhedra. We call the *wedge* of an edge E (of $\mathcal{A}$ or $\mathcal{B}$) the intersection of the two (closed) inner half-spaces of the two faces intersecting at E. Show that $\mathcal{A}$ and $\mathcal{B}$ have a null intersection if and only if all edges of $\mathcal{A}$ (or $\mathcal{B}$) are outside some wedge of an edge of $\mathcal{B}$ (or $\mathcal{A}$). From this result, derive an expression of the predicate CB defined by $\mathbf{q} \in$ CB $\Leftrightarrow$ CB($\mathbf{q}$) [Canny, 1988].

15: Describe how the method of Subsection 1.7 can be adapted to compute the boundary of a C-obstacle when $\mathcal{A}$ is a polyhedron that can only translate and $\mathcal{B}$ is a polyhedron. $\mathcal{A}$ and $\mathcal{B}$ may not be convex. What is the complexity of the adapted method [Avnaim and Boissonnat, 1988]?

16: Let A be a semi-algebraic subset of $\mathbf{R}^n$ with $\phi_A(x_1, ..., x_n)$ as its defining formula. Show that the interior, the boundary and the closure of A are all semi-algebraic subsets of $\mathbf{R}^n$.

Chapter 4

Roadmap Methods

In this chapter we describe a first approach to robot path planning which we name the *roadmap approach.* This approach is based on the following general idea: capture the connectivity of the robot's free space $\mathcal{C}_{free}$ in the form of a network of one-dimensional curves — the *roadmap* — lying in $\mathcal{C}_{free}$ or its closure $cl(\mathcal{C}_{free})$. Once constructed, a roadmap $\mathcal{R}$ is used as a set of standardized paths. Path planning is reduced to connecting the initial and goal configurations to $\mathcal{R}$, and searching $\mathcal{R}$ for a path.

The key issue in this approach is obviously the construction of the roadmap. Methods based on different principles have been proposed, producing various sorts of roadmaps, called "visibility graphs", "Voronoi diagrams", "freeway nets", "silhouettes". In this chapter we present several of these methods.

In Section 1 we describe the *visibility graph* method. This simple method essentially applies to two-dimensional configuration spaces with polygonal C-obstacles. The roadmap is a graph called a "visibility graph" which is constructed by connecting every pair of vertices in $\mathcal{C}_{free}$'s boundary by a straight segment, if this segment does not traverse the interior of a C-obstacle. The method is one of the earliest path planning methods

[Nilsson, 1969] and it has been widely used to implement path planners for mobile robots. It addition, it has been at the origin of a stream of research on the "shortest path problem" in Computational Geometry (e.g. see [Mitchell, 1986] [Akman, 1987]).

In Section 2 we present the *retraction* approach [Ó'Dúnlaing, Sharir and Yap, 1983]. It consists of constructing a roadmap $\mathcal{R}$ by defining a continuous mapping (called a retraction) of $\mathcal{C}_{free}$ onto $\mathcal{R}$. When the configuration space is two-dimensional with polygonal C-obstacles, a simple instantiation of this idea is to retract $\mathcal{C}_{free}$ on its Voronoi diagram. (This diagram is defined as the one-dimensional subset of $\mathcal{C}_{free}$ that maximizes the clearance between the robot and the obstacles.) In principle, the retraction approach is general, but existing results only apply to low-dimensional configuration spaces.

In Section 3 we describe the *freeway* method [Brooks, 1983a], which is loosely related to retraction. It applies to a polygonal robot translating and rotating among polygonal obstacles. The reference point of the robot is constrained to stay on a network of straight line segments which is reminiscent of a Voronoi diagram. The method embeds several empirical assumptions and is not complete, i.e. it may fail to generate a path even if one exists. In practice, however, it often runs efficiently.

Finally, in Section 4 we sketch the principle of a general roadmap method developed by Canny [Canny, 1988]. This method, called the *silhouette* method[1], solves the basic motion planning problem in time exponential in the dimension of the configuration space. Roughly, it consists of constructing the "silhouette" of the robot's free space when it is viewed from a point at infinity, and adding some curve segments linking "critical points" of the silhouette to other curve segments of the silhouette. The silhouette and the linking curves form the roadmap that is subsequently searched for a path.

With the exception of the silhouette method, the methods presented in this chapter are limited to planning problems in configuration spaces of small dimension (2 or 3). However, when it comes to building an actual planner, they are relatively easy to implement and they can be

[1] Canny called the method the "roadmap algorithm". Since we already used the term "roadmap" to name the motion planning approach presented in this chapter, we re-name Canny's method the "silhouette method".

quite efficient. On the other hand, the silhouette method is general, but its details (not presented below) are mathematically involved and its implementation is certainly not an easy task.

1 Visibility Graph Method

1.1 Principle

Let us consider a polygonal object $\mathcal{A}$ that translates at fixed orientation among polygonal obstacles in a two-dimensional workspace $\mathcal{W} = \mathbf{R}^2$. Therefore, $\mathcal{A}$'s configuration space is $\mathcal{C} = \mathbf{R}^2$ and the C-obstacle region $\mathcal{CB}$ (the union of all the C-obstacles) is a polygonal region of $\mathbf{R}^2$. The free space $\mathcal{C}_{free}$ is equal to $\mathcal{C} \setminus \mathcal{CB}$. Techniques described in Subsection 1.6 of Chapter 3 compute the boundary of $\mathcal{CB}$ (hence, of $\mathcal{C}_{free}$) from the description of $\mathcal{A}$ and the obstacles.

The principle of the visibility graph method is to construct a semi-free path as a simple polygonal line connecting the initial configuration $\mathbf{q}_{init}$ to the goal configuration $\mathbf{q}_{goal}$ through vertices of $\mathcal{CB}$. This principle rests on the following proposition:

PROPOSITION 1: *Let the C-obstacle region $\mathcal{CB}$ be a polygonal region of $\mathcal{C} = \mathbf{R}^2$. There exists a semi-free path between any two given configurations $\mathbf{q}_{init}$ and $\mathbf{q}_{goal}$ if and only if there exists a simple polygonal line τ lying in $cl(\mathcal{C}_{free})$ whose endpoints are $\mathbf{q}_{init}$ and $\mathbf{q}_{goal}$, and such that τ's vertices are vertices of $\mathcal{CB}$.*

Proof: If there exists a semi-free path between two configurations of the closed subset $cl(\mathcal{C}_{free}) \subset \mathbf{R}^2$, then there exists a semi-free path τ of minimal Euclidean length between these two configurations. In order to be a shortest path, τ must also locally be a shortest path. This means that any subpath τ' of τ must be the shortest path in $cl(\mathcal{C}_{free})$ between the two configurations that it connects. Hence, at any configuration $\mathbf{q}$ traversed by τ, if $\mathbf{q}$ is not a vertex of $\mathcal{CB}$, τ must have zero curvature. This entails that the vertices of τ be vertices of $\mathcal{CB}$. ∎

This proposition implies that in order to find a semi-free path between any two configurations, it is sufficient for a planner to only consider the polygonal lines in $cl(\mathcal{C}_{free})$ running via vertices of $\mathcal{CB}$. In addition, if a

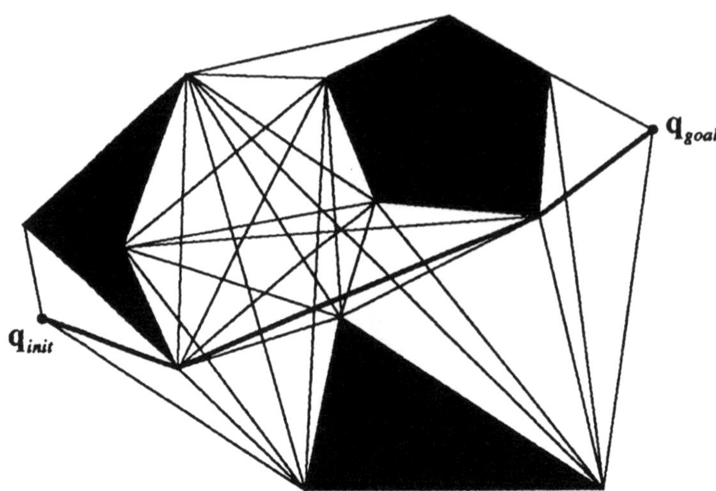

Figure 1. This figure shows the visibility graph G of a simple configuration space in which $\mathcal{CB}$ consists of three disconnected regions. The links of G also include the edges of $\mathcal{CB}$. The bold line is the shortest path in $cl(\mathcal{C}_{free})$ between $\mathbf{q}_{init}$ and $\mathbf{q}_{goal}$.

semi-free path exists between the two configurations, this set of polygonal lines contains the shortest one (for the Euclidean metric in $\mathbf{R}^2$). This set of polygonal lines is called the visibility graph.

DEFINITION 1: *The* **visibility graph** *is the non-directed graph G specified as follows:*

- *G's nodes are $\mathbf{q}_{init}$, $\mathbf{q}_{goal}$, and the vertices of $\mathcal{CB}$ (at finite distance).*
- *Two nodes of G are connected by a link if and only if either the line segment joining them is an edge of $\mathcal{CB}$, or if it lies entirely in $\mathcal{C}_{free}$, except possibly at its two endpoints.*

Figure 1 shows the visibility graph of a simple configuration space. All the edges of $\mathcal{CB}$ are links of G. The subset of G that is restricted to the vertices of $\mathcal{CB}$ can be regarded as the roadmap $\mathcal{R}$ introduced at the beginning of this chapter. $\mathcal{R}$ is independent of the initial and goal configurations. The rest of G contains the links connecting $\mathbf{q}_{init}$ and $\mathbf{q}_{goal}$ to this roadmap.

The algorithm of the visibility graph method is the following:

1. Construct the visibility graph G.

2. Search G for a path from $\mathbf{q}_{init}$ to $\mathbf{q}_{goal}$.

3. If a path is found, return it; otherwise, indicate failure.

A naive algorithm for constructing G consists of considering all pairs of points (X, X'), where X and X' are either $\mathbf{q}_{init}$, $\mathbf{q}_{goal}$, or vertices of $\mathcal{CB}$. If X and X' are the endpoints of the same edge of $\mathcal{CB}$, then the corresponding nodes are connected by a link in G. Otherwise, the intersections with $\mathcal{CB}$ of the line passing through X and X' are computed; the nodes corresponding to X and X' are connected by a link in G if and only if no intersection lies in the open segment joining the two points [Lozano-Pérez and Wesley, 1979]. This algorithm requires time $O(n^3)$, where n is the total number of vertices of $\mathcal{CB}$.

One can improve this algorithm by using a variant of the sweep-line algorithm given in Appendix D:

- For each point X in G, compute the orientation $\alpha_i \in [0, 2\pi)$ of every half-line emanating from X and passing through another point X_i of G. Sort the orientations α_i.

- Rotate a half-line emanating from X, from the orientation 0 to the orientation 2π. During the rotation, stop at every orientation α_i. At each stop, update the intersection of the half-line with $\mathcal{CB}$, keep track of the point in the intersection that is the closest one to X, and check whether the segment connecting X to X_i intersects $\mathcal{CB}$.

This algorithm allows us to construct G in time $O(n^2 \log n)$ [Lee, 1978]. More efficient algorithms performing in time $O(n^2)$ have been proposed (e.g. see [Welzl, 1985] [Asano et al., 1986] [Edelsbrunner, 1987]). An output-sensitive algorithm that takes time $O(k + n \log n)$, where k is the number of edges in G, is given in [Ghosh and Mount, 1987]. In the worst case, k is in $O(n^2)$; in better cases, it may be as small as $O(n)$.

Searching G can be done using various graph searching techniques. When the A^* algorithm is used (Appendix C), with the Euclidean distance to the goal configuration as the heuristic function, the search generates the

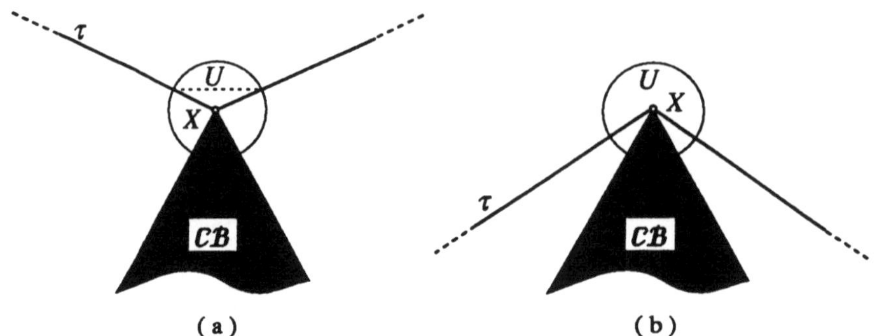

Figure 2. In order to be a shortest path, a polygonal line τ must be such that every vertex X of τ has a neigborhood U such that $\mathcal{CB} \cap U$ is contained in the convex sector of τ at X (Proposition 2). The portion of path shown in Figure a does not satisfy this proposition and can be made shorter (dotted line). The portion of path shown in Figure b verifies this proposition.

shortest path in $cl(\mathcal{C}_{free})$ whenever a path exists and returns failure otherwise. Since the heuristic function is "locally consistent", the search is done in time $O(n^2)$ if the list OPEN is not sorted, and in time $O(k \log n)$ if it is sorted (k being the number of edges of G). Therefore, assuming that the boundary of $\mathcal{C}_{free}$ is given, the visibility graph method can generate a semi-free path in total time $O(n^2)$ in the worst-case where the visibility graph contains $O(n^2)$ nodes. If $cl(\mathcal{C}_{free})$ is a manifold with boundary, the generated semi-free path τ can be transformed into a polygonal free path that can be arbitrarily close to τ.

In the rest of this subsection we assume that $cl(\mathcal{C}_{free})$ is a manifold with boundary. The efficiency of the visibility graph method can be improved by noticing that some links in G are not needed. Let the **convex sector** of a simple polygonal path τ at a vertex X be the closed convex region comprised between the two half-lines drawn from X and containing the straight segments of τ adjacent at X.

PROPOSITION 2: *In order for a polygonal path τ to be a shortest one, each vertex X of τ must have a neighborhood U such that $\mathcal{CB} \cap U$ is completely contained in the convex sector of τ at X.*

Proof: If $\mathcal{CB} \cap U$ is not contained in the convex sector of τ at X, then

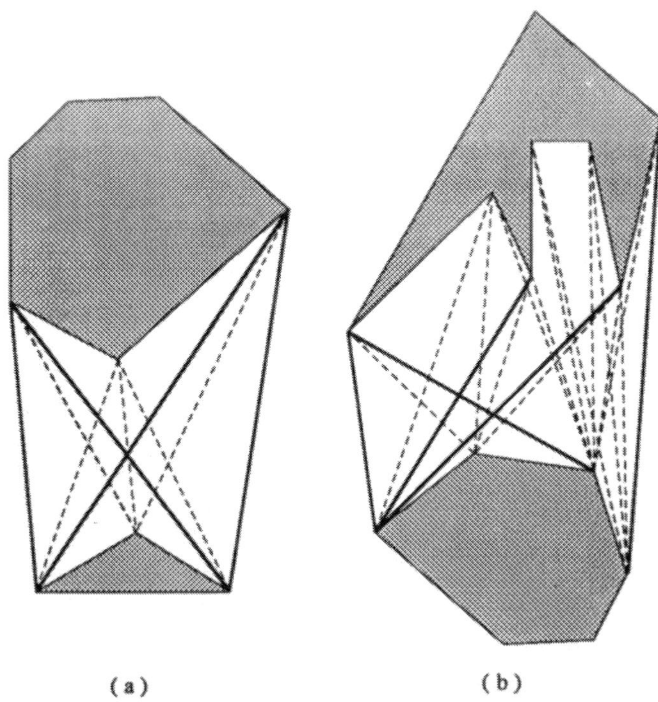

Figure 3. This figure illustrates the concept of a tangent segment between two polygons. The tangent segments are shown in plain lines and the non-tangent ones in dashed lines. There are four and only four tangent segments between any two disjoint convex polygons (Figure a). No tangent segment emanates from a concave vertex (Figure b).

---

one can slightly modify τ into a shorter semi-free path (Figure 2). ∎

Let X be a vertex of $\mathcal{CB}$ and L be a straight line passing through X. L is **tangent** to $\mathcal{CB}$ at X if and only if in a neighborhood U of X the interior of $\mathcal{CB}$ lies entirely on a single side of L.

Let us consider two nodes X and X' of G. Let L be the infinite line passing through X and X'. The segment joining X to X' is said to be a **tangent segment** if and only if:

- if X is a vertex of $\mathcal{CB}$, then L is tangent to $\mathcal{CB}$ at X,

- if X' is a vertex of $\mathcal{CB}$, then L is tangent to $\mathcal{CB}$ at X'.

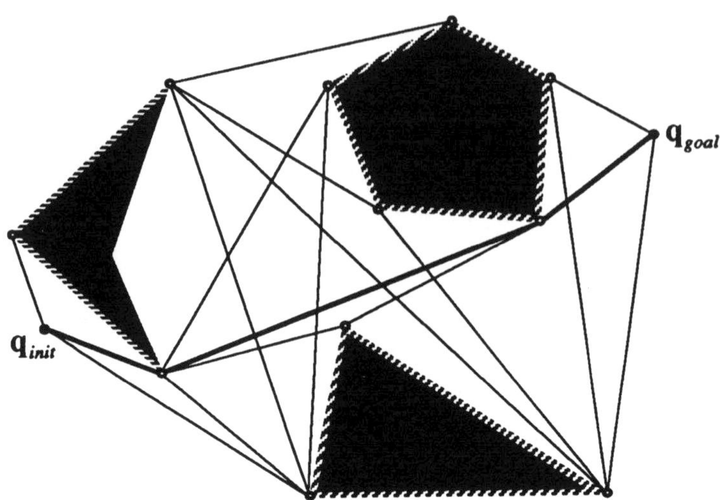

Figure 4. This figure shows the reduced visibility graph for the same configuration space as in Figure 1. The nodes are q_{init}, q_{goal}, and all the convex vertices of $\mathcal{CB}$. The links are the edges of $\mathcal{CB}$ (shown as striped lines) that connect convex vertices and the tangent segments between q_{init}, q_{goal}, and the vertices of $\mathcal{CB}$ that lie in free space (except possibly at their endpoints).

Figure 3 shows the tangent (plain lines) and non-tangent (dashed lines) segments between the vertices of two polygons. The figure also illustrates two simple properties of tangent segments:

- Between two disjoint convex polygons there are exactly four tangent segments. Two are "supporting" tangent segments (the two polygons lie on the same side of the line containing the segment) and two are "separating" tangent segments (the two polygons lie on different sides of the line containing the segment).

- If a vertex X is concave (i.e. the inner angle at X is greater than π), then no tangent segment ends at X.

Using Proposition 2, we can simplify G as follows: If a segment between X and X' is not a tangent segment, it need not be included in the search graph[2] [Rohnert, 1986]. Indeed, any path τ between q_{init} and q_{goal} that contains this segment is such that $\mathcal{CB}$ does not locally lie in the convex

[2] Hence, the concave vertices of $\mathcal{CB}$ need not be included in the graph either.

sector τ of either X or X' (or both). The subgraph G' of G obtained by removing all the non-tangent segments and all the concave vertices is called the **reduced visibility graph.** Whenever there exists a semi-free path between $\mathbf{q}_{init}$ and $\mathbf{q}_{goal}$, G' contains the shortest path in $cl(\mathcal{C}_{free})$ between these two configurations. Figure 4 shows the reduced visibility graph for the same configuration space as in Figure 1.

Rohnert [Rohnert, 1986] presented an algorithm for computing the tangent segments between two convex polygons in $O(\log n_1 + \log n_2)$ time, where n_1 and n_2 are the respective numbers of vertices of the two polygons. Based on this algorithm, he proposed a method that computes the shortest path between two input configurations when $\mathcal{CB}$ consists of c disjoint convex polygons. This method runs in $O(c^2 + n \log n)$ time after $O(n + c^2 \log n)$ preprocessing time.

Reif and Storer proposed an algorithm that computes the shortest path between two input configurations in $O(nk + n \log n)$ time, where k is the number of convex parts of $\mathcal{CB}$ [Reif and Storer, 1985].

1.2 Extension to Generalized Polygons

The visibility graph method can be extended to the case where the C-obstacles are "generalized polygons", i.e. regions bounded by straight segments and/or circular arcs [Laumond, 1987a]. Such C-obstacles occur when $\mathcal{A}$ is a generalized polygon translating at fixed orientation among obstacles which are also modeled as generalized polygons. This extension is particularly useful when the objects in the workspace are best modeled as generalized polygons, which is often the case with mobile robot problems. It is also useful when the polygonal C-obstacles are isotropically grown into generalized polygons by a disc of radius ε in a preprocessing phase, in order to guarantee a minimal clearance of ε between the generated path and the C-obstacles.

We first present a technique to compute generalized polygonal C-obstacles. Next, we describe the visibility graph method with such C-obstacles.

1.2.1 Generalized Polygonal Configuration Space

DEFINITION 2: *A* **generalized polygon** *is a subset of* $\mathbf{R}^2$ *that is*

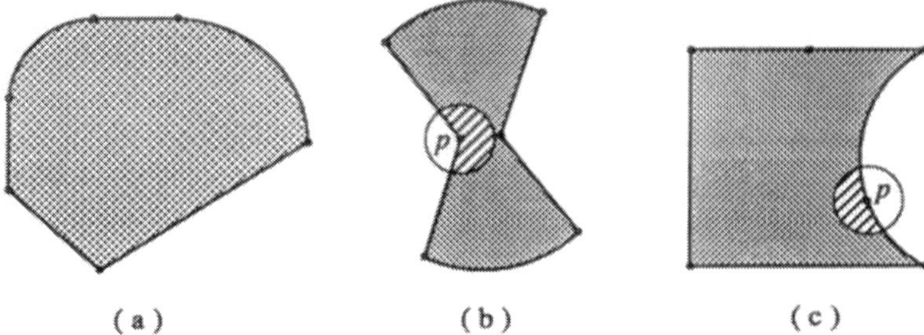

Figure 5. This figure shows three generalized polygons. The generalized polygon shown in Figure a is convex, while those shown in Figures b and c are non-convex. The generalized polygon in Figure b is locally non-convex, because there only exists a finite number of points p (two in the example) such that the intersection of the generalized polygon with a convex neighborhood of p (e.g. a disc) is non-convex. The generalized polygon in Figure c is not locally non-convex.

homeomorphic to the closed unit disc and whose boundary is a closed-loop sequence of straight line segments and circular arcs.

Hence, a generalized polygon is compact, regular, and simply-connected. Each maximal line segment and circular arc in its boundary is called an **edge**. Every endpoint of an edge is called a **vertex**.

Figure 5 shows three generalized polygons, a convex one (Figure 5.a) and two non-convex ones (Figures 5.b and 5.c). The generalized polygon in 5.b is said to be **locally non-convex**, because there only exists a finite number of points p (all of them being vertices) such that the intersection of the generalized polygon with a sufficiently small convex neighborhood of p is non-convex. Any locally non-convex generalized polygon can be decomposed into a finite number of convex generalized polygons. The non-convex generalized polygon in Figure 5.c is not locally non-convex. Such a generalized polygon cannot be decomposed into a finite number of convex parts.

In the following, we assume that a generalized polygon is described by the sequence of its vertices and edges in counterclockwise order. The description of every circular edge specifies the position of the circle center and the position of the edge relative to the straight line passing through the two endpoints.

Let $\mathcal{A}$ and $\mathcal{B}$ be two convex generalized polygons. $\mathcal{A}$ can only translate at fixed orientation[3]. We denote the vertices and the edges of $\mathcal{A}$ (listed in counterclockwise order) by a_i and $E_i^{\mathcal{A}} = a_i a_{i+1}$, respectively, with $i = 1, ..., n_{\mathcal{A}}$, mod $n_{\mathcal{A}}$. We denote the vertices and the edges of $\mathcal{B}$ (in counterclockwise order) by b_j and $E_j^{\mathcal{B}} = b_j b_{j+1}$, respectively, with $j = 1, ..., n_{\mathcal{B}}$, mod $n_{\mathcal{B}}$. We denote the outgoing normal vectors of $E_i^{\mathcal{A}}$ by $\vec{\nu}_i^{\mathcal{A}}$. If $E_i^{\mathcal{A}}$ is a circular edge, these vectors span an angular sector comprised between the vectors $\vec{\nu}_i^{\mathcal{A}}$ at a_i and a_{i+1}. We denote these two extreme vectors by $\vec{\nu}_i^{\mathcal{A}}(-)$ and $\vec{\nu}_i^{\mathcal{A}}(+)$, respectively. If $E_i^{\mathcal{A}}$ is a straight edge, the vector $\vec{\nu}_i^{\mathcal{A}}$ is unique. In the same way, we denote the outgoing normal vectors of $E_j^{\mathcal{B}}$ by $\vec{\nu}_j^{\mathcal{B}}$. If $E_j^{\mathcal{B}}$ is a circular edge, we denote the two extreme vectors at b_j and b_{j+1} by $\vec{\nu}_j^{\mathcal{B}}(-)$ and $\vec{\nu}_j^{\mathcal{B}}(+)$, respectively.

One may verify that the C-obstacle $\mathcal{CB} = \mathcal{B} \ominus \mathcal{A}(0)$ is another generalized convex polygon [Laumond, 1987a]. It can be computed by a simple adaptation of the linear-time algorithm used in the case where both $\mathcal{A}$ and $\mathcal{B}$ are convex polygons (see Subsection 1.6 of Chapter 3).

We illustrate the computation of $\mathcal{CB}$ with the example shown in Figure 6. The two objects $\mathcal{A}$ and $\mathcal{B}$ are depicted in Figures 6.a and 6.b, respectively. $\mathcal{A}$'s boundary consists of three edges, including a circular one ($E_3^{\mathcal{A}}$). $\mathcal{B}$'s boundary consists of four edges, including a circular one ($E_3^{\mathcal{B}}$). The various normals $-\vec{\nu}_i^{\mathcal{A}}$ and $\vec{\nu}_j^{\mathcal{B}}$ (the extreme vectors in the case of the circular edges) are affixed at the center of a unit circle S^1 shown in Figure 6.c. (Recall from Subsection 1.6 of Chapter 3 that the algorithm uses the ingoing normal vectors $-\vec{\nu}_i^{\mathcal{A}}$. This results from the fact that $\mathcal{CB}$ is the Minkowski *difference* $\mathcal{B} \ominus \mathcal{A}(0)$.) The bold arcs in the graphic representation of S^1 represent the angular sectors corresponding to the circular edges.

S^1 is scanned in counterclockwise order and the edges of $\mathcal{CB}$ are generated in the same order (see Figure 6.d). Fictitious vertices are introduced during the construction of $\mathcal{CB}$ when a qualitative change occurs in the contact between $\mathcal{A}$ and $\mathcal{B}$. One fictitious vertex, $a_{3'}$, is introduced in $E_3^{\mathcal{A}}$. It corresponds to the transition from the contact between $E_3^{\mathcal{A}}$ and

[3]For simplification, we denote the robot at the fixed orientation θ_0 by $\mathcal{A}$, rather than by $\mathcal{A}_{\theta_0}$. $\mathcal{A}(0)$ denotes the robot in its fixed orientation at the reference position, i.e. where the origins of the coordinate frames $\mathcal{F}_{\mathcal{A}}$ and $\mathcal{F}_{\mathcal{W}}$ coincide.

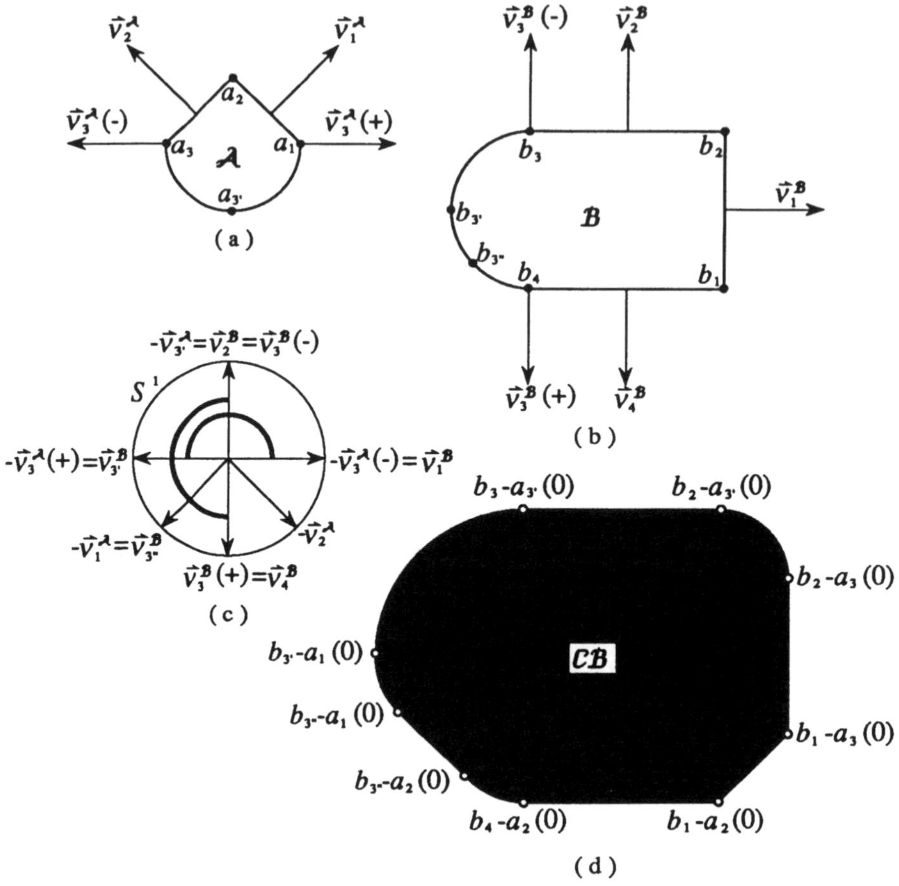

Figure 6. $\mathcal{A}$ and $\mathcal{B}$ are convex generalized polygons. The C-obstacle $\mathcal{CB}$ is also a convex generalized polygon. The figure illustrates the construction of $\mathcal{CB}$.

b_2 to the contact between $E_3^{\mathcal{A}}$ and $E_2^{\mathcal{B}}$. In the same fashion, two fictitious vertices, $b_{3'}$ and $b_{3''}$, are introduced in $E_3^{\mathcal{B}}$. They correspond to the transition from the contact between $E_3^{\mathcal{A}}$ and $E_3^{\mathcal{B}}$ to the contact between a_1 and $E_3^{\mathcal{B}}$, and to the transition from the contact between a_1 and $E_3^{\mathcal{B}}$ to the contact between $E_1^{\mathcal{A}}$ and $E_3^{\mathcal{B}}$. When these fictitious vertices are introduced, the corresponding normal vectors $-\vec{\nu}_{3'}^{\mathcal{A}}$, $\vec{\nu}_{3'}^{\mathcal{B}}$ and $\vec{\nu}_{3''}^{\mathcal{B}}$ are installed in S^1 (see Figure 6.c).

Suppose that we start the scanning process of S^1 with $-\vec{\nu}_2^{\mathcal{A}}$:

. $-\vec{\nu}_2^{\mathcal{A}}$ points between $\vec{\nu}_4^{\mathcal{B}}$ and $\vec{\nu}_1^{\mathcal{B}}$. The straight edge of $\mathcal{CB}$ joining $b_1 - a_2(0)$ to $b_1 - a_3(0)$ is generated.

. Next in S^1, the two vectors $-\vec{\nu}_3^{\mathcal{A}}(-)$ and $\vec{\nu}_1^{\mathcal{B}}$ coincide. This means that a_3 can slide in contact with $E_1^{\mathcal{B}}$. The straight edge of $\mathcal{CB}$ joining $b_1 - a_3(0)$ to $b_2 - a_3(0)$ is generated.

. We now encounter the beginning of the angular sector spanned by the vectors $-\vec{\nu}_3^{\mathcal{A}}$. A subset of this sector points between $\vec{\nu}_1^{\mathcal{B}}$ and $\vec{\nu}_2^{\mathcal{B}}$. The fictitious vertex $a_{3'}$ is introduced here and we install the vector $\vec{\nu}_{3'}^{\mathcal{A}}$ in S^1. The circular edge of $\mathcal{CB}$ joining $b_2 - a_3(0)$ to $b_2 - a_{3'}(0)$ is generated.

. Next in S^1, the three vectors $\vec{\nu}_{3'}^{\mathcal{A}}$, $\vec{\nu}_2^{\mathcal{B}}$, and $\vec{\nu}_3^{\mathcal{B}}(-)$ coincide. This means that $a_{3'}$ can slide in contact with $E_2^{\mathcal{B}}$. The straight edge of $\mathcal{CB}$ joining $b_2 - a_{3'}(0)$ to $b_3 - a_{3'}(0)$ is generated.

. We now encounter the second part of the angular sector spanned by the vectors $-\vec{\nu}_3^{\mathcal{A}}$ and the beginning of the angular sector spanned by the vectors $\vec{\nu}_3^{\mathcal{B}}$. The intersection of these two sectors yields the fictitious vertex $b_{3'}$ and the corresponding vector $\vec{\nu}_{3'}^{\mathcal{B}}$ in S^1. The circular edge of $\mathcal{CB}$ joining $b_3 - a_{3'}(0)$ to $b_{3'} - a_1(0)$ is generated.

. Etc ...

The algorithm illustrated above computes $\mathcal{CB}$'s boundary in $O(n_{\mathcal{A}} + n_{\mathcal{B}})$ time. It can be adapted to handle the case where $\mathcal{B}$ is a convex, non-bounded generalized polygonal region of $\mathcal{W}$.

If $\mathcal{A}$ and $\mathcal{B}$ are locally non-convex generalized polygons, one can apply the same algorithm after having decomposed the non-convex objects into a finite collection of convex generalized polygons. For example, a sweep-line algorithm decomposes a locally non-convex generalized polygon having n vertices into $O(n)$ convex generalized polygons in $O(n \log n)$ time (see Section 1 of Chapter 5). The above algorithm applies to each pair of convex generalized polygons resulting from the decomposition of $\mathcal{A}$ and $\mathcal{B}$.

If there are several obstacles $\mathcal{B}_i$, $i = 1, ..., q$, and if both $\mathcal{A}$ and the obstacles are modeled as convex and/or locally non-convex generalized polygons, the technique described at the end of Subsection 1.6 of Chapter 3 allows us to compute the boundary of $\mathcal{C}_{free}$ in $O(n_{\mathcal{A}}^2 n_{\mathcal{B}}^2 \log n_{\mathcal{A}} n_{\mathcal{B}})$ time, where $n_{\mathcal{B}}$ is the total number of vertices of the obstacles. The total number of vertices (resp. edges) of this boundary is $O(n_{\mathcal{A}}^2 n_{\mathcal{B}}^2)$.

If $\mathcal{A}$ and/or $\mathcal{B}$ are non-convex and not locally non-convex, the above algorithm does not apply. However, in this case, $\mathcal{CB}$'s boundary still consists of straight and circular edges. It can be computed as follows:

- For every $i \in [1, n_{\mathcal{A}}]$ and $j \in [1, n_{\mathcal{B}}]$, the region $E_j^{\mathcal{B}} \ominus E_i^{\mathcal{A}}(0)$ is computed. Its boundary consists of at most four edges.

- $\mathcal{CB}$'s boundary is a subset of the boundary of $\bigcup_{i=1}^{n_{\mathcal{A}}} \bigcup_{j=1}^{n_{\mathcal{B}}} E_j^{\mathcal{B}} \ominus E_i^{\mathcal{A}}(0)$. It can be computed using a sweep-line technique that traces out the boundary of $\bigcup_{i=1}^{n_{\mathcal{A}}} \bigcup_{j=1}^{n_{\mathcal{B}}} E_j^{\mathcal{B}} \ominus E_i^{\mathcal{A}}(0)$, while keeping track of the position of $\mathcal{CB}$ relative to this boundary.

This computation can be done in $O((n_{\mathcal{A}} n_{\mathcal{B}} + c) \log n_{\mathcal{A}} n_{\mathcal{B}})$ time, where c is the number of intersections between the boundaries of the regions $E_j^{\mathcal{B}} \ominus E_i^{\mathcal{A}}(0)$. In the worst case, $c \in O(n_{\mathcal{A}}^2 n_{\mathcal{B}}^2)$.

1.2.2 Extension of the Visibility Graph Method

Let the configuration space $\mathcal{C} = \mathbf{R}^2$ be populated by generalized polygonal C-obstacles whose union is denoted by $\mathcal{CB}$.

The (reduced) **generalized visibility graph**, denoted by G, is defined as follows:

1. $\mathbf{q}_{init}$, $\mathbf{q}_{goal}$ and the convex vertices of $\mathcal{CB}$ are nodes of G.

2. Let X and X' be two nodes among those defined above. If the open segment joining X and X' lies entirely in free space and is a tangent segment to $\mathcal{CB}$, then it is a link of G.

3. Let X be a node among those defined above and E' be a circular edge of $\mathcal{CB}$. If there exists a point X' in E' such that the open segment joining X and X' lies entirely in free space and its supporting line is tangent to $\mathcal{CB}$ at X and X', then the point X' is a node of G and the segment joining X and X' is a link of G.

4. Let E and E' be two circular edges of $\mathcal{CB}$. If there exist a point X in E and a point X' in E' such that the open segment joining X and X' lies entirely in free space and its supporting line is tangent to $\mathcal{CB}$

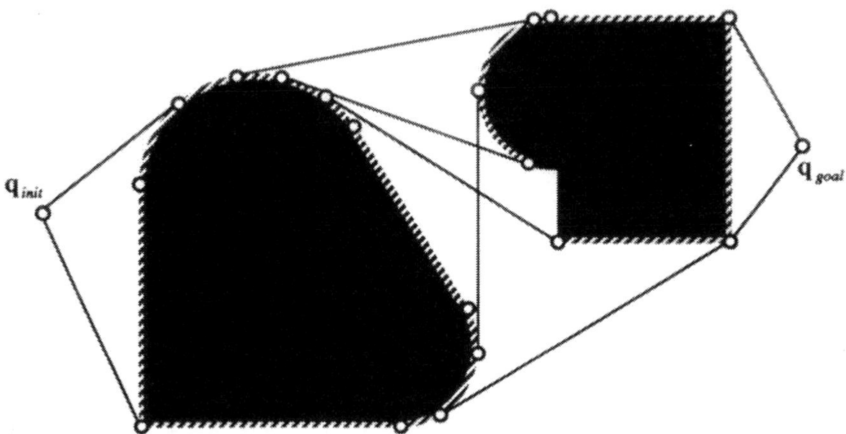

Figure 7. This figure shows a simple generalized visibility graph G in a configuration space with two C-obstacles. The points marked as small circles are the nodes of G. Only those edges of $\mathcal{CB}$ which are depicted as striped lines are links of G.

at X and X', then the two points X and X' are nodes of G and the segment joining them is a link in G.

5. Every straight edge of $\mathcal{CB}$ connecting two convex vertices is a link of G.

6. Every two nodes X and X' of G contained in the same circular edge E of $\mathcal{CB}$ are connected by a link of G, if there exists no other node X'' of G which lies in E between X and X'.

Figure 7 shows the generalized visibility graph in a configuration space with two generalized polygonal C-obstacles.

Using the same arguments as in the previous subsection — that any subpath in a shortest path is also a shortest path — one can show that there exists a semi-free path between $\mathbf{q}_{init}$ and $\mathbf{q}_{goal}$ if and only if there exists a path between these two nodes in G. In addition, if there exists a semi-free path, then G contains the shortest one.

Let n be the total number of vertices of $\mathcal{CB}$. There are two lines at most which are tangent to a circular edge while passing through a given point. There are four lines at most which are simultaneously tangent to

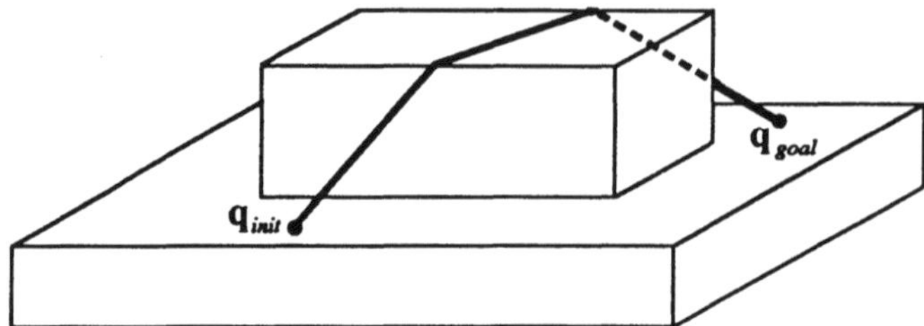

Figure 8. The shortest path in a three-dimensional configuration space with polyhedral C-obstacles may not connect q_{init} to q_{goal} through vertices of the C-obstacle region. It is a polygonal line whose vertices are located in the closed edges of the C-obstacle region.

two circular edges. For every pair (point,edge) and (edge,edge), these lines can be computed in constant time. Thus, the number of nodes in G is $O(n^2)$ and the number of links is also $O(n^2)$. G can easily be constructed in $O(n^3)$ time and its search performed in $O(n^2)$ time. The shortest path can be found in $O(n^2 \log n)$ time using the A^* algorithm with the Euclidean distance to the goal as the heuristic function.

1.3 Higher-Dimensional Configuration Spaces

Consider the case where $\mathcal{C} = \mathbf{R}^3$, with polyhedral C-obstacles. It occurs when the object $\mathcal{A}$ is a polyhedron that translates at fixed orientation in $\mathcal{W} = \mathbf{R}^3$ among obstacles which are also modeled as polyhedra. The visibility graph method can still be applied, but the generated path may not be the shortest path in $cl(\mathcal{C}_{free})$. Indeed, the shortest path is only guaranteed to be a polygonal line whose vertices are contained in the edges of the C-obstacle region, as is illustrated in Figure 8.

Canny [Canny, 1988] showed that the problem of generating a shortest semi-free path in a three-dimensional space populated by polyhedral obstacles is NP-hard, and Reif and Storer [Reif and Storer, 1985] gave a $2^{n^{O(1)}}$-time algorithm, where n is the total number of vertices, by reducing the problem to a decision problem in the theory of the real numbers. Papadimitriou [Papadimitriou, 1985] gave an algorithm that finds a path which is at most $(1+\varepsilon)$ times the length of the shortest path in time that is polynomial in n and $1/\varepsilon$. This algorithm breaks edges into short segments and searches a graph with these segments as nodes.

When the visibility graph method is applied in a polyhedral configuration space, one may get shorter paths by adding fictitious vertices in the edges of the C-obstacle region. Lozano-Pérez and Wesley [Lozano-Pérez and Wesley, 1979] suggested adding these fictitious vertices in such a way that no resulting edge is longer than a prespecified maximum length (the idea is similar to that of Papadimitriou).

The visibility graph method does not generalize to the case where the robot is a polygon translating and rotating among polygonal obstacles in a two-dimensional workspace. Then, $\mathcal{C} = \mathbf{R}^2 \times S^1$, and in $\mathbf{R}^2 \times [0, 2\pi)$ the C-obstacles are three-dimensional regions bounded by curved ruled surfaces. However, the visibility graph method can be combined with other techniques in order to handle rotation. For instance, we can "slice" the orientation axis into a finite number of intervals $[\theta_k, \theta_{k+1}]$, $k = 1, ..., p$, mod p, and compute the projection in $\mathbf{R}^2$ of the C-obstacle region within each slice. Let $\mathcal{C}_k$ be the copy of $\mathbf{R}^2$ containing the projection $\mathcal{CB}_k$ of the subset of $\mathcal{CB}$ contained in the slice $[\theta_k, \theta_{k+1}]$. One can verify that $\mathcal{CB}_k$ is a generalized polygonal region (see Section 3 of Chapter 6). Hence, a generalized visibility graph G_k can be built for every interval $[\theta_k, \theta_{k+1}]$. The graphs G_k can then be combined into a larger graph G by linking any two nodes $X \in G_k$ and $X' \in G_{k+1}$, $k = 1, ..., p$ mod p, whenever the open segment joining the two configurations X and X' does not intersect $\mathcal{CB}_k$ in $\mathcal{C}_k$, nor $\mathcal{CB}_{k+1}$ in $\mathcal{C}_{k+1}$. A path can be generated by searching G. This approach, however, rests on conservative approximations; hence, it is not complete, i.e. it may fail to find a path even if one exists.

The visibility graph method has rarely been used for planning motions in configuration spaces of dimensions higher than two. Other methods are then usually more suitable.

2 Retraction Approach

The retraction approach consists of defining a continuous mapping of the robot's free space $\mathcal{C}_{free}$ onto a one-dimensional network of curves $\mathcal{R}$ lying in $\mathcal{C}_{free}$. In Subsection 2.1 we develop the theoretical principle of this approach. In Subsection 2.2 we illustrate this principle with a simple planning method applicable to the case where $\mathcal{A}$ is a polygonal object that translates at fixed orientation among polygonal obstacles.

2.1 Principle

The notion of retraction is a classical one in Topology [Jänich, 1984]:

DEFINITION 3: *Let $\mathcal{X}$ be a topological space and $\mathcal{Y}$ be a subset of $\mathcal{X}$. A surjective map $\mathcal{X} \rightarrow \mathcal{Y}$ is called a* **retraction** *from $\mathcal{X}$ onto $\mathcal{Y}$ if and only if it is continuous and its restriction to $\mathcal{Y}$ is the identity map.*

In the following, we consider retractions refined as follows:

DEFINITION 4: *Let ρ be a retraction of a topological space $\mathcal{X}$ onto $\mathcal{Y}$. ρ is said to* **preserve the connectivity** *of $\mathcal{X}$ if and only if, for all $x \in \mathcal{X}$, x and $\rho(x)$ belong to the same path-connected component of $\mathcal{X}$.*

Let ρ be a connectivity-preserving retraction of $\mathcal{C}_{free}$ onto a one-dimensional subset $\mathcal{R}$ of $\mathcal{C}_{free}$ (hence, a network of curves). The following proposition reduces motion planning in $\mathcal{C}_{free}$ to motion planning in $\mathcal{R}$.

PROPOSITION 3: *Let $\rho : \mathcal{C}_{free} \rightarrow \mathcal{R}$, where $\mathcal{R} \subset \mathcal{C}_{free}$ is a network of one-dimensional curves, be a connectivity-preserving retraction. There exists a free path between two free configurations $\mathbf{q}_{init}$ and $\mathbf{q}_{goal}$ if and only if there exists a path in $\mathcal{R}$ between $\rho(\mathbf{q}_{init})$ and $\rho(\mathbf{q}_{goal})$.*

Proof: Let $\tau : [0,1] \rightarrow \mathcal{C}_{free}$ be a free path between $\mathbf{q}_{init}$ and $\mathbf{q}_{goal}$. We can apply ρ to τ. Thanks to the continuity of ρ, we obtain another free path $\rho \circ \tau : [0,1] \rightarrow \mathcal{R}$ between $\rho(\mathbf{q}_{init})$ and $\rho(\mathbf{q}_{goal})$. Conversely, if there is a path in $\mathcal{R}$ between $\rho(\mathbf{q}_{init})$ and $\rho(\mathbf{q}_{goal})$, then $\mathbf{q}_{init}$ and $\mathbf{q}_{goal}$ are connected by a free path which is the product of three paths: a free path from $\mathbf{q}_{init}$ to $\rho(\mathbf{q}_{init})$, a path from $\rho(\mathbf{q}_{init})$ to $\rho(\mathbf{q}_{goal})$ in $\mathcal{R}$, and a free path from $\rho(\mathbf{q}_{goal})$ to $\mathbf{q}_{goal}$. The first and the third of these paths exist since ρ preserves the connectivity of $\mathcal{C}_{free}$. ■

A retraction planning method is based on this proposition. It is determined by the choice of the map ρ. In order to make the method effective, we also need algorithms to construct the graph-like representation of $\mathcal{R}$, to compute the maps $\rho(\mathbf{q}_{init})$ and $\rho(\mathbf{q}_{goal})$, and to generate free paths from $\mathbf{q}_{init}$ to $\rho(\mathbf{q}_{init})$ and from $\rho(\mathbf{q}_{goal})$ to $\mathbf{q}_{goal}$.

We now describe an effective retraction method in $\mathcal{C} = \mathbf{R}^2$.

2.2 Polygonal Configuration Space

The retraction method described below is due to Ó'Dúnlaing and Yap [Ó'Dúnlaing and Yap, 1982]. It is applicable when $\mathcal{C} = \mathbf{R}^2$ and $\mathcal{C}_{free}$ is the interior of a bounded polygonal region[4]. It consists of retracting $\mathcal{C}_{free}$ onto its *Voronoi diagram*. This diagram has the interesting property of maximizing the clearance between the robot and the obstacles.

DEFINITION 5: *Let $\beta = \partial\mathcal{C}_{free}$. For any $\mathbf{q} \in \mathcal{C}_{free}$, let:*

$$clearance(\mathbf{q}) = \min_{p \in \beta} \|\mathbf{q} - p\|,$$

where $\|\mathbf{q} - p\|$ is the Euclidean distance between $\mathbf{q}$ and p, and let:

$$near(\mathbf{q}) = \{p \in \beta \;/\; \|\mathbf{q} - p\| = clearance(\mathbf{q})\}.$$

The **Voronoi diagram**[5] *of $\mathcal{C}_{free}$ is the set:*

$$Vor(\mathcal{C}_{free}) = \{\mathbf{q} \in \mathcal{C}_{free} \;/\; card(near(\mathbf{q})) > 1\},$$

where $card(E)$ denotes the cardinality of the set E.

The Voronoi diagram of $\mathcal{C}_{free}$ is the roadmap $\mathcal{R}$ used by the retraction method described in this subsection. Figure 9 shows a simple Voronoi diagram.

One can verify that $Vor(\mathcal{C}_{free})$ consists of a finite collection of straight and parabolic curve segments, which we call **arcs**. A straight arc in $Vor(\mathcal{C}_{free})$ is the set of configurations that are closest to the same pair of edges, or the same pair of vertices. A parabolic arc in $Vor(\mathcal{C}_{free})$ is the set of configurations that are closest to the same pair consisting of an edge and a vertex. The pair (edge,edge), (vertex,vertex), or (edge,vertex) of elements closest to any configuration in a straight or parabolic arc

[4]Actually, the method described in [Ó'Dúnlaing and Yap, 1982] is for the case where $\mathcal{A}$ is a disc moving among polygonal obstacles. Leven and Sharir [Leven and Sharir, 1985b] studied the case where $\mathcal{A}$ is a convex polygon translating among polygonal obstacles.

[5]The Voronoi diagram was originally defined for a set of points. The Voronoi diagram as defined here is often called the "generalized Voronoi diagram".

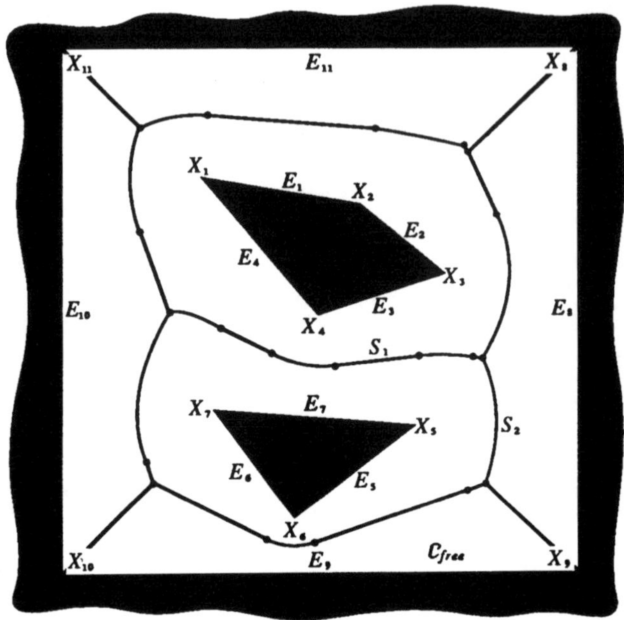

Figure 9. The Voronoi diagram of $\mathcal{C}_{free}$ is the set of points whose minimal distance to $\partial\mathcal{C}_{free}$ is achieved with more than one point in $\partial\mathcal{C}_{free}$. When $cl(\mathcal{C}_{free})$ is a bounded polygonal region, as is the case in the above figure, the Voronoi diagram consists of straight segments and parabolic arcs. A straight segment in the diagram is the set of configurations that are closest to the same pair of edges or the same pair of vertices. For example, the straight segment S_1 is closest to edges E_3 and E_7. A parabolic arc is the set of configurations that are closest to the same pair consisting of an edge and a vertex. For example, the arc S_2 is closest to edge E_8 and vertex X_5.

determines the equation of the curve supporting the arc. The arcs of $Vor(\mathcal{C}_{free})$ and the edges of $\partial\mathcal{C}_{free}$ delimit open regions in $\mathcal{C}_{free}$. For any point $\mathbf{q}$ in such a region, $card(near(\mathbf{q})) = 1$.

Let n be the number of vertices of $\partial\mathcal{C}_{free}$. A naive algorithm constructs the Voronoi diagram of $\mathcal{C}_{free}$ in $O(n^4)$ time by considering the $O(n^2)$ pairs (edge,edge), (vertex,edge) and (vertex,vertex) and computing the intersections of the corresponding equidistant curves. It is not hard, however, to show that the total number of arcs of $Vor(\mathcal{C}_{free})$ is $O(n)$, and from this result to derive an algorithm that computes $Vor(\mathcal{C}_{free})$ in

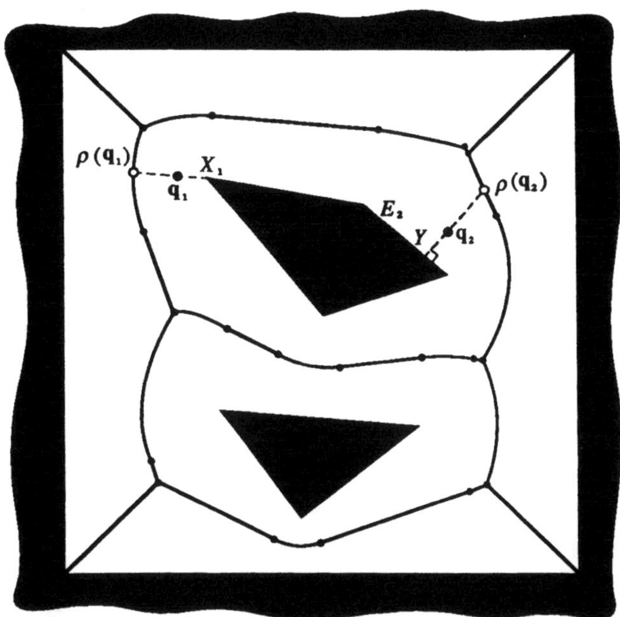

Figure 10. This figure illustrates the construction of the images $\rho(\mathbf{q}_1)$ and $\rho(\mathbf{q}_2)$ of two configurations $\mathbf{q}_1$ and $\mathbf{q}_2$ by the retraction $\rho : \mathcal{C}_{free} \rightarrow Vor(\mathcal{C}_{free})$ defined in the text.

$O(n^2)$ time. There also exist algorithms running in $O(n \log^2 n)$ time [Lee and Drysdale, 1981] [Leven and Sharir, 1987] and in optimal $O(n \log n)$ time [Kirkpatrick, 1979] [Yap, 1985] [Fortune, 1986].

Let us now consider a free configuration $\mathbf{q} \notin Vor(\mathcal{C}_{free})$. There exists a unique vertex or open edge Z of $\partial\mathcal{C}_{free}$ which is closer to $\mathbf{q}$ than any other edge or vertex of $\partial\mathcal{C}_{free}$. Let $p \in \partial\mathcal{C}_{free}$ be the point such that $\|\mathbf{q}-p\| = clearance(\mathbf{q})$. If Z is an open edge, then $p \in Z$; if it is a vertex, then $p = Z$. Consider the half-line L drawn from p and passing through $\mathbf{q}$. The straight segment connecting p to the closest intersection of L with $Vor(\mathcal{C}_{free})$ follows the steepest ascent of the function *clearance*, i.e. the vector $\vec{\nabla} clearance(\mathbf{q}')$ points along L at any $\mathbf{q}'$ in the segment. The first intersection of L with $Vor(\mathcal{C}_{free})$, denoted by $\rho(\mathbf{q})$, is the image of $\mathbf{q}$ in the Voronoi diagram. Figure 10 illustrates the construction of $\rho(\mathbf{q})$. Beyond $\rho(\mathbf{q})$, the function *clearance* decreases or the direction of its gradient changes. The function ρ can be extended to all of $\mathcal{C}_{free}$ by setting $\rho(\mathbf{q}) = \mathbf{q}$ for all $\mathbf{q} \in Vor(\mathcal{C}_{free})$.

One can show the following proposition:

PROPOSITION 4: *The application $\rho : \mathcal{C}_{free} \rightarrow Vor(\mathcal{C}_{free})$ is continuous.*

Hence, ρ is a retraction of $\mathcal{C}_{free}$ onto $Vor(\mathcal{C}_{free})$. By construction, it preserves the connectivity of $\mathcal{C}_{free}$.

The path planning method based on the retraction ρ uses the Voronoi diagram $Vor(\mathcal{C}_{free})$ as the roadmap. It proceeds as follows:

1. Compute the Voronoi diagram $Vor(\mathcal{C}_{free})$.
2. Compute the points $\rho(\mathbf{q}_{init})$ and $\rho(\mathbf{q}_{goal})$ and identify the arcs of $Vor(\mathcal{C}_{free})$ containing these two points.
3. Search $Vor(\mathcal{C}_{free})$ for a sequence of arcs $A_1, ..., A_p$, such that $\rho(\mathbf{q}_{init}) \in A_1$, $\rho(\mathbf{q}_{goal}) \in A_p$, and, for all $i \in [1, p-1]$, A_i and A_{i+1} share an endpoint.
4. If the search terminates successfully return $\rho(\mathbf{q}_{init})$, $\rho(\mathbf{q}_{goal})$, and the sequence of arcs connecting them; otherwise, return failure.

Step 1 requires $O(n \log n)$ time. At step 2, one can:

- first, compute both the distance of $\mathbf{q}_{init}$ (resp. $\mathbf{q}_{goal}$) to every edge and every vertex of $\partial\mathcal{C}_{free}$, and the point p_{init} (resp. p_{goal}) of $\partial\mathcal{C}_{free}$ to which it is closest;
- second, determine the intersections of the ray drawn from p_{init} (resp. p_{goal}) and passing through $\mathbf{q}_{init}$ (resp. $\mathbf{q}_{goal}$) with $Vor(\mathcal{C}_{free})$, and select the intersection which is closest to p_{init} (resp. p_{goal}).

Since there are $O(n)$ arcs in $Vor(\mathcal{C}_{free})$, all these computations require $O(n)$ time. The search at Step 3 also takes $O(n)$ time. Thus, the overall time complexity of this path planning method is $O(n \log n)$. The most expensive step is Step 1, but it depends on $\mathcal{C}_{free}$ only, not on $\mathbf{q}_{init}$ and $\mathbf{q}_{goal}$. Planning other paths in the same free space between other pairs of configurations can be done by omitting Step 1 and takes $O(n)$ time.

In addition, there exist linear techniques to update the Voronoi diagram when a few C-obstacles are changed [Gowda et al., 1983].

This method is also applicable when the C-obstacles are generalized polygons.

2.3 "Retraction-Like" Methods

Other retraction methods have been proposed to solve more involved instances of the basic motion planning problem. For instance, Ó'Dúnlaing, Sharir and Yap [Ó'Dúnlaing, Sharir and Yap, 1983, 1984a and 1984b] described a retraction method for planning the motion of a segment ("ladder") among polygonal obstacles. In this method, the three-dimensional free space is first retracted onto a two-dimensional variant of the Voronoi diagram. In a second step, this diagram is itself retracted onto a network of one-dimensional curves of $\mathcal{C}_{free}$. The whole computation takes $O(n^2 \log n \log\log n)$ time. Canny and Donald proposed a "simplified Voronoi diagram" which is easier to extend to higher-dimensional space than the classical generalized Voronoi diagram [Canny and Donald, 1988].

A retraction method in such a higher-dimensional configuration space typically uses an application ρ that is "retraction-like" in the sense that it is only continuous almost everywhere, i.e. it is continuous except in a subset of $\mathcal{C}_{free}$ having zero measure. A "retraction-like" application $\rho : \mathcal{C}_{free} \rightarrow \mathcal{R}$ preserves connectivity if and only if:

- $\forall \mathbf{q} \in \mathcal{C}_{free}$, $\rho(\mathbf{q})$ and $\mathbf{q}$ belong to the same connected component of $\mathcal{C}_{free}$, and

- each connected component of $\mathcal{C}_{free}$ contains exactly one connected component of $\mathcal{R}$.

The second condition makes Proposition 3, with "retraction-like application" substituted for "retraction", still valid, despite the fact that ρ is not continuous everywhere. Yap [Yap, 1987a] proved the existence of a connectivity-preserving retraction-like application of any semi-algebraic m-dimensional manifold $\mathcal{C}_{free}$ onto an algebraic one-dimensional subset $\mathcal{R} \subset \mathcal{C}_{free}$.

Several authors suggested using the Voronoi diagram of the empty subset of a two-dimensional workspace as a heuristic guideline to plan free paths

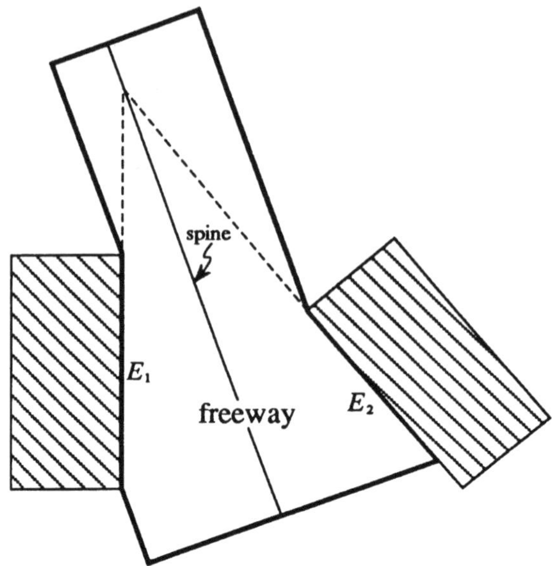

Figure 11. A freeway is a straight linear generalized cylinder whose spine is annotated with a description of the free orientations of the robot.

for a robot that can both translate and rotate in the plane (e.g. see [Takahashi and Schilling, 1989]). The method described in the next section is of the same vein.

3 Freeway Method

The freeway method [Brooks, 1983a] presented in this section specifically applies to a polygonal object moving both in translation and rotation in a two-dimensional polygonal workspace. The intuition behind it is similar to the intuition behind retracting free space onto its Voronoi diagram, i.e. keep the robot as far away as possible from the obstacles.

The freeway method consists of extracting geometric figures called *freeways* from the workspace, connecting them into a graph called the freeway net, and searching this graph. A freeway is a straight linear generalized cylinder [Binford, 1971] [Ponce and Chelberg, 1987] whose straight axis, the *spine*, is annotated with a conservative description of the free orientations of $\mathcal{A}$ when the reference point $O_{\mathcal{A}}$ moves along it. Figure

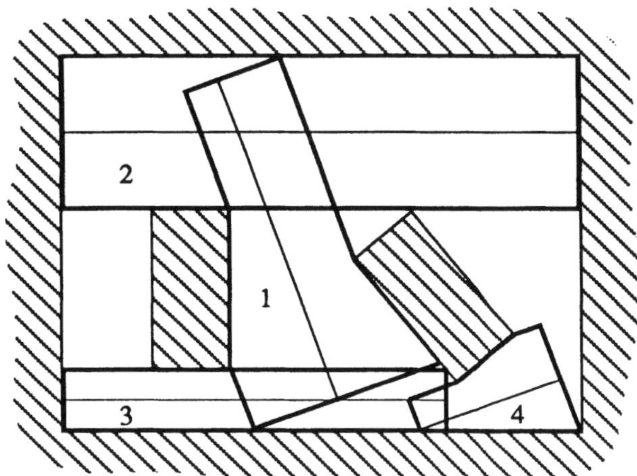

Figure 12. Multiple overlapping freeways can usually be extracted from a workspace. This figure shows four freeways, but many others could be constructed.

11 illustrates the geometry of a freeway. Figure 12 shows four freeways in a bounded workspace.

$\mathcal{A}$ can move along a freeway (or part of it) if there is a connected set of free orientations of $\mathcal{A}$ along the spine. In addition, whenever two spines intersect, $\mathcal{A}$ can transfer from one freeway to the other if the ranges of free orientations of $\mathcal{A}$ along both spines have a non-empty intersection at the intersection point. The freeway net is a representation of the possible motions of $\mathcal{A}$ along spines and between spines. However, its construction rests on some empirical conservative assumptions, and the description it gives of the connectivity of free space is in general incomplete.

Let $\mathcal{B}$ be the obstacle region (union of all the obstacles). We assume that the empty subset of the workspace, $\mathcal{E} = \mathcal{W} \setminus \mathcal{B}$, is the interior of a bounded polygonal region.

3.1 Extraction of Freeways

The terminology introduced in the following definition is illustrated in Figure 13.

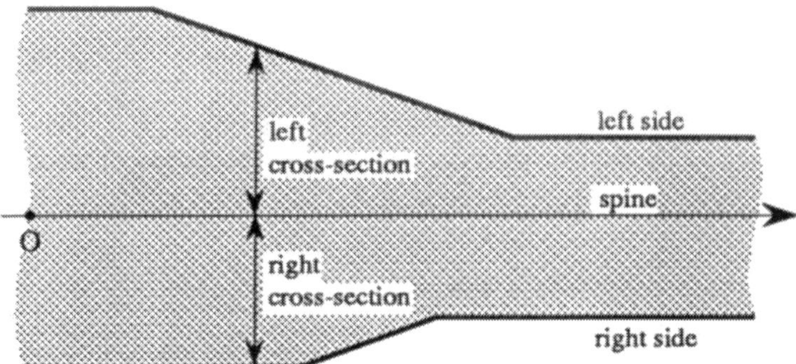

Figure 13. This figure shows a two-dimensional straight linear generalized cylinder. It illustrates the terminology introduced in Definition 6.

DEFINITION 6: *A* **two-dimensional straight linear generalized cylinder** *is a region of* $\mathbf{R}^2$ *obtained by sweeping a straight line segment, the* **cross-section**, *along a straight line, the* **spine**. *An origin and an orientation are defined on the spine. The cross-section stays perpendicular to the spine. It is partitioned by the spine into two segments, the* **right** *and the* **left** *cross-sections. The lengths of the right and left cross-sections are independent, continuous, piecewise linear functions of the abscissa along the spine. The two lines drawn by the extremities of the cross-section are called the right and left* **sides** *of the cylinder.*

Freeways are extracted from $\mathcal{E} = \mathcal{W} \backslash \mathcal{B}$ by considering all pairs of edges of $\mathcal{B}$. A pair of edges (E_1, E_2) produces a generalized cylinder if and only if it satisfies the following two conditions:

1. For $i, j \in \{1,2\}$, $i \neq j$, both extremities of E_i are in the outer half-plane of E_j.

2. The inner product of the outgoing normal vectors of E_1 and E_2 is negative.

The satisfaction of these two conditions means that E_1 and E_2 "face" each other. Figure 14 shows two examples of pairs of edges that do not satisfy these conditions.

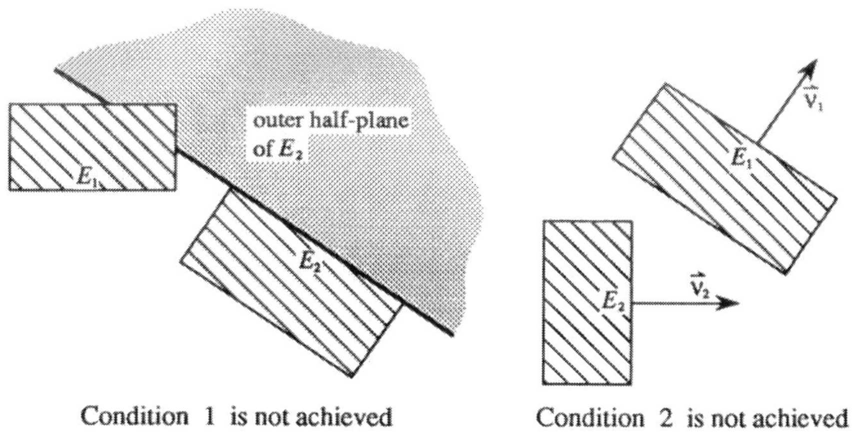

Figure 14. A freeway is constructed from a pair of obstacle edges. These edges must satisfy two conditions implying that they "face" each other. This figure shows two cases where one of these conditions is not satisfied.

Given any pair of edges (E_1, E_2) satisfying the above two conditions, a generalized cylinder (from now on, we omit the "straight" and "linear" qualifiers) *GC* is constructed as follows (see Figure 15). *GC*'s spine is the bisector of the angle formed by the lines containing E_1 and E_2. (If these lines are parallel, the spine is parallel and equidistant from both of them.) Each side of *GC* is made of one edge (E_1 or E_2) extended at each extremity by a half-line parallel to the spine. The choice of the bisector as the spine of the freeway is reminiscent of the construction of a Voronoi diagram (in the workspace). In fact, the spines of the extracted generalized cylinders contain some of the straight segments of the Voronoi diagram of $\mathcal{E}$.

GC lies partly outside $\mathcal{E}$ and we now select those pieces of *GC* which completely lie within $\mathcal{E}$ as illustrated by Figure 16. To that purpose, *GC* is intersected with the obstacle region $\mathcal{B}$. The intersection is normally projected on the spine, and the corresponding slices are removed from the cylinder. *GC* is thus sliced into one or several truncated generalized cylinders. Truncated cylinders whose sides do not include portions of both E_1 and E_2 are discarded. In the example of Figure 16, this only leaves one slice denoted by Φ. This way of removing slices from generalized cylinders is empirical and may seem a bit radical. However, we

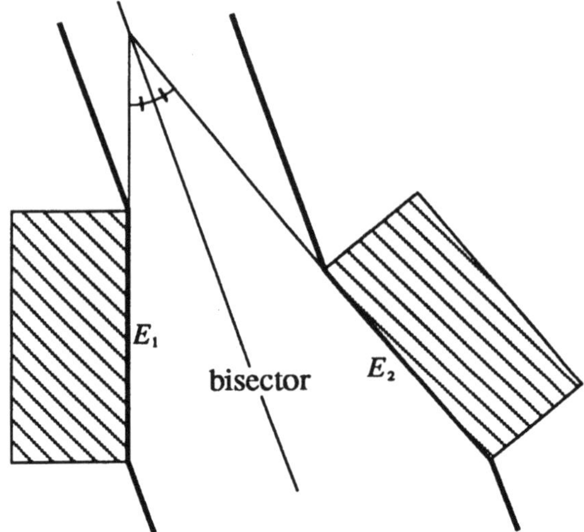

Figure 15. Given two edges E_1 and E_2 of $\mathcal{B}$ that face each other, a generalized cylinder is constructed. Its spine is taken as the bisector of the angle formed by the two lines containing E_1 and E_2. Its sides are made of the two edges extended at each extremity by a half-line parallel to the spine.

should keep in mind that there are usually many generalized cylinders leading to many overlapping freeways. In particular, some of the edges bounding the obstacles intersecting GC can usually be paired with either E_1 or E_2 to produce other generalized cylinders overlapping GC. The success of the freeway method derives from this multiplicity of freeways.

When all the pairs of edges of $\mathcal{B}$ have been considered, the remaining truncated generalized cylinders are the freeways to be used for path planning.

Each freeway is an instance of the general case shown in Figure 17. By convention, the spine of a freeway is oriented from the "big" end toward the "small" end; if the freeway has parallel sides, an arbitrary orientation is selected. The origin of spine abscissae is taken at the freeway's end (when it exists, the big end) such that any point in the freeway projects on the spine at a positive abscissa. The geometry of every freeway is completely specified by 7 parameters illustrated in Figure 17: L, B_l, B_r,

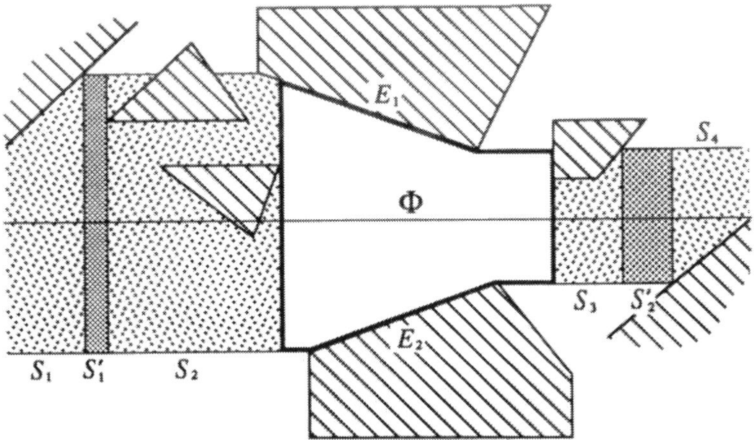

Figure 16. After a generalized cylinder has been constructed using two edges E_1 and E_2, it is intersected with the obstacles. The intersection regions are normally projected into the spine, and the corresponding slices (in the figure, S_1, S_2, S_3, and S_4), are removed. There remains a finite set of truncated generalized cylinders. Those cylinders whose sides no longer include portions of both E_1 and E_2 are discarded (in the figure, S'_1 and S'_2). In the above figure, only Φ remains.

S_l, S_r, ϕ and W. A freeway is a polygon with at most 8 edges.

The computation of the initial non-truncated set of generalized cylinders requires $O(n_{\mathcal{B}}^2)$ time, where $n_{\mathcal{B}}$ is the number of edges of $\mathcal{B}$. Intersecting a generalized cylinder with the obstacle region can be done in time $O(n_{\mathcal{B}})$. If the intersection is not empty, the intersection can be projected on the spine in time $O(n_{\mathcal{B}})$. Therefore, the time complexity of the complete operation is $O(n_{\mathcal{B}}^3)$.

3.2 Freeways as Cross-Sections in Configuration Space

After having extracted freeways, the next step of the freeway method is to compute the range of free orientations of $\mathcal{A}$ when it moves along the spine of a freeway. Before describing how this is done, we define more accurately what it means for $\mathcal{A}$ to "move along" the spine.

We represent a configuration $\mathbf{q}$ of $\mathcal{A}$ by (x, y, θ), where x and y are

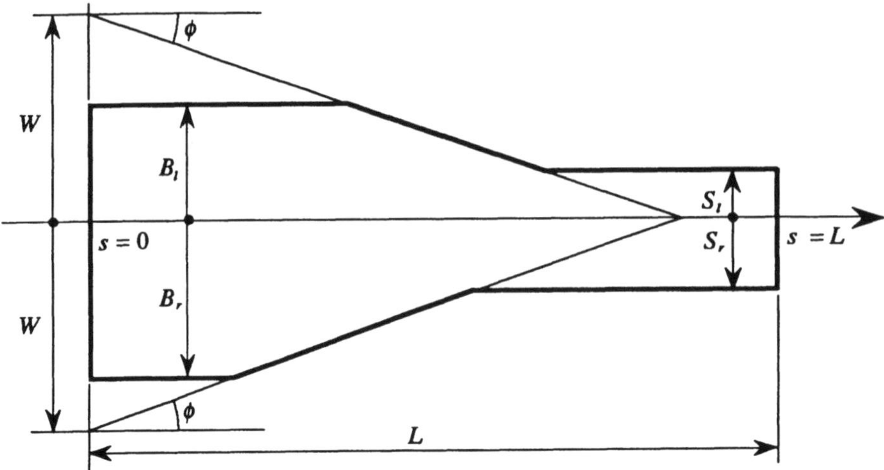

Figure 17. Each freeway is an instance of the general figure shown above. It is specified by 7 parameters: L, B_l, B_r, S_l, S_r, ϕ and W. Particular cases are included in this general specification. For example, if the freeway is a traditional cylinder, then $B_r = B_l = S_r = S_l = W$ and $\phi = 0$.

the coordinates of the reference point $O_{\mathcal{A}}$ in the coordinate frame $\mathcal{F}_{\mathcal{W}}$ attached to the workspace, and $\theta \in [0, 2\pi)$ is the angle (mod 2π) between the x-axes of $\mathcal{F}_{\mathcal{W}}$ and $\mathcal{F}_{\mathcal{A}}$ (the frame attached to $\mathcal{A}$).

When $\mathcal{A}$ "moves along" the spine of a freeway Φ, we impose that $O_{\mathcal{A}}$ stays on Φ's spine. In other words, we constrain paths in the robot's configuration space $\mathcal{C}$ to be contained in planes which are perpendicular to the xy-plane and which project on this plane along the spines of the extracted freeways. This is not absurd since each spine is equally distant from the two supporting lines of the obstacle edges that were used to generate the corresponding freeway. Determining the range of free orientations of $\mathcal{A}$ along the spine of a freeway is thus equivalent to computing a cross-section through the C-obstacle region by a plane perpendicular to the xy-plane.

In order to maximize the range of free orientations of $\mathcal{A}$ when it moves along a spine, it is preferable to select $O_{\mathcal{A}}$ such that the maximum distance from $O_{\mathcal{A}}$ to the points on the boundary of $\mathcal{A}$ is the smallest possible. This is achieved by taking $O_{\mathcal{A}}$ at the center of the minimum spanning

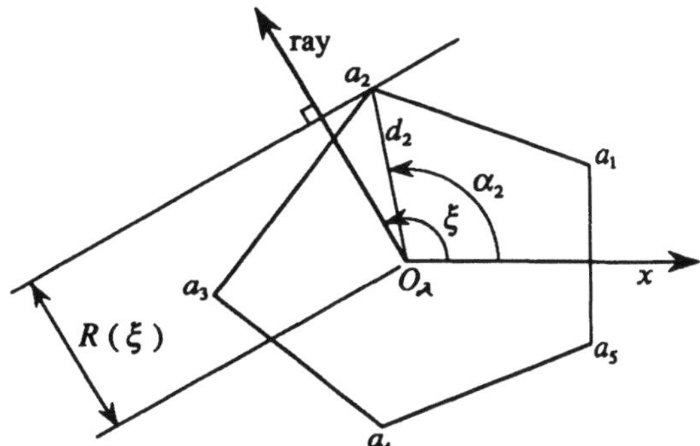

Figure 18. The radius function $R(\xi)$ of $\mathcal{A}$ is defined as the infimum of the distances from $O_{\mathcal{A}}$ to the lines not intersecting $\mathcal{A}$ and perpendicular to the ray erected from $O_{\mathcal{A}}$ making an angle ξ with the x-axis of $\mathcal{F}_{\mathcal{A}}$.

circle[6] of $\mathcal{A}$.

It is interesting to notice that Brooks' intention, when he developed the freeway method, was to capture the effects of both translating and rotating an object into a more intuitive representation than configuration space [Brooks, 1983a]. Since configuration space is a general representational tool, it is not surprising that the "new" representation can easily be related to configuration space. As will become clear later, the actual contribution of the freeway method is that it provides a means for constructing a fairly comprehensive description of the connectivity of $\mathcal{C}_{free}$ without having to exhaustively compute and represent the C-obstacles.

3.3 Determination of Free Orientations

We now describe an efficient technique for approximating the range of free orientations of $\mathcal{A}$ along the spine of a freeway. This technique makes

[6]The *minimum spanning circle* of a polygon is the smallest circle that contains the polygon. An obvious algorithm computes this circle in $O(n^4)$ time, where n is the number of vertices of the polygon. This algorithm may be sufficient here, since in most practical cases, the polygon $\mathcal{A}$ has a small number of vertices, and furthermore the computation has to be done only once. Nevertheless, the circle can be more efficiently computed in $O(n^2)$ time by noticing that it passes through two diametrically opposite points or through three points that define a triangle with acute angles. An optimal algorithm runs in $O(n \log n)$ time [Preparata and Shamos, 1985]. If the polygon is convex, it can be computed in $O(n)$ time [Castells and Melville, 1985].

use of a function, called the radius function of $\mathcal{A}$, and of its inverse. We assume that $\mathcal{A}$ is a convex polygon; otherwise, we approximate it by its convex hull.

DEFINITION 7: *A half-line issued from $O_\mathcal{A}$ is called a* **ray**. *The angle $\xi \in [0, 2\pi)$ between the x-axis of $\mathcal{F}_\mathcal{A}$ and the ray is called the* **angle** *of the ray. The* **radius function** *$R(\xi)$ of $\mathcal{A}$ is defined as the infimum of the distances from $O_\mathcal{A}$ to the lines which are normal to a ray of angle ξ and do not intersect $\mathcal{A}$.*

We have (see Figure 18):

$$R(\xi) = \max_{i \in [1, n_\mathcal{A}]} \{d_i \cos(\xi - \alpha_i)\}$$

where:
- $n_\mathcal{A}$ is the number of vertices of $\mathcal{A}$,
- d_i is the distance from $O_\mathcal{A}$ to the ith vertex of $\mathcal{A}$,
- α_i is the angle of the ray passing through the ith vertex of $\mathcal{A}$.

$R(\xi)$ specifies how close a line perpendicular to the ray of angle ξ can be to $O_\mathcal{A}$ without intersecting the interior of $\mathcal{A}$. The inverse image of an interval $[0, r]$ by R is the function $R^{-1}(r)$ defined by:

$$R^{-1}(r) = \{\xi \in [0, 2\pi) \; / \; R(\xi) < r\}.$$

A line perpendicular to a ray of any angle $\xi \in R^{-1}(r)$ and distant from $O_\mathcal{A}$ by more than r is guaranteed to have no intersection with $\mathcal{A}$.

$R^{-1}(r)$ is computed as follows:

$$R^{-1}(r) \;=\; [0, 2\pi) \setminus \bigcup_{i=1}^{n_\mathcal{A}} \mathcal{I}_i,$$

where:
- $\mathcal{I}_i = \emptyset$ if $r > d_i$,
- $\mathcal{I}_i = [\alpha_i - |\arccos \frac{r}{d_i}|, \alpha_i + |\arccos \frac{r}{d_i}|] \mod 2\pi$, if $r \leq d_i$.

We can use the inverse radius function for computing the free orientations of $\mathcal{A}$ along the spine of a freeway Φ. In order to simplify computation, we approximate $\mathcal{A}$ by a bounding rectangle. By doing so, we obtain a subrange of the free orientations of $\mathcal{A}$.

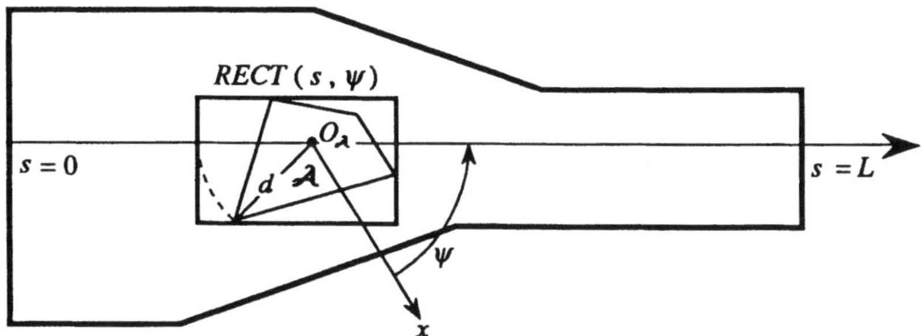

Figure 19. In order to conservatively approximate the range of free orientations of $\mathcal{A}$ when $O_{\mathcal{A}}$ moves along the spine of a freeway, the object $\mathcal{A}$ is enclosed in a rectangle $RECT(s, \psi)$ which depends on the orientation ψ of $\mathcal{A}$ relative to the spine of the freeway.

Let s denote the spine abscissa of $O_{\mathcal{A}}$ and ψ the angle between the x-axis of $\mathcal{F}_{\mathcal{A}}$ and the spine of Φ. We enclose $\mathcal{A}$ in a rectangle $RECT(s, \psi)$ defined as follows (see Figure 19). The *right* and *left* sides of $RECT(s, \psi)$ are parallel to the spine. The distances from $O_{\mathcal{A}}$ to these sides are $R(\psi - \pi/2)$ and $R(\psi + \pi/2)$, respectively. The other two sides are the *front* and *rear* sides. The distance from $O_{\mathcal{A}}$ to the front side is $R(\psi)$. The distance from $O_{\mathcal{A}}$ to the rear side is $d = \max_{i \in [1, n_{\mathcal{A}}]}\{d_i\}$. The later choice is quite conservative, but the independence of d from ψ presents practical advantages as we will see soon. Notice that the bounding rectangle $RECT(s, \psi)$ depends on the orientation of $\mathcal{A}$ relative to the spine of Φ.

Let $F_{\Phi}(s)$, with $0 \leq s \leq L$, denote the set of orientations ψ such that $RECT(s, \psi)$ lies in the interior of Φ. Since $RECT(s, \psi)$ is a bounding approximation of $\mathcal{A}$, $F_{\Phi}(s)$ is a subset of the range of free orientations of $\mathcal{A}$ when $O_{\mathcal{A}}$'s spine abscissa is s. $F_{\Phi}(s)$ can be computed using the formulas given in the following proposition.

PROPOSITION 5: *For a freeway Φ constructed from two non-parallel edges, we have:*

- *if $0 \leq s \leq d$, then:*

$$F_{\Phi}(s) = \emptyset; \tag{1}$$

- *if $d < s \leq L$, then:*

$$\begin{array}{rcl} F_{\Phi}(s) & = & R^{-1}(L-s) \\ & & \bigcap([(R^{-1}(S_r)+\frac{\pi}{2}) \cap (R^{-1}(S_l)-\frac{\pi}{2})] \\ & & \quad \bigcup[(R^{-1}(B_r)+\frac{\pi}{2}) \cap (R^{-1}(B_l)-\frac{\pi}{2}) \\ & & \quad\quad \cap R^{-1}_{-\frac{\pi}{2},\alpha}(\alpha s+W) \cap R^{-1}_{\frac{\pi}{2},\alpha}(\alpha s+W)]), \end{array} \qquad (2)$$

where:

$$\begin{array}{rcl} \alpha & = & -\tan\phi, \\ R_{\zeta,\alpha} & = & R(\psi+\zeta)-\alpha R(\psi), \\ R^{-1}_{\zeta,\alpha}(u) & = & \{\psi \;/\; R_{\zeta,\alpha}(\psi) < u\}. \end{array}$$

In the particular case where Φ is constructed from two parallel edges, the relation (1) *is unchanged, but the relation* (2) *simplifies to:*

$$F_{\Phi}(s) = R^{-1}(L-s) \cap (R^{-1}(W)+\frac{\pi}{2}) \cap (R^{-1}(W)-\frac{\pi}{2}).$$

Relation (1) is obvious, since for $0 \leq s \leq d$, the rear side of $RECT(s,\psi)$ lies outside Φ or in its boundary, for any value of ψ. (Notice here the simplification brought about by the conservative definition of d.) The best way to understand the relation (2) is to parse it:

- The first part of the main conjunct, $R^{-1}(L-s)$, expresses the constraint that the abscissa of the front side of the rectangle must be less than L.

- The first sub-expression in the second part of the main conjunct:

$$(R^{-1}(S_r)+\frac{\pi}{2}) \cap (R^{-1}(S_l)-\frac{\pi}{2}),$$

  expresses that the distance from the spine to the right (resp. left) side of Φ is nowhere smaller that S_r (resp. S_l). The conjunction of $R^{-1}(L-s)$ and this sub-expression is a sufficient condition insuring that $RECT(s,\psi)$ lies in Φ's interior (for $s > d$). However, it is too strong a condition.

- The next two elements in relation (2):

$$(R^{-1}(B_r)+\frac{\pi}{2}) \cap (R^{-1}(B_l)-\frac{\pi}{2}),$$

express the fact that the distance from the spine to the right and left sides of Φ cannot be greater than B_r and B_l, respectively. These two conditions are too weak and remain to be composed with conditions considering the non-parallel sides of Φ.

- The last two elements in relation (2):

$$R^{-1}_{-\frac{\pi}{2},\alpha}(\alpha s + W) \cap R^{-1}_{\frac{\pi}{2},\alpha}(\alpha s + W),$$

express a necessary and sufficient condition that the bounding rectangle be between the two lines containing the non-parallel sides of Φ. Indeed, for each line, we can write:

$$R(\psi - \frac{\pi}{2}) < \alpha(s + R(\psi)) + W,$$

$$R(\psi + \frac{\pi}{2}) < \alpha(s + R(\psi)) + W,$$

where $\alpha = -\tan\phi$. The last two elements in relation (2) directly derive from these inequalities.

Using formula (2) to compute $F_\Phi(s)$ requires being able to compute $R^{-1}_{\zeta,\alpha}(u)$, where $R_{\zeta,\alpha}(\psi) = R(\psi + \zeta) - \alpha R(\psi)$. This may be done by noticing that the sum of two radius functions can be expressed in the same form as a radius function, whose inverse can be computed as shown above. Indeed, if:

$$R(\xi) = \max_{i\in[1,n_A]}\{d_i \cos(\xi - \alpha_i)\}$$

and

$$S(\xi) = \max_{j\in[1,n_A]}\{d'_j \cos(\xi - \alpha'_j)\},$$

then:

$$R(\xi) + S(\xi) = \max_{i,j\in[1,n_A]}\{d_i \cos(\xi - \alpha_i) + d'_j \cos(\xi - \alpha'_j)\}$$

where each term

$$d_i \cos(\xi - \alpha_i) + d'_j \cos(\xi - \alpha'_j)$$

can be re-written in the form:

$$e_{ij} \cos(\xi - \nu_{ij}),$$

where e_{ij} and ν_{ij} depend on d_i, d'_j, α_i, and α'_j.

Since both the right and left radii of Φ are non-increasing functions of the spine abscissa, we have that:

PROPOSITION 6: $\forall s_1, s_2 \in [d, L] : s_1 \leq s_2 \Rightarrow F_\Phi(s_2) \subseteq F_\Phi(s_1)$.

This proposition is central to the construction of the freeway net.

$F_\Phi(s)$ may consist of several disconnected intervals. We assume below that $F_\Phi(s)$ is described as a set of maximal connected intervals.

3.4 Freeway Net

At this point, we have built a collection of freeways. For each freeway Φ, $F_\Phi(s)$ describes the complement of a bounding approximation of the cross-section of the C-obstacle region by a plane containing the spine of Φ and perpendicular to the xy-plane. We thus have a partial representation of free space. We now use this representation to construct a non-directed graph G called the **freeway net**. G determines all the possible ways for $\mathcal{A}$ to move along spines.

Let $\mathcal{X}$ be the set containing every point P such that the spines of two freeways intersect at P inside both freeways. Since there are $O(n_\mathcal{B}^2)$ lines supporting the freeway spines ($n_\mathcal{B}$ being the number of edges of $\mathcal{B}$), a sweep-line algorithm computes $\mathcal{X}$ in $O((n_\mathcal{B}^2 + c)\log n_\mathcal{B})$ time, where $c \in O(n_\mathcal{B}^4)$ is the cardinality of $\mathcal{X}$.

The nodes of G are created as follows:

for every freeway Φ do:
  for every point P of $\mathcal{X}$ in Φ's spine do:
    - let s be the spine abscissa of P;
    - create one node of G for every interval in $F_\Phi(s)$.

The corresponding interval of $F_\Phi(s)$ is associated with each node of G. Since each interval is originally defined with respect to a spine's orientation, an appropriate angle has to be added to both extremities of the interval so that all the intervals are defined with respect to the same reference orientation, say the x-axis of $\mathcal{F}_\mathcal{W}$.

In addition, the initial and goal positions of $O_\mathcal{A}$ are used to create the *initial* and *goal* nodes of G. For simplification, we assume that the initial (resp. goal) position lies in the spine of a freeway Φ_1 (resp. Φ_2) inside the freeway at abscissa s_{init} (resp. s_{goal}). The interval of $F_{\Phi_1}(s_{init})$ (resp. $F_{\Phi_2}(s_{goal})$) containing the initial (resp. goal) orientation of $\mathcal{A}$ is associated with the corresponding node of G. If the initial and goal positions of $\mathcal{A}$ are not located on freeway spines, then a technique has to be devised for moving $\mathcal{A}$ from its initial position to a spine, and from a spine to its goal position.

The links connecting the nodes of G are created as follows:

1. Let N_1 and N_2 be two nodes corresponding to two different points in the same freeway spine having abscissae s_1 and s_2, respectively. Let I_1 and I_2 be the two intervals of free orientations of $\mathcal{A}$ associated with these nodes. If $I_1 \cap I_2 \neq \emptyset$, then N_1 and N_2 are connected by a link in G. Indeed, Proposition 6 entails that at any orientation in $I_1 \cap I_2$, $\mathcal{A}$ can translate in free space from abscissa s_1 to abscissa s_2, and vice-versa.

2. Let N_1 and N_2 be two nodes corresponding to the same point P in the xy-plane for two different freeways. Let I_1 and I_2 be the intervals of free orientations associated with these nodes. If $I_1 \cap I_2 \neq \emptyset$, then N_1 and N_2 are connected by a link in G. Indeed, with any orientation in $I_1 \cap I_2$, $\mathcal{A}$ can transfer in free space from one of the freeways to the other at P.

Hence, it is sufficient for constructing G to compute $F_\Phi(s)$ at a finite number of abscissae along the spine of each freeway Φ.

The constructed freeway net is searched for a path between the initial and the goal node. A path in the net determines a class of free paths in $\mathcal{C}_{free}$. A particular free path is obtained by rotating $\mathcal{A}$, if necessary, only at the intersections between spines. Figure 20 shows such a path.

3.5 Discussion

Experiments with the freeway method have shown that it works fast in a "relatively uncluttered" workspace [Brooks, 1983a]. However, the method is not complete, and hence may not always find a free path even if one exists. Indeed, it embodies several empirical choices whose justification is merely intuitive, for instance: the spine of each freeway is

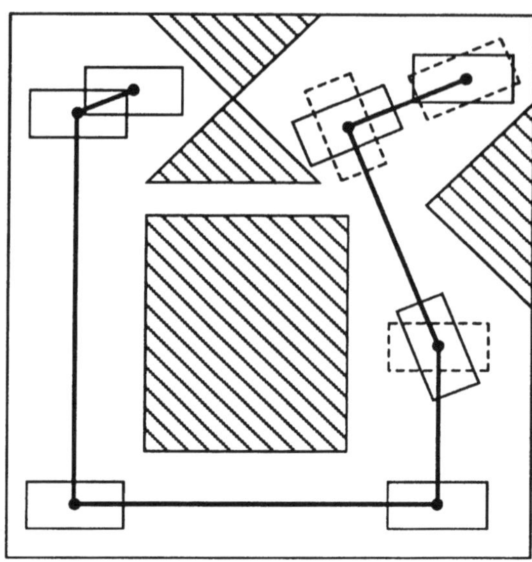

Figure 20. This figure shows a path generated using the freeway method. The robot translates along the spines and rotates (if necessary) at the intersections between spines.

the bisector of the supporting lines of two obstacle edges; each freeway is extended at the endpoints of these edges by lines parallel to the spine; slices containing subsets of the obstacle region are removed; $O_{\mathcal{A}}$ is constrained to navigate along spines; etc... One of the major drawbacks when the workspace is densely occupied by obstacles is that freeways are often too short to contain the rectangle enclosing $\mathcal{A}$.

The example shown in Figure 21 illustrates a limitation of the method [Gouzènes, 1984]. $\mathcal{A}$ is a segment that must move from the initial configuration $\mathbf{q}_{init}$ to the goal configuration $\mathbf{q}_{goal}$ through the right corner made by two corridors. Only three freeways can be built. If we take $O_{\mathcal{A}}$ at the mid-point of $\mathcal{A}$, which is the best choice, the freeway method finds a path only if the length of $\mathcal{A}$ is less than $\sqrt{2}U$, where U is the width of the corridors. A free path nevertheless exists if $\mathcal{A}$'s length is comprised between $\sqrt{2}U$ and $2\sqrt{2}U$.

Some of the limitations of the freeway method could possibly be alleviated. For example, it is possible to construct additional freeways by introducing fictitious edges in the workspace and/or by considering spines

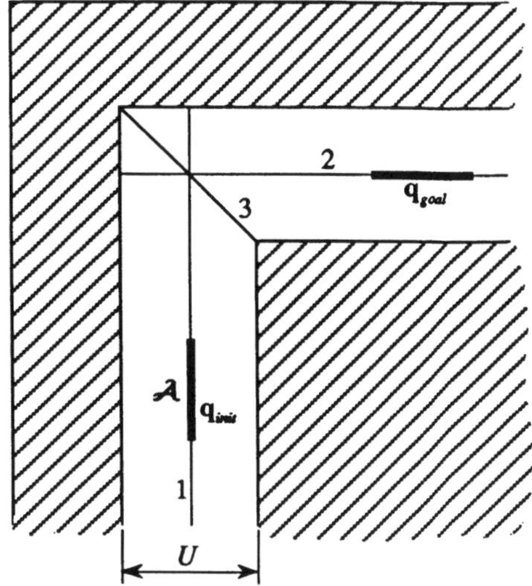

Figure 21. Only three freeways can be extracted from this workspace subset made of two corridors and a corner. The robot $\mathcal{A}$ is a segment and $O_{\mathcal{A}}$ is at the midpoint of this segment. The freeway method succeeds in planning a path for $\mathcal{A}$ only if its length is less than $\sqrt{2}U$, where U is the width of the two corridors, while a free path actually exists as long as the length of the robot is smaller than $2\sqrt{2}U$.

which are not the bisectors of lines supporting obstacle edges. But, the size of the freeway net and the cost of searching it would eventually be significantly increased. Perhaps it is better to take the freeway method as is, i.e. relatively fast but incomplete, and not try to make it more complicated, slower and still incomplete.

4 Silhouette Method

In this section we sketch the principle of a general roadmap method, the *silhouette* method, developed by Canny [Canny, 1988]. This method generates semi-free paths, i.e. paths in $cl(\mathcal{C}_{free})$. It is complete, i.e. guaranteed to find a semi-free path if one exists and to return failure otherwise. The closure of free space is input as a compact semi-algebraic set.

The silhouette method can be used to solve the basic motion planning

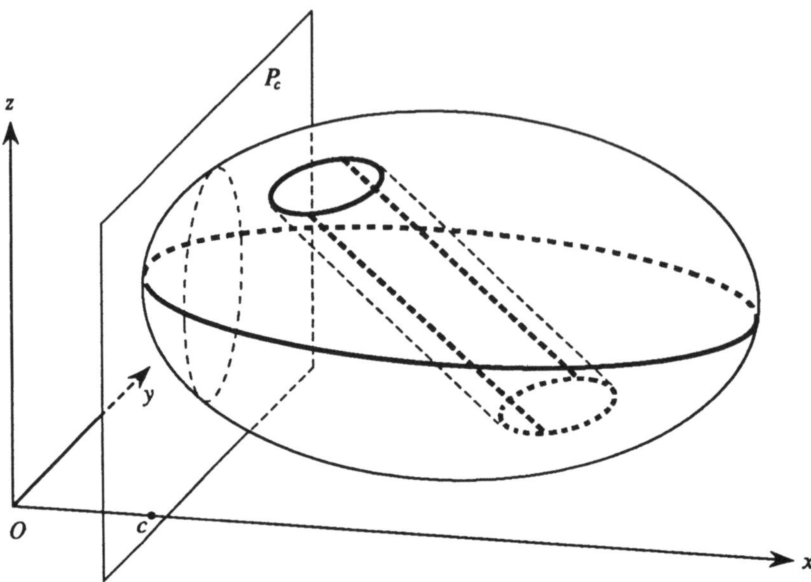

Figure 22. This figure shows the three-dimensional set $\mathcal{S} = cl(\mathcal{C}_{free})$ used to illustrate the silhouette method. It is an ellipsoid pierced through by a cylindrical hole. The set is described in a coordinate system $Oxyz$. The plane P_c is the sweep-plane at $x = c$. This plane is swept along the x-axis to construct the roadmap. The silhouette curves are the curves shown in bold, plain and dashed lines which are traced out by the extremal points of $\mathcal{S} \cap P_c$ in the y-direction during the sweep process.

problem with an appropriate parameterization of the configuration space $\mathcal{C}$. However, the method does not require $\mathcal{C}$ to be of the form $\mathbf{R}^N \times SO(N)$. As a matter of fact, it is applicable to any problem that can be reduced to planning a path in a compact semi-algebraic set. This includes problems involving multiple robots and/or articulated robots (see Chapter 8). It is the only known complete path planning algorithm which runs in single-exponential time in $\mathcal{C}$'s dimension. Involved tools from Differential Geometry ("stratifications") and Elimination Theory ("multivariate resultant") are used by the method to achieve this single exponential bound. Here we only outline the general principle of the method independently of these tools. We illustrate this principle with an example drawn from [Canny, 1988].

Let $\mathcal{S} = cl(\mathcal{C}_{free})$ be a compact subset of $\mathbf{R}^m$, where m is the dimension of $\mathcal{C}$. For simplification, we assume that $\mathcal{C}$ is covered by a single chart. In the example, $m = 3$ and $\mathcal{S}$ is the three-dimensional set (an ellipsoid pierced through by a cylindrical hole) depicted in Figure 22.

The first step of the silhouette method is to construct a roadmap $\mathcal{R}$ of $\mathcal{S}$. It is applied recursively to sets of decreasing dimensions. At the recursion level zero, a hyperplane of dimension $m - 1$ is swept across $\mathcal{S}$. This plane is taken perpendicular to one of the axes of the coordinate system in which $\mathcal{S}$ is represented, say the x-axis (this choice is arbitrary), and it is translated along this axis. At every position $x = c$, the plane is denoted by P_c. In the example, P_c is a two-dimensional plane as shown in Figure 22.

We choose arbitrarily a direction in the sweep-plane, say the y-axis. $\mathcal{R}$ contains all the locally extremal points (maxima, minima and inflection points) of $\mathcal{S} \cap P_c$ in this direction. Since $\mathcal{S}$ is compact, it is guaranteed that such an extremal point exists in every connected component of $\mathcal{S} \cap P_c$. Hence, any point in $\mathcal{S} \cap P_c$ may be connected to $\mathcal{R}$ by a path lying in $\mathcal{S} \cap P_c$. We assume that the directions have been carefully chosen so that there is a finite number of extremal points in every slice (technical "details" of the method not described here deal with this issue).

When c varies, i.e. when the plane is swept across $\mathcal{S}$, the extremal points trace out piecewise algebraic curves called **silhouette curves**. These curves are contained in the boundary of $\mathcal{S}$, as shown in Figure 22. When $m = 3$, these curves are the silhouette contours (hence their name) that a viewer would see from a point at infinity in the z direction if $\mathcal{S}$ was made of a translucent material. In the example, they consist of two connected components. One is the external silhouette of the ellipsoid, i.e. the set of points where the tangent plane to the ellipsoid is parallel to the z-axis. The other consists of the two closed-loop curves at the ends of the cylindrical hole connected by two straight lines parallel to the cylinder's axis.

The silhouette curves extracted above form only a subset of the roadmap $\mathcal{R}$. We want the final roadmap $\mathcal{R}$ to be connected in every connected component of $\mathcal{S}$ so that path planning can be reduced to searching $\mathcal{R}$. Obviously, the curves extracted so far may not satisfy this requirement, as illustrated in Figure 22. We must add some additional curves.

The above silhouette curves are constructed during the sweep process.

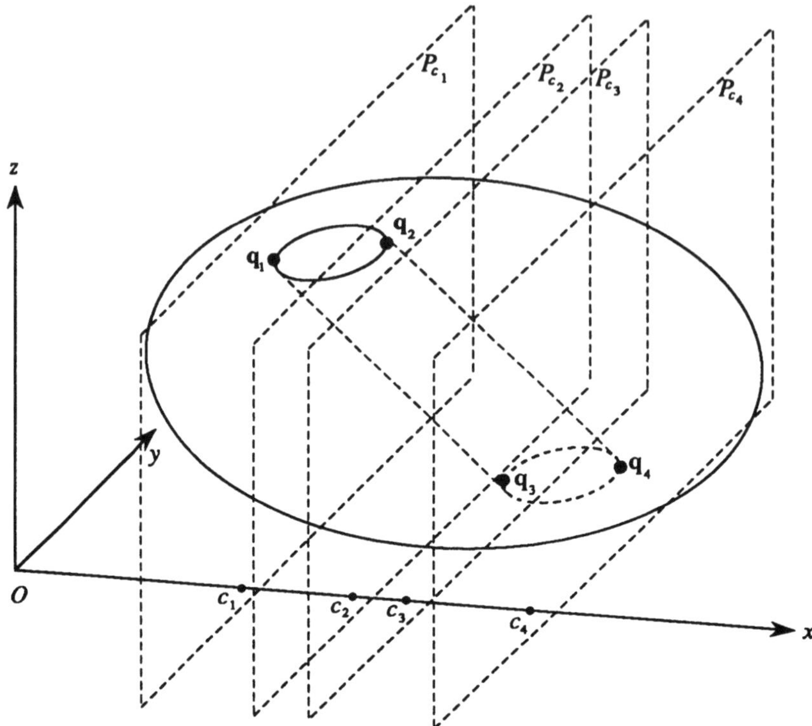

Figure 23. During plane sweeping, new extremals may appear. Such extremals are called critical points. This figure shows the critical points (and the associated critical abscissae) in the set $\mathcal{S}$ of Figure 22, when a plane is swept from left to right along the x-axis with the extremal points extracted in the y-axis direction.

The process starts at some abscissa $c = c_0$ such that $\mathcal{S} \subset \{x > c_0\}$. From there, c is continuously increased. For most values of c, if the curves traced out in $x \leq c$ were connected, they would remain connected when c is slightly increased to $c+\delta c$. The connectivity of the curves may change only at a finite number of values of c, called **critical values**, where new extrema appear. These new extrema are called **critical points**. In the example, there are 4 such critical values denoted by c_1 through c_4 (in addition to the first critical value occurring when the sweep-plane starts intersecting with $\mathcal{S}$). They are shown in Figure 23.

When it passes through a critical point $\mathbf{q}_i$ at $c = c_i$, the sweep process

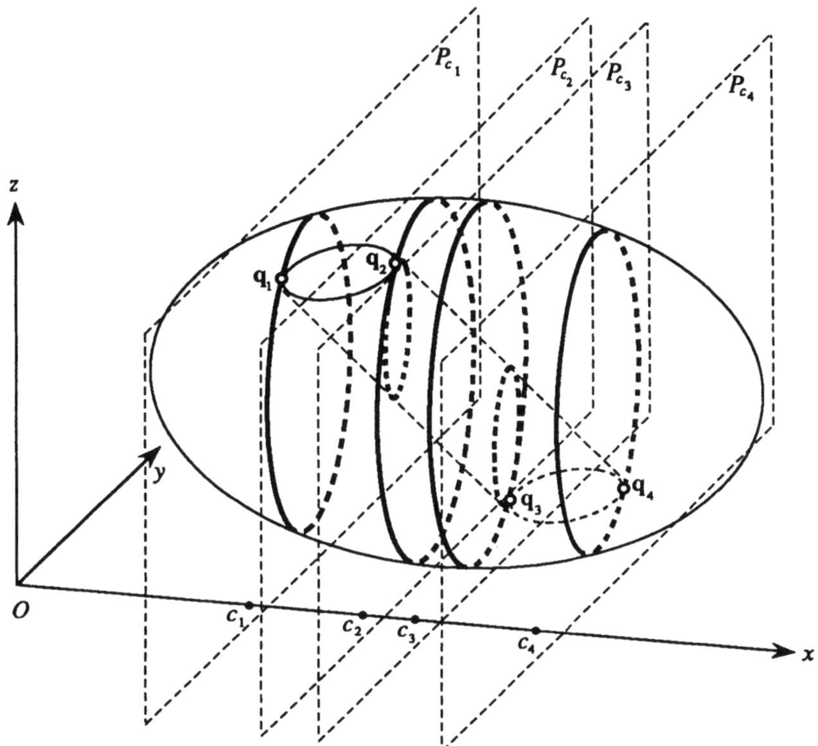

Figure 24. This figure shows the curves (bold lines) built by the recursive calls to the silhouette algorithm when the sweep-plane passes through each of the critical points shown in Figure 23.

connects[7] $\mathbf{q}_i$ to the rest of the silhouette curves by a set of curves contained in $\mathcal{S} \cap P_{c_i}$. This is always possible since any connected component of $\mathcal{S} \cap P_{c_i}$ contains at least one point of the silhouette curves. The way it is done is by calling the silhouette algorithm recursively on the $(m-1)$-dimensional set $\mathcal{S} \cap P_{c_i}$. This recursive call causes the method to sweep another hyperplane of dimension $m-2$ perpendicularly to, say, the y-axis across $\mathcal{S} \cap P_{c_i}$ and to extract the silhouette curves traced out by the extremal points of $\mathcal{S} \cap P_{c_i}$ in a third direction, say the z-axis direction. The new extracted silhouette curves are included in the roadmap. Figure 24 shows the silhouette curves built in the planes P_{c_i}, $i = 2, ..., 5$, in our three-dimensional example. Since each P_{c_i} is a two-dimensional

[7]We assume that there is a unique critical point at every critical value of c.

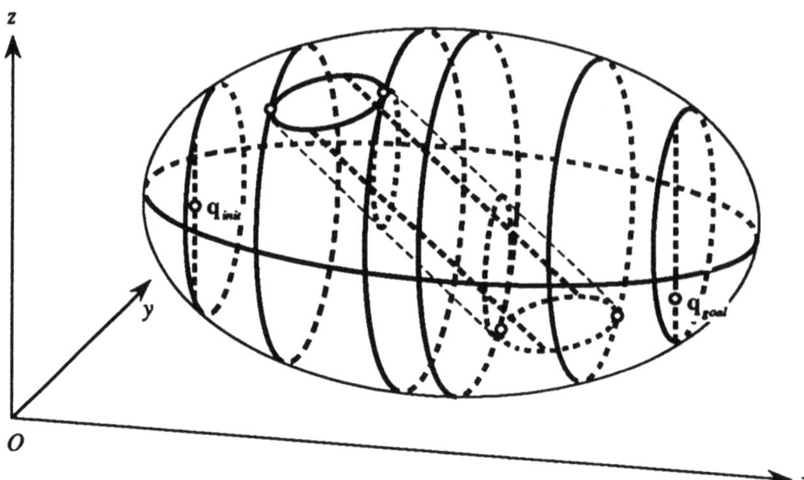

Figure 25. This figure shows in bold lines the complete roadmap built for the example of Figure 22 and the links connecting the initial and the goal configurations to this roadmap.

plane, the construction of the curves is done by sweeping a line parallel to the z-axis along the y-axis. The total set of curves generated by the algorithm after all the recursive calls have returned is the roadmap built by the silhouette method.

Therefore, the algorithm recursively reduces the problem of constructing a roadmap in $\mathcal{S}$ to several subproblems of constructing roadmaps in sets of dimension $m - 1$. The recursion ends when there is no critical point to connect to the silhouette curves in the set that is currently swept out, or when this set has dimension two. Indeed, in dimension two, if the sweep-line crosses a critical point, this point can simply be linked to the silhouette curves by straight segments. This is equivalent to saying that the roadmap of a one-dimensional set is this set itself.

If the recursive calls on the $(m-1)$-dimensional sets $\mathcal{S} \cap P_{c_i}$ correctly connect the critical points to the current roadmap, then the final roadmap $\mathcal{R}$ is connected in each connected component of $\mathcal{S}$. It is easy to see that in dimension two, the algorithm works correctly. Hence, by induction on dimension, the algorithm correctly computes $\mathcal{R}$.

It only remains to connect the initial configuration $\mathbf{q}_{init}$ and the goal configuration $\mathbf{q}_{goal}$ to the roadmap. This can simply be done by treating these configurations as critical points and calling the algorithm recursively when the sweep-plane passes through them. Figure 25 shows the complete roadmap $\mathcal{R}$ for the set $\mathcal{S}$ of Figure 22 and the links connecting two configurations $\mathbf{q}_{init}$ and $\mathbf{q}_{goal}$ to it.

The roadmap can be represented as a graph whose links are algebraic curve segments and whose nodes are the endpoints of these segments. The second step of the silhouette method is to search this graph for a path connecting $\mathbf{q}_{init}$ to $\mathbf{q}_{goal}$.

Exercises

1: Let the C-obstacle region $\mathcal{CB}$ consist of c disjoint convex polygons. Let n be the number of vertices of $\mathcal{CB}$. What is the worst-case number of links in the reduced visibility graph?

2: Implement the visibility graph method for a polygonal configuration space. (Use the sweep-line algorithm to construct the links of the reduced visibility graph.)

3: Consider a two-dimensional configuration space with polygonal C-obstacles. Assume that $cl(\mathcal{C}_{free})$ is a manifold with boundary. Propose a simple method for transforming a semi-free path generated by the visibility graph method into a free path.

4: Let C_1 and C_2 be two circular arcs of radius ρ_1 and ρ_2, respectively. Show that the Minkowski sum $C_1 \oplus C_2$ is a circular arc. What is its radius? Prove that the Minkowski sum of two convex generalized polygons is a convex generalized polygon.

5: Show that the Voronoi Diagram of an open bounded region whose boundary is made of straight and circular edges is a network of curves made of straight and parabolic curve segments.

6: Let $\mathcal{C}_{free}$ be the interior of a bounded polygonal region. Let $near(\mathbf{q})$ be the set of points of $\partial\mathcal{C}_{free}$ that achieve minimal distance with $\mathbf{q} \in \mathcal{C}_{free}$. Let Z be a vertex or an open edge of $\partial\mathcal{C}_{free}$; the set $V(Z)$ of points in $\mathcal{C}_{free}$ that are closer to Z than to any other vertex or edge of $\partial\mathcal{C}_{free}$ is called a *Voronoi cell.*

a. Show that $V(Z)$ is an open connected subset of $\mathcal{C}_{free}$ bounded by arcs of the Voronoi diagram and straight segments joining nodes of the diagram of degree 2 to their respective closest points in $\partial\mathcal{C}_{free}$.

b. Let $\mathbf{q}$ be any point in $V(Z)$. Show that the segment that joins $\mathbf{q}$ to its closest point in Z (Z itself if it is a vertex) is entirely contained in $V(Z)$.

c. Let $\mathbf{q}_1$ and $\mathbf{q}_2$ be two configurations in $V(Z)$. Show that the two line segments joining $\mathbf{q}_1$ and $\mathbf{q}_2$ to their respective closest points in Z do not intersect, except possibly at their endpoint in Z, or else if one segment is contained in the other.

7: Derive from the previous exercise (question a) the fact that $Vor(\mathcal{C}_{free})$ has $O(n)$ arcs, where n is the number of vertices of $\partial\mathcal{C}_{free}$. Design an $O(n^2)$ time algorithm for computing $Vor(\mathcal{C}_{free})$.

8: Prove Proposition 4. [Hint: Consider a sequence $\{\mathbf{q}_i\}_{i=1,2,\ldots}$ of configurations in $\mathcal{C}_{free}$ that converges toward $\mathbf{q} \in \mathcal{C}_{free}$. Show that $\lim_{i\to\infty} \rho(\mathbf{q}_i) = \rho(\mathbf{q})$ (by decomposing the sequence into several subsequences).]

9: Implement a retraction-based path planner for a point in a bounded workspace with polygonal obstacles. (Use an existing algorithm or design your own to compute the Voronoi diagram. This planner can usefully be combined with the program proposed in Exercise 7.a of Chapter 3 in order to plan paths for a polygonal object $\mathcal{A}$ translating among polygonal obstacles.)

10: Let $\mathcal{E}$ be the interior of a bounded polygonal region. Suppose that the Voronoi diagram of $\mathcal{E}$ has been computed. We now want to plan the path of a disc of given radius between two positions in $\mathcal{E}$ so that the disc remains wholly contained in $\mathcal{E}$. How can the Voronoi diagram of $\mathcal{E}$ be used for planning this path?

11: Compare the advantages and the drawbacks of the visibility graph method and the retraction method (with the Voronoi diagram) for planning the paths of an object that translates at fixed orientation among polygonal obstacles.

12: Design an $O(n^2)$ time algorithm for computing the minimum spanning circle of a polygon with n vertices.

13: Discuss various heuristic functions for searching a freeway net using the A^* algorithm.

14: What are the most interesting properties of the freeway net as a representation of the connectivity of free space?

15: Let $\mathcal{S}$ be a toric three-dimensional volume containing two configurations $\mathbf{q}_{init}$ and $\mathbf{q}_{goal}$. Construct the roadmap $\mathcal{R}$ of $\mathcal{S}$ and the connections of $\mathbf{q}_{init}$ and $\mathbf{q}_{goal}$ to $\mathcal{R}$ that would be computed by the silhouette method.

Chapter 5

Exact Cell Decomposition

In this chapter we describe a second approach to motion planning, *exact cell decomposition.* The principle of this approach is to first decompose the robot's free space $\mathcal{C}_{free}$ into a collection of non-overlapping regions, called *cells*, whose union is exactly[1] $\mathcal{C}_{free}$ (or its closure). Next, the *connectivity graph* which represents the adjacency relation among the cells is constructed and searched. If successful, the outcome of the search is a sequence of cells, called a *channel*, connecting the cell containing the initial configuration to the cell containing the goal configuration. A path is finally extracted from this sequence.

Not all exact cell decompositions of free space are appropriate, however. For example, one extreme view would be to consider the whole set $\mathcal{C}_{free}$ as a single cell, but extracting a path from this cell would bring us back to the original problem. As a matter of fact, the cells generated by the decomposition should have the following two characteristics:

1. The geometry of each cell should be "simple" enough to make it easy

[1] As mentioned in Chapter 1, there are two cell decomposition approaches to motion planning: the exact one and the approximate one. The approximate cell decomposition approach will be described in the next chapter.

to compute a path between any two configurations in the cell.

2. It should not be difficult to test the adjacency of any two cells and to find a path crossing the portion of boundary shared by two adjacent cells.

It follows from these requirements that the boundary of a cell corresponds to a "criticality" of some sort, i.e. something significant which changes when the boundary is crossed. Conversely, cells are "noncritical" regions, such that all paths lying in the same cell between any two configurations are qualitatively equivalent.

We present three planning methods based on the exact cell decomposition approach and a fourth method based on a variant of this approach.

The first method (Section 1) applies to problems where the configuration space $\mathcal{C}$ is two-dimensional and the C-obstacles are polygonal. This allows us to introduce the approach in a very simple fashion. We propose a planning method based on the "trapezoidal decomposition" of free space (this kind of decomposition is used in other places of the book).

The second method (Section 2) is a method due to Schwartz and Sharir [Schwartz and Sharir, 1983a]. It plans the motion of a robot modeled as a straight segment (sometimes called a "ladder") translating and rotating among polygonal obstacles. In this case, $\mathcal{C} = \mathbf{R}^2 \times S^1$ and the C-obstacles are three-dimensional regions bounded by ruled surfaces. The method presented is not the most efficient, but it is a good illustration of the exact cell decomposition approach and it is conceptually rather simple.

The third planning method (Section 3) is the general path planning method due to Schwartz and Sharir [Schwartz and Sharir, 1983b] which is based on the Collins decomposition of the robot's free space. In theory, this method can plan the paths of a point in any smooth semi-algebraic manifold. Hence, it is applicable not only to the basic motion planning problem in its most general form, but also to various extensions of this problem. However, quoting their authors, the method is also "catastrophically inefficient". It mainly serves as a proof of existence of a general path planning method. It also illustrates the application of decision methods in the theory of the reals to robot motion planning.

Finally, in Section 4 we describe a method based on a variant of the exact cell decomposition approach. This method, due to Avnaim, Boissonnat and Faverjon [Avnaim, Boissonnat and Faverjon, 1988], plans the paths of a (possibly non-convex) polygon $\mathcal{A}$ that can both translate and rotate among polygonal obstacles. Rather than decomposing free space into three-dimensional cells, this method only decomposes $\mathcal{C}_{free}$'s boundary (and some well-chosen subsets of $\mathcal{C}_{free}$) into two-dimensional cells. The method is relatively simple to implement and experiments with it showed that it is also quite efficient.

Notice that the connectivity graph representing the adjacency relationship among cells resulting from the decomposition of $\mathcal{C}_{free}$ implicitly determines a network of one-dimensional curves. These curves could be obtained by selecting an arbitrary "sample point" in each cell and joining by a path the sample point of every cell κ to the sample point of every cell κ' adjacent to κ. This construction relates the exact cell decomposition approach to the roadmap approach of the previous chapter[2]. However, the explicit construction of a roadmap from the connectivity graph does not seem to bring anything useful. Furthermore, the notion of a channel is interesting; it is less constraining than a path and may provide the robot with useful information for dealing with dynamic constraints and unexpected obstacles at execution time.

1 Polygonal Configuration Space

In this section we introduce the exact cell decomposition approach in the simple case where $\mathcal{C} = \mathbf{R}^2$ and the C-obstacle region $\mathcal{CB}$ (the union of the C-obstacles) forms a polygonal region in $\mathcal{C}$. For simplifying the presentation, we assume that the robot's free space $\mathcal{C}_{free} = \mathcal{C} \backslash \mathcal{CB}$ is bounded. Figure 1 depicts such a configuration space. The assumption that $\mathcal{C}_{free}$ is bounded is not actually needed and will be retracted later.

We define the decomposition of $\mathcal{C}_{free}$ and the associated connectivity graph as follows:

DEFINITION 1: *A* **convex polygonal decomposition** $\mathcal{K}$ *of* $\mathcal{C}_{free}$ *is a finite collection of convex polygons, called* cells, *such that the interiors*

[2]Yap's proof of the existence of a connectivity-preserving retraction-like application of any semi-algebraic manifold $\mathcal{C}_{free}$ onto an algebraic one-dimensional subset $\mathcal{R} \subset \mathcal{C}_{free}$ (see Subsection 2.3 of Chapter 4) makes use of such a construct [Yap, 1987a].

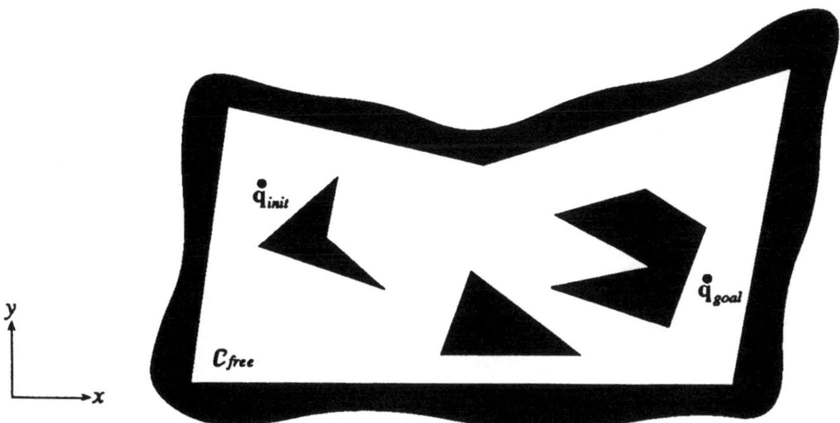

Figure 1. This figure depicts the two-dimensional configuration space used to illustrate the exact cell decomposition method presented in Section 1.

of any two cells do not intersect and the union of all the cells is equal to $cl(C_{free})$. Two cells κ and κ' in $\mathcal{K}$ are **adjacent** *if only if $\kappa \cap \kappa'$ is a line segment of non-zero length.*

DEFINITION 2: *The* **connectivity graph** *G associated with a convex polygonal decomposition $\mathcal{K}$ of C_{free} is the non-directed graph specified as follows:*

- *G's nodes are the cells in $\mathcal{K}$.*
- *Two nodes in G are connected by a link if and only if the corresponding cells are adjacent.*

Figure 2 shows a convex polygonal decomposition of the robot's free space shown in Figure 1 and the connectivity graph associated with this decomposition.

Consider an initial configuration $\mathbf{q}_{init}$ and a goal configuration $\mathbf{q}_{goal}$ in C_{free}. The exact cell decomposition algorithm for planning a free path connecting the two configurations is the following:

1. Generate a convex polygonal decomposition $\mathcal{K}$ of C_{free}.

2. Construct the connectivity graph G associated with $\mathcal{K}$.

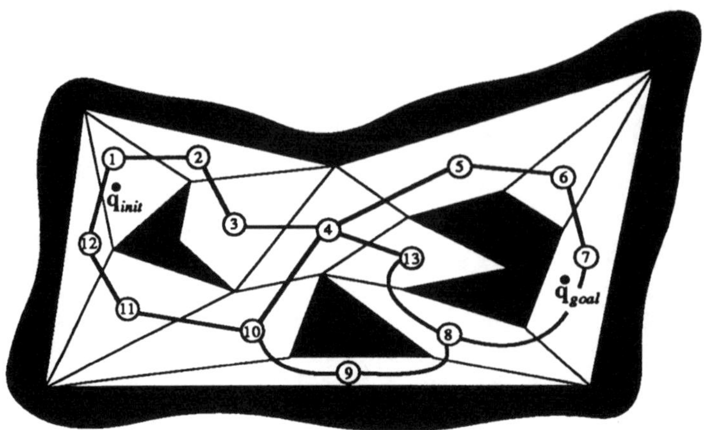

Figure 2. Free space has been decomposed into a collection of convex polygonal cells whose interiors do not intersect. Each cell is labeled by a distinct integer and the associated connectivity graph is displayed (bold lines) with each node placed in the corresponding cell.

3. Search G for a sequence of adjacent cells between $\mathbf{q}_{init}$ and $\mathbf{q}_{goal}$.

4. If the search terminates successfully, return the generated sequence of cells; otherwise, return failure.

The output of the algorithm is a sequence $\kappa_1, \ldots, \kappa_p$ of cells such that $\mathbf{q}_{init} \in \kappa_1$, $\mathbf{q}_{goal} \in \kappa_p$ and for every $j \in [1, p-1]$, κ_j and κ_{j+1} are adjacent. This sequence is called a **channel**. For instance, in Figure 2, $\mathbf{q}_{init}$ is in cell 1 and $\mathbf{q}_{goal}$ is in cell 7; a possible channel is the sequence of cells labeled as 1, 2, 3, 4, 13, 8, and 7.

The interior of a channel, i.e. :

$$int\left(\bigcup_{j=1}^{p} \kappa_j\right)$$

lies entirely in free space. Let $\beta_j = \partial\kappa_j \cap \partial\kappa_{j+1}$, where $\partial\kappa_j$ (resp. $\partial\kappa_{j+1}$) denotes the boundary of the cell κ_j (resp. κ_{j+1}). One simple way to generate a free path contained in the interior of the channel produced by the search of G is to consider the midpoint Q_j of every segment β_j, and

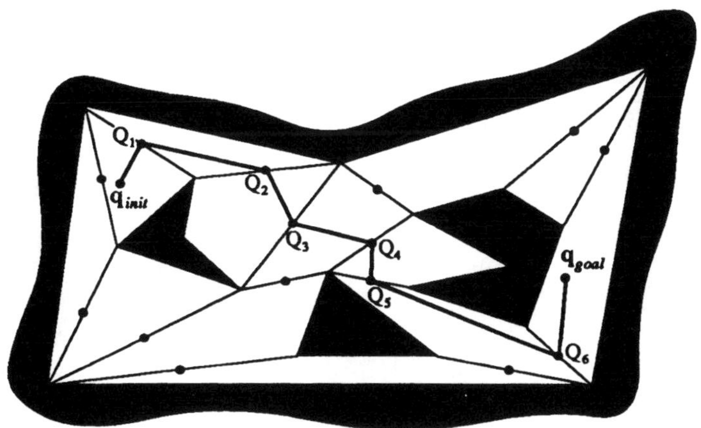

Figure 3. Once a channel has been constructed, a free path can easily be generated as the polygonal line joining $\mathbf{q}_{init}$ to $\mathbf{q}_{goal}$ through the midpoints of the portions of boundary shared by every two successive cells in the channel.

to connect $\mathbf{q}_{init}$ to $\mathbf{q}_{goal}$ by a polygonal line whose successive vertices are $Q_1, ..., Q_{p-1}$. This construction[3] is illustrated in Figure 3. If needed, a smooth path can be obtained by applying spline curve fitting techniques (e.g. see [Kant and Zucker, 1986b]).

As mentioned in Subsection 1.3 of Chapter 3, the optimal convex decomposition of a polygon is computable in time polynomial in its number n of vertices. Unfortunately, the presence of holes in the polygon makes this problem NP-hard [Lingas, 1982]. Nevertheless, a non-optimal decomposition can be generated rather efficiently as shown below.

A non-optimal convex polygonal decomposition of $\mathcal{C}_{free}$ is generated by sweeping a line L parallel to the, say, y-axis across $\mathcal{C}_{free}$. The sweep process is interrupted whenever L encounters a vertex X of $\mathcal{CB}$. A maximum of two vertical line segments are created, connecting X to the edges of $\mathcal{CB}$ that are immediately above and immediately below X, as shown in Figure 4. The boundary of $\mathcal{CB}$ and the erected vertical line segments determine a **trapezoidal decomposition** of $\mathcal{C}_{free}$ (also called a "vertical decomposition" [Chazelle, 1987]). Each cell of the decomposition is

[3] Actually, one additional via point Q'_j must be introduced in every cell κ_j whenever the segment $Q_{j-1}Q_j$ lies in $\partial\kappa_j$.

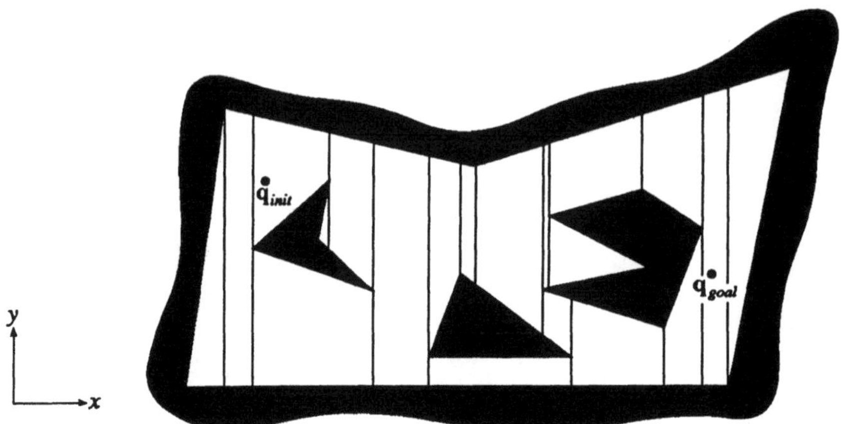

Figure 4. This figure illustrates the trapezoidal decomposition of free space. A vertical line L parallel to the y-axis is swept across C_{free} from left to right. At every vertex X of $\mathcal{CB}$, a maximum of two vertical line segments are erected. They connect X to the edges of $\mathcal{CB}$ immediately above and below X.

either a trapezoid or a triangle. Two cells are adjacent if and only if their boundaries share a vertical segment. When such a segment is crossed the vertical structure of the constraints imposed by $\mathcal{CB}$ on the motion of $\mathcal{A}$ changes discontinuously. Notice that the same algorithm also applies when C_{free} is not bounded; the generated decomposition then includes cells that extend infinitely in the y-axis direction.

During line sweeping, it is possible to concurrently create the vertical line segments emanating from $\mathcal{CB}$'s vertices, generate the connectivity graph G, and identify the cells containing the initial and goal configurations. The whole treatment takes $O(n \log n)$ time, where n is the total number of vertices in $\mathcal{CB}$. The number of generated cells and the number of links in G are both in $O(n)$. Notice that the decomposition of C_{free} can also be generated during the sweep process used to construct $\mathcal{CB}$ (see Subsection 1.6 of Chapter 3).

Figure 6 in Chapter 1 illustrates the path planning method described above, with the trapezoidal decomposition technique.

The search of G can be done in various ways. A simple breadth-first search takes $O(n)$ time. But the produced channel may be far from optimal with respect to the Euclidean lengths of the paths it contains. One possible alternative is to use the A^* search algorithm, with a different search graph G' specified as follows:

- The nodes of G' are $\mathbf{q}_{init}$, $\mathbf{q}_{goal}$, and the midpoints Q_j of all the portions of boundary shared by adjacent cells.
- Two nodes are connected by a link if and only if they belong to the same cell (they can be joined by a straight line segment in the cell).

Each link is weighted by the Euclidean length of the straight line segment joining the two nodes. A^* can be applied with the Euclidean distance to the goal as the heuristic function. If a free path exists, the search produces the shortest free path contained in G'. It also determines a sequence of cells from which another path can be extracted (e.g. a smooth path). The search takes $O(n \log n)$ time. From a practical point of view, the paths generated using this search method are usually significantly better than those generated using the visibility graph method. Indeed, they avoid obstacles more widely, while not being significantly longer.

The above method can be extended to the case where $\mathcal{C} = \mathbf{R}^3$, with polyhedral C-obstacles. One can use a sweep-plane algorithm to construct a decomposition of $\mathcal{C}_{free}$ into parallelepipedic cells. Two cells are adjacent if and only if two of their faces share a trapezoid of non-zero area. In order to apply the A^* algorithm, one may build a search graph G' defined as above, with every point Q_j selected in the interior of the trapezoidal intersection of the boundaries of two adjacent cells.

2 Translation and Rotation in the Plane

We now illustrate the exact cell decomposition approach in a more difficult, but still particular case. The robot $\mathcal{A}$ is modeled as a line segment, called a *ladder*[4], that can both translate and rotate in a two-dimensional workspace $\mathcal{W} = \mathbf{R}^2$. $\mathcal{W}$ is populated by obstacles whose union forms a polygonal region denoted by $\mathcal{B}$. We assume that $\mathcal{B}$ is a manifold with boundary.

$\mathcal{A}$'s endpoints are denoted by P and Q, respectively. The configuration space $\mathcal{C}$ of $\mathcal{A}$ is $\mathbf{R}^2 \times S^1$. We parameterize each configuration by (x, y, θ),

[4]Such a robot is also called a *rod* or a *bar* in the literature.

where x and y are the coordinates of P in the coordinate frame $\mathcal{F}_\mathcal{W}$ attached to $\mathcal{W}$, and $\theta \in [0, 2\pi)$ is the angle (mod 2π) between the x-axis of $\mathcal{F}_\mathcal{W}$ and the vector $\vec{PQ}$. The pair (x, y) of coordinates of P is called the *position* of $\mathcal{A}$ and the angle θ is called the *orientation* of $\mathcal{A}$. $\mathcal{A}(x, y, \theta)$ denotes $\mathcal{A}$ at position (x, y) and orientation θ. We use the same copy of $\mathbf{R}^2$ to represent both the workspace $\mathcal{W}$ and the xy-plane of $\mathcal{C}$. The length of the segment $\mathcal{A}$ is d.

The method presented below is due[5] to Schwartz and Sharir [Schwartz and Sharir, 1983a]. The basic idea is to decompose the set of positions of $\mathcal{A}$ into two-dimensional *noncritical regions*, to "lift" these regions into three-dimensional *cells*, and to represent the adjacency relation among the cells in a *connectivity graph*. The noncritical regions are such that the C-obstacles maintain a constant "structure" in the cylinders above these regions. The cells contained in such a cylinder make this structure explicit. The boundaries of the noncritical regions are called *critical curves*.

The method described below is a good illustration of the exact cell decomposition approach and it is conceptually rather simple. But it is not the only one, nor the most efficient. Other exact cell decomposition methods have been proposed for planning the motion of a ladder or a polygon, with lower asymptotic time complexity bounds. For example, the method described by Leven and Sharir [Leven and Sharir, 1985a] only requires $O(n^2 \log n)$ time, where n is the complexity (number of edges and vertices) of the obstacle region $\mathcal{B}$, while the method studied below requires $O(n^5)$ time. However, most of these methods are quite involved and difficult to implement. In Section 4 we will describe a method based on a variant of the exact cell decomposition approach, which plans the paths of a (possibly non-convex) polygon moving freely among polygonal obstacles. This method is relatively efficient and easy to implement.

2.1 Critical Curves and Noncritical Regions

The C-obstacle region $\mathcal{CB}$ is a three-dimensional region in $\mathbf{R}^2 \times [0, 2\pi)$ which is bounded by faces which are portions of ruled C-surfaces of types A and B. A type A face corresponds to the situation when the segment PQ contains an obstacle vertex. A type B face corresponds to

[5] A similar method was independently presented in [Chien, Zhang and Zhang, 1984].

the situation where either P or Q is contained in an obstacle edge.

The **critical curves** are defined as the projections of the following curves on the xy-plane:

- the boundaries of the faces of the C-obstacle region $\mathcal{CB}$, and
- the curves in $\mathcal{CB}$'s faces where the tangent plane to $\mathcal{CB}$ is perpendicular to the xy-plane.

These curves are the set of points where the structure of the C-obstacle region above the xy-plane undergoes a qualitative change. Indeed, when such a curve is crossed, either the set of $\mathcal{CB}$'s faces which are intersected by a line perpendicular to the xy-plane at the current position changes, or the number of intersection points changes.

Let us illustrate the notion of critical curve on a simple example. Assume first that there is no obstacle in $\mathcal{W}$. Let $\mathcal{A}$ be at position (x,y). Any orientation θ is free, i.e. $(x,y,\theta) \in \mathcal{C}_{free}$. Now, bring an obstacle into the workspace. Let this obstacle be a closed half-plane bounded by a straight line E (the edge of the obstacle). A configuration (x,y,θ) of $\mathcal{A}$ may be in free space only if (x,y) is not in the obstacle. Thus, the edge E is one critical curve: in the inner half-plane of E no orientation is free, while in the outer half-plane of E some orientations are free. Now, let us assume that the distance of P to E is greater than d (the length of $\mathcal{A}$). All the orientations of $\mathcal{A}$ are free. If P gets closer to E and crosses the line K parallel to E at distance d, some orientations of $\mathcal{A}$ are no longer free. K is another critical curve. For any (x,y) on one side of K, the line perpendicular to the xy-plane at (x,y) traverses no C-obstacle, while on the other side it traverses one. Notice that in this particular example there are two type B faces corresponding to the contact situations where P and Q are in the edge E, respectively. Both E and K are the projection of curves where the tangent plane to these faces is perpendicular to the xy-plane.

The critical curves can be constructed from the representation of the boundary of $\mathcal{CB}$ computed as in Subsection 1.7 of Chapter 3. They can also be constructed as all the edges bounding the obstacles[6] and all the curves listed below. Obstacle edges are said to be critical curves of **type**

[6] We use the same copy of $\mathbf{R}^2$ to represent both $\mathcal{W}$ and the xy-plane of $\mathcal{C}$.

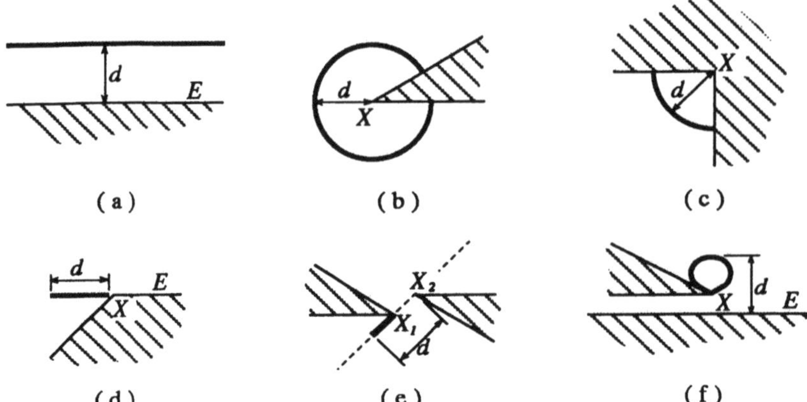

Figure 5. This figure illustrates the various types of critical curves other than the obstacle edges. The critical curves (shown in bold lines) are the set of positions of $\mathcal{A}$ where the structure of the C-obstacle region along the θ direction undergoes a qualitative change.

0. The other critical curves are classified as types 1 through 5 (Figure 5):

- Let E be an obstacle edge. The line segment at distance d from E is a critical curve of **type 1** (Figure 5.a). It has the same length as E.

- Let X be an obstacle vertex. The circular arc of radius d, centered at X and bounded by the two half-lines containing the edges abutting at X is a critical curve of **Type 2** (Figures 5.b and 5.c).

- Let E be an obstacle edge and X be a convex vertex at one endpoint of E. The line segment traced by P, as $\mathcal{A}$ slides in a way such that PQ is contained in the line containing E, and Q is in E, is a critical curve of **type 3** (Figure 5.d).

- Let X_1 and X_2 be two convex obstacle vertices such that the line passing through them is tangent to $\mathcal{B}$ at both X_1 and X_2. The line segment traced by P, as $\mathcal{A}$ translates while touching both X_1 and X_2, is a critical curve of **type 4** (Figure 5.e).

- Let E be an obstacle edge and X be a convex obstacle vertex that is not an endpoint of E. Let h be the distance from X to E. If $h < d$

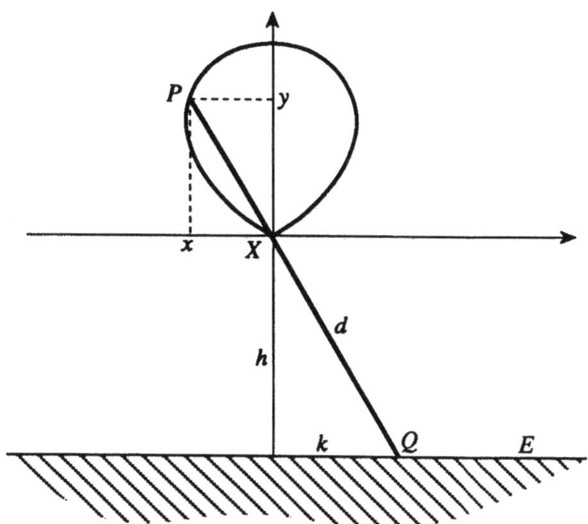

Figure 6. A conchoid of Nicomedes (see text) is an algebraic curve of degree 4. We have: $d^2 = (y+h)^2 + (x+k)^2$ and $\frac{y}{x} = \frac{h}{k}$. Thus: $x^2 = y^2(\frac{d^2}{(y+h)^2} - 1)$.

the curve traced by P, as $\mathcal{A}$ moves in a way such that it touches both E and X while being tangent to $\mathcal{B}$ at X, is a critical curve of **type 5** (Figure 5.f).

A critical curve of type 5 is a subset of a conchoid of Nicomedes. Selecting the coordinates' origin at X, the x-axis parallel to E, and the y-axis pointing along the outgoing normal of E, it is described by the following algebraic equation of degree 4 (see Figure 6):

$$x^2 = y^2 \left(\frac{d^2}{(y+h)^2} - 1 \right)$$

with $y \geq 0$.

Example: Figure 7 shows the critical curves in a particular polygonal workspace[7]. There are 21 critical curves other than the obstacle edges:

- β_1 through β_6 are curves of type 1 generated by the edges E_1 through E_6, respectively.

[7]This example is drawn from [Yap, 1987a].

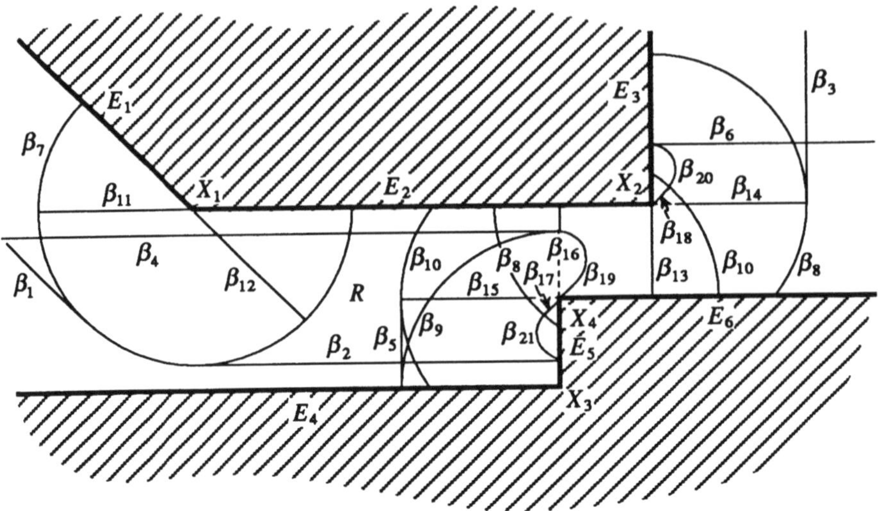

Figure 7. This figure illustrates the concepts of a critical curve and a noncritical region in a polygonal workspace. The boundaries of the obstacles are shown in bold lines. The critical curves of types 1 through 5 are shown in plain thin lines. A subset of the critical curve β_{16} is shown in dashed lines and corresponds to a "redundant" critical curve section. Any open region enclosed by critical curves, e.g. R, is a noncritical region.

- β_7 through β_{10} are curves of type 2 generated by the vertices X_1 through X_4, respectively.

- β_{11} through β_{16} are curves of type 3 generated by E_2 and X_1, E_1 and X_1, E_3 and X_2, E_2 and X_2, E_6 and X_4, and E_5 and X_4, respectively.

- β_{17} and β_{18} are curves of type 4 generated by the vertices X_2 and X_4.

- β_{19} through β_{21} are curves of type 5. β_{19} is generated by the edge E_4 and the vertex X_4, β_{20} by E_6 and X_2, and β_{21} by E_2 and X_4. ■

Remark: The curves defined as above form a superset of the set of critical curves. Indeed, some of them are generated by configurations of $\mathcal{A}$ where it intersects with the interior of $\mathcal{B}$. Such curves are said to be **redundant**. A subset of the critical curve β_{16} in Figure 7 is an example of a redundant curve. Figure 8 shows two redundant curves in another workspace. Redundant curves result from the fact that we

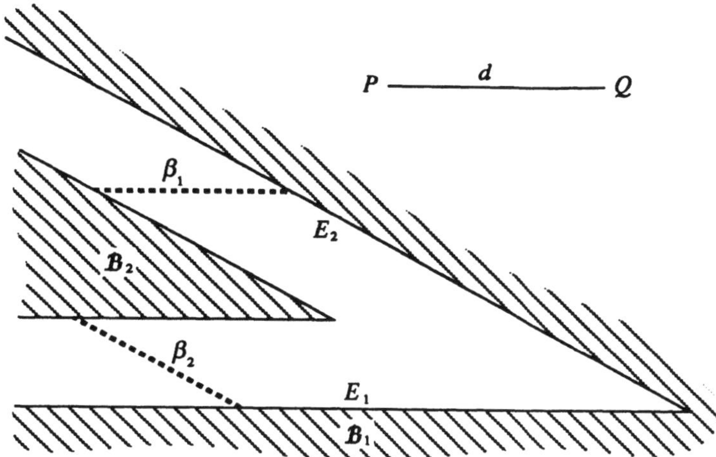

Figure 8. This figure illustrates the notion of a redundant critical curve section in a subset of $\mathcal{W}$ that contains two obstacles $\mathcal{B}_1$ and $\mathcal{B}_2$ forming a "corner". Two redundant critical curves of type 1, β_1 and β_2, are shown in dashed lines. β_1 (resp. β_2) is the locus of P when Q is in E_1 (resp. E_2) and PQ is perpendicular to E_1 (resp. E_2). However, in both cases, PQ intersects with $\mathcal{B}_2$, implying that β_1 and β_2 are the projections of curves contained in the C-obstacle $C\mathcal{B}_2$.

constructed them by considering vertices and edges, and pairs of them, independently of the rest of the obstacle region. They are the projection of curves contained in the interior of the C-obstacle region. We will see that they can easily be removed later. ∎

The set of critical curves is finite. Each endpoint of a critical curve is located either in another critical curve or at infinity. In addition, every critical curve is a smooth algebraic curve of degree 1 (types 0, 1, 3, and 4), 2 (type 2), or 4 (type 5). Thus, every two critical curves can have only finitely many intersections, except if they coincide. Figure 9 shows cases where several critical curves coincide along sections of non-zero length. Under the assumption that $\mathcal{B}$ is a manifold with boundary, every coincidence can be eliminated by an infinitesimal change in $\mathcal{B}$ without changing its topology. Therefore, we assume below that there is no such coincidence. A **section** of a critical curve is any maximal closed connected subset of the curve which is intersected by no critical curve,

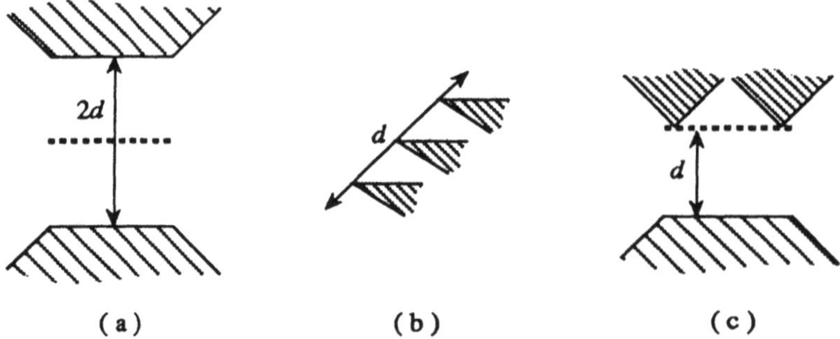

Figure 9. This figure illustrates cases where several critical curves coincide over a section of non-zero length. In our presentation we assume that there is no such coincidence. In theory, a coincidence can be eliminated by changing the position of an obstacle by an arbitrarily small amount.

except at its two endpoints.

A position (x, y) of $\mathcal{A}$ is **admissible** if there exists at least one orientation θ such that $(x, y, \theta) \in \mathcal{C}_{free}$. A **noncritical region** is a maximal subset of admissible positions of $\mathcal{A}$ which intersects no critical curve. The finite set of critical curves thus determines a finite collection of noncritical regions. Every noncritical region is an open subset of the xy-plane. For example, in Figure 7, R denotes a noncritical region bounded by sections of the critical curves designated by β_4, β_7, β_2, β_5, and β_{10}. The boundary of any noncritical region is a finite union of sections of critical curves.

An admissible position (x, y) of $\mathcal{A}$ is **noncritical** if and only if it lies in a noncritical region, otherwise it is **critical.**

2.2 Decomposition of Free Space

Let (x, y) be a noncritical position and $F(x, y)$ be the set of all the free orientations of $\mathcal{A}$ at this position, i.e. $F(x, y) = \{\theta \mid (x, y, \theta) \in \mathcal{C}_{free}\}$. If all the orientations of $\mathcal{A}$ at (x, y) are free, then $F(x, y) = [0, 2\pi)$. Otherwise, $F(x, y)$ consists of a finite number of open maximal connected intervals. Each interval endpoint θ_c is such that $\mathcal{A}(x, y, \theta_c)$ touches $\mathcal{B}$'s boundary; it is called a **limit orientation** of $\mathcal{A}$ at (x, y).

Let us assume that $F(x, y) \neq [0, 2\pi)$. For each maximally connected

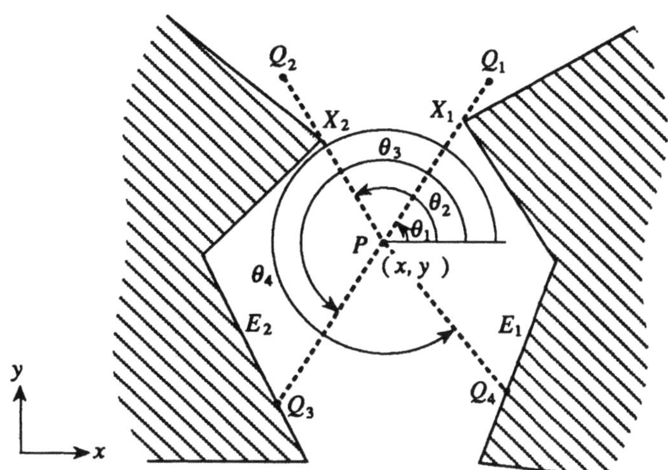

Figure 10. $\mathcal{A}$ is at a noncritical position (x, y). The obstacle edges and vertices it can touch without intersecting the interior of $\mathcal{B}$ when it rotates about P are called stops. A stop is (counter)clockwise if it can be reached from a free configuration by a (counter)clockwise rotation. X_1 and E_2 are clockwise stops, while X_2 and E_1 are counterclockwise stops. The limit orientations corresponding to the stops X_1, X_2, E_2, and E_1 are θ_1, θ_2, θ_3, and θ_4, respectively.

interval $(\theta_1, \theta_2) \subseteq F(x, y)$, let s_1 (resp. s_2) be the vertex or open edge of $\mathcal{B}$ that is touched by $\mathcal{A}(x, y, \theta_1)$ (resp. $\mathcal{A}(x, y, \theta_2)$). Both s_1 and s_2 are uniquely defined, since otherwise the position (x, y) would be a critical one. s_1 (resp. s_2) is called a **clockwise stop** (resp. a **counterclockwise stop**) of $\mathcal{A}$ at (x, y). Clockwise and counterclockwise stops are illustrated in Figure 10.

Let θ_c be a limit orientation at the noncritical position (x, y). We denote by $s(x, y, \theta_c)$ the unique stop touched by $\mathcal{A}(x, y, \theta_c)$ and by $\sigma(x, y)$ the set of all the pairs $[s(x, y, \theta_c), s(x, y, \theta_c')]$ such that $s(x, y, \theta_c)$ (resp. $s(x, y, \theta_c')$) is a clockwise (resp. counterclockwise) stop at (x, y) and the interval $(\theta_c, \theta_c') \subseteq F(x, y)$. In the example of Figure 10, we have:

$$\begin{aligned} s(x, y, \theta_1) &= X_1, \\ s(x, y, \theta_2) &= X_2, \\ s(x, y, \theta_3) &= E_2, \\ s(x, y, \theta_4) &= E_1, \end{aligned}$$

and:

$$\sigma(x,y) = \{[X_1, X_2], [E_2, E_1]\}.$$

If $F(x,y) = [0, 2\pi)$, then we write $\sigma(x,y) = \{[\Omega, \Omega]\}$, where Ω designates a "non-existent" stop. Given a pair $[s_1, s_2] \neq [\Omega, \Omega]$ in $\sigma(x,y)$, we denote by $\lambda_1(x,y,s_1)$ the unique orientation such that $\mathcal{A}(x,y,\lambda_1(x,y,s_1))$ touches the clockwise stop s_1. Similarly, we denote by $\lambda_2(x,y,s_2)$ the unique orientation such that $\mathcal{A}(x,y,\lambda_2(x,y,s_2))$ touches the counterclockwise stop s_2. For example, in Figure 10:

$$\begin{aligned}
\lambda_1(x,y,X_1) &= \theta_1, \\
\lambda_2(x,y,X_2) &= \theta_2, \\
\lambda_1(x,y,E_2) &= \theta_3, \\
\lambda_2(x,y,E_1) &= \theta_4.
\end{aligned}$$

By convention: $\lambda_1(x,y,\Omega) = 0$ and $\lambda_2(x,y,\Omega) = 2\pi$. Thus, if $[s_1, s_2] \in \sigma(x,y)$, $(\lambda_1(x,y,s_1), \lambda_2(x,y,s_2))$ is a maximal connected interval in $F(x,y)$.

Let R be a noncritical region. By construction, we have:

$$\forall (x,y), (x',y') \in R : \sigma(x,y) = \sigma(x',y').$$

We write $\sigma(R) = \sigma(x,y)$, where (x,y) is any position in R.

DEFINITION 3: *Let R be a noncritical region and $[s_1, s_2]$ be a pair in $\sigma(R)$. The three-dimensional region:*

$$cell(R, s_1, s_2) = \{(x,y,\theta) \;/\; (x,y) \in R \text{ and } \theta \in (\lambda_1(x,y,s_1), \lambda_2(x,y,s_2))\}$$

is called a **cell**[8]. *It is an open connected subset of $\mathcal{C}_{free}$.*

For every cell $\kappa = cell(R, s_1, s_2)$ and every position (x,y) in R, we write: $\phi_\kappa(x,y) = \lambda_1(x,y,s_1)$ and $\psi_\kappa(x,y) = \lambda_2(x,y,s_2)$. Thus:

$$(x,y,\theta) \in \kappa \Leftrightarrow \theta \in (\phi_\kappa(x,y), \psi_\kappa(x,y)).$$

$\phi_\kappa(x,y)$ and $\psi_\kappa(x,y)$ are continuous functions $R \to \mathbf{R}$.

[8] In the various sections of this chapter, the word "cell" is given several (though conceptually similar) formal definitions. Within each section, however, the word keeps the same definition.

The set of all the cells defined as above forms a decomposition of C_{free}. The cells are disjoint and the closure of their union is equal to $cl(C_{free})$. The cells belong to "cylinders" that project in the xy-plane onto the noncritical regions.

It is possible that two noncritical regions R and R' whose boundaries share a section β of a critical curve are such that $\sigma(R) = \sigma(R')$. This corresponds to the case where β is redundant.

2.3 Connectivity Graph

At this point, we know how to decompose the free space into open connected cells. We now define the connectivity graph associated with this decomposition.

Let $\kappa = cell(R, s_1, s_2)$ be a cell. We extend by continuity the domain of ϕ_κ and ψ_κ to the closure $cl(R)$ of R.

DEFINITION 4: *Two cells $\kappa = cell(R, s_1, s_2)$ and $\kappa' = cell(R', s_1', s_2')$ are* **adjacent** *if and only if:*

- *the boundaries ∂R and $\partial R'$ of R and R' share a critical curve section β, and*

- $\forall (x, y) \in int(\beta) : (\phi_\kappa(x, y), \psi_\kappa(x, y)) \cap (\phi_{\kappa'}(x, y), \psi_{\kappa'}(x, y)) \neq \emptyset$.

If two cells κ and κ' are adjacent, any configuration in κ can be connected to any configuration in κ' by a free path whose projection on the xy-plane crosses β transversely, with a constant orientation in some neighborhood of the crossing point.

Let R and R' be two noncritical regions such that β is a section of critical curve contained in $\partial R \cap \partial R'$. The **crossing rule** of β is the description of the adjacency relation among the cells projecting on R and those projecting on R'. Below, we first express the crossing rule in two particular cases. Then, we give a general formulation of the rule.

1. Let β be a section of a critical curve of type 1, hence a line segment parallel to an obstacle edge E at distance d. Let R' be the region between β and E, and R be the region lying on the other side of β. The crossing rule is:

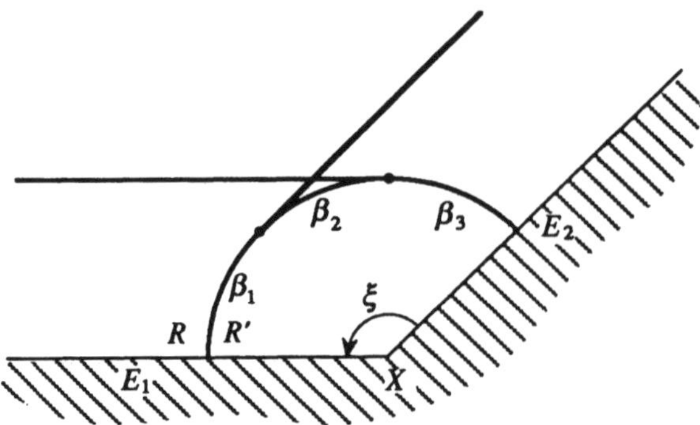

Figure 11. This figure illustrates the notations used to establish the crossing rule of a section of critical curve of type 2.

The two cells $cell(R, s_1, s_2)$ *and* $cell(R', s'_1, s'_2)$ *are adjacent if and only if* $[s'_1, s'_2] = [s_1, s_2]$, *or* $[s'_1, s'_2] = [s_1, E]$, *or* $[s'_1, s'_2] = [E, s_2]$, *or* $[s_1, s_2] = [\Omega, \Omega]$ *and* $[s'_1, s'_2] = [E, E]$.

2. Let β be a section of critical curve of type 2 generated by a concave vertex X at the extremity of two edges E_1 and E_2, such that the outer angle ξ between the two edges is greater than $\frac{\pi}{2}$ (see Figure 11). The critical curve generated by X is a circular arc which intersects the critical curves of type 1 generated by E_1 and E_2. The two intersection points partition the critical curve of type 2 into three sections denoted by β_1, β_2, and β_3, as shown in the figure. We consider the case where $\beta = \beta_1$. Let R' be the region lying on the side of β which is closer than d to X, and R the region lying on the other side of β. The crossing rule is:

   The two cells $cell(R, s_1, s_2)$ *and* $cell(R', s'_1, s'_2)$ *are adjacent if and only if* $[s'_1, s'_2] = [s_1, s_2]$, *or* $s_1 = E_1$, $s'_1 = E_2$ *and* $s_2 = s'_2$.

It is straightforward to analyze in the same fashion all the other cases. Such an analysis shows that whenever $int(\beta)$ is crossed transversely, if β is not a redundant section of critical curve and is not contained in two coincident critical curves, the set $\sigma(x, y)$ is changing in one of the following ways:

- One pair in $\sigma(x, y)$ disappears, or one new pair appears.

- One element in one pair in $\sigma(x, y)$ changes.

If β is a redundant section of a critical curve, then $\sigma(x, y)$ is unchanged when β is crossed.

Thus, the crossing rules for the different types of critical curves can be generalized in the single following rule, which is valid for any critical curve section β, if no two critical curves coincide along β:

Connect $cell(R, s_1, s_2)$ *to* $cell(R', s_1, s_2)$ *for each* $[s_1, s_2] \in \sigma(R) \cap \sigma(R')$, *and connect each* $cell(R, s_1, s_2)$, $[s_1, s_2] \in \sigma(R) \setminus \sigma(R')$, *if any, to each* $cell(R', s'_1, s'_2)$, $[s'_1, s'_2] \in \sigma(R') \setminus \sigma(R)$, *if any.*

Now, we can define and build the connectivity graph:

DEFINITION 5: *The* **connectivity graph** G *is the non-directed graph whose nodes are all the cells* $cell(R, s_1, s_2)$, *where* R *is a non-critical region and* $[s_1, s_2] \in \sigma(R)$. *A link of* G *connects any two nodes if and only if the corresponding cells are adjacent.*

Given two configurations:

$$\begin{aligned} \mathbf{q}_{init} &= (x_{init}, y_{init}, \theta_{init}) \\ \mathbf{q}_{goal} &= (x_{goal}, y_{goal}, \theta_{goal}) \end{aligned}$$

in $\mathcal{C}_{free}$ such that neither (x_{init}, y_{init}) nor (x_{goal}, y_{goal}) lie in a critical curve, there exists a free path between them if and only if the two cells containing $\mathbf{q}_{init}$ and $\mathbf{q}_{goal}$ are connected by a path in the connectivity graph G.

Example: Consider the "corner example" shown in Figure 12. The non-redundant critical curves sections determine 13 noncritical regions denoted by R_1 through R_{13}. We have:

$$\begin{aligned} \sigma(R_1) &= \{[E_1, E_3], [E_3, E_1]\} \\ \sigma(R_2) &= \{[E_1, E_3], [E_3, X_1]\} \\ \sigma(R_3) &= \{[E_1, E_3], [E_3, E_4]\} \\ \sigma(R_4) &= \{[E_1, E_3], [E_3, E_4], [E_4, X_1]\} \\ \sigma(R_5) &= \{[E_1, E_3], [E_4, X_1]\} \\ \sigma(R_6) &= \{[E_1, E_3], [E_4, E_2]\} \end{aligned}$$

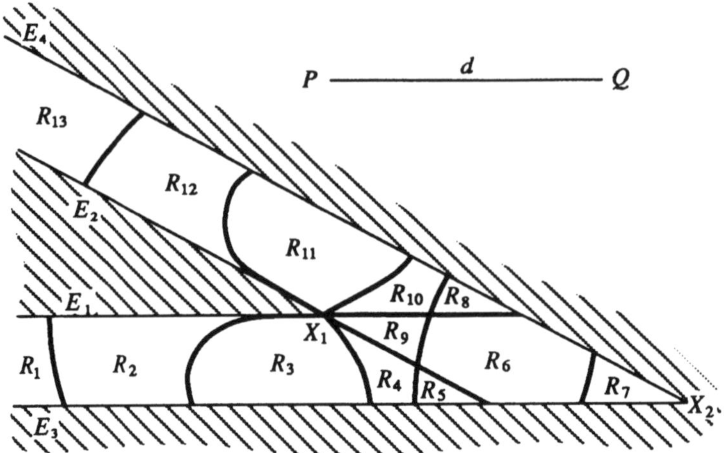

Figure 12. This figure shows the decomposition of a polygonal workspace into 13 noncritical regions. The robot can move from one end of the "corner" to the other, but it cannot make a full rotation in the corner. Thus, when the robot exits from the corner at one end, its orientation is determined by its orientation when it entered the corner at the other end.

$$
\begin{aligned}
\sigma(R_7) &= \{[E_4, E_3]\} \\
\sigma(R_8) &= \{[E_4, E_2], [X_1, E_3]\} \\
\sigma(R_9) &= \{[E_4, E_2], [E_1, E_3], [E_3, E_4]\} \\
\sigma(R_{10}) &= \{[E_4, E_2], [X_1, E_3], [E_3, E_4]\} \\
\sigma(R_{11}) &= \{[E_4, E_2], [E_3, E_4]\} \\
\sigma(R_{12}) &= \{[E_4, E_2], [X_1, E_4]\} \\
\sigma(R_{13}) &= \{[E_4, E_2], [E_2, E_4]\}
\end{aligned}
$$

Figure 13 shows the resulting connectivity graph. This graph contains two connected components, which means that $\mathcal{C}_{free}$ also consists of two connected subsets. The first component of G (at the top of Figure 13) corresponds to the orientation of $\mathcal{A}$ in which P is closer to the corner vertex X_2. The existence of two connected components means that the robot cannot make a full rotation inside the corner. But it can move from one end of the corner to the other with a "forward - backward" or "backward - forward" motion. ∎

We now sketch the overall planning algorithm that is derived from the

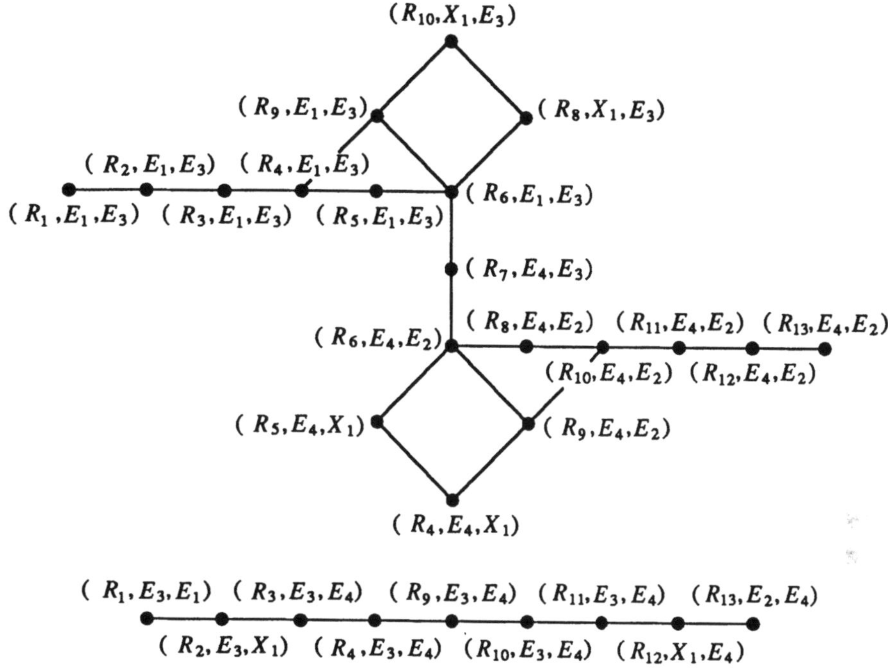

Figure 13. This figure shows the connectivity graph for the example of Figure 12. It consists of two connected components, hence verifying the legend of Figure 12.

notions presented above.

2.4 Sketch of the Algorithm

Given a polygonal obstacle region $\mathcal{B}$, $\mathcal{A}$'s length d, and $\mathcal{A}$'s initial and goal configurations $\mathbf{q}_{init}$ and $\mathbf{q}_{goal}$ in $\mathcal{C}_{free}$, such that neither configuration projects on the xy-plane in a critical curve, the algorithm sketched below generates a free path from $\mathbf{q}_{init}$ to $\mathbf{q}_{goal}$, whenever one exists, and returns failure otherwise.

In this algorithm, we do not build an explicit representation of the noncritical regions, say, in the form of a list of critical curve sections forming their boundaries. Instead, we give an arbitrary orientation to every section β and we designate the two regions separated by β by $right(\beta)$ and $left(\beta)$. As a consequence, the algorithm searches a connectivity graph

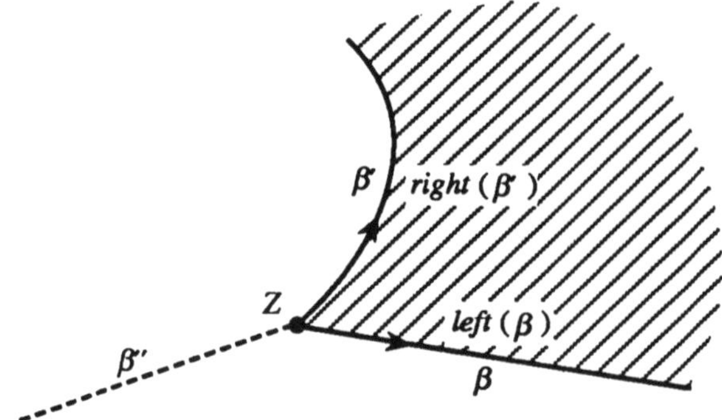

Figure 14. β and β' are two critical curve sections sharing a common endpoint Z. We assume that both sections are oriented from Z toward their other endpoints. If there is no other critical curve section β'' having Z as one of its endpoints and lying in the sector between β and β' depicted as a striped area in the figure, then $R = left(\beta)$ and $R' = right(\beta')$ are the same noncritical region.

G' which is slightly different from the graph G defined in the previous subsection. The nodes of G' are all the cells $cell(R, s_1, s_2)$, where R is identified as one side of a curve section β, i.e. $right(\beta)$ or $left(\beta)$. Thus, each actual cell appears in G' as many times as there are curve sections forming the boundary of the noncritical region onto which the cell projects. Two nodes $cell(R, s_1, s_2)$ and $cell(R', s'_1, s'_2)$ are connected by a link as follows:

- Let $R = left(\beta)$ and $R' = right(\beta')$, with β and β' sharing a common endpoint Z. We can assume that β and β' are both oriented from Z toward their other endpoints. If there is no other critical curve section β'' having Z as one of its endpoints and lying in the sector between β and β' considered in counterclockwise order, then R and R' are the same noncritical region (see Figure 14). The cells $cell(R, s_1, s_2)$ and $cell(R', s'_1, s'_2)$ are connected by a link of G' if and only if $[s_1, s_2] = [s'_1, s'_2]$, i.e. $cell(R, s_1, s_2)$ and $cell(R', s'_1, s'_2)$ are the same actual cell.

- Let $R = right(\beta)$ and $R' = left(\beta)$. The cells $cell(R, s_1, s_2)$ and $cell(R', s'_1, s'_2)$ are connected by a link of G' if and only if they are adjacent according to the crossing rule stated in Subsection 2.3.

The path planning algorithm is the following:

1. **Compute the critical curve sections.** Compute the set of all the critical curves and their intersections. For each curve, sort the intersection points according to their order on the curve (this is done by defining a suitable parameterization for each type of curve). Decompose each curve into sections, each lying between two consecutive intersection points.

2. **Compute the adjacency relation among cells.** For every critical curve section β which is not part of an obstacle edge, compute[9] $\sigma(right(\beta))$ and $\sigma(left(\beta))$ and build the adjacency relation among the cells above $right(\beta)$ and those above $left(\beta)$ (using the general crossing rule). If $\sigma(right(\beta)) = \sigma(left(\beta))$, β is redundant and may be removed from further consideration. For every critical curve section β of type 0 (i.e. part of an obstacle edge), there is no adjacency relation to build.

3. **Form the clusters of adjacent curve sections.** At every intersection point Z of critical curves, form the circular list of all the curve sections incident at Z and sort this list in, say, counterclockwise order (the sorted list is called a *cluster*). To this end, compute and sort the outgoing tangential directions of the curve sections at Z. Whenever two such directions are equal, compute and sort higher-order derivative directions.

4. **Identify the initial and goal cells.** Find[10] a curve section β_{init} (resp. β_{goal}) such that $(x_{init}, y_{init}) \in right(\beta_{init})$ or $left(\beta_{init})$ (resp. $(x_{goal}, y_{goal}) \in right(\beta_{goal})$ or $left(\beta_{goal})$). Let κ_{init} (resp. κ_{goal}) be the cell containing $(x_{init}, y_{init}, \theta_{init})$ (resp. $(x_{goal}, y_{goal}, \theta_{goal})$).

---

[9]One may pick a point X in the interior of β, draw the line L perpendicular to β at X, and compute the intersections of L with the critical curves. Then, it is easy to select a point X' (resp. X'') in $right(\beta)$ (resp. $left(\beta)$) located in L. The last step is to compute σ at X' and at X''.

[10]To this end, draw the line segment between (x_{init}, y_{init}) and (x_{goal}, y_{goal}); compute its intersections with all the critical curve sections (if there is no intersection, (x_{init}, y_{init}) and (x_{goal}, y_{goal}) lie in the same noncritical region); find the curve sections β_{init} and β_{goal} whose intersections with the line segment are nearest to (x_{init}, y_{init}) and (x_{goal}, y_{goal}), respectively; identify the side of β_{init} (resp. β_{goal}) that contains (x_{init}, y_{init}) (resp. (x_{goal}, y_{goal})); and find the pair $[s_1^{init}, s_2^{init}]$ (resp. $[s_1^{goal}, s_2^{goal}]$) such that $\theta_{init} \in (\lambda_1(x_{init}, y_{init}, s_1^{init}), \lambda_2(x_{init}, y_{init}, s_2^{init}))$ (resp. $\theta_{goal} \in (\lambda_1(x_{goal}, y_{goal}, s_1^{goal}), \lambda_2(x_{goal}, y_{goal}, s_2^{goal}))$).

5. **Search the connectivity graph.** Initialize the graph G' with the node κ_{init}. Then, iteratively expand a node that has not been expanded yet. Stop either when the node κ_{goal} has been generated, or when there are no more nodes to expand. In the first case, return the sequence of cells on the path connecting κ_{init} to κ_{goal} in the graph. In the second case, return failure.

The output of the algorithm is a sequence of cells $(\kappa_1, ..., \kappa_p)$, with $\kappa_j = right(\beta_j)$ or $left(\beta_j)$. This sequence determines a channel $\bigcup_{j=1}^{p} cl(\kappa_p)$. Every path between $\mathbf{q}_{init}$ and $\mathbf{q}_{goal}$ lying in the interior of this channel is a free path. One can construct such a semi-free path from the sequence $(\beta_1, ..., \beta_p)$. Since the obstacle region $\mathcal{B}$ is a manifold with boundary, an arbitrarily small deformation is sufficient to transform this path into a free path.

Let n be the complexity of $\mathcal{B}$. There are $O(n)$ critical curves of types 0, 1, 2 and 3, and $O(n^2)$ critical curves of type 4 and 5. Since all critical curves are algebraic of degree 4 or less, the number of intersection points of a pair of critical curves is bounded by a fixed constant. Hence, there are $O(n^2)$ intersection points on each curve and $O(n^4)$ intersection points in all. The number of pairs in $\sigma(R)$ is $O(n)$. Therefore, there are $O(n^5)$ nodes in the graph searched at step 5. The crossing rule determines $O(n^5)$ links between these nodes. The connectivity graph specified in the previous subsection also has $O(n^5)$ nodes and links.

The time complexity of the algorithm depends significantly on the implementation of some of the steps, in particular step 2. With some modifications not discussed here[11], it is $O(n^5)$ [Schwartz and Sharir, 1983a]. This estimate, however, derives from the number $O(n^4)$ of intersections among critical curves. In many practical cases, this number is much lower and the actual complexity of the algorithm is more reasonable.

The above algorithm assumes that the curve sections in a connected subset of free space form a connected graph. This is not true if some noncritical regions, collectively denoted by R', form a hole in an enclosing noncritical region R, as is illustrated in Figure 15. Then, the curve sections delimiting the regions forming R' are not connected to the curve sections in the external boundary of R. This is a problem, however,

[11] In particular, the naive computation of $\sigma(right(\beta))$ and $\sigma(left(\beta))$ proposed in footnote 9 would lead to a much higher time complexity bound.

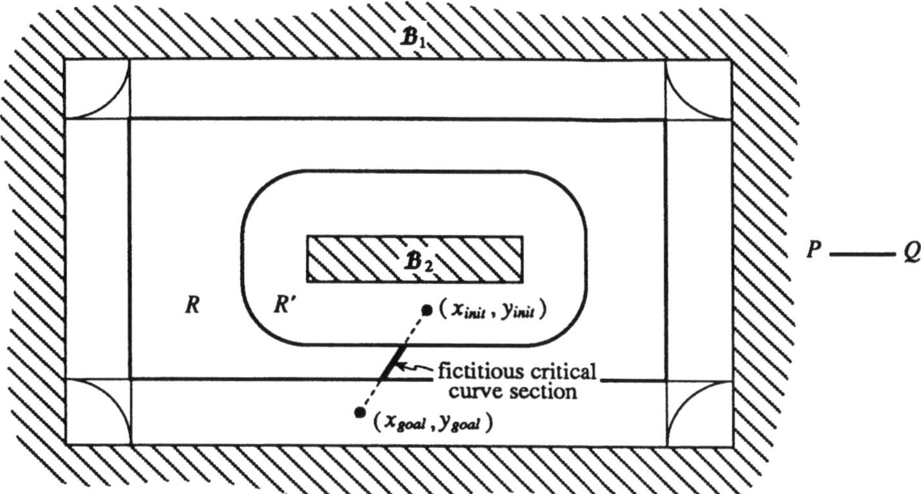

Figure 15. This workspace contains two obstacles, $\mathcal{B}_1$ and $\mathcal{B}_2$. The critical curves determined by $\mathcal{B}_2$ delimit several noncritical regions which we collectively denote by R'. R' forms a hole in another noncritical region, R. Hence, the critical curve sections form two connected subgraphs. This problem is fixed by generating a straight segment between the initial and goal positions of $\mathcal{A}$, and joining every two consecutive intersections of this segment with the critical curves by a fictitious critical curve section.

only if (x_{init}, y_{init}) or (x_{goal}, y_{goal}), but not both, is contained in R'. The problem can be fixed by joining the initial and goal positions by a segment, and creating a fictitious critical curve section between every two consecutive intersections of this segment with the critical curves [Bañon, 1990].

An implementation of the above planning method is described in [Bañon, 1990]. Technical details underlying such an implementation (e.g. the separation of the intersection points of the critical curves requires some care), as well as experimental results, are provided in this publication.

3 General Case

In the previous two sections we have illustrated the exact cell decomposition approach with two methods applying to two particular cases of the basic motion planning problem. We now describe a method which solves this problem in its full generality. As we will see in Chapter 8, the same method is also applicable to several extensions of the basic problem, including motion planning for multi-robot systems and articulated robots.

The only constraint imposed by the method is that both the robot $\mathcal{A}$ and the obstacle region $\mathcal{B}$ be described as semi-algebraic sets. Then, as shown in Subsection 3.2 of Chapter 3, using an adequate parameterization of the m-dimensional configuration space $\mathcal{C}$, the C-obstacle region $\mathcal{CB}$, as well as the free space $\mathcal{C}_{free}$, can also be described as semi-algebraic sets.

The method partitions $\mathcal{C}_{free}$ into a finite collection of semi-algebraic cells. As in the previous section, the cells are obtained by first computing their projection in a lower-dimensional space and then lifting them in $\mathcal{C}$. Since the dimension of $\mathcal{C}$ is in general greater than 3, the projection/lift mechanism is recursive, that is, it produces projections of cells on spaces of dimensions $m-1, m-2, ..., 1$, before lifting them back through the reverse sequence. Due to this recursive computation, the method is sometimes called a "projection" method.

The method is due to Schwartz and Sharir [Schwartz and Sharir, 1983b]. It makes use of a well-known result established by Collins [Collins, 1975] for deciding the satisfiability of Tarski sentences (see Subsection 3.1 of Chapter 3). We discuss this result first. The motion planning algorithm is sketched next. Our presentation concentrates on the main ideas of the method. The actual algorithm requires us to pay attention to many details, including exact computation with algebraic numbers, which are not covered here. We refer the reader to the original paper for more details [Schwartz and Sharir, 1983b].

3.1 Cylindrical Algebraic Decomposition

DEFINITION 6: *An* **algebraic decomposition** *of* $\mathbf{R}^n$ *is a finite collection* $\mathcal{K}_n$ *of disjoint semi-algebraic subsets of* $\mathbf{R}^n$*, each homeomorphic to* $\mathbf{R}^j$ *for some* $j \in [0, n]$*, whose union is* $\mathbf{R}^n$*. Each element* κ *of* $\mathcal{K}_n$ *is called a* **cell**. *If* $j = 0$*, then the cell is a single algebraic point.*[12]

[12] An algebraic point is a point whose coordinates are all algebraic numbers. The set of algebraic numbers is the set of real roots of polynomials in $Q[X]$, i.e. polynomials in one real variable with rational coefficient.

The fact that each cell in an algebraic decomposition is homeomorphic to some $\mathbf{R}^j$ implies that it is connected and contains no holes.

A **sample** of an algebraic decomposition $\mathcal{K}_n$ of $\mathbf{R}^n$ is a function $\sigma : \mathcal{K}_n \rightarrow \mathbf{R}^n$ such that for all $\kappa \in \mathcal{K}_n$, $\sigma(\kappa) \in \kappa$ and $\sigma(\kappa)$ is an algebraic point. $\sigma(\kappa)$ is called the **sample point** of κ.

The **projection** of a point $\mathbf{x}_n = (x_1, ..., x_n) \in \mathbf{R}^n$ is the point $\mathbf{x}_{n-1} = (x_1, ..., x_{n-1}) \in \mathbf{R}^{n-1}$, obtained by omitting the nth component of the point. The projection of a subset of $\mathbf{R}^n$ is the set of projections of its members.

DEFINITION 7: *A* **cylindrical algebraic decomposition** $\mathcal{K}_n$ *of* $\mathbf{R}^n$ $(n > 0)$ *is an algebraic decomposition defined recursively as follows:*

- *For $n > 1$, there exists a cylindrical algebraic decomposition $\mathcal{K}_{n-1}$ of $\mathbf{R}^{n-1}$ such that for each cell κ in $\mathcal{K}_n$ there is a cell κ' in $\mathcal{K}_{n-1}$ that is the projection of κ on $\mathbf{R}^{n-1}$. $\mathcal{K}_{n-1}$ is called the* **base decomposition** *of $\mathcal{K}_n$, and κ is said to be* **based** *on κ'.*

- *For $n = 1$, $\mathcal{K}_1$ is a partitioning of $\mathbf{R}$ into a finite set of algebraic numbers and the finite and infinite open intervals bounded by these numbers.*

The collection of all the cells κ in $\mathcal{K}_n$ which have the same projection κ' in $\mathcal{K}_{n-1}$ is called the **cylinder** over κ'.

A **cylindrical sample** of a cylindrical algebraic decomposition of $\mathbf{R}^n$, $\mathcal{K}_n$, is a sample $\sigma : \mathcal{K}_n \rightarrow \mathbf{R}^n$, such that:

- if κ_1 and κ_2 are cells of $\mathcal{K}_n$ having the same base cell then $\sigma(\kappa_1)$ and $\sigma(\kappa_2)$ have the same projection in $\mathbf{R}^{n-1}$,

- if $n > 1$, σ recursively induces a cylindrical sample of the base decomposition $\mathcal{K}_{n-1}$ of $\mathcal{K}_n$.

Let $sign(x)$ be -1, 0, or $+1$ whenever $x < 0$, $x = 0$, or $x > 0$, respectively. Let $\mathcal{F}$ be a set of functions of n real variables. A decomposition $\mathcal{K}_n$ of $\mathbf{R}^n$ is said to be $\mathcal{F}$**-invariant** if and only if, for each cell κ in $\mathcal{K}_n$ and each function f in $\mathcal{F}$, $sign(f(\mathbf{x}))$ is constant as $\mathbf{x} = (x_1, ..., x_n)$ varies over κ.

THEOREM 1 (Collins): *Given any set $\mathcal{P}$ of polynomials in $Q[X_1; ...; X_n]$ there is an effective algorithm which constructs a $\mathcal{P}$-invariant cylindrical algebraic decomposition $\mathcal{K}_n$ of $\mathbf{R}^n$ together with a cylindrical sample of $\mathcal{K}_n$. The algorithm takes polynomial time in the number of polynomials in $\mathcal{P}$* (geometric *complexity*) *and in their maximum degree* (algebraic *complexity*)[13], *with double exponential dependence on n. The number of cells generated by the algorithm has the same order of magnitude.*[14]

The constructive proof of this theorem is the Collins decomposition algorithm which constructs the $\mathcal{P}$-invariant decomposition $\mathcal{K}_n$. This algorithm is sketched in Subsection 3.4. Each of the cells produced by the algorithm is described by a quantifier-free polynomial expression over $\mathbf{R}^n$. The algorithm also produces the base decomposition $\mathcal{K}_{n-1}$ of $\mathcal{K}_n$, the base decomposition $\mathcal{K}_{n-2}$ of $\mathcal{K}_{n-1}$, etc..., and a cylindrical sample[15] of $\mathcal{K}_n$. $\mathcal{K}_n$ is called the $\mathcal{P}$-invariant **Collins decomposition** of $\mathbf{R}^n$.

3.2 Application to the First-Order Theory of Reals

Two straightforward (and related) applications of the Collins decomposition algorithm are truth decision and quantifier elimination for Tarski sentences:

- **Truth decision:** The problem is to determine the truth value of a Tarski sentence that contains no free variables. Consider, for example, the sentence $\forall x, \exists y : \phi(x, y)$. Let $\mathcal{P}$ be the set of polynomials appearing in ϕ. Using Collins algorithm, we can build a $\mathcal{P}$-invariant cylindrical algebraic decomposition $\mathcal{K}_2$ of $\mathbf{R}^2$, a base decomposition $\mathcal{K}_1$ of $\mathcal{K}_2$, and a cylindrical sample $\sigma : \mathcal{K}_2 \rightarrow \mathbf{R}^2$. The above sentence is true if and only if, for all $\kappa' \in \mathcal{K}_1$, there exists a cell $\kappa \in \mathcal{K}_2$ contained in the cylinder over κ' such that $\phi(x_\kappa, y_\kappa)$ holds with $\sigma(\kappa) = (x_\kappa, y_\kappa)$.

- **Quantifier elimination:** The problem is to transform a given Tarski sentence with quantifiers into a quantifier-free polynomial expression.

[13] The algorithm is also polynomial in the magnitude of the integers used to describe the rational coefficients in $\mathcal{P}$.

[14] The number of cells is independent of the magnitude of the coefficients.

[15] Every coordinate x_i of every sample point is "exactly" represented by a polynomial of $Q[X]$ together with an interval that isolates a unique root of the polynomial equal to x_i.

For instance, consider the sentence $\exists y : \phi(x,y)$. Variable x is free, so that the sentence defines a semi-algebraic subset of $\mathbf{R}$. We wish to define this subset by a quantifier-free polynomial expression over $\mathbf{R}$. Let $\mathcal{K}_2$, $\mathcal{K}_1$, and σ be as above. Let ψ_κ denote the quantifier-free polynomial expression defining the cell κ, for any cell κ in $\mathcal{K}_2$ and $\mathcal{K}_1$. Let $\Psi(x)$ be the disjunction of all the $\psi_{\kappa'}$, where κ' is the base cell of at least one $\kappa \in \mathcal{K}_2$ such that $\phi(x_\kappa, y_\kappa)$ holds with $\sigma(\kappa) = (x_\kappa, y_\kappa)$. The two Tarski sentences $\exists y : \phi(x,y)$ and $\Psi(x)$ define the same semi-algebraic subset of $\mathbf{R}$. In order to build a general quantifier-elimination procedure, an order has to be imposed on the variables such that the quantified variables — those which will be eliminated by projection — appear after the non-quantified ones.

The principles illustrated in the above two examples can be embedded in general procedures. In the next subsection we will use quantifier elimination to construct a cylindrical algebraic decomposition of $\mathcal{C}_{free}$ and truth decision to test adjacency among the cells of this decomposition.

3.3 General Approach to Path Planning

We now sketch how the Collins decomposition can be used for path planning. The robot $\mathcal{A}$ and the obstacle region $\mathcal{B}$ are represented as semi-algebraic subsets of $\mathbf{R}^N$, and their defining formulas are $\Lambda_\mathcal{A}(a)$ and $\Lambda_\mathcal{B}(b)$, respectively, where a and b are points in $\mathbf{R}^N$.

As shown in Section 3 of Chapter 3, the C-obstacle region $\mathcal{CB}$ can be represented as a semi-algebraic set in $\mathbf{R}^m$, where m is the dimension of $\mathcal{A}$'s configuration space. We have:

$$\mathcal{CB} = \{\mathbf{q} \in \mathbf{R}^m \ / \ \mathsf{CB}(\mathbf{q})\}$$

with:

$$\mathsf{CB}(\mathbf{q}) \equiv \exists a \in \mathbf{R}^N, \exists b \in \mathbf{R}^N : \ \Lambda_\mathcal{A}(a) \wedge \Lambda_\mathcal{B}(b) \wedge \Upsilon'(\mathbf{q}, a, b)$$

where $\mathbf{q}$ represents a configuration as a point in $\mathbf{R}^m$ using a parameterization of $\mathcal{A}$'s configuration space such that $\Upsilon'(\mathbf{q}, a, b)$, which expresses that $a(\mathbf{q}) = b$, is algebraic.

We assume for a moment that $\mathcal{C}$ is covered by a single chart which maps every configuration of $\mathcal{C}$ into $\mathbf{R}^m$, so that $\mathcal{C}_{free}$ is fully represented as:

$$C_{free} = \{\mathbf{q} \in \mathbf{R}^m \ / \ \neg\texttt{CB}(\mathbf{q})\}.$$

We can construct a Collins decomposition $\mathcal{K}_{m+2N}$ with the polynomials appearing in the expression

$$\exists a \in \mathbf{R}^N, \exists b \in \mathbf{R}^N : \Lambda_{\mathcal{A}}(a) \wedge \Lambda_{\mathcal{B}}(b) \wedge \Upsilon'(\mathbf{q}, a, b)$$

and eliminate the quantified N-dimensional variables a and b. This yields a cylindrical algebraic decomposition $\mathcal{K}_m$ of $\mathbf{R}^m$, the successive base decompositions of $\mathcal{K}_m$, and a cylindrical sample σ of $\mathcal{K}_m$. A cell $\kappa \in \mathcal{K}_m$ is **free** if it is contained in C_{free}. It is the projection of a cylinder in $\mathcal{K}_{m+2N}$ that contains no cell κ' whose sample point satisfies the above expression. We denote by $\mathcal{K}_m^{free}$ the set of all the free cells of $\mathcal{K}_m$.

In the following, we assume that the Collins decomposition $\mathcal{K}_m$ is "well-based", meaning that the closure of every cell κ in $\mathcal{K}_l$, for all $l \in [1, m]$, is a union of cells of $\mathcal{K}_l$. In general, a Collins decomposition is well-based, but this may not be true for certain unfavorable orientations of the axes in $\mathbf{R}^m$. In that case, one can always restore this property by an appropriate rotation of the coordinate axes (see [Schwartz and Sharir, 1983b]).

Let $\mathbf{q}_{init}$ and $\mathbf{q}_{goal}$ be the initial and goal free configurations, respectively, represented as points of $\mathbf{R}^m$. By evaluating the polynomial expression defining each cell in C_{free} with each of these configurations, we can identify the cell of $\mathcal{K}_m^{free}$ containing $\mathbf{q}_{init}$, called the *initial cell*, and the cell of $\mathcal{K}_m^{free}$ containing $\mathbf{q}_{goal}$, called the *goal cell*. We assume that both the initial and goal cells are m-dimensional cells. (This assumption can always be made true by an infinitesimal shift of the two configurations.)

The search for a free path between the two configurations can be confined to the m- and $(m-1)$-dimensional cells of $\mathcal{K}_m^{free}$. Indeed, C_{free} is an open manifold of dimension m; thus no submanifold of dimension $m-2$ or less can disconnect a maximal connected component of C_{free}.

DEFINITION 8: *Assuming that the decomposition $\mathcal{K}_m$ is well-based, two cells κ and κ' in $\mathcal{K}_m^{free}$ are* **adjacent** *if and only if:*

- *one of the cell, say κ, is m-dimensional and the other, κ', is $(m-1)$-dimensional,*

- *$\kappa' \subset \partial\kappa$ (κ's boundary).*

The **connectivity graph** *G is the non-directed graph whose nodes are all the m- and $(m-1)$-dimensional cells of $\mathcal{K}_m^{free}$. Two nodes are connected by a link if and only if the corresponding cells are adjacent.*

According to the above definition, two m-dimensional free cells κ_1 and κ_2 cannot be adjacent. However, if they are both adjacent to the same $(m-1)$-dimensional free cell, the latter cell is in the intersection of the boundaries $\partial\kappa_1$ and $\partial\kappa_2$, so that any two configurations in κ_1 and κ_2 can be connected by a free path. Hence, G contains a path between the initial and the goal cells if and only if there exists a free path between the initial and the goal configurations.

The adjacency condition between an m-dimensional cell κ and an $(m-1)$-dimensional cell κ' can be expressed with the following Tarski sentence [Arnon, 1979]:

$$\forall \mathbf{q}', \forall \varepsilon, \exists \mathbf{q} : (\varepsilon > 0) \wedge \psi_{\kappa'}(\mathbf{q}') \Rightarrow \psi_\kappa(\mathbf{q}) \wedge (\|\mathbf{q} - \mathbf{q}'\| < \varepsilon).$$

(Read: For every configuration $\mathbf{q}'$ in κ' and every $\varepsilon > 0$, there exists a configuration $\mathbf{q}$ in κ that is distant from $\mathbf{q}'$ by less than ε.)

Thus, testing adjacency between the two cells and consequently building the connectivity graph G are decidable problems that can be solved by applying the Collins decomposition algorithm to the polynomials in the above Tarski sentence for all pairs (κ, κ') of cells of $\mathcal{K}_m^{free}$ of respective dimensions m and $m-1$ and checking the truth value of the sentence.

Once constructed, G is searched for a path connecting the initial cell to the goal cell. The output of the search, if successful, is a sequence of cells, $(\kappa_1, \kappa_2, ..., \kappa_{2p+1})$, $p \geq 0$, such that κ_{2j+1}, with $j \in [0, p]$, is m-dimensional and κ_{2j}, with $j \in [1, p]$, is $(m-1)$-dimensional. The union $\bigcup_{j=1}^{2p+1} \kappa_j$ forms an open connected region in $\mathcal{C}_{free}$, called a **channel**. The final step is to construct a free path τ from $\mathbf{q}_{init}$ to $\mathbf{q}_{goal}$ lying in this channel.

We have assumed above that every configuration $\mathbf{q}$ is represented as a point $(x_1, ..., x_m)$ in $\mathbf{R}^m$. In addition, the cells in the sequence $(\kappa_1, ..., \kappa_{2p+1})$ are alternately m-dimensional and $(m-1)$-dimensional cells, with $\kappa_{2j} \subseteq \partial\kappa_{2j-1} \cap \partial\kappa_{2j+1}$, for any $j \in [1, p]$. We can construct τ

as the product of $p+1$ paths τ_j, $j = 1, 3, ..., 2p+1$, respectively connecting $\mathbf{q}_{init}$ to $\sigma(\kappa_2)$, $\sigma(\kappa_2)$ to $\sigma(\kappa_4)$, ..., $\sigma(\kappa_{2p-2})$ to $\sigma(\kappa_{2p})$, and $\sigma(\kappa_{2p})$ to $\mathbf{q}_{goal}$.

Assume that the projection $\tau' = \tau_1' \bullet \tau_3' \bullet ... \bullet \tau_{2p+1}'$ of $\tau = \tau_1 \bullet \tau_3 \bullet ... \bullet \tau_{2p+1}$, in $\mathbf{R}^{m-1}$, has already been generated within $\kappa_1', ..., \kappa_{2p+1}'$, the base cells of $\kappa_1, ..., \kappa_{2p+1}$. Each path τ_j' in τ' can be lifted in $\kappa_1, ..., \kappa_{2p+1}$ as follows. The x_m coordinate along the path τ_j in the m-dimensional cell κ_j is obtained by linearly interpolating between the values of x_m at the beginning and the end of τ_j, using the proportion of the distances to the boundary of κ_j in the $+x_m$ and $-x_m$ directions and by extending the interpolation to the closure of κ_j, which includes κ_{j-1} (if $j > 1$) and κ_{j+1} (if $j < 2p+1$). In order to construct the interpolation, one must distinguish between the case where κ_{j-1} (resp. κ_{j+1}) and κ_j have the same base cell and the case where they do not have the same base cell. In both cases, if κ_j is a semi-infinite cell, an artificial boundary has to be generated by adding (resp. subtracting) a fixed appropriate amount to the cell boundary below (resp. above) the path in order to make the interpolation possible. The projected path τ' can be generated in the same way by lifting in $\mathcal{K}_{m-1}$ a path τ'' generated in the base decomposition $\mathcal{K}_{m-2}$ of $\mathcal{K}_{m-1}$. Thus, the problem is recursively reduced to the generation of a path in a base decomposition in $\mathbf{R}$, which is trivial.

As mentioned in Subsection 3.1, the time complexity for generating the cell decomposition of $\mathcal{C}_{free}$ is polynomial in the number of polynomials in the expression defining the C-obstacle region and in their maximum degree, with a double exponential dependence on m. For a free-flying rigid object, m is fixed. Therefore, the decomposition of $\mathcal{C}_{free}$ is generated in polynomial time. The number of produced cells has the same order of magnitude. The number of polynomials in the expression ψ_κ defining each cell in the decomposition of $\mathcal{C}_{free}$ and their maximum degree are polynomial since the algorithm that constructs these expressions is itself polynomial (for more detail see Subsection 3.4). Therefore, since the number of variables appearing in the Tarski sentences representing the adjacency condition between two cells is $2m+1$, and hence fixed, the construction of the connectivity graph G is polynomial. However, it has a multiple exponential dependence on m. A more efficient test of cell adjacency than the one above is proposed in [Schwartz and Sharir, 1983b]. Using this improved test, the construction and search of the connectivity graph takes double-exponential time in m.

Let us now remove the assumption that a single chart suffices to cover all of $\mathcal{C}$:

- If $\mathcal{C} = \mathbf{R}^3 \times SO(3)$, one can build an algebraic application of $\mathcal{C}$ onto $\mathbf{R}^6$ by mapping the 3-sphere S^3 onto $\mathbf{R}^3$ using the stereographic projection (remember from Subsection 4.3 of Chapter 2 that S^3 represents $SO(3)$ in a bivalent fashion). This application maps all but one of the points in S^3. However, the removal of one point in S^3 cannot change the connectivity of $\mathcal{C}_{free}$, since $\mathcal{C}_{free}$ is a manifold of dimension 6, while the removed point determines a submanifold of $\mathcal{C}$ of dimension 3, i.e. of codimension greater than 2. Since S^3 is a bivalent representation of $SO(3)$, two points represent the initial configuration in $\mathbf{R}^m$ and two points represent the goal configuration. Thus, the connectivity graph contains two initial cells and two goal cells. Any path joining an initial cell to a goal cell is appropriate.

- If $\mathcal{C} = \mathbf{R}^2 \times S^1$, one can build an algebraic application of $\mathcal{C}$ onto $\mathbf{R}^3$ by mapping the unit circle S^1 onto $\mathbf{R}$ using the stereographic projection. This application maps all but one of the points in S^1. This point determines a submanifold of dimension 2 in $\mathcal{C}$, and its removal may affect the connectivity of $\mathcal{C}_{free}$. Hence, it is necessary to consider two charts, built with two stereographic projections from two different points P_1 and P_2 in S^1. This yields two representations of $\mathcal{C}$, two Collins decompositions $\mathcal{K}_m^{free,1}$ and $\mathcal{K}_m^{free,2}$ of $\mathcal{C}_{free}$, and two connectivity graphs G_1 and G_2. The two graphs have to be combined into one by connecting the cells in G_1 which intersect with cells in G_2. This can be done by including the polynomials in the equations defining $\mathbf{R}^2 \times \{P_2\}$ (resp. $\mathbf{R}^2 \times \{P_1\}$) among the polynomials from which the decomposition $\mathcal{K}_m^{free,1}$ (resp. $\mathcal{K}_m^{free,2}$) is computed. Then a cell $\kappa_1 \in \mathcal{K}_m^{free,1}$ and a cell $\kappa_2 \in \mathcal{K}_m^{free,2}$ intersect if the sample point $\sigma(\kappa_1)$ is the image in $\mathcal{K}_m^{free,1}$ of a configuration whose image in $\mathcal{K}_m^{free,2}$ belongs to κ_2.

Combining all the above results, we get the following theorem:

THEOREM 2 (Schwartz and Sharir): *The problem of planning a free path of a semi-algebraic rigid free-flying object $\mathcal{A}$ among finitely many semi-algebraic fixed rigid obstacles $\mathcal{B}_i$, $i = 1, ..., q$, can be solved in* *polynomial time in the number of polynomials defining the objects in the workspace and in the maximal degree of these polynomials.*

For example, if both $\mathcal{A}$ and the obstacles are polygonal (resp. polyhedral) objects in a two-dimensional (resp. three-dimensional) workspace, the time complexity of the above method is polynomial in the complexity of the objects.

The above general approach to motion planning requires performing exact computations with algebraic numbers. This is a delicate issue not treated here. The paper by Schwartz and Sharir provides many details on exact computation with algebraic numbers. It also analyzes the effect of the growth in size of the integers involved in the computations on the total time complexity of the algorithm. It turns out that the algorithm still requires time polynomial in the number of polynomials defining the object and in the maximal degree of these polynomials. However, it is also polynomial in the size of the integers used to describe the rational coefficients of the polynomials.

The Collins decomposition algorithm generates many cells, allowing its application to the general problem of deciding the satisfiability of Tarski sentences. However, it is probably too general for motion planning. The more specific methods presented in the first two sections of this chapter, although based on the same general approach, generate fewer cells than the Collins decomposition would do in the same cases. This suggests that decompositions coarser than the Collins one may still be appropriate for motion planning and that general algorithms based on exact cell decomposition, but with smaller time complexity than the algorithm presented above, may exist.

3.4 Sketch of the Collins Decomposition Algorithm

Let $\mathcal{P} = \{P_i(x_1, ..., x_n)\}$ be any set of polynomials in $Q[X_1; ...; X_n]$ and $P(x_1, ..., x_n)$ be the product of all the non-constant polynomials $P_i(x_1, ..., x_n)$ in this set. We do not have to worry about the constant polynomials (i.e. the polynomials having degree 0) since their signs remain constant. We consider $P(x_1, ..., x_n)$ as a polynomial in x_n whose coefficients are functions of $\mathbf{x}_{n-1} = (x_1, ..., x_{n-1})$; hence, we also denote it by $P_{\mathbf{x}_{n-1}}(x_n)$.

Let κ be a region of $\mathbf{R}^{n-1}$ such that $P_{\mathbf{x}_{n-1}}(x_n)$ has a constant number of distinct real roots when $\mathbf{x}_{n-1}$ varies over κ. This implies that the number of distinct real roots of every polynomial P_i (considered as a polynomial in x_n) is also constant over κ. It can be shown that these roots are continuous functions of $\mathbf{x}_{n-1}$ [Schwartz and Sharir, 1983b].

Suppose that we know a technique (we will present one below) for constructing a cylindrical algebraic decomposition $\mathcal{K}_{n-1}$ of $\mathbf{R}^{n-1}$ such that in any cell $\kappa \in \mathcal{K}_{n-1}$ the number of distinct real roots of $P_{\mathbf{x}_{n-1}}(x_n)$ is constant, say j (j may vary from one cell to another). Let $f^1_\kappa(\mathbf{x}_{n-1})$, ..., $f^j_\kappa(\mathbf{x}_{n-1})$ be these roots and $y_1, ..., y_j$ their algebraic values computed at $\sigma(\kappa)$. We can always write the indices in such a way that $y_1 < y_2 < ... < y_j$. Then, for all $\mathbf{x}_{n-1} \in \kappa$, we have: $f^1_\kappa(\mathbf{x}_{n-1}) < f^2_\kappa(\mathbf{x}_{n-1}) < ... < f^j_\kappa(\mathbf{x}_{n-1})$, since otherwise the number of distinct real roots of $P_{\mathbf{x}_{n-1}}(x_n)$ would not remain constant over κ. Thus, $f^i_\kappa(\mathbf{x}_{n-1})$ can be regarded as the ith root of $P_{\mathbf{x}_{n-1}}(x_n)$ in κ. It is a continuous function of $\mathbf{x}_{n-1}$ over κ. A $\mathcal{P}$-invariant cylindrical algebraic decomposition $\mathcal{K}_n$ of $\mathbf{R}^n$ is obtained from $\mathcal{K}_{n-1}$ by decomposing every cylinder $\kappa \times \mathbf{R}$, $\kappa \in \mathcal{K}_{n-1}$, into the following cells, where $(\mathbf{x}_{n-1}, x_n)$ stands for $(x_1, ..., x_{n-1}, x_n)$:

- $\{(\mathbf{x}_{n-1}, x_n)\}$, with $\mathbf{x}_{n-1} \in \kappa$ and $x_n < f^1_\kappa(\mathbf{x}_{n-1})$;
- $\{(\mathbf{x}_{n-1}, x_n)\}$, with $\mathbf{x}_{n-1} \in \kappa$ and $x_n = f^i_\kappa(\mathbf{x}_{n-1})$, for $i \in [1, j]$;
- $\{(\mathbf{x}_{n-1}, x_n)\}$, with $\mathbf{x}_{n-1} \in \kappa$ and $f^i_\kappa(\mathbf{x}_{n-1}) < x_n < f^{i+1}_\kappa(\mathbf{x}_{n-1})$, for $i \in [1, j-1]$;
- $\{(\mathbf{x}_{n-1}, x_n)\}$, with $\mathbf{x}_{n-1} \in \kappa$ and $f^j_\kappa(\mathbf{x}_{n-1}) < x_n$.

In addition, let σ_{n-1} be a cylindrical sample of $\mathcal{K}_{n-1}$. For any $\kappa \in \mathcal{K}_{n-1}$, the points:

$$(\mathbf{s}_{n-1}, y_1 - 1), (\mathbf{s}_{n-1}, y_j + 1), (\mathbf{s}_{n-1}, y_1), ..., (\mathbf{s}_{n-1}, y_j), (\mathbf{s}_{n-1}, \tfrac{y_1+y_2}{2}), ..., (\mathbf{s}_{n-1}, \tfrac{y_{j-1}+y_j}{2})$$

with $\mathbf{s}_{n-1} = \sigma_{n-1}(\kappa)$, are algebraic points which define a cylindrical sample of $\mathcal{K}_n$.

So, the remaining problem is to construct $\mathcal{K}_{n-1}$, i.e. a cylindrical algebraic decomposition of $\mathbf{R}^{n-1}$ such that in any cell κ of this decomposition the number of distinct real roots of $P_{\mathbf{x}_{n-1}}(x_n)$ is constant.

It can be shown that if the number of distinct (both complex and real) roots of $P_{\mathbf{x}_{n-1}}(x_n)$ is constant over a certain region of $\mathbf{R}^{n-1}$, then the number of distinct real roots is also constant over the same region. In addition, it is well-known that the number of distinct roots of a polynomial $P(x)$ of degree k is equal to $k-h$, where h is the degree of the greatest common divisor (GCD) of P and P', the derivative of P. Thus, one way to construct a cylindrical decomposition $\mathcal{K}_{n-1}$ that can serve as a base decomposition of the decomposition of $\mathbf{R}^n$ is to insure that both the degree of $P_{\mathbf{x}_{n-1}}(x_n)$ and the degree of the GCD of $P_{\mathbf{x}_{n-1}}(x_n)$ and the derivative of $P_{\mathbf{x}_{n-1}}(x_n)$ relative to x_n, $P'_{\mathbf{x}_{n-1}}(x_n)$, stay constant in each cell of $\mathcal{K}_n$. We now show how this can be done.

Let $A(x)$ and $B(x)$ be two polynomials in x having degrees a and b ($a>0$, $b>0$), respectively. Consider the equation:

$$A(x)U_j(x) - B(x)V_j(x) = 0 \tag{1}$$

where U_j and V_j are two polynomials having degrees $b-j-1$ and $a-j-1$ ($0 \leq j \leq \min\{a,b\}-1$), respectively. Let us treat the coefficients of U_j and V_j as unknowns. The identification to zero of the coefficients of the same powers of x in the above equation produces a set of $a+b-j$ linear equations involving $a+b-2j$ unknowns. Let $\psi_j(A,B)$ be the determinant of the $a+b-2j$ equations corresponding to the highest $(a+b-2j)$ powers of x.

PROPOSITION 1: *The number of common roots of two polynomials $A(x)$ and $B(x)$ having non-zero degrees is the smallest integer j such that $\psi_j(A,B) \neq 0$.*

(We include the proof of this proposition since it provides part of the decomposition algorithm.)

Proof: Let us write:

$$\begin{array}{rcl} A(x) & = & (x-\alpha_1)...(x-\alpha_a), \\ B(x) & = & (x-\beta_1)...(x-\beta_b), \\ U_j(x) & = & (x-\mu_1)...(x-\mu_{b-j-1}), \\ V_j(x) & = & (x-\nu_1)...(x-\nu_{a-j-1}). \end{array}$$

Assume that j is the smallest integer such that equation (1) admits a non-zero solution. Then, $\forall r \in [1, b-j-1]$ and $\forall s \in [1, a-j-1]$, we have $\mu_r \neq \nu_s$, otherwise the assumption is contradicted. In addition, the unique factorization theorem for polynomials implies that every root of U_j is also a root of B, every root of V_j is also a root of A, and consequently, every remaining root of A (if any) is also a root of B. Thus, A and B have $j+1$ common roots. Inversely, if A and B have $j+1$ roots in common, there exist non-zero polynomials U_j and V_j satisfying equation (1).

If $j = 0$, equation (1) determines a square system of $a+b$ linear equations with the $a+b$ coefficients of U_0 and V_0 as unknowns. If the determinant of this system, i.e. $\psi_0(A, B)$, is non-zero, the only solution to this system is the trivial solution, and A and B have no root in common. If $\psi_0(A, B) = 0$, there exist non-zero U_0 and V_0 satisfying equation (1), and A and B have at least one root in common.

Let us now proceed by recurrence. Suppose that we already know that equation (1) admits a non-zero solution for all $i = 0, ..., j-1$. Then we know that A and B have at least j common roots. Now, consider the weaker equation:

$$A(x)U_j(x) - B(x)V_j(x) = C_j(x) \tag{2}$$

where U_j and V_j are defined as above, and C_j is an arbitrary polynomial whose degree is at most $j-1$. The system Σ of linear equations in the coefficients of U_j, V_j, and C_j, obtained by identifying the coefficients of the same powers of x in both sides of equation (2), involves $a+b-j$ equations with $a+b-j$ unknowns; hence, it is a square system. The $a+b-2j$ equations representing the condition that every coefficient of every power of x in the interval $[j, a+b-j-1]$ is zero is a square $(a+b-2j)\times(a+b-2j)$ homogeneous subsystem Σ' of Σ. Σ has a non-zero solution if and only if Σ' has a non-zero solution, i.e. if $\psi_j(A, B) = 0$. If $\psi_j(A, B) \neq 0$, then equation (2), and hence also equation (1), has only trivial solutions, so that A and B have exactly j common roots. However, if $\psi_j(A, B) = 0$, then there exist U_j, V_j, and C_j satisfying equation (2). As we already know that A and B have at least j common roots, $C_j(x)$ must be divisible by their product. But, since the degree of C_j is at most $j-1$, C_j must be identically zero, so that U_j and V_j also satisfy equation (1). Hence, A and B have at least $j+1$ roots in common. ∎

Thus, given two polynomials A and B of $Q[X]$ having non-zero degrees, we can determine exactly how many roots they have in common by computing $\psi_j(A, B)$ for increasing values of j. This computation requires, however, that we know the degrees of both A and B. Let us assume, for example, that we consider A as a polynomial of degree a, while the coefficient of the power a is zero. Then, for any j, including $j = 0$, the polynomials $U_j(x) \equiv \lambda x^{b-j-1}$, with $\lambda \neq 0$, and $V_j(x) \equiv 0$ form a non-zero solution of equation (1), even if A and B have one or several roots in common (one can also easily verify that $\psi_j(A, B) = 0$ for all j).

Since the degree of the GCD of A and B is equal to the number of common roots of the two polynomials, we have:

PROPOSITION 2: *Let $P_{\mathbf{x}_{n-1}}(x_n)$ be of degree k in x_n. For each $j = 1, ..., k$, let $Q_j(x_1, ..., x_{n-1})$ denote the coefficient of the term in $P_{\mathbf{x}_{n-1}}(x_n)$ having degree j in x_n, and $P_{j,\mathbf{x}_{n-1}}(x_n)$ denote the sum of the terms of $P_{\mathbf{x}_{n-1}}(x_n)$ whose degrees in x_n are less than or equal to j. Let $R_{jl}(x_1, ..., x_{n-1}) = \psi_l(P_{j,\mathbf{x}_{n-1}}, P'_{j,\mathbf{x}_{n-1}})$, with $l = 0, ..., j-2$. Let $\mathcal{Q}$ be the collection of all polynomials $Q_j(x_1, ..., x_{n-1})$ and $R_{jl}(x_1, ..., x_{n-1})$. The number of distinct real roots of $P_{\mathbf{x}_{n-1}}(x_n)$ is constant over each connected subset of $\mathbf{R}^{n-1}$ on which all polynomials in $\mathcal{Q}$ maintain a constant sign $(-, 0, +)$.*

This proposition may look over-sophisticated, but keep in mind that any coefficient of the polynomials $P_{\mathbf{x}_{n-1}}$ may become zero over a certain region of $\mathbf{R}^{n-1}$. Including every coefficient $Q_j(x_1, ..., x_{n-1})$ of $P_{\mathbf{x}_{n-1}}(x_n)$ in $\mathcal{Q}$ insures that the degree of $P_{\mathbf{x}_{n-1}}$ stays constant as $\mathbf{x}_{n-1}$ varies over a cell of $\mathcal{K}_{n-1}$; since several leading coefficients of $P_{\mathbf{x}_{n-1}}$ may be identically zero over a cell, we must conservatively include all the coefficients in $\mathcal{Q}$. There is no need to include the constant term of $P_{\mathbf{x}_{n-1}}$. Similarly, because the degree of $P_{\mathbf{x}_{n-1}}$ may be less than the full degree k, in order to insure that the degree of $GCD(P_{\mathbf{x}_{n-1}}, P'_{\mathbf{x}_{n-1}})$ stays constant over any cell of $\mathcal{K}_{n-1}$, we have to conservatively include in $\mathcal{Q}$ all polynomials $P_{j,\mathbf{x}_{n-1}}$ obtained from $P_{\mathbf{x}_{n-1}}$ by deleting one or more leading high order terms.

Thus, a $\mathcal{Q}$-invariant cylindrical algebraic decomposition $\mathcal{K}_{n-1}$ of $\mathbf{R}^{n-1}$ is an appropriate base decomposition to construct a $\mathcal{P}$-invariant cylindrical algebraic decomposition of $\mathbf{R}^n$. Hence, the original problem is reduced to a problem of the same form for a new set of polynomials in fewer variables, so that the same method as above can be applied recursively. When only one variable remains, the decomposition problem is trivial and the recursion can be terminated.

3.5 Example

We now illustrate the Collins decomposition on a very simple example[16] where $\mathcal{C} = \mathbf{R}^3$. The C-obstacle region is a ball of radius 1 centered at configuration $(0,0,0)$. $\mathcal{C}_{free}$ is thus defined by:

$$\mathcal{C}_{free} = \{(x,y,z) \in \mathbf{R}^3 \;/\; x^2 + y^2 + z^2 - 1 > 0\}.$$

A single polynomial appears in this definition, so that $P(x,y,z) \equiv x^2 + y^2 + z^2 - 1$.

Let us consider $P(x,y,z)$ as a polynomial in z denoted by $P_{(x,y)}(z)$. We have:
- $P_{(x,y)}(z) = z^2 + (x^2 + y^2 - 1)$,
- $P'_{(x,y)}(z) = 2z$.

The degree of $P_{(x,y)}$ is constant over $\mathbf{R}^2$ and equal to 2. Since $\psi_0(P_{(x,y)}, P'_{(x,y)}) = 4(x^2 + y^2 - 1)$, the number of common roots of $P_{(x,y)}$ and $P'_{(x,y)}$ is 0 if $x^2 + y^2 - 1 \neq 0$, and 1 otherwise. Thus, the degree of $GCD(P_{(x,y)}, P'_{(x,y)})$ is constant when the sign of $x^2 + y^2 - 1$ remains constant.

Let us now proceed recursively in $\mathbf{R}^2$, and consider $Q(x,y) = x^2 + y^2 - 1$ as a polynomial in y, i.e. as $Q_x(y) = y^2 + (x^2 - 1)$. The degree of this polynomial is constant over $\mathbf{R}$ and equal to 2. The degree of the greatest common divisor of Q_x and Q'_x, i.e. the number of common roots of Q_x and Q'_x, is constant in any region of $\mathbf{R}$ where $x^2 - 1$ keeps a constant sign.

The following cells in $\mathbf{R}$:

$$(-\infty, -1)\;[-1]\;(-1, +1)\;[+1]\;(+1, +\infty)$$

form a decomposition $\mathcal{K}_1$ of $\mathbf{R}$ such that $x^2 - 1$ keeps a constant sign over any cell of this decomposition. Thus, this decomposition can be used as the base decomposition of a Q-invariant cylindrical algebraic decomposition $\mathcal{K}_2$ of $\mathbf{R}^2$. This decomposition is obtained by lifting every cell κ

[16] This example is drawn from [Whitesides, 1985].

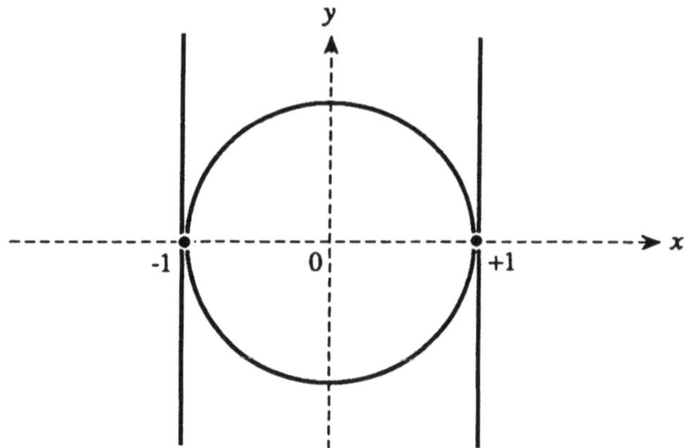

Figure 16. This figure shows the Collins decomposition of $\mathbf{R}^2$ for the polynomial $x^2 + y^2 - 1$. It consists of 13 cells of dimensions 2, 1, and 0.

of $\mathcal{K}_1$ in $\mathbf{R}^2$ and by decomposing the cylinder above κ into cells according to the distinct real roots of $Q_x(y)$, i.e. $\{-\sqrt{1-x^2}, +\sqrt{1-x^2}\}$ for $x \in (-1,+1)$, $\{0\}$ for $x = -1$ and $x = +1$, and $\emptyset$ otherwise. Therefore, $\mathcal{K}_2$ consists of the 13 cells shown in Figure 16:

- the two open half-planes $x < -1$ and $x > +1$,
- the four open half-lines $x = -1 \wedge y > 0$, $x = -1 \wedge y < 0$, $x = +1 \wedge y > 0$, $x = +1 \wedge y < 0$,
- the two points $(-1,0)$ and $(+1,0)$,
- the two open half-circles $-1 < x < +1 \wedge x^2 + y^2 - 1 = 0 \wedge y > 0$ and $-1 < x < +1 \wedge x^2 + y^2 - 1 = 0 \wedge y < 0$,
- the open disk $x^2 + y^2 - 1 < 0$,
- the two two-dimensional regions above and below the unit circle $-1 < x < +1 \wedge x^2+y^2-1 > 0 \wedge y > 0$ and $-1 < x < +1 \wedge x^2+y^2-1 > 0 \wedge y < 0$.

This decomposition can be used in turn as the base decomposition of a P-invariant cylindrical algebraic decomposition $\mathcal{K}_3$ of $\mathbf{R}^3$. For that purpose, every cell $\kappa \in \mathcal{K}_2$ is lifted in $\mathbf{R}^3$, using the distinct real roots of $P_{(x,y)}(z)$ to decompose the cylinder above κ, i.e. $\{-\sqrt{1-(x^2+y^2)}, +\sqrt{1-(x^2+y^2)}\}$ if (x,y) is in the disk $x^2+y^2-1 <$

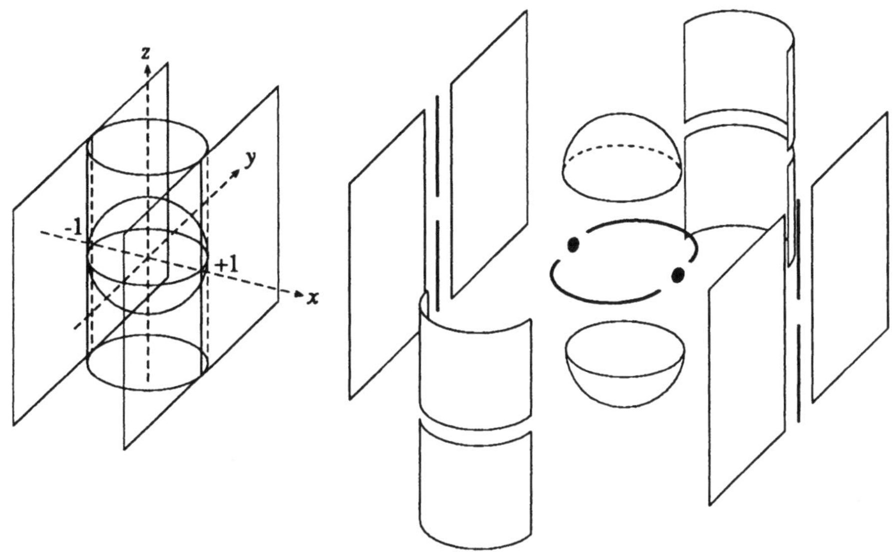

Figure 17. This figure shows the Collins decomposition of $\mathbf{R}^3$ for the polynomial $x^2 + y^2 + z^2 - 1$. It consists of 25 cells of dimensions 3, 2, 1, and 0.

0, $\{0\}$ if (x, y) is $(-1, 0)$ or $(+1, 0)$ or is in one of the two half-circles $y = \pm\sqrt{1 - x^2} \wedge -1 < x < +1$, and $\emptyset$ otherwise. Thus, $\mathcal{K}_3$ consists of the 25 cells shown in Figure 17:

- the two half-spaces $x < -1$ and $x > +1$,
- the four half-lines $x = -1 \wedge y = 0 \wedge z > 0$, $x = -1 \wedge y = 0 \wedge z < 0$, $x = +1 \wedge y = 0 \wedge z > 0$, $x = +1 \wedge y = 0 \wedge z < 0$,
- the two points $(-1, 0, 0)$ and $(+1, 0, 0)$,
- the four half-planes $x = -1 \wedge y > 0$, $x = -1 \wedge y < 0$, $x = +1 \wedge y > 0$, $x = +1 \wedge y < 0$,
- the two half-circles $-1 < x < +1 \wedge x^2 + y^2 - 1 = 0 \wedge y > 0 \wedge z = 0$ and $-1 < x < +1 \wedge x^2 + y^2 - 1 = 0 \wedge y < 0 \wedge z = 0$,
- the two half-spheres $-1 < x < +1 \wedge -1 < y < +1 \wedge x^2 + y^2 + z^2 - 1 = 0 \wedge z > 0$ and $-1 < x < +1 \wedge -1 < y < +1 \wedge x^2 + y^2 + z^2 - 1 = 0 \wedge z < 0$,
- the open ball $x^2 + y^2 + z^2 - 1 < 0$,

- the four cylindrical surfaces $-1 < x < +1 \wedge x^2+y^2-1 = 0 \wedge y > 0 \wedge z > 0$, $-1 < x < +1 \wedge x^2+y^2-1 = 0 \wedge y < 0 \wedge z > 0$, $-1 < x < +1 \wedge x^2+y^2-1 = 0 \wedge y > 0 \wedge z < 0$, and $-1 < x < +1 \wedge x^2+y^2-1 = 0 \wedge y < 0 \wedge z < 0$.
- the two three-dimensional volumes above and below the unit sphere $x^2 + y^2 - 1 < 0 \wedge x^2 + y^2 + z^2 - 1 > 0 \wedge z > 0$ and $x^2 + y^2 - 1 < 0 \wedge x^2 + y^2 + z^2 - 1 > 0 \wedge z < 0$.
- the two three-dimensional volumes $-1 < x < +1 \wedge x^2+y^2-1 > 0 \wedge y > 0$ and $-1 < x < +1 \wedge x^2 + y^2 - 1 > 0 \wedge y < 0$.

Remark: In this example, we used the fact that the analytical expression of the roots of a polynomial of degree 2 are easily computable. These expressions for a polynomial of arbitrary degree may not be computable. The Collins algorithm provides an effective way to deal with this difficulty. ■

4 Exact Decomposition Variant

In this section we describe a variant of the exact cell decomposition approach. Rather than decomposing the m-dimensional free space into cells, this variant consists of first decomposing the free space boundary $\partial \mathcal{C}_{free}$ into $(m-1)$-dimensional cells and next searching the connectivity graph representing the adjacency relation among these cells. Since the boundary of a connected component of $\mathcal{C}_{free}$ may not be connected, additional $(m-1)$-dimensional cells are included in the search graph. These cells are selected so that one of them contains the initial configuration and another one contains the goal configuration. This approach generates semi-free paths.

We illustrate below the approach with a specific method originally presented in [Avnaim, Boissonnat and Faverjon, 1988]. The robot $\mathcal{A}$ is modeled as a polygon which can translate and rotate freely in a two-dimensional workspace $\mathcal{W}$ among obstacles whose union is modeled as a regular polygonal region $\mathcal{B}$. $\mathcal{CB}$ denotes the C-obstacle region in $\mathcal{C} = \mathbf{R}^2 \times S^1$. We assume that $\mathcal{CB}$ is a manifold with boundary. We assume further that the complement of $\mathcal{B}$ in $\mathcal{W}$ is bounded, so that $\mathcal{C} \backslash \mathcal{CB}$ is also bounded.

We represent a configuration $\mathbf{q}$ as a triple $(x, y, \theta) \in \mathbf{R}^2 \times [0, 2\pi)$, where x and y are the coordinates of $\mathcal{A}$'s reference point in $\mathcal{F}_{\mathcal{W}}$ and θ is the angle (mod 2π) between the x-axis of $\mathcal{F}_\mathcal{W}$ and the x-axis of $\mathcal{F}_\mathcal{A}$. The method described below makes use of the explicit representation of the boundary $\partial\mathcal{C}_{free} = \partial\mathcal{CB}$ computed as in Subsection 1.7 of Chapter 3 in $O(n_\mathcal{A}^3 n_\mathcal{B}^3 \log n_\mathcal{A} n_\mathcal{B})$ time, where $n_\mathcal{A}$ and $n_\mathcal{B}$ are the number of vertices of $\mathcal{A}$ and $\mathcal{B}$, respectively. This representation describes $\mathcal{C}_{free}$'s boundary as a collection of $O(n_\mathcal{A}^3 n_\mathcal{B}^3)$ C-surface patches called faces and their adjacency relation. Each face is a portion of a ruled surface generated by a line segment $L(\theta)$ whose endpoints are of the form $(f(\theta), g(\theta), \theta)$, where f and g are functions of known analytic forms. Each face is bounded by at most four edges. Two edges are line segments parallel to the xy-plane (these edges may possibly degenerate to points). The other two edges are curve segments traced by the endpoints of $L(\theta)$ when θ varies over some interval of non-zero length. Two faces are adjacent if and only if the intersection of their boundaries is a curve segment of non-zero length.

Let S be a connected component of $cl(\mathcal{C}_{free})$. The boundary ∂S of S may not be connected. However, since $\mathcal{C}\backslash\mathcal{CB}$ is bounded, S is also bounded. Hence, in all cases, there exists a connected component of ∂S, called the *external boundary* of S and denoted by ∂S_{ext}, which encloses S and consequently all the other connected components of ∂S. Consider any configuration $\mathbf{q} = (x, y, \theta)$ in S and let S_θ denote the cross-section of S by a plane at constant orientation θ. There necessarily exists a path in $S_\theta \cup \partial S$ that connects $\mathbf{q}$ to a configuration in ∂S_{ext}. From Chapter 3 we know that S_θ is a bounded polygonal region with $O(n_\mathcal{A}^2 n_\mathcal{B}^2)$ vertices.

Let $\mathbf{q}_{init}$ and $\mathbf{q}_{goal}$ be the initial and goal configurations, respectively. They are connected by a semi-free path if and only if they belong to the same connected component S of $cl(\mathcal{C}_{free})$. If this is the case, then they are connected by a semi-free path contained in $S_{\theta_{init}} \cup \partial S \cup S_{\theta_{goal}}$. One way to construct such a path is to construct and search the graph G defined as follows:

1. Construct a graph G_1 by associating a node of G_1 with every face in the boundary of $\mathcal{C}_{free}$. Connect every two such nodes by a link if the corresponding faces are adjacent.

2. Compute a trapezoidal decomposition T_{init} of $S_{\theta_{init}}$. Construct a graph G_2 by associating a node of G_2 with every cell of this decomposition. Connect every two such nodes by a link if the corresponding cells are adjacent.

3. Compute a trapezoidal decomposition T_{goal} of $S_{\theta_{goal}}$. Construct a graph G_3 by associating a node of G_3 with every cell of this decomposition. Connect every two such nodes by a link if the corresponding cells are adjacent.

4. Merge G_1, G_2 and G_3 into a graph G by connecting every node associated with a face of $\partial\mathcal{C}_{free}$ to every node associated with a cell of T_{init} or T_{goal} if the boundaries of the face and the cell share a straight segment of non-zero length.

The initial cell containing $\mathbf{q}_{init}$ and the goal cell containing $\mathbf{q}_{goal}$ are identified during the construction of T_{init} and T_{goal}. G is searched for a path connecting the initial cell to the goal cell.

G_1 is generated by the algorithm described in Subsection 1.7 of Chapter 3. Let $n = n_{\mathcal{A}}^3 n_{\mathcal{B}}^3$ be the number of faces of $\partial\mathcal{C}_{free}$. Since each face is bounded by four edges at most, the number of links in G_1 is $O(n)$. The polygonal region $S_{\theta_{init}}$ and $S_{\theta_{goal}}$ can be computed from the description of the boundary of $\partial\mathcal{C}_{free}$ in $O(n)$ time. Their boundaries contain $O(n^{2/3})$ vertices. The trapezoidal decompositions T_{init} and T_{goal} can be computed in $O(n^{2/3}\log n)$ time and contain $O(n^{2/3})$ cells. Merging the three graphs G_1, G_2 and G_3 into G takes $O(n^{2/3})$ time (the inclusion of the edges of $S_{\theta_{init}}$ and $S_{\theta_{goal}}$ in the faces of $\partial\mathcal{C}_{free}$ is determined when $S_{\theta_{init}}$ and $S_{\theta_{goal}}$ are computed). Hence, G is constructed in $O(n)$ time. It contains $O(n)$ nodes and $O(n)$ links. G can be searched in $O(n)$ time.

The outcome of the search of G is a sequence of two-dimensional cells/faces. It remains to extract a path from that sequence. We can construct this path by connecting $\mathbf{q}_{init}$ to $\mathbf{q}_{goal}$ through the midpoint of every portion of boundary shared by two successive cells/faces of T_{init}, $\partial\mathcal{C}_{free}$ and T_{goal} in the output sequence. In the case of two successive faces sharing a curved edge ranging between θ_1 and θ_2, we can take the midpoint as the point in the edge corresponding to the orientation $(\theta_1 + \theta_2)/2$. The subpath between any two configurations in a trapezoidal or triangular cell of T_{init} or T_{goal} is the straight segment joining them. The subpath between any two configurations $\mathbf{q}_1 = (x_1, y_1, \theta_1)$ and $\mathbf{q}_2 = (x_2, y_2, \theta_2)$ in a face F of $\partial\mathcal{C}_{free}$ can be computed as follows. Let $L(\theta)$ be the line segment that sweeps out F, and $P(\theta) = (x_P(\theta), y_P(\theta), \theta)$ and $Q(\theta) = (x_Q(\theta), y_Q(\theta), \theta)$ be its two endpoints (the functions x_P, y_P, x_Q and y_Q have known analytical forms). If $\theta_1 = \theta_2$, the subpath is the

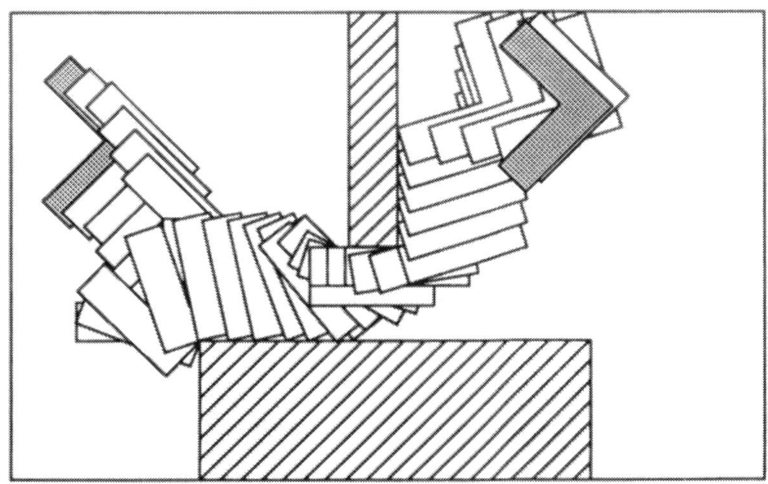

Figure 18. This figure shows a semi-free path generated by an implementation of the method described in Section 4 [Avnaim, Boissonnat and Faverjon, 1988].

line segment joining $\mathbf{q}_1$ to $\mathbf{q}_2$ (it is a subset of $L(\theta_1)$). If $\theta_1 \neq \theta_2$, the subpath is the curve segment $\gamma : \theta \in [\theta_1, \theta_2] \mapsto (x(\theta), y(\theta), \theta) \in F$, with:

$$x(\theta) = \left[\lambda_1 + (\lambda_2 - \lambda_1)\tfrac{\theta - \theta_1}{\theta_2 - \theta_1}\right](x_Q(\theta) - x_P(\theta)) + x_P(\theta),$$

$$y(\theta) = \left[\lambda_1 + (\lambda_2 - \lambda_1)\tfrac{\theta - \theta_1}{\theta_2 - \theta_1}\right](y_Q(\theta) - y_P(\theta)) + y_P(\theta),$$

where:

$$\lambda_1 = \tfrac{x_1 - x_P(\theta_1)}{x_Q(\theta_1) - x_P(\theta_1)} \quad \text{and} \quad \lambda_2 = \tfrac{x_2 - x_P(\theta_2)}{x_Q(\theta_2) - x_P(\theta_2)}.$$

Experimental results obtained with this method are described in [Avnaim, Boissonnat and Faverjon, 1988]. They show that the method is quite efficient and compares well with approximate cell decomposition methods (see Chapter 6). Figure 18 shows a path generated by the method. In this particular example, the search graph consists of approximately 270 nodes.

Exercises

1: Implement a path planner applying the exact cell decomposition method described in Section 1. If you previously implemented a program that computes the C-obstacle region $\mathcal{CB}$ for a polygon $\mathcal{A}$ that can only translate and a polygonal obstacle region $\mathcal{B}$ (Chapter 3), use this program as a front-end to the planner. Display the path generated by the planner by showing $\mathcal{A}$ in its workspace at successive configurations.

2: The method of Section 1 generates a channel consisting of a sequence of trapezoidal and triangular cells. We suggested connecting $\mathbf{q}_{init}$ to $\mathbf{q}_{goal}$ through the midpoints of the segments shared by every two successive cells in the channel. Comment on this technique, propose other techniques, and compare their respective advantages and drawbacks.

3: Consider a channel generated by the planning method of Section 1 between two configurations $\mathbf{q}_{init}$ and $\mathbf{q}_{goal}$. Explain how B-splines can be used to compute a path of class C^k lying entirely in the channel and joining $\mathbf{q}_{init}$ to $\mathbf{q}_{goal}$ [Kant and Zucker, 1986b]. Write a program based on this method that computes a C^2 path.

4: Extend the planning method of Section 1 to the case where the C-obstacles are generalized polygonal regions bounded by straight and circular edges.

5: Extend the planning method of Section 1 to the case where $\mathcal{C} = \mathbf{R}^3$ and the C-obstacle region is polyhedral.

6: Propose a method for computing the critical curves defined in Subsection 2.1 using the explicit representation of $\mathcal{CB}$'s boundary computed as in Subsection 1.7 of Chapter 3.

7: Assume that $\mathcal{B}$ is a manifold with boundary. List all the cases where two critical curves constructed as proposed in Subsection 2.1 coincide. Show that all coincidences can be eliminated by infinitesimal changes of $\mathcal{B}$ without changing its topology.

8: In Subsection 2.2 we defined two functions λ_1 and λ_2. Why do we have to differentiate between these two functions?

9: Explain in detail the two specific crossing rules given in Subsection 2.3.

10: Identify all the critical curves shown in Figure 12.

11: Discuss why in the planning approach described in Subsection 3.3 it is important that the decomposition of $\mathcal{C}_{free}$ be cylindrical.

12: Compare the decompositions of $\mathcal{C}_{free}$ into cells which are described in Sections 1 and 2 with the Collins decomposition.

13: Assume that the graph G constructed by the planning method described in Section 4 is searched using an A^* algorithm. Propose and discuss several possible heuristic functions.

Chapter 6

Approximate Cell Decomposition

In this chapter we investigate another cell decomposition approach to path planning which is known as the *approximate cell decomposition* approach. It consists again of representing the robot's free space $\mathcal{C}_{free}$ as a collection of cells. But it differs from the exact cell decomposition approach in that the cells are now required to have a simple prespecified shape, e.g. a rectangloid shape. Such cells do not in general allow us to represent free space exactly. Instead, a conservative approximation of this space is constructed, hence the name of the approach. As with the exact cell decomposition approach, a connectivity graph representing the adjacency relation among the cells is built and searched for a path.

The rationales for the standardization of the shape of the cells are (1) to achieve space decomposition by iterating the same simple computation, and (2) to be relatively insensitive to numerically approximate computations. It indeed results in planning methods which are usually much easier to implement than exact cell decomposition ones. As a matter of fact, more planners have been implemented using the approach described in this chapter than using exact cell decomposition methods. In addition, with the approximate cell decomposition approach, one can directly control the "amount of free space" around a generated path by setting a minimal size for the cells. This may be an important advantage when the errors in the geometric models and/or in robot control are significant.

On the other hand, the boundaries of the cells generated by an approximate cell decomposition method are somewhat arbitrary. They no longer characterize discontinuities of the motion constraints, as is the case with exact cell decomposition. Consequently, the methods presented below provide less insight into the mathematical structure of the basic motion planning problem. Moreover, since they represent the robot's free space conservatively, they may fail to find a free path, even if one exists. Under some simple assumptions, this drawback can be fixed, i.e. the methods can be made complete, but at the expense of an unbounded time complexity.

Most approximate cell decomposition methods allow the size of the cells to be locally adapted to the geometry of the C-obstacle region. In fact, presetting the size of the cells could result in significant difficulties: a large cell size would prevent free paths from being found in too many cases, while a small size would systematically require increased computation times. Therefore, most methods operate in a hierarchical fashion, by generating an initial coarse decomposition and then locally refining this decomposition until a free path is found or the decomposition becomes too fine.

The principle of the approximate cell decomposition approach is general and can be applied (in theory) to the basic motion planning problem in its full generality, as well as to several of its extensions (see Chapter 8). However, in practice, the time and space complexity of the methods based on this approach grows quickly with the dimension m of the configuration space. These methods are realistically applicable only when this dimension is small enough (say, $m \leq 4$). In these cases, for instance when the robot is a polygon that translates and rotates freely among polygonal obstacles in a two-dimensional workspace, they often compare favorably to the other approaches. Specific tricks exploiting the structure of a particular task domain may possibly be used to develop planners working in higher-dimensional spaces.

Section 1 gives a general description of the approximate cell decomposition approach. Sections 2 and 3 describe two techniques for approximating free space as a collection of rectangloid cells. Section 4 proposes hierarchical search techniques for exploring the connectivity graph at multiple levels of cell decomposition. Finally, Section 5 describes variants of the approach.

The approximate cell decomposition approach was first introduced by Lozano-Pérez and Brooks [Lozano-Pérez, 1981] [Brooks and Lozano-Pérez, 1983]. It was subsequently developed by other researchers (e.g. [Laugier and Germain, 1985] [Faverjon, 1986] [Kambhampati and Davis, 1986] [Zhu and Latombe, 1989]). Large subsets of the material presented in this chapter are derived from [Zhu and Latombe, 1989].

1 General Description

In the sequel a **rectangloid** designates a closed region of the following form in a Cartesian space $\mathbf{R}^n$:

$$\{(x_1, ..., x_n) \ / \ x_1 \in [x'_1, x''_1], ..., x_n \in [x'_n, x''_n]\}.$$

The differences $x''_i - x'_i$, $i = 1, ..., n$, are called the **dimensions** of the rectangloid. None of these dimensions is zero.

Let $\mathcal{A}$ be a robot whose configuration space $\mathcal{C}$ is $\mathbf{R}^N$ or $\mathbf{R}^N \times SO(N)$, with $N = 2$ or 3. If $\mathcal{C} = \mathbf{R}^N$, a configuration $\mathbf{q}$ is represented by the coordinates of $\mathcal{A}$'s reference point $O_\mathcal{A}$ in the frame $\mathcal{F}_\mathcal{W}$ attached to the workspace. If $\mathcal{C} = \mathbf{R}^2 \times SO(2) = \mathbf{R}^2 \times S^1$, a configuration is represented as (x, y, θ), where θ is the angle between the x-axes of $\mathcal{F}_\mathcal{W}$ and $\mathcal{F}_\mathcal{A}$ (the frame attached to $\mathcal{A}$). If $\mathcal{C} = \mathbf{R}^3 \times SO(3)$, a configuration is represented as $(x, y, z, \phi, \theta, \psi)$, where ϕ, θ and ψ are the Euler angles defined as in Subsection 4.2 of Chapter 2.

Without practical loss of generality, we assume that the set of possible positions of $\mathcal{A}$ is contained in a rectangloid $D \subset \mathbf{R}^N$. We represent $\mathcal{C}_{free}$ as:

$$\mathcal{C}_{free} = R \backslash \mathcal{CB},$$

where $\mathcal{CB}$ is the C-obstacle region and:

- $R = int(D)$, if $\mathcal{C} = \mathbf{R}^N$,

- $R = int(D) \times [0, 2\pi]$, if $\mathcal{C} = \mathbf{R}^2 \times S^1$,
- $R = int(D) \times [0, 2\pi] \times [0, \pi] \times [0, 2\pi]$, if $\mathcal{C} = \mathbf{R}^3 \times SO(3)$.

Let $\Omega = cl(R)$. It is a rectangloid of $\mathbf{R}^m$, where m is the dimension of the configuration space $\mathcal{C}$.

A **rectangloid decomposition** $\mathcal{P}$ of Ω is a finite collection of rectangloids $\{\kappa_i\}_{i=1,\ldots,r}$ such that:

- Ω is equal to the union of the κ_i, i.e. :

$$\Omega = \bigcup_{i=1}^{r} \kappa_i.$$

- The interiors of the κ_i's do not intersect, i.e. :

$$\forall i_1, i_2 \in [1, r], i_1 \neq i_2 : \; int(\kappa_{i_1}) \cap int(\kappa_{i_2}) = \emptyset.$$

Each rectangloid κ_i is called a **cell** of the decomposition $\mathcal{P}$ of Ω.

Two cells are **adjacent** if and only if their intersection is a set of non-zero measure in $\mathbf{R}^{m-1}$. The intersection is computed by taking into account that:

- if $\mathcal{C} = \mathbf{R}^2 \times S^1$, $(x, y, 2\pi)$ is identified with $(x, y, 0)$,
- if $\mathcal{C} = \mathbf{R}^3 \times SO(3)$, $(x, y, z, 2\pi, \theta, \psi)$, $(x, y, z, \phi, \pi, \psi)$, and $(x, y, z, \phi, \theta, 2\pi)$ are identified with $(x, y, z, 0, \theta, \psi)$, $(x, y, z, \phi, 0, 2\pi - \psi)$, and $(x, y, z, \phi, \theta, 0)$, respectively.

A cell κ_i is classified as:

- EMPTY, if and only if its interior does not intersect the C-obstacle region, i.e. $int(\kappa_i) \cap \mathcal{CB} = \emptyset$.
- FULL, if and only if κ_i is entirely contained in the C-obstacle region, i.e. $\kappa_i \subseteq \mathcal{CB}$.
- MIXED, otherwise.

The **connectivity graph** associated with a decomposition $\mathcal{P}$ of Ω is the non-directed graph G defined as follows:

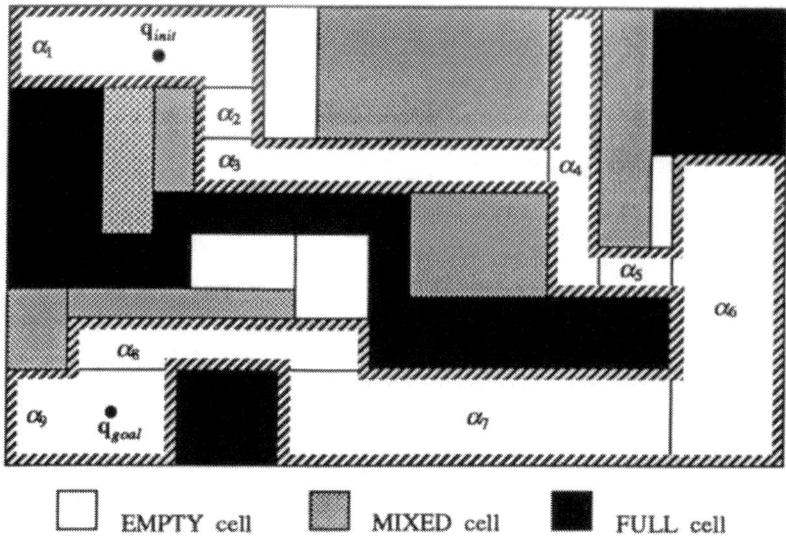

Figure 1. A channel is a sequence of adjacent cells which are either EMPTY or MIXED. If all the cells are EMPTY the channel is said to be an E-channel, otherwise it is said to be an M-channel. This figure shows an E-channel (striped contour) in a two-dimensional space. It connects the two cells that contain the initial configuration $\mathbf{q}_{init}$ and the goal configuration $\mathbf{q}_{goal}$.

- The nodes of G are the EMPTY and MIXED cells of $\mathcal{P}$.

- Two nodes of G are connected by a link if and only if the corresponding cells are adjacent.

Given a rectangloid decomposition $\mathcal{P}$ of Ω, a **channel** is defined as a sequence $(\kappa_{\alpha_j})_{j=1,\ldots,p}$ of EMPTY and/or MIXED cells such that any two consecutive cells κ_{α_j} and $\kappa_{\alpha_{j+1}}$, $j \in [1, p-1]$, are adjacent. A channel that only contains EMPTY cells is called an **E-channel** (see Figure 1). A channel that contains at least one MIXED cell is called an **M-channel**. If $(\kappa_{\alpha_j})_{j=1,\ldots,p}$ is an E-channel, then any path connecting any configuration in κ_{α_1} to any configuration in κ_{α_p} and lying in $int(\cup_{j=1}^{p} \kappa_{\alpha_j})$ is a free path. If $(\kappa_{\alpha_j})_{j=1,\ldots,p}$ is an M-channel, there may exist a free path connecting two configurations in κ_{α_1} and κ_{α_p}, and lying in $int(\cup_{j=1}^{p} \kappa_{\alpha_j})$, but there is no guarantee that this is the case.

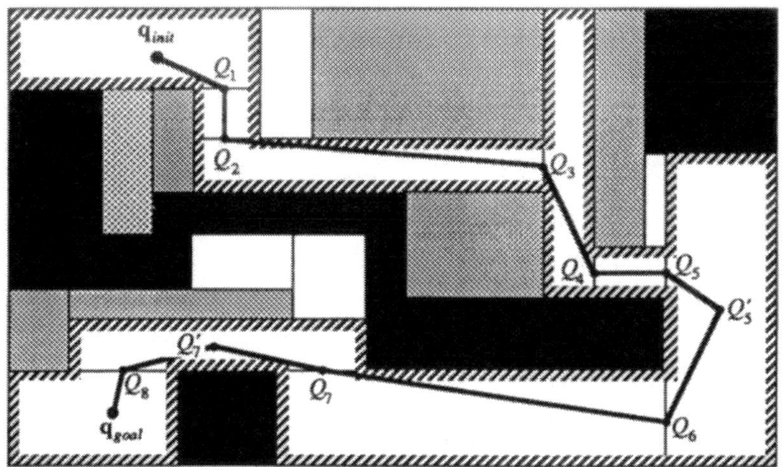

Figure 2. A path can be extracted from an E-channel by joining $\mathbf{q}_{init}$ to $\mathbf{q}_{goal}$ by a polygonal line.

Given an initial configuration $\mathbf{q}_{init} \in \mathcal{C}_{free}$ and a goal configuration $\mathbf{q}_{goal} \in \mathcal{C}_{free}$, the problem is to generate an E-channel $(\kappa_{\alpha_j})_{j=1,\dots,p}$, such that $\mathbf{q}_{init} \in \kappa_{\alpha_1}$ and $\mathbf{q}_{goal} \in \kappa_{\alpha_p}$. If such a channel is generated, let $\beta_j = \partial\kappa_{\alpha_j} \cap \partial\kappa_{\alpha_{j+1}}$, $j = 1, \dots, p-1$, be the intersection of the boundaries of two successive cells. A free path joining the initial to the goal configuration can be extracted from the E-channel by linking $\mathbf{q}_{init}$ to $\mathbf{q}_{goal}$ by a polygonal line whose vertices are points $Q_j \in int(\beta_j)$. For every j such that β_{j-1} and β_j are subsets of the same face of κ_{α_j}, an additional point Q'_{j-1} located in the interior of κ_{α_j} should be included among the path's vertices, since in this case the line segment $Q_{j-1}Q_j$ is not guaranteed to lie entirely in the robot's free space (see Figure 2). If necessary, the polygonal path can be smoothed by fitting spline curves.

Hierarchical path planning consists of generating an E-channel by constructing successive rectangloid decompositions of Ω and searching the associated connectivity graphs. (We assume that Ω is a MIXED cell, otherwise there is nothing to do. This assumption can be checked in a preprocessing step.) Let $\mathcal{P}_i$, $i = 1, 2, \dots$, denote the successive decompositions of Ω. Each decomposition $\mathcal{P}_i$ is obtained from the previous one, $\mathcal{P}_{i-1}$ (with $\mathcal{P}_0 = \{\Omega\}$), by decomposing one or several MIXED cells, the other cells being unchanged. Whenever a decomposition $\mathcal{P}_i$ is computed, the associated connectivity graph, denoted by G_i, is searched for a channel connecting $\mathbf{q}_{init}$ to $\mathbf{q}_{goal}$.

A simple first-cut planning algorithm is the following:

1. Compute a rectangloid decomposition $\mathcal{P}_1$ of Ω. Set i to 1.

2. Search the connectivity graph G_i associated with the decomposition $\mathcal{P}_i$ for a channel connecting the initial cell containing $\mathbf{q}_{init}$ to the goal cell containing $\mathbf{q}_{goal}$. If the outcome of the search is an E-channel, return success. If it is an M-channel, proceed to the next step. Otherwise, return failure.

3. Let Π_i be the M-channel generated at Step 2. Set $\mathcal{P}_{i+1}$ to $\mathcal{P}_i$. For every MIXED cell κ in Π_i, compute a rectangloid decomposition $\mathcal{P}^\kappa$ of κ and set $\mathcal{P}_{i+1}$ to $[\mathcal{P}_{i+1}\backslash\{\kappa\}] \cup \mathcal{P}^\kappa$. Set i to $i+1$. Go to Step 2.

The search of G_i at Step 2 can be guided by various heuristics. In particular, one could search for an E-channel before searching for an M-channel. But, although the heuristic function should put an extra cost on MIXED cells in order to generate an E-channel quicker, it may also be appropriate to prefer short channels over long ones (according to some metrics). Thus, although an E-channel may exist in a graph G_i, it may nevertheless be preferable to generate a significantly shorter M-channel instead, and refine G_i accordingly. Notice that any E-channel existing in G_i will continue to exist in all the graphs G_j, $j > i$.

Let us assume that the region $cl(C_{free})$ is a manifold with boundary. Then the algorithm outlined above can be made complete — i.e. guaranteed to (1) terminate and (2) return an E-channel whenever $\mathbf{q}_{init}$ and $\mathbf{q}_{goal}$ lie in the same connected component of C_{free} — by working out some details appropriately, for instance:

- the search of the connectivity graph at Step 2 should be complete and it should output an E-channel whenever one exists[1], and
- all the dimensions of every MIXED cell in $\mathcal{P}_i$ should tend toward 0 when $i \to \infty$.

[1]This constraint can easily be made weaker in order to accommodate a heuristic function that favors short channels over long ones, as suggested above.

However, for an unknown region C_{free}, there is no upper bound on the worst-case computation time.

The computing time can be bounded at the expense of completeness by imposing constraints on the minimal dimensions of the cells. For example, one possible constraint is that the total volume of the EMPTY and FULL cells in the decomposition $\mathcal{P}^{\kappa}$ of κ be greater than a predefined ratio $\lambda \in (0,1)$ of the volume of κ; in addition, every MIXED cell whose volume is smaller than a prespecified value ε is re-labeled as FULL. If such a constraint is imposed, the algorithm is no longer guaranteed to output an E-channel whenever one exists. However, if one exists, the algorithm will find one provided that both λ and ε are selected small enough. For this reason, the planning method is said to be *resolution-complete.*

Figure 3 shows paths generated by a planner based on the approximate cell decomposition approach. In all the input problems, the robot is a (possibly non-convex) polygon that can translate and rotate freely in the plane among polygonal obstacles. For this class of problems, the approach is attractive. Compared to other path planning approaches, it is relatively easy to implement and reasonably efficient to run.

2 Divide-and-Label

In this section we describe a first method for computing a rectangloid decomposition of a MIXED cell. This is a basic computation in Steps 1 and 3 of the algorithm outlined in the previous section. We call the method *divide-and-label.* It consists of first dividing a cell into smaller cells (Subsection 2.1) and next labeling each newly created cell according to its intersection with the C-obstacle region (Subsections 2.2 and 2.3).

2.1 2^m-Tree Decomposition

There are many ways to decompose a rectangloid into smaller rectangloids. Perhaps the most widely used technique is to compute a 2^m-tree decomposition, where m is the dimension of the configuration space.

A 2^m-**tree** decomposition of Ω is a tree of degree 2^m (each node which is not a leaf has exactly 2^m children). Each node of the tree is a rectangloid cell which is labeled as EMPTY, FULL, or MIXED. The root of the tree is Ω. Only those nodes which are MIXED cells may have children and

Figure 3. This figure shows paths generated by a planner based on the approximate cell decomposition approach for various input problems [Zhu and Latombe, 1989]. In every problem the robot is a polygon that can translate and rotate freely in the plane among polygonal obstacles.

then they each have 2^m of them. All the children of a cell κ have the same dimensions and are obtained by cutting each edge of κ into two segments of equal length.

If $m = 2$, the tree is called a **quadtree**. If $m = 3$, it is called an **octree**. Figure 4 shows a quadtree decomposition of a two-dimensional

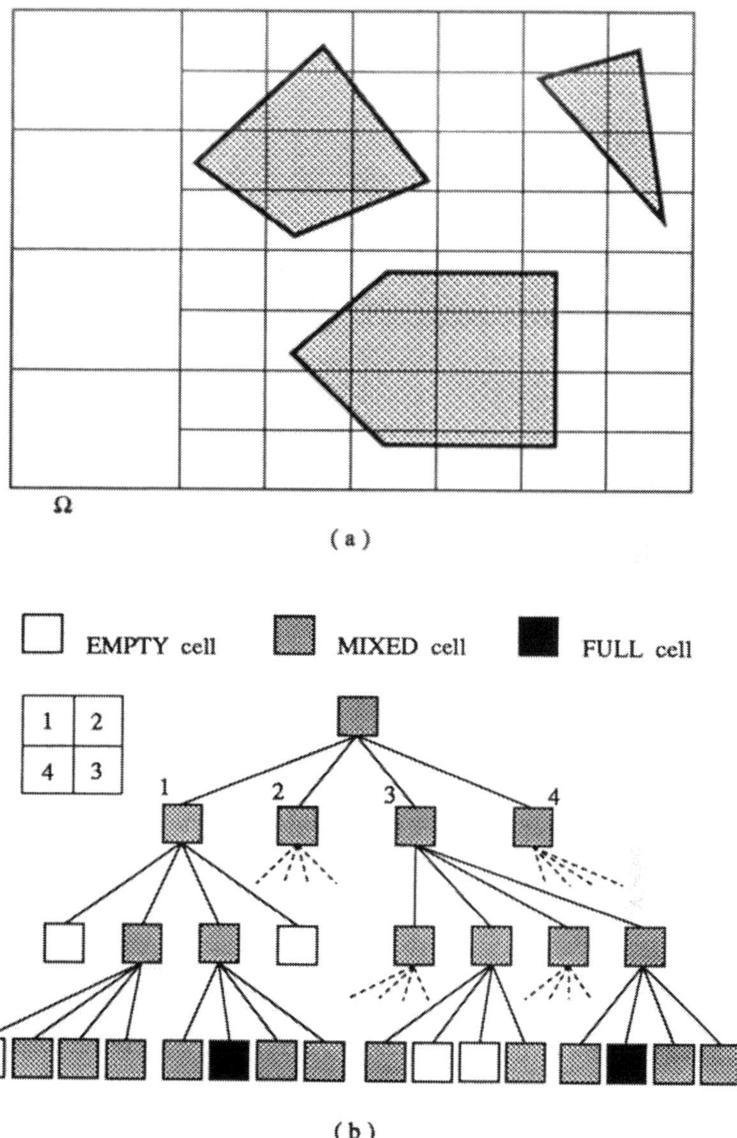

Figure 4. A quadtree decomposition of Ω is obtained by recursively dividing Ω and the generated MIXED cells into smaller cells. The division of a cell creates four new rectangloid cells of equal dimensions. Figure a shows the quadtree decomposition at depth 3 of a simple configuration space. Figure b shows a subset of the corresponding tree.

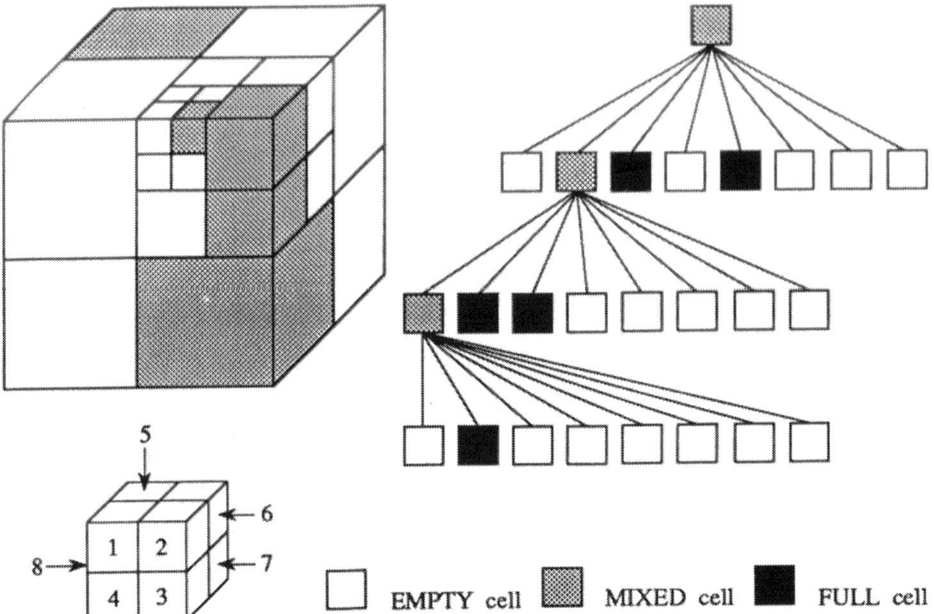

Figure 5. The concept of octree is similar to the concept of quadtree, but it applies to a three-dimensional space. Each MIXED cell is divided into eight rectangloid cells of equal size, thus generating a tree of degree 8.

configuration space and displays a portion of the quadtree graph. Figure 5 illustrates the notion of octree.

The notion of 2^m-tree has been first developed in Solid Modeling, Computer Graphics and Computer Vision (see [Jackins and Tanimoto, 1980] [Samet, 1980] [Meagher, 1982] and [Ayala et al., 1985]). Its use in collision avoidance and motion planning is more recent and reported in several papers[2], including [Ahuja et al., 1980] [Faverjon, 1984] [Boaz and Roach, 1985] [Hayward, 1986] [Herman, 1986] [Kambhampati and Davis, 1986] [Fujimura and Samet, 1989].

The depth of a node (i.e. the number of arcs joining the root to the node) in a 2^m-tree representing Ω determines the dimensions of the corresponding cell relative to Ω. The height h of the tree (i.e. the depth of

[2]Not all of them use 2^m-trees in the context of the approximate cell decomposition approach presented in this chapter.

the deepest node) determines the resolution of the decomposition of Ω, hence the size of the smallest cells in the decomposition. In order to run the planning algorithm outlined in Section 1, we must specify the height h_1 of the initial decomposition $\mathcal{P}_1$ generated at Step 1. All the MIXED cells whose depths are less than h_1 are decomposed at this step. Later on, only those MIXED cells which are located in an M-channel found at Step 2 are decomposed further. Usually, a maximal height h_{max} of the tree is also specified in order to bound the iterative process carried out by Steps 2 and 3. Every MIXED cell whose depth is equal to h_{max} is re-labeled as FULL, yielding a "truncated" 2^m-tree.

In the worst-case, the number of leaves in a 2^m-tree of height h is 2^{mh}; hence, it increases exponentially with both the dimension of $\mathcal{C}$ and the depth of the decomposition. In practice, this number is usually significantly smaller, since the tree is pruned at every EMPTY and FULL nodes. Nevertheless, the number of leaves still tends to increase exponentially with both m and h.

In the following two subsections we describe a technique for labeling the newly created cells. We limit our presentation to the cases where $\mathcal{C} = \mathbf{R}^N$ and $\mathcal{C} = \mathbf{R}^2 \times S^1$.

2.2 Cell labeling in $\mathbf{R}^N$

Let $\Omega \subset \mathbf{R}^2$. We assume that the C-obstacle region $\mathcal{CB}$ is described by the following disjunctive expression, called a **C-sentence**:

$$\bigvee_i \bigwedge_j e_{ij}$$

where e_{ij} is a C-constraint of the form $a_{ij}x + b_{ij}y + c_{ij} \leq 0$. This corresponds to the case where both the robot and the obstacles are polygonal objects, the robot only being allowed to translate. The above C-sentence is obtained by decomposing both $\mathcal{A}$ and the obstacles into convex polygons. Every conjunct $\wedge_j e_{ij}$ in the above C-sentence describes the C-obstacle that corresponds to a pair made from a convex polygon issued from the decomposition of $\mathcal{A}$ and a convex polygon issued from the decomposition of an obstacle (see Subsection 1.3 of Chapter 3). The C-sentence describing the whole C-obstacle region is associated with Ω and we denote it by S_Ω.

In a similar fashion, we associate a C-sentence S_κ with every MIXED cell κ. S_κ consists of the C-constraints which describe the intersection of κ with $\mathcal{CB}$. S_κ is iteratively derived from S_Ω as explained below. When a MIXED cell is divided into smaller cells (this occurs at Steps 1 and 3 of the algorithm given in Section 1), the C-sentence associated with it is used to compute the label (EMPTY, FULL, or MIXED) of each newly created cell. During the labeling process (described below), the C-sentence may get simplified — this occurs whenever a C-constraint e_{ij} is either satisfied or contradicted everywhere in the new cell — and the simplified C-sentence obtained at the end of the labeling process is associated with the new cell, if it is MIXED. In general, the association of simplified C-sentences with MIXED cells considerably reduces the amount of computation spent for labeling the forthcoming cells, since the number of C-constraints in S_κ tends to decrease rapidly with the size of κ, hence with its depth in the quadtree.

A cell κ is said to be **inside** the C-constraint e_{ij} if all the points $(x, y) \in \kappa$ satisfy e_{ij}. It is **outside** e_{ij} if all the points $(x, y) \in int(\kappa)$ contradict e_{ij}. It is **cut** by the C-constraint if it is neither inside nor outside e_{ij}. Testing whether a cell κ is inside or outside a C-constraint e_{ij}, or cut by it, is simple:

- κ is inside e_{ij} if each of its four vertices verifies: $a_{ij}x + b_{ij}y + c_{ij} \leq 0$.

- κ is outside e_{ij} if each of its four vertices verifies: $a_{ij}x + b_{ij}y + c_{ij} \geq 0$.

- κ is cut by e_{ij} otherwise.

The algorithm for labeling a new cell and simplifying the C-sentence is given below[3]. Its input parameters are κ and S. κ denotes the new cell to be labeled. S denotes a C-sentence initialized to the C-sentence associated with the parent cell of κ. S is represented as a set of conjuncts Λ, and each conjunct Λ is represented as a set of C-constraints e.

```
procedure LABEL(κ,S);
begin
    for each conjunct Λ ∈ S do
        begin
            for each C-constraint e ∈ Λ do
```

[3]This algorithm was originally given in [Brooks and Lozano-Pérez, 1983].

```
                    if κ is inside e
                        then Λ ← Λ\{e};
                        else if κ is outside e then S ← S\{Λ};
                if Λ ∈ S and Λ = ∅ then
                    begin
                        label κ with FULL;
                        exit from procedure;
                    end;
            end;
        if S = ∅
            then label κ with EMPTY;
            else
                begin
                    label κ with MIXED;
                    associate S with κ;
                end;
end;
```

This procedure takes $O(n)$ time to label a cell, where n is the number of C-constraints in the input C-sentence S.

The operations performed by the procedure LABEL are illustrated by the example shown in Figure 6. In this example, a MIXED cell κ' with the C-sentence $S_{\kappa'}$ attached to it is cut by three constraints e_1, e_2, and e_3. κ' is decomposed into four cells, κ_1 through κ_4. κ_1, κ_2 and κ_3 get labeled as MIXED and the simplified C-sentences S_{κ_1}, S_{κ_2}, and S_{κ_3} are attached to them, respectively. κ_4 gets labeled as EMPTY.

The procedure is slightly conservative. Indeed, it treats each C-constraint $a_{ij}x + b_{ij}y + c_{ij} \leq 0$, independently of the other C-constraints. Therefore, some cells may get labeled as MIXED, while they do not actually intersect any C-obstacle. This is illustrated in Figure 7, where the cell κ is incorrectly (but conservatively) labeled as MIXED because it is outside none of the C-constraints e_1 and e_2, although no point of κ is inside *both* e_1 and e_2. However, such an "error", which never yields incorrect paths, is remedied later if the cell κ is decomposed further. For instance, in Figure 7, the decomposition of κ generates four smaller cells which are all correctly labeled as EMPTY by the above procedure.

Remark: The C-sentences associated with the MIXED cells can be used

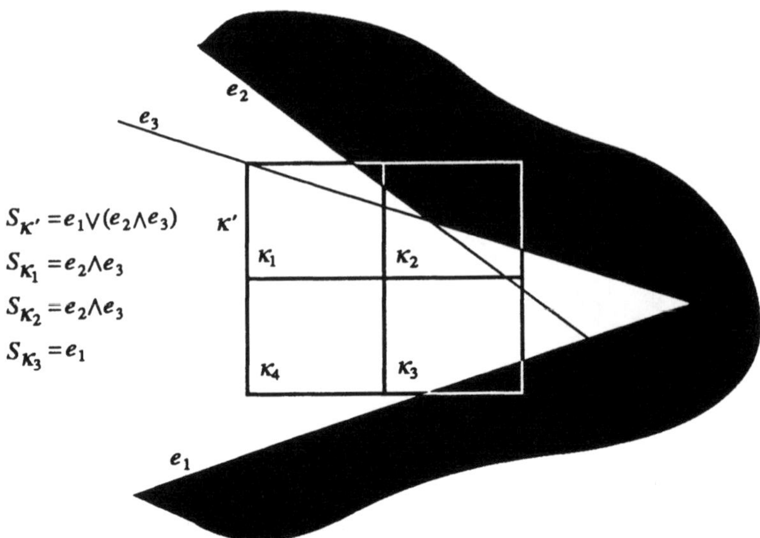

Figure 6. This figure illustrates the simplification of a C-sentence when new cells are created and labeled. The sentence $S_{\kappa'} = e_1 \vee (e_2 \wedge e_3)$ is associated with the MIXED cell κ'. When this cell is decomposed (in a quadtree fashion), four new cells denoted by κ_1 through κ_4 are generated. Both κ_1 and κ_2 are outside e_1, but are cut by e_2 and e_3. Hence, they are labeled as MIXED. The C-sentence $e_2 \wedge e_3$, is associated with κ_1 and κ_2. κ_3 is cut by e_1 and outside e_3. It is labeled as MIXED, and the C-sentence e_1 is associated with it. Finally, κ_4 is outside all three C-constraints and is labeled as EMPTY.

as follows to refine the definition of the adjacency relation between two cells. Let κ_1 and κ_2 be two cells such that one of them is MIXED and the other is either MIXED or EMPTY. The two cells are *adjacent* if and only if they satisfy the following two conditions:

- $\beta = \partial\kappa_1 \cap \partial\kappa_2$ is a line segment of non-zero length.

- β considered as a degenerate one-dimensional cell is not FULL.

The second condition can easily be tested by running the procedure LABEL on β with $S = S_{\kappa_1}$ (if κ_1 is MIXED) and $S = S_{\kappa_2}$ (if κ_2 is MIXED). The second condition is satisfied if none of the calls to LABEL labels β as FULL. ∎

The above procedure is also applicable when $\Omega \subset \mathbf{R}^3$ with polyhedral

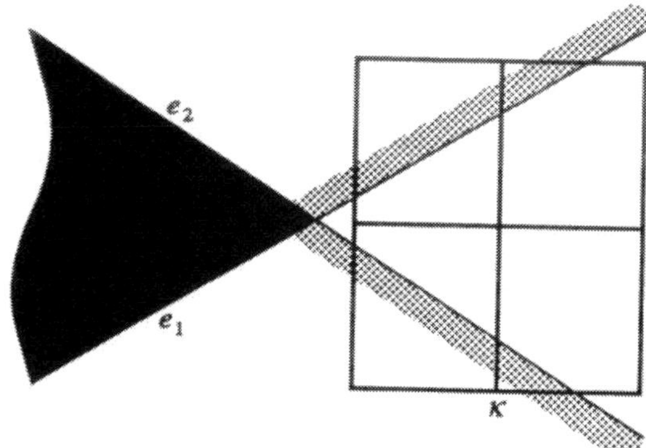

Figure 7. This figure illustrates how a cell κ gets labeled as MIXED though it has no intersection with the C-obstacle region. Assume that the C-sentence associated with the parent cell of κ is $e_1 \wedge e_2$. Since κ is cut by both e_1 and e_2, κ is labeled as MIXED and the C-sentence $e_1 \wedge e_2$ is associated with it. However, no point in κ satisfies e_1 and e_2 *simultaneously*. This incorrect (but conservative) labeling results from the fact that each C-constraint is individually considered as a half-plane. The error is eventually corrected at a deeper level in the quadtree decomposition. (The "inside" side of a C-constraint line is shown grey.)

C-obstacles. Each C-constraint e_{ij} in a C-sentence is of the form $a_{ij}x + b_{ij}y + c_{ij}z + d_{ij} \leq 0$. Determining whether a rectangloid cell κ is inside or outside e_{ij}, or cut by e_{ij}, requires the 8 vertices of κ to be checked against the C-constraint.

A drawback of the decomposition of Ω into a quadtree or an octree is that it does not take advantage of the information contained in the C-sentence associated with a MIXED cell κ in order to decide where to cut its edges. Hence, relatively many of the created cells may get labeled as MIXED, which may ultimately result in producing a large number of cells. Another decomposition technique described in [Brooks and Lozano-Pérez, 1982] makes some limited use of the information contained in the C-sentence. It consists of considering potential cuts of the cell, scoring them, and choosing the "best". The potential cuts are chosen wherever a C-surface would go through a vertex of one of the new cells generated by the decomposition. The scoring function favors cuts which do not generate small cells, i.e. cuts that are closer to the midpoints of each edge.

The scoring function also attempts to minimize the number of C-surfaces intersected by each new cell in order to reduce future computations.

2.3 Cell labeling in $\mathbf{R}^2 \times S^1$

We now extend the labeling technique of the previous subsection to the case where $\mathcal{C} = \mathbf{R}^2 \times S^1$, with $\Omega = R \times [0, 2\pi]$. We assume that the C-obstacle region $\mathcal{CB}$ is described as *a collection* of C-sentences σ, each being of the form:

$$\bigvee_i \bigwedge_j e_{ij}$$

where e_{ij} is a C-constraint $a_{ij}(\theta)x + b_{ij}(\theta)y + c_{ij}(\theta) \leq 0$. Each C-sentence σ only applies over a certain closed interval δ_σ of orientations θ of $\mathcal{A}$; hence, it only describes the C-obstacle region within this angular interval. This corresponds to the case where both the robot and the obstacles are polygonal objects, the robot being allowed to translate and rotate freely in the plane (see Section 1 of Chapter 3). Both endpoints of δ_σ are critical orientations of $\mathcal{A}$ where the nature of the possible contacts between $\mathcal{A}$ and the obstacles changes, i.e. an edge of $\mathcal{A}$ is parallel to an obstacle edge. The intervals δ_σ do not overlap and their union is equal to $[0, 2\pi]$. The collection $\mathcal{S}_\Omega$ of C-sentences describing $\mathcal{CB}$ is associated with Ω (together with the intervals δ_σ corresponding to these sentences).

Let $\kappa = [x_\kappa, x'_\kappa] \times [y_\kappa, y'_\kappa] \times [\theta_\kappa, \theta'_\kappa]$ be a cell and $\mathcal{S}$ be the collection of C-sentences associated with its parent cell. (If $\kappa = \Omega$, $\mathcal{S} = \mathcal{S}_\Omega$.) The labeling of κ is made more difficult than in the previous subsection for two reasons:

1. The applicable C-sentence may vary over the orientation range of κ.

2. C-constraints are non-linear in the orientation θ of $\mathcal{A}$.

Multiplicity of C-sentences. Let e_{ij} be a C-constraint in a C-sentence of $\mathcal{S}$. If it does not apply to the whole cell κ, we cannot say that κ is "inside" or "outside" e_{ij}, or "cut" by it, in the fashion defined in Subsection 2.2.

One way to solve this difficulty is to build a separate octree for the interval δ_σ associated with every C-sentence in $\mathcal{S}_\Omega$. However, this technique yields $O(n_{\mathcal{A}}n_{\mathcal{B}})$ octrees, where $n_{\mathcal{A}}$ and $n_{\mathcal{B}}$ are the respective complexity of $\mathcal{A}$ and the obstacle region $\mathcal{B}$. In addition, these octrees would in general have different angular thicknesses.

A better technique which constructs a single octree is the following:

- We partition the interval $[\theta_\kappa, \theta'_\kappa]$ into r maximal closed subintervals I_l, $l \in [1, r]$, such that a single C-sentence of $\mathcal{S}$, denoted by S_l, applies over each I_l.

- For every $l \in [1, r]$, we consider the subcell κ_l of κ defined as:

$$\kappa_l = [x_\kappa, x'_\kappa] \times [y_\kappa, y'_\kappa] \times I_l.$$

  Let e be any C-constraint in S_l. We say that κ_l is **inside** e if all the points $(x, y, \theta) \in \kappa_l$ satisfy e. We say that it is **outside** e if all the points $(x, y, \theta) \in int(\kappa_l)$ contradict e. If κ_l is neither inside nor outside e, we say that it is **cut** by e. In order to label κ_l as EMPTY, FULL, or MIXED, we apply the procedure LABEL(κ_l, S_l) given in Subsection 2.2, with these definitions. (As we will see below, a C-sentence S_l may be equal to `true` or `false`. If S_l = `true` (resp. `false`), then κ_l is directly labeled as FULL (resp. EMPTY) without running LABEL.)

- After the procedure LABEL has been applied to all the subcells κ_l, $l = 1, ..., r$, we label κ as:
  - EMPTY, if all the subcells κ_l have been labeled as EMPTY,
  - FULL, if all the subcells κ_l have been labeled as FULL,
  - MIXED, otherwise.

- While it was applied, the procedure LABEL simplified every C-sentence S_l into S'_l. If κ has been labeled as MIXED, the collection of simplified C-sentences $\mathcal{S}_\kappa = \{S'_1, ..., S'_{r'}\}$, is associated with κ, with S'_l set to `true` (resp. `false`) if κ_l has been labeled as FULL (resp. EMPTY).

Non-linearity of C-constraints. We now need to compute whether a subcell κ_l is inside or outside a C-constraint e, or cut by it. This subcell is a rectangloid of the form:

$$[x_\kappa, x'_\kappa] \times [y_\kappa, y'_\kappa] \times [\theta_l, \theta'_l]$$

where $I_l = [\theta_l, \theta'_l]$ is contained in the applicability domain of e. The C-constraint e is of the form:

$$a(\theta)x + b(\theta)y + c(\theta) \leq 0.$$

Consider the projection of the portion of the C-surface $a(\theta)x + b(\theta)y + c(\theta) = 0$ comprised in the interval $[\theta_l, \theta'_l]$ on the xy-plane. Let (x, y) be a point outside this projection. The segment connecting (x, y, θ_l) to (x, y, θ'_l) lies on a single side of the C-surface, i.e. it is either completely inside or completely outside e. Inversely, let (x, y) be a point inside the projection of the surface. The segment connecting (x, y, θ_l) to (x, y, θ'_l) crosses the C-surface. Hence, part of it lies inside e and part of it lies outside. Thus, we can check whether κ_l is inside, outside, or cut by e by first projecting the $[\theta_l, \theta'_l]$ slice of the C-surface into the xy-plane and next comparing the position of the rectangle $[x_\kappa, x'_\kappa] \times [y_\kappa, y'_\kappa]$ to this projection. We detail this computation below.

Recall that there are two types of C-constraints:

- A type A C-constraint is caused by the possible contact between an edge $E_i^{\mathcal{A}}$ connecting vertices a_i and a_{i+1} of $\mathcal{A}$ and an obstacle vertex b_j. It is of the form (see Subsection 1.4 of Chapter 3):

$$-x\cos(\phi_i + \theta) - y\sin(\phi_i + \theta) + \|b_j\|\cos(\phi_i + \theta - \beta_j) - \|a_i\|\cos(\phi_i - \alpha_i) \leq 0.$$

- A type B C-constraint is caused by the possible contact between a vertex a_i of $\mathcal{A}$ and an obstacle edge $E_j^{\mathcal{B}}$ connecting vertices b_j and b_{j+1}. It is of the form:

$$x\cos\xi_j + y\sin\xi_j + \|a_i\|\cos(\alpha_i + \theta - \xi_j) - \|b_j\|\cos(\beta_j - \xi_j) \leq 0.$$

The intersection of a C-surface with a plane at constant θ is a straight line $L(\theta)$. When $\mathcal{A}$ rotates slightly, if the constraint is of type A, this line rotates by the same angle around the contact vertex b_j. If the constraint is of type B, the line may translate (but does not rotate) and remains parallel to the contact edge $E_j^{\mathcal{B}}$. Thus, the projection of the $[\theta_l, \theta'_l]$ slice of a type A C-surface into the xy-plane is the region swept out by a line (the projection of $L(\theta)$) rotating around the vertex b_j. This projection is the union of the area comprised between two extreme lines which are the projections of $L(\theta_l)$ and $L(\theta'_l)$ and the area comprised between these two

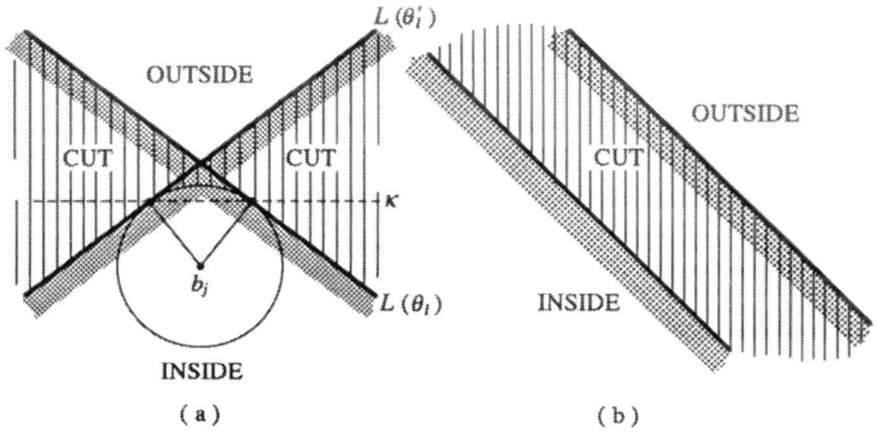

Figure 8. This figure illustrates the projection of the portion of the C-surface $a(\theta)x + b(\theta)y + c(\theta) = 0$ which is comprised in the angular interval $[\theta_l, \theta_l']$ into the xy-plane. When the C-surface is of type A, the projection is the region swept out by a line rotating around an obstacle vertex b_j (Figure a). When the C-surface is of type B, the projection is the region swept out by a translating line parallel to an obstacle edge $E_j^{\mathcal{B}}$ (Figure b). In both cases, the projection divides the plane into three regions designated by OUTSIDE, INSIDE, and CUT. For every (x, y) in the OUTSIDE (resp. INSIDE) region, the segment joining (x, y, θ_l) to (x, y, θ_l') is completely outside (resp. inside) the C-constraint $a(\theta)x + b(\theta)y + c(\theta) \leq 0$. In both Figures a and b, the CUT regions are the striped regions.

extreme lines on the one hand and the circle centered at b_j and tangent to these two lines on the other hand (see Figure 8.a). The projection of the $[\theta_l, \theta_l']$ slice of C-surface of type B is the region swept out by a translating line (the projection of $L(\theta)$) parallel to $E_j^{\mathcal{B}}$. This projection is a ribbon bounded by two parallel lines as shown in Figure 8.b.

The projection of the slice of C-surface comprised between θ_l and θ_l' is referred to as the CUT region in Figures 8.a and 8.b. It separates two other regions of the plane which are referred to as the OUTSIDE and INSIDE regions. For every (x, y) in the OUTSIDE (resp. INSIDE) region, the segment joining (x, y, θ_l) to (x, y, θ_l') is completely outside (resp. inside) the C-constraint $a(\theta)x + b(\theta)y + c(\theta) \leq 0$.

Notice that in the case of the projection of a C-surface of type A, the vertex b_j can be either in the INSIDE or the OUTSIDE region of the xy-plane. It is in the INSIDE region (as in Figure 8.a) if $\cos(\phi_i - \alpha_i) > 0$, which is the most typical situation. It is in the OUTSIDE region if $\cos(\phi_i - \alpha_i) < 0$, which implies that $\mathcal{A}$'s reference point $O_{\mathcal{A}}$ is in the outer half-plane of the edge $E_i^{\mathcal{A}}$, a situation which may occur when $\mathcal{A}$ is not convex. It is exactly at the intersection of $L(\theta_l)$ and $L(\theta_l')$ if $\cos(\phi_i - \alpha_i) = 0$.

The cell κ_l is outside the C-constraint e if the four points (x_κ, y_κ), (x_κ, y_κ'), (x_κ', y_κ), and (x_κ', y_κ') lie in the closure of the OUTSIDE region. It is inside e if they all lie in the closure of the INSIDE region. It is cut by e otherwise.

It is easy to determine the position of the four points relative to the projection of a type B C-surface. It suffices to compare them to the two lines that bound the projection. The equations of these lines are:

$$x \cos \xi_j + y \sin \xi_j + \|a_i\| \cos(\alpha_i + \theta_a - \xi_j) - \|b_j\| \cos(\beta_j - \xi_j) = 0$$

and:

$$x \cos \xi_j + y \sin \xi_j + \|a_i\| \cos(\alpha_i + \theta_b - \xi_j) - \|b_j\| \cos(\beta_j - \xi_j) = 0$$

where θ_a and θ_b provide the extreme values for $\cos(\alpha_i + \theta - \xi_j)$ over $\theta \in [\theta_l, \theta_l']$. Thus, θ_a and $\theta_b \in \{\theta_l, \theta_l', \xi_j - \alpha_i\}$ if $\xi_j - \alpha_i \in [\theta_l, \theta_l']$; θ_a and $\theta_b \in \{\theta_l, \theta_l'\}$ otherwise.

Determining the position of the four points (x_κ, y_κ), (x_κ, y_κ'), (x_κ', y_κ), and (x_κ', y_κ') relative to the projection of a type A C-surface is hardly more involved. Assume that b_j is in the INSIDE region (the symmetrical case is treated in a similar fashion). Each of the four points is in the OUTSIDE region if it lies on the "outside" side of both $L(\theta_l)$ and $L(\theta_l')$. It is in the INSIDE region if it lies on the "inside" side of both $L(\theta_l)$ and $L(\theta_l')$ and not in the small area between the two lines and the circle. Let K be the line supporting the chord joining the two points where the circle is tangent to the extreme two lines (the dashed line in Figure 8.a). Testing that a point is not in the small area can be done by first checking that it is not distant from b_j by more than the radius of the circle, and if it is, by checking that it is on the same side of K as b_j.

In the same fashion as in Section 2.2, a cell may be labeled as MIXED, while it actually intersects no C-obstacle. Again, this conservative error will ultimately be remedied if the cell is decomposed further.

3 Approximate-and-Decompose

In this section we describe a second method for computing a rectangloid decomposition of a MIXED cell κ. The goal of this method, proposed in [Zhu and Latombe, 1989], is to reduce the total volume of the newly created MIXED cells relative to the volume of κ, while producing a reasonably small total number of cells. We call the method *approximate-and-decompose*. It decomposes a cell by first analyzing and approximating the space occupied by the C-obstacle region. The labeling of the generated cells is a by-product of the decomposition process. We present the method in the case where $\mathcal{C} = \mathbf{R}^2 \times S^1$.

3.1 Outline

The approximate-and-decompose method consists of first approximating the intersection of the C-obstacle region $\mathcal{CB}$ with the cell κ to be decomposed as a collection of non-overlapping rectangloids. The complement of a union of rectangloids within a rectangloid can easily be computed as a union of rectangloids. In order to generate EMPTY and FULL cells, as well as MIXED cells, two types of approximation are computed: a *bounding* approximation and a *bounded* approximation.

Let $\mathcal{CB}[\kappa] = \mathcal{CB} \cap \kappa$. The two types of approximation are defined as follows:

- A **bounding rectangloid approximation** of $\mathcal{CB}[\kappa]$ is a collection of non-overlapping rectangloids R_i, $i = 1, ..., p$, whose union contains $\mathcal{CB}[\kappa]$, i.e. :
  - for every $i \in [1,p]$, $R_i \subseteq \kappa$,
  - for every $i, i' \in [1,p]$, $i \neq i'$, $int(R_i) \cap int(R_{i'}) = \emptyset$, and
  - $\mathcal{CB}[\kappa] \subseteq \bigcup_{i=1,...,p} R_i$.
- A **bounded rectangloid approximation** of $\mathcal{CB}[\kappa]$ is a collection of non-overlapping rectangloids R'_i, $i = 1, ..., p'$, whose union is contained in $\mathcal{CB}[\kappa]$, i.e. :
  - for every $i \in [1,p']$, $R'_i \subseteq \kappa$,
  - for every $i, i' \in [1,p']$, $i \neq i'$, $int(R'_i) \cap int(R'_{i'}) = \emptyset$, and
  - $\bigcup_{i=1,...,p'} R'_i \subseteq \mathcal{CB}[\kappa]$.

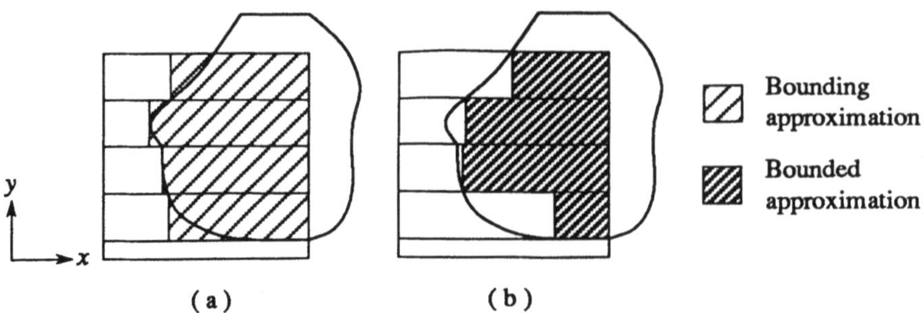

Figure 9. This figure illustrates the notions of bounding (Figure a) and bounded (Figure b) approximations in a rectangular cell. In this example, both approximations are built by slicing the y-axis into the same intervals.

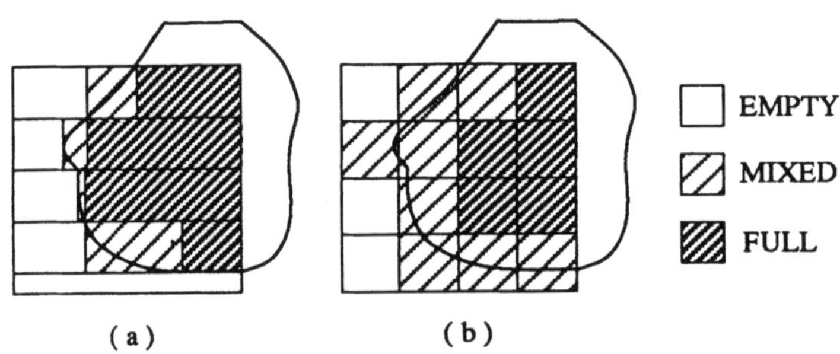

Figure 10. Figure a illustrates the decomposition of a rectangular cell into a collection of EMPTY, FULL and MIXED cells built using the bounding and bounded approximations of Figure 9. Figure b shows the quadtree decomposition of depth 2 of the same cell. The area covered by the MIXED cells is smaller in Figure a than in Figure b, though the total number of cells is smaller.

The EMPTY cells of the decomposition of κ are obtained by computing the complement of $\cup_i R_i$ in κ. The FULL cells of the decomposition are the R'_i's. The MIXED cells are obtained by computing the complement of $\cup_{i'} R'_{i'}$ in every R_i.

Figure 9 illustrates the notion of bounding and bounded approximations of $\mathcal{CB}[\kappa]$ in a two-dimensional cell κ. In this illustration, both decompositions are built by slicing the y-axis into the same intervals, which makes the subsequent computation of the MIXED cells even simpler. Figure 10.a illustrates the rectangloid decomposition of κ into EMPTY, FULL and MIXED cells which is derived from these two approximations. Figure 10.b shows the quadtree decomposition (at depth 2) of the same cell.

There are many ways of generating bounding and bounded approximations of $\mathcal{CB}[\kappa]$. The method given below computes "outer" and "inner" projections of $\mathcal{CB}$ into the xy-plane, next into the x- or y-axis. Then, it lifts back the projections into rectangloid cells. Let:

$$\kappa = [x_\kappa, x'_\kappa] \times [y_\kappa, y'_\kappa] \times [\theta_\kappa, \theta'_\kappa].$$

The method consists of the following two steps:

1. **Decomposition of** $[\theta_\kappa, \theta'_\kappa]$: The range of orientations $[\theta_\kappa, \theta'_\kappa]$ is cut into non-overlapping intervals $[\gamma_u, \gamma_{u+1}]$, $u = 1, ..., r$, $r \geq 1$, with $\gamma_1 = \theta_\kappa$ and $\gamma_{r+1} = \theta'_\kappa$. We denote the rectangloid $[x_\kappa, x'_\kappa] \times [y_\kappa, y'_\kappa] \times [\gamma_u, \gamma_{u+1}]$ by κ^u.

   For every $u \in [1, r]$, we compute the **outer projection** and the **inner projection** of $\mathcal{CB}[\kappa^u]$ into the xy-plane (see Subsection 3.2). These two projections, which we respectively denote by $\mathcal{OCB}_{xy}[\kappa^u]$ and $\mathcal{ICB}_{xy}[\kappa^u]$, are defined by:

   $$\mathcal{OCB}_{xy}[\kappa^u] = \{(x, y) \,/\, \exists\theta \in [\gamma_u, \gamma_{u+1}] : (x, y, \theta) \in \mathcal{CB}[\kappa^u]\},$$

   $$\mathcal{ICB}_{xy}[\kappa^u] = \{(x, y) \,/\, \forall\theta \in [\gamma_u, \gamma_{u+1}] : (x, y, \theta) \in \mathcal{CB}[\kappa^u]\}.$$

   Clearly, we have: $\mathcal{ICB}_{xy}[\kappa^u] \subseteq \mathcal{OCB}_{xy}[\kappa^u]$.

2. **Decomposition of** $[x_\kappa, x'_\kappa]$ **and** $[y_\kappa, y'_\kappa]$: For every $u \in [1, r]$, the interval $[x_\kappa, x'_\kappa]$ or $[y_\kappa, y'_\kappa]$, whichever is longer, is cut into non-overlapping subintervals.

   Let us assume that $[x_\kappa, x'_\kappa]$ is subdivided (a similar presentation would be made if $[y_\kappa, y'_\kappa]$ was decomposed instead). The generated subintervals are $[x_v, x_{v+1}]$, $v = 1, ..., s$, $s \geq 1$ with $x_1 = x_\kappa$ and $x_{s+1} = x'_\kappa$. We denote the rectangloid $[x_v, x_{v+1}] \times [y_\kappa, y'_\kappa] \times [\gamma_u, \gamma_{u+1}]$ by κ^{uv}.

   For every $v \in [1, s]$, we compute the outer projection of $\mathcal{OCB}_{xy}[\kappa^u]$ into the y-axis, i.e. :

   $$\mathcal{OCB}_y[\kappa^{uv}] = \{y \,/\, \exists x \in [x_v, x_{v+1}] : (x, y) \in \mathcal{OCB}_{xy}[\kappa^u]\}$$

and the inner projection of $\mathcal{ICB}_{xy}[\kappa^u]$ into the y-axis, i.e. :

$$\mathcal{ICB}_y[\kappa^{uv}] = \{y \;/\; \forall x \in [x_v, x_{v+1}] : (x,y) \in \mathcal{ICB}_{xy}[\kappa^u]\}.$$

We have: $\mathcal{ICB}_y[\kappa^{uv}] \subseteq \mathcal{OCB}_y[\kappa^{uv}]$.

Each rectangloid $[x_v, x_{v+1}] \times [c, c'] \times [\gamma_u, \gamma_{u+1}]$, where $[c, c']$ is a maximal connected interval in $\mathcal{OCB}_y[\kappa^{uv}]$ (resp. $\mathcal{ICB}_y[\kappa^{uv}]$) is a rectangloid R_i (resp. R'_i) in the bounding (resp. bounded) approximation of $\mathcal{CB}[\kappa]$.

The choice of the γ_u's and the x_v's is essentially empirical. One should, however, try to keep the three dimensions of every MIXED cell approximately "homogeneous". Let δx, δy and $\delta\theta$ be these dimensions. We say they are "homogeneous" if $\delta x \approx \delta y \approx \rho\delta\theta$, where ρ is the maximal distance between $\mathcal{A}$'s reference point and the points in $\mathcal{A}$'s boundary.

Experimental results with this decomposition method are given in [Zhu and Latombe, 1989]. Comparison to the divide-and-label method shows that the it usually produces a much smaller number of cells.

3.2 Computation of the Outer and Inner Projections

In this subsection we describe a method for computing the outer and inner projections of $\mathcal{CB}[\kappa^u]$.

Principle. Let us first assume that both $\mathcal{A}$ and $\mathcal{B}$ are convex polygons and that $\mathcal{A}$'s reference point lies in $int(\mathcal{A})$. In $\mathbf{R}^2 \times [0, 2\pi]$, $\mathcal{CB}$ is a volume bounded by faces formed by patches of ruled surfaces.

Consider a point (x_0, y_0) in the xy-plane and the segment $\{(x_0, y_0, \theta) \;/\; \theta \in [\gamma_u, \gamma_{u+1}]\}$ above this point. If the segment pierces a face of $\mathcal{CB}$, then (x_0, y_0) is in $\mathcal{OCB}_{xy}[\kappa^u]$, but not in $\mathcal{ICB}_{xy}[\kappa^u]$. If it pierces no face, then either (x_0, y_0) is not in $\mathcal{OCB}_{xy}[\kappa^u]$, or it is in $\mathcal{ICB}_{xy}[\kappa^u]$. The segment $\{(x_0, y_0, \theta) \;/\; \theta \in [\gamma_u, \gamma_{u+1}]\}$ pierces a face F if and only if (x_0, y_0) lies in the projection of F into the xy-plane.

We show below that the intersection of every face with the interval $[\gamma_u, \gamma_{u+1}]$ projects into the xy-plane according to a generalized polygon. Therefore, the projection of the boundary of $\mathcal{CB}$ that is comprised between γ_u and γ_{u+1} is the union of generalized polygons. Under the

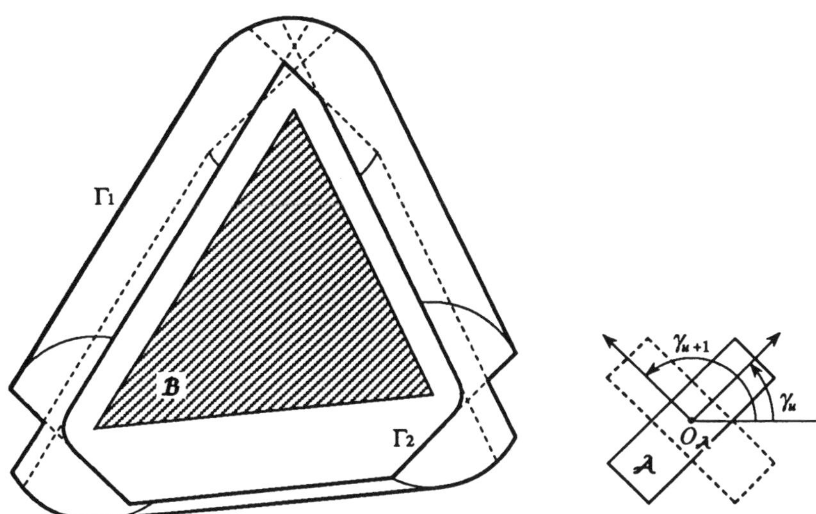

Figure 11. $\mathcal{A}$ and $\mathcal{B}$ are convex polygons. The reference point of $\mathcal{A}$ is selected in the interior of $\mathcal{A}$. The faces of $\mathcal{CB}$ which are comprised in an angular interval $[\gamma_u, \gamma_{u+1}]$ project into the xy-plane according to generalized polygons. The union of these generalized polygons is a "donut" shaped region bounded by an external boundary Γ_1 and an internal boundary Γ_2.

assumptions made above, this union is a "donut" shaped region (see Figure 11), with an external boundary Γ_1 and an internal boundary Γ_2. The generalized polygon bounded by Γ_1 is $\mathcal{OCB}_{xy}[\kappa^u]$. The generalized polygon bounded by Γ_2 is $\mathcal{ICB}_{xy}[\kappa^u]$.

The outer and the inner projections of $\mathcal{CB}[\kappa^u]$ are computed by: (1) projecting all the faces of $\mathcal{CB}$ (to the extent they are contained in the interval $[\gamma_u, \gamma_{u+1}]$) into the xy-plane; (2) tracking Γ_1 and Γ_2 and clipping them by the rectangle $[x_\kappa, x'_\kappa] \times [y_\kappa, y'_\kappa]$. The sets $\mathcal{OCB}_y[\kappa^{uv}]$ and $\mathcal{ICB}_y[\kappa^{uv}]$ are easily derived from these results.

Projection of Type A Face. Let F_A be a face of type A comprised between the limit orientations ϕ_1 and ϕ_2. The contact that generates F_A is illustrated in Figure 12.a. It occurs between an edge $E^{\mathcal{A}}$ of $\mathcal{A}$ and a vertex b of $\mathcal{B}$. $E^{\mathcal{A}}$ connects two vertices of $\mathcal{A}$, denoted by a_1 and a_2. The vertex b is the extremity of two edges of $\mathcal{B}$, $E_1^{\mathcal{B}}$ and $E_2^{\mathcal{B}}$. The reference

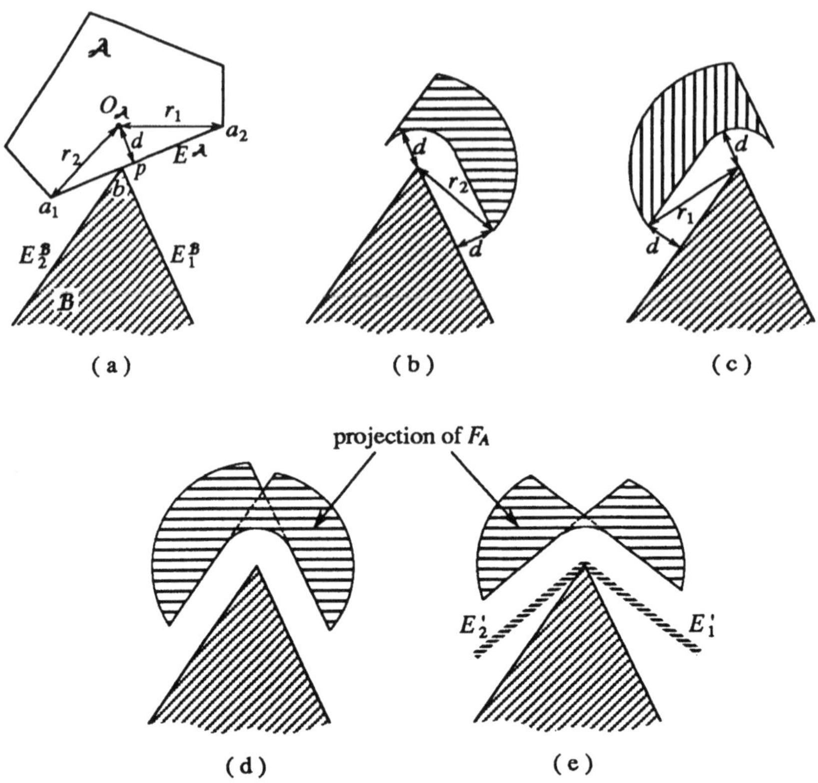

Figure 12. This figure illustrates the computation of the projection of a face of type A into the xy-plane.

point $O_{\mathcal{A}}$ projects into the supporting line of $E^{\mathcal{A}}$ at the point p. We assume below that p is located between a_1 and a_2. (The case where p is outside the segment a_1a_2 would be treated in a similar fashion.) The orientation ϕ_1 (resp. ϕ_2) is achieved when the edge $E^{\mathcal{A}}$ is colinear with the edge $E_1^{\mathcal{B}}$ (resp. $E_2^{\mathcal{B}}$). Assuming that $[\phi_1, \phi_2] \subseteq [\gamma_u, \gamma_{u+1}]$, the projection of F_A into the xy-plane is shown in Figure 12.d. It is obtained as the union of two regions shown in Figures 12.b and 12.c.

The region in Figure 12.b is the locus of $O_{\mathcal{A}}$ when $\mathcal{A}$ translates and rotates while the edge segment a_1p stays in contact with the vertex b. This region is bounded by two circular arcs and two straight segments. The two arcs are centered at b. The smaller one is the locus of $O_{\mathcal{A}}$

when p coincides with b and $\mathcal{A}$ rotates from ϕ_1 to ϕ_2. The larger arc is the locus of $O_{\mathcal{A}}$ when a_1 coincides with b and $\mathcal{A}$ rotates from ϕ_1 to ϕ_2. The straight segments are the loci of $O_{\mathcal{A}}$ when $\mathcal{A}$ translates at fixed orientations ϕ_1 and ϕ_2 from the position where p and b coincide to the position where a_1 and b coincide. The region in Figure 12.c is the locus of $O_{\mathcal{A}}$ when $\mathcal{A}$ translates and rotates while the edge segment pa_2 stays in contact with the vertex b.

The case where $[\gamma_u, \gamma_{u+1}] \subset [\phi_1, \phi_2]$ is treated in the same way, after having drawn fictitious edges E'_1 and E'_2 from vertex b (see Figure 12.e). These edges are such that $\mathcal{A}$'s orientation is γ_u (resp. γ_{u+1}) when $E^{\mathcal{A}}$ is colinear with E'_1 (resp. E'_2).

Projection of a Type B C-Face. Let F_B be a face of type B comprised between the orientations ϕ_1 and ϕ_2. The contact that generates F_B is illustrated in Figure 13.a. It occurs between a vertex a of $\mathcal{A}$ and an edge $E^{\mathcal{B}}$ of $\mathcal{B}$. a is the extremity of two edges of $\mathcal{A}$, $E_1^{\mathcal{A}}$ and $E_2^{\mathcal{A}}$. $E^{\mathcal{B}}$ connects two vertices of $\mathcal{B}$, b_1 and b_2. The orientation ϕ_1 (resp. ϕ_2) is achieved when the edge $E_2^{\mathcal{A}}$ (resp. $E_1^{\mathcal{A}}$) is colinear with the edge $E^{\mathcal{B}}$. Assuming that $[\phi_1, \phi_2] \subseteq [\gamma_u, \gamma_{u+1}]$, the projection of F_B into the xy-plane is shown in Figure 13.d. It is obtained as the union of two regions shown in Figures 13.b and 13.c.

Under the assumptions made above, $O_{\mathcal{A}}$ lies within the convex angular sector bounded by the two half-lines supporting $E_1^{\mathcal{A}}$ and $E_2^{\mathcal{B}}$ and erected from a. Let ψ be the orientation of $\mathcal{A}$ when the segment $aO_{\mathcal{A}}$ is perpendicular to $E^{\mathcal{B}}$. The region in Figure 13.b is the locus of $O_{\mathcal{A}}$, when $\mathcal{A}$ translates and rotates with the vertex a staying in $E^{\mathcal{B}}$ and the orientation θ ranging over $[\phi_1, \psi]$. (If $\psi < \phi_1$, then this region is empty.) This region is bounded by two circular arcs and two straight segments. The two arcs are centered at b_1 and b_2, respectively, and have the same radius which is equal to the distance r between a and $O_{\mathcal{A}}$. The region in Figure 13.c is the locus of $O_{\mathcal{A}}$ when $\mathcal{A}$ translates and rotates with a staying in $E^{\mathcal{B}}$ and the orientation θ ranging over $[\psi, \phi_2]$. (If $\psi > \phi_2$, this region is empty.)

The case where $[\gamma_u, \gamma_{u+1}] \subset [\phi_1, \phi_2]$ is treated in the same way, after having drawn two fictitious edges E'_1 and E'_2 from vertex a (see Figure 13.e and 13.f). These edges are such that $\mathcal{A}$'s orientation is γ_u (resp. γ_{u+1}) when E'_1 (resp E'_2) is colinear with $E^{\mathcal{B}}$.

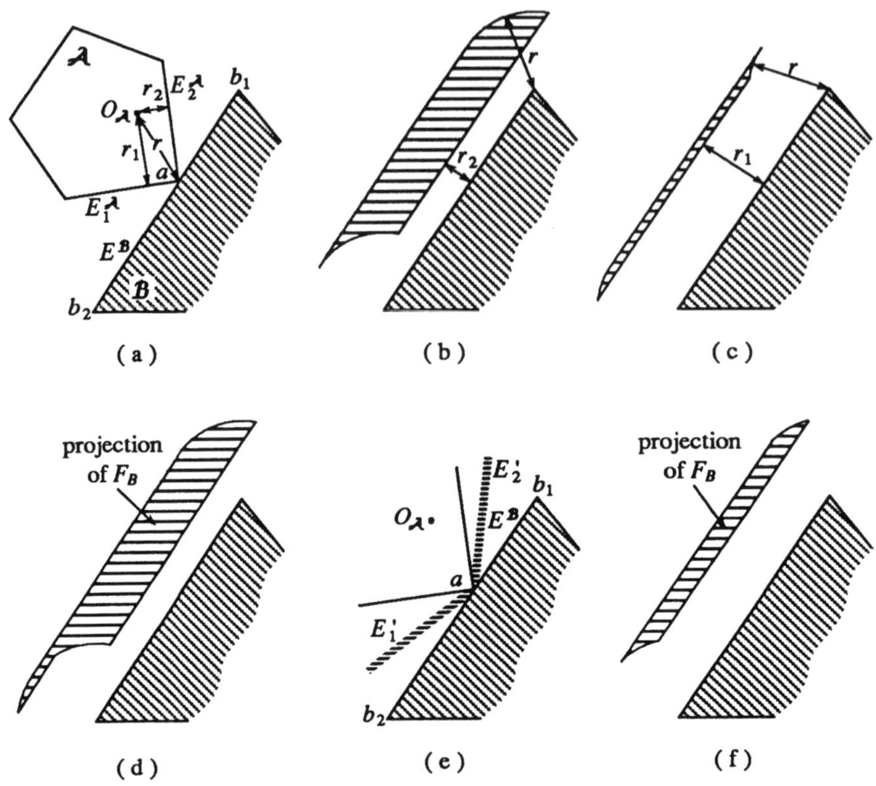

Figure 13. This figure illustrates the computation of the projection of a face of type B into the xy-plane.

Computation of $\mathcal{OCB}_{xy}[\kappa^u]$ and $\mathcal{ICB}_{xy}[\kappa^u]$. The projection of every face is a generalized polygon with a bounded number of edges.

The contours Γ_1 and Γ_2 (see Figure 11) are computed by sweeping a line across the xy-plane. Assuming that the $[x_\kappa, x'_\kappa]$ interval will be decomposed next in order to compute $\mathcal{OCB}_y[\kappa^{uv}]$ and $\mathcal{ICB}_y[\kappa^{uv}]$, the sweep-line is chosen parallel to the x-axis. The sweep starts at the smallest ordinate of all the points in the generalized polygons. During the sweep, the contours Γ_1 and Γ_2 are tracked by labeling the intervals comprised between the abscissae listed in the sweep-line status as being outside $\mathcal{OCB}_{xy}[\kappa^u]$, inside $\mathcal{OCB}_{xy}[\kappa^u]$ but outside $\mathcal{ICB}_{xy}[\kappa^u]$, or inside $\mathcal{ICB}_{xy}[\kappa^u]$.

Once Γ_1 and Γ_2 have been extracted in the form of sequences of straight and circular edges, they are clipped by the rectangle $[x_\kappa, x'_\kappa] \times [y_\kappa, y'_\kappa]$. The projections $\mathcal{OCB}_y[\kappa^{uv}]$ and $\mathcal{ICB}_y[\kappa^{uv}]$ are easily computed by determining both the points of Γ_1 and Γ_2 at the abscissae x_v, $v = 1, ..., s+1$, and the extremal points (minima and maxima) of Γ_1 and Γ_2 within each interval $[x_v, x_{v+1}]$, $v = 1, ..., s$. All these computations can be done during line sweeping. The sweep process is stopped as soon as it leaves the $[x_\kappa, x'_\kappa] \times [y_\kappa, y'_\kappa]$ rectangle.

Assuming s is constant, the overall sweep process takes time $O((n + c)\log n)$, where n is the number of faces that intersect the interval $[\gamma_u, \gamma_{u+1}]$ and c is the number of intersections of the generalized polygons. We know that $n \in O(n_\mathcal{A} n_\mathcal{B})$, where $n_\mathcal{A}$ and $n_\mathcal{B}$ are the respective numbers of edges of $\mathcal{A}$ and $\mathcal{B}$. On the other hand, $c \in O(n^2)$, but it is usually much smaller.

The computation can be made faster by associating the set of faces intersected by κ with every MIXED cell κ' generated by the decomposition of κ. If κ' is decomposed later, then only the faces associated with it have to be considered.

Generalization. Let us now consider the case where $\mathcal{A}$ and $\mathcal{B}$ are non-convex polygons. Both regions can be decomposed into finite collections $\{\mathcal{A}_i\}$ and $\{\mathcal{B}_j\}$ of convex polygons.

If the interiors of the convex polygons $\mathcal{A}_i$ have a non-empty intersection, then the reference point can be chosen within this intersection and the above method directly applies by considering the faces of the C-obstacles $\mathcal{CB}_{ij}$ corresponding to all the pairs $(\mathcal{A}_i, \mathcal{B}_j)$.

If the interiors of the convex polygons $\mathcal{A}_i$ do not overlap (this is the case, for instance, if $\mathcal{A}$ is a ⨆-shaped object), the above method can be applied, but with some changes. Indeed, since the reference point is now outside some of the convex components of $\mathcal{A}$, the projections of the faces of the corresponding C-obstacles have shapes that are different from those described above, but they are not difficult to establish.

An alternative to decomposing $\mathcal{A}$ and $\mathcal{B}$ into convex parts is to first compute the boundary of the C-obstacle region $\mathcal{CB}$ with the method described in Subsection 1.7 of Chapter 3 and next project every face of this boundary into the xy-plane.

4 Hierarchical Graph Searching

The first-cut planning algorithm of Section 1 searches successive connectivity graphs G_i, $i = 1, 2, ...$, until either it finds an E-channel connecting $\mathbf{q}_{init}$ to $\mathbf{q}_{goal}$ (success), or it does not find any channel connecting these two configurations (failure). Each graph G_i, $i > 0$, is obtained from the previous one, G_{i-1}, by decomposing the MIXED cells contained in the M-channel found in G_{i-1}. The major drawback of the algorithm is that the search work done in G_{i-1} is not used to help the search of G_i, although large subsets of these graphs may be identical. In this section we discuss an improved algorithm that remedies this drawback.

4.1 Improved Algorithm

The improved algorithm constructs a hierarchy of cg's (connectivity graphs). The cg at the root corresponds to the first decomposition $\mathcal{P}_1$ of Ω (as is computed in Step 1 of the first-cut algorithm) and we now denote this cg by G^{Ω} (previously it was denoted by G_1). Every other cg in the hierarchy corresponds to the decomposition of a MIXED cell, say κ, and we denote it by G^{κ}.

The graph G^{Ω} is searched for a channel connecting $\mathbf{q}_{init}$ to $\mathbf{q}_{goal}$. Assume that the outcome of the search is an M-channel Π_1. (If the outcome is an E-channel, the problem is solved; if no channel is found, the planner returns failure.) Every MIXED cell κ in Π_1 is recursively decomposed and searched for a channel that "connects appropriately" to the rest of Π_1. This is achieved as explained below, by constructing a complete channel Π_i, $i = 2, 3, ...$, connecting $\mathbf{q}_{init}$ to $\mathbf{q}_{goal}$ at each new level in the hierarchy of cg's.

Once the M-channel Π_1 has been found in G^{Ω}, Π_2 is initialized to the empty sequence of cells and the planner considers every successive cell κ in Π_1, starting with the cell that contains $\mathbf{q}_{init}$. If κ is EMPTY, it is simply appended at the end of the current Π_2. If κ is MIXED, it is decomposed into a set $\mathcal{P}^{\kappa}$ of smaller cells. Let G^{κ} be the cg associated with this decomposition. G^{κ} is searched for a channel Π^{κ} that verifies the following four conditions:

1. If κ is the first cell in Π_1 (hence, it contains $\mathbf{q}_{init}$), then the first cell in Π^{κ} also contains $\mathbf{q}_{init}$.

2. If κ is the last cell in Π_1 (hence, it contains $\mathbf{q}_{goal}$), then the last cell in Π^κ also contains $\mathbf{q}_{goal}$.

3. If κ is not the first cell in Π_1, the first cell of Π^κ is adjacent to the last cell of the current Π_2.

4. If κ is not the last cell in Π_1, the last cell of Π^κ is adjacent to the cell following κ in Π_1.

In the following, Π^κ is called a **subchannel**.

If the planner succeeds in generating Π^κ, Π_2 is modified by appending Π^κ to it. The case where the planner fails to generate Π^κ typically corresponds to the situation where the free space in the M-channel Π_1 is disconnected by the C-obstacle region. (Remember that the fact that Π_1 is an M-channel does not guarantee that $\mathbf{q}_{init}$ and $\mathbf{q}_{goal}$ can be connected by a free path lying in Π_1.) The treatment of this situation will be described in Subsection 4.3.

If all the cells in Π_1 have been considered successfully, we ultimately obtain a new E- or M-channel Π_2 that is a refinement of Π_1. If Π_2 is an E-channel, the problem is solved and the planner returns Π_2. Otherwise, Π_2 is refined in turn into a new channel Π_3 by decomposing the MIXED cells it contains, and so on.

4.2 Cell Occurrences in a Channel

An E-channel is now eventually generated by refining an M-channel without reconsidering a global cg. Therefore, it is important that we allow an M-channel to contain the same MIXED cell several times. Indeed, different occurrences of the same MIXED cell may yield different subchannels. This is best explainea using an example.

Consider the two-dimensional example of Figure 14. In Figure 14.a the M-channel $\Pi_1 = (\kappa_1, \kappa_2, \kappa_3, \kappa_2, \kappa_4)$ contains κ_2 twice. (The striped areas represent C-obstacles, so that κ_2 and κ_3 are MIXED, whereas κ_1 and κ_4 are EMPTY.) We assume that κ_2 and κ_3 are then decomposed as shown in Figure 14.b. κ_2 is divided into three smaller cells, two EMPTY ones, κ_{21} and κ_{22}, and one FULL one, κ_{23}. κ_3 is divided into four cells, three EMPTY ones, κ_{31}, κ_{32}, and κ_{33}, and one FULL one, κ_{34}. When the two occurrences of κ_2 are refined, two different subchannels are produced

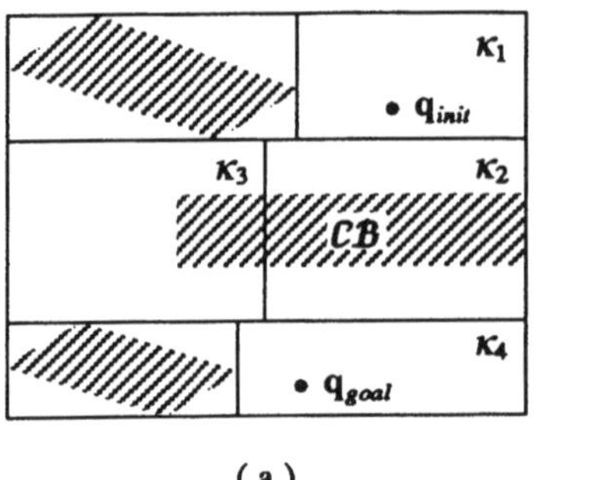

(a)

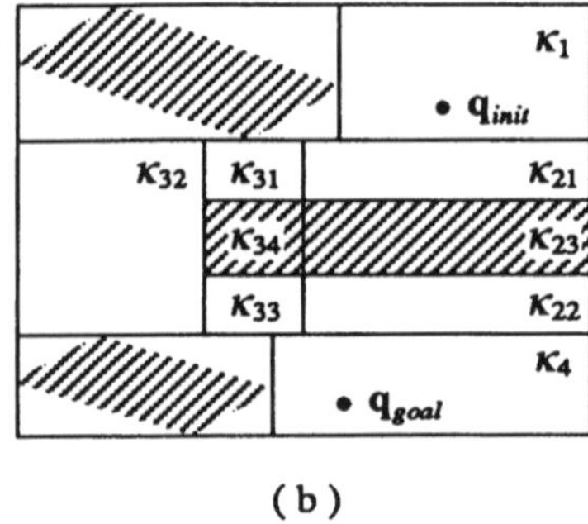

(b)

Figure 14. Figure a represents a first-level rectangloid decomposition of a rectangle (the striped areas are C-obstacles). In order to ultimately construct an E-channel at a deeper level of decomposition, the planner must consider the sequence $(\kappa_1, \kappa_2, \kappa_3, \kappa_2, \kappa_4)$, which contains the cell κ_2 twice, as an M-channel. This leads the planner to decompose the two MIXED cells κ_2 and κ_3 into smaller cells (for example, as shown in Figure b) and to generate the E-channel $(\kappa_1, \kappa_{21}, \kappa_{31}, \kappa_{32}, \kappa_{33}, \kappa_{22}, \kappa_4)$.

---

using the *same* decomposition of κ_2. The first occurrence of κ_2 yields a one-cell subchannel (κ_{21}), while the second occurrence yields another one-cell subchannel (κ_{22}). On the other hand, the occurrence of κ_3 is refined into the subchannel $(\kappa_{31}, \kappa_{32}, \kappa_{33})$. This yields the E-channel $\Pi_2 = (\kappa_1, \kappa_{21}, \kappa_{31}, \kappa_{32}, \kappa_{33}, \kappa_{22}, \kappa_4)$. This E-channel could not have been constructed if Π_1 had not contained two occurrences of κ_2.

Since the same cell may appear several times in a channel Π_i, we now refer to the elements in a channel as **cell occurrences**, rather than just cells. We denote the jth cell occurrence in the channel Π_i by ω_i^j and the cell of which ω is an occurrence by $cell(\omega)$. A cell occurrence ω is said to be MIXED (resp. EMPTY) if and only if the cell $cell(\omega)$ is MIXED (resp. EMPTY). A subchannel generated as a refinement of a MIXED cell occurrence ω is denoted by Π^ω; it is constructed by searching the cg G^κ where $\kappa = cell(\omega)$.

If an M-channel Π_i contains multiple occurrences of the same MIXED cell κ, a *single* rectangloid decomposition of κ is computed in order to construct the channel Π_{i+1}. However, as illustrated above, every occurrence of κ may yield a different subchannel.

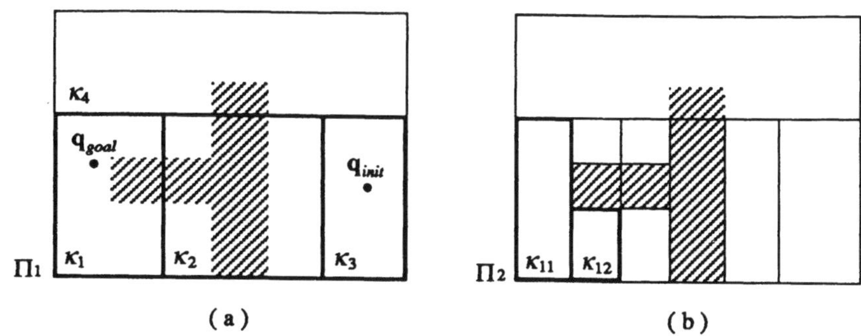

Figure 15. Figure a shows an M-channel $\Pi_1 = (\kappa_1, \kappa_2, \kappa_3)$ in bold lines. Assume that κ_1 and κ_2 are decomposed as shown in Figure b. The subchannel $(\kappa_{11}, \kappa_{12})$ is constructed within κ_1. But then the planner fails to find a subchannel within κ_2 because a C-obstacle obstructs the passage between κ_1 and κ_3 through κ_2.

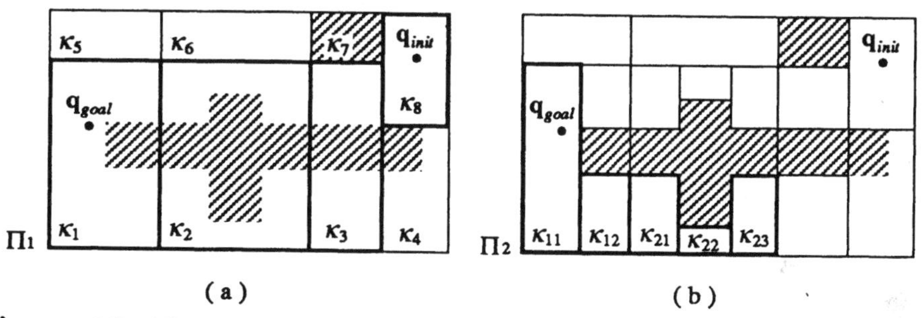

Figure 16. Figure a shows an M-channel $\Pi_1 = (\kappa_1, \kappa_2, \kappa_3, \kappa_8)$. Assume that κ_1, κ_2 and κ_3 are decomposed as shown in Figure b, and that the partial channel $\Pi_2 = (\kappa_{11}, \kappa_{12}, \kappa_{21}, \kappa_{22}, \kappa_{23})$ has been generated. Then, the planner fails to find a subchannel in the decomposition of κ_3.

Allowing a channel to contain several occurrences of the same MIXED cell affects the way the cg's should be searched for (sub)channels. This issue will be studied in Subsection 4.5.

4.3 Failure Recovery

Consider now the situation where the planner is constructing the channel Π_{i+1} and fails to refine a MIXED cell occurrence ω_i^j contained in Π_i into a subchannel. We have explicitly mentioned this situation in Subsection 4.1. It may be caused either by the fact that Π_i does not contain an E-channel as illustrated in Figure 15, or by the fact that the partial channel Π_{i+1} generated so far is a blind alley as illustrated in Figure 16.

Of course, the planner may have no way to directly determine the cause of the failure. Instead, when it fails to refine the MIXED cell occurrence ω_i^j into a subchannel, it distinguishes among the three following cases, which are mutually exclusive:

Case (1): $j = 1$.

$\omega_i^j = \omega_i^1$ is the first cell occurrence in Π_i. The failure typically means that there is no way to connect $\mathbf{q}_{init}$ to the second cell occurrence in Π_i. The only alternative for the planner is to attempt to generate a new channel Π_i. If no new channel Π_i can be generated and $i > 1$, the planner recursively attempts to generate a new channel Π_{i-1}. If no new channel Π_i can be generated and $i = 1$, the planner returns failure.

Case (2): $j > 1$ and ω_i^{j-1} is EMPTY.

The failure typically means that there is no way to connect any configuration in the EMPTY cell occurrence ω_i^{j-1} to a configuration in ω_i^{j+1} through ω_i^j. The planner then proceeds as described in case (1). (Some differences justifying the distinction between cases (1) and (2) will, however, appear in the next subsection.)

Case (3): $j > 1$ and ω_i^{j-1} is MIXED.

The failure typically means that there is no way to connect any configuration in the last cell occurrence of the current Π_{i+1} (which is a subset of ω_i^{j-1}) to a configuration in ω_i^{j+1} through ω_i^j. It may be possible to recover from the failure by finding another subchannel in ω_i^{j-1}. Hence, the planner attempts to generate another subchannel $\Pi^{\omega_i^{j-1}}$ in the cg of $cell(\omega_i^{j-1})$. If it finds another subchannel, the planner substitutes it for the previous one in Π_{i+1}, and tries again to refine ω_i^j into a subchannel by searching the cg of $cell(\omega_i^j)$. Otherwise, it iteratively treats ω_i^{j-1} as it just treated ω_i^j, i.e. it sets j to $j-1$ and applies the treatment of case (1), (2), or (3), whichever is applicable.

4.4 Recording Failure Conditions

Within the framework of the improved algorithm, the same cg may be searched multiple times for (sub)channels. The planner can be significantly more efficient if it annotates the cells of the cg's with its "mistakes" and later use the annotations in order to avoid making the same mistakes again. Cell annotation is performed whenever the planner recovers from the failure to refine a MIXED cell occurrence into a subchannel.

Annotations may become quite involved and costly to use when there are too many of them. For these reasons, we will restrict the generation of annotations to the cases (1) and (2) presented in the previous subsection, when these cases do not occur after an iteration on case (3). Then the annotations are in reasonable number and they are very simple to use. Below, n_i denotes the number of cell occurrences in Π_i and λ stands for the inexistent cell.

Case (1):

Let $\xi = cell(\omega_i^1)$ and $\psi = cell(\omega_i^2)$ be the first two cells in the current Π_i ($\psi = \lambda$ if $n_i = 1$). The cell ξ is annotated with: (ψ).

This annotation is later used as follows: If a new channel Π_i is generated and if it contains an occurrence ω_ξ of ξ in the first position, then an occurrence of ψ should not be considered a valid successor of ω_ξ.

Notice that the first cell occurrence in the new Π_i is necessarily an occurrence of ξ, since ξ contains $\mathbf{q}_{init}$. But there may be other occurrences of ξ appearing in this new Π_i to which the above annotation does not apply.

Case (2):

Let $\xi = cell(\omega_i^j)$ $(j > 1)$, $\varphi = cell(\omega_i^{j-1})$, and $\psi = cell(\omega_i^{j+1})$ in the current Π_i (see Figure 17). If $j = n_i$, then $\psi = \lambda$. ξ is annotated with: (φ, ψ).

This annotation is later used as follows: If a new channel Π_i is generated and if it contains an occurrence ω_ξ of ξ preceded by an occurrence of φ, then an occurrence of ψ should not be considered a valid successor of ω_ξ. In the particular case where $\psi = \lambda$, this means that ω_ξ must not be the

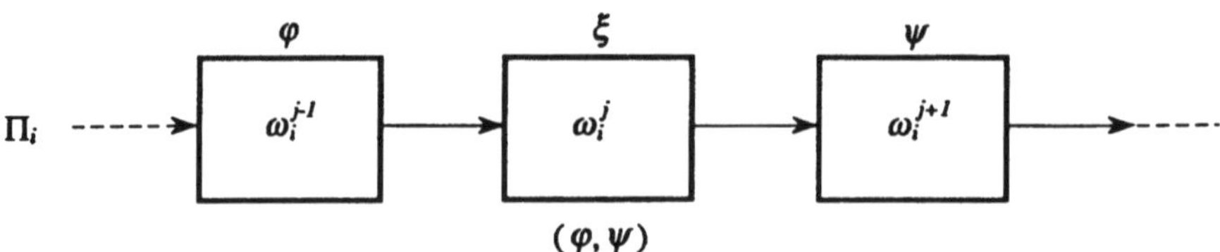

Figure 17. This figure illustrates the notations used in the text to explain the construction of an annotation in case (2). Each box represents a cell occurrence. The symbol above the box designates the corresponding cell.

last cell occurrence of the new Π_i.

Notice that if $\psi \neq \lambda$ the above annotation is commutative, i.e. if ξ cannot be traversed by a subchannel connecting φ to ψ, it can neither be traversed by a subchannel connecting ψ to φ. Therefore, the annotation (φ, ψ) is used commutatively whenever $\psi \neq \lambda$.

4.5 Search of a Connectivity Graph

Below we regard Ω as a cell and we denote the channel (ω_0^1), with $cell(\omega_0^1) = \Omega$, by Π_0. (We assume that Π_0 is an M-channel.) The channel Π_1 can be regarded as a refinement of Π_0.

Let us consider the situation where the planner refines the MIXED cell occurrence $\omega_i^j \in \Pi_i$. Let $\kappa = cell(\omega_i^j)$. If κ has already been decomposed, the planner just reuses the cg G^κ that was previously generated; otherwise, it decomposes κ into a set $\mathcal{P}^\kappa$ of smaller rectangloid cells and builds G^κ. It then searches G^κ for a subchannel refining ω_i^j that will be appended to the tail of the current Π_{i+1}.

The search of G^κ requires that the possible initial and goal cells be first determined:

- If $j = 1$, ω_i^j is the first cell of Π_i. Then the only possible initial cell in G^κ is the cell of $\mathcal{P}^\kappa$ that contains $\mathbf{q}_{init}$. Otherwise, the possible initial cells of G^κ are all the cells of $\mathcal{P}^\kappa$ which are adjacent to the cell — call it φ — of the last cell occurrence ω_φ in the current Π_{i+1} and whose insertion in Π_{i+1} as cell occurrences succeeding ω_φ does not violate any annotation.

- If $j = n_i$, ω_i^j is the last cell occurrence of Π_i. The only possible goal cell in G^κ is the cell of $\mathcal{P}^\kappa$ that contains $\mathbf{q}_{goal}$. Otherwise, the possible goal cells of G^κ are all the cells of $\mathcal{P}^\kappa$ which are adjacent to $cell(\omega_i^{j+1})$.

If there is no possible initial or goal cell in G^κ, the planner considers this situation as a failure to refine ω_i^j and applies the treatments described in Subsections 4.3 and 4.4.

If at least one initial and one goal cell are established in G^κ, the planner searches G^κ for a subchannel $\Pi^{\omega_i^j}$ linking any of the possible initial cells to any of the possible goal cells. It does that in a way such that no annotation is violated.

As mentioned earlier, the subchannel $\Pi^{\omega_i^j}$ to be generated in G^κ should be allowed to contain several occurrences of the same MIXED cells in $\mathcal{P}^\kappa$. This means that there may be an infinity of potential subchannels to consider. A reasonable way to deal with this difficulty in most practical cases is by limiting the number of occurrences of the same cell in a subchannel to a small one.

5 Variant Methods

In this section, we present two variant methods based on the approximate cell decomposition approach. They apply to the case where the robot $\mathcal{A}$ is a polygon that can translate and rotate freely among polygonal obstacles. In both variants, the robot's free space is approximated as a collection of *prismatic* cells whose union is contained in free space. However, the complement of this approximation is not decomposed into MIXED and FULL cells. One variant — *orientation slicing* — consists of decomposing the range $[0, 2\pi]$ of orientations of $\mathcal{A}$ into intervals and approximating the C-obstacles within each interval. The other variant — *position partitioning* — consists of decomposing the set of positions of $\mathcal{A}$ into small polygonal regions and approximating the C-obstacles above each region.

5.1 Orientation Slicing

This method [Lozano-Pérez, 1981 and 1983] consists of decomposing the range of the orientations of $\mathcal{A}$ into a finite number of intervals. In every

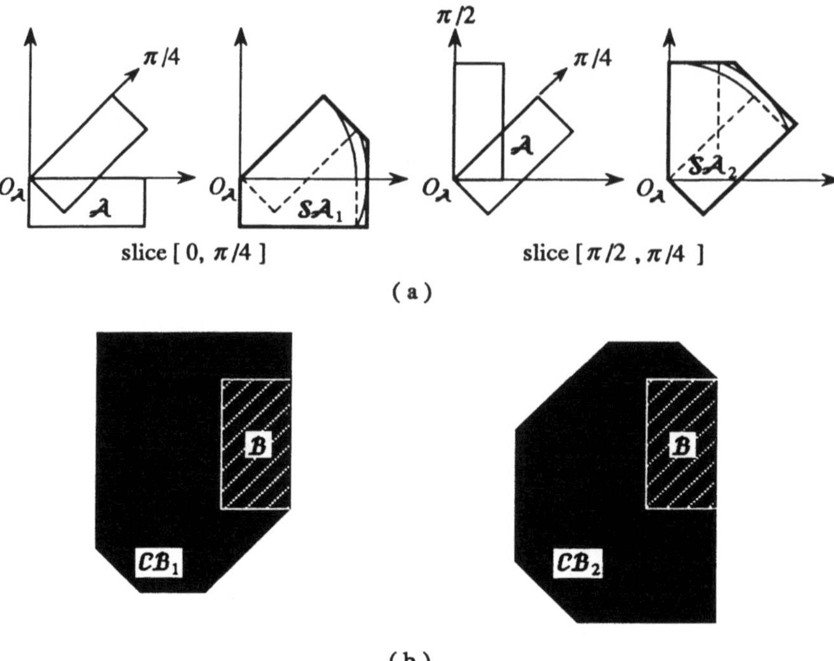

Figure 18. In the orientation slicing method, the range of orientations of $\mathcal{A}$ is sliced into intervals. In each interval — here $[0, \pi/4]$ and $[\pi/4, \pi/2]$ — the area swept out by $\mathcal{A}$, when it rotates around the reference point $O_{\mathcal{A}}$ is computed and approximated by a polygon (Figure a). This polygon is then treated as a translating object. The C-obstacle region corresponding to each such polygon is computed (Figure b).

---

interval, the area swept out by $\mathcal{A}$ as it rotates around its reference point $O_{\mathcal{A}}$ is computed and approximated by a polygon. This polygon is then considered as a translating object.

Let $\mathbf{q}_{init} = (x_{init}, y_{init}, \theta_{init})$ and $\mathbf{q}_{goal} = (x_{goal}, y_{goal}, \theta_{goal})$. The overall algorithm is the following:

1. Decompose the range of orientations of $\mathcal{A}$ into a finite number of consecutive closed intervals $[\theta_k, \theta'_k]$, $k = 1, ..., p$, such that $\theta_1 = 0$, $\theta_{k+1} = \theta'_k$ (for all $k \in [1, p-1]$), and $\theta'_p = 2\pi$.

2. For every $k \in [1, p]$, compute the area swept out by $\mathcal{A}$ when it rotates

around $O_{\mathcal{A}}$ from the orientation $\theta = \theta_k$ to the orientation $\theta = \theta'_k$. Approximate this area by a bounding polygon. This polygon is denoted by $\mathcal{SA}_k$.

(This step is illustrated in Figure 18, where we consider two intervals, $[0, \pi/4]$ and $[\pi/4, \pi/2]$.)

3. For every $k \in [1,p]$, treat the polygon $\mathcal{SA}_k$ as a robot that is only allowed to translate with $O_{\mathcal{A}}$ as its reference point. Compute the C-obstacle region $\mathcal{CB}_k$ in $\mathcal{SA}_k$'s configuration space $\mathcal{C}_k = \mathbf{R}^2$. Set $\mathcal{C}_{k,free}$ to $\mathcal{C}_k \backslash \mathcal{CB}_k$.

4. For every $k \in [1,p]$, decompose $\mathcal{C}_{k,free}$ into convex polygonal cells, using, say, the exact trapezoidal decomposition method[4] described in Section 1 of Chapter 5. Construct the connectivity graph G_k representing the adjacency relation between these cells. If θ_{init} (resp. θ_{goal}) $\in [\theta_k, \theta'_k]$, then label the cell containing (x_{init}, y_{init}) (resp. (x_{goal}, y_{goal})) as the *initial* (resp. *goal*) cell.[5]

5. Merge the graphs G_k into a graph G by linking any two nodes $X_1 \in G_{k_1}$ and $X_2 \in G_{k_2}$, with $k_2 = k_1 + 1 \bmod p$, whenever the two cells corresponding to X_1 and X_2 intersect at a subset of non-zero area in $\mathbf{R}^2$ when $\mathcal{C}_{k_1}$ and $\mathcal{C}_{k_2}$ are represented as the same copy of $\mathbf{R}^2$.

6. Search G for a path joining the initial cell to the goal cell. If the search terminates successfully, then return the sequence of cells along the path that has been found. Otherwise, return failure.

We can regard each $\mathcal{CB}_k$ as a bounding polygonal approximation of the projection of the slice of the C-obstacle region $\mathcal{CB}$ that is comprised between θ_k and θ'_k. Thus, $\mathcal{CB}_k \times [\theta_k, \theta'_k]$ is a union of prisms bounding $\mathcal{CB}$ within the $[\theta_k, \theta'_k]$ interval. Similarly, every cell in G_k is a vertical prism. Any two cells in G_k can only be adjacent by their vertical faces. Two cells in G that are not contained in the same angular slice are adjacent if and only if they belong to two adjacent slices and the intersection of the top face of one with the bottom face of the other is a region of non-zero area in $\mathbf{R}^2$.

[4]One may also use an approximate cell decomposition method.

[5]If the decomposition of $[0, 2\pi]$ is not fine enough, it may happen that no cell in the p decompositions contain (x_{init}, y_{init}) or (x_{goal}, y_{goal}).

In Step 2 of the algorithm, the area swept out by $\mathcal{A}$ when it rotates within an angular interval $[\theta_k, \theta'_k]$ around its reference point is computed and approximated by a polygon. The computation of the swept area is simple and is based on the observation that its contour consists of portions of edges of $\mathcal{A}$ at orientations θ_k and θ'_k, and by circular arcs traced by vertices of $\mathcal{A}$ during the rotation. The swept area is a generalized polygon.

The above method is conservative. If it generates a path, it is actually a free path; but it may fail to find a free path, even when one exists, if the intervals $[\theta_k, \theta'_k]$ are too large. Then, it is possible to decompose these intervals into smaller ones and run the method again. However, in the absence of MIXED cells, there is no obvious information that can be used to select which intervals have to be refined. In addition, there is no simple way to capitalize on the computations performed for one interval in order to simplify future computations when this interval is decomposed into subintervals.

5.2 Position Partitioning

This method [Germain, 1984] consists of decomposing the set of positions of $\mathcal{A}$ — i.e. the xy-plane — into small polygonal regions. In every polygon, the range of free orientations of $\mathcal{A}$ is computed, yielding a conservative approximation of free space in the form of a collection of vertical prisms. (The method has some similarities to the exact cell decomposition method presented in Section 2 of Chapter 5.)

In order to simplify our description, we assume that the set of possible positions of $\mathcal{A}$'s reference point $O_{\mathcal{A}}$ is restricted to the interior of a rectangle $D \subset \mathbf{R}^2$. If we select $O_{\mathcal{A}}$ in $\mathcal{A}$, the set of admissible positions of $\mathcal{A}$ — i.e. the set of positions which admit free orientations of $\mathcal{A}$ — is included in $\mathcal{E} = int(D) \backslash \mathcal{B}$, where $\mathcal{B}$ is the obstacle region.

The position partitioning method first decomposes $\mathcal{E}$ into a finite collection of polygons denoted by R_k, $k = 1, ..., p$, such that $\cup_k R_k \subseteq cl(\mathcal{E})$ and for all $k \neq k'$, $int(R_k) \cap int(R_{k'}) = \emptyset$. For each R_k, we compute the range $S(R_k)$ of orientations of $\mathcal{A}$ which are free or semi-free for all positions of $\mathcal{A}$ in R_k, i.e. :

$$S(R_k) = \bigcap_{(x,y) \in R_k} \{\theta \;/\; (x, y, \theta) \in cl(C_{free})\}.$$

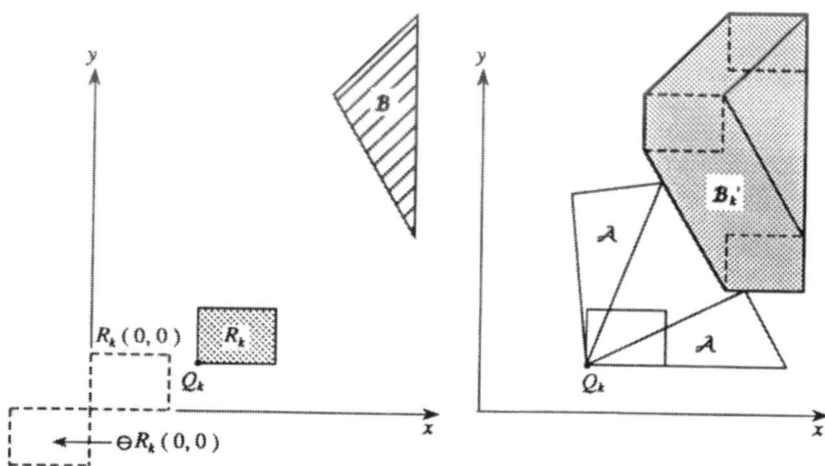

Figure 19. The range of free and semi-free orientations of $\mathcal{A}$, when its position varies over a closed region R_k, can be computed as the range of free and semi-free orientations of $\mathcal{A}$ at a single position $Q_k = (x_k, y_k)$ after having "grown" the obstacle region $\mathcal{B}$ by $\ominus R_k(0,0)$ into $\mathcal{B}'_k$.

$S(R_k)$ is a (possibly empty) union of closed intervals of non-zero lengths. Next, we approximate $cl(\mathcal{C}_{free})$ by prismatic cells of the form $R_k \times I_{kl}$, where $I_{kl} = [\theta_{kl}, \theta'_{kl}]$ is a maximal connected interval in $S(R_k)$. Finally, we construct the connectivity graph among these cells and we search the graph. If the search terminates successfully, the outcome is a channel whose interior lies entirely in $\mathcal{C}_{free}$. These computations are described in more detail below.

The decomposition of $\mathcal{E}$ into convex polygons may be an exact decomposition (in which case $\bigcup_k R_k = cl(\mathcal{E})$) or it may be an approximate decomposition (in which case $\bigcup_k R_k \subset cl(\mathcal{E})$). Approximate decomposition makes it possible using a single type of polygon, e.g. rectangles.

Let $S_{\mathcal{R}}(x,y) = \{\theta \;/\; int(\mathcal{A}(x,y,\theta)) \cap \mathcal{R} = \emptyset\}$, where $\mathcal{R}$ is a closed region of $\mathbf{R}^2$. The following proposition reduces the computation of $S(R_k)$ to the computation of the set of free and semi-free orientations of $\mathcal{A}$ at a single position with respect to a transformed obstacle region. More precisely, it reduces the computation of $S(R_k)$ to the computation of $S_{\mathcal{B}'_k}(x_k, y_k)$ at a single point (x_k, y_k), with $\mathcal{B}'_k$ obtained by "growing" $\mathcal{B}$ according to R_k, as illustrated in Figure 19.

PROPOSITION 1: *Let Q_k be an arbitrary point selected in R_k, and x_k and y_k its coordinates in $\mathcal{F}_\mathcal{W}$. Let $R_k(0,0) = \{(x - x_k, y - y_k) \,/\, (x,y) \in R_k\}$. We have:*

$$S(R_k) = S_{\mathcal{B}\ominus R_k(0,0)}(x_k, y_k),$$

where $\ominus$ is the Minkowski operator for affine set difference.

Proof: (1) Consider an orientation $\theta \in S(R_k)$, and let us prove that $\theta \in S_{\mathcal{B}\ominus R_k(0,0)}(x_k, y_k)$.

Assume that $\theta \notin S_{\mathcal{B}\ominus R_k(0,0)}(x_k, y_k)$. Then, there exists a point $a \in int(\mathcal{A})$ such that $a(x_k, y_k, \theta) \in \mathcal{B} \ominus R_k(0,0)$. This implies that there exists a point $b \in \mathcal{B}$ and a point $c = (x_c, y_c) \in R_k(0,0)$ such that $a(x_k, y_k, \theta) = b - c$, hence $a(x_k, y_k, \theta) + c = b$. Since $a(x_k, y_k, \theta) + c \in \mathcal{A}(x_k + x_c, y_k + y_c, \theta)$ and $(x_k + x_c, y_k + y_c) \in R_k$, it contradicts $\theta \in S(R_k)$. Therefore: $\theta \in S_{\mathcal{B}\ominus R_k(0,0)}(x_k, y_k)$.

Hence: $S(R_k) \subseteq S_{\mathcal{B}\ominus R_k(0,0)}(x_k, y_k)$.

(2) Consider an orientation θ such that $\theta \in S_{\mathcal{B}\ominus R_k(0,0)}(x_k, y_k)$, and let us prove that $\theta \in S(R_k)$.

Assume that $\theta \notin S(R_k)$. Then, there exists a point (x,y) in R_k and a point a in $\mathcal{A}$ such that $a(x,y,\theta) \in \mathcal{B}$. We can write $(x,y) = (x_k + x_c, y_k + y_c) = (x_k, y_k) + (x_c, y_c)$, with $c = (x_c, y_c) \in R_k(0,0)$. Thus, the assumption implies that there exists $b \in \mathcal{B}$ such that $a(x_k + x_c, y_k + y_c, \theta) = b$, or $a(x_k, y_k, \theta) = b - c \in \mathcal{B} \ominus R_k(0,0)$. This contradicts $\theta \in S_{\mathcal{B}\ominus R_k(0,0)}(x_k, y_k)$.

Hence: $S_{\mathcal{B}\ominus R_k(0,0)}(x_k, y_k) \subseteq S(R_k)$.

Therefore: $S(R_k) = S_{\mathcal{B}\ominus R_k(0,0)}(x_k, y_k)$. ∎

$S_{\mathcal{B}'_k}(x_k, y_k)$, with $\mathcal{B}'_k = \mathcal{B} \ominus R_k(0,0)$, can be computed as follows. First, the intersections of the circles centered at $Q_k = (x_k, y_k)$ and passing through the vertices of $\mathcal{A}$ (resp. $\mathcal{B}'_k$), with the edges of $\mathcal{B}'_k$ (resp. $\mathcal{A}$) are computed. Next, each intersection is classified as a candidate contact if the two edges of $\mathcal{A}$ (resp. $\mathcal{B}'_k$) abutting the contact vertex lie on the outer side of the contact edge of $\mathcal{B}'_k$ (resp. $\mathcal{A}$). The intersections that do not satisfy this condition are not considered further. Each candidate contact is classified as clockwise (resp. counterclockwise) if it is attained by a clockwise (counterclockwise) rotation of $\mathcal{A}$. The set of candidate contacts is sorted along the unit circle S^1 in counterclockwise order of the corresponding orientations of $\mathcal{A}$. The intervals of $S(R_k) = S_{\mathcal{B}'_k}(x_k, y_k)$ are all the angular intervals between two successive candidate contacts such that the first one is a clockwise contact and the second one is a counterclockwise contact.

At this point, we have $S(R_k) = \{I_{k1}, ..., I_{ks}\}$, where I_{kl}, for any $l \in [1, s]$, is a closed angular interval of non-zero length. We approximate $cl(\mathcal{C}_{free})$ as a collection of prismatic **cells** of the form $R_k \times I_{kl}$. Two cells $R_k \times I_{kl}$ and $R_{k'} \times I_{k'l'}$ are **adjacent** if and only if:

- $\partial R_k \cap \partial R_{k'}$ is a segment of non-zero length, and

- $I_{kl} \cap I_{k'l'}$ is an interval of non-zero length.

It is easy to verify that any two configurations in the interiors of two adjacent cells can be connected by a free path.

Path finding consists of searching the connectivity graph G representing the adjacency relation among these prismatic cells.

The method is conservative and it may fail to generate a free path even if one exists. In case of failure, one may attempt to refine the decomposition of $\mathcal{E}$ and run the method again. However, as with the orientation slicing method, no simple information is available in order to determine which part of the decomposition should be refined.

Exercises

1: Let Ω be a rectangloid in $\mathbf{R}^m$. Let $\mathcal{CB}_1$ and $\mathcal{CB}_2$ be two C-obstacles in $\mathbf{R}^m$. Let T_1 (resp. T_2) be the 2^m-tree decomposition of Ω when it contains $\mathcal{CB}_1$ (resp. $\mathcal{CB}_2$). The height of the two decompositions is limited to the same maximal height by imposing that under a specified size any cell that would normally be labeled as MIXED be labeled as FULL.

Give an algorithm that takes the two trees T_1 and T_2 as input and computes the 2^m-tree decomposition T of Ω when it contains both C-obstacles $\mathcal{CB}_1$ and $\mathcal{CB}_2$.

2: Let a cell at depth d in a quadtree be identified by a list of d integers in $\{1,2,3,4\}$ describing the path from the root to the cell. Propose an algorithm that determines whether two input cells in the quadtree are adjacent or not.

3: Implement a path planner for a point moving among polygonal obstacles using the quadtree decomposition. If you have already implemented a program that computes the C-obstacle region $\mathcal{CB}$ for a polygon $\mathcal{A}$ translating among polygonal obstacles, use this program as a front-end to the planner.

4: Consider the decomposition method presented in Section 3. Discuss the impact of the choice of the position of the reference point $O_{\mathcal{A}}$ relative to $\mathcal{A}$ on the efficiency of the method. Where would you choose the reference point?

5: Implement a planner based on the approximate cell decomposition approach for a polygonal robot that can translate and rotate freely among polygonal obstacles. Run the planner on examples such as those shown in Figure 3. (This exercise requires a substantial amount of programming.)

6: Let the robot $\mathcal{A}$ be a polygon that translates and rotates in the plane. A configuration of $\mathcal{A}$ is represented as (x, y, θ), where (x, y) is the position of the reference point $O_{\mathcal{A}}$. The *outer swept area* and the *inner swept area* are the regions $\mathcal{OSA}[\theta_1, \theta_2]$ and $\mathcal{ISA}[\theta_1, \theta_2]$ of the xy-plane which are defined as follows:

$$\begin{aligned} \mathcal{OSA}[\theta_1, \theta_2] &= \{(x, y) \,/\, \exists\theta \in [\theta_1, \theta_2] : (x, y) \in \mathcal{A}(0, 0, \theta)\}, \\ \mathcal{ISA}[\theta_1, \theta_2] &= \{(x, y) \,/\, \forall\theta \in [\theta_1, \theta_2] : (x, y) \in \mathcal{A}(0, 0, \theta)\}. \end{aligned}$$

a. Show that both $\mathcal{OSA}[\theta_1, \theta_2]$ and $\mathcal{ISA}[\theta_1, \theta_2]$ are generalized polygons. Show that if $\mathcal{A}$ is convex and $O_{\mathcal{A}} \in \mathcal{A}$, $\mathcal{OSA}[\theta_1, \theta_2]$ and $\mathcal{ISA}[\theta_1, \theta_2]$ are convex or locally non-convex.

b. Give an algorithm for computing $\mathcal{OSA}[\theta_1, \theta_2]$ and an algorithm for computing $\mathcal{ISA}[\theta_1, \theta_2]$.

c. Let $\mathcal{B}$ be the obstacle region in $\mathcal{A}$'s workspace. We define $\mathcal{OCB}[\theta_1, \theta_2]$ and $\mathcal{ICB}[\theta_1, \theta_2]$ as follows:

$$\mathcal{OCB}[\theta_1, \theta_2] = \{(x, y) \,/\, \exists\theta \in [\theta_1, \theta_2] : (x, y, \theta) \in \mathcal{CB}\},$$

$$\mathcal{ICB}[\theta_1, \theta_2] = \{(x, y) \,/\, \forall\theta \in [\theta_1, \theta_2] : (x, y, \theta) \in \mathcal{CB}\}.$$

Show that:

- $\mathcal{B} \ominus \mathcal{OSA}[\theta_1, \theta_2] = \mathcal{OCB}[\theta_1, \theta_2]$.

- $\mathcal{B} \ominus \mathcal{ISA}[\theta_1, \theta_2] \subseteq \mathcal{ICB}[\theta_1, \theta_2]$.

d. Derive from the previous result an alternative to the method presented in Subsection 3.2 for computing the bounding and bounded approximations of $\mathcal{CB}[\kappa]$. Discuss the conditions under which this alternative may be more efficient than the method of Subsection 3.2 [Zhu and Latombe, 1989].

7: In Subsection 4.4 we described a way to remember failure conditions by annotating cells with these conditions. The annotations are used during the search to prune the search of the connectivity graphs. Describe another type of connectivity graph in which the annotations of types 1 and 2 are no longer needed and the corresponding failure conditions are more simply remembered by removing links of the graph. [Hint: A node of the new graph corresponds to the intersection of the boundaries of two adjacent cells. A link corresponds to a cell.] Describe in detail how this new type of graph would be used for planning. Discuss the advantages and the drawbacks of the new graph relative to the connectivity graph used throughout Section 4 [Zhu and Latombe, 1989].

8: Consider the orientation slicing method presented in Subsection 5.1. In our presentation we approximated the area swept out by $\mathcal{A}$ when its orientation varies in an interval $[\theta_k, \theta'_k]$ by a polygon. However, this is not necessary. Explain how the method could proceed without such an approximation. [Hint: Compute cells as generalized prisms, i.e. prisms whose cross-sections are generalized polygons.] Discuss the advantages and the drawbacks of avoiding the approximation.

9: In the orientation slicing method of Subsection 5.1 we decomposed the range $[0, 2\pi]$ of orientations of $\mathcal{A}$ into non-overlapping intervals. In the position partitioning method of Subsection 5.2 we decomposed the empty subset of the workspace, $\mathcal{E}$, into non-overlapping polygons. What would be the advantages and the drawbacks of decomposing $[0, 2\pi]$ into overlapping intervals (for the orientation slicing method) and $\mathcal{E}$ into overlapping polygons (for the position partitioning method)?

10: Implement the position partitioning method presented in Subsection 5.2. Propose and discuss heuristics for refining the decomposition of $\mathcal{E}$ (the empty subset of the workspace) when the planner fails to find a path.

11: Implement a planner based on the following variant of the approximate cell decomposition approach, for a polygonal robot $\mathcal{A}$ that translates and rotates freely among polygonal obstacles. The rectangloid Ω is decomposed into a fixed three-dimensional array (e.g. a 128^3 array) of small rectangloids having all the same size. Decompose $\mathcal{A}$ and the obstacle region $\mathcal{B}$ into convex components $\{\mathcal{A}_i\}$ and $\{\mathcal{B}_j\}$. Let $\mathcal{CB}_{ij}$ be the C-obstacle corresponding to $\mathcal{A}_i$ and $\mathcal{B}_j$. Represent the decomposition of Ω by an array $\mathcal{G}$ of integers (or bits) initialized to 0. Set the elements of $\mathcal{G}$ representing cells intersecting the boundary of $\mathcal{CB}_{ij}$, for some i and j, to 1. Search the decomposition of Ω for a path joining $\mathbf{q}_{init}$ to $\mathbf{q}_{goal}$, and crossing no C-obstacle boundary. Describe in detail the components of the planner that are the most critical to its efficiency. Run the planner on the examples of Figure 3.

Chapter 7

Potential Field Methods

The planning methods described in the previous three chapters aim at capturing the global connectivity of the robot's free space into a condensed graph that is subsequently searched for a path. The approach presented in this chapter proceeds from a different idea. It treats the robot represented as a point in configuration space as a particle under the influence of an *artificial potential field* $\mathbf{U}$ whose local variations are expected to reflect the "structure" of the free space. The potential function is typically (but not necessarily) defined over free space as the sum of an *attractive* potential pulling the robot toward the goal configuration and a *repulsive* potential pushing the robot away from the obstacles. Motion planning is performed in an iterative fashion. At each iteration, the artificial force $\vec{F}(\mathbf{q}) = -\vec{\nabla}\mathbf{U}(\mathbf{q})$ induced by the potential function at the current configuration is regarded as the most promising direction of motion, and path generation proceeds along this direction by some increment.

Potential field was originally developed as an on-line collision avoidance approach, applicable when the robot does not have a prior model of the obstacles, but senses them during motion execution [Khatib, 1980 and 1986]. Emphasis was put on real-time efficiency, rather than on guaranteeing the attainment of the goal. In particular, since an on-line potential field method essentially acts as a fastest descent optimization procedure, it may get stuck at a local minimum of the potential function other than the goal configuration. However, the idea underlying potential field can be combined with graph searching techniques. Then, using a prior model of the workspace, it can be turned into a systematic motion planning approach. This is the approach described in this chapter.

Even when graph searching techniques are used, local minima remain an important cause of inefficiency for potential field methods. Hence, dealing with local minima is *the* major issue that one has to face in designing a planner based on this approach. This issue can be addressed at two levels: (1) in the definition of the potential function, by attempting to specify a function with no or few local minima, and (2) in the design of the search algorithm, by including appropriate techniques for escaping from local minima. A large portion of this chapter describes results and techniques developed at these two levels.

Potential field methods are often referred to as "local methods". This comes from the fact that most potential functions are defined in such a way that their values at any configuration do not depend on the distribution and shapes of the C-obstacles beyond some limited neighborhood around the configuration. However, as mentioned in Chapter 1, if one could construct an "ideal" potential function with a single minimum (at the goal configuration) in the connected subset of the free space containing the goal configuration, this function could be regarded as some kind of "global" information about free space, and the expression "local method" would then be much less relevant.

Most planning methods based on the potential field approach have a strong empirical flavor. They are usually incomplete, i.e. they may fail to find a free path, even if one exists. But, the counterpart is that some of them are particularly fast in a wide range of situations. The strength of the approach is that, with some limited engineering, it makes it possible to construct motion planners which are both quite efficient and reasonably reliable. This explains why they are increasingly popular for implementing practical motion planners.

In Sections 1 and 2 we give a classical definition of a potential function and the induced field of forces. We first consider the case where the robot can only translate (Section 1); then, we treat the case where it can also rotate (Section 2). In Section 3 we describe three simple potential-guided path planning techniques: depth-first, best-first and variational planning. In Section 4 we introduce the notion of *navigation function* (local-minimum-free potential) and we define potential functions that may generate less or smaller local minima than the function of Sections 1 and 2. In Section 5 we combine results presented in previous sections with randomized search techniques for escaping from local minima; we describe a powerful potential-field-based planner that is able to plan free paths in high-dimensional configuration spaces. In Section 6 we describe a simple learning technique that may be used to improve the ability of a potential-field-based planning method to avoid local minima when the geometry of free space remains unchanged over many successive tasks.

1 Potential Field in Translational Case

Most proposed potential functions are based upon the following general idea: the robot should be attracted toward its goal configuration, while being repulsed by the obstacles. In this section we illustrate this idea with the definition of one possible potential function, in the case where $\mathcal{A}$ translates freely at fixed orientation in $\mathcal{W} = \mathbf{R}^N$, with $N = 2$ or 3, i.e. $\mathcal{C} = \mathbf{R}^N$. In the next section we will adapt this definition to the case where $\mathcal{A}$ can both translate and rotate. The function defined below is used in several implemented systems (e.g. see [Khatib, 1986]).

1.1 General Structure

The field of artificial forces $\vec{F}(\mathbf{q})$ in $\mathcal{C}$ is produced by a differentiable potential function $\mathbf{U} : \mathcal{C}_{free} \rightarrow \mathbf{R}$, with:

$$\vec{F}(\mathbf{q}) = -\vec{\nabla}\mathbf{U}(\mathbf{q})$$

where $\vec{\nabla}\mathbf{U}(\mathbf{q})$ denotes the gradient vector of $\mathbf{U}$ at $\mathbf{q}$. In $\mathcal{C} = \mathbf{R}^N$ ($N = 2$ or 3), we can write $\mathbf{q} = (x, y)$ or (x, y, z), and:

$$\vec{\nabla}\mathbf{U} = \begin{pmatrix} \partial\mathbf{U}/\partial x \\ \\ \partial\mathbf{U}/\partial y \end{pmatrix} \quad \text{or} \quad \begin{pmatrix} \partial\mathbf{U}/\partial x \\ \\ \partial\mathbf{U}/\partial y \\ \\ \partial\mathbf{U}/\partial z \end{pmatrix}.$$

In order to make the robot be attracted toward its goal configuration, while being repulsed from the obstacles, $\mathbf{U}$ is constructed as the sum of two more elementary potential functions:

$$\mathbf{U}(\mathbf{q}) = \mathbf{U}_{att}(\mathbf{q}) + \mathbf{U}_{rep}(\mathbf{q})$$

where $\mathbf{U}_{att}$ is the **attractive potential** associated with the goal configuration $\mathbf{q}_{goal}$ and $\mathbf{U}_{rep}$ is the **repulsive potential** associated with the C-obstacle region. $\mathbf{U}_{att}$ is independent of the C-obstacle region, while $\mathbf{U}_{rep}$ is independent of the goal configuration.

With these conventions, $\vec{F}$ is the sum of two vectors:

$$\vec{F}_{att} = -\vec{\nabla}\mathbf{U}_{att} \quad \text{and} \quad \vec{F}_{rep} = -\vec{\nabla}\mathbf{U}_{rep}$$

which are called the **attractive** and the **repulsive forces**, respectively.

1.2 Attractive Potential

The attractive potential field $\mathbf{U}_{att}$ can be simply defined as a parabolic well, i.e. :

$$\mathbf{U}_{att}(\mathbf{q}) = \frac{1}{2}\,\xi\,\rho_{goal}^2(\mathbf{q})$$

where ξ is a positive scaling factor and $\rho_{goal}(\mathbf{q})$ denotes the Euclidean distance $\|\mathbf{q} - \mathbf{q}_{goal}\|$. The function $\mathbf{U}_{att}$ is positive or null, and attains its minimum at $\mathbf{q}_{goal}$, where $\mathbf{U}_{att}(\mathbf{q}_{goal}) = 0$.

The function ρ_{goal} is differentiable everywhere in $\mathcal{C}$. At every configuration $\mathbf{q}$, the artificial attractive force $\vec{F}_{att}$ deriving from $\mathbf{U}_{att}$ is:

$$\begin{aligned} \vec{F}_{att}(\mathbf{q}) &= -\vec{\nabla}\mathbf{U}_{att}(\mathbf{q}) \\ &= -\xi\,\rho_{goal}(\mathbf{q})\vec{\nabla}\rho_{goal}(\mathbf{q}) \\ &= -\xi\,(\mathbf{q} - \mathbf{q}_{goal}). \end{aligned}$$

When used for on-line collision avoidance, the parabolic well has good stabilizing characteristics [Khatib, 1986] [Khosla and Volpe, 1988], since it generates a force $\vec{F}_{att}$ that converges linearly toward 0 when the robot's configuration gets closer to the goal configuration. (Asymptotic stabilization of the robot can be achieved by adding dissipative forces proportional to the velocity $\dot{\mathbf{q}}$.)

However, the parabolic-well attractive potential generates a force that augments with the distance to the goal configuration and tends toward infinity when $\rho_{goal}(\mathbf{q}) \rightarrow \infty$. Alternatively, one may define $\mathbf{U}_{att}$ as a conic well, i.e. :

$$\mathbf{U}_{att}(\mathbf{q}) = \xi \rho_{goal}(\mathbf{q}).$$

Then, the resulting attractive force is:

$$\begin{aligned} \vec{F}_{att}(\mathbf{q}) &= -\xi \vec{\nabla} \rho_{goal}(\mathbf{q}) \\ &= -\xi(\mathbf{q} - \mathbf{q}_{goal}) \;/\; \|\mathbf{q} - \mathbf{q}_{goal}\|. \end{aligned}$$

The amplitude of $\vec{F}_{att}(\mathbf{q})$ is constant over $\mathcal{C}$, except at $\mathbf{q}_{goal}$, where $\mathbf{U}_{att}$ is singular. Since the amplitude of the force does not tend toward 0 when $\mathbf{q} \rightarrow \mathbf{q}_{goal}$, the conic-well potential does not have the stabilizing characteristics of the parabolic-well function.

One way to combine the advantages of both the parabolic and the conic wells is to define the attractive potential as a parabolic well within some distance from the goal configuration and a conic well beyond that distance, with a defined derivative at their juncture.

In the rest of the chapter we will exclusively use parabolic-well attractive potentials (whenever we construct the potential as the sum of an attractive and a repulsive potential).

1.3 Repulsive Potential

The main idea underlying the definition of the repulsive potential is to create a potential barrier around the C-obstacle region that cannot be traversed by the robot's configuration. In addition, it is usually desirable that the repulsive potential not affect the motion of the robot when it is sufficiently far away from the C-obstacles. One way to achieve these constraints is to define the repulsive potential function as follows:

$$\mathbf{U}_{rep}(\mathbf{q}) = \begin{cases} \frac{1}{2}\eta\left(\frac{1}{\rho(\mathbf{q})} - \frac{1}{\rho_0}\right)^2 & \text{if } \rho(\mathbf{q}) \leq \rho_0, \\ 0 & \text{if } \rho(\mathbf{q}) > \rho_0, \end{cases}$$

where η is a positive scaling factor, $\rho(\mathbf{q})$ denotes the distance from $\mathbf{q}$ to the C-obstacle region $\mathcal{CB}$, i.e. :

$$\rho(\mathbf{q}) = \min_{\mathbf{q}' \in \mathcal{CB}} \|\mathbf{q} - \mathbf{q}'\|,$$

and ρ_0 is a positive constant called the **distance of influence** of the C-obstacles. The function $\mathbf{U}_{rep}$ is positive or null, tends to infinity as $\mathbf{q}$ gets closer to the C-obstacle region, and is null when the distance of the robot's configuration to the C-obstacle region is greater than ρ_0.

If $\mathcal{CB}$ is a convex region with a piecewise differentiable boundary, ρ is differentiable everywhere in $\mathcal{C}_{free}$. The artificial repulsive force deriving from $\mathbf{U}_{rep}$ is:

$$\begin{aligned} \vec{F}_{rep}(\mathbf{q}) &= -\vec{\nabla}\mathbf{U}_{rep}(\mathbf{q}) \\ &= \begin{cases} \eta(\frac{1}{\rho(\mathbf{q})} - \frac{1}{\rho_0})\frac{1}{\rho^2(\mathbf{q})}\vec{\nabla}\rho(\mathbf{q}) & \text{if } \rho(\mathbf{q}) \leq \rho_0, \\ 0 & \text{if } \rho(\mathbf{q}) > \rho_0. \end{cases} \end{aligned}$$

Let $\mathbf{q}_c$ be the unique configuration in $\mathcal{CB}$ that is closest to $\mathbf{q}$, i.e. that achieves $\|\mathbf{q} - \mathbf{q}_c\| = \rho(\mathbf{q})$. The gradient $\vec{\nabla}\rho(\mathbf{q})$ is a unit vector pointing away from $\mathcal{CB}$ and supported by the line passing through $\mathbf{q}_c$ and $\mathbf{q}$.

If we retract the (unrealistic) assumption that $\mathcal{CB}$ is convex, $\rho(\mathbf{q})$ remains differentiable everywhere in $\mathcal{C}_{free}$, except at those configurations $\mathbf{q}$ for which there exist several $\mathbf{q}_c \in \mathcal{CB}$ verifying $\|\mathbf{q} - \mathbf{q}_c\| = \rho(\mathbf{q})$. These configurations form a set of measure zero in $\mathcal{C}$ which is in general locally $(N-1)$-dimensional. The force field $\vec{F}_{rep}$ is defined on both sides of this set, but with differently oriented vector values. This may result in producing paths that oscillate between the two sides of the set.

One way to eliminate this difficulty is to decompose $\mathcal{CB}$ into (possibly overlapping) convex components $\mathcal{CB}_k$, $k = 1, ..., r$, and to associate a repulsive potential $\mathbf{U}_{\mathcal{CB}_k}$ with each component. Then, the repulsive potential is the sum of the repulsive potential fields created by each individual C-obstacle, i.e. :

$$\mathbf{U}_{rep}(\mathbf{q}) = \sum_{k=1}^{r} \mathbf{U}_{\mathcal{CB}_k}(\mathbf{q})$$

with:

$$\mathbf{U}_{\mathcal{CB}_k}(\mathbf{q}) = \begin{cases} \frac{1}{2}\eta(\frac{1}{\rho_k(\mathbf{q})} - \frac{1}{\rho_0})^2 & \text{if } \rho_k(\mathbf{q}) \leq \rho_0, \\ 0 & \text{if } \rho_k(\mathbf{q}) > \rho_0, \end{cases}$$

where $\rho_k(\mathbf{q})$ denotes the distance from $\mathbf{q}$ to $\mathcal{CB}_k$.

The artificial repulsive force deriving from $\mathbf{U}_{rep}$ is:

$$\vec{F}_{rep}(\mathbf{q}) = \sum_{k=1}^{r} \vec{F}_{\mathcal{CB}_k}(\mathbf{q})$$

with:

$$\vec{F}_{\mathcal{CB}_k}(\mathbf{q}) = -\vec{\nabla}\mathbf{U}_{\mathcal{CB}_k}(\mathbf{q}).$$

The $\mathcal{CB}_k$'s may be constructed by first decomposing $\mathcal{A}$ and the workspace obstacles into convex components, and then computing the C-obstacle corresponding to every pair consisting of a component of $\mathcal{A}$ and a component of an obstacle.

The drawback of decomposing the C-obstacle region is that several small components that are close to each other arbitrarily produce a combined repulsive force which is larger than the one produced by a single bigger component. One empirical way to deal with this drawback is to weigh each individual potential by a coefficient depending on the "size" of the C-obstacle component that generates it.

Figure 1 illustrates the above definitions in a two-dimensional configuration space containing two C-obstacles (Figure 1.a). The attractive potential is a parabolic well with its minimum at the goal configuration $\mathbf{q}_{goal}$ (Figure 1.b). The repulsive potential is displayed in Figure 1.c. Since the two C-obstacles are sufficiently far apart from each other, both definitions given above yield the same function. This function tends toward infinity near the C-obstacles' boundary; hence, it is truncated in the figure. The total potential is displayed in Figure 1.d and some of its equipotential contours are shown in Figure 1.e together with a path connecting $\mathbf{q}_{init}$ to $\mathbf{q}_{goal}$. A matrix showing the discretized negated gradient vector of the total potential over free space is shown in Figure 1.f.

In the above definitions of $\mathbf{U}_{rep}(\mathbf{q})$, we use a unique scaling factor η and a unique distance of influence ρ_0. However, there is no difficulty in using different parameters η and ρ_0 for various subsets of the C-obstacle region. In particular, if the goal configuration $\mathbf{q}_{goal}$ is close to $\mathcal{CB}$, the influence distance ρ_0 should be set to a value smaller than the distance from $\mathbf{q}_{goal}$ to $\mathcal{CB}$ for the subset of $\mathcal{CB}$ that is located in the surrounding of $\mathbf{q}_{goal}$, so that the robot is not prevented from achieving $\mathbf{q}_{goal}$ by the

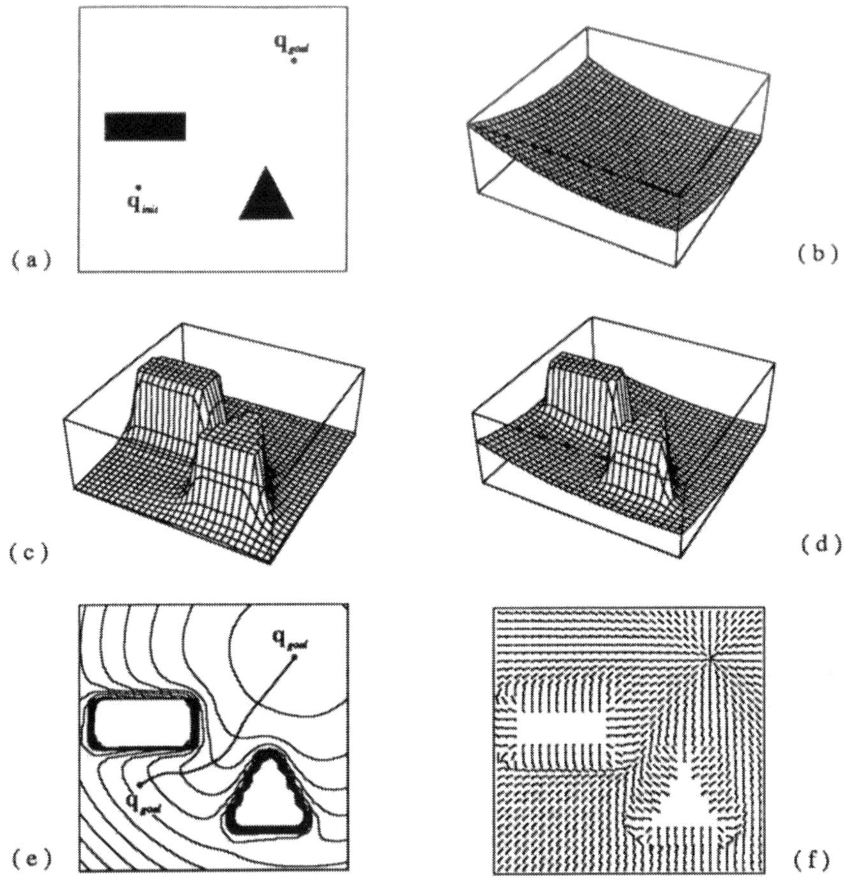

Figure 1. This figure shows an attractive potential field (Figure b), a repulsive potential field (Figure c) and the sum of the two (Figure d) in a two-dimensional configuration space containing two C-obstacles (Figure a). Figure e displays both several equipotential contours of the total potential and a path generated by following the negated gradient of this function. Figure f shows a matrix of the negated gradient vector orientations over free space.

repulsive potential field. (If the repulsive potential is non-zero at $\mathbf{q}_{goal}$, then in general the global minimum of the total potential function is not $\mathbf{q}_{goal}$.) It can be set to a greater value for the rest of $\mathcal{CB}$. From another perspective, if a specific workspace obstacle is known to be more "threatening" than others, it may be appropriate to use larger values of

η and ρ_0 for the corresponding C-obstacle.

The repulsive potential $\mathbf{U}_{\mathcal{CB}}(\mathbf{q})$ and the associated force field $\vec{F}_{\mathcal{CB}}(\mathbf{q})$ can easily be computed when the C-obstacle region is a polygonal or a polyhedral region. Then, if an explicit representation of $\mathcal{CB}$'s boundary has been previously computed, the computation can be done in time $O(n)$, with n being the complexity of $\mathcal{CB}$. Repulsive potential and forces are also easy to compute for some other C-obstacles' shapes (e.g. spheres). In more general cases, however, there may exist no straightforward algorithm. When the C-obstacle region can be decomposed into convex components $\mathcal{CB}_k$, each described as a conjunction of inequalities $f_j^{(k)}(\mathbf{q}) \leq 0$, $j = 1, 2, \ldots$, the problem of computing $\rho_k(\mathbf{q})$ is a convex optimization problem which can be solved using iterative numerical techniques.

Remark: In two dimensions, the potential field approach with the potential defined as above can be loosely related to retraction (see Section 2 of Chapter 4). The valleys of the repulsive potential form the Voronoi diagram of the free space. The attractive potential acts as a heuristic function guiding the search of the diagram. However, since the two potentials are added in the total potential function, the potential field approach does not restrict the robot to moving along the standardized paths formed by the arcs of the Voronoi diagram. ∎

2 Potential Field in General Case

This section considers the case where $\mathcal{C}$ is the manifold $\mathbf{R}^N \times SO(N)$. The content of the section is not essential to the understanding of the rest of the chapter and may be skipped in a first reading. In particular, the reader who wishes to get an immediate look at potential-guided planning methods can jump to the next section, and later come back to the present section.

2.1 Extension of Previous Notions

The notions introduced in the previous section can be extended to the general case where $\mathcal{C} = \mathbf{R}^N \times SO(N)$. We must however pay some attention to the gradient vector $\vec{\nabla}\mathbf{U}$ whose definition requires us to specify an inner product in the tangent spaces of $\mathcal{C}$.

Let (U, ϕ) be a chart of $\mathcal{C}$, with $\phi : \mathbf{q} \in U \mapsto (x_1(\mathbf{q}), \ldots, x_m(\mathbf{q})) \in \mathbf{R}^m$.

Let $\vec{t}$ be a vector of the tangent space $T_{\mathbf{q}}(\mathcal{C})$, $[\vec{t}\,]_\beta$ be the m-vector of its components in the base induced by the chart (U, ϕ), and $\mathbf{q}_{\alpha\vec{t}}$ be the configuration defined by $\phi^{-1}((x_1(\mathbf{q}), ..., x_m(\mathbf{q})) + \alpha[\vec{t}\,]_\beta)$. We denote the directional derivative of $\mathbf{U}$ along the direction $\vec{t}$, taken at $\mathbf{q}$, by $D_{\vec{t}}\mathbf{U}(\mathbf{q})$. It is defined as the following linear functional on $T_{\mathbf{q}}(\mathcal{C})$:

$$D_{\vec{t}}\mathbf{U}(\mathbf{q}) = \lim_{\alpha \to 0} \frac{\mathbf{U}(\mathbf{q}_{\alpha\vec{t}}) - \mathbf{U}(\mathbf{q})}{\alpha}.$$

Let $<,>$ designate the inner product in $T_{\mathbf{q}}(\mathcal{C})$. By definition, the gradient vector $\vec{\nabla}\mathbf{U}(\mathbf{q})$ is the vector of $T_{\mathbf{q}}(\mathcal{C})$ such that for any vector $\vec{t} \in T_{\mathbf{q}}(\mathcal{C})$, we have:

$$D_{\vec{t}}\mathbf{U}(\mathbf{q}) \;\; = \;\; <\vec{\nabla}\mathbf{U}, \vec{t}> .$$

Notice the similarity with the definition of a force vector in Section 10 of Chapter 2: while the directional derivative of $\mathbf{U}$ (resp. the work of a force) is independent of the choice of the inner product, the representation of the gradient vector (resp. the force vector) depends on that choice.

Let us select the inner product in $T_{\mathbf{q}}(\mathcal{C})$ such that β is an orthonormal basis. Let t_i, $i = 1$ to m, be the components of the vector $\vec{t}$ in β. We have:

$$\mathbf{U}(\mathbf{q}) = \mathbf{U}(x_1(\mathbf{q}), ..., x_m(\mathbf{q}))$$

and:

$$D_{\vec{t}}\mathbf{U}(\mathbf{q}) = \sum_{i=1}^{m} \frac{\partial \mathbf{U}}{\partial x_i}(x_1(\mathbf{q}), ..., x_m(\mathbf{q}))\, t_i.$$

Hence:

$$[\vec{\nabla}\mathbf{U}(\mathbf{q})\,]_\beta = \left(\frac{\partial \mathbf{U}}{\partial x_1} \;\; \cdots \;\; \frac{\partial \mathbf{U}}{\partial x_m}\right)^T_{|(x_1(\mathbf{q}),...,x_m(\mathbf{q}))}$$

which coincides with the conventional definition of a gradient vector in the Euclidean space $\mathbf{R}^m$.

Let $\vec{\nabla}_1\mathbf{U}(\mathbf{q})$ and $\vec{\nabla}_2\mathbf{U}(\mathbf{q})$ denote the gradient vectors of $\mathbf{U}$ for two inner products designated by $<,>_1$ and $<,>_2$, respectively. Let β_1 and β_2 be two bases of $T_{\mathbf{q}}(\mathcal{C})$ such that β_i, for any $i \in \{1, 2\}$, is an orthonormal basis when $<,>_i$ is used. Let J be the $m \times m$ matrix of the linear application that maps the m-vector $[\vec{t}\,]_1$ of the components of a vector

$\vec{t} \in T_{\mathbf{q}}(\mathcal{C})$ in β_1 to the m-vector $[\vec{t}]_2$ of the components of $\vec{t}$ in β_2. For any $\vec{t} \in T_{\mathbf{q}}(\mathcal{C})$, we have:

$$\begin{aligned}[\vec{\nabla}_1 \mathbf{U}(\mathbf{q})]_1^T\,[\vec{t}]_1 &= [\vec{\nabla}_2 \mathbf{U}(\mathbf{q})]_2^T\,[\vec{t}]_2 \\ &= [\vec{\nabla}_2 \mathbf{U}(\mathbf{q})]_2^T\,J\,[\vec{t}]_1.\end{aligned}$$

Thus:

$$[\vec{\nabla}_1 \mathbf{U}(\mathbf{q})]_1 = J^T\,[\vec{\nabla}_2 \mathbf{U}(\mathbf{q})]_2.$$

(Notice the similarity of this relation with the relation established for forces in Section 10 of Chapter 2.)

Given a metric d in $\mathcal{C}$, the attractive potential function and the associated field of forces can be defined as in the previous section by:

$$\begin{aligned}\mathbf{U}_{att}(\mathbf{q}) &= \tfrac{1}{2}\xi\rho_{goal}^2(\mathbf{q}) \\ \vec{F}_{att}(\mathbf{q}) &= -\vec{\nabla}\mathbf{U}_{att}(\mathbf{q}) \\ &= -\xi\rho_{goal}(\mathbf{q})\vec{\nabla}\rho_{goal}(\mathbf{q})\end{aligned}$$

where $\rho_{goal}(\mathbf{q}) = d(\mathbf{q}, \mathbf{q}_{goal})$. In the same way, the repulsive potential and the associated field of forces are:

$$\mathbf{U}_{rep}(\mathbf{q}) = \begin{cases} \frac{1}{2}\eta(\frac{1}{\rho(\mathbf{q})} - \frac{1}{\rho_0})^2 & \text{if } \rho(\mathbf{q}) \leq \rho_0, \\ 0 & \text{if } \rho(\mathbf{q}) > \rho_0, \end{cases}$$

and:

$$\begin{aligned}\vec{F}_{rep}(\mathbf{q}) &= -\vec{\nabla}\mathbf{U}_{rep}(\mathbf{q}) \\ &= \begin{cases} \eta(\frac{1}{\rho(\mathbf{q})} - \frac{1}{\rho_0})\frac{1}{\rho^2(\mathbf{q})}\vec{\nabla}\rho(\mathbf{q}) & \text{if } \rho(\mathbf{q}) \leq \rho_0, \\ 0 & \text{if } \rho(\mathbf{q}) > \rho_0, \end{cases}\end{aligned}$$

where $\rho(\mathbf{q}) = \min_{\mathbf{q}' \in \mathcal{CB}} d(\mathbf{q}, \mathbf{q}')$.

However, this extension yields computational difficulties. Indeed, for a given metric d there is in general no simple and efficient method for computing both the distance $\rho(\mathbf{q})$ from a configuration $\mathbf{q}$ to the C-obstacle region $\mathcal{CB}$ and its gradient $\vec{\nabla}\rho(\mathbf{q})$.

One practical way to proceed is to define the attractive and repulsive potential functions in $\mathcal{W} = \mathbf{R}^N$ and combine their effects at several points of $\mathcal{A}$. This technique is described in the next two subsections.

2.2 Attractive Potential

Let a_j, $j = 1, ..., N$, be points selected in $\mathcal{A}$, with N being the dimension of the workspace. (If $N = 3$, the 3 points a_j should not be aligned in order to uniquely determine the configuration of $\mathcal{A}$.) The points a_j are called the *control points subject to the attractive potential field.*

For each point a_j, $j \in [1, N]$, we define a distinct attractive potential function $\mathbf{V}^j_{att} : \mathcal{W} \to \mathbf{R}$ with:

$$\mathbf{V}^j_{att}(\mathbf{x}) = \frac{1}{2}\xi\|\mathbf{x} - a_j(\mathbf{q}_{goal})\|^2$$

where ξ is a positive scaling factor.

Each potential $\mathbf{V}^j_{att}$ induces a field of forces $-\vec{\nabla}\mathbf{V}^j_{att}$ over the workspace. Only the control point a_j is "sensitive" to that field. At each configuration $\mathbf{q}$ of $\mathcal{A}$, the artificial force $\vec{\mathbf{F}}^j_{att}(\mathbf{q}) \in \mathbf{R}^N$ defined by:

$$\begin{aligned}\vec{\mathbf{F}}^j_{att}(\mathbf{q}) &= -\left(\vec{\nabla}\mathbf{V}^j_{att}(\mathbf{x})\right)_{|\mathbf{x}=a_j(\mathbf{q})} \\ &= -\xi(a_j(\mathbf{q}) - a_j(\mathbf{q}_{goal}))\end{aligned}$$

is exerted on $\mathcal{A}$ at point a_j. From Section 10 of Chapter 2, and given an inner product in $T_{\mathbf{q}}(\mathcal{C})$, we know how to map this force into a vector $\vec{F}^j_{att}(\mathbf{q})$ of $T_{\mathbf{q}}(\mathcal{C})$.

The total attractive force in $T_{\mathbf{q}}(\mathcal{C})$ is:

$$\vec{F}_{att}(\mathbf{q}) = \sum_{j=1}^{N} \vec{F}^j_{att}(\mathbf{q}).$$

Let:

$$\mathbf{U}^j_{att}(\mathbf{q}) = \mathbf{V}^j_{att}(a_j(\mathbf{q})).$$

One can easily verify that:

$$\vec{F}^j_{att}(\mathbf{q}) = -\vec{\nabla}\mathbf{U}^j_{att}(\mathbf{q}).$$

Thus:

$$\mathbf{U}_{att}(\mathbf{q}) = \sum_{j=1}^{N} \mathbf{V}^j_{att}(a_j(\mathbf{q}))$$

is the attractive potential in $\mathcal{C}$ which produces the attractive force $\vec{F}_{att}(\mathbf{q})$, i.e. :

$$\vec{F}_{att}(\mathbf{q}) = -\vec{\nabla}\mathbf{U}_{att}(\mathbf{q}).$$

Remark 1: The number of points a_j may be smaller than N. Then, specifying the goal positions of these points only partially determines a goal configuration of $\mathcal{A}$, i.e. the goal configurations form a region of $\mathcal{C}$. ■

Remark 2: If there is more than one control point a_j subject to the attractive potential field, the above definition of the potential $\mathbf{U}_{att}$ makes the control points compete in trying to attain their respective goal positions. This competition may produce local minima of the total potential in narrow passages and cluttered areas of the workspace. A better definition might be:

$$\mathbf{U}_{att}(\mathbf{q}) = \mathbf{V}^1_{att}(a_1(\mathbf{q})) + \varepsilon \sum_{j>1} \mathbf{V}^j_{att}(a_j(\mathbf{q}))$$

where ε is a small positive constant (say, $\varepsilon = 0.1$). This corresponds to choosing a "leader" among the control points (a_1 in the above definition). Along the generated path, this leading point is vigorously pulled toward its goal position, while the small ε coefficient lets the orientation of $\mathcal{A}$ be mainly under the influence of the repulsive potential (defined below). The second term of the attractive potential nevertheless tends to make the robot ultimately achieve a goal orientation. The above function $\mathbf{U}_{att}$ gives good results when the goal configuration is rather far from the C-obstacle region, but on the other hand it may cause a deep local minimum to exist near the goal configuration if this is not the case. ■

2.3 Repulsive Potential

We can define and compute the repulsive potential in a similar fashion.

For all $\mathbf{x} \in \mathcal{W}\backslash\mathcal{B}$, where $\mathcal{B}$ is the obstacle region in $\mathcal{W}$, we define the repulsive potential in the workspace by:

$$\mathbf{V}_{rep}(\mathbf{x}) = \begin{cases} \frac{1}{2}\eta(\frac{1}{\rho(\mathbf{x})} - \frac{1}{\rho_0})^2 & \text{if } \rho(\mathbf{x}) \leq \rho_0, \\ 0 & \text{if } \rho(\mathbf{x}) > \rho_0. \end{cases}$$

η is a positive scaling factor, $\rho(\mathbf{x})$ denotes the Euclidean distance from the point $\mathbf{x}$ to $\mathcal{B}$, and ρ_0 is the distance of influence of $\mathcal{B}$.

Let a_j, $j = 1, \ldots, Q$, be points selected in the boundary of $\mathcal{A}$. These Q points are called the *control points subject to the repulsive potential field.* The control points used in the definition of the attractive forces do not have to be among these Q points, although they may.

When $\mathcal{A}$ is at configuration $\mathbf{q}$, $\mathbf{V}_{rep}$ exerts a force $\vec{\mathbf{F}}^j_{rep}(\mathbf{q})$ on $\mathcal{A}$ which is applied at point a_j:

$$\vec{\mathbf{F}}^j_{rep}(\mathbf{q}) = \begin{cases} \eta \left(\frac{1}{\rho(a_j(\mathbf{q}))} - \frac{1}{\rho_0}\right) \frac{1}{\rho^2(a_j(\mathbf{q}))} \vec{\nabla}\rho(\mathbf{x})_{|\mathbf{x}=a_j(\mathbf{q})} & \text{if } \rho(a_j(\mathbf{q})) \leq \rho_0, \\ 0 & \text{if } \rho(a_j(\mathbf{q})) > \rho_0. \end{cases}$$

If $\mathcal{B}$ is a convex region with a piecewise differentiable boundary, ρ is differentiable everywhere in $\mathcal{C}_{free}$. If $\mathcal{B}$ is non-convex, we may decompose it into convex components and associate a repulsive potential with each component (see Subsection 1.3).

Given an inner product in $T_{\mathbf{q}}(\mathcal{C})$, we know how to map $\vec{\mathbf{F}}^j_{rep}(\mathbf{q})$ to a vector $\vec{F}^j_{rep}(\mathbf{q})$ of $T_{\mathbf{q}}(\mathcal{C})$. The resulting repulsive force in $T_{\mathbf{q}}(\mathcal{C})$ is:

$$\vec{F}_{rep}(\mathbf{q}) = \sum_{j=1}^{Q} \vec{F}^j_{rep}(\mathbf{q}).$$

Let:

$$\mathbf{U}^j_{rep}(\mathbf{q}) = \mathbf{V}_{rep}(a_j(\mathbf{q})).$$

One can verify that:

$$\vec{F}^j_{rep}(\mathbf{q}) = -\vec{\nabla}\mathbf{U}^j_{rep}(\mathbf{q}).$$

Thus:

$$\mathbf{U}_{rep}(\mathbf{q}) = \sum_{j=1}^{Q} \mathbf{V}_{rep}(a_j(\mathbf{q}))$$

is the repulsive potential in $\mathcal{C}$ that produces the repulsive force $\vec{F}_{rep}(\mathbf{q})$, i.e. :

$$\vec{F}_{rep}(\mathbf{q}) = -\vec{\nabla}\mathbf{U}_{rep}(\mathbf{q}).$$

In order to make sure that $\mathcal{A}$ cannot come close to $\mathcal{B}$ without being repulsed, we may combine a small number of "fixed" control points and a "variable" control point which depends on $\mathcal{A}$'s configuration. The variable control point is the point a' in $\mathcal{A}$'s boundary that is closest to $\mathcal{B}$ at the current configuration $\mathbf{q}$ of $\mathcal{A}$. Hence, it is the solution of:

$$\min_{b \in \mathcal{B}} \|a'(\mathbf{q}) - b\| = \min_{a \in \mathcal{A}, b \in \mathcal{B}} \|a(\mathbf{q}) - b\|.$$

The repulsive force that is applied at a' is:

$$\vec{\mathbf{F}}'_{rep}(\mathbf{q}) = \begin{cases} \eta \left(\frac{1}{\rho(a'(\mathbf{q}))} - \frac{1}{\rho_0} \right) \frac{1}{\rho^2(a'(\mathbf{q}))} \vec{\nabla} \rho(\mathbf{x})_{|\mathbf{x}=a'(\mathbf{q})} & \text{if } \rho(a'(\mathbf{q})) \leq \rho_0, \\ 0 & \text{if } \rho(a'(\mathbf{q})) > \rho_0. \end{cases}$$

If both $\mathcal{A}$ and $\mathcal{B}$ are convex polygons or polyhedra, the computation of $\min_{a \in \mathcal{A}, b \in \mathcal{B}} \|a(\mathbf{q}) - b\|$, together with a pair of points achieving this minimal distance, can be made in time $O(n_{\mathcal{A}} + n_{\mathcal{B}})$, where $n_{\mathcal{A}}$ and $n_{\mathcal{B}}$ are the number of vertices of $\mathcal{A}$ and $\mathcal{B}$, respectively [Dobkin and Kirkpatrick, 1985]. However, for some critical configurations of $\mathcal{A}$, where an edge or a face of $\mathcal{A}$ is parallel to an edge or a face of $\mathcal{B}$, the minimal distance is achieved by more than a single pair of points. At these configurations, the variable control point a' may change discontinuously in $\mathcal{A}$'s boundary, resulting in a discontinuity of the repulsive force $\vec{F}'_{rep}(\mathbf{q})$. This might cause the generation of an oscillating path. One way to reduce the discontinuity of the repulsive force at these configurations is to distribute the repulsive force over several points of $\mathcal{A}(\mathbf{q})$ among those which are closest to $\mathcal{B}$. Let S be the set of points of $\mathcal{A}(\mathbf{q})$ which are closest to $\mathcal{B}$. If S is not a singleton, it is either a line segment or a polygon. The repulsive force $\vec{F}'_{rep}(\mathbf{q})$ can be distributed over the extreme points of S, i.e. the extremities of the line segment or the vertices of the polygon.

If $\mathcal{A}$ and $\mathcal{B}$ are non-convex polygons, they can be decomposed into two sets of convex polygons $\{\mathcal{A}_k\}$ and $\{\mathcal{B}_l\}$. For each pair $(\mathcal{A}_k, \mathcal{B}_l)$ we can define a variable control point a'_{kl} defined as the point in $\mathcal{A}_k$'s boundary that is the closest to $\mathcal{B}_l$. All such points a'_{kl} are simultaneously taken as variable control points subject to the repulsive potential field.

If $\mathcal{A}$ and $\mathcal{B}$ have more complex shapes, but can be decomposed into convex components $\{\mathcal{A}_k\}$ and $\{\mathcal{B}_l\}$, iterative numerical techniques can be used to compute the distance between $\mathcal{A}_k$ and $\mathcal{B}_l$, together with a pair of points achieving this distance (e.g. see [Gilbert, Johnson and Keerthi, 1988]). We can define the repulsive potential field by using several variable control points a'_{kl}, one for each pair $(\mathcal{A}_k, \mathcal{B}_l)$. If either

$\mathcal{A}_k$ or $\mathcal{B}_l$ is strictly convex[1], the other being convex, and if both have piecewise differentiable boundaries, then a'_{kl} is uniquely defined and the repulsive force exerted at a'_{kl} varies continuously.

3 Potential-Guided Path Planning

We now describe simple potential-guided path planning techniques. In principle, these techniques do not assume any specific potential function. Hence, they are applicable with the potential function defined in the previous two sections, and with other potential functions as well.

In its original conception, the potential field approach to motion generation consists of regarding the robot in configuration space as a unit mass particle moving under the influence of the force field $\vec{F} = -\vec{\nabla}\mathbf{U}$. At every configuration $\mathbf{q}$ the artificial force $\vec{F}(\mathbf{q})$ determines the acceleration of the particle. Knowing the dynamic equation of $\mathcal{A}$ and assuming no limitation on the power of the actuators, one can compute the forces/torques that should be delivered by the actuators at each instant so that the robot actually behaves according to the particle metaphor. These forces/torques are the commands sent to the robot's servo controllers [Hogan, 1985] [Khatib, 1986] [Koditschek, 1989].

This way of using the potential function is applicable for generating paths on-line. It is well-suited when the obstacles are not known in advance, but sensed during motion execution. If a prior model of the obstacles is available, the same method can be used to plan a path by simulating the motion of the particle. However, in this case, there exist simpler and more efficient path planning techniques using potential field.

In the following subsections, we present several such techniques. The first of them generates a path in a "depth-first" fashion, without backtracking. Like on-line path generation, it may be very fast in favorable cases, but it may also get stuck at local minima of the potential function. The second technique operates in a "best-first" mode. It deals with local minima by "filling" them up. The third technique consists of optimizing a functional constructed by integrating the potential along a complete

[1] A closed set $S \subset \mathbf{R}^N$ is *strictly convex* if and only if it is homeomorphic to the closed unit ball in $\mathbf{R}^N$ and for every point p in the boundary of S, there exists a line passing through p and intersecting S at p only.

path between the initial and the goal configurations.

3.1 Depth-First Planning

This technique constructs a path as the product of successive path segments starting at the initial configuration $\mathbf{q}_{init}$. Each segment is oriented along the negated gradient of the potential function computed at the configuration attained by the previous segment. The amplitude of the segment is chosen so that the segment lies in free space.

Let $\mathbf{q}_i$ and $\mathbf{q}_{i+1}$ be the origin and end extremities of the ith segment in the path. Let $x_j(\mathbf{q}_i)$, $j = 1, ..., m$, be the coordinates of $\mathbf{q}_i$ in some chart (U, ϕ). We define the inner product in the tangent space $T_{\mathbf{q}}(\mathcal{C})$ so that the basis β induced by this chart in $T_{\mathbf{q}}(\mathcal{C})$ is orthonormal. We then have:

$$[\vec{F}]_\beta = -[\vec{\nabla}\mathrm{U}]_\beta = (-\partial\mathrm{U}/\partial x_1 \ \ ... \ \ -\partial\mathrm{U}/\partial x_m)^T.$$

We denote the components of the unit vector $\vec{t}(\mathbf{q}_i) = \vec{F}(\mathbf{q}_i)/\|\vec{F}(\mathbf{q}_i)\|$ in β by $t_j(\mathbf{q}_i)$. The coordinates of the configuration $\mathbf{q}_{i+1}$ attained at the ith iteration, in (U, ϕ), are:

$$x_j(\mathbf{q}_{i+1}) = x_j(\mathbf{q}_i) + \delta_i t_j(\mathbf{q}_i), \ j = 1, ..., m,$$

with δ_i denoting the length of the ith increment (measured with the Euclidean metric of $\mathbf{R}^m$). The segment $\mathbf{q}_i\mathbf{q}_{i+1}$ is the inverse image in $\mathcal{C}$ of the straight line segment joining $\phi(\mathbf{q}_i)$ to $\phi(\mathbf{q}_{i+1})$ in $\mathbf{R}^m$.

For example, if $\mathcal{A}$ is a planar object moving in $\mathcal{W} = \mathbf{R}^2$, we can parameterize any $\mathbf{q}$ by $(x_1, x_2, x_3) = (x, y, \theta) \in \mathbf{R}^2 \times [0, 2\pi)$, with x and y being the coordinates of $\mathcal{A}$'s reference point $O_\mathcal{A}$ at $\mathbf{q}$, and θ being the angle (modulo 2π) between the x-axes of the frames $\mathcal{F}_\mathcal{W}$ and $\mathcal{F}_\mathcal{A}$ attached to $\mathcal{W}$ and $\mathcal{A}$, respectively. Then:

$$\begin{aligned} x(\mathbf{q}_{i+1}) &= x(\mathbf{q}_i) + \delta_i \tfrac{\partial \mathrm{U}}{\partial x}(x, y, \theta), \\ y(\mathbf{q}_{i+1}) &= y(\mathbf{q}_i) + \delta_i \tfrac{\partial \mathrm{U}}{\partial y}(x, y, \theta), \\ \theta(\mathbf{q}_{i+1}) &= \theta(\mathbf{q}_i) + \delta_i \tfrac{\partial \mathrm{U}}{\partial \theta}(x, y, \theta) \mod 2\pi. \end{aligned}$$

In order to "normalize" the displacements along the θ-axis relative to displacements along the x- and y-axes, one may parameterize $\mathbf{q}$ by

$(x, y, \phi) \in \mathbf{R}^2 \times [0, 2\pi R)$, by posing $\phi = \theta R$, with $R = \max_{a \in \partial\mathcal{A}} \|O_\mathcal{A} - a\|$ being the maximal distance between the reference point $O_\mathcal{A}$ and $\mathcal{A}$'s boundary. This yields:

$$\phi(\mathbf{q}_{i+1}) = \phi(\mathbf{q}_i) + \delta_i \frac{\partial \mathbf{U}}{\partial \phi}(x, y, \phi) \mod 2\pi R,$$

or, equivalently:

$$\theta(\mathbf{q}_{i+1}) = \theta(\mathbf{q}_i) + \frac{\delta_i}{R} \frac{\partial \mathbf{U}}{\partial \phi}(x, y, \theta R) \mod 2\pi.$$

Several considerations intervene in the choice of a value for δ_i. First, δ_i should be taken small enough so that the direction of the force and the mapping of that direction in the local coordinate system keeps some meaning along the segment $\mathbf{q}_i\mathbf{q}_{i+1}$. Typically, δ_i is taken equal to some small prespecified increment δ, unless other considerations examined below require δ_i to be smaller.

The increment δ_i should also be chosen small enough so that no collision happens along the segment $\mathbf{q}_i\mathbf{q}_{i+1}$. Suppose that δ_i is equal to δ. The motion from $\mathbf{q}_i$ to $\mathbf{q}_{i+1}$ can be expressed as a small rotation about $O_\mathcal{A}$, followed by a small translation. The point of $\mathcal{A}$ that moves along the longest distance during the rotation is the point $a \in \partial\mathcal{A}$ that is the most distant from $O_\mathcal{A}$. Let l_1 be the length of this displacement and l_2 be the length of the translation. The sum $l = l_1 + l_2$ is an upper bound on the maximal displacement of a point of $\mathcal{A}$ during the motion from $\mathbf{q}_i$ to $\mathbf{q}_{i+1}$. If l is smaller than the minimum distance of $\mathcal{A}$ to the obstacles, then the segment $\mathbf{q}_i\mathbf{q}_{i+1}$ is guaranteed to lie in free space. Otherwise, a smaller value of δ_i, say $\delta/2$, may be selected and checked again for collision.

Finally, the value of the increment should not lead the path to go beyond the goal configuration. Assume that the chart (U, ϕ) used at $\mathbf{q}_i$ also allows us to map $\mathbf{q}_{goal}$. We should take δ_i smaller than the Euclidean distance (in $\mathbf{R}^m$) between the mappings $\phi(\mathbf{q}_i)$ and $\phi(\mathbf{q}_{goal})$.

This technique simply follows the steepest descent of the potential function until the goal configuration is attained. However, it may get stuck at a minimum of the potential other than the goal configuration (Figure 2). Dealing with local minima within depth-first planning is not simple. First, the fact that a local minimum has been attained must be recognized. Since motions are discretized, the planner usually does not stop

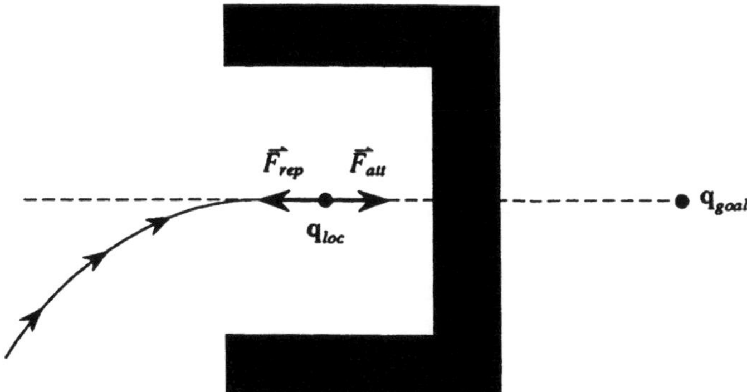

Figure 2. The attractive potential guides the robot's path into a C-obstacle concavity. At some configuration $\mathbf{q}_{loc}$, the repulsive force generated by the C-obstacle cancels exactly the attractive force generated by the goal. This stable zero-force configuration is a local minimum of the total potential function.

exactly at the zero-force configuration; instead, it typically generates segments "looping" around this configuration. This may be detected by checking that several successive configurations $\mathbf{q}_i$'s are not too close to each other. Second, the planner must escape the local minimum. A simple technique consists of moving along a certain direction by some distance before resuming depth-first path generation. A possible direction of motion is any direction in the tangent plane of the repulsive equipotential surface at the local minimum. This corresponds to "moving around" the combination of obstacles that creates the problematic minimum.[2] A more systematic way of dealing with local minima consists of seeing the generated path as a path in a search graph whose nodes are the $\mathbf{q}_i$'s. This leads to presenting best-first planning.

3.2 Best-First Planning

Let us throw a fine regular grid of configurations across $\mathcal{C}$. We denote this grid by $\mathcal{GC}$. $\mathcal{GC}$ can be defined by considering a single chart over

[2] Donald [Donald, 1984 and 1987b] described a variety of heuristics, which he called "local experts", for "moving around" C-obstacles. The principle of these heuristics is to slide over level C-surfaces and/or their intersections. (If $f(\mathbf{q}) = 0$ is the equation of a C-surface (see Chapter 3), $f(\mathbf{q}) = K$, with K constant, is the equation of a level C-surface.) Although Donald did not explicitly use potential functions, his heuristics are directly relevant here.

$\mathcal{C}$ and discretizing each of the m corresponding coordinate axes. For instance, if $\mathcal{A}$ is a free-flying object in $\mathcal{W} = \mathbf{R}^2$, the grid consists of the configurations $(k_x\delta x, k_y\delta y, k_\theta\delta\theta)$, with $k_x, k_y, k_\theta \in \mathbf{Z}$ and modulo 2π arithmetic on θ. The angular increment $\delta\theta$ is an integer fraction of 2π. (As in the previous section, we may use $\phi = \theta R$ rather than θ for the third coordinate, in order to normalize the scale of the angular axis relative to the translational axes.)

Given a configuration $\mathbf{q}$ in the m-dimensional grid $\mathcal{GC}$, its **p-neighbors** $(1 \leq p \leq m)$ are defined as all the configurations in $\mathcal{GC}$ having at most p coordinates differing from those of $\mathbf{q}$, the amount of the difference being exactly one increment in absolute value (with appropriate modular arithmetic applied to the angular parameters). There are $2m$ 1-neighbors, $2m^2$ 2-neighbors, ..., and $3^m - 1$ m-neighbors. In this subsection we consider that two configurations in $\mathcal{GC}$ are neighbors if and only if they are p-neighbors for a predefined $p \in [1, m]$. In practice, we can take $p = 1$ or 2. In addition, for simplifying the presentation, we make the following assumptions:

- Both $\mathbf{q}_{init}$ and $\mathbf{q}_{goal}$ are configurations in $\mathcal{GC}$.
- If two neighbors in $\mathcal{GC}$ are in free space, the straight line segment connecting them in $\mathbf{R}^m$ also lies in free space.
- The grid $\mathcal{GC}$ is bounded and forms a rectangloid (this is achieved by bounding the set of possible positions of $\mathcal{A}$ by a rectangloid).

Best-first planning consists of iteratively constructing a tree T whose nodes are configurations in $\mathcal{GC}$. The root of T is $\mathbf{q}_{init}$. At every iteration, the algorithm examines the neighbors of the leaf of T that has the smallest potential value, retains the neighbors not already in T at which the potential function is less than some large threshold[3] M, and installs the retained neighbors in T as successors of the currently considered leaf. The algorithm terminates when the goal configuration $\mathbf{q}_{goal}$ has been attained (success) or when the free subset of $\mathcal{GC}$ accessible from $\mathbf{q}_{init}$ has been fully explored (failure). Each node in T has a pointer toward its parent. If $\mathbf{q}_{goal}$ is attained, a path is generated by tracing the pointers from $\mathbf{q}_{goal}$ to $\mathbf{q}_{init}$.

[3]We assume here that the potential $U(\mathbf{q})$ is defined such that it grows toward infinity when $\mathbf{q}$ gets closer to the C-obstacle region and that it is infinite when $\mathbf{q} \in \mathcal{CB}$.

The procedure BFP given below is a formal expression of the best-first planning algorithm. In addition to the tree T, BFP uses a list OPEN that contains the leaves of T sorted by increasing values of the potential function. OPEN supports the following three operations: FIRST(OPEN), which removes the configuration of OPEN having the smallest potential value and returns it; INSERT($\mathbf{q}$,OPEN), which inserts the configuration $\mathbf{q}$ in OPEN; and EMPTY(OPEN), which returns `true` (resp. `false`) if the list OPEN is empty (resp. non-empty).

```
procedure BFP;
begin
    install q_init in T; [initially, T is the empty tree]
    INSERT(q_init,OPEN); mark q_init visited;
    [initially, all the configurations in GC are marked "unvisited"]
    SUCCESS ← false;
    while ¬ EMPTY(OPEN) and ¬ SUCCESS do
        begin
            q ← FIRST(OPEN);
            for every neighbor q' of q in GC do
                if U(q') < M and q' is not visited then
                    begin
                        install q' in T with a pointer toward q;
                        INSERT(q',OPEN); mark q' visited;
                        if q' = q_goal then SUCCESS ← true;
                    end;
        end;
    if SUCCESS then
        return the constructed path by tracing the pointers in T
        from q_goal back to q_init;
    else return failure;
end;
```

This procedure follows a discrete approximation of the negated gradient of the potential function until it reaches a local minimum. When this happens, it operates in a way that corresponds to filling the well of this local minimum until a saddle point is attained.

The algorithm is guaranteed to return a free path whenever there exists one in the free subset of the grid $\mathcal{GC}$ and to report failure otherwise. Let $O(r)$ be the number of discretization points along each coordinate axis. The size of $\mathcal{GC}$ is $O(r^m)$. We can represent the list OPEN as a balanced tree [Aho, Hopcroft and Ullman, 1983] so that each operation FIRST and INSERT takes logarithmic time in the size of OPEN, which is $O(r^m)$ in the worst case. The time complexity of the procedure BFP given above is thus $O(mr^m \log r)$.

Most of the time, the procedure BFP will only explore a small subset of $\mathcal{GC}$. Rather than representing $\mathcal{GC}$ explicitly in a large array and marking its configurations *visited* or *unvisited*, we may only represent the configurations that are installed in T. Then, instead of checking whether a configuration $\mathbf{q}'$ is marked *visited*, we test if it is in T. By storing the configurations in T in a balanced-tree data structure sorted according to the configurations' coordinates, the test takes logarithmic time in the number of configurations in T, which is $O(r^m)$ in the worst case. This modification slightly increases the running time of the procedure (without changing its asymptotic time complexity), but reduces the amount of memory space it needs. It also makes it possible to run the algorithm with a grid $\mathcal{GC}$ of unbounded size (but then the running time is no longer bounded).

BFP is practical only when m is small, say less than 5. In the case of a free-flying object in a two-dimensional workspace ($m = 3$), it provides a means for implementing a very fast and reliable planner with grid resolutions of the order of 256^3 (see [Barraquand and Latombe, 1989a]). It may even be made faster (in average time) by using a pyramid of grids at various resolutions. However, when m becomes larger, filling up local minimum wells is no longer tractable (the number of discretized configurations in a well tends to grow exponentially with the dimension m). Other techniques, e.g. the "randomized planning" technique presented in Section 5, must then be used.

3.3 Variational Planning

Another approach to path planning using a potential field consists of constructing a functional J of a path τ and optimizing this functional over all possible paths. For example, if $\mathbf{U}$ is a potential function that is defined over the whole configuration space, with large values in the C-obstacle region and small values in free space, a possible definition of J is the following:

$$J(\tau) = \int_0^1 \mathbf{U}(\tau(s))ds + \int_0^1 \|\frac{d\vec{\tau}}{ds}\|ds.$$

The purpose of the first term in J is to make the optimization produce a free path. The second term in J is optional and is aimed at producing shorter paths. In fact, other objectives could be encoded in this form. The optimization of the functional J can be done using standard variational calculus methods. We report the reader to publications describing implementations of this approach for more detail about the construction of both an appropriate potential $\mathbf{U}$ and an appropriate objective functional J, and about the optimization of this functional (e.g. [Buckley, 1985] [Gilbert and Johnson, 1985] [Hwang and Ahuja, 1988] [Suh and Shin, 1988]).

Variational path planning suffers from the same sort of drawback as depth-first planning. Indeed, since it essentially consists of minimizing a functional J by following its negated gradient, it may get stuck at a local minimum of J that does not correspond to a free path. In addition, the optimization of J is conducted over a space of much larger dimension (the number of configurations in the path) and can be quite costly. The advantage of variational path planning is that it allows additional objective criteria to be encoded in J.

4 Other Potential Functions

A variety of potential functions other than those presented in Sections 1 and 2 have been proposed in the literature. The most interesting of them are aimed at either one of these two goals: (1) improving the "local dynamic behavior" of the robot along the generated paths; (2) reducing the number of local minima and/or the size of their attractive wells.

The first goal is important when the potential field method is used to generate a path on-line or in a depth-first fashion (although in the second case, the path can be improved in a postprocessing step) [Tilove, 1990]. One of the proposed function is the "generalized potential field" [Krogh, 1984] which is a function of *both* the robot's configuration and velocity. It is constructed in such a way that the robot is repelled by an obstacle only if it is close to the obstacle *and* its velocity points toward the obstacle. If the robot moves parallel to the obstacle, it does not have to be repulsed by it. Using a similar idea, Faverjon and Tournassoud [Faverjon and Tournassoud, 1987] described a planning method that consists of minimizing a quadratic criterion under linear constraints. The minimization pulls the robot toward the goal, while each constraint acts as a damper pushing the robot away from an obstacle when the robot is close to it and moves toward it. Such potential functions are interesting, but they address a secondary issue, the main one being the presence of local minima. They may even make it more difficult to deal with local minima.

In the following subsections we present theoretical and empirical results related to the definition of "good" potential functions with no, few, or small local minima[4]. The definition of the potential function is indeed a good place to start solving the local minima problem.

4.1 Notion of a Navigation Function

Since local minima other than the goal are a major cause of inefficiency for potential field methods, and since the potential function is not something imposed on us in the original formulation of the basic motion planning problem, an important question is the following: Is it possible to construct a potential function $\mathbf{U} : \mathcal{C}_{free} \rightarrow \mathbf{R}$ with a minimum located at $\mathbf{q}_{goal}$ whose domain of attraction includes the entire subset of $\mathcal{C}_{free}$ that is connected to $\mathbf{q}_{goal}$? Such a function, if it exists, is called a "global navigation function". If it could be constructed, then depth-first path generation would be guaranteed to reach $\mathbf{q}_{goal}$ whenever $\mathbf{q}_{init}$ and $\mathbf{q}_{goal}$ belong to the same connected subset of $\mathcal{C}_{free}$, and to fail recognizably otherwise.

Using the Poincaré-Hopf Index Theorem [Guillemin and Pollack, 1974], Koditschek [Koditschek, 1987] showed that in the presence of C-obstacles (i.e. when $\mathcal{C}_{free}$ is strictly included in $\mathcal{C}$) a global navigation function does not exist in general. More specifically, if $\mathcal{C} = \mathbf{R}^2$ and if there are q ($q > 0$) disjoint C-obstacles, each homeomorphic to the closed unit disc, then a potential function $\mathbf{U}$ must possess at least q saddle points.

However, there seems to be no topological obstruction to the existence of

[4]Indeed, the number of local minima is not the only measure of the quality of a potential function. Another critical factor is the size of the attraction domain of these minima. A search technique may possibly escape from several small local minima without difficulty, but may fail to escape from a single, big one.

an analytical potential function with a minimum located at $\mathbf{q}_{goal}$ whose domain of attraction includes the whole subset of $\mathcal{C}_{free}$ that is connected to $\mathbf{q}_{goal}$, except a set of points of measure zero which are saddles of the potential function. These points are unstable equilibrium configurations, and any slight random deviation allows the planner to evade them. Such a potential function is called an "almost global navigation function". In the following, we will simply call it a **navigation function.** If $\mathcal{C} = \mathbf{R}^N$ and all the C-obstacles are disjoint spherical regions, then one can define a navigation function $\mathbf{U}$ using the formulae given in Section 1, as the sum of an attractive quadratic-well potential and repulsive potentials, each generated by a C-obstacle, with the distance of influence ρ_0 set to less than the minimum of half the minimal distance between any two C-obstacles and the distance between $\mathbf{q}_{goal}$ and the nearest C-obstacle.

Of course, the problem is to generalize this specific construct to C-obstacles with other shapes. One way to proceed is to construct a smooth "deformation" of these more complex C-obstacles into spherical ones. Indeed, if $\mathbf{U}$ is a navigation function over $\mathbf{R}^N$ and $\vartheta : \mathbf{R}^N \rightarrow \mathbf{R}^N$ is a diffeomorphism, then $\tilde{\mathbf{U}} = \mathbf{U} \circ \vartheta$ is also a navigation function [Rimon and Koditschek, 1988]. Hence, the problem is to construct a diffeomorphism ϑ that deforms a given set of C-obstacles into a set of spherical regions. Rimon and Koditschek [Rimon and Koditschek, 1989] constructed such a diffeomorphism in the particular case where the configuration space is a star-shaped subset[5] of $\mathbf{R}^N$ punctured by a finite number of disjoint N-dimensional star-shaped C-obstacles. They also extended the construction to C-obstacles of more complicated shapes defined as finite unions of intersecting star-shaped regions whose connectivity graph is a tree [Rimon and Koditschek, 1990].

4.2 Grid Potentials

Although it seems difficult to construct an analytical navigation function over a free space of arbitrary geometry, the computation of a numerical navigation function over a representation of the configuration space in the form of a grid turns out to be a much easier problem. Below we describe algorithms originally proposed in [Barraquand and Latombe,

[5] A subset S of $\mathbf{R}^N$ is *star shaped* if and only if it is homeomorphic to the closed unit ball in $\mathbf{R}^N$ and there exists a point $z \in S$ such that for all points $p \in S$, the line segment zp is contained in S.

1989a] for computing two such navigation functions (Subsections 4.2.1 and 4.2.2). These algorithms are efficient when the dimension of the configuration space is small, i.e. $m = 2$ and 3, and their time complexity is independent of the geometry of the free space. But they rapidly become impractical when m grows larger. Nevertheless, when m is large, they can still be used to construct potential fields with "good" properties (Subsection 4.2.3).

4.2.1 Simple Numerical Navigation Function

Let us discretize $\mathcal{C}$ into a rectangloid grid $\mathcal{GC}$ as we already did in Subsection 3.2. Each configuration in $\mathcal{GC}$ is labeled "0" if it lies in free space and "1" otherwise. The subset of configurations labeled "0" is denoted by $\mathcal{GC}_{free}$. We assume that both $\mathbf{q}_{init}$ and $\mathbf{q}_{goal}$ belong to $\mathcal{GC}_{free}$.

A simple navigation function $\mathbf{U}$ is the L^1 distance (also called "Manhattan distance") to the goal in $\mathcal{GC}_{free}$. It can be easily computed using the following "wavefront expansion" algorithm [Barraquand, Langlois and Latombe, 1989b]. First, the value of $\mathbf{U}$ is set to 0 at $\mathbf{q}_{goal}$. Next, it is set to 1 at every 1-neighbor[6] of $\mathbf{q}_{goal}$ in $\mathcal{GC}$ (we assume that the distance between two 1-neighbors in $\mathcal{GC}$ is normalized to 1); to 2 at every 1-neighbor of these new configurations (if it has not been computed yet); etc. The algorithm terminates when the entire subset of $\mathcal{GC}_{free}$ accessible from $\mathbf{q}_{goal}$ has been fully explored.

Figures 3.1 and 3.2 illustrate the operations of the algorithms in a two-dimensional configuration space[7] represented by a 64^2 grid[8]. The goal configuration $\mathbf{q}_{goal}$ is located in the upper left-hand corner of the grid. Figures 3.a through 3.g show various stages of the wavefront expansion. The displayed front lines are equipotential contours of the potential $\mathbf{U}$ computed by the algorithm. Figure 3.h shows a path following the steepest descent of $\mathbf{U}$ from an initial configuration $\mathbf{q}_{init}$. This path is a minimum-length path (for the L^1 distance in $\mathcal{GC}$) between $\mathbf{q}_{init}$ and $\mathbf{q}_{goal}$.

The algorithm computes $\mathbf{U}$ only in the connected subset of $\mathcal{GC}_{free}$ that

[6] Neighborhood in a grid is defined in Subsection 3.2.

[7] This space is treated here as a configuration space example. Later, we will also use the same space as a two-dimensional workspace example.

[8] This low resolution is intended to facilitate the interpretation of the figures.

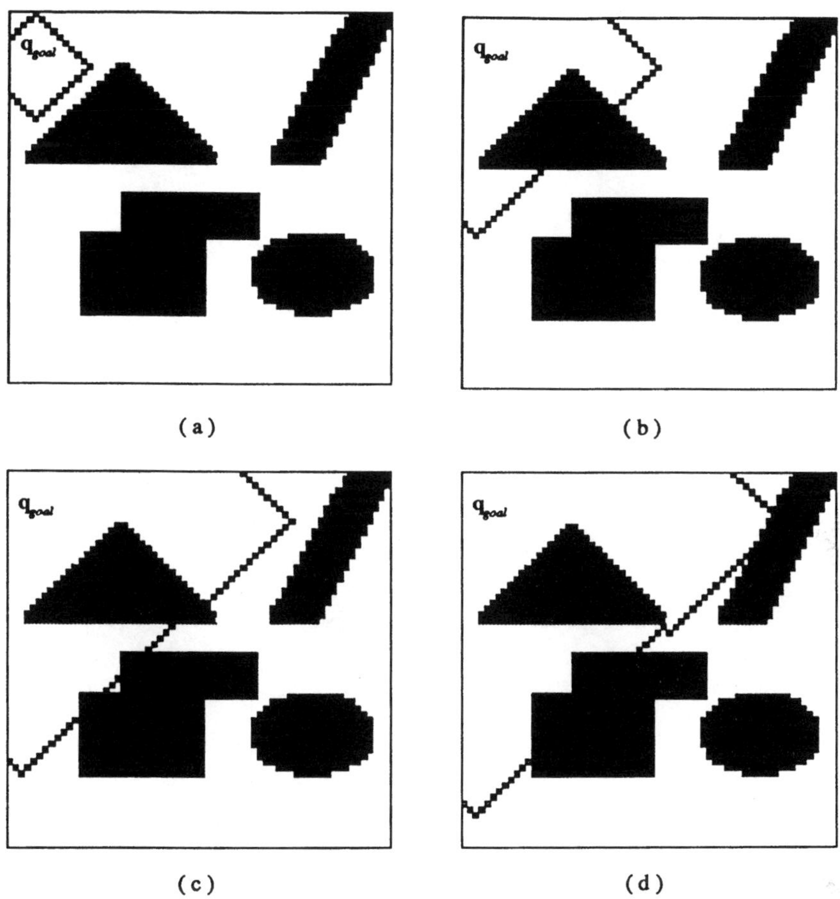

Figure 3.1. This figure and the next one illustrate the computation of a navigation function by the wavefront expansion procedure NF1 in a two-dimensional configuration space represented as a grid of 64^2 configurations. Figures a through g show the expansion front at various stages of the computation. The displayed front lines are equipotential contours of the computed navigation function. Figure h displays a path following the steepest descent of the function from an initial configuration $\mathbf{q}_{init}$.

contains $\mathbf{q}_{goal}$. Hence, if $\mathbf{U}(\mathbf{q}_{init})$ has not been computed by the algorithm, we immediately know that no free path connects $\mathbf{q}_{init}$ to $\mathbf{q}_{goal}$ at the resolution of the grid.

A more formal expression of the above algorithm is the following:

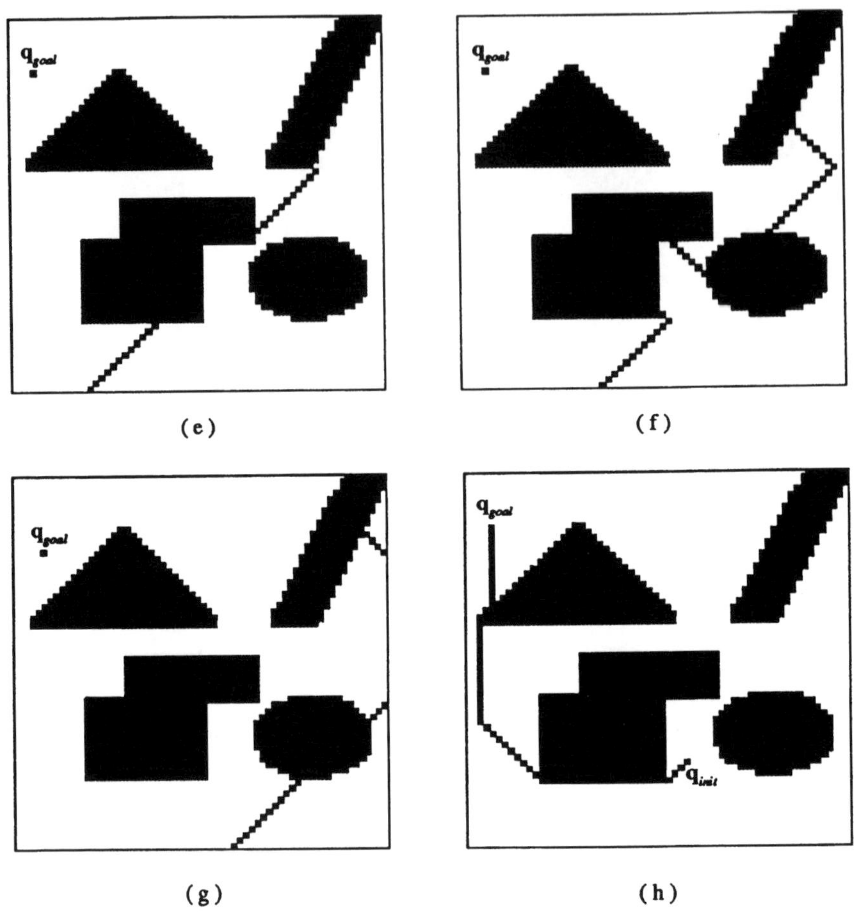

Figure 3.2. This figure is the continuation of Figure 3.1.

---

procedure NF1;
begin
  for every $\mathbf{q}$ in $\mathcal{GC}_{free}$ do $\mathbf{U}(\mathbf{q}) \leftarrow$ M; *[M is a large number]*
  $\mathbf{U}(\mathbf{q}_{goal}) \leftarrow 0$; insert $\mathbf{q}_{goal}$ in L_0;
  [L_i, $i = 0, 1, \ldots$, is a list of configurations; it is initially empty]
  for $i = 0, 1, \ldots$, until L_i is empty do
    for every $\mathbf{q}$ in L_i do
      for every 1-neighbor $\mathbf{q}'$ of $\mathbf{q}$ in $\mathcal{GC}_{free}$ do
        if $\mathbf{U}(\mathbf{q}') = $ M then

```
                    begin
                        U(q') ← i + 1;
                        insert q' at the end of L_{i+1};
                    end;
    end;
```

The time complexity of this algorithm is linear in the number of configurations in the grid $\mathcal{GC}$. It is independent of the number and the shape of the C-obstacles.

An implemented planner using the navigation function computed by NF1 with a best-first search of the grid $\mathcal{GC}$ is described in [Lengyel et al., 1990] for a polygonal robot that both translates and rotates in $\mathcal{W} = \mathbf{R}^2$ among polygonal obstacles. The polygonal cross-sections at each discretized orientation in the grid $\mathcal{GC}$ are computed and passed to a rasterizing Computer Graphics hardware that gives the label "1" to the configurations of $\mathcal{GC}$ located in these cross-sections.

4.2.2 Improved Numerical Navigation Function

A drawback of the navigation function computed by the procedure NF1 given in the previous subsection is that it induces paths which in general graze C-obstacles. In contrast, the improved navigation function computed below induces paths that lie as far as possible away from the C-obstacles. But the L^1 lengths of these paths are no longer minimal.

The improved navigation function is computed in three steps. First, an $(m-1)$-dimensional subset $\mathcal{S}$ of $\mathcal{GC}_{free}$ called the **skeleton** is extracted. Second, the potential function $\mathbf{U}$ is computed in $\mathcal{S}$. Third, $\mathbf{U}$ is computed in the rest of $\mathcal{GC}_{free}$. We develop these three steps below:

1. Computation of the skeleton. The L^1 distance $d_1(\mathbf{q})$ from every configuration $\mathbf{q} \in \mathcal{GC}_{free}$ to the C-obstacle region is computed using a wavefront expansion algorithm with the expansion starting from the boundary of $\mathcal{GC}_{free}$. This boundary is constructed in lines 5-10 of the procedure L1 given below.

The following procedure computes the map d_1 over $\mathcal{GC}_{free}$:

```
1   procedure L1;
2   begin
```

```
3        for every q in GC_free do U(q) ← M; [M is a large number]
4        [L_i, i = 0,1,..., is a list of configurations; it is initially empty]
5        for every q in GC\GC_free do
6            if there exists a 1-neighbor q' of q in GC_free then
7                begin
8                    d1(q) ← 0;
9                    insert q at the end of L0;
10               end;
11       for i = 0,1,..., until L_i is empty do
12           for every q in L_i do
13               for every 1-neighbor q' of q in GC_free do
14                   if d1(q') = M then
15                       begin
16                           d1(q') ← i + 1;
17                           insert q' at the end of L_{i+1};
18                       end;
19   end;
```

Concurrently with the computation of the map d_1, an $(m-1)$-dimensional subset[9] $\mathcal{S}$ of $\mathcal{GC}_{free}$ is constructed as the set of configurations where the "waves" (the lists L_i in the above procedure) issued from the boundary of $\mathcal{GC}_{free}$ meet. This is done by propagating not only the values of d_1, but also the points in the discretized boundary of $\mathcal{GC}_{free}$ that are at the origin of the propagation. For each configuration $\mathbf{q}$ in $\mathcal{GC}_{free}$ such that $d_1(\mathbf{q})$ has been computed, let $O(\mathbf{q})$ denote the point computed as follows. At line 8 of the algorithm, we set $O(\mathbf{q})$ to $\mathbf{q}$. At line 16, we set $O(\mathbf{q}')$ to $O(\mathbf{q})$. The "if" statement at lines 14-18 is extended by an "else" part that considers the case where $d_1(\mathbf{q}') \neq \mathrm{M}$, i.e. the case where two waves possibly meet, and builds $\mathcal{S}$: if the L^1 distance $\mathrm{D1}(O(\mathbf{q}'), O(\mathbf{q}))$ between $O(\mathbf{q}')$ and $O(\mathbf{q})$ is greater than some threshold α, we include $\mathbf{q}$ in $\mathcal{S}$.

The resulting complete algorithm which computes both the map d_1 and the skeleton $\mathcal{S}$ is the following (the set $\mathcal{S}$ is initially empty):

```
procedure SKELETON;
begin
    for every q in GC_free do U(q) ← M; [M is a large number]
```

[9] Actually, $\mathcal{S}$ is a discretized approximation of an $(m-1)$-dimensional subset of $\mathcal{C}$.

```
[L_i, i = 0,1,..., is a list of configurations; it is initially empty]
for every q in GC\GC_free do
    if there exists a 1-neighbor q' of q in GC_free then
        begin
            d1(q) ← 0; O(q) = q;
            insert q at the end of L0;
        end;
for i = 0,1,..., until L_i is empty do
    for every q in L_i do
        for every 1-neighbor q' of q in GC_free do
            if d1(q') = M then
                begin
                    d1(q') ← i + 1; O(q') ← O(q);
                    insert q' at the end of L_{i+1};
                end;
            else if D1(O(q'), O(q)) > α then
            [α is an integer constant set to a value typically
            comprised between 2 and 6]
                if q ∉ S then insert q' in S;
                [this test avoids generating a double skeleton (i.e. a
                skeleton of thickness 2) resulting from the fact that
                two waves are meeting]
end;
```

Figure 4 displays the skeleton computed in the same two-dimensional space as in Figure 3. In two dimensions, the skeleton is a kind of Voronoi diagram (see Section 2 of Chapter 4) for the L^1 distance. It is also similar to the "skeleton" extracted from a region in a digitized image using techniques from Mathematical Morphology [Serra, 1982].

2. Computation of U in the skeleton. The goal configuration $\mathbf{q}_{goal}$ is first connected to $\mathcal{S}$ by a path σ following the steepest ascent of the map d_1 in $C\mathcal{G}_{free}$. This path σ is included in $\mathcal{S}$ (lines 6-12 of the procedure NF2 given below). Then, **U** is computed in $\mathcal{S}$ by a wavefront expansion algorithm restricted to $\mathcal{S}$ and starting at $\mathbf{q}_{goal}$. The algorithm uses the d_1 map previously computed in order to guide the expansion. **U** is first given the value 0 at $\mathbf{q}_{goal}$, and $\mathbf{q}_{goal}$ is inserted in a list Q of configuration sorted by decreasing value of d_1. Next, until Q is empty, the first element

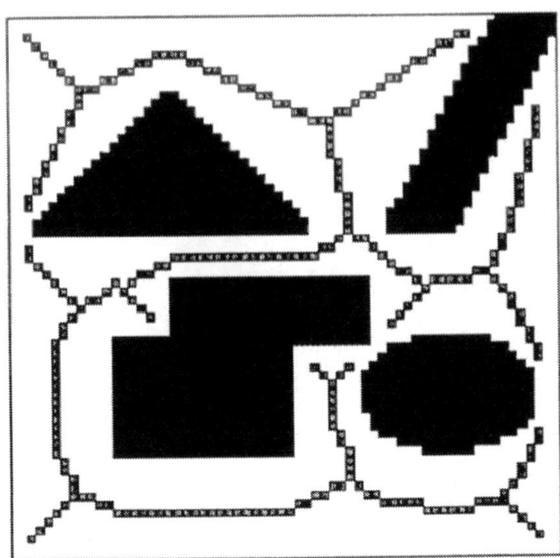

Figure 4. This figure shows the skeleton computed in the same two-dimensional space as in Figure 3 (with the parameter α equal to 4).

of Q — call it $\mathbf{q}$ — is removed from Q; every m-neighbor[10] $\mathbf{q}'$ of $\mathbf{q}$ in $\mathcal{S}$ whose potential has not been computed yet receives a potential value equal to $\mathbf{U}(\mathbf{q})+1$ and is inserted in Q. The algorithm terminates when Q is empty, i.e. when all the configurations in $\mathcal{S}$ accessible from $\mathbf{q}_{goal}$ have been given a potential value. A formal expression of the algorithm computing $\mathbf{U}$ in $\mathcal{S}$ consists of the lines 14-29 of the procedure NF2 given below. The outcome of the algorithm is a list L_0 of all the configurations of $\mathcal{S}$ accessible from $\mathbf{q}_{goal}$.

Figure 5 illustrates various stages of the wavefront propagation in the skeleton of Figure 4. The elements of $\mathcal{S}$ shown black are those which have been attained so far. $\mathcal{S}$ includes the path σ that connects $\mathbf{q}_{goal}$ to the original skeleton.

3. Computation of U in the rest of free space. The potential U in the rest of the subset of $\mathcal{GC}_{free}$ accessible from $\mathbf{q}_{goal}$ is computed

[10] $\mathcal{S}$ is a subset of $\mathcal{GC}$ of dimension lower than m. In order to track it reliably, we must use the m-neighborhood.

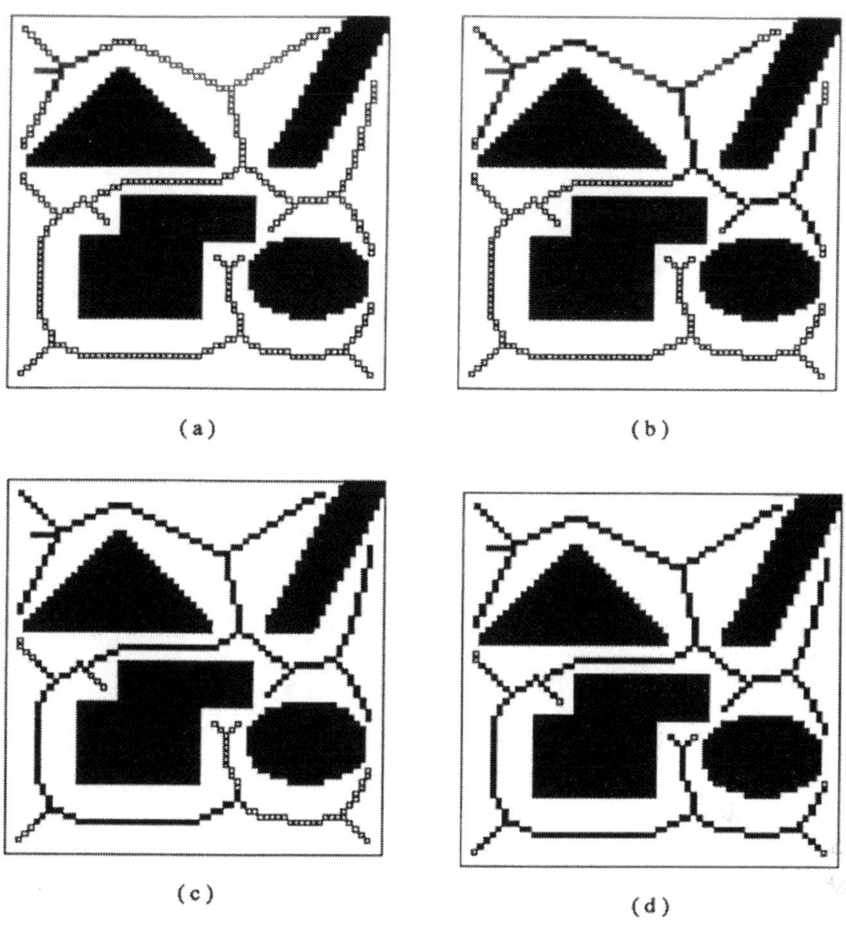

Figure 5. This figure illustrates four stages of the wavefront expansion carried out by the procedure NF2 in the skeleton of Figure 4. The skeleton elements shown black are those attained at the current stage of the propagation.

by a wavefront expansion algorithm starting from the configurations in L_0. The potential at each 1-neighbor $\mathbf{q}'$ of every configuration $\mathbf{q}$ in L_0 is set to $\mathbf{U}(\mathbf{q}) + 1$. Next, the potential at every neighbor $\mathbf{q}''$ of a configuration $\mathbf{q}'$ whose potential has been previously computed is set to $\mathbf{U}(\mathbf{q}') + 1$, etc. The algorithm (lines 34-41 of NF2) terminates when all the configurations of $\mathcal{GC}_{free}$ accessible from $\mathbf{q}_{goal}$ have been considered.

Figure 6 shows two stages of the wavefront expansion from the skeleton

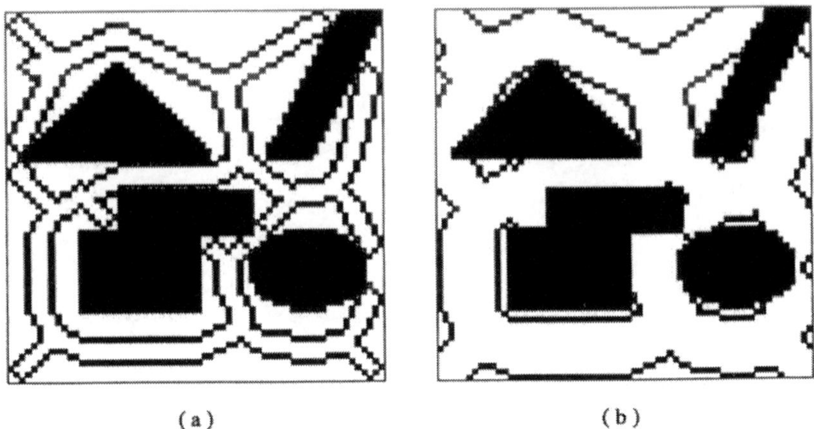

Figure 6. This figure shows two successive stages of the wavefront expansion carried out by the procedure NF2 from the skeleton of Figure 4.

of Figure 4.

The procedure NF2 given below is a more formal expression of the algorithm computing $\mathbf{U}$ in the subset of $\mathcal{GC}_{free}$ accessible from $\mathbf{q}_{goal}$:

```
procedure NF2;
begin
    for every q in GC_free do U(q) ← M; [M is a large number]
    [the following sequence of instructions connects q_goal to S]
    [σ is a list of configurations; it is initially empty]
    insert q_goal in σ; q ← q_goal;
    while q ∉ S do
        begin
            select a neighbor q' having the largest value of d_1;
            insert q' in σ; q ← q';
        end;
    S ← σ ∪ S;
    [the following sequence of instructions computes U in S]
    U(q_goal) ← 0; INSERT(q_goal, Q);
    [Q is a list of configurations sorted by decreasing values of d_1;
    it supports the operations FIRST, INSERT, and EMPTY (see
    Subsection 3.2); it is initially empty]
```

```
    [L_i, i = 0, 1, ..., is a list of configurations; it is initially empty]
    while ¬ EMPTY(Q) do
        begin
            q ← FIRST(Q);
            insert q at the end of L_0;
            for every m-neighbors q' of q in S do
                if U(q') = M then
                    begin
                        U(q') ← U(q) + 1;
                        INSERT(q', Q);
                    end;
        end;
    [at the end of this loop, L_0 contains all the configurations in S
    accessible from q_goal]
    [the following instructions compute U in the rest of the subset
    of GC_free accessible from q_goal]
    for i = 0, 1, ..., until L_i is empty do
        for every q in L_i do
            for every neighbor q' of q in GC_free do
                if U(q') = M then
                    begin
                        U(q') ← U(q) + 1;
                        insert q' at the end of L_{i+1};
                    end;
end;
```

Figure 7 shows the potential function computed by NF2 in the two-dimensional space introduced in Figure 3. Figure 7.a displays equipotential contours of the function and Figures 7.b and c show three-dimensional views of the function. The valleys of the function are the skeleton $\mathcal{S}$ shown in Figure 4. This potential generates no stable equilibrium configuration other than $\mathbf{q}_{goal}$. However, it is usually not continuous and it is not suitable to explicitly compute its gradient. A best-first search (see Subsection 3.2) starting from any initial configuration $\mathbf{q}_{init}$ produces a path that first connects $\mathbf{q}_{init}$ to $\mathcal{S}$, then follows the curve in $\mathcal{S}$ that lies the furthest away from the C-obstacles, and finally connects $\mathcal{S}$ to $\mathbf{q}_{goal}$ (see Figure 7.d).

The time complexity of the above algorithm is slightly higher than that

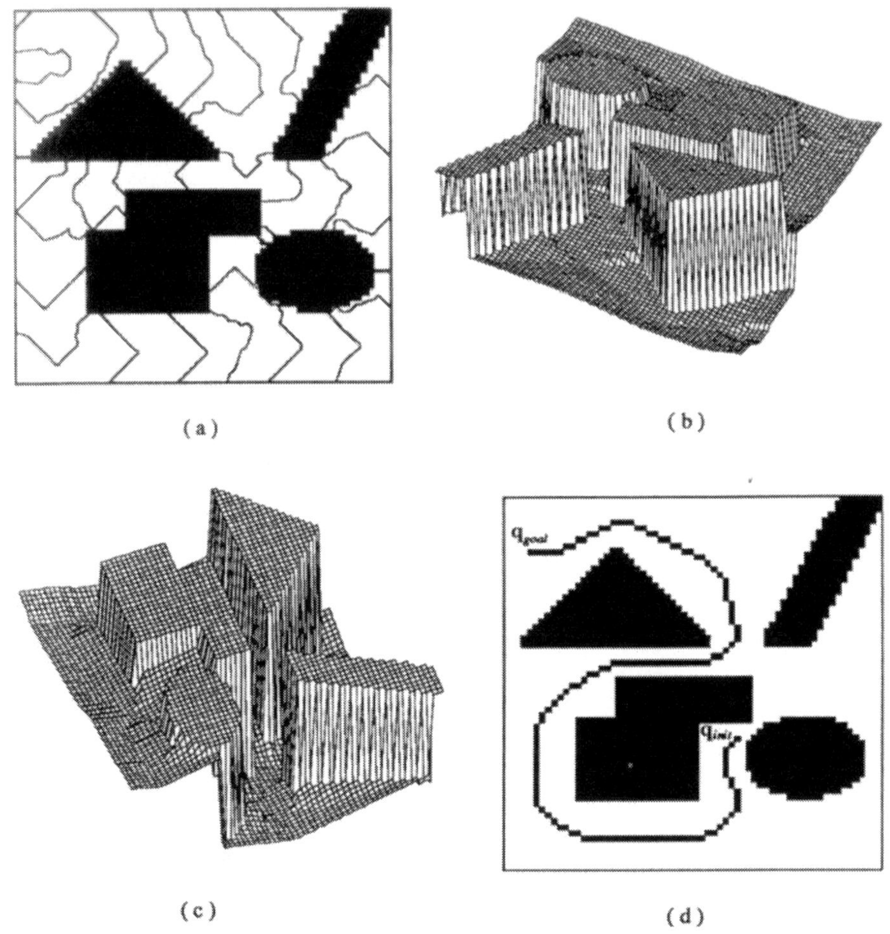

Figure 7. This figure illustrates the potential function **U** computed by the procedure NF2 in the same two-dimensional space as in Figure 3. Figure a displays equipotential contours of **U**. Figures b and c show two three-dimensional views of the function. Figure d shows a path generated by a best-first search of the grid starting at an initial configuration $\mathbf{q}_{init}$.

of the algorithm NF1 (Subsection 4.2.1). Let n be the number of configurations in $\mathcal{GC}$ and s (resp. s') the number of configurations in $\mathcal{S}$ (resp. in $\mathcal{S}$ augmented by σ). By representing $\mathcal{S}$ in a balanced tree, the complexity of SKELETON is $O(n + s \log s)$. By representing the list Q in a balanced tree, the complexity of NF2 is $O(n + s' \log s')$. In general, we can expect s and s' to be proportional to $n^{\frac{m-1}{m}}$ since $\mathcal{S}$ and σ are $(m-1)$-dimensional subsets of $\mathcal{GC}$. Under this condition, the complexity of both SKELETON and NF2 is $O(n + n^{\frac{m-1}{m}} \log n)$, and hence is linear in n.

Variants of the above potential are easy to define. For example, at line 39 of NF2, we could compute $\mathbf{U}(\mathbf{q}') = \mathbf{U}(\mathbf{q}) + 1/d_1(\mathbf{q})$, which yields a potential $\mathbf{U}$ that becomes infinite on the boundary of $\mathcal{CG}_{free}$.

4.2.3 Application to High-Dimensional Spaces

The procedures NF1 and NF2 are general in theory. However, in practice, they can only be applied to configuration spaces of small dimension. Indeed, the size of the grid $\mathcal{GC}$ increases exponentially with the dimension m of the robot's configuration space. Hence, both NF1 and NF2 require exponential space and time in m. Nevertheless, in high-dimensional spaces, we can apply a technique similar to the one presented in Subsections 2.2 and 2.3. It consists of using NF1 and NF2 to compute "workspace potentials" associated with *control points* in the robot and combining them into a "configuration space potential", as described below. Since the workspace dimension is 2 or 3, the cost of running NF1 or NF2 remains reasonable.

Let $\mathcal{GW}$ be a regular grid of points thrown across $\mathcal{A}$'s workspace $\mathcal{W}$. This grid is constructed by discretizing the axes of the frame $\mathcal{F}_{\mathcal{W}}$ embedded in $\mathcal{W}$. We assume that the grid is bounded and forms a rectangloid. Each point in $\mathcal{GW}$ is labeled "1" if it lies in the obstacle region $\mathcal{B}$ or if it lies in the boundary of $\mathcal{GW}$; it is labeled "0" otherwise. $\mathcal{GW}_{free}$ denotes the subset of points of $\mathcal{GW}$ that are labeled "0".

Let a_j, $j = 1, \ldots, Q$, be the control points selected in $\mathcal{A}$. With each point a_j we associate a potential function $\mathbf{V}_j$ defined over $\mathcal{GW}_{free}$. $\mathbf{V}_j$ is computed using either NF1 or NF2, by substituting $\mathcal{GW}$, $\mathcal{GW}_{free}$, $a_j(\mathbf{q}_{goal})$, $\mathbf{x}$, and $\mathbf{x}'$ for $\mathcal{GC}$, $\mathcal{GC}_{free}$, $\mathbf{q}_{goal}$, $\mathbf{q}$, and $\mathbf{q}'$, respectively, in the text of the procedures.

Once the potentials $\mathbf{V}_j$, $j = 1, \ldots, Q$, have been computed over $\mathcal{GW}_{free}$, the potential function $\mathbf{U}$ is computed at a configuration $\mathbf{q} \in \mathcal{CG}_{free}$ as:

$$\mathbf{U}(\mathbf{q}) = G(\mathbf{V}_1(a_1(\mathbf{q})), \ldots, \mathbf{V}_Q(a_Q(\mathbf{q})))$$

where G is called the **arbitration function** (we will give examples of G below). This potential function is not computed at every configuration in $\mathcal{GC}$ (as mentioned above, this would be impractical), but only at the configurations where the planner needs to know the value of $\mathbf{U}$ (this is in general a very small subset of $\mathcal{GC}$ relative to the total size of the grid). In order to compute $\mathbf{U}$ at a given configuration $\mathbf{q}$, the position $a_j(\mathbf{q})$ of every control point is first computed and the value of $\mathbf{V}_j$ at the closest position of $a_j(\mathbf{q})$ in $\mathcal{GW}$ is retrieved and used to compute $\mathbf{U}(\mathbf{q})$. With this definition, $\mathbf{U}$ is positive or null over free space and becomes null at the goal configurations. If $Q = N$, the goal configuration is usually uniquely defined. If $Q \leq N$, there are multiple goal configurations.

The above computation requires that the resolutions of the grids $\mathcal{GW}$ and $\mathcal{GC}$ be related to each other appropriately. Let Δ_i, $i = 1, ..., m$, be the increment between two discretization points along the ith axis of $\mathcal{GC}$. We assume that the scales along the m axes of $\mathcal{GC}$ have been normalized so that a change of the robot's configuration by Δ_i along one axis of $\mathcal{GC}$ causes the longest Euclidean displacement of the points in $\mathcal{A}$ to be of the same order of magnitude as the one caused by a change by Δ_j along another axis (see Subsection 3.2). Let δ be the increment between two discretization points along an axis of the grid $\mathcal{GW}$. In the same way as the Δ_i's have been normalized, the increment δ should be of the same order of magnitude as the length of the longest Euclidean displacement of the points in $\mathcal{A}$ when its configuration is changed by Δ_i along any of its axes. Hence, δ is equal to the increment Δ_i along any of the translational axes of $\mathcal{GC}$.

The rationale behind the two-step construction of $\mathbf{U}$ is to avoid computing $\mathbf{U}$ systematically over an enormous grid, while using the workspace as a source of inspiration for generating a "good" function $\mathbf{U}$. The first goal is achieved since the only potentials which we exhaustively compute are the workspace potentials. The second goal is hopefully achieved by combining workspace potentials that are free of local minima. Their combination certainly allows us to construct a potential $\mathbf{U}$ that prevents the robot from getting trapped in simple concavities formed by the obstacles. However, the combination usually have local minima. Typically, these minima result from the fact that the various control points, which have fixed relative positions, are concurrently attracted toward their respective goal positions. Hence, they are competing among themselves to attain their goal positions. This competition creates unwanted minima and the role of the arbitration function G is to arbitrate in this competition. Various arbitration functions are possible, resulting in more or less numerous, more or less deep local minima.

A typical arbitration function consists of adding the various potentials $\mathbf{V}_j$, i.e. :

$$\mathbf{U}(\mathbf{q}) = \sum_{j=1}^{j=Q} \mathbf{V}_j(a_j(\mathbf{q})).$$

However, this choice does not favor any control point over any other. For this reason, it tends to increase the number of conflicts and to produce numerous minima.

Another choice for G is:

$$\mathbf{U}(\mathbf{q}) = \min_{j=1}^{j=Q} \mathbf{V}_j(a_j(\mathbf{q})) + \varepsilon \max_{j=1}^{j=Q} \mathbf{V}_j(a_j(\mathbf{q})) \quad (1)$$

where ε is a small constant (typically, $\varepsilon = 0.1$). This arbitration function favors the attraction of the point in $\mathcal{A}$ that is the closest to its goal position. The second term of G makes the robot rotate (less vigorously) toward a goal configuration. Experience shows that this arbitration function tends to generate fewer local minima than the previous one. However, it may create a deep local minimum if there is not enough space for the robot to rotate when it is close to its goal position.

Another choice for G is:

$$\mathbf{U}(\mathbf{q}) = \max_{j=1}^{j=Q} \mathbf{V}_j(a_j(\mathbf{q})).$$

This choice tends to increase the number of competitions among the control points and, therefore, the number of local minima. But experimental results show that it also tends to reduce the size of the attraction domains of these minima.

In general, NF2 computes workspace potentials $\mathbf{V}_j$ that can be combined into a "better" potential function $\mathbf{U}$ than those computed by NF1. This results from the fact that, unlike NF1, NF2 tends to keep the control points as far away as possible from the obstacles. Hence, it increases the maneuvering space for the robot.

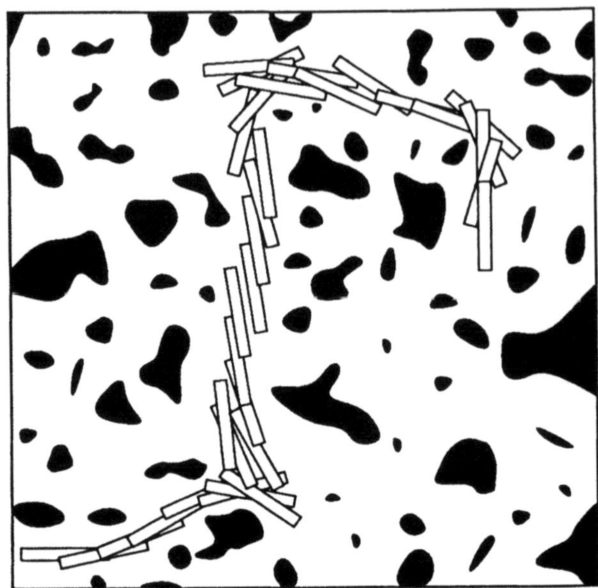

Figure 8. This figure illustrates a path for a rectangular robot moving among many obstacles. The path has been generated by a best-first planner using a potential field defined by equation (1) with two potentials $\mathbf{V}_j$ computed by NF2 for two control points located at the small ends of the robot.

The potential **U** defined in equation (1) has been used in a very efficient path planner using a best-first search technique (see Subsection 3.2) for a robot that can both translate and rotate in the plane [Barraquand and Latombe, 1989a]. Since the function **U** does not grow to infinity near the boundary of the C-obstacle region, best-first search using this function does not guarantee avoiding collision with obstacles. Collisions have to be checked separately in the workspace (as will be explained in Subsection 5.2). Figure 8 shows a path generated by this planner with two control points located at the small ends of the robot.

4.3 Elliptical Potentials

Another approach to reducing the number and the size of the local minima of the potential function **U** consists of defining **U** as the combination of an attractive potential having a revolute symmetry around the goal configuration $\mathbf{q}_{goal}$ (e.g. a parabolic well) and elliptical repulsive

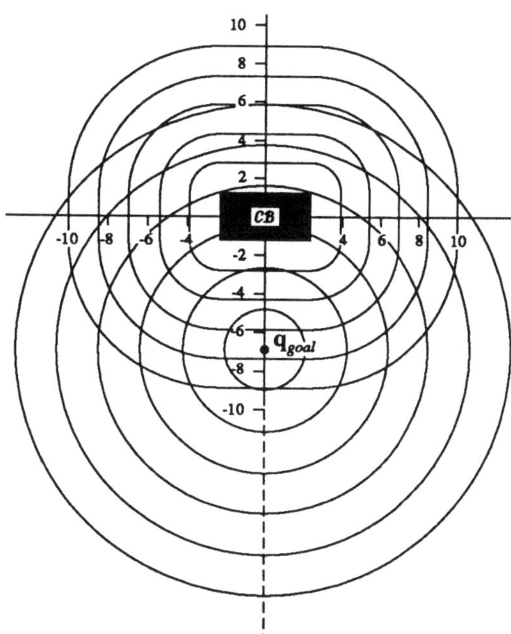

Figure 9. The equipotential contours of the attractive potential field (parabolic well) are circles centered at $\mathbf{q}_{goal}$. The equipotential contours of the repulsive potential field are of the form $\max_{\mathbf{q}' \in \mathcal{CB}} \|\mathbf{q} - \mathbf{q}'\| = K$, with K constant. They are "rectangles with circular corners".

potentials. The basic idea is to define the repulsive potential around a C-obstacle in such a way that the equipotential contours converge toward the boundary of the C-obstacle when the distance to the C-obstacle tends toward zero, and toward spherical surfaces when the distance increases [Volpe and Khosla, 1987]. We develop this idea in a two-dimensional configuration space $\mathcal{C}$.

Consider the case where there is a single rectangular C-obstacle $\mathcal{CB}$ in $\mathcal{C}$ and the goal configuration $\mathbf{q}_{goal}$ is located on a symmetry axis of $\mathcal{CB}$ (Figure 9). Let the potential in $\mathcal{C}_{free}$ be the function $\mathbf{U} = \mathbf{U}_{att} + \mathbf{U}_{rep}$ defined in Section 1. The equipotential contours of $\mathbf{U}_{att}$ are circles centered at $\mathbf{q}_{goal}$. Within the influence distance ρ_0 from $\mathcal{CB}$, the equipotential contours of $\mathbf{U}_{rep}$ are "rectangles with circular corners". Figure 10 illustrates how the composition of $\mathbf{U}_{att}$ and $\mathbf{U}_{rep}$ creates a local minimum of $\mathbf{U}$ along the symmetry axis of $\mathcal{CB}$ containing $\mathbf{q}_{goal}$:

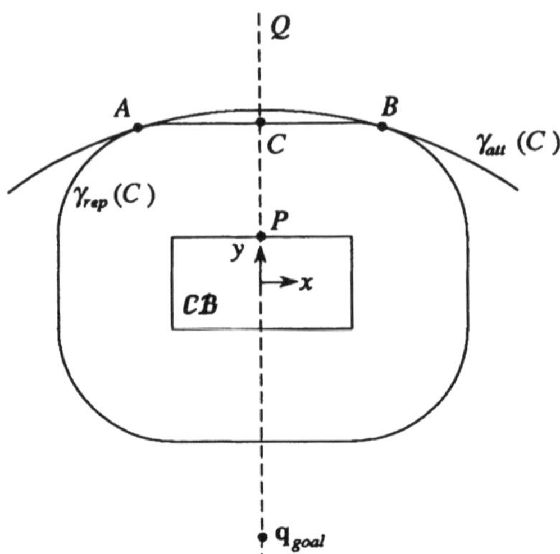

Figure 10. A local minimum is located on the symmetry axis of $\mathcal{CB}$, which contains $\mathbf{q}_{goal}$, "uphill" from $\mathcal{CB}$ with respect to $\mathbf{q}_{goal}$. More generally, a local minimum may appear whenever the radius of curvature of a repulsive equipotential curve at a configuration is greater than the radius of curvature of the attractive equipotential curve at the same configuration [Volpe and Khosla, 1987].

- Consider the half-line PQ. P is on the boundary of $\mathcal{CB}$, while Q lies at infinity. Let $\mathbf{q}$ be a configuration in PQ at a finite non-null distance from P. The potential $\mathbf{U}(\mathbf{q})$ is finite. When $\mathbf{q}$ tends toward either P or Q, $\mathbf{U}(\mathbf{q})$ tends toward $+\infty$.

- Let C be a point in PQ at a finite non-null distance from P, $\gamma_{rep}(C)$ be the equipotential contour of $\mathbf{U}_{rep}$ that contains C, and $\gamma_{att}(C)$ be the equipotential contour of $\mathbf{U}_{att}$ that is tangent to $\gamma_{rep}(C)$ at two distinct points A and B. As $\mathbf{q}$ moves along $\gamma_{rep}(C)$ from A to B, the potential $\mathbf{U}$ first decreases until C is reached, then increases until B is reached.

Therefore, the potential function $\mathbf{U}$ has a local minimum at a configuration $\mathbf{q}_{loc}$ located on PQ between P and Q. This minimum results from the fact that the radius of curvature of the repulsive equipotential curve at $\mathbf{q}_{loc}$ is greater than the radius of curvature of the attractive equipotential at $\mathbf{q}_{loc}$. For this reason, the total potential is non-convex despite the fact that both the attractive potential and the repulsive potential are convex.

This remark suggests that we should approximate the C-obstacle region by a collection of disjoint discs. If $\mathbf{q}_{goal}$ is outside any of these discs, we can select the distance of influence ρ_0 to be smaller than the minimum of half the minimal distance between any two discs and the distance between $\mathbf{q}_{goal}$ and the nearest disc. Then, it is guaranteed that the radius of the repulsive equipotential curve at any configuration $\mathbf{q}$ is smaller than the radius of the attractive equipotential at $\mathbf{q}$. The total potential is a navigation function previously proposed in Subsection 4.1. However, the drawback of approximating the C-obstacle region by disjoint discs is that it may prevent the robot from accessing to a large subset of the free space. This is in particular the case if the C-obstacle region is made of elongated objects. An elliptical repulsive potential, as defined below, is a compromise between the repulsive potential of Subsection 1.3, which tends to create large local minima, and the circular repulsive potential presented above, which prevents access to a large subset of the free space.

Consider again the simple case where the C-obstacle region $\mathcal{CB}$ is a rectangle. Let L and H denote the lengths of the long and small sides of $\mathcal{CB}$. We select the coordinate frame in $\mathcal{C} = \mathbf{R}^2$ with its origin at the center of $\mathcal{CB}$ and its x-axis pointing along the long side of $\mathcal{CB}$. The elliptical repulsive potential generated by $\mathcal{CB}$ is defined in two steps [Volpe and Khosla, 1987]. First, the equipotential contours are specified. Second, the value of the potential on each equipotential contour is defined.

An n-ellipse ($n \geq 1$) is a curve with the following equation:

$$\left(\frac{x}{a}\right)^{2n} + \left(\frac{y}{b}\right)^{2n} = 1 \tag{2}$$

where a is the semi-major axis and b is the semi-minor axis. If we set:

$$a = \frac{L}{2}\left(2^{\frac{1}{2n}}\right) \quad \text{and} \quad b = \frac{H}{2}\left(2^{\frac{1}{2n}}\right) \tag{3}$$

we get an n-ellipse that touches $\mathcal{CB}$ at its four corners. In addition, the surface of the area between the rectangle and the n-ellipse is minimal (over all the n-ellipses passing through the four corners of $\mathcal{CB}$). When $n \rightarrow \infty$, the n-ellipse converges toward the boundary of $\mathcal{CB}$.

In order to make the n-ellipse become a circle when $n = 1$, we multiply

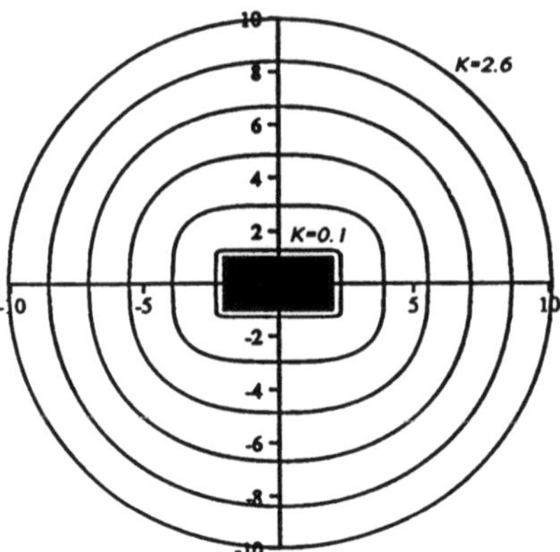

Figure 11. This figure shows elliptical equipotential contours for values of K between 0.1 and 2.6, and $\alpha = 1.5$. The contours are close to the rectangle when K is small and almost circular when K is large [Volpe and Khosla, 1987].

the term in y by $(b/a)^2$. The n-ellipse becomes:

$$\left(\frac{x}{a}\right)^{2n} + \left(\frac{b}{a}\right)^2 \left(\frac{y}{b}\right)^{2n} = 1 \quad (n \geq 1).$$

This n-ellipse is a circle if $n = 1$. For every $n \geq 1$, it contains $\mathcal{CB}$, and it converges toward the boundary of $\mathcal{CB}$ when $n \to \infty$.

For a given n, we treat the quantity:

$$\xi_n(x,y) = \left[\left(\frac{x}{a}\right)^{2n} + \left(\frac{b}{a}\right)^2 \left(\frac{y}{b}\right)^{2n}\right]^{\frac{1}{2n}} - 1$$

as a pseudo-distance between the configuration (x, y) and $\mathcal{CB}$. It is zero in the n-ellipse curve and grows to infinity for configurations moving away from this curve. For every given constant value K, the curve $\xi_n(x,y) = K$ is taken as a repulsive equipotential, for some n determined by:

$$n = \frac{1}{1 - e^{-\alpha K}}$$

where α is an adjustable parameter. Thus, when $K \to \infty$, $n \to 1$, and the equipotential tends toward a circle. When $K \to 0$, $n \to \infty$, and the equipotential tends toward the boundary of $\mathcal{CB}$. The parameter α determines how quickly the "n-ness" of the ellipse increases when $K \to 0$.

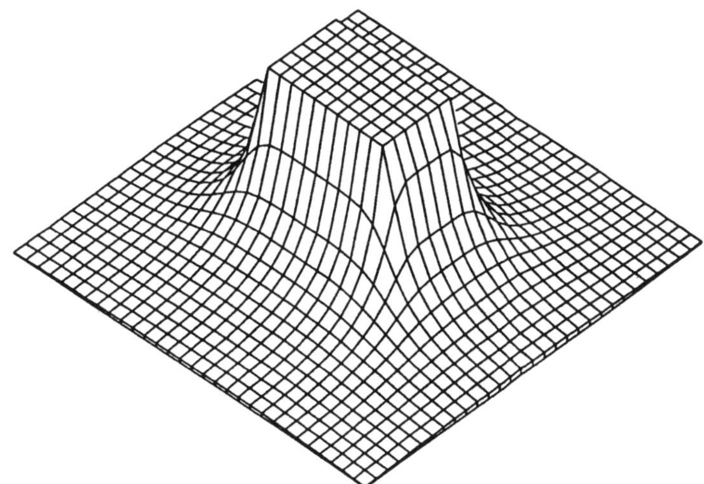

Figure 12. This figure shows the elliptical repulsive potential generated by a rectangular C-obstacle with $\alpha = \beta = 1$ and $\eta = 1$.

Figure 11 shows equipotential contours around $\mathcal{CB}$ for various values of K. Let $R(K)$ be the repulsive equipotential contour[11] corresponding to the pseudo-distance K.

Now that the form of the repulsive equipotential contours has been specified, we define the value of the repulsive potential in each of them. It is desirable that the repulsive potential of $\mathcal{CB}$ have a limited range of influence, with an inverse dependence on the distance to the C-obstacle. This can (almost) be achieved by taking:

$$\forall (x,y) \in R(K): \quad \mathbf{U}_{rep}(x,y) = \eta \frac{e^{-\beta K}}{K}$$

where η is a scaling factor. The exponential term makes the potential drop toward zero faster than K^{-1}, when K increases. Figure 12 shows the repulsive potential $\mathbf{U}_{rep}$ generated by the rectangular C-obstacle $\mathcal{CB}$.

When the above repulsive potential is added to the attractive potential (with $\mathbf{q}_{goal}$ located as in Figure 10), a local minima may still be created

[11]The use of n-ellipsoids in the definition of repulsive potentials was first proposed in [Khatib, 1986]. Khatib suggested that a rectangular C-obstacle be approximated by an n-ellipsoid, for some fixed n. In contrast, here n varies from 1 to infinity.

"uphill" from $\mathcal{CB}$. However, the size of the domain of attraction of this minimum can be tuned by setting η, α and β appropriately. It is possible to avoid the local minimum by making the minimum of the total potential along the y-axis happen at a point $(0, y_0)$ where the radius of curvature of the repulsive equipotential contour is smaller than or equal to the radius of the attractive equipotential contour. (Then, in the plane, $(0, y_0)$ is a saddle point.) Since this happens with a non-circular repulsive equipotential contour, less free space is wasted than with circular repulsive potentials. The wasted space is minimum when the parameters are set in so that y_0 is as close as possible to $\mathcal{CB}$. The wasted free space may still be too large (e.g. it may include the goal configuration), and the parameters may be set differently. Then, the domain of attraction of the created local minimum is typically smaller than for the local minimum created by the repulsive potential function of Subsection 1.3.

Nevertheless, elliptical potentials have several drawbacks which make their application to practical problems difficult. First, they can only be generalized to simple convex shapes (e.g. trapezoid) in two- and three-dimensional spaces, using superquadric formulation with non-constant scaling parameters (see [Khosla and Volpe, 1988]). Second, they require C-obstacles to be sufficiently distant from each other, so that no C-obstacle is within the distance of influence of another C-obstacle. Third, their definition does not provide a direct way to compute the potential at a given configuration; it requires interpolating among precomputed values at nearby configurations.

One can also compute the elliptical repulsive potential in the workspace, apply it at selected control points in the robot, and thus construct a configuration space potential (see Subsection 2.3).

5 Randomized Planning

In Section 3 we have described simple potential-field-based planning techniques. These techniques, however, have a limited ability to deal with local minima of the potential function. On the other hand, in Section 4 we have seen that the construction of a navigation function, i.e. a potential field with no local minimum other than the goal configuration, is a difficult problem that has a known solution only when the C-obstacles have simple shapes and/or when the dimension of the configuration space is small ($m = 2$ or 3). In this section we present a powerful planner, which we call RPP (for Randomized Path Planner), that combines some of the techniques described above with randomized search techniques. This planner has been implemented and successfully experimented with difficult planning problems [Barraquand and Latombe, 1989a and 1990a].

This section serves three purposes. First, it introduces a new technique — randomized search — within the framework of the potential field approach. Second, it presents an integrated view of what a planner based on the potential field approach looks like. Indeed, so far we have only described separate components of such a planner. Third, RPP will also be applicable to problems involving high-dimensional configuration spaces studied in the next chapter (multiple robots, articulated robots).

5.1 Overview

The techniques presented below do not require any specific potential function to be used. However, in order to make our presentation more precise, we assume that this function $\mathbf{U}$ is a grid potential computed as described in Subsection 4.2.3 and we use the same notations. We denote the coordinate axes in the grid $\mathcal{GC}$ by x_1, ..., x_m, and the grid increment along any x_i-axis by Δ_i. Since the dimension m of the grid may be large, the planner has no explicit representation of $\mathcal{GC}$, e.g. in the form of an array. We assume that $\mathbf{U} \geq 0$ and that any configuration $\mathbf{q}$ such that $\mathbf{U}(\mathbf{q}) = 0$ is a goal configuration.

In a first approximation, we may imagine that RPP constructs and searches a graph G whose nodes are local minima of the potential function $\mathbf{U}$. Two minima are connected by a link in the graph if the planner has constructed a path joining them.

The planner starts at the input initial configuration $\mathbf{q}_{init}$. From there, it executes a **best-first motion**, i.e. it follows the steepest descent of the potential $\mathbf{U}$. The best-first motion stops when it reaches a minimum $\mathbf{q}_{loc}$. If $\mathbf{U}(\mathbf{q}_{loc}) = 0$, the problem is solved and the algorithm returns the constructed path. Otherwise, it attempts to escape from the local minimum by executing a series of **random motions** issued from $\mathbf{q}_{loc}$. Each random motion is immediately followed by a new best-first motion that attains a minimum. If this minimum is different from $\mathbf{q}_{loc}$, it is installed as a successor of $\mathbf{q}_{loc}$ in the graph G. Hence, two adjacent minima in G are connected by a path obtained by composing a random motion and a best-first motion. G is built incrementally until a goal configuration is attained or the planner gives up. If the search terminates successfully, the constructed path is transformed into a smoother path.

In the following subsections, we describe in more detail the various components of RPP. Our description is directly drawn from [Barraquand and Latombe, 1989a].

5.2 Best-First Motion

The planner operates in the configuration space grid $\mathcal{GC}$ on which $\mathbf{U}$ is defined. In principle, a best-first motion consists of searching this grid in a best-first fashion as described in Subsection 3.2. Assume that it starts at a configuration $\mathbf{q}_0$. From there, it moves to the best neighbor $\mathbf{q}_1$ of $\mathbf{q}_0$, i.e. the neighbor which has the smallest value of $\mathbf{U}$, then to the best neighbor $\mathbf{q}_2$ of $\mathbf{q}_1$, and so on, until it reaches a configuration $\mathbf{q}_{loc}$, whose neighbors all have a higher potential than $\mathbf{U}(\mathbf{q}_{loc})$.

The potential $\mathbf{U}$ computed as described in Subsection 4.2.3 does not tend toward infinity when the robot's configuration gets closer to the C-obstacle region. Hence, a best-first motion is not guaranteed to be collision-free. Before selecting a configuration $\mathbf{q}_i$ as a successor of a configuration $\mathbf{q}_{i-1}$, the planner must verify that $\mathbf{q}_i$ lies in free space. Various collision checking techniques can be used for that purpose. For example, one may compute the distance from the robot to the obstacle region; the tested configuration is free if and only if this distance is positive. A divide-and-conquer technique [Barraquand and Latombe, 1989a] can be used to accelerate the test.

Remark: One may argue that the need for collision checking would be eliminated by the use of a potential that becomes infinite in the boundary of the C-obstacle region. However, it must be noticed that the computation of such a potential at any configuration $\mathbf{q}$ either requires the C-obstacle region to be explicitly computed (which we definitively want to avoid in a high-dimensional configuration space), or contains the computation of the shortest distance between the robot at $\mathbf{q}$ and the obstacle region (and it then includes the computational cost of making the collision test). ∎

When m is small enough (e.g. $m = 3$), best-first motions can be generated by using the m-neighborhood in $\mathcal{GC}$. However, when m becomes larger (e.g. $m = 6$), the m-neighborhood is too large to be fully explored at each step. One may then use a smaller neighborhood. Another technique consists of using the m-neighborhood and checking only a small number of configurations randomly selected in this neighborhood [Barraquand and Latombe, 1990a]. The first considered configuration which both lies in $\mathcal{GC}_{free}$ and has a smaller potential configuration than the current configuration is taken as the next configuration along the best-first motion. If none of the randomly selected configurations is a possible successor of the current configuration, the latter is considered a local minimum.

5.3 Random Motion

Random motions are executed whenever a local minimum $\mathbf{q}_{loc}$ is encountered. They are defined as random walks [Papoulis, 1965] consisting of a series of t motion steps such that the projection of every step along each axis x_i, $i = 1, ..., m$ is randomly Δ_i or $-\Delta_i$. Each step is selected independently of the previously executed steps. Such a random walk is known to converge almost surely toward a Brownian motion when every Δ_i tends toward 0 [Papoulis, 1965].

We may fictitiously consider that each step takes a unit of time, so that t is the duration of the random walk. Without loss of generality, let us take the local minimum $\mathbf{q}_{loc}$ as the coordinates' origin. The configuration attained by a random walk of duration t is a random variable $\mathbf{Q}(t) = (Q_1(t), ..., Q_m(t))$ verifying the following two properties [Papoulis, 1965]:

- The density of $Q_i(t)$ is:

$$p_i(q_i) = \frac{1}{\Delta_i\sqrt{2\pi t}} \exp(-\frac{q_i^2}{2\Delta_i^2 t}).$$

- The standard deviation of $Q_i(t)$ is:

$$D_i = \Delta_i\sqrt{t}.$$

A random walk may hit a C-obstacle. In fact, as in the case of a best-first motion, whenever a new configuration in the random motion is generated, the planner must check that this configuration lies in free space. If the new configuration is not free, the planner randomly selects another step that it substitutes for the previous one. This substitution is an approximation of the classical adaptation of a Brownian motion in a bounded space that consists of reflecting the motion that would have taken place if there were no boundary, symmetrically to the tangent hyperplane of the boundary at the collision configuration [Anderson and Orey, 1976].

The duration t of a random walk still has to be chosen. If it is too short, the motion has little chance of escaping from the local minimum. If it is too long, the planner may waste time and loose the opportunity of using the information provided by the potential when this information becomes pertinent again.

Let us define the **attraction radius** $R_i(\mathbf{q}_{loc})$ of any local minimum $\mathbf{q}_{loc}$ of $\mathbf{U}$ along any x_i-axis as the distance along this axis between $\mathbf{q}_{loc}$ and the nearest saddle point of $\mathbf{U}$ in that direction. The minimum distance that the robot must travel along any x_i-axis in order to escape from the local minimum is precisely $R_i(\mathbf{q}_{loc})$. If we were able to estimate the statistics of R_i, the property $D_i = \Delta_i\sqrt{t}$ would give us a clue for choosing t. The duration of the random walk could then be defined by:

$$t = \max_{i\in[1,m]} \left(\frac{R_i(\mathbf{q})}{\Delta_i}\right)^2.$$

But, as we make no assumption on the obstacle distribution, we cannot infer any strong statistical property about $\mathbf{U}$ and the R_i's. Nonetheless, in general, we may consider that a modification of the robot's configuration by R_i along the x_i-axis, for any $i \in [1, m]$, would not cause any point in $\mathcal{A}$ to be displaced by more than the length L of the sides of $\mathcal{GW}$ (we assume that $\mathcal{GW}$ is square or cubic). Assuming that the angular increments Δ_i have been normalized with respect to the translational ones, this consideration yields an estimate of R_i equal to L for any local minimum $\mathbf{q}_{loc}$, hence an estimate of t equal to $t = L/\delta^2$, with δ being the increment length along the axes of $\mathcal{GW}$.

One could take t equal to this estimate. However, this choice would implicitly assume that all the attraction radii are the same. Instead, we can consider that R_i is a random variable with a truncated Laplace distribution having L for the expected value (this distribution maximizes entropy, and hence is the less informed distribution for a positive random variable of given expected value). This leads us to choose t as the value of a random variable T with the following density:

$$p(t) = \frac{\delta}{2\sqrt{Lt}} \exp(-\delta\sqrt{Lt}). \tag{4}$$

One can verify that the expected value of T is indeed L/δ^2.

In fact, the value of T gives the maximal duration of the random walk. At each time unit (i.e. each step), the planner checks the value of $\mathbf{U}$ at the current configuration against $\mathbf{U}(\mathbf{q}_{loc})$. If it is smaller, the planner terminates the random walk.

5.4 Graph Searching

By combining best-first motions and random motions, RPP can incrementally construct a graph of the local minima of the potential function $\mathbf{U}$. The search of this graph can be done using a best-first strategy. This simply consists of iteratively generating the successors of the pending local minimum having the smallest potential value and limiting the number of random motions issued from the same minimum to a prespecified number K. A drawback of this strategy is that the same minimum may be attained several times, which is difficult and costly to detect. The strategy may also waste time exploring a local-minimum well containing smaller local-minimum wells imbricated in one another.

Another search technique is the following. When a local minimum $\mathbf{q}_{loc}$ is attained, a maximum of K random motions are generated. Each random motion is immediately followed by a best-first motion that attains a local minimum $\mathbf{q}'_{loc}$. If $\mathbf{U}(\mathbf{q}'_{loc}) > \mathbf{U}(\mathbf{q}_{loc})$, then the planner forgets $\mathbf{q}'_{loc}$; otherwise, if it is not a goal configuration (in which case the problem is solved), it generates its successors in the same fashion. If none of the K motions issued from $\mathbf{q}_{loc}$ allow the planner to attain a lower local minimum than $\mathbf{q}_{loc}$, the latter is considered to be a dead-end and the search is resumed at the most recently considered local minimum whose K successors have not all been generated yet.

This second search strategy gives better experimental results than the first, but it may still waste time exploring imbricated minima. Moreover, if a low local minimum has been attained, it may have difficulty attaining a lower one from it. (It is not difficult to construct examples where all the free paths between a very low local minimum and a goal configuration are quite long and unlikely to be generated by a single random motion.)

The above strategy can be slightly modified as follows. Rather than memorizing the whole graph G, the planner only memorizes the constructed path τ connecting the initial configuration to the current one. At every local minimum, the planner iteratively generates a maximum of K (typically $K \approx 20$) random/best-first motions as described above. If one of them reaches a lower minimum $\mathbf{q}'_{loc}$, it inserts the path from $\mathbf{q}_{loc}$ to $\mathbf{q}'_{loc}$ at the end of the current path τ and continues the search from $\mathbf{q}'_{loc}$. If none of the motions executed from a local minimum attains a lower minimum, the planner randomly selects a configuration $\mathbf{q}_{back}$ in the subset of τ formed by random motions, using a uniform distribution law over that subset, and backtracks to $\mathbf{q}_{back}$. The search is resumed at $\mathbf{q}_{back}$ by executing a best-first motion. Since $\mathbf{q}_{back}$ belongs to the path of a random motion, this best-first motion may terminate at a new local minimum that has not been attained so far. (If the first local minimum $\mathbf{q}_{loc}$ encountered by the planner turns out to be a dead-end, the above backtracking mechanism cannot be applied. Then, RPP randomly selects one of the local minima $\mathbf{q}'_{loc}$ attained from $\mathbf{q}_{loc}$, and resumes the search with a best-first motion starting at a configuration randomly selected along the path of the random motion leading to $\mathbf{q}'_{loc}$.)

A more formal expression of this search is given by the procedure RPP below:

```
procedure RPP;
begin
    τ ← BEST-FIRST-PATH(q_init); q_loc ← LAST(τ);
    while q_loc ≠ q_goal do
        begin
            ESCAPE ← false;
            for i = 1 to K until ESCAPE do
                begin
                    t ← RANDOM-TIME;
                    τ_i ← RANDOM-PATH(q_loc,t);
                    q_rand ← LAST(τ_i);
                    τ_i ← PRODUCT(τ_i,BEST-FIRST-PATH(q_rand));
                    q'_loc ← LAST(τ_i);
                    if U(q'_loc) < U(q_loc) then
```

```
                    begin
                        ESCAPE ← true;
                        τ ← PRODUCT(τ,τ_i);
                    end;
            end;
        if ¬ ESCAPE then
            begin
                τ ← BACKTRACK(τ, τ_1, ..., τ_K);
                q_back ← LAST(τ);
                τ ← PRODUCT(τ,BEST-FIRST-PATH(q_back));
            end;
        q_loc ← LAST(τ);
    end;
end;
```

where:

- τ is a list of configurations representing the path constructed so far;
- LAST(τ) returns the last configuration in a path τ;
- PRODUCT(τ_1, τ_2) returns the list of configurations representing the path $\tau_1 \bullet \tau_2$, i.e. the product of τ_1 and τ_2;
- BEST-FIRST-PATH($\mathbf{q}$) returns the path generated by a best-first motion starting at $\mathbf{q}$;
- RANDOM-PATH($\mathbf{q}$,t) returns the path generated by a random walk of duration t starting at $\mathbf{q}$;
- RANDOM-TIME returns a random duration computed with the distribution defined in formula (4);
- BACKTRACK($\tau, \tau_1, ..., \tau_K$) selects a backtracking configuration and returns the path from $\mathbf{q}_{init}$ to this configuration; if τ includes a subpath generated by a random motion, then the returned path is a subpath of τ; otherwise it is a subpath of $\tau \bullet \tau_i$, with i randomly chosen in $[1, K]$.

The above search techniques have been implemented, and the technique described in the procedure RPP gave the best experimental results [Barraquand and Latombe, 1990a].

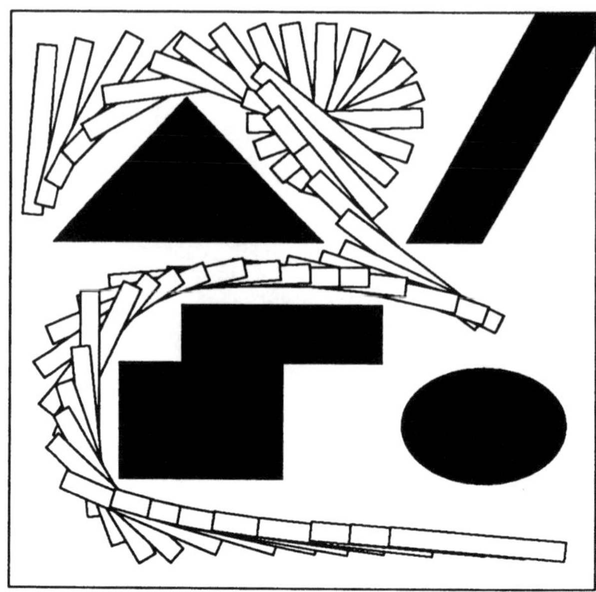

Figure 13. This figure illustrates a path generated by RPP for a rectangular robot moving among several obstacles.

5.5 Path Smoothing

A path τ produced by the randomized planning technique presented above has to be smoothed before it is executed. One simple and relatively fast way to do this is to iteratively replace subpaths of τ of decreasing lengths with straight line segments in the space $\mathbf{R}^m$ containing the grid $\mathcal{GC}$. Each new straight segment is discretized at the resolution of the grid $\mathcal{GC}$ and checked for collision before it is inserted in the new path. The algorithm initially attempts to replace subpaths whose lengths are of the order of the total path length. Then, it considers shorter and shorter subpaths until the resolution of $\mathcal{GC}$ is attained. More sophisticated variational calculus techniques may be applied instead, in order to optimize more involved objective criteria.

5.6 Discussion

Experiments with RPP have shown that randomized planning is both efficient and reliable. Figure 13 illustrates a path generated by RPP

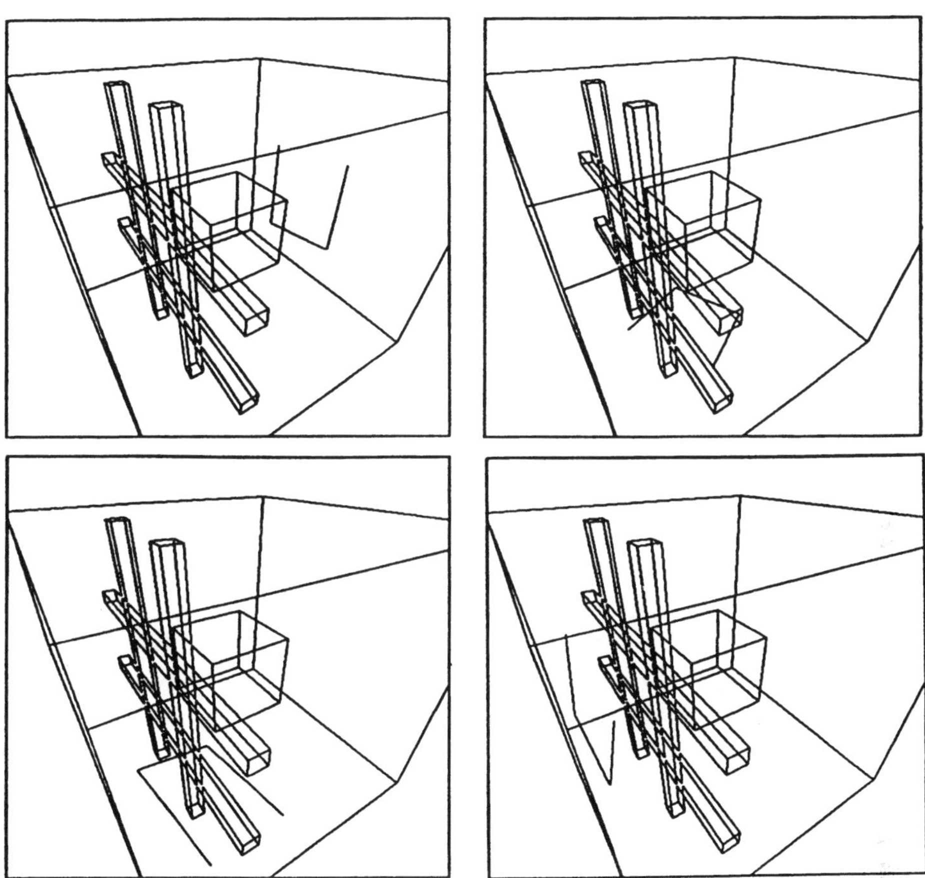

Figure 14. This figure shows a sequence of configurations of a ⊔-shaped object along a path generated by RPP in a three-dimensional workspace containing a grid-shaped obstacle and a parallelepipedic obstacle.

(after smoothing) for a rectangular robot that translates and rotates in a two-dimensional workspace. Figure 14 shows a sequence of configurations of a ⊔-shaped object moving in a three-dimensional workspace along a path generated by RPP. In chapter 8 we will give other examples solved by RPP involving multiple robots and articulated robots, with high-dimensional configuration spaces.

The efficiency of RPP results from the fact that a typical path planning problem has many solutions, so that a globally random search procedure can find one if it is well-informed most of the time (by the potential function). In fact, similar randomized techniques (e.g. "simulated annealing") have proven to be useful[12] for solving other NP-hard problems (e.g. traveling salesman problem [Cerny, 1985], VLSI placement and routing [Kirkpatrick, Gelatt and Vecchi, 1983] [Sechen, 1988]). In these problems the very large search space is associated with a large number of "good" sub-optimal solutions.

Randomized planning has some drawbacks, however. The planner typically generates different paths if it is run several times with the same problem and the running time varies from one run to another. Furthermore, if the input path planning problem admits no solution, the planner has no way to recognize it even after a large amount of computation. Hence, a limit on the running time of the algorithm has to be imposed. But, if this limit is attained and no path has been generated yet, there is no guarantee that no paths exist. Experiments show, however, that it is not difficult, for a given class of problem (e.g. an object moving in a three-dimensional workspace) and a given size of the grid $\mathcal{GC}$, to determine (through a series of trials) a time limit such that if no path has been found yet, there is little chance that one actually exists. In [Barraquand and Latombe, 1989a] it is shown that randomized planning is *probabilistically resolution-complete*. This means that if there exists a free path at the resolution of the grid $\mathcal{GC}$ the probability of finding one converges toward 1 when the running time grows toward infinity. But no formal result has been established so far on the rate of convergence, i.e. on the probability distribution of finding a path (if one exists) as a function of the running time.

6 Global Learning

Another approach for dealing with local minima, which has been proposed in [Faverjon and Tournassoud, 1989], is to accumulate statistics through a series of experiments. In this section we outline the principle of this approach.

We assume that the range of possible configurations is a rectangloid of dimension m. This rectangloid is decomposed into an array of cells of

[12]See [Barraquand and Latombe, 1989a] for a comparison of RPP with simulated annealing.

equal size, called the **state array**. The size of the cells is relatively large so that the memory space required for representing the state array remains reasonable, even when $m = 6$. Let κ_i and κ_j be two adjacent cells in the array. A number $p_{i,j} \in [0,1]$ is attached to κ_i. It measures the "probability" that the potential-field-based planner will succeed in finding a path from any configuration in κ_i to any configuration in κ_j. Typically, the numbers $p_{i,j}$ are initially set to 0.5 and then modified through a learning process.

Planning using the above state array proceeds according to the following two steps.

First, the state array is searched for a sequence $\kappa_1, ..., \kappa_p$ of adjacent cells connecting the initial configuration $\mathbf{q}_{init}$ to the goal configuration $\mathbf{q}_{goal}$ that maximizes the product $\prod_{k=1}^{k=p-1} p_{k,k+1}$ (i.e. the probability for the planner to successfully traverse the cells κ_1 through κ_p). Equivalently, we can minimize the quantity $K = -\sum_{k=1}^{k=p-1} \log p_{k,k+1}$, which is called the *cost* of the sequence. From the sequence $\kappa_1, ..., \kappa_p$ produced by the search, the planner extracts intermediate goal configurations, $\mathbf{q}_2, ..., \mathbf{q}_{p-1}$, defined, for example, as the centers of the cells $\kappa_2, ..., \kappa_{p-1}$.

Secondly, the planner uses a potential-field-based method to plan a path τ_1 from $\mathbf{q}_{init} \in \kappa_1$ to $\mathbf{q}_2 \in \kappa_2$, a path τ_2 from $\mathbf{q}_2$ to $\mathbf{q}_3 \in \kappa_3$, ..., and a path τ_{p-1} from $\mathbf{q}_{p-1} \in \kappa_{p-1}$ to $\mathbf{q}_{goal} \in \kappa_p$. Each path τ_k, $k \in [1, p-1]$, is constructed with a potential field whose global minimum is $\mathbf{q}_{k+1}$ (with $\mathbf{q}_p = \mathbf{q}_{goal}$). If the planner succeeds in generating τ_1 through τ_{p-1}, it returns the product $\tau_1 \bullet \tau_2 \bullet ... \bullet \tau_{p-1}$ as the constructed path. Otherwise, it searches for the next best sequence of cells in the state array; if the cost K of this sequence is lower than some prespecified threshold, the planner treats it as it treated the previous sequence, otherwise it exits with failure. (The planner is considered as failing to find a path τ_k from $\mathbf{q}_k \in \kappa_k$ to $\mathbf{q}_{k+1} \in \kappa_{k+1}$ if either the planner attains a local minimum before reaching $\mathbf{q}_{k+1}$, or the path τ_k leaves $\kappa_k \cup \kappa_{k+1}$, typically because there is an obstacle along the way.)

The probability $p_{i,j}$ associated with a cell κ_i is updated whenever the planner succeeds or fails to find a path from κ_i to κ_j. This is done by computing $p_{i,j}$ as follows [Faverjon and Tournassoud, 1989]:

$$p_{i,j} = \frac{s+1}{e+2}$$

where e is the number of attempts by the planner to plan a path from κ_i to κ_j and s is the number of successful attempts. Initially, $e = s = 0$; hence, $p_{i,j} = 0.5$.

Experiments described in [Faverjon and Tournassoud, 1989] show that after a reasonable period of training, a planner based on the above approach is capable of solving quite intricate problems. The learning approach is interesting only if the obstacle arrangement in the robot's workspace remains the same over many successive tasks.

Exercises

1: Consider the potential field $\mathbf{U}$ defined in Section 1. What would happen if the repulsive potential $\mathbf{U}_{rep}$ is non-zero at the goal configuration?

2: Design an algorithm for computing both the distance $\rho(x, y)$ between a point $(x, y) \in \mathbf{R}^2$ and a line segment L, and the gradient $\vec{\nabla}\rho(x, y)$ of this distance at (x, y). Extend this algorithm to the case where the line segment L is replaced by a convex polygon P.

3: Let $\mathcal{CB}$ be a convex polygon in a two-dimensional configuration space $\mathcal{C} = \mathbf{R}^2$. Let:

$$\mathbf{U}_{rep}(\mathbf{q}) = \begin{cases} \frac{1}{2}\eta(\frac{1}{\rho(\mathbf{q})} - \frac{1}{\rho_0})^2 & \text{if } \rho(\mathbf{q}) \leq \rho_0, \\ 0 & \text{if } \rho(\mathbf{q}) > \rho_0. \end{cases}$$

be the repulsive potential generated by $\mathcal{CB}$ (see Subsection 1.3). Implement an algorithm that computes the artificial force $\vec{F}_{rep}(\mathbf{q}) = -\vec{\nabla}\mathbf{U}_{rep}(\mathbf{q})$ at any given configuration $\mathbf{q} = (x, y)$ (see previous exercise).

4: Same exercise as above with $\mathcal{CB}$ being a convex polyhedron in a three-dimensional configuration space $\mathcal{C} = \mathbf{R}^3$ and $\mathbf{q} = (x, y, z)$.

5: Design an algorithm for computing both the distance $\rho(x, y, z)$ between a point $(x, y, z) \in \mathbf{R}^3$ and a given cylinder, and the gradient $\vec{\nabla}\rho(x, y, z)$ of this distance at (x, y, z).

6: Same exercise as above with a cone instead of a cylinder.

7: Let $\mathcal{A}$ and $\mathcal{B}$ be two convex polygons in $\mathcal{W} = \mathbf{R}^2$. Design an algorithm for computing a pair of points (a_0, b_0), $a_0 \in \mathcal{A}$ and $b_0 \in \mathcal{B}$, such that:

$$\|a_0 - b_0\| = \min_{a \in \mathcal{A}, b \in \mathcal{B}} \|a - b\|.$$

8: Verify the following expression given in Subsection 2.2:

$$\vec{F}^j_{att}(\mathbf{q}) = -\vec{\nabla}\mathbf{U}^j_{att}(\mathbf{q})$$

with $\mathbf{U}^j_{att}(\mathbf{q}) = \mathbf{V}^j_{att}(a_j(\mathbf{q}))$.

9: Comment on Remark 2 in Subsection 2.2. Compare the attractive potential:

$$\mathbf{U}_{att}(\mathbf{q}) = \sum_j \mathbf{V}^j_{att}(a_j(\mathbf{q}))$$

with:

$$\mathbf{U}_{att}(\mathbf{q}) = \mathbf{V}^1_{att}(a_1(\mathbf{q})) + \varepsilon \sum_{j>1} \mathbf{V}^j_{att}(a_j(\mathbf{q})).$$

Illustrate your answer with figures.

10: Let $\mathcal{C}$ be the configuration space of a free-flying object $\mathcal{A}$ in $\mathbf{R}^3$. We represent any configuration in $\mathcal{C}$ by $(x, y, z, \phi, \theta, \psi)$, where x, y, z are the coordinates of the reference point of $\mathcal{A}$ and (ϕ, θ, ψ) are the Euler angles defined as in Subsection 4.2 of Chapter 2. We assume that the range of possible positions (x, y, z) for $\mathcal{A}$ is bounded by a parallelepiped and we discretize $\mathcal{C}$ into a six-dimensional rectangloid grid $\mathcal{GC}$ as described in Subsection 3.2.

a. Define the 1-neighborhood of a configuration $\mathbf{q}$ in $\mathcal{GC}$.

b. How should the scale of the angular axes (ϕ, θ, ψ) be normalized with respect to the scale of the translational axes?

11: Let $\mathcal{C} = \mathbf{R}^2$. Assume that the C-obstacles are disjoint discs. Construct an analytical navigation function over $\mathcal{C}_{free}$, i.e. a local-minimum-free potential function with a global minimum at the goal configuration.

12: Adapt the procedure NF1 given in Subsection 4.2.1 to the case where several configurations in $\mathcal{GC}$ are specified as goal configurations.

13: Same exercise as above with the procedure NF2 given in Subsection 4.2.2.

14: Implement the best-first planning procedure BFP (Subsection 3.2) with a potential function computed as in Subsection 4.2.3. Experiment with the planner using a rectangular robot on examples such as those shown in Figures 8 and 13. (Collision can be checked at every new configuration by "drawing" the four sides of the rectangle on the grid representing the workspace. If the map d_1 is first computed as indicated in Subsection 4.2.2, a more efficient "divide-and-conquer" technique can be used [Barraquand and Latombe, 1989a].)

15: Consider Figure 10. Assume that the attractive potential is a parabolic well and that the repulsive potential is defined as in Subsection 1.3. Give the analytical expression of the total potential in the vertical strip above $\mathcal{CB}$. Show that it admits a local minimum and determine the coordinates of this minimum.

16: Implement the planner RPP (Section 5) and experiment with it. [This exercise requires a substantial amount of programming and experimentation.]

17: Suppose that a semi-free path τ has been generated between two configurations using a roadmap planning method or the exact cell decomposition method described in Section 4 of Chapter 5. Describe, in general terms, how a potential field method could be used in a postprocessing step to transform this path τ into a free path.

18: Let $\mathcal{A}$ be a circular robot of radius r in $\mathbf{R}^2$ (the reference point is selected at the center of the robot). Assume that an E-channel of rectangloid cells has been generated for $\mathcal{A}$ using an approximate cell decomposition method (Chapter 6). However, there may be a few small obstacles lying in the workspace that the planner did not know about when it produced the channel. While the robot is moving, a sensor detects and recognizes the obstacles that enter a circular area of some radius $R > r$ centered at the current position of the reference point. Propose an on-line potential-field-based method for guiding the robot's motion through the channel, from $\mathbf{q}_{init}$ to $\mathbf{q}_{goal}$, while avoiding collisions with the unexpected obstacles. Define the various potential functions used by the controller [Choi, Zhu and Latombe, 1989]. Discuss the choices you made.

19: Suppose that a path τ has been generated for a robot from an initial configuration $\mathbf{q}_{init}$ to a goal configuration $\mathbf{q}_{goal}$. How can this path be reformulated into a potential function $\mathbf{U}_\tau$ so that: (1) if the robot is controlled to move along the steepest descent of $\mathbf{U}_\tau$, it exactly follows the planned path; and (2) if there are unexpected obstacles along τ, we can add a repulsive potential $\mathbf{U}_{unexp}$ such that $\mathbf{U}_\tau + \mathbf{U}_{unexp}$ keeps the actual path of the robot close to τ while avoiding collisions with the unexpected obstacles? Compare this approach to dealing with unexpected obstacles with the approach of the previous exercise.

Chapter 8

Multiple Moving Objects

In this chapter as well as the next three, we investigate various extensions of the basic motion planning problem. In the present chapter we begin this investigation with a series of extensions which all involve multiple moving objects.

We first consider the case where some (or all) of the obstacles in the workspace are moving (Section 1). The resulting problem is called the *dynamic motion planning problem.* Unlike the (static) basic problem, it cannot be solved by merely constructing a geometric path. Instead, a continuous function of time specifying the robot's configuration at each instant has to be generated. The general approach taken in this chapter is to add a dimension — time — to the robot's configuration space. The robot maps in this *configuration-time space* to a point moving among stationary obstacles. But the time dimension is irreversible and hence is significantly different from the other dimensions. Several path planning methods presented in the previous chapters are still applicable to the dynamic motion planning problem, but with some modifications and extensions, which are aimed at taking the specificity of the time dimension into account.

The dynamic motion planning problem can be made more realistic by imposing constraints on the robot's velocity and acceleration. In our presentation, however, we will only consider the two "simple" cases where (1) the robot's velocity modulus is not upper-bounded, and (2) where it is upper-bounded (with no other constraint on the velocity and the acceleration). As we will see, from a computational point of view, the problem with bounded velocity is substantially harder than the problem without bounded velocity.

As a second extension, we then consider the *multi-robot path planning problem* where several robots move independently in the same workspace among stationary obstacles (Section 2). One conceptually simple approach to this problem is to consider the various robots as the components of a composite robot and plan a free path in the configuration space of this composite robot (*centralized planning*). A difficulty, of course, is that the dimension of this configuration space is usually large and we know that the time complexity of path planning methods is exponential in this dimension. Another approach consists of planning a path for each robot more or less independently of the other robots and then considering the interactions among the paths (*decoupled planning*). By doing so, we significantly reduce the amount of computation, but we also loose completeness.

In the third extension, we consider *articulated robots*, i.e. robots which are made of several moving rigid objects connected by joints (e.g. hinges, sliding joints) that constrain the relative movements of the objects (Section 3). We focus our presentation on the case where the robots contain no kinematic loop. This case includes the problem of planning collision-free paths for manipulator arms.

1 Moving Obstacles

1.1 Configuration-Time Space

Let $\mathcal{A}$ be a rigid object moving in a workspace $\mathcal{W} = \mathbf{R}^N$, with $N = 2$ or 3, populated by obstacles $\mathcal{B}_i$ (rigid objects), $i = 1$ to q, which may possibly be in motion. $\mathcal{B}_i(t)$ designates the region of $\mathcal{W}$ occupied by the object $\mathcal{B}_i$ at time t ($t \geq 0$). We assume that the regions $\mathcal{B}_1(t)$ through $\mathcal{B}_q(t)$ are known in advance. This assumption requires that the motions of the obstacles not be influenced by the motion of $\mathcal{A}$.

Let $\mathcal{C}$ be the configuration space of $\mathcal{A}$ as defined in Chapter 2. At every instant t, the obstacle $\mathcal{B}_i(t)$ maps in $\mathcal{C}$ to the C-obstacle:

$$\mathcal{CB}_i(t) = \{\mathbf{q} \in \mathcal{C} \mid \mathcal{A}(\mathbf{q}) \cap \mathcal{B}_i(t) \neq \emptyset\}.$$

The shape of $\mathcal{CB}_i(t)$ does not depend on t, but its location in $\mathcal{C}$ does.

The dynamic motion planning problem is to plan a collision-free motion of $\mathcal{A}$ from the initial configuration $\mathbf{q}_{init}$ at time 0 to the goal configuration $\mathbf{q}_{goal}$ at time $T \geq 0$. (We assume without loss of generality that the motion of $\mathcal{A}$ starts at time 0.) The time T is called the **arrival time.** The planning problem requires that a function of time be generated which specifies the configuration of $\mathcal{A}$ at every instant in the interval $[0, T]$. More formally, a solution of the problem is a continuous function:

$$t \in [0, T] \quad \mapsto \mathbf{q}(t) \; \in \; \mathcal{C} \setminus \bigcup_{i=1}^{q} \mathcal{CB}_i(t)$$

with t representing time, $\mathbf{q}(0) = \mathbf{q}_{init}$ and $\mathbf{q}(T) = \mathbf{q}_{goal}$. This function is called a **trajectory** of $\mathcal{A}$.

Since the locations of the C-obstacles may vary with t, we cannot represent them in $\mathcal{C}$ in such a way that we can still reduce the motion of $\mathcal{A}$ to the motion of a point among fixed constraints. This difficulty is solved by simply adding another dimension — namely, time — to $\mathcal{C}$. We obtain a new space, $\mathcal{CT} = \mathcal{C} \times [0, +\infty)$, which is called the **configuration-time space**[1] of $\mathcal{A}$. $\mathcal{CT}$ is a smooth manifold of dimension $m + 1$, where m is the dimension of $\mathcal{C}$. A continuous map $\xi : [0, 1] \rightarrow \mathcal{CT}$ is called a **CT-path.**

The robot $\mathcal{A}$ maps in $\mathcal{CT}$ to a **configuration-time** $(\mathbf{q}, t)$, meaning that $\mathcal{A}$'s configuration at instant t is $\mathbf{q}$. The cross-section through $\mathcal{CT}$ at every instant t is $\mathcal{C}$ populated by the C-obstacles $\mathcal{CB}_i(t)$, $i = 1$ to q. Thus, every obstacle $\mathcal{B}_i$ maps in $\mathcal{CT}$ to a stationary region $\mathcal{CTB}_i$, called a **CT-obstacle**, defined by:

$$\mathcal{CTB}_i = \{(\mathbf{q}, t) \mid \mathcal{A}(\mathbf{q}) \cap \mathcal{B}_i(t) \neq \emptyset\}.$$

---

[1]In the literature, this space is often called *configuration space-time.*

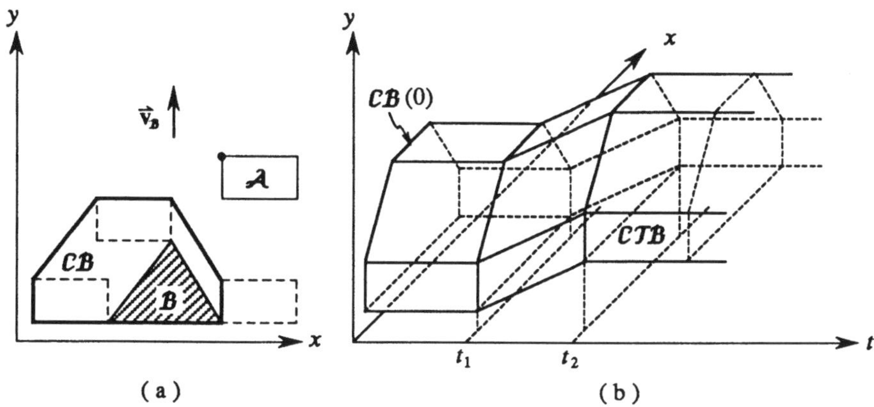

Figure 1. The robot $\mathcal{A}$ translates freely without rotation in the plane (Figure a). The obstacle $\mathcal{B}$ translates at fixed orientation, at constant linear velocity $\vec{\mathbf{v}}_{\mathcal{B}}$, between the instants t_1 and t_2. $\vec{\mathbf{v}}_{\mathcal{B}}$ is parallel to the y-axis and points upward. At any time $t \notin [t_1, t_2]$, $\mathcal{B}$ is motionless. The CT-obstacle $\mathcal{CTB}$ is obtained (see Figure b) by sweeping $\mathcal{CB}$ along a line that is perpendicular to the xy-plane, when $t \notin [t_1, t_2]$, and oriented along the vector $(0, \|\vec{\mathbf{v}}_{\mathcal{B}}\|, 1)$, when $t \in [t_1, t_2]$.

Example: Let $\mathcal{W} = \mathbf{R}^2$ and assume that $\mathcal{A}$ can only translate at fixed orientation. Hence, $\mathcal{C} = \mathbf{R}^2$ and $\mathcal{CT} = \mathbf{R}^2 \times [0, +\infty)$. Let the workspace contain a moving obstacle $\mathcal{B}$ that translates at fixed orientation, at constant linear velocity $\vec{\mathbf{v}}_{\mathcal{B}}$, between the instants t_1 and t_2. The rest of the time, $\mathcal{B}$ is stationary. Let $\mathcal{CB}(0)$ be the C-obstacle in $\mathcal{C}$ at instant 0. For all $t \in [t_1, t_2]$, $\mathcal{CB}(t)$ is obtained by translating $\mathcal{CB}(0)$ by $\vec{\mathbf{v}}_{\mathcal{B}}(t - t_1)$ in $\mathcal{C}$. Thus:

$$\begin{aligned} \mathcal{CTB} = & \{(\mathbf{q}, t) \;/\; \mathbf{q} \in \mathcal{CB}(0) \text{ and } t \in [0, t_1]\} \\ & \cup\{(\mathbf{q} + \vec{\mathbf{v}}_{\mathcal{B}}(t - t_1), t) \;/\; \mathbf{q} \in \mathcal{CB}(0) \text{ and } t \in [t_1, t_2]\} \\ & \cup\{(\mathbf{q} + \vec{\mathbf{v}}_{\mathcal{B}}(t_2 - t_1), t) \;/\; \mathbf{q} \in \mathcal{CB}(0) \text{ and } t \geq t_2\}. \end{aligned}$$

Figure 1 illustrates the construction of $\mathcal{CTB}$. ∎

We define the **free space** in $\mathcal{CT}$ as:

$$\mathcal{CT}_{free} = \mathcal{CT} \setminus \bigcup_{i=1}^{q} \mathcal{CTB}_i.$$

The closure of $\mathcal{CT}_{free}$ is called the **semi-free space** in $\mathcal{CT}$. A continuous map $\xi : [0,1] \rightarrow \mathcal{CT}_{free}$ (resp. $cl(\mathcal{CT}_{free})$) is called a **free** (resp.

semi-free) CT-path. The dynamic motion planning problem can now be regarded as that of constructing a free CT-path between two given configuration-times $(\mathbf{q}_{init}, 0)$ and $(\mathbf{q}_{goal}, T)$. Hence, it is reduced to the problem of finding a path among stationary obstacles, i.e. the $\mathcal{CTB}_i$'s.

With respect to the basic motion planning problem, there is a major difference. Since no object can move back in time, a CT-path ξ should (at least) be *strictly monotone in time*, i.e. satisfy the following condition:

$$\forall s, s' \in [0,1], \text{let } \xi(s) = (\mathbf{q}, t) \text{ and } \xi(s') = (\mathbf{q}', t');\ s < s' \ \Rightarrow\ t < t'.$$

Then ξ determines a trajectory $\mathbf{q}(t)$ such that: $\forall s \in [0,1] : \xi(s) = (\mathbf{q}(t), t)$.

In practice, the condition that a CT-path ξ be strictly time-monotone may not be sufficient, since it allows for unbounded velocity and acceleration. A variant of the dynamic motion planning problem consists of imposing constraints on $\mathcal{A}$'s velocity and acceleration, that is, on the slope and the curvature of ξ relative to the time axis. These constraints make the planning problem considerably harder. In the following, we will only consider the two cases where the robot's velocity modulus is/is not bounded. We will not examine the case where the acceleration modulus is bounded.[2]

Another classical variant of the dynamic motion planning problem is to leave the arrival time unspecified. This means that we only want $\mathcal{A}$ to reach a goal configuration $\mathbf{q}_{goal}$ at some time $T \geq 0$. Most of the methods solving the dynamic problem with specified arrival time T can solve the problem with unspecified T, with minor adaptations.

1.2 Complexity Results

It has been shown in [Reif and Sharir, 1985] that the dynamic motion planning problem in a three-dimensional workspace, with arbitrarily many rotating obstacles, is PSPACE-hard if the velocity modulus is bounded and NP-hard otherwise. (Notice that rotating obstacles may generate non-algebraic constraints in the configuration-time space, even if the objects are semi-algebraic.) It has also been shown in [Canny,

[2]Ó'Dúnlaing [Ó'Dúnlaing, 1987] is one of the rare publications considering the case where the robot's acceleration is bounded.

1988] that dynamic motion planning for a point in the plane, with a bounded velocity modulus and arbitrarily many obstacles, is NP-hard, even when the moving obstacles are convex polygons moving without rotation at constant linear velocities.

On the other hand, the dynamic motion planning problem for a polygonal robot $\mathcal{A}$ translating at fixed orientation in the plane with bounded velocity modulus among polygonal obstacles that translate without rotation at fixed linear velocity can be solved by an algorithm that is polynomial in the total number of vertices of $\mathcal{A}$ and the $\mathcal{B}_i$'s, if the number of obstacles is bounded (see Subsection 1.4.1). However, as one could suspect from the results cited above, the algorithm takes exponential time in the number of obstacles. When all the constraints can be expressed semi-algebraically, the dynamic motion planning problem with no velocity constraints and arbitrarily many obstacles can be solved in polynomial time in the number and the degree of the semi-algebraic constraints (see Subsection 1.3.1).

1.3 Planning Without Velocity Bound

1.3.1 Exact Cell Decomposition

When the modulus of $\mathcal{A}$'s velocity is not bounded and the obstacles' trajectories are expressed in algebraic form, the general path planning algorithm based on the Collins decomposition presented in Section 3 of Chapter 5 is applicable with minor changes. The various CT-obstacles are represented as semi-algebraic sets in $\mathcal{CT}$ and planning is performed with the additional constraint that time is irreversible.

This constraint can be taken into account by representing a configuration-time as a list $(t, q_1, ..., q_m) \in \mathbf{R}^{m+1}$ (m is the dimension of $\mathcal{A}$'s configuration space), where the time t stands as the first coordinate in the charts put on $\mathcal{CT}$. Hence, all the cells resulting from the Collins decomposition of $\mathcal{CT}$ ultimately project on the time axis. In other words, they are contained in "cylinders" above intervals or points along the time axis. Therefore, two cells κ and κ' are in the same cylinder, i.e. "occur" in the same interval of time, if and only if the sample points $\sigma(\kappa)$ and $\sigma(\kappa')$ project at the same time coordinate. If, instead, the sample points project at different time coordinates, one cell "occurs" before the other. The connectivity graph CG is now defined as a *directed graph* whose nodes are the $(m+1)$- and m-dimensional cells of the Collins decomposition of $\mathcal{CT}$. A cell κ is connected to a cell κ' by an arc if and only if κ and κ' are adjacent and either κ occurs before κ', or κ and κ' occur simultaneously, i.e. the time coordinate of $\sigma(\kappa)$ is less than or equal to the time coordinate of $\sigma(\kappa')$.

In order to construct a free motion between $(\mathbf{q}_{init}, 0)$ and $(\mathbf{q}_{goal}, T)$, the planner searches CG for a sequence of adjacent cells $\kappa_1, ..., \kappa_p$ such that $(\mathbf{q}_{init}, 0) \in \kappa_1$ and $(\mathbf{q}_{goal}, T) \in \kappa_p$. From this sequence, if one exists, it generates a CT-path from $(\mathbf{q}_{init}, 0)$ to $(\mathbf{q}_{goal}, T)$ in the same way as for the static problem. Clearly, however, the slope of such a path with respect to the time axis is not bounded.

If the arrival time T is unspecified, the same method still applies. The goal locus in $\mathcal{CT}$ is the half-line $\{(\mathbf{q}_{goal}, T) \ / \ T \geq 0\}$ rather than a single configuration-time. The graph CG has to be searched for a sequence of cells $\kappa_1, ..., \kappa_p$ such that κ_p intersects this half-line. A search of CG with time as the cost function can generate the CT-path that attains the goal configuration at the earliest possible time, if a free CT-path exists.

This general method is complete, but its computational complexity makes it impractical, even for small values of m.

1.3.2 Approximate Cell Decomposition

The approximate cell decomposition approach presented in Chapter 6 also applies to the dynamic motion planning problem with no velocity constraint. It only requires the definition of the connectivity graph to be slightly modified.

Assume that the set of possible configuration-times in $\mathcal{CT}$ is a rectangloid which has been decomposed into smaller rectangloid cells, each classified as EMPTY (if its interior intersects no CT-obstacle), FULL (if it is entirely contained in the union of the CT-obstacles), or MIXED (otherwise). The connectivity graph CG corresponding to this decomposition is a directed graph whose nodes are the EMPTY cells (or both the EMPTY and MIXED cells) of the decomposition. A cell κ is connected to a cell κ' by an oriented arc if and only if:

- either they are adjacent by a face that is parallel to the time axis,

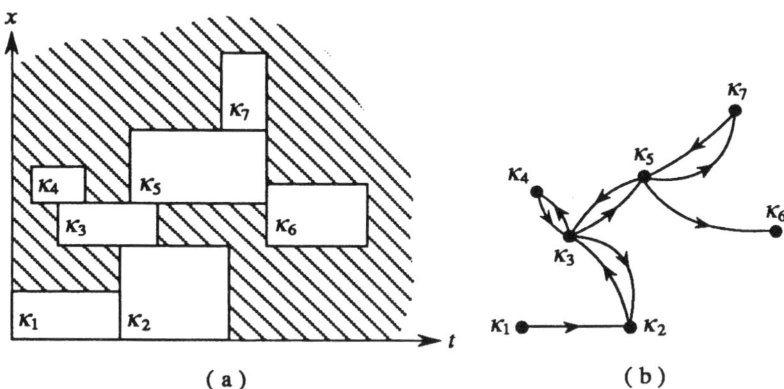

Figure 2. Figure a depicts a two-dimensional configuration-time space with seven EMPTY cells. Figure b shows the corresponding connectivity graph. Although no path from κ_2 to κ_4 can be time-monotone, this fact is not represented in the graph. Therefore, the search must be appropriately constrained.

- or they are adjacent by a face that is perpendicular to the time axis, and κ projects on the time axis before κ'.

Planning is accomplished by searching *CG* for a sequence of adjacent cells forming an EMPTY or MIXED channel such that the first cell of the sequence contains the initial configuration-time and the last cell contains the goal configuration-time.

Figure 2 illustrates this refinement with a two-dimensional configuration-time space. Seven EMPTY cells, κ_1 through κ_7, and the corresponding connectivity graph are depicted in the figure.

This refinement of the connectivity graph is not sufficient, however, to guarantee that the constructed channel contains a strictly time-monotone CT-path. It is necessary to put an additional constraint on the search of the graph. This need is illustrated in the graph of Figure 2. Although there exists a strictly time-monotone path between κ_2 and κ_3 and a strictly time-monotone path between κ_3 and κ_4, there is no such path between κ_2 and κ_4. This fact is not represented in the connectivity graph because it depends on which cells were traversed before κ_4 was entered, and hence involves more than two cells. The planner can deal with this difficulty during graph searching by maintaining the earliest time $t_{earliest}$ at which it can be in the current cell [Erdmann and

Lozano-Pérez, 1986]. Let $[t_{min}(\kappa), t_{max}(\kappa)]$ be the time interval on which the cell κ projects itself on the time axis. $t_{earliest}$ is simply computed as follows:

- when the planner starts the search and is in the initial cell:

$$t_{earliest} \leftarrow 0,$$

- when the planner decides to move from a cell κ to a cell κ':

$$t_{earliest} \leftarrow \max\{t_{earliest}, t_{min}(\kappa')\}.$$

Suppose that the planner has attained κ. It can choose κ' as the successor of κ in the search if and only if there is an arc from κ to κ' in CG and $t_{earliest} < t_{max}(\kappa')$. Consider Figure 2 and suppose that the search starts at κ_1. The earliest time in κ_1 is 0. The search can proceed to κ_2 with $t_{earliest} = t_{min}(\kappa_2)$, then to κ_3 with $t_{earliest}$ still equal to $t_{min}(\kappa_2)$. κ_3 has two successors, κ_4 and κ_5, in CG. However, κ_4 cannot be selected as a successor of κ_3 in the search, since $t_{earliest} > t_{max}(\kappa_4)$.

If an EMPTY channel is constructed, a free strictly time-monotone CT-path can easily be extracted from it.

For a configuration space of given dimension m, the time complexity of the approach is increased by the fact that it is now applied in the $(m+1)$-dimensional configuration-time space. But the constraint on the search does not increase the complexity further. As a matter of fact, it may only reduce the size of the search space.

1.4 Planning With Velocity Bound

In this subsection we consider the case where the modulus of the robot's velocity is upper-bounded. There is no constraint, however, on the robot's acceleration.

1.4.1 Asteroid Avoidance Problem

We first consider a rather specific type of dynamic motion planning problem called the two-dimensional *asteroid avoidance problem* [Reif and Sharir, 1985] and we describe a complete planning method for this problem. Roughly speaking, the method is a kind of visibility graph method extended in the robot's configuration-time space.

The robot is a polygon that can translate freely in $\mathbf{R}^2$ without rotation. Hence, $\mathcal{CT} = \mathbf{R}^2 \times [0, +\infty)$. The obstacles $\mathcal{B}_i$, $i = 1, ..., q$, are polygons that translate without rotation at fixed velocities (which may vary from one obstacle to another). The modulus $\|\vec{\mathbf{v}}_\mathcal{A}\|$ of the linear velocity of $\mathcal{A}$ is upper-bounded by V_{max}. In $\mathcal{A}$'s configuration space, the problem is equivalent to planning the motion of a point among polygonal C-obstacles that translate at constant linear velocities. We assume that the C-obstacles never intersect themselves. As mentioned in Subsection 1.2, this problem is NP-hard when the number of obstacles is not bounded.

Each CT-obstacle is obtained by sweeping the corresponding C-obstacle along a straight line whose slope relative to the time axis is determined by the constant linear velocity of the obstacle. The straight lines traced in $\mathcal{CT}$ by the C-obstacle's vertices are called the **vertex lines**.

We use the following terminology [Reif and Sharir, 1985]:

- A **direct motion** is a motion of $\mathcal{A}$ at constant velocity (possibly zero) such that $\mathcal{A}$ makes no contact with the obstacles, except possibly at the endpoints of the motion. In addition, at every endpoint where the motion makes a contact with an obstacle $\mathcal{B}_i$, the straight CT-path representing the direct motion in $\mathcal{CT}$ is supported by a line that is tangent at the contact point to the corresponding CT-obstacle $\mathcal{CTB}_i$.

- A **contact motion** is a motion of $\mathcal{A}$ along a convex subset (possibly a single point) of the boundary of a C-obstacle $\mathcal{CB}_i$, starting and ending at vertices of $\mathcal{CB}_i$. In $\mathcal{CT}$, the motion is thus described by a CT-path (possibly of length 0) which is contained in the boundary of $\mathcal{CTB}_i$ and begins and ends in a vertex line.

- A **fundamental motion** is a motion of $\mathcal{A}$ that consists of a direct motion possibly followed by a contact motion.

- A **normal motion** is a motion of $\mathcal{A}$ that consists of a sequence of fundamental motions such that any two direct motions are separated by a contact motion and no two contact motions are performed within the same maximal convex subset of the boundary of a C-obstacle. (Hence, if a C-obstacle is convex, a normal motion can visit it at most once.)

PROPOSITION 1: *There exists a time-monotone semi-free CT-path of $\mathcal{A}$ from $(\mathbf{q}_{init}, 0)$ to $(\mathbf{q}_{goal}, T)$ satisfying the velocity constraint*

$\|\vec{\mathbf{v}}_{\mathcal{A}}\| \leq V_{max}$ *if and only if there exists a normal motion from* $(\mathbf{q}_{init}, 0)$ *to* $(\mathbf{q}_{goal}, T)$ *satisfying the velocity constraint.*

Proof: Obviously, a normal motion satisfying the velocity constraint produces a strictly time-monotone semi-free CT-path.

For the converse part, let the metric in $\mathcal{CT} \subset \mathbf{R}^3$ be the Euclidean metric in $\mathbf{R}^3$. If there exists a time-monotone semi-free CT-path ξ connecting two configuration-times that satisfies the velocity constraint, then the shortest CT-path ξ_{min} homotopic to ξ in $cl(\mathcal{CT}_{free})$ also satisfies the velocity constraint. We show below that ξ_{min} describes a normal motion.

The CT-path ξ_{min} must be such that any subpath ξ' of ξ_{min} is also the shortest path (in its homotopy class) between the two configuration-times that ξ' connects. Hence, at any configuration-time in $cl(\mathcal{CT}_{free})$ that is not in a vertex line, ξ_{min} must have zero curvature, since otherwise there is locally a shorter subpath. At any configuration-time contained in a vertex line, ξ_{min} must locally (i.e. in a sufficiently small neighborhood) be either a straight line or a polygonal line with a single vertex (the considered configuration-time); if it is a polygonal line, then the CT-obstacle must lie in its convex sector. (In particular, this implies that ξ_{min} does not traverse any vertex line generated by a C-obstacle's concave vertex.) Finally, consider a subpath ξ' of ξ_{min} whose endpoints lie in the same maximal convex subset S of the boundary of a CT-obstacle; ξ' must lie in S, be contained in a plane, and thus consist of successive straight segments. Hence, ξ_{min} cannot visit S twice.

Therefore, ξ_{min} describes a normal motion. ∎

The shortest semi-free CT-path in some homotopy class between two free configuration-times may lie partly in the boundary of $\mathcal{CT}_{free}$. Since the CT-obstacles are disjoint, one can obtain a free CT-path by an arbitrarily small perturbation of the shortest semi-free CT-path. But such a free CT-path may violate the velocity constraint by an arbitrarily small offset. Indeed, assume that the straight segment connecting $(\mathbf{q}_{init}, 0)$ to $(\mathbf{q}_{goal}, T)$ lies in $cl(\mathcal{CT}_{free})$ and that its slope relative to the time axis is exactly V_{max}. Assume further that it makes a single contact with a CT-obstacle. Then, in order to avoid the contact, the robot would have to go faster, either before or after it passes near the obstacle. An alternative to violating the constraint is to slightly shift either the initial or the goal configuration-time.

From the above proposition, we can infer the planning method presented below, which is due to Reif and Sharir [Reif and Sharir, 1985]. We first outline the algorithm. Then we analyze some of its components in more detail. For simplifying the presentation, we assume that all the C-obstacles are convex polygons. However, it is not difficult to extend the algorithm to non-convex polygonal C-obstacles.

Let q be the number of obstacles in the environment, Q_i the set of vertices of $\mathcal{CB}_i$, $Q_0 = \{\mathbf{q}_{init}\}$, $Q_{q+1} = \{\mathbf{q}_{goal}\}$, and $Q = \bigcup_{i=0}^{q+1} Q_i$. For any point b in Q, we denote by $b(t)$ its position in $\mathcal{C}$ at time t, by $\overline{b}$ the (half-)line $\{b(t) \mid t \in [0, +\infty)\}$ to which b maps in $\mathcal{CT}$, and by $\overline{Q}$ the set $\{\overline{b} \mid b \in Q\}$. Since $\mathbf{q}_{init}$ and $\mathbf{q}_{goal}$ are stationary points in $\mathcal{C}$, $\overline{\mathbf{q}}_{init}$ and $\overline{\mathbf{q}}_{goal}$ are lines parallel to the time axis of $\mathcal{CT}$. In the following, we only consider motions that satisfy the velocity constraint.

Let $I_b^{(k)}$ denote the set of all times[3] $t \geq 0$ at which the line $\overline{b} \in \overline{Q}$ can be attained from the initial configuration-time $(\mathbf{q}_{init}, 0)$ by a motion consisting of at most k fundamental motions. We have:

$$I_{\mathbf{q}_{init}}^{(0)} = [0, +\infty) \quad \text{and} \quad I_b^{(0)} = \emptyset$$

for every $b \in Q \setminus \{\mathbf{q}_{init}\}$.

Let $DM_{b,b'}(I)$ be the set of all times $t' \geq 0$ such that $\overline{b'} \in \overline{Q}$ can be reached at time t' by a single direct motion starting at the configuration-time $(b(t), t)$, with $t \in I$. For every $b' \in Q$,

$$J_{b'}^{(k)} = \bigcup_{b \in Q} DM_{b,b'}(I_b^{(k)})$$

is the set of all times at which $\overline{b'}$ can be reached from $(\mathbf{q}_{init}, 0)$ by a motion consisting of at most k fundamental motions followed by a direct motion.

Let $CM_{b',b''}(I)$, with b' and b'' being in the same set Q_j for some $j \in [1, q]$, be the set of all times $t'' \geq 0$ for which $\overline{b''}$ can be attained at time t'' by

[3]It would be slightly more efficient to restrict t to the interval $[0, T]$ rather than $[0, +\infty)$. However, by making the presentation more general, the same algorithm applies to the case where the arrival time T is unspecified.

a contact motion in the boundary of $\mathcal{CTB}_j$ starting at $(b'(t'), t')$, with $t' \in I$. For every $b'' \in Q_j$ ($j \in [1, q]$), we have:

$$I_{b''}^{(k+1)} = \bigcup_{b' \in Q_j} CM_{b',b''}(J_{b'}^{(k)}).$$

Assume that we know how to compute the sets $DM_{b,b'}(I)$ and $CM_{b',b''}(I)$. Then, for $k = 1, 2, ...$, we can successively compute the sets $I_b^{(k)}$ for all $b \in Q \backslash \{\mathbf{q}_{init}\}$, until either $T \in I_{\mathbf{q}_{goal}}^{(k)}$, or $k > q + 1$. In the first case, the problem is solved and a semi-free CT-path can be generated by moving backward along the time axis from $(\mathbf{q}_{goal}, T)$. In the second case, the problem admits no solution; indeed, according to the above proposition, if a semi-free CT-path exists, there exists a normal motion that visits each CT-obstacle at most once, and hence consists of at most $q + 1$ fundamental motions. If the arrival time T is not specified, the same algorithm applies; it terminates when either $I_{\mathbf{q}_{goal}}^{(k)} \neq \emptyset$ (success) or $k > q + 1$ (failure).

The computation of $CM_{b,b'}(I)$ is simple and not described here. The number of disjoint intervals in $CM_{b,b'}(I)$ is less than or equal to the number of disjoint intervals in I.

The computation of $DM_{b,b'}(I)$ is more involved. Consider $DM_{b,b'}(\{t\})$, i.e. the set of all times t' such that $\overline{b'}$ can be reached by a single direct motion starting at $(b(t), t)$. The set of CT-paths representing the direct motions starting at $(b(t), t)$ spans a one-sided cone of angle $\tan^{-1}(V_{max})$. This cone is anchored at $(b(t), t)$ and its axis points along the time axis. Its intersection with the line $\overline{b'}$ is a segment (possibly semi-infinite or empty) joining two points $(b'(t_1), t_1)$ and $(b'(t_2), t_2)$, with t_1 and t_2 depending on t. We denote by $L(t)$ the interval $[t_1, t_2]$. We have that $DM_{b,b'}(\{t\}) \subseteq L(t)$, since for some $t' \in L(t)$ the straight CT-path between $(b(t), t)$ and $(b(t'), t')$ may traverse the interior of a CT-obstacle. Let us consider the triangle Δ in $\mathcal{CT}$, whose vertices are $(b(t), t)$, $(b(t_1), t_1)$ and $(b(t_2), t_2)$. Each CT-obstacle $\mathcal{CTB}_i$ ($i \in [1, q]$) intersects Δ at a (possibly empty) convex polygon. The two rays drawn from $(b(t), t)$ which are tangent to this polygon cut a segment $S_i(t)$ off $L(t)$. $S_i(t)$ is the set of configuration-times in $\overline{b'}$ that are not reachable from $(b(t), t)$ by a direct motion due to $\mathcal{CTB}_i$. Let $U_i(t)$ be the projection of $S_i(t)$ on the time axis of $\mathcal{CT}$. Let $F_{b,b'}$ denote the set of all pairs of times (t, t') such that $(b'(t'), t')$ can be reached from $(b(t), t)$ by a single direct motion. From the above discussion, we have:

$$F_{b,b'} = \{(t, t') \in [0, +\infty) \times [0, +\infty) \;/\; t' \in L(t) \setminus \bigcup_{i=1}^{q} U_i(t)\}.$$

It is a two-dimensional region and its intersection by a line at constant t consists of $q+1$ intervals at most. Suppose it has been computed. Then:

$$DM_{b,b'}(I) = \{t' \in [0, +\infty) \;/\; \exists t \in I \text{ such that } (t, t') \in F_{b,b'}\}.$$

We now sketch the computation of $F_{b,b'}$. In the space $\{(t, t') \in [0, +\infty) \times [0, +\infty)\}$, the endpoints of the interval $L(t)$ trace an algebraic curve of maximal degree 2, whose equation is obtained by intersecting the fixed line $\overline{b'}$ with the surface of the cone spanned by the CT-paths originating at $b(t)$. This curve bounds a domain D which includes $F_{b,b'}$. Now for each C-obstacle vertex b'', we compute the time coordinate $e_{b''}(t)$ of the intersection[4] of the plane containing both $(b(t), t)$ and $\overline{b''}$ with the line $\overline{b'}$. The curve $t' = e_{b''}(t)$ is an algebraic curve of maximal degree 2. We can compute the intersections of the various curves $t' = e_{b''}(t)$ inside D, determine the regions bounded by these curves and the boundary of D, and label each of these regions according to whether it is contained in $F_{b,b'}$ or not. (A sweep-line technique can perform these computations and that of $DM_{b,b'}(I)$ concurrently.)

Let n be the total number of vertices in the C-obstacles. There are $O(n^2)$ intersections among the curves $t' = e_{b''}(t)$, and between these curves and the boundary of D. Therefore, there are at most $O(n^2)$ regions in D, and $F_{b,b'}$ can be computed in polynomial time. It follows that the computation of every set $I_b^{(k)}$ also takes time polynomial in n, but exponential in k (since $I_b^{(k)}$ contains $O(q^k)$ disjoint intervals in the worst case). Hence, in the worst case where $I_{\mathbf{q}_{goal}}^{(q+1)}$ has to be computed, the overall algorithm requires time polynomial in n and exponential in q. If q is fixed, the algorithm is polynomial.

1.4.2 Approximate Cell Decomposition

In Subsection 1.3.2 we described a modification of the approximate cell decomposition approach that solves the dynamic motion planning prob-

[4] Pathological cases can be avoided by imposing that all obstacles move at different non-zero velocities.

lem with no velocity constraint. This approach can be adapted further in order to deal with bounded velocity.

The principle of the adaptation is the following[5]. Concurrent with the generation of an EMPTY (or MIXED) channel, the planner constructs a time-monotone CT-path contained in this channel that satisfies the velocity constraint. One way to construct this path is to discretize each rectangloid cell in the channel into a grid of configuration-times and search the composite grid contained in the channel generated so far. Hence, the planner concurrently searches (1) the cell connectivity graph for a channel, and (2) the grid contained in the channel generated so far for a CT-path satisfying the velocity constraint. Whenever the planner considers the inclusion of a new cell at the end of the current channel, it verifies that a path can be extended to this cell without violating the velocity constraint. If this is not the case, the cell is not included in the channel.

Due to the discretization in each cell, the planning method is not guaranteed to find a CT-path satisfying the velocity constraint in a channel, even if one exists. Hence, it may lead to reject a solution channel and fail to solve the planning problem at hand, although a solution exists. The method can be made more reliable by using a finer discretization of the cells, but this also increases the computational cost of the method.

1.4.3 Velocity Tuning

Velocity tuning is a two-phase approach to the dynamic motion planning problem. This approach, which was introduced in [Kant and Zucker, 1986a], consists of first planning a path for $\mathcal{A}$ from $\mathbf{q}_{init}$ to $\mathbf{q}_{goal}$ among the stationary obstacles lying in $\mathcal{W}$, and then tuning the velocity of $\mathcal{A}$ along the path in order to avoid collisions with the moving obstacles. Hence, the original problem is broken down into two subproblems. The first subproblem is a basic motion planning problem that can be solved using a method chosen among those presented in Chapters 4 through 7. The second subproblem is a *velocity planning* problem. As we will see below, it can be formulated as a dynamic motion planning problem in a path-time space similar to a two-dimensional configuration-time space.

[5]This adaptation is derived from [Fujimura and Samet, 1989], where an implemented planner is described in detail.

Let $\tau : l \in [0, L] \mapsto \tau(l) \in \mathcal{C}'_{free}$ be the path planned in the first phase. $\mathcal{C}'_{free}$ denotes the set of configurations where $\mathcal{A}$ has no intersection with the stationary obstacles. The parameter l measures the length of the path between $\tau(0)$ and $\tau(l)$ (in the metric of $\mathcal{C}$). Hence, L denotes the total length of the path. The velocity planning problem is to generate a continuous function $\lambda : t \in [0, T] \mapsto \lambda(t) \in [0, L]$, with $\lambda(0) = 0$ and $\lambda(T) = L$, such that the composed map $\tau \circ \lambda : t \in [0, T] \mapsto \tau(\lambda(t))$ verifies $(\tau(\lambda(t)), t) \in \mathcal{CT}_{free}$ for all $t \in [0, T]$. The map λ must also be such that $\mathcal{A}$'s velocity be bounded by V_{max}. It may not be bijective, so that $\mathcal{A}$ may traverse some configurations in τ more than once.

Let the lt-space $[0, L] \times [0, T]$ be denoted by $\mathcal{PT}$. $\mathcal{PT}$ is called the **path-time space**. For a given path τ of $\mathcal{A}$ in $\mathcal{C}$, a moving obstacle $\mathcal{B}(t)$ maps in this space to a region $\mathcal{PTB}$, called a **PT-obstacle**, defined as follows:

$$\mathcal{PTB} = \{(l, t) \in [0, L] \times [0, T] \;/\; \tau(l) \in \mathcal{CB}(t)\}.$$

Let:

$$\mathcal{PT}_{free} = [0, L] \times [0, T] \backslash \bigcup_i \mathcal{PTB}_i$$

where the $\mathcal{PTB}_i$'s are the PT-obstacles corresponding to the moving obstacles in $\mathcal{W}$. The velocity planning problem is thus reduced to finding a strictly time-monotone continuous path in $\mathcal{PT}_{free}$, called a **free PT-path**, between $(0, 0)$ and (L, T). Indeed, this free PT-path determines a unique function $\lambda(t)$ such that $\forall t \in [0, T]$, $(\tau(\lambda(t)), t) \in \mathcal{CT}_{free}$. The slope of the PT-path with respect to the time axis must also satisfy the velocity constraint, i.e. $|\frac{dl}{dt}| \leq V_{max}$. (Hence, the velocity tuning approach corresponds to constraining the solution CT-path to lie in a two-dimensional surface swept out in $\mathcal{CT}$ by a line parallel to the time axis and following the path τ generated in $\mathcal{C}$. $\mathcal{PT}$ is the space obtained by "flattening" this surface on a plane.)

The PT-obstacles may have complicated shapes. A connected C-obstacle $\mathcal{CB}$ can even map in $\mathcal{PT}$ to a region made of several connected components. In the following, we will assume that the PT-obstacles are polygons. This representation is exact when the path τ is a sequence of straight segments in $\mathcal{C} = \mathbf{R}^2$ or $\mathbf{R}^3$ and the moving C-obstacles are polygons/polyhedra translating at constant linear velocities without rotation (asteroid avoidance problem). In other cases, it is obtained by approximating the PT-obstacles.

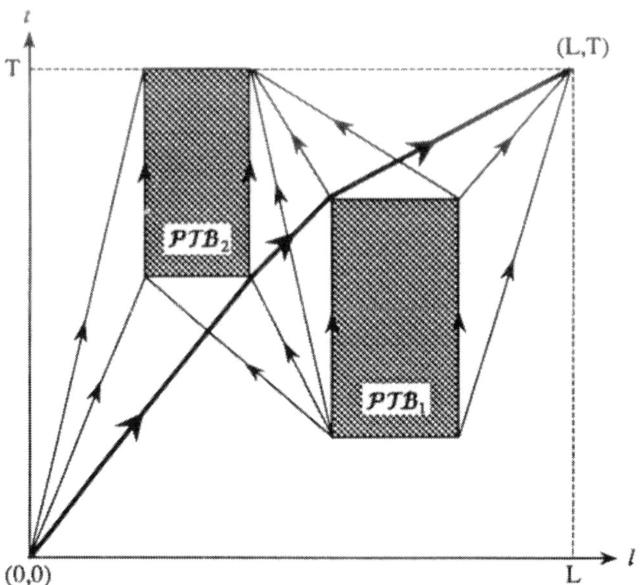

Figure 3. This figure displays a simple polygonal path-time space with two rectangular PT-obstacles and the corresponding search graph. This graph is a refinement of the reduced visibility graph that incorporates the irreversible nature of time and the upper-bound constraint on the robot's velocity modulus. The path connecting $(0,0)$ to (L,T) is shown in bold line.

With polygonal PT-obstacles, the velocity planning problem can easily be solved by using a variant of the visibility graph method presented in Section 1 of Chapter 4. The variant consists of searching a refinement of the reduced visibility graph. The new graph — let us denote it by G — is an oriented graph defined as follows. The nodes of G are the convex vertices of the PT-obstacles and the two points $(0,0)$ and (L,T). An arc of G connects the node (l,t) to (l',t') if and only if:

- $t < t'$, $|\frac{l'-l}{t'-t}| \leq V_{max}$, and
- either the line segment joining the two points is a PT-obstacle's edge, or it lies entirely in $\mathcal{PT}_{free}$, except possibly at each of the two endpoints, where it is tangent to a PT-obstacle.

One can show that there exists a semi-free PT-path connecting $(0,0)$ to (L,T) in $cl(\mathcal{PT}_{free})$ if and only if there exists a path in G between the two corresponding nodes (see proof of Proposition 1 in Subsection 1.4.1). If a path in G entails a contact with some obstacles, an arbitrarily small perturbation makes it into a free PT-path (which may slightly violate the velocity constraint). Figure 3 shows a simple path-time space and the corresponding graph G. If $(0,0)$ and (L,T) are in the same connected component of G, the search of G allows us to construct a solution PT-path (shown in a bold line in Figure 3). Otherwise, the search terminates with failure.

The above method can be easily extended to the case where the arrival time T is unspecified.

The velocity tuning approach is inherently incomplete. Given a path of $\mathcal{A}$ that avoids collisions with the stationary obstacles, there may exist no velocity profile along that path that allows $\mathcal{A}$ to avoid collisions with the moving obstacles. This happens, for example, when an obstacle comes at the robot on approximately the same geometric path (for instance, in a narrow corridor). Perhaps another path might have made it possible to avoid collision with both the stationary and the moving obstacles. Whenever the velocity planning phase fails, the planner may try to generate another path among the stationary obstacles. But the approach does not provide a systematic way of choosing this other path.

The velocity tuning approach, which can nevertheless solve many practical problems, is substantially more time-efficient than systematic methods. It is polynomial (rather than exponential) in the number of C-obstacles and the total number of their vertices. In addition, it substitutes a twofold search in an m-dimensional configuration space and in a two-dimensional path-time space for a search in an $(m+1)$-dimensional configuration-time space.

2 Multiple Robots

In this section we consider a collection of p robots, $\mathcal{A}_i$, $i = 1, ..., p$, moving in the same workspace $\mathcal{W} = \mathbf{R}^N$, $N = 2$ or 3, among stationary obstacles $\mathcal{B}_j$, $j = 1, ..., q$. The robots move independently of each other, but they cannot occupy the same space at the same time. The problem is to plan a collision-free motion of the robots from an initial configuration to a goal configuration.

2.1 Composite Configuration Space

Let $\mathcal{C}_1$ through $\mathcal{C}_p$ denote the individual configuration spaces of the p robots $\mathcal{A}_1$ through $\mathcal{A}_p$. Each space $\mathcal{C}_i$, $i \in [1,p]$, contains two sorts of C-obstacles:

- C-obstacles corresponding to the collision of $\mathcal{A}_i$ with the stationary obstacles $\mathcal{B}_1$ through $\mathcal{B}_q$,

- C-obstacles corresponding to the collision of $\mathcal{A}_i$ with the other robots.

C-obstacles of the second sort dependent on the motion of the other robots and cannot be represented as fixed regions in $\mathcal{C}_i$.

The most direct way to extend the notion of configuration space to multiple robots is to consider the objects $\mathcal{A}_1$ through $\mathcal{A}_p$ as the components of the same composite robot $\mathcal{A} = \{\mathcal{A}_1, ..., \mathcal{A}_p\}$. Let $\mathcal{F}_{\mathcal{A}_i}$ be a Cartesian frame attached to $\mathcal{A}_i$. A **configuration q** of $\mathcal{A}$ is a specification of the position and orientation of every frame $\mathcal{F}_{\mathcal{A}_i}$ ($i = 1$ to p) with respect to the workspace frame $\mathcal{F}_{\mathcal{W}}$. Since the robots move independently of each other, a configuration of $\mathcal{A}$ is of the form $\mathbf{q} = (\mathbf{q}_1, ..., \mathbf{q}_p)$, where $\mathbf{q}_i \in \mathcal{C}_i$ denotes a configuration of $\mathcal{A}_i$, and we have: $\mathcal{A}(\mathbf{q}) = \mathcal{A}_1(\mathbf{q}_1) \cup ... \cup \mathcal{A}_p(\mathbf{q}_p)$. $\mathcal{A}$'s configuration space, $\mathcal{C} = \mathcal{C}_1 \times ... \times \mathcal{C}_p$, is called the **composite configuration space** of $\mathcal{A}_1$ through $\mathcal{A}_p$. It is a smooth manifold of dimension $m = \sum_{i=1}^{p} m_i$, where m_i denotes the dimension of $\mathcal{C}_i$.

There are still two sorts of C-obstacles in $\mathcal{C}$, but they are now both time independent:

- The C-obstacle due to the interaction of $\mathcal{A}_i$ with the obstacle $\mathcal{B}_j$ is denoted by $\mathcal{CB}_{ij}$ and defined by:

$$\mathcal{CB}_{ij} = \{\mathbf{q} = (\mathbf{q}_1, ..., \mathbf{q}_i, ..., \mathbf{q}_p) \in \mathcal{C} \ / \ \mathcal{A}_i(\mathbf{q}_i) \cap \mathcal{B}_j \neq \emptyset\}.$$

- The C-obstacle due to the interaction of $\mathcal{A}_i$ and $\mathcal{A}_j$, $i < j$, is denoted by $\mathcal{CA}_{ij}$ and defined by:

$$\mathcal{CA}_{ij} = \{\mathbf{q} = (\mathbf{q}_1, ..., \mathbf{q}_i, ..., \mathbf{q}_j, ..., \mathbf{q}_p) \in \mathcal{C} \ / \ \mathcal{A}(\mathbf{q}_i) \cap \mathcal{A}(\mathbf{q}_j) \neq \emptyset\}.$$

The computation of the $\mathcal{CA}_{ij}$'s can be done by a slight adaptation of the methods presented in Chapter 3.

The **free space** in $\mathcal{C}$ is defined by:

$$\mathcal{C}_{free} = \mathcal{C} \setminus \left(\bigcup_{i\in[1,p],j\in[1,q]} \mathcal{CB}_{ij} \right) \cup \left(\bigcup_{i\in[1,p],j\in[i+1,p]} \mathcal{CA}_{ij} \right).$$

A **free path** between two configurations $\mathbf{q}_{init}$ and $\mathbf{q}_{goal}$ is a continuous map $\tau : [0,1] \rightarrow \mathcal{C}_{free}$, with $\tau(0) = \mathbf{q}_{init}$ and $\tau(1) = \mathbf{q}_{goal}$.

Hence, the problem of planning a free path for multiple robots moving in the same workspace is formulated as the problem of planning a free path τ for a single composite robot. The projection of τ on every space $\mathcal{C}_i$, $i \in [1,p]$, is the corresponding path of $\mathcal{A}_i$. Although time does not appear explicitly, the execution of τ requires that the individual robots follow their own paths in a coordinated fashion. Changing the velocity of one robot requires the velocities of the other robots to be changed in the same ratio.

2.2 Centralized Planning

We call *centralized planning* the approach which consists of planning the coordinated paths of multiple robots as a path in their composite configuration space. This approach allows us to use general path planning methods described in Chapters 4 through 7.

As an illustration, Figure 4 shows several configurations along a path for two rectangular robots in a workspace with three narrow corridors. The two robots cannot pass each other in the same corridor. The planning problem is essentially that of interchanging the two robots in the central corridor. The path was generated using the randomized potential-field-based method (RPP) described in Section 5 of Chapter 7 in the six-dimensional composite configuration space of the two robots.

Several specific methods based on the approaches described in Chapters 4 through 7 have also been developed to solve particular multi-robot motion planning problems. For instance, Schwartz and Sharir [Schwartz and Sharir, 1983c] presented an exact cell decomposition method for planning the motion of two discs in the plane among polygonal obstacles. The method is based on the definition of critical curves (in the individual two-dimensional configuration spaces of the discs) where the possible contacts among the discs and the obstacles change qualitatively.

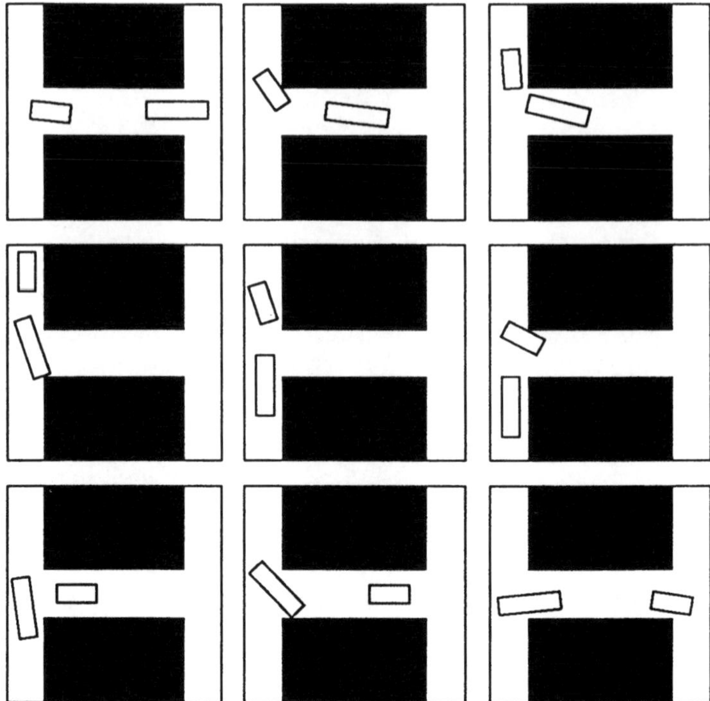

Figure 4. The planning problem is to interchange two robots in a workspace made of narrow corridors. The two robots cannot pass each other in the same corridor. The figure shows several snapshots along a path planned using a centralized planning approach with the randomized potential-field-based method described in Section 5 of Chapter 7. Both robots have to move away to intermediate configurations before actually moving to their goal configurations.

The method runs in time $O(n^3)$, where n is the number of obstacle edges. The same method can be extended to more than two discs, but its complexity rapidly increases with the number of discs. An $O(n^2)$ exact cell decomposition algorithm for two discs was also presented in [Sharir and Sifrony, 1988].

Barraquand, Langlois and Latombe [Barraquand, Langlois and Latombe, 1989b] described a potential field method for planning the motion of many discs in an environment including narrow corridors in which two discs cannot pass each other. The method concurrently builds and searches a local-minima graph of the potential function defined over

$\mathcal{C} = \mathbf{R}^{2p}$, where p is the number of discs. The specificity of the method rests on the technique used to escape local minima. At each encountered local minimum the method simulates a series of *constrained motions.* Each constrained motion is generated by forcing one configuration coordinate to increase or decrease until a saddle point of the potential function is attained. During the constrained motion, the other $m - 1$ coordinates are selected along the negative slope of the gradient of the potential. As an illustration, consider the case where a gradient motion leads two discs to move in opposite directions toward each other, in the same narrow corridor. At some point, the two discs cannot progress further toward their respective goals without colliding with each other. Then, a local minimum of the potential function has been attained by the gradient motion. A constrained motion corresponds to one of the discs moving backward, with the other disc still proceeding forward along the negated gradient until there is sufficient free room for the two discs to pass each other (saddle point of the potential). This planning method interweaving gradient and constrained motions was successfully experimented with up to 10 discs.

Tournassoud [Tournassoud, 1986] proposed a variant of the potential field approach where motion coordination is expressed as a local optimization problem. The basic idea is to force the moving objects to slide along tangent separating hyperplanes.

The practical difficulty with centralized planning is that it yields methods whose time complexity is exponential in the dimension of the composite configuration space, i.e. $m = \sum_{i=1}^{p} m_i$. Indeed, as mentioned in Section 6 of Chapter 1, path planning in an m-dimensional configuration space is PSPACE-hard when m is not bounded. Even the apparently simpler problem of coordinating the motion of arbitrarily many rectangular objects in an empty rectangular workspace is PSPACE-complete.

2.3 Decoupled Planning

One way to reduce the computational complexity of the multi-robot path planning problem is to first plan a path for each robot more or less independently of the other robots and then consider the interactions among the paths. We call such an approach *decoupled planning.* Roughly speaking, while centralized planning is exponential in $m = \sum_{i=1}^{p} m_i$, decoupled planning is exponential in $\max_{i\in[1,p]}\{m_i\}$. However, this gain results in a loss of completeness. A decoupled planning approach may fail to plan a path, while a centralized planning approach using the same sort of path planning method would have succeeded (perhaps at the expense of a long computation).

Below, we describe two decoupled planning approaches, which we call *prioritized planning* and *path coordination.*

2.3.1 Prioritized Planning

Prioritized planning consists of considering the motions of the $\mathcal{A}_i$'s, one robot at a time, say from $i = 1$ to $i = p$ [Erdmann and Lozano-Pérez, 1986]. At each iteration, a path of $\mathcal{A}_i$ is generated in the configuration-time space $\mathcal{CT}_i$ of $\mathcal{A}_i$, avoiding collisions with both the obstacles $\mathcal{B}_j$, $j = 1, ..., q$, and the robots $\mathcal{A}_1, ..., \mathcal{A}_{i-1}$. Hence, the motion of $\mathcal{A}_i$ is planned as if the robot was moving among stationary obstacles and $i - 1$ moving obstacles. The path of $\mathcal{A}_1$ is planned in $\mathcal{C}_1$, rather than in $\mathcal{CT}_1$, and an arbitrary velocity is selected in order to map $\mathcal{A}_1$ into the configuration-time spaces of the other robots. This velocity merely determines a scale factor along the time axes of these spaces. Multiplying the velocity of $\mathcal{A}_1$ by a constant factor results in multiplying the other robots' velocities by the same factor. The path of $\mathcal{A}_1$ can be generated using an available path planning method. The motion of every $\mathcal{A}_i$, $i > 1$, can be generated using a planning method described in Section 1.

The ordering on the robots requires a priority to be assigned to each robot. Priorities may be assigned randomly. However, a better way is to infer priorities from motion constraints. For instance, a simple rule could be to plan the motions of the bigger robots before those of the smaller ones. Buckley [Buckley, 1989a] proposed to assign priorities by attempting to maximize the number of robots which can move in straight line from their initial configuration to their goal configuration (in Buckley's application the robots are translating, two-dimensional objects).

Note that the prioritized planning approach would have great difficulty in solving the planning problem of Figure 4. Indeed, the path of the first robot (say, the longest one) is likely to be the most direct one, i.e. along this path, the robot stays in the central corridor. This makes it impossible to find a path for the second robot, since the corridor is too narrow for the two robots to pass each other. The planner may try to generate another path for the first robot, but the prioritizing planning approach offers no systematic way of choosing it. In addition, by allowing backtracking, the time complexity of the prioritized planning approach is substantially augmented. Thus, when the approach fails, rather than backtracking, it might be more suitable to call another algorithm based on the centralized planning approach, or to give up.

2.3.2 Path Coordination

Path coordination is a decoupled planning approach proposed in [O'Donnell and Lozano-Pérez, 1989]. It is based on a scheduling technique for dealing with limited resources which was originally developed for concurrent access to a database by several users [Yannakakis, Papadimitriou and Kung, 1979]. Here, space is the shared resource. The approach is applicable when the planning problem involves only two robots $\mathcal{A}_1$ and $\mathcal{A}_2$. It consists of first generating a free path for each of the two robots, independently of the other robot, and then coordinating the two paths so that the robots do not collide with each other:

Let:

$$\tau_1 : s_1 \in [0,1] \mapsto \tau_1(s_1) \in \mathcal{C}_{1,free}$$

$$\tau_2 : s_2 \in [0,1] \mapsto \tau_2(s_2) \in \mathcal{C}_{2,free}$$

be the two free paths generated for $\mathcal{A}_1$ and $\mathcal{A}_2$. $\mathcal{C}_{1,free}$ (resp. $\mathcal{C}_{2,free}$) denotes the free space in the configuration space of $\mathcal{A}_1$ (resp. $\mathcal{A}_2$) obtained by ignoring the other robot. Consider the s_1s_2-space $[0,1]\times[0,1]$. Any continuous path in this space connecting $(0,0)$ to $(1,1)$ is called a **schedule** and determines a coordinated execution of the two paths. For instance, the schedule that first connects $(0,0)$ to $(1,0)$ by a straight segment and then $(1,0)$ to $(1,1)$ by another straight segment corresponds to the case where $\mathcal{A}_1$ moves first and completely executes the path τ_1 before $\mathcal{A}_2$ starts moving along τ_2. If s_1 (resp. s_2) is proportional to the length of the subpath from $\tau_1(0)$ to $\tau_1(s_1)$ (resp. from $\tau_2(0)$ to $\tau_2(s_2)$), then the schedule that connects $(0,0)$ to $(1,1)$ by a single straight segment — a diagonal of the rectangle $[0,1]\times[0,1]$ — corresponds to the case where the two robots move simultaneously, with a constant ratio of their velocity moduli.

Let us define the obstacle region in the s_1s_2-space as the set of pairs (s_1,s_2) such that $\mathcal{A}(\tau_1(s_1)) \cap \mathcal{A}_2(\tau_2(s_2)) \neq \emptyset$. The second phase of the path coordination approach is to construct a **free schedule**, i.e. a schedule that does not traverse this obstacle region. Hence, the path coordination approach corresponds to constraining the path of $\mathcal{A} = \{\mathcal{A}_1, \mathcal{A}_2\}$ in the composite configuration space $\mathcal{C} = \mathcal{C}_1 \times \mathcal{C}_2$ to lie in the intersection of two subspaces whose projections into $\mathcal{C}_1$ and $\mathcal{C}_2$ are τ_1 and τ_2, respectively. The $s_1 s_2$-space is obtained by "flattening" this intersection, which has dimension 2, on a plane.

Notice that the path coordination approach puts an even stronger constraint on the search of a solution than the prioritizing planning approach. Therefore, it may fail more often. When the path coordination approach fails, it seems appropriate to execute an algorithm based on the prioritizing planning approach. This algorithm may use the path τ_1 already generated and plan a CT-path of $\mathcal{A}_2$ in $\mathcal{CT}_2$, treating $\mathcal{A}_1$ as a moving obstacle. If this algorithm fails again, a centralized planning algorithm may then be executed.

In general, the obstacle region in the $s_1 s_2$-space consists of several connected subsets whose shapes are arbitrarily complicated. Let us subdivide each path τ_i, $i = 1$ or 2, into a sequence of w_i path segments determined by the intervals $\delta_{i,k_i} = [s_{i,k_i}, s_{i,k_i+1}]$, with $k_i = 0, \ldots, w_i - 1$, $s_{i,0} = 0$ and $s_{i,w_i} = 1$. Typically, all the intervals $\delta_{i,k}$ along the same path τ_i determine subpaths of τ_i of equal length. This subdivision transforms the continuous $s_1 s_2$-space into an array of $w_1 \times w_2$ closed rectangular cells $\delta_{1,k_1} \times \delta_{2,k_2}$. In the following, we denote by κ_{k_1,k_2} the cell $\delta_{1,k_1} \times \delta_{2,k_2}$. Each cell κ_{k_1,k_2} is classified as EMPTY if the regions swept out by $\mathcal{A}_1$ and $\mathcal{A}_2$ when (s_1, s_2) varies in $\delta_{1,k_1} \times \delta_{2,k_2}$ do not intersect, i.e. if:

$$\{\mathcal{A}_1(\tau_1(s_1)) \ / \ s_1 \in \delta_{1,k_1}\} \cap \{\mathcal{A}_2(\tau_2(s_2)) \ / \ s_2 \in \delta_{2,k_2}\} = \emptyset.$$

The cell is classified as FULL otherwise. Hence, the two robots are guaranteed to not collide in an EMPTY cell (even in its boundary), while they may collide in a FULL cell. The classification of the cells may be more or less complicated depending on the shape of the two robots, but it is often reasonable to approximate the shape of the area swept out by a robot when its configuration varies in a path segment.

The discretized $s_1 s_2$-space with labeled cells is called a **coordination diagram**. Figure 5 shows an example of such a diagram, with EMPTY cells shown white. Let us restrict the possible paths in the coordination diagram to the network defined as follows. The nodes of the network

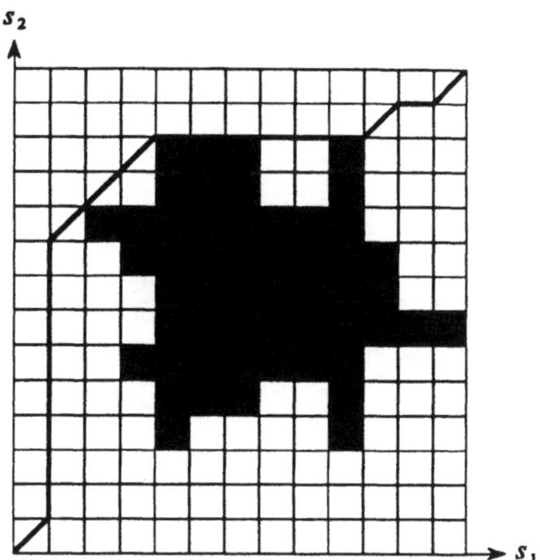

Figure 5. This figure shows the coordination diagram of two robots. Each rectangular cell determines segments of configurations of the two robots along two paths τ_1 and τ_2. Each cell is labeled as EMPTY (white cells) if the robots intersect with each other at none of these configurations; otherwise, it is labeled FULL (dark cells). The bold line shows a free schedule that is generated by searching the network formed by the edges bounding the EMPTY cells and their diagonals. A free schedule determines a class of coordinated motions along which the two robots do not collide with each other.

are all the vertices of the EMPTY cells. Its links are all the edges and the diagonals of these cells. We denote each node by its discretized coordinates (i, j), with $i \in [0, w_1]$ and $j \in [0, w_2]$. A free schedule can be generated by searching this network for a path connecting $(0, 0)$ to (w_1, w_2). The bold line in Figure 5 shows such a possible free schedule. (Remember that the boundary of an EMPTY cell is collision-free.)

We now describe a technique that allows us to generate a free schedule without search [O'Donnell and Lozano-Pérez, 1989]. This technique can only generate schedules that are nondecreasing with respect to both s_1 and s_2. It succeeds whenever one such free schedule exists. Prior to the description, we introduce the notion of SW-closure [Preparata and Shamos, 1985] [Yannakakis, Papadimitriou and Kung, 1979].

Let (s'_1, s'_2) and (s''_1, s''_2) be two points in the s_1s_2-plane. They are **incomparable** if and only if $(s'_1 - s''_1)(s'_2 - s''_2) < 0$. Let us assume that

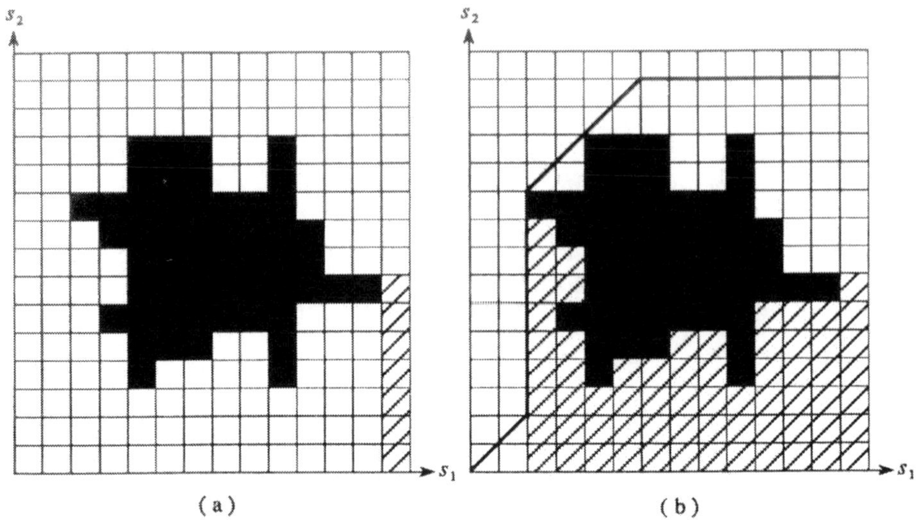

Figure 6. The extended coordination diagram is obtained by adding a fictitious path segment at the end of each of the two paths τ_1 and τ_2 (Figure a). Along this path segment the corresponding robot is motionless. The obstacle region is extended as shown in Figure a (striped cells). The SW-closure of the extended obstacle region is shown in Figure b (dark and striped cells). If there exists a nondecreasing free schedule, it only traverses cells that are not contained in the SW-closure of the extended obstacle region. The bold line in Figure b shows the free schedule constructed by the algorithm given in the text.

(s'_1, s'_2) and (s''_1, s''_2) are incomparable and (without loss of generality) that $s'_1 < s''_1$ (hence: $s'_2 > s''_2$). The **SW-conjugate** of (s'_1, s'_2) and (s''_1, s''_2) is the point (s'_1, s''_2). A connected region R of the s_1s_2-space is **SW-closed** if and only if the SW-conjugate of any two incomparable points in R is also in R. The **SW-closure** of a connected region S is the smallest SW-closed region R containing S. The SW-closure of a non-connected region is defined as the union of the SW-closures of all its connected components.

Now let us consider the coordination diagram. Let S be the obstacle region. We extend the diagram by adding a row and a column of cells at the top and the right of the diagram, respectively. This extension corresponds to adding a fictitious path segment at the end of each of the two paths τ_1 and τ_2. Along this path segment the corresponding robot is motionless. We extend the region S as follows:

- If for any $j \in [0, w_2 - 1]$ the cell $\kappa_{w_1-1,j}$ is FULL, then all the cells $\kappa_{w_1,k}$, for $k = 0, ..., j$, are included in S.
- If for any $i \in [0, w_1 - 1]$ the cell κ_{i,w_2-1} is FULL, then all the cells κ_{k,w_2}, for $k = 0, ..., i$, are included in S.

Figure 6.a shows the extended diagram corresponding to the coordination diagram of Figure 5.

Let R be the SW-closure of S, as illustrated in Figure 6.b. It can be computed in time $O(w_1 w_2(\log w_1 + \log w_2))$ [Preparata and Shamos, 1985]. None of the nondecreasing free schedules in the coordination diagram traverse cells contained in R. In addition, a nondecreasing free schedule exists if and only if neither of the two cells $\kappa_{0,0}$ and κ_{w_1-1,w_2-1} are contained in R. More generally, a nondecreasing free schedule exists between (i,j), $0 \leq i \leq w_1$, $0 \leq j \leq w_2$, and (ω_1, ω_2) if and only if neither of the two cells $\kappa_{i,j}$ and κ_{w_1-1,w_2-1} are contained in R.

Therefore, if a nondecreasing free schedule exists, one such schedule is constructed (without searching any graph) by the following algorithm:

```
i ← 0; j ← 0;
while i < w1 or j < w2 do
    begin
        if κ(i,j) ∉ R then
            if i < w1 and j < w2 then
                    i ← i + 1; j ← j + 1; move to (i, j);
            else
                if i < w1 then
                        i ← i + 1; move to (i, j);
                else
                        j ← j + 1; move to (i, j);
        else
            if i < w1 and κ(i,j-1) ∉ R then
                    i ← i + 1; move to (i, j);
            else
                    j ← j + 1; move to (i, j);
    end;
```

A free schedule constructed by this algorithm is shown as a bold line in Figure 6.b.

3 Articulated Robots

In this section we consider the case where the robot is made of several moving rigid objects connected by mechanical joints which constrain their relative motion. A manipulator arm is a typical example of such a robot. It consists of a sequence of rigid objects connected in a chain by joints.

3.1 Configuration Space

Consider a robot $\mathcal{A}$ made of p rigid objects $\mathcal{A}_1, ..., \mathcal{A}_p$. Any two objects $\mathcal{A}_i$ and $\mathcal{A}_j$ may possibly be connected by a joint. For simplification, we will assume that the joint is either a **revolute joint** or a **prismatic joint**. A revolute joint is a hinge that constrains the relative motion of $\mathcal{A}_i$ and $\mathcal{A}_j$ to be a rotation around an axis fixed with respect to both objects. A prismatic joint is a sliding connection that constrains the relative motion to be a translation along an axis fixed with respect to both objects. These are the most usual joints, but other types of joint are possible. Many of them would be treated in much the same way as we will treat the revolute and prismatic joints.

Let $\mathcal{F}_{\mathcal{A}_i}$ be the frame attached to $\mathcal{A}_i$, $i \in [1,p]$. As defined in Section 2, the configuration of $\mathcal{A}$ is a specification of the position and orientation of every frame $\mathcal{F}_{\mathcal{A}_i}$, $i = 1$ to p, with respect to $\mathcal{F}_{\mathcal{W}}$. If the various objects were independent free-flying objects in $\mathcal{W} = \mathbf{R}^N$, the configuration space of $\mathcal{A}$ would be:

$$\mathcal{C}' = \underbrace{(\mathbf{R}^N \times SO(N)) \times ... \times (\mathbf{R}^N \times SO(N))}_{p}.$$

However, the various joints impose constraints[6] on the feasible configurations in $\mathcal{C}'$. These constraints select out a subspace $\mathcal{C}$ of $\mathcal{C}'$ of lower dimension which is the actual configuration space of $\mathcal{A}$.

[6]In the next chapter, each such constraint will be called a *holonomic* kinematic constraint.

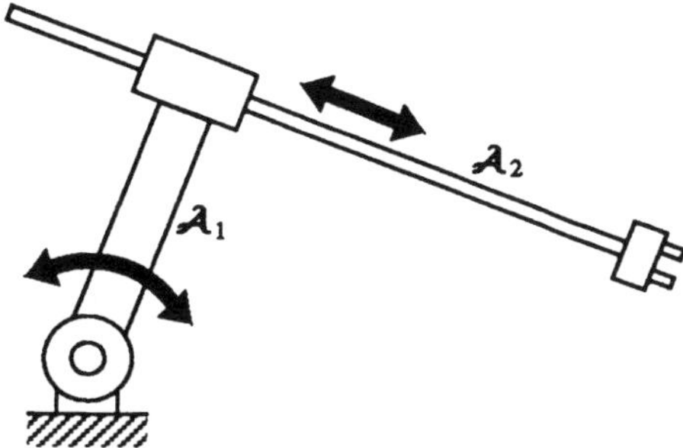

Figure 7. This planar manipulator arm $\mathcal{A}$ consists of two objects $\mathcal{A}_1$ and $\mathcal{A}_2$. $\mathcal{A}_1$ is connected to the workspace by a revolute joint, and $\mathcal{A}_2$ to $\mathcal{A}_1$ by a prismatic joint. The configuration space of $\mathcal{A}$ is a two-dimensional subspace of $(\mathbf{R}^2 \times SO(2))^2$.

Example: Consider the planar two-joint manipulator arm shown in Figure 7. It consists of two objects $\mathcal{A}_1$ and $\mathcal{A}_2$. $\mathcal{A}_1$ is connected to the workspace by a revolute joint; $\mathcal{A}_2$ is connected to $\mathcal{A}_1$ by a prismatic joint. The workspace is $\mathcal{W} = \mathbf{R}^2$. In this example, $\mathcal{C}' = (\mathbf{R}^2 \times SO(2))^2$ is a six-dimensional space. The first joint imposes two constraints expressing that the point about which $\mathcal{A}_1$ can rotate is fixed in the workspace. The second joint imposes two additional constraints on the position and orientation of $\mathcal{A}_2$ relative to $\mathcal{A}_1$. The four constraints are independent and determine a subspace $\mathcal{C}$ of dimension 2 in $\mathcal{C}'$. ∎

It is in general suitable to consider $\mathcal{C}$ independently of $\mathcal{C}'$, since its dimension is usually much smaller.

In the rest of this section, we assume that $\mathcal{A}$ is an articulated robot containing no kinematic loop, i.e. no closed sequence $\mathcal{A}'_1, ..., \mathcal{A}'_r, \mathcal{A}'_{r+1}$, with $\mathcal{A}'_{r+1} = \mathcal{A}'_1$ and $\mathcal{A}'_{i+1}$ being connected to $\mathcal{A}'_i$ by a joint, for every $i \in [1, r]$. Then the total number of joints in $\mathcal{A}$ is exactly p.

Under the previous assumption, let us consider the workspace as a fictitious object denoted $\mathcal{A}_0$. Every object $\mathcal{A}_j$ ($j \in [1,p]$) is connected to a *single* object $\mathcal{A}_i$, $i \in [0,p]$, $i \neq j$. Hence, the whole robot $\mathcal{A}$ can be represented by a tree structure, whose nodes are the objects $\mathcal{A}_0$ through $\mathcal{A}_p$ and whose arcs are the joints. The root of the tree is $\mathcal{A}_0$. For any two objects $\mathcal{A}_i$ and $\mathcal{A}_j$, such that $\mathcal{A}_j$ is a child of $\mathcal{A}_i$ in the tree, we can define the configuration of $\mathcal{A}_j$ relative to $\mathcal{A}_i$ as a specification of the

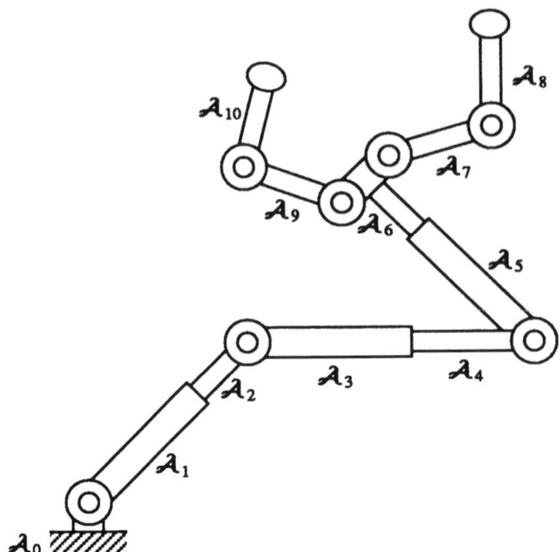

Figure 8. This figure depicts an articulated robot with three telescopic prismatic joints and seven revolute joints. Assuming no mechanical stops in the joints, the configuration space of this robot is $\mathcal{C} = (S^1)^7 \times \mathbf{R}^3$.

position and orientation of $\mathcal{F}_{\mathcal{A}_j}$ with respect to $\mathcal{F}_{\mathcal{A}_i}$. Let $\mathcal{C}_j^{(i)}$ denote the configuration space of $\mathcal{A}_j$ relative to $\mathcal{A}_i$. Assume for a moment that there is no mechanical stop in the joints. If the two objects are connected by a revolute joint, we have that: $\mathcal{C}_j^{(i)} = S^1$. If the two objects are connected by a prismatic joint, we have that: $\mathcal{C}_j^{(i)} = \mathbf{R}$. It follows that the configuration space $\mathcal{C}$ of an articulated object $\mathcal{A}$ with no loop, made of p rigid objects connected by p_1 revolute joints and p_2 prismatic joints ($p_1 + p_2 = p$), is [Burdick, 1988]:

$$\mathcal{C} = \underbrace{S^1 \times \ldots \times S^1}_{p_1} \times \underbrace{\mathbf{R} \times \ldots \times \mathbf{R}}_{p_2}.$$

Hence, the configuration space of the two-joint manipulator arm of Figure 7 is $S^1 \times \mathbf{R}$. In the same way, the configuration space of the ten-joint robot shown in Figure 8 is $(S^1)^7 \times \mathbf{R}^3$.

Therefore, the configuration space of an articulated robot with p revolute and prismatic joints and no kinematic loop is a p-dimensional smooth manifold. A chart of this manifold can be simply defined by associating

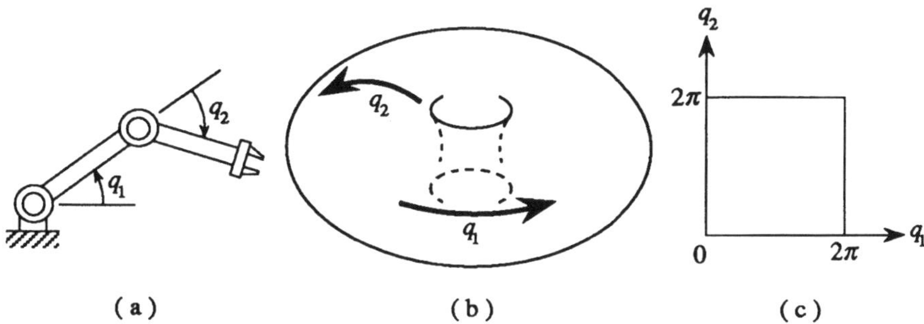

Figure 9. The configuration space of a planar two-revolute-joint manipulator arm (Figure a) is a torus in $\mathbf{R}^3$ (Figure b). A parameterization of this space is obtained by representing every configuration as $(q_1, q_2) \in [0, 2\pi) \times [0, 2\pi)$ (see Figure c), where q_1 and q_2 are angles associated with the joints as illustrated in Figure a. This parameterization corresponds to cutting the torus along two generators. Opposite edges of the square shown in Figure c must be procedurally identified through modulo 2π arithmetic in order to capture the multiple connectedness of the torus.

an angle, say in $(0, 2\pi)$, with each revolute joint and a real number with each prismatic joint. Then, a configuration $\mathbf{q}$ is represented by a list $(q_1, ..., q_p)$ of coordinates, each one describing the relative position and orientation of two joint-connected objects in $\mathcal{A}$. Robots with revolute joints have a multiply-connected configuration space and several charts are necessary to form an atlas. However, in practice, it is often sufficient to consider a single chart, with each angular coordinate belonging to, say, $[0, 2\pi)$, and applying modulo 2π arithmetic to the coordinate.

Example: Consider the planar two-revolute-joint manipulator arm of Figure 9.a. The configuration space of this robot is $S^1 \times S^1$, hence a torus in a three-dimensional Euclidean space (Figure 9.b). A chart on this manifold can be defined by associating an angle in $(0, 2\pi)$ with each of the two joints. Figure 9.a illustrates two such angles q_1 and q_2. The torus is multiply-connected and several charts are required to form an atlas. Nevertheless, in most cases, it is appropriate to consider a single chart and let the two angles q_1 and q_2 vary in $[0, 2\pi)$, with modulo 2π arithmetic. This simplification corresponds to representing the torus by a square $[0, 2\pi] \times [0, 2\pi]$ (see Figure 9.c) whose opposite edges are procedurally identified. ∎

The notions of **C-obstacle, free space** and **free path** defined for multiple objects extend to an articulated robot without modification. There are also two sorts of C-obstacles: those due to the interaction of an object $\mathcal{A}_i$ and an obstacle $\mathcal{B}_j$, and those due to the interaction between two objects $\mathcal{A}_i$ and $\mathcal{A}_j$. Figure 10 illustrates the construction of C-obstacles of the first sort for a two-revolute-joint planar manipulator. The computation of the exact equations of a C-obstacle boundary for an articulated robot has been recently investigated in [Hwang, 1990], [Dooley and McCarthy, 1990] and [Ge and McCarthy, 1990].

In general, the relative motion of two objects connected by a joint is further limited by a pair of *mechanical stops*, which restrict the range of values of the corresponding configuration coordinate to a connected interval. The mechanical stops thus act like "invisible obstacles". The set of configurations where no joint has reached a mechanical stop is an open subset of $\mathcal{C}$. The complement of this subset is the "C-obstacle region" to which the invisible obstacles due to the mechanical stops map. The mechanical stops in every joint are usually independent of the configuration of the robot. Then the subset of configurations that are left feasible by the mechanical stops is simply a cross-product of intervals in the Cartesian space (chart) representing $\mathcal{C}$.

Consider a revolute joint connecting $\mathcal{A}_i$ to $\mathcal{A}_j$. If the mechanical stops in this joint prevent $\mathcal{A}_j$ from making a full rotation of 2π with respect to $\mathcal{A}_i$, the limitation eliminates the need for applying modulo 2π arithmetic to the corresponding configuration coordinate. If, instead, the mechanical stops allow $\mathcal{A}_i$ to perform r ($r > 1$) full rotations[7], then it is appropriate to let the corresponding configuration coordinate vary in an interval of length $2r\pi$ (r does not have to be an integer). In this way, the configuration is explicitly represented with respect to both mechanical stops. One drawback of this representation, however, is that the same obstacle may map to several C-obstacle regions of identical shape.

3.2 Path Planning Methods

The configuration space $\mathcal{C}$ of an articulated robot is just another smooth manifold. As long as the constraints imposed by the joints on the relative

[7]This is often the case with some revolute joints located in the "wrist" of a manipulator arm.

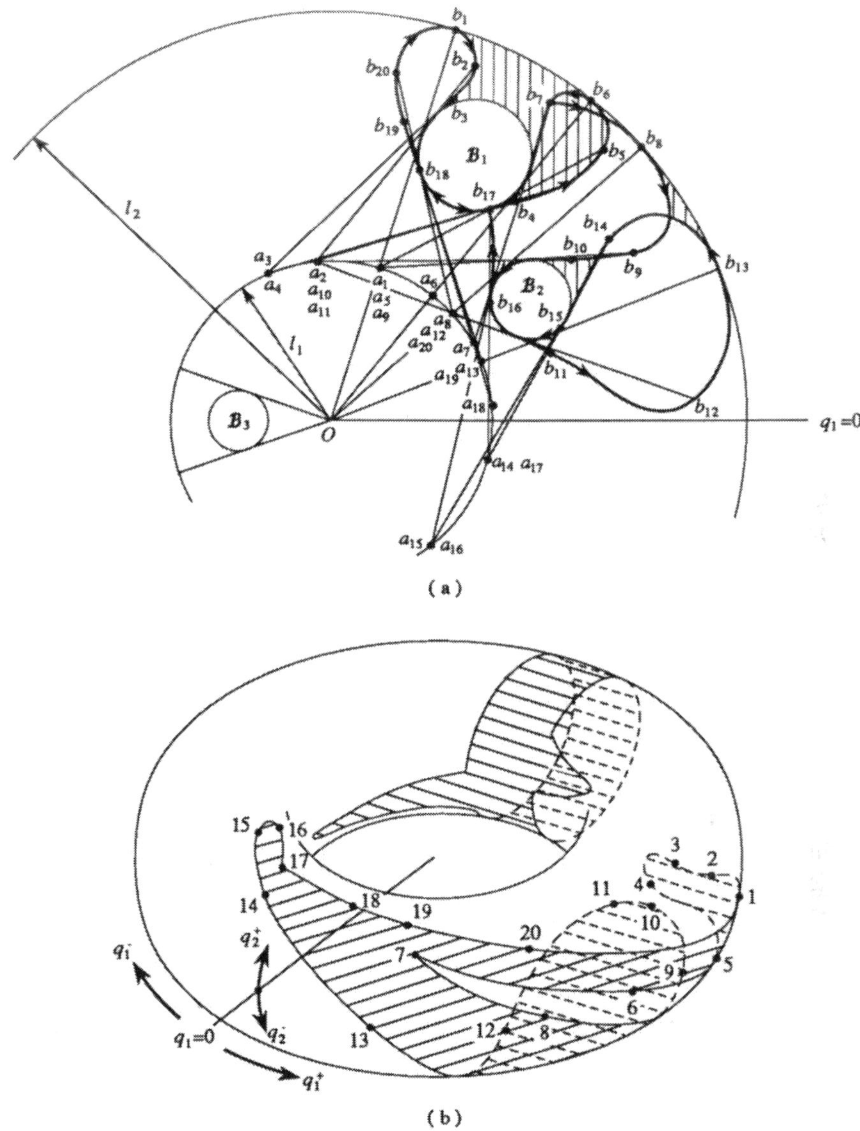

Figure 10. The robot is a two-revolute-joint planar manipulator moving among three circular obstacles $\mathcal{B}_1$, $\mathcal{B}_2$ and $\mathcal{B}_3$ (Figure a). Its two links are line segments of lengths l_1 and l_2. The first joint is located at O. The second moves in the circle of radius l_1 centered at O. The configuration space is the torus $S^1 \times S^1$ shown in Figure b. The striped areas in Figure b depict the C-obstacles. This construction is due to Lumelsky [Lumelsky, 1987b].

motion of the objects they connect can be expressed in algebraic form (which is the case with revolute and prismatic joints) the general methods described in Section 4 of Chapter 4 (silhouette method) and in Section 3 of Chapter 5 (Collins decomposition) are applicable. These methods take simple and double exponential time, respectively, in the dimension of the configuration space $\mathcal{C}$.

An extension of the freeway method presented in Section 3 of Chapter 4 is described in [Brooks, 1983b] for a specific manipulator arm with revolute joints only.

The approximate cell decomposition approach is also applicable, although the mapping of the obstacles in the configuration space is more difficult than for the basic motion planning problem with polygonal/polyhedral objects. We study this mapping in detail in the next subsection, in the case of a manipulator arm. Several path planners based upon the approximate cell decomposition approach have been implemented for manipulator arms (e.g. [Gouzènes, 1984b] [Laugier and Germain, 1985] [Faverjon, 1986] [Lozano-Pérez, 1987]). However, the approach is limited to robots with relatively few joints.

The potential field approach can be applied to articulated objects with minor adaptations. These adaptations are presented below in Subsection 3.4. So far, the best practical results for robots with many joints have been obtained with this approach. For instance, the randomized planner described in Section 5 of Chapter 7 has been successfully experimented with problems involving up to a few tens of joints in two- and three-dimensional workspace [Barraquand and Latombe, 1989a]. The learning technique introduced in Section 6 of Chapter 7 has been successfully applied to plan the coordinated motion of two six-joint manipulator arms [Faverjon and Tournassoud, 1989].

Because the motions of the rigid objects in an articulated robot are strongly interdependent, the "decoupled planning" techniques described in Subsection 2.3 do not apply.

3.3 Approximate Cell Decomposition

The principle of the approximate cell decomposition approach is general enough to be applicable to an articulated robot. The main issue is the construction of both the free space and the C-obstacle region in the Cartesian space representing $\mathcal{C}$. In Chapter 6 we used the explicit equations defining the boundaries of the C-obstacles to label the rectangloid cells generated by the decomposition of the configuration space. These equations were established in Chapter 3 for a polygonal/polyhedral object moving freely among polygonal/polyhedral obstacles. We could use a similar approach here, but obviously, the explicit equations of the boundaries of the C-obstacles are now considerably more difficult to establish and to exploit. Instead, we describe a method that generates a conservative approximation of the free space (i.e. a subset of the actual free space) by discretizing the motion of each joint into small intervals. This approximate representation is then used to construct a connectivity graph, which is searched for a channel connecting the initial configuration to the goal configuration. With minor variations, this is the method proposed in [Gouzènes, 1984b], [Laugier and Germain, 1985], [Faverjon, 1986] and [Lozano-Pérez, 1987]. Various implementations and experimental results of the method are presented in these publications.

In order to simplify our presentation, we make a few assumptions which are not difficult to remove. First, we assume that $\mathcal{A}$ is a sequence of p objects $\mathcal{A}_1$ through $\mathcal{A}_p$, with every two successive objects $\mathcal{A}_i$ and $\mathcal{A}_{i+1}$ being connected by either a revolute or a prismatic joint. $\mathcal{A}_1$ is the first object in the sequence and is connected to the workspace by a joint. As introduced in Subsection 3.1, we represent a configuration $\mathbf{q}$ of $\mathcal{A}$ by a list of p parameters, $(q_1, ..., q_p)$. Each parameter q_i ($i \in [1,p]$) determines the position and orientation of $\mathcal{F}_{\mathcal{A}_i}$ relative to $\mathcal{F}_{\mathcal{A}_{i-1}}$, with $\mathcal{F}_{\mathcal{A}_0} = \mathcal{F}_{\mathcal{W}}$. The second assumption is that each q_i varies in a bounded interval $I_i = (q_i^-, q_i^+)$. If q_i corresponds to a revolute joint with no mechanical stop, the interval is $I_i = [0, 2\pi)$ and modulo 2π arithmetic is used. The third assumption is that no two objects $\mathcal{A}_i$ and $\mathcal{A}_j$ can ever collide with each other. Hence, the only collisions we consider are those between the objects $\mathcal{A}_i$ ($i \in [1,p]$) and the obstacles $\mathcal{B}_j$ ($j \in [1,q]$).

The position and orientation of $\mathcal{F}_{\mathcal{A}_i}$ relative to $\mathcal{F}_{\mathcal{W}}$, for any $i \in [1,p]$, is completely determined by the parameters q_1 through q_i. Hence, we denote by $\mathcal{A}_i(q_1, ..., q_i)$ the region of $\mathcal{W}$ occupied by the object $\mathcal{A}_i$.

It is possible to construct an approximation of the C-obstacle region in $I_1 \times ... \times I_p \subset \mathbf{R}^p$ as follows. Each interval I_i, $i = 1$ to p, is partitioned into non-overlapping smaller intervals δ_{i,k_i}, $k_i = 1, 2, ...$, at some prespecified resolution. If the region swept out by $\mathcal{A}$ in the workspace,

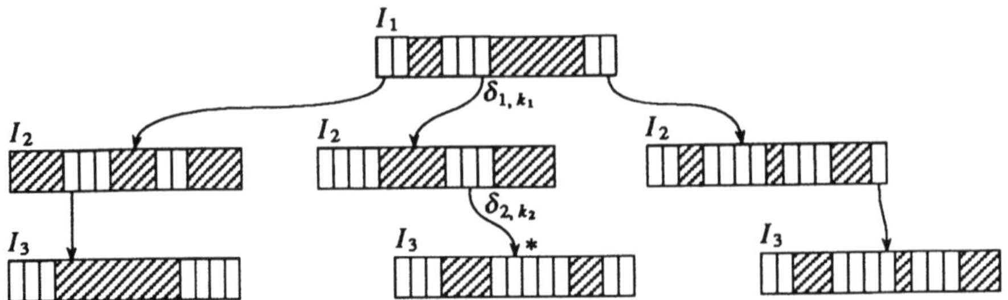

Figure 11. The configuration space of a manipulator arm is represented in the form of a p-level tree structure [Lozano-Pérez, 1987]. The nodes at depth $i-1$ ($i \in [1,p]$) represent decompositions of the range I_i into EMPTY and FULL intervals. Except at depth $p-1$, the EMPTY intervals are decomposed into smaller intervals δ_{i,k_i}. Consider the node at depth 2 marked by "*". The EMPTY intervals of this node are formed by all the values of q_3 such that $\mathcal{A}_3(q_1, q_2, q_3)$ intersects with no obstacle when q_1 varies over δ_{1,k_1} and q_2 varies over δ_{2,k_2}.

when $(q_1, \ldots, q_p)$ varies over the rectangloid cell $\delta_{1,k_1} \times \ldots \times \delta_{p,k_p}$, does not intersect with any obstacle, then $\delta_{1,k_1} \times \ldots \times \delta_{p,k_p}$ is part of the free space $\mathcal{C}_{free}$; otherwise, it is considered to be in the C-obstacle region. By considering all the cells, we construct a conservative approximation of the free space as a collection of small rectangloid cells. The connectivity graph representing the adjacency relation among these cells can then be generated and searched for a channel. As such, however, the overall planning method is not likely to be very efficient in terms of computing time and memory space. Fortunately, in most cases, simple modifications allow for a considerable reduction in the number of cells that have to be explicitly checked by this method. These modifications are described below.

Let us consider the first object $\mathcal{A}_1$ in $\mathcal{A}$. Its position and orientation in $\mathcal{W}$ depend only on the value of the coordinate q_1. If, for a given value of this coordinate, it intersects with an obstacle, then there is no need to check $\mathcal{A}_2$ through $\mathcal{A}_p$ for collisions with obstacles. Indeed, the whole $(p-1)$-dimensional slice of $\mathcal{C}$ projecting at the given value of q_1 is part of the C-obstacle region. In the same way, if for given values of $q_1, \ldots, q_i$, $\mathcal{A}_i$ intersects with an obstacle, the whole $(p-i)$-dimensional slice of $\mathcal{C}$ projecting at the given values of q_1 through q_i is part of the C-obstacle region. This suggests constructing a representation of the configuration space in the form of a p-level tree, as shown in Figure 11 [Gouzènes, 1984b] [Lozano-Pérez, 1987].

Each node at depth $i-1$ ($i \in [1,p]$) in this tree represents a decomposition of the range I_i of values of q_i into intervals classified as EMPTY or FULL. The root of the tree, i.e. the node at depth 0, represents the range I_1. The (open) EMPTY intervals of the decomposition are formed by all the values of q_1 for which $\mathcal{A}_1(q_1)$ intersects with no obstacle. The (closed) FULL intervals are the complement of the EMPTY intervals in I_1. The EMPTY intervals are subdivided into smaller intervals δ_{1,k_1}, $k_1 = 1, 2, \ldots$, at the specified resolution. The tree contains one node at depth 1 for each such interval. Consider any interval δ_{1,k_1}. The corresponding successor represents a decomposition of the range I_2 into EMPTY and FULL intervals. The EMPTY intervals are formed by all the values of q_2 for which $\mathcal{A}_2(q_1, q_2)$ intersects with no obstacle when q_1 varies over δ_{1,k_1}. In turn, these intervals are decomposed into smaller intervals δ_{2,k_2}, and so on. Since entire slices may get classified as part of the C-obstacle region at early stages of the computation, the time and memory required to build the tree structure is usually much smaller than with the more simplistic version of the method presented above.

When the tree is fully computed, we have an approximate representation of the free space and the C-obstacle region. In particular, given a configuration $(q_1, \ldots, q_p)$, we can easily check whether it belongs to the approximate free space or not. Indeed, the configuration determines a unique path in the tree. If this path terminates at a FULL interval (this may happen at any depth in the tree), the configuration is considered to be in the C-obstacle region. Instead, if the path terminates in an EMPTY interval (this can only happen at depth $p-1$), the configuration is in free space.

Figure 12 illustrates the representation of the configuration space of a two-revolute-joint arm computed by the above algorithm. The C-obstacle region (shown dark in the figure) is approximated as a set of slices parallel to the q_2-axis.

The key computation for constructing the tree of Figure 11 is to determine the range of values of q_i ($i \in [1,p]$) at which $\mathcal{A}_i(q_1, \ldots, q_{i-1}, q_i)$ intersects no obstacle when $q_1, \ldots, q_{i-1}$ vary over intervals $\delta_{1,k_1}, \ldots, \delta_{i-1,k_{i-1}}$.

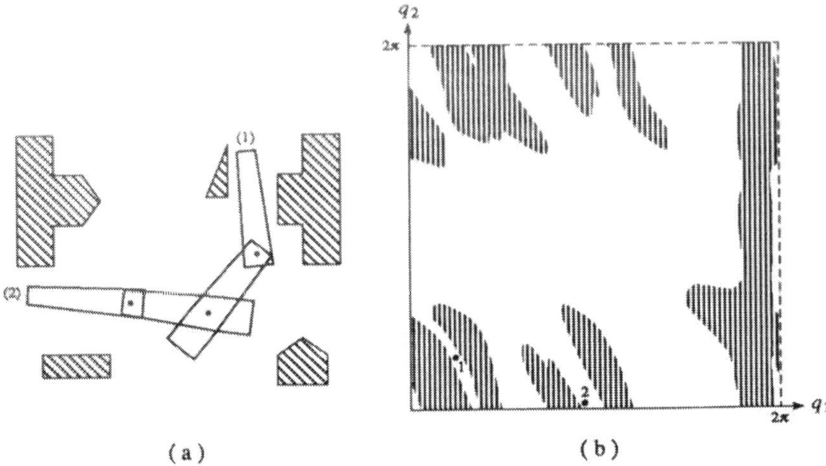

Figure 12. Figure a shows a planar two-revolute-joint arm among obstacles, at two free configurations marked (1) and (2). We assume that there is no mechanical stop in the joints. Figure b shows the representation of the robot's configuration space constructed by the algorithm described in the text. The C-obstacle region (shown dark) is approximated by a set of slices parallel to the q_2-axis. The points designated by 1 and 2 represent the arm's configurations marked (1) and (2) in Figure a.

A simple technique proposed in [Lozano-Pérez, 1987] consists of fixing the values of $q_1, ..., q_{i-1}$ to, say, the midpoints $q_{1,k_1}, ..., q_{i-1,k_{i-1}}$ of the intervals $\delta_{1,k_1}, ..., \delta_{i-1,k_{i-1}}$. Then the computation reduces to determining the intervals of values of q_i for which $\mathcal{A}_i(q_{1,k_1}, ...q_{i-1,k_{i-1}}, q_i)$ intersects no obstacle when it rotates around a fixed axis (in the case of a revolute joint) or translates along a fixed axis (in the case of a prismatic joint). If $\mathcal{A}_i$ and the obstacles are modeled by polygons or polyhedral, this computation is not difficult. (The case where $\mathcal{A}_i$ is a polygon rotating about a fixed point among polygonal obstacles has already been studied in Subsection 5.2 of Chapter 6.)

However, fixing the values of $q_1,...,q_{i-1}$, rather than letting them vary over intervals, has the drawback of generating an approximation of the free space that is no longer guaranteed to be a conservative one. One way to deal with this drawback is to compute the largest Euclidean displacement ρ_i of any point in $\mathcal{A}_i$ for any $q_i \in I_i$, when $q_1, ..., q_{i-1}$ vary over $\delta_{1,k_1}, ..., \delta_{i-1,k_{i-1}}$. The object $\mathcal{A}_i$ is grown isotropically by ρ_i and the grown object (rather than $\mathcal{A}_i$) is used to decompose the range I_i into EMPTY and FULL intervals.

The construction of the tree representing configuration space can be improved further as follows [Gouzènes, 1984b] [Faverjon, 1986]. Let $\mathcal{S}_i(q_1, ..., q_i)$, $i \in [1, p]$, be the region swept out by the objects $\mathcal{A}_{i+1}, ..., \mathcal{A}_p$ when $q_{i+1}, ..., q_p$ vary over $I_{i+1}, ..., I_p$, and the previous i objects $\mathcal{A}_1, ..., \mathcal{A}_i$ are at configuration $(q_1, ..., q_i)$. Note that only the position and orientation of $\mathcal{S}_i(q_1, ..., q_i)$, not its shape, depend on the values of $q_1, ..., q_i$. When the range I_i labeling a node at depth $i - 1$ in the tree is decomposed, the generated intervals are labeled as FULL, C/EMPTY (for "certainly empty") and P/EMPTY (for "possibly empty"):

- The FULL intervals are the same as before, i.e. they are formed by all the values of q_i such that $\mathcal{A}_i(q_1, ..., q_{i-1}, q_i)$ intersects with an obstacle when $q_1, ..., q_{i-1}$ vary over $\delta_{1,k_1}, ..., \delta_{i-1,k_{i-1}}$.
- The C/EMPTY intervals are formed by all the values of q_i such that $\mathcal{A}_i(q_1, ..., q_{i-1}, q_i) \cup \mathcal{S}_i(q_1, ..., q_{i-1}, q_i)$ intersects with no obstacle when $q_1, ..., q_{i-1}$ vary over $\delta_{1,k_1}, ..., \delta_{i-1,k_{i-1}}$.
- The P/EMPTY intervals are the complementary of the FULL and C/EMPTY intervals in I_i.

Only the P/EMPTY intervals of I_i are subdivided into smaller intervals δ_{i,k_i}. With each P/EMPTY interval in a node of the tree, one can associate the list of obstacles with which $\mathcal{S}_i$ collides. When the successors of this node (corresponding to this interval) are considered, only the obstacles in this list have to be considered for checking the possible collisions of $\mathcal{A}_{i+1}$. Then, in each P/EMPTY interval of I_{i+1}, if any, the list is pruned further by removing the obstacles that do not collide with $\mathcal{S}_{i+1}$ [Faverjon and Tournassoud, 1989].

Once the tree representation of the robot's configuration space has been built, we still have the problem of constructing a free path between the initial configuration and the goal one. The tree directly provides a representation of the free space in the form of a collection of rectangloid cells. The connectivity graph representing the adjacency relation between these cells can be constructed and searched for a channel. However, the cells derived from the tree representation are very thin along one or several dimensions and usually there are many of them, resulting in a large connectivity graph. It may be preferable to group these cells into larger rectangloids. One way[8] to do this is to consider the tree representation as the exact representation of the configuration space and to apply an approximate cell decomposition method (as described in Section 1 of Chapter 6) to this representation. Using the information contained in the tree, the labeling of the new cells generated by this method as EMPTY, FULL and MIXED is particularly simple.

3.4 Potential Field

The potential field approach described in Chapter 7 applies to articulated robots. As is the case for a single rigid object, there are many possible methods based on this approach.

Typically, one considers a collection of control points $a_j \in \mathcal{A}$, $j = 1, ..., Q$, which may be distributed over the various objects $\mathcal{A}_i$. Each point a_j is subject to a potential field $\mathbf{V}_j(\mathbf{x})$ defined in the workspace. This potential may be either an attractive potential depending on the goal position of a_j in $\mathcal{W}$ (see Subsection 2.2, Chapter 7), a repulsive potential depending on the obstacles in the workspace (see Subsection 2.3, Chapter 7), or a mixed potential depending on both the goal position of a_j and the obstacles (see Subsection 4.2.3, Chapter 7).

The potential function $\mathbf{U}(\mathbf{q})$ in the robot's configuration space is obtained by adding these potentials:

$$\mathbf{U}(\mathbf{q}) = \sum_{j=1}^{Q} \mathbf{V}_j(a_j(\mathbf{q})).$$

It induces the attractive force defined by:

$$\vec{F}(\mathbf{q}) = -\vec{\nabla}\mathbf{U}(\mathbf{q}).$$

Let us consider a chart in the robot's configuration space $\mathcal{C}$ such that a configuration $\mathbf{q}$ is parameterized by $(q_1, ..., q_p)$, with each q_i defining the relative position of two objects connected by a joint, as described in Subsection 3.1. The components of $\vec{F}(\mathbf{q})$ in the basis of $T_{\mathbf{q}}(\mathcal{C})$ (tangent space of $\mathcal{C}$ at $\mathbf{q}$) induced by this chart are the forces/torques applied at every prismatic/revolute joint of the robot. Assume further that $\mathcal{W} =$

[8] See [Lozano-Pérez, 1987] [Siméon, 1989] for other techniques.

$\mathbf{R}^3$, with x, y, z being the coordinates of a point $\mathbf{x} \in \mathcal{W}$ in $\mathcal{F}_{\mathcal{W}}$. Let $X_j(q_1, ..., q_p)$, $Y_j(q_1, ..., q_p)$ and $Z_j(q_1, ..., q_p)$ be the coordinates of $a_j(\mathbf{q})$ in $\mathcal{F}_{\mathcal{W}}$. The ith component of $\vec{F}(\mathbf{q})$ is:

$$-\frac{\partial}{\partial q_i}\mathbf{U}(q_1, ..., q_p) = -\sum_{j=1}^{Q}\left(\frac{\partial \mathbf{V}_j}{\partial x}\frac{\partial X_j}{\partial q_i} + \frac{\partial \mathbf{V}_j}{\partial y}\frac{\partial Y_j}{\partial q_i} + \frac{\partial \mathbf{V}_j}{\partial z}\frac{\partial Z_j}{\partial q_i}\right).$$

Hence, we have that:

$$\vec{F}(\mathbf{q}) = -\sum_{j=1}^{Q} J_j^T(\mathbf{q})\vec{\nabla}\mathbf{V}_j(\mathbf{x})_{|\mathbf{x}=a_j(\mathbf{q})}$$

where J_j^T is the transpose of the Jacobian matrix J_j of the map:

$$(q_1, ..., q_p) \in \mathbf{R}^p \mapsto (X_j(q_1, ..., q_p), Y_j(q_1, ..., q_p), Z_j(q_1, ..., q_p)) \in \mathbf{R}^3.$$

The same result also holds with $\mathcal{W} = \mathbf{R}^2$.

If the motion at a joint is limited by mechanical stops, these stops can be taken into account in the potential function. One way to proceed is as follows [Khatib, 1986]. Let a configuration coordinate q_i be constrained to be in an interval (q_i^-, q_i^+) by the mechanical stops in the corresponding joint. Both endpoints of the interval can be treated as obstacles generating repulsive potential fields $\mathbf{W}^-(q_i)$ and $\mathbf{W}^+(q_i)$. Using the same general form as for the repulsive potential functions described in Subsection 2.3 of Chapter 7, we define $\mathbf{W}^-(q_i)$ and $\mathbf{W}^+(q_i)$ by:

$$\mathbf{W}^-(q_i) = \begin{cases} \frac{1}{2}\eta(\frac{1}{q_i - q_i^-} - \frac{1}{\rho_0})^2 & \text{if } q_i - q_i^- \leq \rho_0, \\ 0 & \text{if } q_i - q_i^- > \rho_0 \end{cases}$$

and:

$$\mathbf{W}^+(q_i) = \begin{cases} \frac{1}{2}\eta(\frac{1}{q_i^+ - q_i} - \frac{1}{\rho_0})^2 & \text{if } q_i^+ - q_i \leq \rho_0, \\ 0 & \text{if } q_i^+ - q_i > \rho_0. \end{cases}$$

where η is a positive scaling factor and ρ_0 is the "distance of influence" of the mechanical stops. The various functions $\mathbf{W}_i^-(q_i)$ and $\mathbf{W}_i^+(q_i)$ are added to the function $\mathbf{U}(q_1, ..., q_p)$ in order to get a potential function

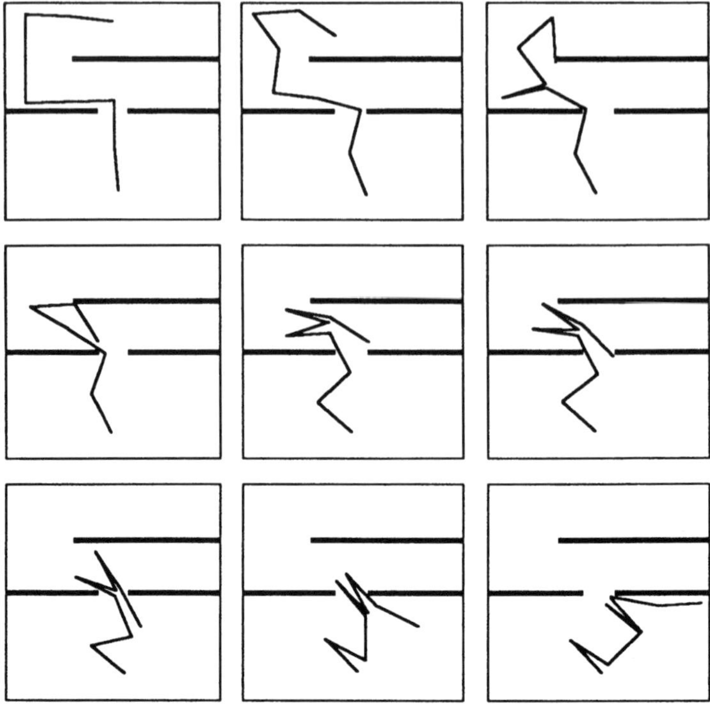

Figure 13. This figure shows snapshots along a path generated for an eight-revolute-joint robot by an implementation of the randomized potential-field-based method described in Section 5 of Chapter 7.

that takes the mechanical stops into account. Notice that all the components of the forces induced by $\mathbf{W}_i^-(q_i)$ and $\mathbf{W}_i^+(q_i)$, except for the ith ones, are zero. The ith components of these forces are:

$$-\frac{d\mathbf{W}_i^-(q_i)}{dq_i} = \begin{cases} \eta(\frac{1}{q_i - q_i^-} - \frac{1}{\rho_0}) & \text{if } q_i - q_i^- \leq \rho_0, \\ 0 & \text{if } q_i - q_i^- > \rho_0 \end{cases}$$

and:

$$-\frac{d\mathbf{W}_i^+(q_i)}{dq_i} = \begin{cases} \eta(\frac{1}{q_i^+ - q_i} - \frac{1}{\rho_0}) & \text{if } q_i^+ - q_i \leq \rho_0, \\ 0 & \text{if } q_i^+ - q_i > \rho_0. \end{cases}$$

These are forces/torques applying to the ith joint of the robot.

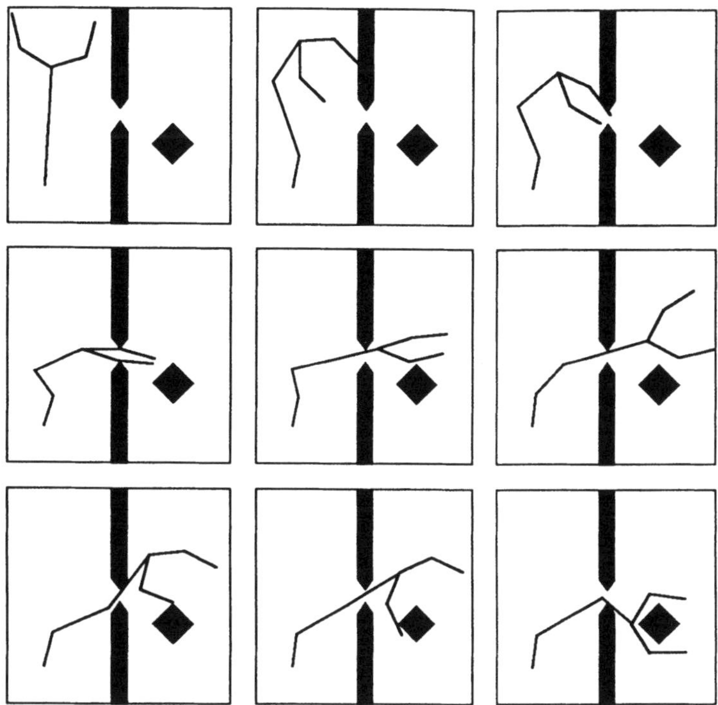

Figure 14. This figure shows snapshots along a path generated for a ten-joint robot (see Figure 8) by an implementation of the randomized potential-field-based method described in Section 5 of Chapter 7.

The various potential field methods described in Chapter 7 can be used to plan free paths for manipulator arms. As mentioned in Subsection 3.2, several potential-field-based planners have been implemented. As an illustration of the power of this approach, Figures 13 and 14 show snapshots along two paths generated by the randomized planner described in Section 5 of Chapter 7 [Barraquand, Langlois and Latombe, 1989b]. Figure 13 illustrates a path for an eight-revolute-joint robot. Figure 14 illustrates a path for the ten-joint robot of Figure 8. The example of Figure 13 was produced using a single control point located at the endpoint of the arm. The example of Figure 14 was produced using two control points located at the endpoints of the two kinematic chains. In both examples, collisions of the various links among themselves were checked explicitly in the same way as the collisions between the robot and the obstacles. Mechanical stops were imposed on each joint and the collisions with each stop were also checked.

Exercises

1: (One-dimensional asteroid problem) Let $\mathcal{A}$ be a polygonal object moving in translation along a single line L (locus of the origin of the frame $\mathcal{F}_{\mathcal{A}}$ attached to $\mathcal{A}$) in the plane $\mathbf{R}^2$. The plane contains moving obstacles $\mathcal{B}_i$, $i = 1, ..., q$. Each obstacle $\mathcal{B}_i$ is a convex polygon translating without rotation at constant linear velocity $\vec{\mathbf{v}}_{\mathcal{B}_i}$. No two obstacles ever collide. The geometry, the position at time 0, and the velocity of each obstacle are given. The problem is to plan a motion of $\mathcal{A}$ from an initial position x_{init} in L at time 0 to a goal position x_{goal} at time T. The velocity $\vec{\mathbf{v}}_{\mathcal{A}}$ of $\mathcal{A}$ is bounded, i.e. $\|\vec{\mathbf{v}}_{\mathcal{A}}\| \leq V_{max}$. There is no constraint on its acceleration.

a. Characterize the shape of the CT-obstacle in the configuration-time space of $\mathcal{A}$.

b. Adapt the visibility graph method to plan the motion of $\mathcal{A}$ in the configuration-time space of $\mathcal{A}$.

c. Assuming that the search graph is constructed using a sweep-line technique, what would be the time complexity of the method?

2: Show that the planning algorithm of Subsection 1.4.1 still applies if $\mathbf{q}_{goal}$ moves at constant velocity.

3: Modify the velocity tuning method (with polygonal PT-obstacles) presented in Subsection 1.4.3 to handle the dynamic motion planning problem with unspecified arrival time.

4: Consider two convex polygonal objects $\mathcal{A}_1$ and $\mathcal{A}_2$ which translate and rotate freely in $\mathcal{W} = \mathbf{R}^2$. Let $\mathcal{C}$ be their composite configuration space and $\mathcal{CA}_{12}$ be the C-obstacle defined by:

$$\mathcal{CA}_{12} = \{\mathbf{q} \in \mathcal{C} \;/\; \mathcal{A}_1(\mathbf{q}) \cap \mathcal{A}_2(\mathbf{q}) \neq \emptyset\}.$$

Using the approach of Section 2 of Chapter 3, construct a computable expression (in the form of a conjunction of C-constraints) of the predicate

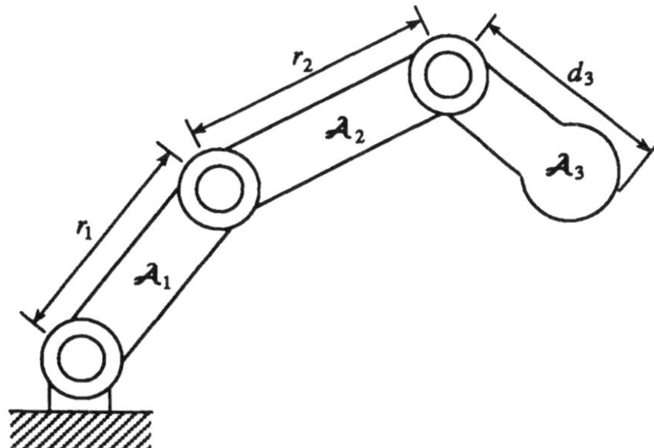

Figure 15. This planar robot arm has three revolute joints (Exercise 9).

$\mathtt{CA}_{12}$ such that:

$$\mathbf{q} \in \mathcal{CA}_{12} \Leftrightarrow \mathtt{CA}_{12}(\mathbf{q}).$$

5: What is the configuration space of a robot that consists of a six-revolute-joint manipulator arm mounted on a platform? (The arm operates in a three-dimensional workspace. The platform translates and rotates freely in a plane.) Discuss the application of various path planning methods to this robot.

6: Let $\mathcal{A}_1$ and $\mathcal{A}_2$ be two polygons that can translate freely at fixed orientation among fixed obstacles. Let τ_1 and τ_2 be two paths defined as in Subsection 2.3.2. Hence, τ_1 (resp. τ_2) is a path of $\mathcal{A}_1$ (resp. $\mathcal{A}_2$) that avoids the fixed obstacles, but ignores the other robot. Assume that both τ_1 and τ_2 are sequences of straight segments. Give a method for constructing the coordination diagram.

7: Write a program based on the techniques described in Subsection 3.3 for computing an approximation of the free space of a robot arm with two revolute joints. Assume that the two links are modeled by straight line segments and that there are no mechanical stops in the joints.

8: Complete the techniques described in Subsection 3.3 to: (1) handle the possible collisions among the various objects composing an articulated robot; (2) to choose the resolution of the decomposition of the EMPTY intervals into small intervals δ_{i,k_i}.

9: Consider a planar manipulator $\mathcal{A} = \{\mathcal{A}_1, \mathcal{A}_2, \mathcal{A}_3\}$ with three revolute joints (see Figure 15). Its configuration is parameterized by three angles q_1, q_2 and q_3, each varying over $[0, 2\pi)$ (we assume no mechanical stop). The distance between the first two joints (resp. the last two joints) is r_1 (resp. r_2). The maximal distance of any point in $\mathcal{A}_3$ to joint 3 is d_3. Write the formula that computes the largest Euclidean displacement of any point in $\mathcal{A}_3$ when q_1 and q_2 vary over any small intervals of amplitudes δ_1 and δ_2, with q_3 being anywhere in $[0, 2\pi)$.

10: Discuss in general terms a method for solving the problem of planning the coordinated motion of two robot arms carrying the same object (closed-loop kinematic structure) among obstacles.

Chapter 9

Kinematic Constraints

In this chapter we study the path planning problem for a robot subject to kinematic constraints, i.e. an object (or a collection of objects) that cannot translate and rotate freely in the workspace. Indeed, so far, we have only investigated the case of free-flying robots. The only exception occurred in the last section of the previous chapter, when we considered articulated robots made of several objects connected by joints. The constraint imposed by a joint on the relative motion of the two objects it connects is a simple example of kinematic constraint.

The constraints which affect the motion of a car-like mobile robot are another important example of kinematic constraints. Like a car, such a robot can move forward and backward, and can make turns. However, it cannot move sidewise (hence, is does not translate freely) and its turns are usually limited by mechanical stops in the steering gear. Nevertheless, it can take any position and orientation in the plane.

These two examples (articulated robot and car-like robot) illustrate the two different types of kinematic constraints which may affect a robot's motion. The constraints applying to an articulated robot can be expressed as equations relating the configuration parameters. These equations can be used to eliminate some parameters and reduce the dimension of the configuration space. They are called *holonomic constraints*. On the other hand, as we will see in more detail later, the constraints affecting the motion of a car-like robot translate into an equation and an inequation involving the velocity parameters of the robot. They do not reduce the dimension of the robot's configuration space, but that of its velocity space. They are said to be *nonholonomic*. Both holonomic and nonholonomic constraints will be more formally defined in Section 1.

Holonomic constraints do not fundamentally change the path planning problem. They will only be examined briefly in Section 2.

Nonholonomic constraints are much harder to deal with and they are the topic of most of this chapter. They raise three main issues:

- First, given a possibly nonholonomic constraint, how can we be sure that it is actually a nonholonomic one? Indeed, if it is integrable, an equation involving velocity parameters reduces to a holonomic constraint. Classical results from Differential Geometry will allow us to precisely characterize nonholonomic constraints (Section 3).

- The second issue is to determine whether nonholonomic constraints restrict the set of configurations reachable from a given configuration. Using tools from Control Theory, we will establish simple conditions under which nonholonomic constraints have no effect on the range of achievable configurations (Sections 4 and 5).

- The third issue is to build an effective planner capable of generating feasible free paths for a robot subject to given nonholonomic constraints. We will describe in detail two implemented planning approaches in Sections 6 and 7.

Motion planning with nonholonomic constraints is a relatively new area of research. Following the work of Laumond [Laumond, 1986], it has recently attracted considerable interest, mainly in connection to its application to mobile robots. Throughout this chapter we will use this application as a major example, both to illustrate issues and solutions. We will also use the tractor-trailer example to analyze the case where multiple nonholonomic constraints apply simultaneously. A large subset of the material presented in this chapter is derived from [Barraquand and Latombe, 1989b].

1 Types of Kinematic Constraints

Let us consider a robot $\mathcal{A}$ having an m-dimensional configuration space $\mathcal{C}$. An atlas has been defined on $\mathcal{C}$ and any configuration $\mathbf{q}$ of $\mathcal{A}$ is represented by a list of m coordinates $(q_1, ..., q_m)$ in some chart of the atlas. Throughout the chapter we will consider a single chart. Dealing with several charts does not significantly modify any of the definitions and results given below, but requires additional administration for appropriately switching between charts.

1.1 Holonomic Constraints

Let us now suppose that at any time t, we impose an additional scalar constraint of the following form to the configurations of the robot:

$$F(\mathbf{q}, t) = F(q_1, ..., q_m, t) = 0 \tag{1}$$

where F is a smooth function with non-zero derivative. This constraint selects a subset of configurations of $\mathcal{C}$ (those which satisfy the constraint) where the robot is allowed to be.

Using the Implicit Function Theorem we can use the equation (1) to solve for one of the coordinates, say q_m, by expressing it as a function g of the $m-1$ remaining coordinates and time, i.e. $g(q_1, ..., q_{m-1}, t)$. The function g is smooth so that the equation (1) defines a $(m-1)$-dimensional smooth submanifold of $\mathcal{C}$. This submanifold is in fact the actual configuration space[1] of $\mathcal{A}$ and the $m-1$ remaining parameters are the coordinates of the configuration $\mathbf{q}$ in some chart. (Notice that the number of charts needed to form an atlas of the new configuration space is in general not equal to the number of charts needed to form an atlas of $\mathcal{C}$.)

DEFINITION 1: *A scalar constraint of the form $F(\mathbf{q}, t) = 0$, where F is a smooth function with non-zero derivative, is called a* **holonomic** *equality constraint.*

More generally, there may be k holonomic equality constraints ($k \leq m$).

[1] If constraint (1) depends on t, this space is time-dependent, otherwise it is time-independent. Many usual holonomic constraints, e.g. those generated by the prismatic and revolute joints of a manipulator arm, are time-independent.

If they are independent (i.e. their Jacobian matrix has rank k) they determine a $(m-k)$-dimensional submanifold of $\mathcal{C}$, which is the actual configuration space of $\mathcal{A}$.

Example: Typical holonomic constraints are those imposed by the prismatic and revolute joints of a manipulator arm (see Section 3 in Chapter 8). A planar arm with two revolute joints is made of two moving objects $\mathcal{A}_1$ and $\mathcal{A}_2$. The composite configuration space of these objects is the six-dimensional manifold $\mathbf{R}^2 \times S^1 \times \mathbf{R}^2 \times S^1$. Each joint applies two scalar constraints of the form $F(\mathbf{q}) = 0$ to the configurations of the arm. The first joint requires that a point of $\mathcal{A}_1$ (the rotation center of the first joint) be fixed with respect to the workspace. The second joint requires that a point of $\mathcal{A}_2$ (the rotation center of the second joint) be fixed with respect to $\mathcal{A}_1$. With these four constraints, the actual configuration space of the robot is the two-dimensional manifold $S^1 \times S^1$. ■

A constraint of the form:

$$F(\mathbf{q}, t) < 0 \quad \text{or} \quad F(\mathbf{q}, t) \leq 0$$

typically corresponds to a mechanical stop or an obstacle. It determines a subset of $\mathcal{C}$ having in general the same dimension as $\mathcal{C}$. We call it a *holonomic inequality constraint.*

1.2 Nonholonomic Constraints

Let us now consider the robot $\mathcal{A}$ while it is moving. Its configuration $\mathbf{q}$ is a differentiable function of time t. We impose that $\mathcal{A}$'s motion satisfy a scalar constraint of the following form:

$$G(\mathbf{q}, \dot{\mathbf{q}}, t) = G(q_1, \ldots, q_m, \dot{q}_1, \ldots, \dot{q}_m, t) = 0 \tag{2}$$

where G is a smooth function and $\dot{q}_i = \frac{dq_i}{dt}$ for every $i = 1, \ldots, m$. The velocity vector $\dot{\mathbf{q}} = (\dot{q}_1, \ldots, \dot{q}_m)$ is a vector of $T_{\mathbf{q}}(\mathcal{C})$, the tangent space of $\mathcal{C}$ at $\mathbf{q}$. We recall that in the absence of kinematic constraints of the form (2), the tangent space is the space of the velocities of $\mathcal{A}$.

A kinematic constraint of the form (2) is holonomic if it is integrable, i.e. if all the velocity parameters $\dot{q}_1$ through $\dot{q}_m$ can be eliminated and the equation (2) rewritten in the form (1). Otherwise, the constraint is called a *nonholonomic* constraint.

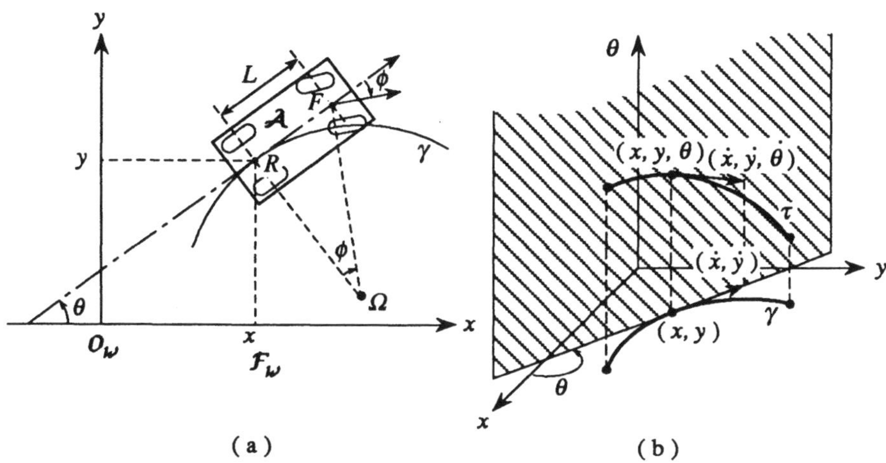

Figure 1. This schematic car-like robot will serve as the main example of nonholonomic robot throughout this chapter. It is modeled as a rectangle $\mathcal{A}$ moving in the plane (Figure a). Its configuration is represented as (x, y, θ). x and y are the coordinates of the midpoint between the two rear wheels (denoted by R). θ is the angle between the x-axis of $\mathcal{F}_{\mathcal{W}}$ and the main axis of the car. At each instant, the velocity $(\dot{x}, \dot{y})$ of R points (forward or backward) along this main axis. Hence, the possible velocity vectors $(\dot{x}, \dot{y}, \dot{\theta})$ at each configuration (x, y, θ) span a plane (shown in Figure b). The angle ϕ between the main axis of the car and the velocity of the midpoint F between the two front wheels is the steering angle. It determines the instantaneous center of rotation Ω of the robot. The curve γ followed by R has an instantaneous curvature radius equal to $L/\tan\phi$, where L is the distance between F and R.

DEFINITION 2: *A non-integrable scalar constraint of the form:*

$$G(q_1, \ldots, q_m, \dot{q}_1, \ldots, \dot{q}_m, t) = 0,$$

where G is a smooth function, is called a **nonholonomic** *equality constraint.*

Example: Consider a car-like robot (i.e. a four-wheel front-wheel-drive vehicle) moving on a flat ground. We model[2] this robot as a rectangle moving in $\mathcal{W} = \mathbf{R}^2$, as illustrated in Figure 1.a. Its configuration space $\mathcal{C}$ is $\mathbf{R}^2 \times S^1$. We represent a configuration as a triple (x, y, θ), where

[2]Other types of mobile robots could be modeled in a similar way.

$(x, y) \in \mathbf{R}^2$ are the coordinates of the midpoint R between the two rear wheels, and $\theta \in [0, 2\pi)$ is the angle (mod 2π) between the x-axis of the frame $\mathcal{F}_{\mathcal{W}}$ attached to the workspace and the main axis of the car. We assume that the contact between each of the wheels and the ground is a pure rolling contact between two perfectly rigid objects. When the robot moves, the point R describes a curve γ that must be tangent to the main axis of the car. Hence, the robot's motion is constrained by:

$$-\dot{x}\sin\theta + \dot{y}\cos\theta = 0$$

which is of the form (2). We will show in Section 3 that this constraint is non-integrable and hence is nonholonomic. At every configuration (x, y, θ), it imposes that the velocity $(\dot{x}, \dot{y}, \dot{\theta})$ lie in a two-dimensional vector subspace of the tangent space. By superposing the $\dot{x}\dot{y}\dot{\theta}$-representation of the robot's tangent space on the $xy\theta$-representation of its configuration space, we can represent the velocity subspace as the plane perpendicular to the xy-plane that projects on this plane along the main axis of the robot (Figure 1.b). The path τ followed by the robot in the $xy\theta$-space is locally an helicoid. ■

Although the car-like robot example will be (with the tractor-trailer example) our main example of nonholonomic robot in the rest of the chapter, it is not unique. Other important examples include levitating and flying robots actuated by thrusters [Alexander, 1987], and dextrous hands when the spherical tips of their fingers perform rolling motions in contact with an object [Cutkosky, 1985] [Cai, 1988] [Li, Canny and Sastry, 1989].

A nonholonomic equality constraint restricts the space of velocities achievable by $\mathcal{A}$ at any configuration $\mathbf{q}$ to an $(m-1)$-dimensional vector subspace of $T_{\mathbf{q}}(\mathcal{C})$. If k independent nonholonomic equality constraints of the form (2) apply to the motion of the robot, the space of achievable velocities is a subspace of $T_{\mathbf{q}}(\mathcal{C})$ of dimension $m - k$.

A nonholonomic equality constraint is often caused by a rolling contact between two rigid objects (the wheels and the ground in the car-like robot example). It expresses that the relative velocity of the two points of contact is zero. When the motion in contact combines rolling and sliding, the expression, which depends on the friction coefficient of the two bodies, is nonlinear. When there is no sliding, the nonholonomic constraint is linear in $\dot{\mathbf{q}}$. In the rest of the chapter we will solely consider nonholonomic equality constraints that are linear in $\dot{\mathbf{q}}$. (See [Barraquand and Latombe, 1990b] for a generalization to both linear and nonlinear constraints.)

An inequality constraint of the form:

$$G(\mathbf{q}, \dot{\mathbf{q}}, t) < 0 \quad \text{or} \quad G(\mathbf{q}, \dot{\mathbf{q}}, t) \leq 0$$

usually restricts the set of velocities achievable by $\mathcal{A}$ without changing its dimension. We call it a *nonholonomic inequality constraint.*

Example: Consider the car-like robot of Figure 1. Let ϕ be the angle between the main axis of the car and the velocity vector of the midpoint F between the two front wheels. ϕ is called the *steering angle.* In general, mechanical stops in the steering gear constrain ϕ in such a way that:

$$|\phi| \leq \phi_{max} < \frac{\pi}{2}.$$

The implication of this constraint on the motion of the robot is easily seen. The midpoint R between the two rear wheels must follow a curve whose curvature (wherever it is defined) is upper-bounded by:

$$\frac{1}{\rho_{min}} = \frac{1}{L} \tan \phi_{max},$$

where L is the distance between F and R. $\rho_{min} = L/\tan\phi_{max}$ is called the *minimal turning radius* of the car. A car-like robot subject to $|\phi| \leq \phi_{max} < \pi/2$ is said to have *limited steering angle* or *lower-bounded turning radius.*

Let v be the velocity of R measured along the main axis of the car. We allow v to vary in some interval $[-v_{max}, +v_{max}]$. If $v > 0$ (resp. $v < 0$) and $\phi = 0$, the car moves straight forward (resp. backward). If $v > 0$ and $\phi > 0$ (resp. $\phi < 0$), the car makes a left (resp. right) forward turn. The curve followed by R is of class C^1 in every interval where v does not change sign.

The constraint $|\phi| \leq \phi_{max} < \pi/2$ can be rewritten as:

$$|\dot{\theta}| \leq \frac{|v|}{\rho_{min}}$$

or:

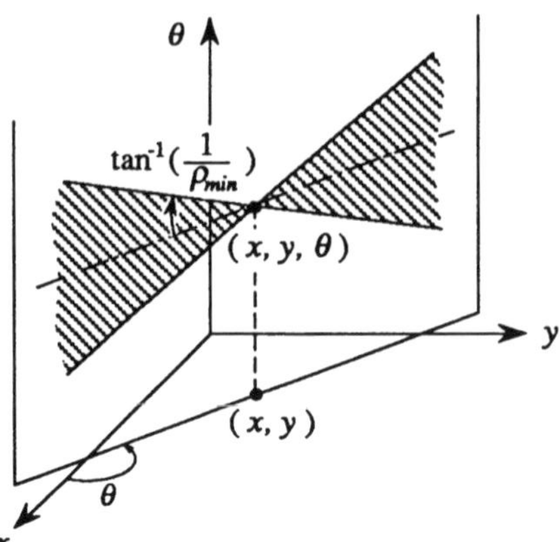

Figure 2. The velocity vector $(\dot{x}, \dot{y}, \dot{\theta})$ of a car-like robot whose turning radius is lower-bounded by $\rho_{min} > 0$ verifies $\dot{x}^2 + \dot{y}^2 - \rho_{min}^2\dot{\theta}^2 \geq 0$. Since the nonholonomic equality constraint $-\dot{x}\sin\theta + \dot{y}\cos\theta = 0$ also imposes that it lie in a plane, it must point into a two-sided cone contained in that plane. This cone is shown in the above figure. Its axis is parallel to the xy-plane and its angle is $2\tan^{-1}(1/\rho_{min})$.

$$\dot{x}^2 + \dot{y}^2 - \rho_{min}^2\dot{\theta}^2 \geq 0.$$

Hence, it is a constraint of the form $G(\mathbf{q}, \dot{\mathbf{q}}) \leq 0$. At any configuration $\mathbf{q}$, it imposes that the velocity vector $(\dot{x}, \dot{y}, \dot{\theta})$ point into a two-sided cone of angle $2\tan^{-1}(1/\rho_{min})$, as illustrated in Figure 2. ■

In the rest of this chapter we will assume for simplification that holonomic and nonholonomic constraints are time-independent.

2 Holonomic Constraints

Path planning with holonomic inequality constraints is what we have studied since the beginning of this book. Indeed, the constraints imposed by the obstacles are exactly of this form. One can imagine other causes for such constraints (for instance, high temperature in a subset of the workspace may not permit the robot to operate properly). The regions of the configuration space determined by these constraints can nevertheless be treated as C-obstacles.

Path planning in the presence of holonomic equality constraints is the same fundamental problem as for a free-flying objects. The constraints may affect the global topology of the configuration space manifold, but the general planning approaches apply to the parameterized representations (charts) of this manifold. Defining convenient charts and managing them may perhaps be slightly more tricky, but that is all. Instead, it is also possible that the holonomic constraints simplify the definition of the charts. For instance, imposing that a rigid object keep a constant orientation may be regarded as a holonomic constraint which reduces the object's configuration space from $\mathbf{R}^N \times SO(N)$, with $N = 2$ or 3, to $\mathbf{R}^N$. The new configuration space is completely covered by a single chart.

Therefore, if m is the dimension of the configuration space after the holonomic inequality constraints have been applied to eliminate parameters, the path planning problem can still be solved in time exponential in m. In the precedent chapter, we have shown how the approximate cell decomposition and the potential field approaches can be applied to plan free paths for articulated objects. Similar adaptations could be done for dealing with other holonomic constraints.

It is worth noticing however that some general topological properties relative to C-obstacles that were established in Chapter 2 may no longer be valid when holonomic constraints apply to the robot. This is the case of Proposition 6 ("If both $\mathcal{A}$ and $\mathcal{B}$ are connected, so is $\mathcal{B}$"), Proposition 7 ("$\mathcal{CB} = cl(\mathcal{CB}_{overlap})$"), and Proposition 8 ("If $\mathcal{A}$ and $\mathcal{B}$ are regular, so is $\mathcal{CB}$"). For example, let $\mathcal{A}$ be a rectangle; take its midpoint as the reference point. If $\mathcal{A}$ is free to rotate, but can only translate so that the reference point stays in a line L, an obstacle $\mathcal{B}$ not intersecting with L but reachable by $\mathcal{A}$ may map in configuration space to a C-obstacle consisting of two disconnected components, so that Proposition 6 does not hold. If $\mathcal{A}$ can only touch $\mathcal{B}$ tangentially without overlapping its interior, Propositions 7 and 8 do not hold.

One interesting case of holonomic constraint is when a robot — say a rigid object — is required to stay in contact with an obstacle. If the obstacle is a polygonal or polyhedral region, the constraint is only piecewise smooth resulting in a piecewise smooth configuration space. The method of Avnaim, Boissonnat and Faverjon [Avnaim, Boissonnat and Faverjon, 1988] described in Section 4 of Chapter 5 can easily be adapted to solve this problem (see also [Koutsou, 1986]). The particular case of finding the shortest path for a point moving in the boundary of a polyhedron (which is equivalent to finding the shortest path of a polyhedron translating in contact with another polyhedron) has been studied in the Computational Geometry literature and is referred to as the *discrete geodesic problem.* An algorithm described in [Mitchell, Mount and Papadimitriou, 1987] solves this problem in time $O(n^2 \log n)$, where n is the complexity of the (possibly non-convex) polyhedron.

3 Characterization of Nonholonomy

Let us now consider constraints of the form:

$$G(\mathbf{q}, \dot{\mathbf{q}}) = G(q_1, \ldots, q_m, \dot{q}_1, \ldots, \dot{q}_m) = 0$$

that are linear in $\dot{\mathbf{q}} = (\dot{q}_1, \ldots, \dot{q}_m)$. The first important question that arises is "Are they integrable?", since, if they are, they simply reduce to holonomic constraints. In this section we focus on this question.

First, we analyze the case where there are several constraints of the above form. The Frobenius Integrability Theorem gives a necessary and sufficient condition for their integrability. Second, we analyze the simpler case where there is a single constraint.

The proof of the results presented in this section are given in textbooks on Differential Geometry [Spivak, 1979]. Below we only give the intuition leading to these results.

3.1 Case of Several Constraints

Any kinematic constraint of the form $G(\mathbf{q}, \dot{\mathbf{q}}) = 0$ that is linear in $\dot{\mathbf{q}}$ can be rewritten as follows:

$$G(\mathbf{q}, \dot{\mathbf{q}}) = \omega(\mathbf{q}) \cdot \dot{\mathbf{q}} = \sum_{i=1}^{i=m} \omega_i(\mathbf{q})\, \dot{q}_i = 0$$

where $\omega_i(\mathbf{q})$, $i = 1, \ldots, m$, are smooth functions of $\mathbf{q}$. By definition, ω is called a *(differential) 1-form* [Spivak, 1979]. Assume that there is no $\mathbf{q}$

where all the ω_i's vanish to zero simultaneously (we say that ω is *non-singular*). For every $\mathbf{q} \in \mathcal{C}$, the velocity vectors $\dot{\mathbf{q}}$ verifying the above equation span an hyperplane $\Delta(\mathbf{q})$ in the tangent space $T_{\mathbf{q}}(\mathcal{C})$. Δ is called the $(m-1)$-*distribution* associated with ω.

Let $\mathcal{A}$'s motions be constrained by the following system of k equations:

$$G^{(l)}(\mathbf{q}, \dot{\mathbf{q}}) = \omega^{(l)}(\mathbf{q}) \cdot \dot{\mathbf{q}} = \sum_{i=1}^{i=m} \omega_i^{(l)}(\mathbf{q})\, \dot{q}_i = 0, \quad l = 1, ..., k. \tag{3}$$

This system is said to be **non-singular** if and only if the determinant $det(\omega^{(1)}, ..., \omega^{(k)})$ is non-zero for every $\mathbf{q} \in \mathcal{C}$, i.e. if the equations are linearly independent for every $\mathbf{q}$. If the system (3) is non-singular, the velocity vectors $\dot{\mathbf{q}}$ verifying the k equations span a $(m-k)$-dimensional linear subspace $\Delta(\mathbf{q})$ of $T_{\mathbf{q}}(\mathcal{C})$. In addition, there exists a set of $m-k$ independent C^∞ vector fields[3] $\vec{X}_1, \ldots, \vec{X}_{m-k}$ spanning Δ [Isidori, 1989]. Hence, at any given configuration $\mathbf{q} \in \mathcal{C}$, the robot's velocity can be expressed as a linear combination of $\vec{X}_1(\mathbf{q}), \ldots, \vec{X}_{m-k}(\mathbf{q})$. Δ is called the $(m-k)$-*distribution* associated with the k 1-forms $\omega^{(1)}, ..., \omega^{(k)}$.

Let $(\vec{X}, \vec{Y})$ be any pair of independent vector fields in Δ. Given any configuration $\mathbf{q}$, let us consider a path of $\mathcal{A}$ starting at $\mathbf{q}$ and obtained by concatenating the four following paths (see Figure 3):

- The first path follows the flow[4] of $\vec{X}$ during δt.
- The second path follows the flow of $\vec{Y}$ during δt.
- The third path follows the flow of $-\vec{X}$ during δt.
- The fourth path follows the flow of $-\vec{Y}$ during δt.

Let $\mathbf{q}'$ be the configuration reached at the end of these four paths. A straightforward Taylor expansion shows that:

---

[3]A *vector field* on $\mathcal{C}$ is a map $\vec{X} : \mathbf{q} \in \mathcal{C} \mapsto \vec{X}(\mathbf{q}) \in T_{\mathbf{q}}(\mathcal{C})$. A C^∞ (or smooth) vector field $\vec{X}(\mathbf{q})$ is such that its components (in any basis) are smooth functions of $q_1, ..., q_m$.

[4]The *integral curve* of a vector field $\vec{X}$ is a curve whose tangent at every $\mathbf{q}$ points along $\vec{X}(\mathbf{q})$. We say that the curve *follows the flow* of $\vec{X}$.

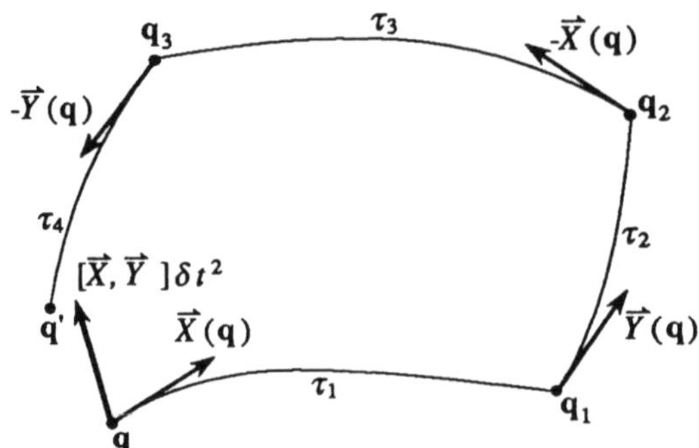

Figure 3. The path from $\mathbf{q}$ to $\mathbf{q}'$ shown in this figure is the concatenation of four paths τ_1, τ_2, τ_3, and τ_4 which follow the flow of the vector fields $\vec{X}$, $\vec{Y}$, $-\vec{X}$, and $-\vec{Y}$, respectively. Each path lasts a short amount of time δt. It can be shown that: $\lim_{\delta t \to 0} \frac{\mathbf{q}-\mathbf{q}'}{\delta t^2} = [\vec{X}, \vec{Y}]$, the Lie bracket of $\vec{X}$ and $\vec{Y}$.

$$\lim_{\delta t \to 0} \frac{\mathbf{q}' - \mathbf{q}}{\delta t^2} = D\vec{Y} \cdot \vec{X} - D\vec{X} \cdot \vec{Y},$$

where $D\vec{Y} \cdot \vec{X}$ and $D\vec{X} \cdot \vec{Y}$ denote the products of the $m \times m$ matrices:

$$D\vec{Y} = \begin{pmatrix} \frac{\partial Y_1}{\partial q_1} & \cdots & \frac{\partial Y_1}{\partial q_m} \\ \cdot & & \cdot \\ \cdot & & \cdot \\ \cdot & & \cdot \\ \frac{\partial Y_m}{\partial q_1} & \cdots & \frac{\partial Y_m}{\partial q_m} \end{pmatrix}; \quad D\vec{X} = \begin{pmatrix} \frac{\partial X_1}{\partial q_1} & \cdots & \frac{\partial X_1}{\partial q_m} \\ \cdot & & \cdot \\ \cdot & & \cdot \\ \cdot & & \cdot \\ \frac{\partial X_m}{\partial q_1} & \cdots & \frac{\partial X_m}{\partial q_m} \end{pmatrix},$$

and the m-vectors:

$$\vec{X} = (X_1 \quad X_2 \quad \ldots \quad X_m)^T; \qquad \vec{Y} = (Y_1 \quad Y_2 \quad \ldots \quad Y_m)^T.$$

$X_1, \ldots, X_m$ and $Y_1, \ldots, Y_m$ denote the components of the vectors $\vec{X}$ and $\vec{Y}$ in the basis of the tangent space induced by the chart of $\mathcal{C}$ (see Subsection 6.2 of Chapter 2). The expression $D\vec{Y} \cdot \vec{X} - D\vec{X} \cdot \vec{Y}$ determines a new vector field, which is commonly denoted by $[\vec{X}, \vec{Y}]$ and called *Lie bracket* of $\vec{X}$ and $\vec{Y}$. The ith component of this vector field can be expressed as:

$$\sum_{j=1}^{j=m} \left(X_j \frac{\partial Y_i}{\partial q_j} - Y_j \frac{\partial X_i}{\partial q_j} \right).$$

The above motion of $\mathcal{A}$ along vectors of the distribution Δ is biased with $\delta t^2[\vec{X},\vec{Y}]$. If the constraints in (3) are integrable, the actual configuration space of $\mathcal{A}$ is a manifold of dimension $m-k$. This space is locally diffeomorphic to $\mathbf{R}^{m-k}$, and thus can locally be treated as the affine space associated with Δ. Hence, if the constraints were integrable, the path defined above would terminate in that space. Therefore, a necessary condition for the integrability of the constraints is that all the Lie brackets of all the vector fields in Δ be in Δ. This condition turns out to be also sufficient, which is precisely the Frobenius Integrability Theorem:

THEOREM 1 (Frobenius): *Let Δ be the $(m-k)$-distribution on the m-dimensional manifold $\mathcal{C}$ associated with a non-singular set of k smooth 1-forms $\omega^{(1)}(\mathbf{q}),...,\omega^{(k)}(\mathbf{q})$. In a neighborhood of any $\mathbf{q}\in\mathcal{C}$, the following two conditions are equivalent:*

1. *The distribution Δ is closed under the Lie bracket operation*[5] *(i.e. for any pair of vector fields $(\vec{X},\vec{Y})$ in Δ, $[\vec{X},\vec{Y}]$ is also in Δ).*
2. *The constraints $\omega^{(l)}(\mathbf{q})\cdot\dot{\mathbf{q}}=0$, $l=1,...,k$, are integrable.*

A proof of this theorem can be found in [Spivak, 1979], where it is also shown (pp. 264-268) that the local result given by the theorem can be globalized to the whole manifold $\mathcal{C}$, which is called the *integral* manifold of Δ.

Unfortunately, this result is stated in terms of vector fields in Δ, not in terms of the $\omega^{(1)},...,\omega^{(k)}$. Thus, it does not provide an effective way to test the nonholonomy of the constraints in (3). In Section 4 we will present a generalization of this result that yields an effective test of nonholonomy.

3.2 Case of a Single Constraint

Consider the case where there is a single non-singular constraint:

$$G(\mathbf{q},\dot{\mathbf{q}})=\omega(\mathbf{q})\cdot\dot{\mathbf{q}}=\sum_{i=1}^{i=m}\omega_i(\mathbf{q})\dot{q}_i=0$$

[5] A distribution that is closed under the Lie bracket operation is also called an *involutive* distribution.

applying to the motions of $\mathcal{A}$.

If there exists a scalar function $\psi : \mathcal{C} \rightarrow \mathbf{R}$ such that for all $i \in [1, m]$:

$$\frac{\partial \psi}{\partial q_i} = \omega_i(\mathbf{q}),$$

then the constraint is integrable. Since G, hence ω, is smooth, we can check the existence of ψ by testing that:

$$\frac{\partial \omega_j}{\partial q_i} = \frac{\partial \omega_i}{\partial q_j}.$$

However, this is only a sufficient condition for the integrability of the constraint. Indeed, when this sufficient condition is not verified, there may nevertheless exist a function ψ such that for all $i \in [1, m]$:

$$\frac{\partial \psi}{\partial q_i} = \lambda(\mathbf{q})\omega_i(\mathbf{q}),$$

where λ is a non-zero scalar function called *integrating factor*. If such a ψ exists, then the kinematic constraint can be integrated.

From the above relation and:

$$\frac{\partial^2 \psi}{\partial q_i \partial q_j} = \frac{\partial^2 \psi}{\partial q_j \partial q_i}, \quad i \neq j,$$

we obtain that:

$$\lambda \left(\frac{\partial \omega_i}{\partial q_j} - \frac{\partial \omega_j}{\partial q_i} \right) = \omega_j \frac{\partial \lambda}{\partial q_i} - \omega_i \frac{\partial \lambda}{\partial q_j}$$

for all $i, j \in [1, m]$, $i \neq j$.

By multiplying this relation by ω_k, with $k \in [1, m]$, $k \neq i, j$, and adding the relations obtained for all the circular permutations of i, j, and k, we get:

$$\omega_i \left(\frac{\partial \omega_k}{\partial q_j} - \frac{\partial \omega_j}{\partial q_k} \right) + \omega_j \left(\frac{\partial \omega_i}{\partial q_k} - \frac{\partial \omega_k}{\partial q_i} \right) + \omega_k \left(\frac{\partial \omega_j}{\partial q_i} - \frac{\partial \omega_i}{\partial q_j} \right) = 0.$$

One such relation can be established for every triple (i, j, k) with $1 \leq i < j < k \leq m$. It can be shown that the conjunction of all these relations form a necessary and sufficient condition for the integrability of the constraint. This result is usually established as a corollary of the Frobenius theorem [Spivak, 1979].

COROLLARY 1: *A non-singular scalar linear kinematic constraint:*

$$G(\mathbf{q}, \dot{\mathbf{q}}) = \omega(\mathbf{q}) \cdot \dot{\mathbf{q}} = \sum_{i=1}^{i=m} \omega_i(\mathbf{q})\dot{\mathbf{q}}_i = 0$$

is holonomic if and only if the following relation holds for any $i, j, k \in [1, m]$ such that $1 \leq i < j < k \leq m$:

$$A_{ijk} = \omega_i \left(\frac{\partial \omega_k}{\partial q_j} - \frac{\partial \omega_j}{\partial q_k} \right) + \omega_j \left(\frac{\partial \omega_i}{\partial q_k} - \frac{\partial \omega_k}{\partial q_i} \right) + \omega_k \left(\frac{\partial \omega_j}{\partial q_i} - \frac{\partial \omega_i}{\partial q_j} \right) = 0.$$

This result provides an effective characterization of holonomy (and hence nonholonomy) for a single scalar linear kinematic constraint.

Example: Let us consider the car-like robot of Figure 1. The kinematic constraint established in Section 1 is:

$$-\dot{x} \sin \theta + \dot{y} \cos \theta = 0.$$

We have:

$$\omega_1 = \omega_x = -\sin \theta, \quad \omega_2 = \omega_y = \cos \theta, \quad \omega_3 = \omega_\theta = 0.$$

Applying Corollary 1, we compute the unique coefficient $A_{123} = A_{xy\theta}$:

$$A_{xy\theta} = \omega_x \left(\frac{\partial \omega_\theta}{\partial y} - \frac{\partial \omega_y}{\partial \theta} \right) + \omega_y \left(\frac{\partial \omega_x}{\partial \theta} - \frac{\partial \omega_\theta}{\partial x} \right) + \omega_\theta \left(\frac{\partial \omega_y}{\partial x} - \frac{\partial \omega_x}{\partial y} \right) = -1.$$

Since $A_{xy\theta} \neq 0$, the kinematic constraint of the car-like robot is nonholonomic. ■

4 Controllability of Nonholonomic Robots

4.1 Notion of Controllability

Once we have established that constraints of the form $G(\mathbf{q}, \dot{\mathbf{q}}) = 0$, are non-integrable, hence nonholonomic, the next important question is "Do they restrict the set of configurations achievable by the robot?". Indeed, we just saw that a non-integrable constraint $G(\mathbf{q}, \dot{\mathbf{q}}) = 0$ restricts the set of the possible velocities of the robot to a vector subspace of the tangent space at $\mathbf{q}$. But we have not determined yet whether it also restricts the set of achievable configurations. (At this point, we only know that it does not reduce the dimension of this set, since otherwise it would be integrable.)

In order to understand the issue, it is convenient to regard the robot's velocity vector $\dot{\mathbf{q}}$ as the control vector, i.e. the vector that creates motion. In the holonomic case, at any configuration $\mathbf{q}$, the control space coincides with the tangent space $T_{\mathbf{q}}(\mathcal{C})$. Hence, the configuration space is locally diffeomorphic to the control space. Every configuration $\mathbf{q}'$ in a small neighborhood of $\mathbf{q}$ can be achieved from $\mathbf{q}$ by selecting a vector $\dot{\mathbf{q}}$ appropriately. In the nonholonomic case, this is no longer true. The dimension of the control space is smaller than that of the tangent space. For example, the control space of the car-like robot is two-dimensional (one dimension corresponds to the linear velocity of the car along its main axis; the other dimension corresponds to its angular velocity, which directly depends on the steering angle). It is no longer certain that there exists a feasible path between a configuration $\mathbf{q}$ and a configuration $\mathbf{q}'$ located in a small neighborhood of $\mathbf{q}$.

However, in the particular case of a car-like robot, if two configurations $\mathbf{q}$ and $\mathbf{q}'$ are located in the same connected component of the robot's free space $\mathcal{C}_{free}$, we know by experience that using the velocity of the rear wheels and the steering angle of the front wheels as the control parameters, we can "drive" the robot from $\mathbf{q}$ to $\mathbf{q}'$. Can this experimental evidence be formally proven? Can it be generalized? These are the questions that we address in this section and the next one.

Let us define the controllability of a nonholonomic robot as follows:

DEFINITION 3: *Let $\mathcal{A}$ be a robot subject to nonholonomic constraints. A path τ of $\mathcal{A}$ is* **feasible** *if it is piecewise of class C^1 and satisfies these constraints*[6]. *We say that the robot is* **fully controllable** *if and only if, for any distribution of obstacles in the workspace, if there exists a free*

[6] If $G(\mathbf{q}, \dot{\mathbf{q}}) = 0$ is one of the nonholonomic constraints, then τ must satisfy $\forall s \in [0,1] : G(\tau(s), \frac{d\tau}{ds}(s)) = 0$.

path between any two configurations $\mathbf{q}_{init}$ *and* $\mathbf{q}_{goal}$, *then there also exists a feasible free path between these two configurations.*

In the following subsection we give general results characterizing the controllability of robots subject to nonholonomic equality constraints. Next, we apply these results to two types of nonholonomic robots, the car-like robot and the tractor-trailer robot. In Section 5 we analyze the effect of the combination of nonholonomic equality and inequality constraints on the controllability of a robot.

4.2 General Results

The Frobenius Integrability Theorem is the theoretical basis of some major results in controllability theory for nonlinear control systems [Isidori, 1989]. The applicability of these results to the analysis of the controllability of nonholonomic robots was first noticed by Li and Canny [Li and Canny, 1989], with subsequent contributions by Laumond and Siméon [Laumond and Siméon, 1989], Barraquand and Latombe [Barraquand and Latombe, 1989b], and Dorst, Mandhyan and Trovato [Dorst, Mandhyan and Trovato, 1989].

We first analyze the controllability of a robot subject to k ($k \geq 1$) independent nonholonomic equality constraints linear in $\dot{\mathbf{q}}$. Next, we consider the particular case where there is a single constraint and we establish a straightforward, but particularly useful, corollary of the general result.

A key concept in controllability theory is the so-called *Control Lie Algebra.* It can be defined as follows. Let Δ be a $(m-k)$-distribution on an m-dimensional manifold generated by a set of independent smooth vector fields $\vec{X}_1, \ldots, \vec{X}_{m-k}$. The Control Lie Algebra associated with Δ, denoted by $CLA(\Delta)$, is the smallest distribution which contains Δ and is closed under the Lie bracket operation. Stated otherwise, $CLA(\Delta)$ is the distribution generated by $\vec{X}_1, \ldots, \vec{X}_{m-k}$ and all their Lie brackets recursively computed. By construction, the dimension p of $CLA(\Delta)$ verifies: $m - k \leq p \leq m$.

Consider the path shown in Figure 3 as the concatenation of four paths along the flow of four successive vector fields, $\vec{X}$, $\vec{Y}$, $-\vec{X}$, $-\vec{Y}$, each path having a small duration δt. In first-order approximation, the configuration reached by this path can also be attained by a motion along the vector field $[\vec{X},\vec{Y}]$. Thus, if $CLA(\Delta)$ has dimension m, then any configuration in a small neighborhood (in $\mathcal{C}$) of a configuration $\mathbf{q}$ can be attained by concatenating a finite number of motions following vector fields of Δ. It turns out that this condition is both necessary and sufficient, and that it can be globalized to the whole integral manifold (see [Sussman and Jurdevic, 1972] [Hermann and Krener, 1977] [Isidori, 1989] [Li and Canny, 1989]). It is expressed in the following theorem:

THEOREM 2: *Let Δ be the $(m-k)$-distribution determined on an m-dimensional configuration space $\mathcal{C}$ by a non-singular system of k nonholonomic equality constraints linear in $\dot{\mathbf{q}}$. Let $CLA(\Delta)$ be the Control Lie Algebra associated with Δ. Any two configurations $\mathbf{q}_{init}$ and $\mathbf{q}_{goal}$ of an open connected subset of $\mathcal{C}$ can be connected by a feasible path lying in this subset and following the flow of a finite sequence of vectors of Δ if and only if the dimension of $CLA(\Delta)$ is equal to m.*

In other words, a nonholonomic robot is fully controllable if and only if its Control Lie Algebra has maximal dimension. Since at each instant the robot's path follows the flow of a vector of Δ, which has dimension lower than m, reaching the goal configuration may require the robot to switch (a finite number of times) between vector fields in order to "create" the necessary motion along the dimensions of $CLA(\Delta)$ not represented in Δ.

There is an immediate corollary of this theorem which is particularly useful for characterizing the controllability of a robot that is subject to a single scalar nonholonomic equality constraint. The constraint determines an $(m-1)$-distribution Δ generated by $\{\vec{X}_1,\ldots,\vec{X}_{m-1}\}$. According to the Frobenius Theorem, for each $\mathbf{q}\in\mathcal{C}$, there must exist at least one pair of integers $i,j\in[1,m-1]$ such that the Lie bracket $[\vec{X}_i,\vec{X}_j]$ does not belong to Δ. Therefore, the Control Lie Algebra has a dimension p strictly greater than $m-1$. Since it cannot be greater than m, it is equal to m. Therefore:

COROLLARY 2: *Any robot that is subject to a single non-singular scalar linear nonholonomic equality constraint is fully controllable.*

From this corollary, we directly derive that the car-like robot with no limitation on the steering angle is fully controllable.

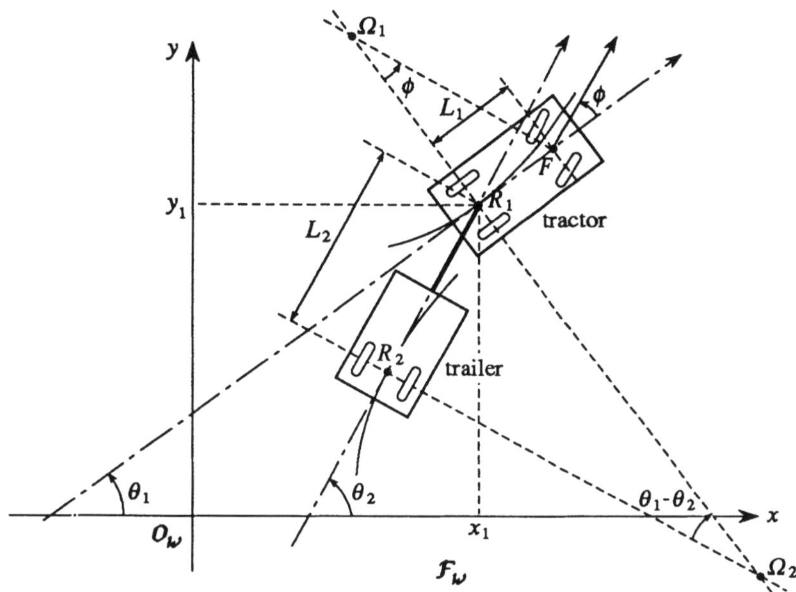

Figure 4. The tractor-trailer robot consists of a tractor (similar to a car-like robot) and a trailer moving on a flat ground. Its configuration space is $\mathbf{R}^2 \times S^1 \times S^1$. We represent a configuration by $(x_1, y_1, \theta_1, \theta_2)$, where x_1 and y_1 are the coordinates of R_1 (the midpoint between the two rear wheels of the tractor), and θ_1 and θ_2 are the respective orientations of the tractor and the trailer. This robot is constrained by two nonholonomic equality constraints. The steering angle ϕ of the tractor determines the instantaneous center of rotation Ω_1 of the tractor. The relative orientation of the tractor and the trailer determines the instantaneous center of rotation Ω_2 of the trailer.

4.3 Application to the Tractor-Trailer Robot

Consider a robot consisting of a tractor towing a two-wheel trailer (Figure 4). We assume that the robot moves in a plane and that the contact between each wheel and the ground is a pure rolling contact. The motion of this robot is constrained by two kinematic constraints:

- The velocity of the midpoint between the rear wheels of the tractor is tangent to the orientation of the tractor.
- The velocity of the midpoint between the two wheels of the trailer is tangent to the orientation of the trailer.

(The kinematics of such a robot is investigated in several papers, including [Alexander, 1984] [Alexander and Maddocks, 1988].)

Let R_1 be the midpoint between the two rear wheels of the tractor, and (x_1, y_1) their coordinates in the frame $\mathcal{F}_\mathcal{W}$ attached to the workspace. Let R_2 be the midpoint between the two wheels of the trailer. We denote by θ_1 (resp. θ_2) the angle between the x-axis of $\mathcal{F}_\mathcal{W}$ and the main axis of the tractor (resp. the trailer). We assume for simplification that the center of rotation of the revolute joint between the tractor and the trailer is R_1. Let $L_2 > 0$ be the distance between R_1 and R_2.

After having eliminated the two holonomic constraints imposed by the revolute joint, the configuration space of the robot is $\mathbf{R}^2 \times S^1 \times S^1$. We represent a configuration as $(x_1, y_1, \theta_1, \theta_2) \in \mathbf{R}^2 \times [0, 2\pi) \times [0, 2\pi)$.

The motion of the robot is subject to the following two constraints:

$$- \dot{x}_1 \sin\theta_1 + \dot{y}_1 \cos\theta_1 = 0 \tag{4}$$

which expresses that the curve followed by R_1 is tangent to the main axis of the tractor, and:

$$-\dot{x}_2 \sin\theta_2 + \dot{y}_2 \cos\theta_2 = 0$$

which expresses that the curve followed by R_2 is tangent to the main axis of the trailer. We have:

$$\begin{array}{rcl} x_2 & = & x_1 - L_2 \cos\theta_2, \\ y_2 & = & y_1 - L_2 \sin\theta_2. \end{array}$$

Hence, the second kinematic constraint can be expressed in terms of the configuration parameters and their derivatives:

$$- \dot{x}_1 \sin\theta_2 + \dot{y}_1 \cos\theta_2 - L_2\dot{\theta}_2 = 0. \tag{5}$$

We now analyze the integrability of the kinematic constraints (4) and (5). Since these two constraints are independent, we cannot use Corollary 1. However, we can compute the dimension p of the Control Lie Algebra associated with the two constraints. We show below that p is maximal, i.e. $p = m = 4$. Combined with Theorem 1 (Frobenius Theorem), this result directly entails that the tractor-trailer robot is nonholonomic. Combined with Theorem 2, it entails that the tractor-trailer robot is fully controllable. This result was established in [Laumond and Siméon, 1989] and [Barraquand and Latombe, 1989b].

PROPOSITION 1: *The Control Lie Algebra associated with the two kinematic constraints of the tractor-trailer robot has maximal dimension* $m = 4$.

Proof: The following two independent vector fields, $\vec{X}_1$ and $\vec{X}_2$, satisfy both constraints (4) and (5):

$$\begin{aligned}\vec{X}_1 &= (L_2\cos\theta_1 \quad L_2\sin\theta_1 \quad 0 \quad -sin(\theta_1-\theta_2))^T,\\ \vec{X}_2 &= (0 \quad 0 \quad 1 \quad 0)^T.\end{aligned}$$

We compute:

$$\begin{aligned}\vec{X}_3 &= [\vec{X}_1,\vec{X}_2] = (-L_2\sin\theta_1 \quad L_2\cos\theta_1 \quad 0 \quad -\cos(\theta_1-\theta_2))^T,\\ \vec{X}_4 &= [\vec{X}_1,\vec{X}_3] = (0 \quad 0 \quad 0 \quad -1)^T.\end{aligned}$$

Finally, we verify that the four above vector fields are independent:

$$det(\vec{X}_1,\vec{X}_2,\vec{X}_3,\vec{X}_4) = L_2^2 \neq 0.$$

Therefore, the Control Lie Algebra has maximal dimension $p = 4$. ∎

Using the same approach, Li and Canny showed that a ball can reach any configuration on a plane by a pure rolling motion. They also showed that a ball can reach any configuration in contact with a fixed ball by rolling, if the radii of the two balls are different [Li and Canny, 1989].

5 Nonholonomic Inequality Constraints

In this section we consider the effect of nonholonomic inequality constraints on the controllability of a robot. We first analyze the particular case of the car-like robot with a lower-bounded turning radius. Then we informally discuss the issue in more generality (see [Barraquand and Latombe, 1990b] for a more formal discussion).

5.1 Car-Like Robot Example

Consider a car-like robot $\mathcal{A}$ (Figure 1). Assume that its turning radius is lower-bounded by the constraint:

$$\dot{x}^2 + \dot{y}^2 - \rho_{min}^2 \dot{\theta}^2 \geq 0$$

established in Section 1. Below, we prove in a constructive way that despite this constraint the robot remains fully controllable. To that purpose, we first define two types of "maneuvers". Each type of maneuver is obtained by concatenating paths separated by configurations where the velocity v changes sign. Next, we show that any (possibly non-feasible) free path can be transformed into a topologically equivalent feasible path between the same two configurations by introducing a finite (but unbounded) number of these maneuvers. The construction is due to Laumond [Laumond, 1986].

To begin with, we notice that there is a bijective correspondence between the paths of $\mathcal{A}$ in the $xy\theta$-space and their projections on the xy-plane. Each path τ projects in the xy-plane onto the curve γ that is followed by the point R. At every point (x, y), the tangent vector of γ makes an angle θ with the x-axis, such that (x, y, θ) is a point of τ. Below, we work with the projected curve γ, rather than with the path τ. By extension, we call γ a path. Each point of γ is designated by a configuration (x, y, θ), where x and y are the coordinates of the point and θ is the orientation of the tangent vector of γ at this point.

A *reversal* is a configuration where the tangent to the path changes sign along the same direction. Hence, it produces a cusp where the path is not differentiable. A *maneuver* is a concatenation of paths separated by reversals.

We define two types of maneuvers which combine turns (with minimal turning radius) and straight segments:

- **Maneuver of type 1:** A maneuver of type 1 allows the robot to perform a sidewise motion. It consists of three subpaths illustrated in Figure 5. The first subpath is a turn from the initial configuration (x_0, y_0, θ_0) to an intermediate configuration $(x_1, y_1, \theta_0 + \delta\theta)$, with $|\delta\theta| < \pi/2$. The second subpath is a straight segment from $(x_1, y_1, \theta_0 + \delta\theta)$ to $(x_2, y_2, \theta_0 + \delta\theta)$. The third subpath is a turn from $(x_2, y_2, \theta_0 + \delta\theta)$

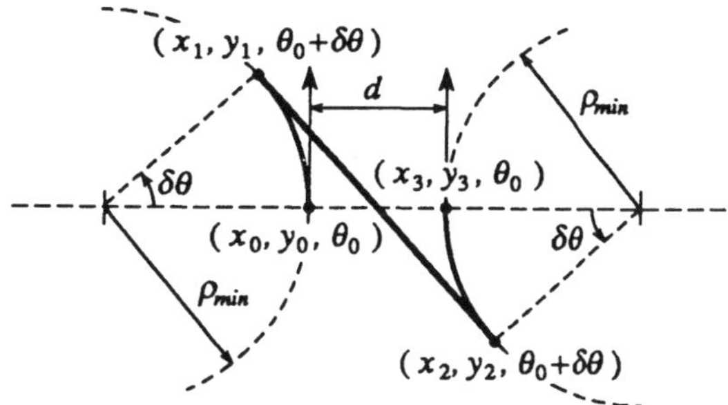

Figure 5. A maneuver of type 1 allows the car-like robot to translate sidewise. Under the condition $|\delta\theta| < \pi/2$, the net length d of the translation is equal to $2\rho_{min}(\frac{1}{\cos\delta\theta} - 1)$, which is strictly positive for any $|\delta\theta| > 0$. The maneuver shown in this figure results in a translation of the robot toward its right. A similar maneuver would produce a leftwise translation.

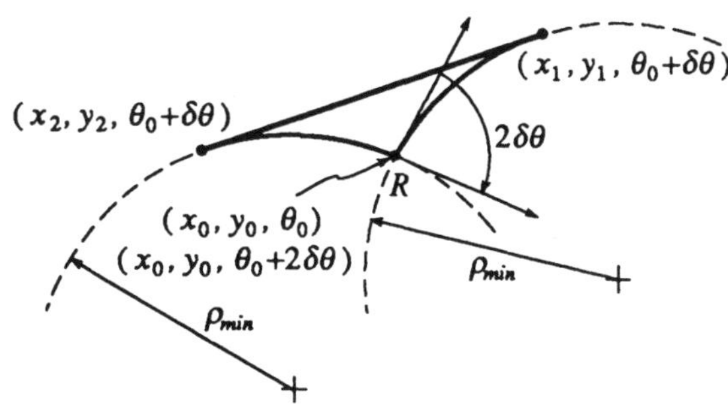

Figure 6. A maneuver of type 2 allows the car-like robot to rotate around the point R, as if it had a zero turning radius. The maneuver shown in the figure makes the robot rotate clockwise. A similar maneuver would result in a counterclockwise rotation.

to the final configuration (x_3, y_3, θ_0). The initial and final orientations of $\mathcal{A}$ are the same.

- **Maneuver of type 2:** A maneuver of type 2 allows the robot to perform a rotation around the point R. It consists of three subpaths illustrated in Figure 6. The first subpath is a turn from the initial configuration (x_0, y_0, θ_0) to an intermediate configuration $(x_1, y_1, \theta_0 + \delta\theta)$, with $|\delta\theta| < \pi/2$. The second subpath is a straight segment from $(x_1, y_1, \theta_0 + \delta\theta)$ to $(x_2, y_2, \theta_0 + \delta\theta)$. The third subpath is a turn from $(x_2, y_2, \theta_0 + \delta\theta)$ to the final configuration $(x_0, y_0, \theta_0 + 2\delta\theta)$. The initial

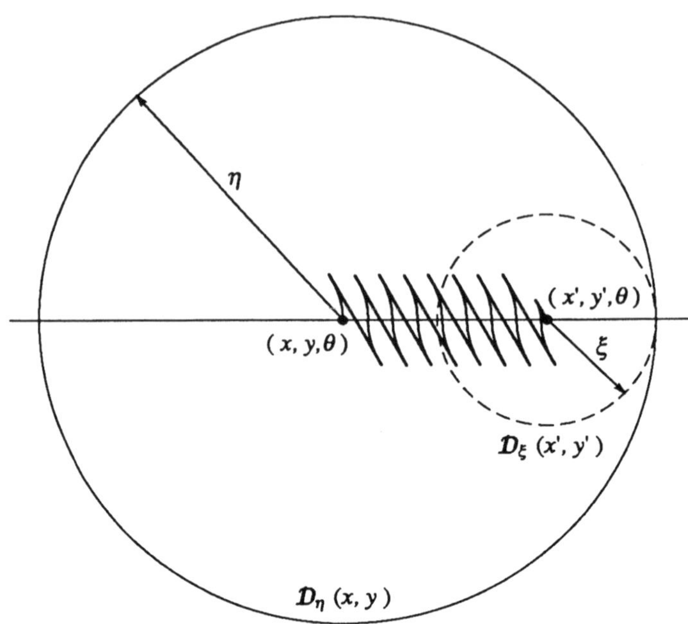

Figure 7. The two configurations (x, y, θ) and (x', y', θ) can be connected by a feasible path completely contained in the cylinder $CYL(x, y, \theta, \eta, \varepsilon)$ by executing a finite number of maneuvers of type 1 (see Lemma 1).

and final positions of R are the same.

Let $CYL(x, y, \theta, \eta, \varepsilon) = \mathcal{D}_\eta(x, y) \times (\theta - \varepsilon, \theta + \varepsilon)$ be an open circular cylinder of height 2ε parallel to the θ-axis, centered at (x, y, θ), and completely contained in $\mathcal{A}$'s free space $\mathcal{C}_{free}$. Its cross-section $\mathcal{D}_\eta(x, y)$ is the open disc of radius η.

LEMMA 1: *Let* $\mathbf{q} = (x, y, \theta)$ *and* $\mathbf{q}' = (x', y', \theta)$ *be two configurations in* $CYL(x, y, \theta, \eta, \varepsilon)$ *such that the line passing through* (x, y) *and* (x', y') *is perpendicular to the main axis of the car. There exists a feasible path from* $\mathbf{q}$ *to* $\mathbf{q}'$ *that consists of a finite sequence of maneuvers of type 1 that are all completely contained in* $CYL(x, y, \theta, \eta, \varepsilon)$.

Proof: The projections of the cylinder $CYL(x, y, \theta, \eta, \varepsilon)$ and the configurations $\mathbf{q}$ and $\mathbf{q}'$ are shown in Figure 7. Let ξ be the distance from the point (x', y') to the boundary of the disc $\mathcal{D}_\eta(x, y)$. If the robot can make a maneuver of type 1 starting at (x, y, θ) that is small enough to be completely contained in the cylinder $CYL(x, y, \theta, \xi, \varepsilon)$, while moving sidewise by a non-zero distance, then it can attain $\mathbf{q}'$ from $\mathbf{q}$ by executing this same maneuver a finite number of times (as illustrated in Figure 7). Eventually, the last maneuver has be made slightly shorter in order to terminate exactly at $\mathbf{q}'$.

The sidewise distance d produced by a maneuver of type 1 is:

$$d = 2\rho_{min}\left(\frac{1}{\cos\delta\theta} - 1\right)$$

(see Figure 5), which is strictly positive for any $0 < |\delta\theta| < \pi/2$. In order for the maneuver to be contained in a cylinder of radius ξ, it is sufficient to impose that the distance between the two intermediate configurations (cusps) of the maneuver be smaller[7] than ξ. This constraint yields:

$$|\tan\delta\theta| < \frac{\xi}{2\rho}.$$

In addition, the maneuver must verify $|\delta\theta| < \varepsilon$. Hence:

$$|\delta\theta| < \min\{\varepsilon, \tan^{-1}(\frac{\xi}{2\rho})\}.$$

Since both arguments of the min are strictly positive, it is possible to select a non-zero angular increment $\delta\theta$ yielding a non-zero distance d. Thus, $\mathbf{q}$ and $\mathbf{q}'$ can be connected by a finite number of maneuvers of type 1. ∎

The following lemma relative to maneuvers of type 2 can be proven in a similar way:

LEMMA 2: *Let* $\mathbf{q} = (x, y, \theta)$ *and* $\mathbf{q}' = (x, y, \theta')$ *be two configurations in* $CYL(x, y, \theta, \eta, \varepsilon)$. *There exists a feasible path from* $\mathbf{q}$ *to* $\mathbf{q}'$ *that consists of a finite sequence of maneuvers of type 2 that are all completely contained in* $CYL(x, y, \theta, \eta, \varepsilon)$.

We can combine the two above results into the following lemma:

[7] This condition is conservative and is certainly not optimal.

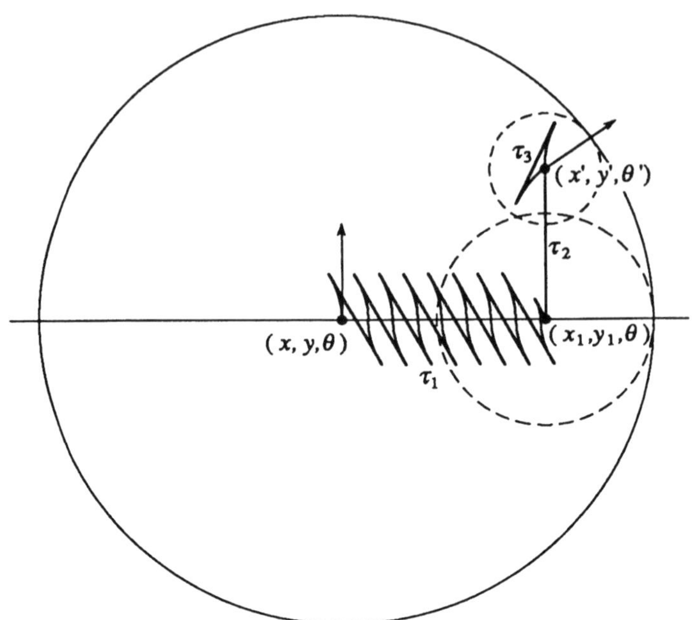

Figure 8. This figure illustrates the proof of Lemma 3. The two configurations (x, y, θ) and (x', y', θ') can be connected by a feasible path τ contained in the cylinder $CYL(x, y, \theta, \eta, \varepsilon)$. This path can be constructed as the product of three paths denoted by τ_1, τ_2 and τ_3. τ_1 is a finite sequence of maneuvers of type 1. τ_2 is a straight path at constant orientation. τ_3 is a finite sequence of maneuvers of type 2 (only one is shown in the figure).

LEMMA 3: *Let* $\mathbf{q}$ *and* $\mathbf{q}'$ *be any two configurations in* $CYL(x, y, \theta, \eta, \varepsilon)$. *There exists a feasible path from* $\mathbf{q}$ *to* $\mathbf{q}'$ *that is completely contained in* $CYL(x, y, \theta, \eta, \varepsilon)$.

Proof: (See Figure 8.) Let (x_1, y_1) be the point obtained by projecting (x', y') on the line of orientation $\theta + \pi/2$ passing through (x, y). From Lemma 1, there exists a feasible path τ_1 contained in the cylinder that connects (x, y, θ) to (x_1, y_1, θ). The straight path τ_2 from (x_1, y_1, θ) to (x', y', θ) is also in the cylinder. Let $CYL(x', y', \theta, \eta', \varepsilon)$, with $\eta' \neq 0$, be an open cylinder centered at (x', y', θ) and contained in $CYL(x, y, \theta, \eta, \varepsilon)$. From Lemma 2, there exists a feasible path τ_3 contained in $CYL(x', y', \theta, \eta', \varepsilon)$ that connects (x', y', θ) to (x', y', θ'). The path $\tau_1 \bullet \tau_2 \bullet \tau_3$ is feasible, it connects $\mathbf{q}$ to $\mathbf{q}'$, and it is contained in $CYL(x, y, \theta, \eta, \varepsilon)$. ∎

The following theorem is a direct consequence of the above lemma:

THEOREM 3: *Let $\mathbf{q}$ and $\mathbf{q}'$ be two configurations in the same connected component of $\mathcal{A}$'s free space $\mathcal{C}_{free}$. There exists a feasible free path of $\mathcal{A}$ from $\mathbf{q}$ to $\mathbf{q}'$.*

Proof: Since $\mathbf{q}$ and $\mathbf{q}'$ are in the same connected component of $\mathcal{C}_{free}$, there exists a (possibly non-feasible) free path τ between them. Since τ is a compact subset of $\mathcal{C}_{free}$, it can be covered by a finite (although non-bounded) number of open cylinders of the same form as above. Hence, we can transform τ into a topologically equivalent feasible free path by introducing a finite number of maneuvers of types 1 and 2 that are contained in these cylinders. ∎

Hence, the problem of deciding whether there exists a feasible path for the car-like robot between any two given configurations can be solved in polynomial time. Unfortunately, the number of maneuvers introduced in a path by the above constructive proof is not bounded. Hence, the time required to solve the planning problem itself is more difficult to analyze.

5.2 Generalization

The above constructive proof essentially consists of combining paths obtained by following the flow of the vector fields $\vec{X}_1$, $\vec{X}_2$ and $\vec{X}_3$, and their negation, with:

$$\begin{aligned}
\vec{X}_1 &= (\cos\theta \quad \sin\theta \quad 0)^T, \\
\vec{X}_2 &= (\cos\theta \quad \sin\theta \quad 1/\rho_{min})^T, \\
\vec{X}_3 &= (\cos\theta \quad \sin\theta \quad -1/\rho_{min})^T.
\end{aligned}$$

These vector fields are not independent and they generate a two-dimensional distribution Δ. It is easy to verify that the Control Lie Algebra of this distribution has (maximal) dimension 3. A maneuver of type 1 allows the robot to move along the third dimension which is not represented in Δ.

If there were no constraint on the turning radius of $\mathcal{A}$, one would rather define $\vec{X}_2$ as $(0\ 0\ 1)^T$, which expresses that the car can rotate about the point R without translation. Instead, the form of the vector fields $\vec{X}_2$ and $\vec{X}_3$ results from the constraint bounding the turning radius. It expresses that the car cannot rotate with R as the instantaneous rotation center. A maneuver of type 2 allows the robot to "remove" the constraint on the turning radius.

More generally, in the absence of nonholonomic inequality constraints, at any configuration $\mathbf{q}$, a path can follow the flow of any vector contained in the vector subspace selected by the nonholonomic equality constraints (if any) in the tangent space $T_{\mathbf{q}}(\mathcal{C})$. Nonholonomic inequality constraints select a subset of these vectors and require the robot to only follow a flow of vectors in this subset. They may also require that the flow of some vectors be followed in a single direction. For example, if the velocity v of R along the main axis of the car-like robot is constrained to be positive or null (i.e. the robot can only move forward), then the robot can move along $\vec{X}_1$, but not along $-\vec{X}_1$.

Let Δ' be the set of vectors selected by the nonholonomic inequality constraints in the $(m-k)$-distribution Δ defined by the k nonholonomic equality constraints. If Δ' contains $m-k$ independent vector fields $\vec{X}_1, \ldots, \vec{X}_{m-k}$, such that for every $i \in [1, m-k]$, $-\vec{X}_i \in \Delta'$, then using these vectors, we can still generate the Control Lie Algebra $CLA(\Delta)$. Therefore, if the robot is fully controllable when the inequality constraints do not apply, it remains fully controllable when these constraints apply. Indeed, there remains a sufficient number of control vectors around a configuration $\mathbf{q}$ to attain any configuration in a small neighborhood of $\mathbf{q}$ by concatenating paths following the flows of a finite sequence of vectors in Δ' (see [Barraquand and Latombe, 1990b] for a formal proof). If τ is a (possibly non-feasible) free path between two configurations $\mathbf{q}_{init}$ and $\mathbf{q}_{goal}$, one can transform τ into a feasible free path τ' that can be made arbitrarily close to τ.

This result applies to car-like and tractor-trailer robots with lower-bounded turning radius[8]. One can also verify that it also applies when the steering angle ϕ is constrained by $0 < \phi_{min} \leq \phi \leq \phi_{max} < \pi/2$. On

[8] The notions of steering angle and turning radius defined for a car-like robot also applies the tractor of the tractor-trailer robot.

the contrary, a car-like or tractor-trailer robot whose steering angle is constrained by $|\phi| \leq \pi/2$ and whose velocity v is constrained to be non-negative is not fully controllable (although the robot has zero minimal turning radius).

In a similar way, we can see that if a car-like or tractor-trailer robot with a lower-bounded turning radius ρ_{min} can move from a configuration $\mathbf{q}_{init}$ to a configuration $\mathbf{q}_{goal}$ without reversal (path τ), it can also move from $\mathbf{q}_{init}$ to $\mathbf{q}_{goal}$ without reversal by only executing turns of radius ρ_{min} (path τ'). Indeed, the fact that the robot can reach $\mathbf{q}_{goal}$ from $\mathbf{q}_{init}$ without reversal means that it follows the flow of a vector field whose projection along the main axis of the car-like robot (or the tractor) keeps a constant sign. This vector field can be expressed as a non-negative linear combination of two vectors, each corresponding to a right or left turn with minimal turning radius. By concatenating a finite number of paths along these two vectors, τ' can be made arbitrarily close to τ.

6 Planning with Nonholonomic Constraints

Planning motions subject to nonholonomic constraints has recently been addressed in several papers. Different planning approaches have been proposed.

One approach consists of using a constructive proof of controllability such as the one given in Subsection 5.1 [Laumond, 1986] [Laumond and Siméon, 1989]. In a first phase, a free path is constructed ignoring the nonholonomic constraints. In a second phase, this path is transformed into a topologically equivalent feasible free path. The drawback of this approach is that the generated path may include a much larger number of reversals than is actually needed. In certain environments, heuristics may possibly be used to help reducing this number.

Another approach has been specifically developed for car-like robots. It consists of planning the motion of the robot in a network of "corridors" [Tournassoud and Jehl, 1988] or "lanes" [Wilfong, 1988b] that are extracted from the workspace in a first phase of processing (using for instance the freeway extraction technique described in Subsection 3.1 of Chapter 4). Local planning techniques are used to generate turns for transferring the robot from a corridor (or lane) to another at a "crossroad" of the network. The difficulty of the approach is that for most workspaces one cannot define an intrinsic set of corridors (or lanes). In addition, the various turns that have to be generated along a path usually interact, and local planning techniques may result in quite inefficient paths.

Below, we present in detail a planning approach that was originally proposed in [Barraquand and Latombe, 1989b]. This approach has been implemented and gives interesting experimental results with car-like and tractor-trailer robots. It consists of decomposing the configuration space into an array of small rectangloids and searching a directed graph whose nodes are these rectangloids. Two rectangloids are adjacent in this graph if there is a feasible free path between a configuration lying in the first rectangloid and a configuration lying in the second. The arcs of the graph are constructed by discretizing the control parameters of the robot according to the results presented in Subsection 5.2. The graph is searched using an un-informed A^* algorithm (Dijkstra's algorithm). Because of the decomposition of the configuration space, the approach is limited to robots with small configuration spaces. Indeed, it requires time $O(mr^m \log r)$ and space $O(r^m)$, where $O(r)$ is the size of the decomposition along each axis of the configuration space and m is the dimension of the configuration space.

We first describe the approach with the car-like robot. Next, we present its application to the tractor-trailer robot.

6.1 Car-Like Robot

Let us parameterize a path of the robot by t, the elapsed time since the beginning of the motion. Given a steering angle ϕ and a velocity v of the rear point R, the velocity parameters of the robot are:

$$\dot{x} = v\cos\theta, \quad \dot{y} = v\sin\theta, \quad \text{and} \quad \dot{\theta} = \frac{v}{L}\tan\phi.$$

We can integrate these equations. We get:

$$\begin{aligned}
\theta(t) &= \theta(0) + \tfrac{v}{L}\tan\phi t,\\
x(t) &= x(0) + \tfrac{L}{\tan\phi}\left(\sin(\theta(0) + \tfrac{v}{L}\tan\phi) - \sin(\theta(0))\right),\\
y(t) &= y(0) - \tfrac{L}{\tan\phi}\left(\cos(\theta(0) + \tfrac{v}{L}\tan\phi) - \cos(\theta(0))\right).
\end{aligned}$$

We assume that the workspace of the robot is enclosed in a rectangular boundary whose interior is denoted by D. The planner decomposes the configuration space $D \times [0, 2\pi]$ into a 3-dimensional array of small rectangloid cells of equal size. In the experimental examples shown below, D is a square and the resolution of the decomposition is of the order[9] of 128^3. Each rectangloid is a node of a search graph G whose construction will be described below. During the search of G, the planner maintains two lists of nodes, CLOSED and OPEN. CLOSED contains all the nodes whose successors have already been generated (i.e. explored). OPEN contains all the attained nodes whose successors have not been generated yet. CLOSED is simply represented by marking the corresponding cells in the array resulting from the decomposition of the configuration space. Hence, the access time to CLOSED is constant. OPEN (which is usually much smaller) is represented as a balanced tree [Aho, Hopcroft and Ullman, 1983]. Every modification and access of OPEN is made in logarithmic time.

Given a configuration $\mathbf{q}$, the planner generates at most six successors of $\mathbf{q}$ by setting the two control parameters v and ϕ to the six values in:

$$\{-v_0, v_0\} \times \{-\phi_{max}, 0, +\phi_{\max}\}$$

and integrating the velocity parameters of the robot along a "short" distance using the expressions established above. By short distance, we mean that the integration time is 1 and v_0 is set to approximately twice the length of an interval along the x-axis. The goal is to avoid staying in the same cell as $\mathbf{q}$, while not moving further than one neighboring cell.

This discretization of the control parameters derives from the discussion of Subsection 5.2. If there exists a feasible free path between two given configurations $\mathbf{q}_{init}$ and $\mathbf{q}_{goal}$, then there exists a feasible free path between these configurations that is only composed of circular arcs of minimal curvature radius ρ_{min} and that minimizes the number of reversals (over the set of all the feasible free paths between $\mathbf{q}_{init}$ and $\mathbf{q}_{goal}$). The inclusion of $\phi = 0$ in the discretization set is aimed at allowing the robot to move along a straight path and hence reducing the total length of the generated path.

[9]The discretization along the θ-axis is normalized relative to the discretization along the x- and y-axes.

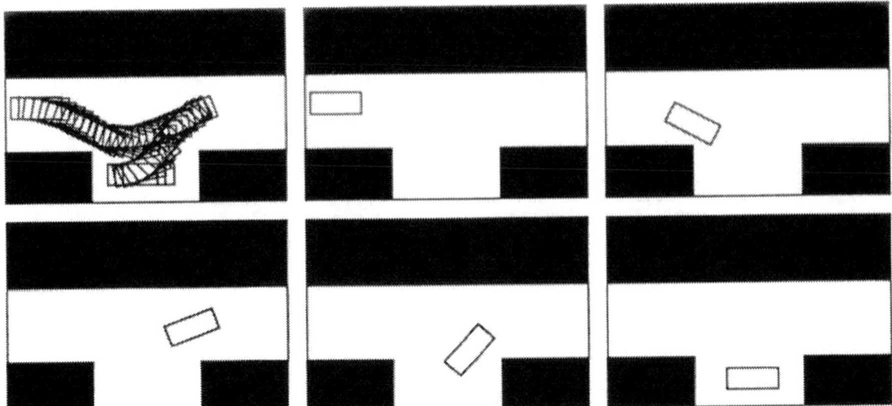

Figure 9. This figure shows a path constructed by the planner described in Subsection 6.1 for the classical parallel parking problem. The steering angle of the car-like robot is limited to 30 degrees.

The planner first places $\mathbf{q}_{init}$ in OPEN. Then, at each iteration, it selects a configuration in OPEN, transfers it to CLOSED, and computes its successors. For each successor, it determines the rectangloid cell to which the configuration belongs. If the cell is marked (hence, is in CLOSED), the configuration is discarded. Otherwise, the configuration is recorded as is (rather than the cell, as indicated before) in OPEN. Hence, the planner never explores twice from the same cell, but it *exactly* records the paths produced by the search. It terminates when the cell containing $\mathbf{q}_{goal}$ is attained (success), or when OPEN is empty (failure). If the goal configuration has to be achieved with a better precision than that provided by the initial decomposition of the configuration space, the planner can decompose the goal cell into an array of smaller cells and repeat the search process in this array.

In the planner described in [Barraquand and Latombe, 1989b], collisions are checked by intersecting the robot at every attained configuration with the obstacles in the workspace (see Subsection 5.2 of Chapter 7). Another way to proceed would be to explicitly compute the C-obstacles and to check the configuration relative to them.

This planner has proven to be able to solve tricky planning problems in reasonable time. We show the paths output by the planner for four

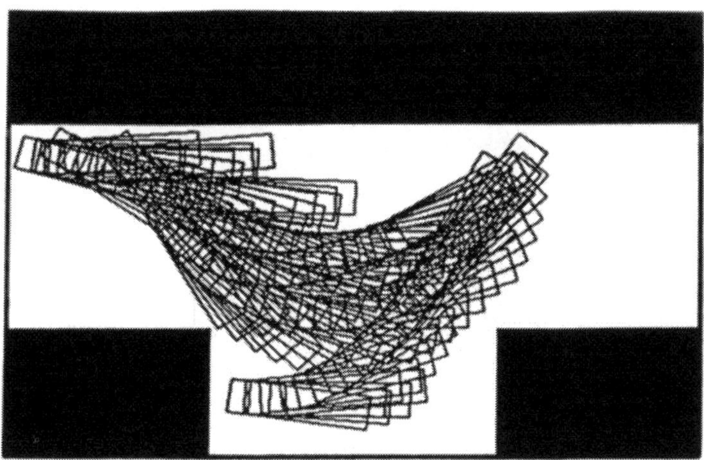

Figure 10. This figure shows a path constructed by the planner described in Subsection 6.1 for the parallel parking problem with a car that can only turn left (22.5 degrees $\leq \phi \leq$ 45 degrees).

problems in Figures 9, 10, 11, and 12. Figure 9 is an illustration of the classical parallel parking problem with a very limited steering angle (ϕ_{max} = 30 degrees). Figure 10 is another parallel parking example, but for a car that can only turn left (22.5 degrees $\leq \phi \leq$ 45 degrees). Figure 11 is an example requiring the robot to maneuver (ten reversals) in a cluttered workspace (ϕ_{max} = 45 degrees). Figure 12 displays a path generated in an "unstructured" workspace (ϕ_{max} = 45 degrees).

The search of a feasible path can be guided by various cost functions. The above examples were generated using the number of reversals as this function. However, in some situations, the minimization of the number of reversals leads the planner to generate paths that are too long. One may avoid this drawback by combining the number of reversals and the length of the path in the cost function.

6.2 Tractor-Trailer Robot

In the case of the tractor-trailer robot, the approach applies in much the same way. Given a steering angle ϕ and a velocity v of the midpoint R_1 between the two rear wheels of the tractor, the velocity parameters of

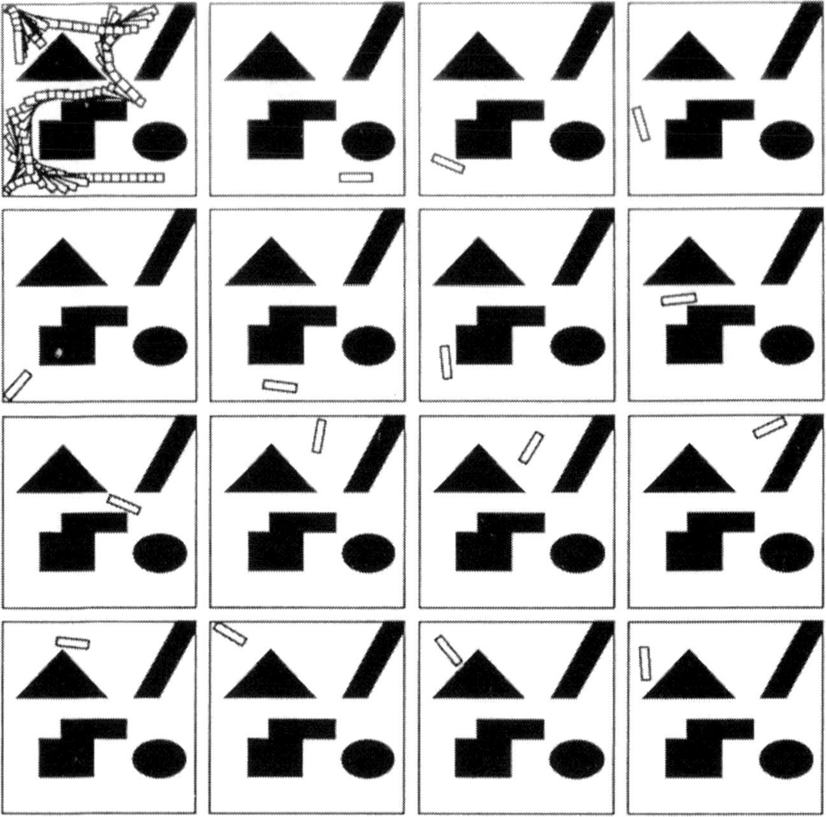

Figure 11. This figure shows another path constructed by the planner of Subsection 6.1. The cluttered environment and the limited steering angle (45 degrees) require the car-like robot to perform multiple reversals.

the robot are:

$$\begin{aligned} \dot{x}_1 &= v\cos\theta_1, & \dot{y}_1 &= v\sin\theta_1, \\ \dot{\theta}_1 &= \tfrac{v}{L_1}\tan\phi, & \dot{\theta}_2 &= \tfrac{v_1}{L_2}\sin(\theta_1-\theta_2). \end{aligned}$$

The generation of the successors of a configuration is slightly more involved than for the car-like robot. While the first three equations in the above system can be integrated analytically when the values of the control parameters v and ϕ remain constant, this is not the case of the fourth equation. The planner can nevertheless solve this equation numerically by using a fourth-order Runge-Kutta method. The paths thus

Figure 12. This figure shows a path constructed by the planner of Subsection 6.1 in a workspace whose obstacles have been randomly generated. The steering angle is limited to 45 degrees.

generated do not exactly satisfy the second nonholonomic constraint.

Figures 13 and 14 shows two paths generated by the planner. Figure 13 illustrates the parallel parking problem for a tractor-trailer with $\phi_{max} =$ 30 degrees. In Figure 14 the robot maneuvers in a cluttered workspace with $\phi_{max} = 45$ degrees. In these examples, the size of the decomposition of the configuration space is of the order of 64^4. The cost function used by the search algorithm is the number of reversals.

7 Point with Bounded Turning Radius

In this section we describe another planning approach applicable when the robot is a point moving among polygonal obstacles along paths whose curvature radius is lower-bounded. The approach, which is adapted from that described in [Jacobs and Canny, 1989], consists of discretizing the robot's contact space and connecting the points resulting from this dis-

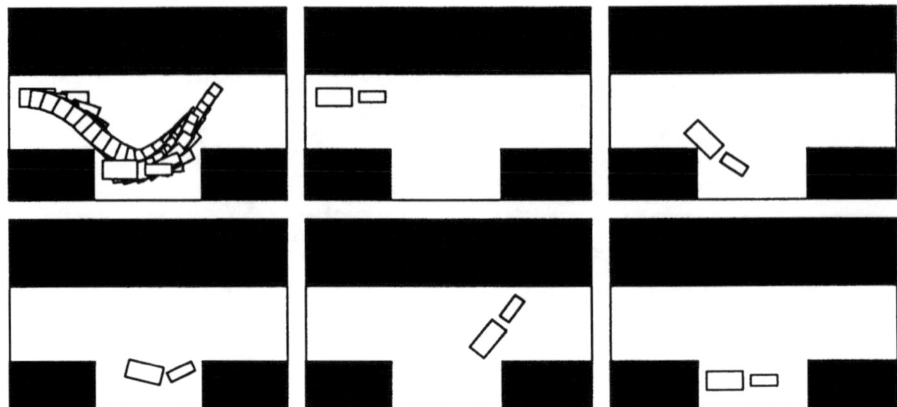

Figure 13. This figure illustrates the parallel parking problem for a tractor-trailer robot. It shows a path generated by the planner described in Subsection 6.2. The steering angle is limited to 30 degrees.

cretization by paths of standard shapes, called *jumps*.

7.1 Problem

The robot $\mathcal{A}$ is a point moving in $\mathcal{W} = \mathbf{R}^2$. The obstacle region $\mathcal{B}$ is a polygonal region such that $\mathcal{W}\backslash\mathcal{B}$ is bounded. In order to be feasible, a path τ of $\mathcal{A}$ must be of class C^1 (no reversals) and piecewise of class C^2; its curvature radius must also be lower-bounded by ρ_{min}.

In the following, we describe a planning method for solving the following problem: Given an initial position (x_{init}, y_{init}) and a goal position (x_{goal}, y_{goal}), compute a feasible semi-free path connecting the initial position to the goal one such that the orientation of the tangent vector to the path at the initial (resp. goal) position is θ_{init} (resp. θ_{goal}). (In order to get a feasible free path, rather than a semi-free one, one would have to slightly grow the obstacles before applying the method.)

The robot can also be regarded as a point with an immaterial arrow of orientation θ attached to it. Under this perspective, its configuration space is three-dimensional and its motion is subject to the following nonholonomic constraints:

$$-\dot{x}\sin\theta + \dot{y}\cos\theta = 0$$

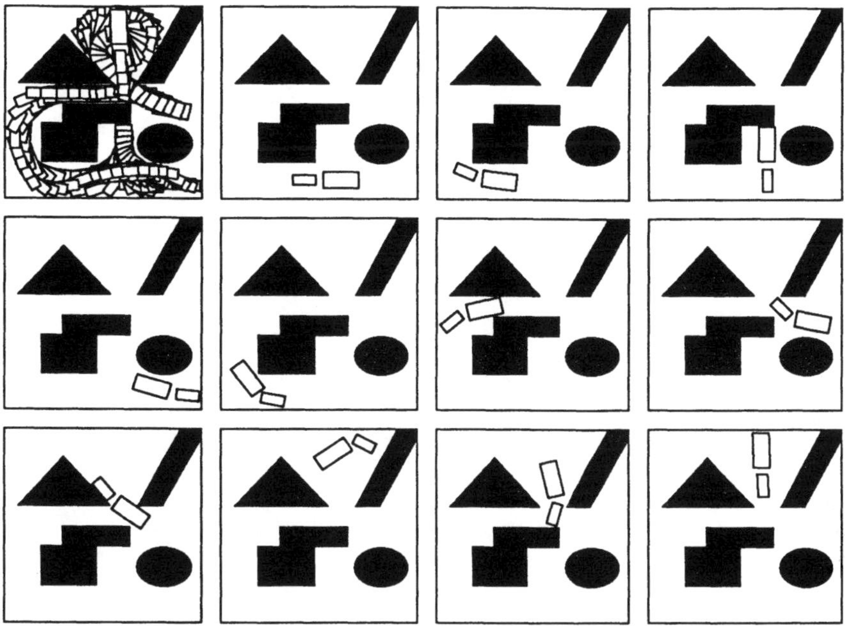

Figure 14. This path was generated by the planner of Subsection 6.2 with a maximal steering angle of 45 degrees.

and:

$$\dot{x}^2 + \dot{y}^2 - \rho_{min}^2 \dot{\theta}^2 \geq 0.$$

A planning problem is specified by the initial configuration (x_{init}, y_{init}, θ_{init}) and the goal configuration (x_{goal}, y_{goal}, θ_{goal}). The C-obstacle region is a cylindrical region parallel to the θ-axis whose projection on the xy-plane is $\mathcal{B}$.

In the following we designate a point in a planar path followed by $\mathcal{A}$ by a configuration (x, y, θ), where x and y are the coordinates of the point robot and θ is the orientation of the tangent vector to the curve at this point.

In practice, the above problem can be obtained by shrinking a nonholonomic mobile robot with a circular perimeter to a point. The polygonal workspace obstacles are then isotropically grown into generalized polygonal C-obstacles. The polygonal obstacle region $\mathcal{B}$ in the above problem may be regarded as an approximation of the generalized polygonal C-

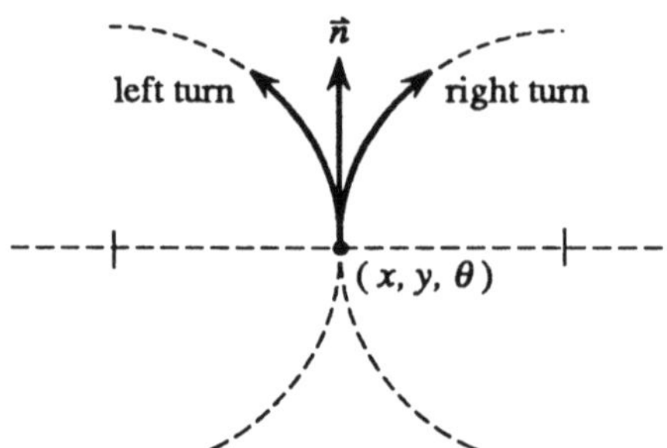

Figure 15. A circular arc issued from (x, y, θ), tangent at this configuration to a vector $\vec{n}$ of orientation θ, and oriented along $\vec{n}$ is a left (resp. right) turn if its orientation is counterclockwise (resp. clockwise).

obstacles. (The problem of planning a reversal-free motion of a point among generalized polygons is investigated in [Laumond, 1987b], where a sketch of a planning algorithm is also presented.)

7.2 Jumps

Let $\mathbf{q} = (x, y, \theta)$ be a configuration of $\mathcal{A}$. A circular arc issued from $\mathbf{q}$, tangent at $\mathbf{q}$ to a vector $\vec{n}$ of orientation θ, and oriented along $\vec{n}$ (see Figure 15) is called a **left turn** (resp. a **right turn**) if its orientation is counterclockwise (resp. clockwise). $C_L(x, y, \theta)$ and $C_R(x, y, \theta)$ denote the two circles of radius ρ_{min} passing through (x, y) and tangent to $\vec{n}$. $C_L(x, y, \theta)$ (resp. $C_R(x, y, \theta)$) corresponds to a left (resp. right) turn.

Let (x_1, y_1, θ_1) and (x_2, y_2, θ_2) be two configurations. A **jump** from (x_1, y_1, θ_1) to (x_2, y_2, θ_2) is any feasible path joining the first configuration to the second, which is obtained by concatenating three subpaths: a circular arc of radius ρ_{min}, a straight line segment, and a circular arc of radius ρ_{min} [Fortune and Wilfong, 1988]. Each of these three subpaths may have zero length.

There are at most four different jumps between two given configurations (see Figure 16), which we respectively denote by LL, LR, RL and RR. The LL (resp. RR) jump contains only left (resp. right) turns. The LR (resp. RL) jump is such that its first turn is a left (resp. right) turn and its second turn is a right (resp. left) turn. A jump may be of several types simultaneously; for example, a jump has both types LL and RL, if

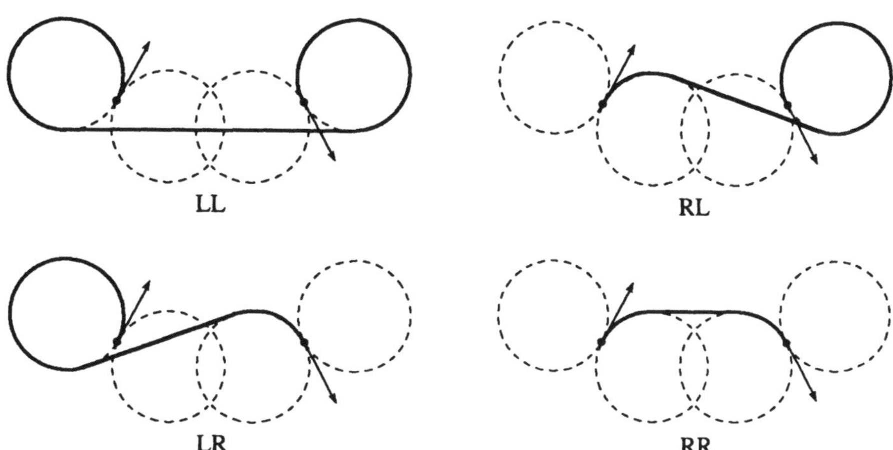

Figure 16. There are at most four jumps between two given configurations. They are denoted by LL, LR, RL, and RR.

its first turn has zero length. The jump of type LR (resp. RL) does not exist if the circles $C_L(x_1, y_1, \theta_1)$ and $C_R(x_2, y_2, \theta_2)$ (resp. $C_R(x_1, y_1, \theta_1)$ and $C_L(x_2, y_2, \theta_2)$) intersect at two distinct points.

The following proposition due to Fortune and Wilfong [Fortune and Wilfong, 1988] states that it is sufficiently general for a planning algorithm to consider paths consisting of a finite sequence of jumps, each taking the robot from the initial configuration or a contact configuration, to the goal configuration or a contact configuration. Each jump lies either in free space (except possibly at its endpoints), or in contact space. A jump in contact space necessarily reduces to a straight line.

PROPOSITION 2: *There exists a feasible semi-free path between two configurations* $(x_{init}, y_{init}, \theta_{init})$ *and* $(x_{goal}, y_{goal}, \theta_{goal})$ *if and only if there exists a feasible semi-free path consisting of either a jump from* $(x_{init}, y_{init}, \theta_{init})$ *to* $(x_{goal}, y_{goal}, \theta_{goal})$ *or the concatenation of a jump from* $(x_{init}, y_{init}, \theta_{init})$ *to a contact configuration, a finite (possibly empty) sequence of jumps between contact configurations, and a final jump to* $(x_{goal}, y_{goal}, \theta_{goal})$.

Proof (informal): This proposition is established in three steps as follows:

- First, we notice that a feasible semi-free path intersects an obstacle at most a finite number of times, i.e. the path can be decomposed into a finite number of subpaths, each one lying either in free space (except at its endpoints) or in contact space. Indeed, a C^1 path has finite length. Furthermore, in order to leave a contact edge and return to it, the robot must follow a feasible path of some non-zero minimum length.

- Second, we decompose each subpath lying in free space (except possibly at its endpoints) into a finite number of small subpaths, so that each of them can be replaced by a free jump. We know from Subsection 5.2 that such a decomposition is possible. It can be effectively constructed as follows. Let τ' be a subpath. If there exists a jump between the initial and the final configuration of τ' that is free (except possibly at its endpoints) we can replace τ' by this jump. Otherwise, we recursively decompose τ' into a sequence of smaller subpaths and we attempt to substitute a jump for each of them. The recursion is necessarily finite since τ' lies in free space (except possibly at its endpoints) and the jumps preserve the tangent orientations at the endpoints.

- Finally, the intermediate circular arcs contained in the subpaths lying in free space are eliminated. An intermediate free circular arc of length $L \geq \pi\rho_{min}$ is eliminated by translating a subarc of length $\pi\rho_{min}$ (thus introducing two new straight segments in the path) until it touches an obstacle (which always happens since we assumed that $\mathcal{W}\backslash\mathcal{B}$ is bounded). This elimination is illustrated in Figure 17. Next, the remaining intermediate free arcs (each has length less than $\pi\rho_{min}$) are eliminated by proceeding from the initial configuration of the subpath toward its final configuration. Eliminating such an arc consists of making it shorter. In the process, the arc may simply disappear or it may hit an obstacle. It may also happen that one of the two free straight subpaths which precede and follow the arc touches an obstacle; in this case, a new turn has to be introduced in the path. Figure 18 illustrates the elimination of an intermediate free arc of length less than $\pi\rho_{min}$. ■

Using the above proposition, Fortune and Wilfong [Fortune and Wilfong, 1988] proposed an exact algorithm for deciding whether there exists a

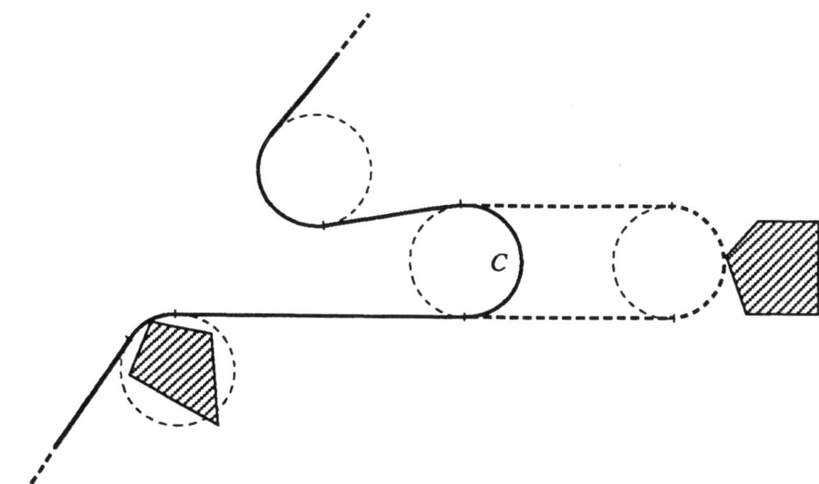

Figure 17. This figure illustrates the third step of the proof of Proposition 2. An intermediate free circular arc (C in the figure) of length $L \geq \pi\rho_{min}$ is eliminated by translating a subarc of length $\pi\rho_{min}$ (thus introducing two new straight segments in the path) until it touches an obstacle. The original subarc of length $\pi\rho_{min}$ is transformed in the dashed subpath shown in the above figure.

feasible semi-free path between two configurations. The time and space complexity of this algorithm is exponential ($2^{O^{poly(n,k)}}$), where n is the number of vertices of $\mathcal{B}$, k is the number of bits needed to specify the coordinates of $\mathcal{B}$'s vertices, and $poly(n,k)$ is a polynomial in n and k.

7.3 Planning Method

Proposition 2 suggests a planning method that consists of discretizing the robot's contact space and building a directed search graph G as follows [Jacobs and Canny, 1989]:

- The initial and goal configurations are nodes of G.

- The range of orientations of the robot is uniformly discretized in $2\pi/\delta$ orientations $\theta_k = k\delta$, with $k = 0, 1, 2, ..., \frac{2\pi}{\delta} - 1$. For each vertex $X = (x, y)$ of $\mathcal{B}$, we generate $2\pi/\delta$ nodes of G, each representing a distinct configuration (x, y, θ_k).

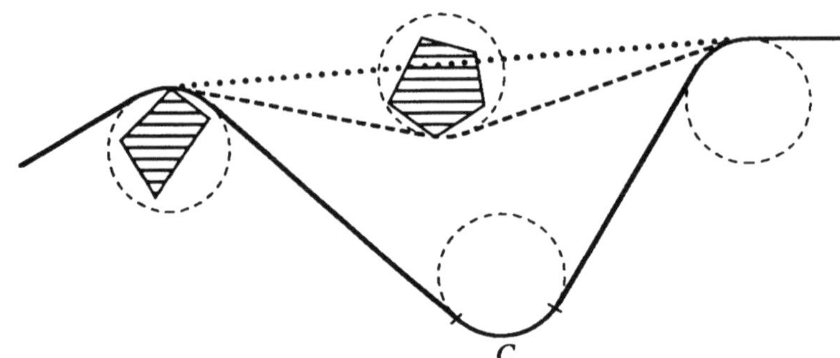

Figure 18. This figure illustrates the elimination of short arcs performed in the third step of the proof of Proposition 2. A circular arc C of length less than $\pi\rho_{min}$ is eliminated by making it shorter. In the process, the arc may simply desappear (the dotted line shows the resulting path if this had happen), it may hit an obstacle (in which case the transformation is completed), or one of the two free straight segments preceding and following the arc (if any) touches an obstacle (this is what happened in the figure). In the later case, a turn is introduced in the path.

- Every edge E of $\mathcal{B}$ is discretized in n_E points $X_l = (x_l, y_l)$, $l = 1, ..., n_E$ (including the endpoints). Let θ_E and $-\theta_E$ be the two possible orientations of $\mathcal{A}$ when it slides along E. For each point X_l, we generate two nodes of G representing the configurations (x_l, y_l, θ_E) and $(x_l, y_l, -\theta_E)$.

A node (x, y, θ) is connected to a node (x', y', θ') by a directed arc of G if and only if there is a jump from (x, y, θ) to (x', y', θ') which lies in free space except possibly at its endpoints. In addition, any two nodes (x, y, θ) and (x', y', θ) are connected by an arc if the two points (x, y) and (x', y') are located in the same contact edge, with (x', y') located beyond (x, y) in the direction of the edge determined by θ.

Path planning can be done by searching this graph for a path connecting the initial configuration to the goal one. Figure 19 shows two paths generated by an implemented planner using this method.

Let us assume that the number of sample points n_E in an edge E is a linear function of $1/\delta$. For example, we may take $n_E \approx \alpha L/\delta$, with L being the length of E and α being a constant independent of E. Then, the number of nodes in G is $O(\frac{n}{\delta})$, where n is the number of vertices

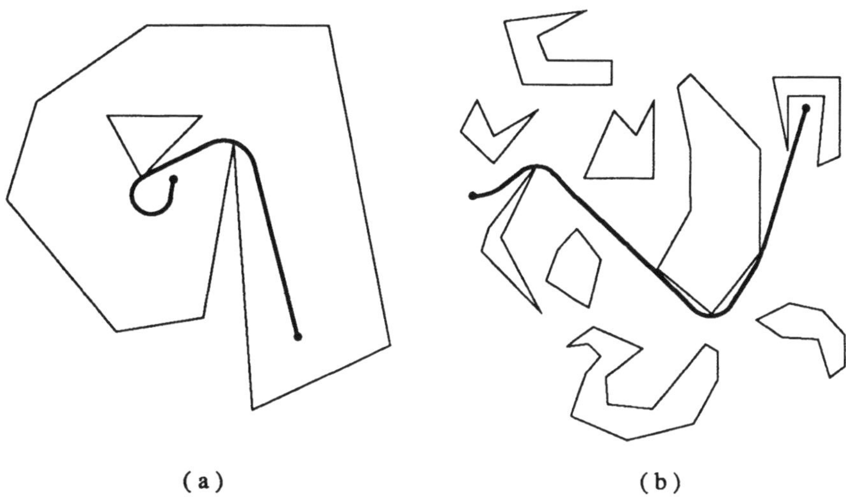

Figure 19. This figure shows two paths generated by the method described in Subsection 7.3 [Jacobs and Canny, 1989].

of $\mathcal{B}$. Each node may have $O(\frac{n}{\delta})$ successors, which are computed in time $O(\frac{n^2}{\delta})$. (Between every two nodes there are at most four possible jumps, which can be computed in constant time. Checking each jump for collision with $\mathcal{B}$ takes time $O(n)$.) Hence, the whole graph G can be computed and searched in time $O(\frac{n^3}{\delta^2})$. Since the number of nodes in G is $O(\frac{n}{\delta})$, the A^* algorithm can be used to search G within this same time complexity (with the Euclidean distance to the goal as the heuristic function). When it terminates successfully, the algorithm then outputs the shortest path among those which are contained in G.

The method can be slightly improved by noticing that only a subset of the discretized configurations (x, y, θ_k), where (x, y) is a contact vertex, can be part of a feasible path. Indeed, if the line of orientation θ_k is not tangent to $\mathcal{B}$ at the contact vertex, the robot either cannot reach the configuration (x, y, θ_k), or cannot leave it. Checking whether a line of orientation θ_k is tangent to $\mathcal{B}$ at a vertex takes constant time. However, this improvement does not modify the worst-case asymptotic time complexity of the overall method.

The method can easily be extended to allow paths to include reversals. Assume now that a path τ is feasible if and only if it is piecewise of class

C^1 and C^2, its curvature radius is lower-bounded by ρ_{min}, and every point where τ is not differentiable is a cusp where $\mathcal{A}$'s velocity changes sign but not direction. It is easily seen that Proposition 2 still applies. The planning method is extended by allowing the robot's path to have a cusp at each contact configuration (hence, between every two nodes there may be as many as eight jumps). If the A^* algorithm is used, an additional cost may be attached to each cusp, in order to avoid producing paths with too many reversals.

7.4 Jump Diagrams

Let us consider the set of jumps of a given type — say, LR jumps — between two points (x_1, y_1) and (x_2, y_2). Call $J_{LR}(x_1, y_1, x_2, y_2)$ this set. Each jump in $J_{LR}(x_1, y_1, x_2, y_2)$ is fully determined by an initial orientation θ_1 and a final one θ_2.

Below, we define critical curves that decompose the $\theta_1\theta_2$-space into regions such that: (1) if a point in a region defines a jump, any other point in the same region defines a jump[10]; (2) any jump in a region can be continuously deformed into any other jump in the same region in such a way that every intermediate paths is a jump belonging to this region; and (3) if a jump in a region intersects with an edge of $\mathcal{B}$, all the jumps in the same region do.

The network of these critical curves is called the **jump diagram** of the set $J_{LR}(x_1, x_2, y_1, y_2)$. The only effective use of this diagram that we present below is the improvement of the worst-case time complexity of the planning method described in the previous subsection. However, jump diagrams (or variants of them) may also play an important role in the future development of exact methods for planning motions subject to nonholonomic constraints.

There are four types of critical curves in the jump diagram of $J_{LR}(x_1, x_2, y_1, y_2)$ [Jacobs and Canny, 1989]:

- A **type 1** critical curve is the set of pairs (θ_1, θ_2) such that $C_L(x_1, y_1, \theta_1)$ (or $C_R(x_2, y_2, \theta_2)$) is tangent to a contact edge or passes through a contact vertex. When such a curve is crossed, the total

[10] We recall that an LR jump may not always exist.

number of intersections of the jump with $\mathcal{B}$'s boundary may change. There are $O(n)$ critical curves of type 1 in the jump diagram.

- A **type 2** critical curve is the set of pairs (θ_1, θ_2) where the initial or the last turn of the jump has zero length. When such a curve is crossed, the deformation of the jump is not continuous. Hence, a jump on one side of the curve may intersect an edge of $\mathcal{B}$, while a jump on the other side may not. There are at most four critical curves of type 2 in the jump diagram.

- A **type 3** critical curve is the set of pairs (θ_1, θ_2) for which the straight segment in the jump passes through a contact vertex. Changing slightly either θ_1 or θ_2 causes the straight segment to intersect $\mathcal{B}$ in a small neighborhood around the vertex or to pass clear of it. There are $O(n)$ critical curves of type 3 in the jump diagram.

- A **type 4** critical curve is the set of pairs (θ_1, θ_2) for which the two circles $C_L(x_1, y_1, \theta_1)$ and $C_R(x_2, y_2, \theta_2)$ are tangent. LR jumps exist on one side of this curve, but not on the other. There are at most two critical curves of type 4 in the jump diagram.

One can easily verify that all these critical curves are algebraic curves of fixed degree when they are represented in the $t_1 t_2$-space, with $t_1 = \tan \frac{\theta_1}{2}$ and $t_2 = \tan \frac{\theta_2}{2}$.

The same types of critical curves can be defined for jumps of type LL, RR, and RL between two given points (for LL and RR jumps there is no critical curve of type 4, since they always exist).

In a similar way, we can parameterize the jumps between a point and an oriented edge by the initial orientation of the jump and the normalized distance of the final point with respect to one of the two endpoints of the edge (the orientation at the final point is that of the oriented edge). We can parameterize as well the jumps between two edges or between an edge and a point. In each case, we can define critical curves decomposing the two-dimensional parameter space in regions with the three properties listed above.

Now, in order to construct the search graph G, instead of considering every pair of discretized configurations in G and checking the jumps between them for collision (as we did in the previous subsection), we can consider every pair consisting of a discretized configuration in G and either a vertex or an oriented edge[11] and proceeds as follows. We only treat the case where the first element of the pair is a configuration (x_1, y_1, θ_1), where (x_1, y_1) is a vertex, and the second element is a vertex (x_2, y_2) (the other cases would be treated in a similar way). For every jump type $jt \in \{\mathrm{LL}, \mathrm{RR}, \mathrm{LR}, \mathrm{RL}\}$, we execute the following algorithm:

1. Set the final orientation θ_2 to 0 and compute the number Q of intersections of the jump of type jt connecting (x_1, y_1, θ_1) to $(x_2, y_2, 0)$ with $\mathcal{B}$'s boundary.

2. Compute the intersections of the vertical line at θ_1 with the critical curves of the jump diagram of $J_{jt}(x_1, y_1, x_2, y_2)$. Sort the intersections in increasing order of θ_2.

3. If $Q = 0$, report (x_2, y_2, θ_2) as a child of (x_1, y_1, θ_1).

4. Increment the final orientation θ_2 by δ. If $\theta_2 = 2\pi$, stop. Otherwise, compute the number of critical curves that have been crossed and update Q accordingly.

5. Go to step 3.

(In the case of a jump of type LR or RL, the algorithm should also check the feasibility of the jump by monitoring the position of the final orientation relative to the critical curves of type 4.)

For a given configuration, this algorithm is executed $O(n)$ times (once for every vertex and every edge of $\mathcal{B}$). Each execution requires time $O(n \log n + \frac{1}{\delta})$. The $O(n \log n)$ part comes from step 2 (there are $O(n)$ critical curves yielding $O(n)$ intersection points which are sorted in time $O(n \log n)$). The $O(\frac{1}{\delta})$ part comes from step 4 which is executed $1/\delta$ times. Notice that when a critical curve of type 2 is crossed, it may be necessary to compute the intersections of a circle with $\mathcal{B}$'s boundary, which requires time $O(n)$, but there is at most a constant number of type 2 critical curve crossings.

Hence, computing all the children of a node of G takes time $O(n^2 \log n + \frac{n}{\delta})$. The time complexity of building and searching the graph is $O(\frac{n^3}{\delta} \log n + \frac{n^2}{\delta^2})$. An A^* algorithm can be used to search the graph

[11] Each edge is considered twice, once for each orientation.

without increasing this asymptotic bound.

Exercises

1: Let $\mathcal{A}$ be a manipulator robot operating in a three-dimensional workspace populated by stationary obstacles. $\mathcal{C}$ is the configuration space of $\mathcal{A}$. Let $\mathcal{A}'$ be the end-effector of $\mathcal{A}$. $\mathcal{A}'$ is a rigid object whose configuration space is $\mathcal{C}' = \mathbf{R}^3 \times SO(3)$. We want $\mathcal{A}'$ to move along a given path τ' in $\mathcal{C}'$. Discuss the problem of planning a free path τ of $\mathcal{A}$ such that when $\mathcal{A}$ performs this path, $\mathcal{A}'$ follows τ'.

2: Consider the path τ defined in Subsection 3.1 as the concatenation of four paths following the flow of four smooth vector fields $\vec{X}$, $\vec{Y}$, $-\vec{X}$, and $-\vec{Y}$, each during a short amount of time δt (see Figure 3). Let $\mathbf{q}$ and $\mathbf{q}'$ be the initial and the final configuration of τ. Show that:

$$\lim_{\delta t \to 0} \frac{\mathbf{q}' - \mathbf{q}}{\delta t^2} = D\vec{Y} \cdot \vec{X} - D\vec{X} \cdot \vec{Y}.$$

3: Let $\mathcal{A}$ be a car-like robot constrained by:

$$-\dot{x} \cos\theta + \dot{y} \sin\theta = 0.$$

Construct two independent vector fields $\vec{X}_1$ and $\vec{X}_2$ that satisfy this constraint. Compute their Lie bracket and show that the Control Lie Algebra associated with $\mathcal{A}$'s nonholonomic constraint has dimension 3.

4: Consider the tractor-trailer robot of Figure 4. F is the midpoint between the two front wheels of the tractor, R_1 is the midpoint between the two rear wheels of the tractor, and R_2 is the midpoint between the wheels of the trailer. Let (x_0, y_0) be the coordinates of F in the frame $\mathcal{F}_\mathcal{W}$ attached to the workspace. $\theta_1 \in [0, 2\pi)$ is the angle between the x-axis of $\mathcal{F}_\mathcal{W}$ and the main axis of the tractor. $\theta_2 \in [0, 2\pi)$ is the angle between the x-axis of $\mathcal{F}_\mathcal{W}$ and the main axis of the trailer. Represent the robot's configurations by $(x_0, y_0, \theta_1, \theta_2)$. Establish the two kinematic constraints with this parameterization and verify that they are nonholonomic.

5: Let $\mathcal{A}$ be a robot consisting of a tractor and a sequence of k trailers. Establish the kinematic constraints of this robot [Barraquand and Latombe, 1989b].

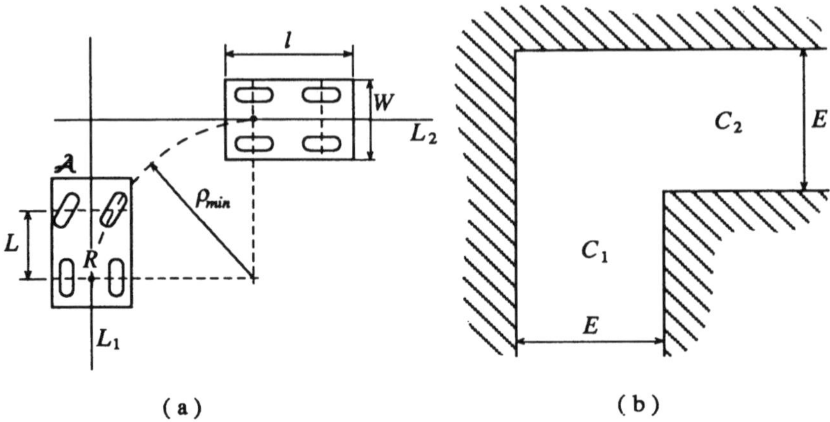

Figure 20. This figure is an illustration for Exercise 9. The purpose of the exercise is to investigate the condition under which the rectangular car-like robot of Figure 1 can turn in the corner made by two corridors without reversal.

6: Prove Lemma 2.

7: Consider a car-like robot $\mathcal{A}$ whose steering angle ϕ is such that $0 < \phi_{min} \leq \phi \leq \phi_{max} < \pi/2$. The car can move both forward and backward. Construct maneuvers that allow $\mathcal{A}$ to move sidewise and to rotate at a given position. Show that a possibly non-feasible free path of $\mathcal{A}$ can be transformed into a topologically equivalent feasible free path between the same configurations by introducing a finite number of these maneuvers.

8: Implement the planning method described in Subsection 6.1 for a rectangular car-like robot with $|\phi| \leq \phi_{max} < \pi/2$.

9: Let $\mathcal{A}$ be the rectangular car-like robot of Figure 1. $\mathcal{A}$'s turning radius is lower-bounded by $\rho_{min} > 0$.

a. Let L_1 and L_2 be two perpendicular lanes. $\mathcal{A}$ transfers from lane L_1 to lane L_2 by a turn of radius ρ_{min} as shown in Figure 20.a. Describe the geometry of the area that is swept out by $\mathcal{A}$ during the transfer.

b. Consider two corridors C_1 and C_2 intersecting at a right angle (Figure 20.b). Assume that each corridor is arbitrary long. Establish the condition under which $\mathcal{A}$ can turn from C_1 to C_2 without reversal.

10: Consider the planning problem described in Subsection 7.1. Assume that the obstacles are generalized polygonal regions. Establish the conditions under which the visibility graph method described in Subsection 1.2 of Chapter 4 can be applied in order to plan a feasible semi-free path.

11: Implement the planning method described in Subsection 7.3.

Chapter 10

Dealing with Uncertainty

In the previous chapters we assumed that the robot was able to perfectly control its motions, i.e. to exactly follow the geometrical paths generated by the planner. This assumption is realistic when the workspace is relatively uncluttered and the goal configuration does not have to be achieved too precisely. In those cases, if necessary, uncertainty in robot's control can be taken into account by slightly "growing" the C-obstacles and planning a free path among the grown C-obstacles. Motions that can be planned in this way are often called *gross motions.*

However, when the workspace is too densely occupied by obstacles, growing the C-obstacles may result in a new free space where the initial and goal configurations are no longer connected by a free path. Moreover, even if they are still connected by a path, executing this path may not allow the robot to attain the goal configuration with the desired precision, if the latter is too tight. In those cases, some sensing must be explicitly incorporated into the motion plans in order to both monitor the effects of the robot's motions and command appropriate corrective motions. A motion plan then consists of a combination of motion commands and sensor readings which interact in order to reduce uncertainty and guide the robot toward the goal. Such a plan is usually called a *motion strategy*, and the motions it produces are called *fine motions*.

In general, sensing is not perfect either and does not provide an exact knowledge of the robot's current configuration during execution. Hence, a fundamental difficulty is that uncertainty exists not only at planning time, but also at execution time. It is thus necessary for the planner to anticipate the knowledge that will be available during execution in order to guarantee that this knowledge will be sufficiently precise to make the robot attain the goal region and stop in this region reliably. The motion strategy should not have to be lucky in order to attain the goal recognizably! This typically requires the planner to generate motion commands that will make some specific features of the workspace ("landmarks") reliably identifiable and located by the sensors. Planning a motion strategy may be complicated further by the fact that the geometry of both the robot and the workspace, and hence of the C-obstacle region in the configuration space, is also imprecisely known (uncertainty in model).

In this chapter we investigate the problem of planning the motion of a rigid object among fixed obstacles in the presence of uncertainty. Introducing uncertainty raises new modeling, computational and combinatorial issues which make the problem significantly more difficult than those considered in the previous chapters. This leads us to simplifying the contents of this chapter by making several assumptions presented in Section 1. These assumptions, however, do not denature the fundamental issues underlying motion planning in the presence of uncertainty. In various places, we will discuss the effect of relaxing some of these assumptions.

Section 1 gives a precise formulation of the motion planning problem with uncertainty in control and sensing [1]. It describes force-compliant motion commands which are less sensitive to errors than pure position control (see Subsection 5.3 of Chapter 1). Section 2 introduces a general approach to motion planning with uncertainty, *preimage backchaining*, which was originally proposed in [Lozano-Pérez, Mason and Taylor, 1984]. This approach is described in detail in the rest of the chapter. The concept of preimage — a region of configuration space from which a

[1] Until the last section of this chapter, we will assume that the geometry of the C-obstacle region is exactly known.

motion command is guaranteed to attain a given goal recognizably — is formally defined in Section 3. Simple methods for computing preimages are presented in Section 4. Using these methods as building blocks, one can build an operational planner. To do so, one must select the search space of the planner by discretizing the range of possible directions of motion of the robot. This issue is addressed in Section 5.

The other sections of the chapter present more advanced concepts. Section 6 explores the construction of conditional strategies, i.e. strategies which include conditional branching statements allowing for different motion commands to be executed depending on the values of parameters sensed at execution time. Section 7 investigates ways of improving the ability of a motion command to recognize goal achievement, in order to make it possible constructing larger preimages. In Section 8 we revisit preimage backchaining and we give a broader view of the approach than in Section 2. Finally, Section 9 extends some of the previous results to the case where there is also uncertainty in the input geometric model of the robot and the workspace.

1 Problem Formulation

1.1 Configuration Space

We assume that the robot $\mathcal{A}$ is a polygon that can translate freely at fixed orientation in the plane among polygonal obstacles. Thus, the configuration space is $\mathcal{C} = \mathbf{R}^2$ and the C-obstacle region is polygonal. A configuration is represented as $\mathbf{q} = (x, y)$, where x and y are the coordinates of $\mathcal{A}$'s reference point in the workspace.

We allow $\mathcal{A}$ to touch obstacles[2], so that the space of the robot's motions is the valid space $\mathcal{C}_{valid}$ (see Subsection 9.5, Chapter 2). To avoid pathological cases, we require this space to be a manifold with boundary. This requirement entails that $\mathcal{C}_{contact} = \partial\mathcal{C}_{free}$ and $\mathcal{C}_{valid} = cl(\mathcal{C}_{free})$; hence, a contact between $\mathcal{A}$ and the obstacle region can be broken by an arbitrarily small displacement of $\mathcal{A}$. It also implies that every vertex in the contact space $\mathcal{C}_{contact}$ is the extremity of two edges only; thus, $\mathcal{C}_{contact}$

[2]Mechanical assembly is typical task domain where uncertainty needs to be taken into account. In this domain, contacts between objects are not only part of the goal; they are also helpful to distinguish among possible geometric situations. Most of the research in motion planning with uncertainty has been done in this context.

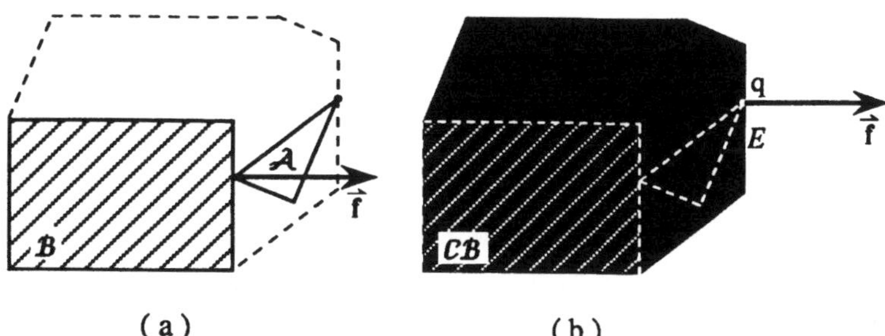

(a) (b)

Figure 1. When the moving object $\mathcal{A}$ can only translate in the plane, the workspace $\mathcal{W}$, the configuration space $\mathcal{C}$ and all the tangent spaces of $\mathcal{C}$ are copies of $\mathbf{R}^2$. We equip all these spaces with the Euclidean metric and we represent them as a single copy of $\mathbf{R}^2$. Then a force acting on $\mathcal{A}$, for instance the reaction force exerted by an obstacle, has the same representation in $\mathcal{W}$ and in $\mathcal{C}$, as illustrated in Figures a and b, respectively.

consists of a finite set of disjoint simple polygonal lines. An edge (resp. a vertex) in $\mathcal{C}_{contact}$ is called a **contact edge** (resp. a **contact vertex**). The **outgoing** normal of a contact edge E is the unit vector $\vec{\nu}(E)$ normal to E pointing toward $\mathcal{C}_{free}$. The vector $-\vec{\nu}(E)$ is the **ingoing** normal of E.

Throughout the chapter, n denotes the number of vertices in $\mathcal{C}_{contact}$.

Allowing $\mathcal{A}$ to move in contact space makes it useful to equip the robot with a force sensor for measuring the reaction forces exerted by the obstacles. Force sensing is then a major source of information for the robot to better determine its position relative to the obstacles. It requires, however, that reaction forces be mapped in configuration space. The rest of the subsection addresses this issue.

Under the above assumptions, the workspace $\mathcal{W}$, the configuration space $\mathcal{C}$ and all the tangent spaces of $\mathcal{C}$ are copies of $\mathbf{R}^2$. We equip all these spaces with the Euclidean metric and we use a single copy of $\mathbf{R}^2$ to represent all of them. Then a force acting on $\mathcal{A}$ has the same vector representation in $\mathcal{W}$ and in $\mathcal{C}$ (see Section 10 of Chapter 2). Indeed, a virtual translation of $\mathcal{A}$ by $(\delta x, \delta y)$ in $\mathcal{W}$ induces the same translation of $\mathcal{A}$'s configuration in $\mathcal{C}$. In particular, if $\mathcal{A}$ pushes on an obstacle, the obstacle exerts a reaction force on $\mathcal{A}$. Assuming that $\mathcal{A}$'s configuration

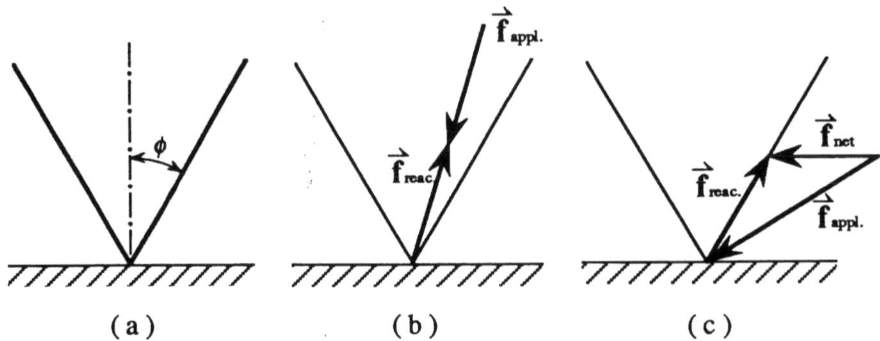

Figure 2. In the presence of friction, the range of possible reaction forces at the contact point between $\mathcal{A}$ and an obstacle consists of all the force vectors applied at the contact point and pointing into a cone, called the friction cone (Figure a). If $\mathcal{A}$ pushes on the obstacle along a line that is contained in the cone, the obstacle produces a reaction force that completely cancels the applied force (Figure b). If $\mathcal{A}$ pushes along a line that is not contained in the cone, the reaction force only partially cancels the applied force and the resulting net force is tangent to the obstacle edge (Figure c).

$\mathbf{q}$ is in the interior of a contact edge E and that the contact between $\mathcal{A}$ and the obstacle is frictionless, the force in configuration space is a vector applied at $\mathbf{q}$ and pointing along the outgoing normal of E, as illustrated in Figure 1.

In practice, contacts are not frictionless. The presence of friction increases the range of reaction forces that the obstacle can produce. Let the contact occur between a vertex of $\mathcal{A}$ and an obstacle edge. The classical Coulomb law of friction represents friction by a cone, called the **friction cone**, which specifies the range of possible reaction forces. The friction cone's axis points along the outgoing normal of the obstacle edge and its sides make an angle $\phi = \tan^{-1}\mu$ $(0 \leq \phi \leq \pi/2)$ with this axis, where μ $(\mu \geq 0)$ is the coefficient of friction between $\mathcal{A}$ and the obstacle at the contact point (see Figure 2.a). Throughout the rest of the chapter, we assume that the static and dynamic friction coefficients have the same value μ and that this value is known and remains fixed for every possible contact point between $\mathcal{A}$ and the obstacles. The range of possible reaction forces at the contact point consists of all the force vectors applied at the contact point and pointing into the friction cone. (The frictionless

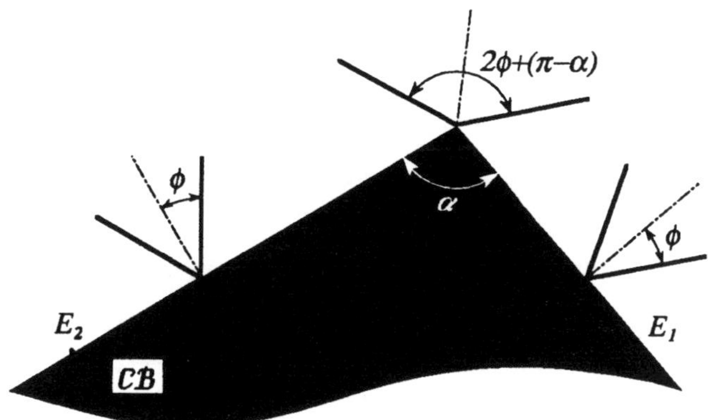

Figure 3. The range of reaction forces at a vertex in contact space can be any non-negative linear combination of the reaction forces that can be generated by the two abutting edges E_1 and E_2. The corresponding friction cone has its sides parallel to the two most extreme rays of the friction cones in E_1 and E_2.

case corresponds to $\phi = 0$.) Hence, if $\mathcal{A}$ pushes on the obstacle along a line that is contained in the cone, the obstacle produces a reaction force that completely cancels[3] the applied force (see Figure 2.b). If, instead, $\mathcal{A}$ pushes along a line that is not contained in the cone, the reaction force only partially cancels the applied force and the resulting net force is tangent to the obstacle edge (see Figure 2.c). The contact between a vertex of $\mathcal{A}$ and an obstacle edge corresponds to a configuration of $\mathcal{A}$ lying in a C-obstacle edge. Since a force has the same representation in $\mathcal{W}$ and $\mathcal{C}$, friction is represented in the C-obstacle edge by the same cone as in the obstacle edge. It is straightforward to extend this result to all the edges of $\mathcal{C}_{contact}$. Since the friction coefficient is constant in the workspace, the friction cone in configuration space has the same angle 2ϕ in every contact edge.

For completeness of the model, we must consider the case when $\mathcal{A}$'s configuration is at a contact vertex X. This case occurs when either a vertex of $\mathcal{A}$ is in contact with an obstacle vertex, or there are two (or more) disconnected contact zones between $\mathcal{A}$ and the obstacles, with different orientations of the friction cones. In the second case, the reaction force at X can be any non-negative linear combination of the reaction forces

[3]We assume that the obstacles are not movable.

that can be generated by the two C-obstacle edges E_1 and E_2 abutting X. In the first case, we assume the same, although the reaction force may not be physically determined (due to the shortcomings of the polygonal representation of the objects). Hence, the friction cone at X is the cone whose sides are respectively parallel to the two most extreme rays of the friction cones in E_1 and E_2 (Figure 3).

The above model of friction can easily be extended to the configuration space of a polyhedral object translating among polyhedral obstacles. But allowing the robot to rotate results in C-obstacles bounded by curved faces. Friction varies in a non-linear fashion over these faces, which makes its description substantially more complicated [Erdmann, 1984].

Throughout this chapter, except in Section 9, we assume that the model of the C-obstacle region is exact.

1.2 Sensing

We assume that the robot $\mathcal{A}$ is equipped with two sensors, a position sensor and a force sensor, which we model below. Other sensors could be modeled in a similar fashion.[4]

The position sensor measures the current configuration of $\mathcal{A}$. The sensed configuration is denoted[5] by $\mathbf{q}^*$, while the current actual configuration is denoted by $\mathbf{q}$. The uncertainty in position sensing is modeled as an open disc $\mathcal{U}_q(\mathbf{q}^*) \subset \mathbf{R}^2$ of fixed radius ρ_q centered at the measured configuration $\mathbf{q}^*$. This means that if the position sensor returns $\mathbf{q}^*$, the current actual configuration $\mathbf{q}$ of $\mathcal{A}$ may be anywhere in $\mathcal{U}_q(\mathbf{q}^*) \cap \mathcal{C}_{valid}$. Notice that although $\mathbf{q}$ can only be in $\mathcal{C}_{valid}$, the sensed configuration $\mathbf{q}^*$ may be in $\mathcal{C} \backslash \mathcal{C}_{valid}$. However, in that case we must have that $\mathcal{U}_q(\mathbf{q}^*) \cap \mathcal{C}_{valid} \neq \emptyset$, since otherwise uncertainty in position sensing would be incorrectly modeled. Since $\mathcal{U}_q$ is a disc, if $\mathbf{q}$ is the actual configuration, then the measured configuration $\mathbf{q}^*$ necessarily lies in $\mathcal{U}_q(\mathbf{q})$.

At this point, our model of uncertainty is worth an explanation. Let the robot be at an actual configuration $\mathbf{q}$ (which of course we do not

---

[4] Using position and force sensors is appropriate for task domains such as mechanical assembly (see footnote 2). It is less appropriate for mobile robot navigation tasks, which are more likely to involve vision and proximity sensors.

[5] Our convention is to denote a measured quantity by a symbol with the superscript "*" and the corresponding actual quantity by the same symbol without superscript.

know) and let the sensed configuration be $\mathbf{q}^*$. The difference $\mathbf{q} - \mathbf{q}^*$ is the sensing *error*, which varies from one measurement to another. Our model of uncertainty bounds the modulus of this error, i.e. $\|\mathbf{q}-\mathbf{q}^*\| < \rho_q$, and accepts all the configurations in $\mathcal{U}_q(\mathbf{q}^*)$ as equally probable ones. Below, we model uncertainty in force sensing and in robot control in a similar fashion. This model is simple, and it is not hard to imagine more involved (and perhaps more realistic) ones. However, it is interesting because it will allow us to neatly define *guaranteed* motion strategies, i.e. strategies which are guaranteed to succeed as long as the errors in control and sensing during execution stay within the bounds specified by the uncertainty regions.

The force sensor measures the force acting on $\mathcal{A}$. The output of the sensor is a vector denoted by $\vec{\mathbf{f}}^*$. The sensor is used to acquire information on whether $\mathcal{A}$ touches obstacles or not and if it touches an obstacle, on the orientation of the outgoing normal of the contact space at the contact configuration. When the sensed force is $\vec{\mathbf{f}}^*$, the actual reaction force acting on $\mathcal{A}$ is a force $\vec{\mathbf{f}} \in \mathcal{U}_f(\vec{\mathbf{f}}^*)$, where $\mathcal{U}_f$ is the force uncertainty region, which is specified as follows. The uncertainty on the sensed force modulus is modeled by $\|\vec{\mathbf{f}}\| \in \Delta_f(\vec{\mathbf{f}}^*)$, where $\Delta_f(\vec{\mathbf{f}}^*)$ is an interval that includes $\|\vec{\mathbf{f}}^*\|$. In addition, the magnitude of the angle between $\vec{\mathbf{f}}$ and $\vec{\mathbf{f}}^*$, if neither is zero, is less than a fixed angle $\varepsilon_f < \pi/2$. Hence, $\vec{\mathbf{f}}$ lies in an open cone of angle $2\varepsilon_f$ whose axis points along $\vec{\mathbf{f}}^*$.

1.3 Robot Control

1.3.1 Motion Command

For a given geometry of $\mathcal{C}_{valid}$, a **motion command** is an application that maps any configuration $\mathbf{q}_s \in \mathcal{C}_{valid}$ (called the **starting configuration** of the motion command) to a nominal motion of the robot. This motion would be the motion of the robot if there was no uncertainty in control.

A motion command $\mathbf{M}$ is described as a pair (**CS**, TC). **CS** is called the **control statement**. Given a starting configuration $\mathbf{q}_s \in \mathcal{C}_{valid}$, it determines a commanded trajectory τ^c of $\mathcal{A}$, i.e. a curve:

$$\tau^c : t \in [0, +\infty) \mapsto \tau^c(t) \in \mathcal{C}_{valid}$$

with $\tau^c(0) = \mathbf{q}_s$ and t denoting time. TC is called the **termination condition**. The most general form of this condition is:[6]

$$\mathtt{TC} = \mathtt{tp}(t, \mathbf{q}^*[0,t], \vec{\mathbf{f}}^*[0,t])$$

where tp is the **termination predicate**, t denotes the current time, and $\mathbf{q}^*[0,t]$ and $\vec{\mathbf{f}}^*[0,t]$ are the position and force sensing history, respectively, since the beginning of the motion, i.e. :

$$\begin{aligned} \mathbf{q}^*[0,t] &= \{(t', \mathbf{q}^*(t')) \;/\; t' \in [0,t]\} \\ \vec{\mathbf{f}}^*[0,t] &= \{(t', \vec{\mathbf{f}}^*(t')) \;/\; t' \in [0,t]\} \end{aligned}$$

where $\mathbf{q}^*(t')$ and $\vec{\mathbf{f}}^*(t')$ are the sensed configuration and force at time t'. TC is continuously evaluated during the execution of the command and $\mathcal{A}$ instantaneously stops moving when TC becomes true. This obviously specifies an ideal behavior which cannot be exactly achieved.

1.3.2 Force-Compliant Control

Throughout the previous chapters, the robot was position-controlled along the planned geometric paths. A drawback of pure position control is that it may be quite sensitive to various kinds of errors. For instance, if the robot is commanded to move along a path close to an obstacle, and if it unexpectedly hits the obstacle (say, because of imperfect control), the normal behavior of a position-controlled robot is to stop. A more interesting behavior would be for the robot to comply with the reaction force exerted by the obstacles (so that it becomes zero), while moving along the projection of the commanded direction in the subspace orthogonal to the reaction force [Mason, 1981]. In the presence of uncertainty, such a compliant behavior is preferable to the previous one, because its ability to attain goals is less sensitive to errors and thus makes it possible to construct motion strategies with fewer motion commands. This behavior requires the robot to be controlled in both position and force.

There exist various models of force-compliant control [Inoue, 1974] [Raibert and Craig, 1981] [Hogan, 1985] [Whitney, 1985]. In order to set up a precise framework for the rest of the chapter, we assume that $\mathcal{A}$'s control follows the so-called **generalized damping** model originally proposed

[6] Less general forms will also be used, for example by not letting tp have access to time and the full sensing history.

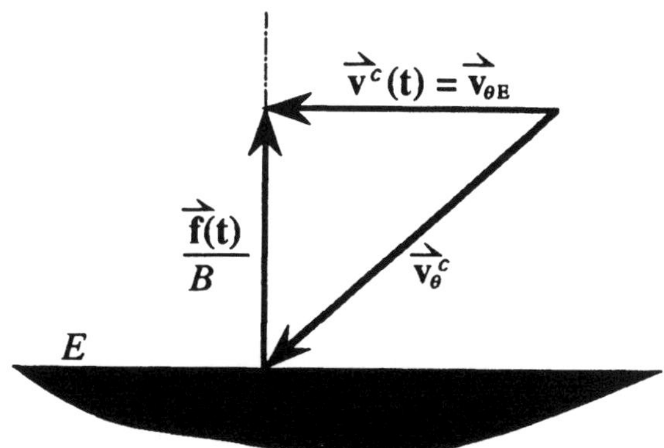

Figure 4. This figure illustrates the robot's nominal behavior in a frictionless contact edge E under the generalized damping control model. The motion is specified by the equation $\vec{\mathbf{f}}(t) = B(\vec{\mathbf{v}}^c(t) - \vec{\mathbf{v}}_\theta^c)$. The robot slides along the edge E at a velocity $\vec{\mathbf{v}}_{\theta \mathrm{E}}^c$ equal to the projection of $\vec{\mathbf{v}}_\theta^c$ on E.

by Whitney [Whitney, 1977]. This means that **CS** is parameterized by a vector $\vec{\mathbf{v}}_\theta^c$ of orientation $\theta \in S^1$, called the **commanded velocity**, and that the robot's commanded trajectory τ_θ^c in $\mathcal{C}$ is specified by the following first-order differential equation:

$$\vec{\mathbf{f}}(t) = B(\vec{\mathbf{v}}^c(t) - \vec{\mathbf{v}}_\theta^c) \tag{1}$$

where $\vec{\mathbf{f}}(t)$ is the actual reaction force exerted by the obstacles on $\mathcal{A}$ at time t, B is a 2×2 matrix called the *damper matrix*, and $\vec{\mathbf{v}}^c(t) = d\vec{\tau}_\theta^c(t)/dt$ is the velocity of $\mathcal{A}$ at t. In the following, we take B as a constant diagonal matrix with equal positive diagonal coefficients. We also take $\vec{\mathbf{v}}_\theta^c$ as a unit vector, so that it is completely specified by its orientation θ.

In equation (1), the vector $B\vec{\mathbf{v}}_\theta^c$ is interpreted as the force exerted by the robot and the vector $B\vec{\mathbf{v}}^c(t)$ as the component of this force which is not canceled by the environment and thus produces motion. At every instant t, given the geometry of $\mathcal{C}_{valid}$ and the friction angle ϕ, the reaction force $\vec{\mathbf{f}}(t)$ is completely determined by the applied force $B\vec{\mathbf{v}}_\theta^c$ and the current configuration. Therefore, for a given starting configuration $\mathbf{q}_s$, the generalized damping equation (1) determines a unique trajectory τ_θ^c, which we describe below.

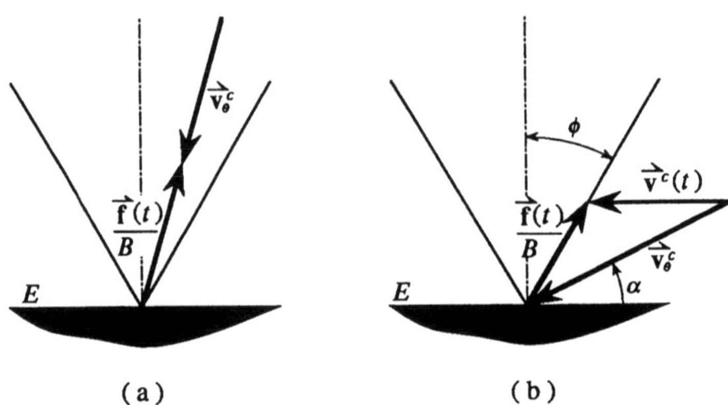

Figure 5. Friction in the contact space increases the range of commanded velocities that lead the robot to stick when it hits a C-obstacle. If the inverted commanded velocity points into the friction cone, the motion sticks to E (Figure a), otherwise it slides along E at velocity $\|\vec{\mathbf{v}}^c(t)\| = \|\vec{\mathbf{v}}^c_\theta\| \cos\alpha[1 - \tan\phi \tan\alpha]$, where $\alpha < \pi/2$ is the angle between $\vec{\mathbf{v}}^c_\theta$ and E (Figure b).

When $\mathcal{A}$ is in free space, it moves at constant velocity $\vec{\mathbf{v}}^c_\theta$, since the reaction force is zero. When $\mathcal{A}$ is in contact space in the interior of an edge E, it may move away from E, slide along E, or stick to E. If the projection of $\vec{\mathbf{v}}^c_\theta$ on the outgoing normal of E is positive or null, $\mathcal{A}$ moves away or tangentially at velocity $\vec{\mathbf{v}}^c_\theta$. Otherwise, let $\vec{\mathbf{v}}^c_{\theta\mathrm{E}}$ be the projection of $\vec{\mathbf{v}}^c_\theta$ on E. If $\vec{\mathbf{v}}^c_{\theta\mathrm{E}}$ is the null vector (i.e. $\vec{\mathbf{v}}^c_\theta$ points perpendicularly to E), then the contact space generates the reaction force $\vec{\mathbf{f}}(t) = B\vec{\mathbf{v}}^c_\theta$ and $\mathcal{A}$ sticks to E, i.e. $\|\vec{\mathbf{v}}^c(t)\| = 0$; otherwise, assuming a frictionless contact space, $\mathcal{A}$ slides along E at velocity $\vec{\mathbf{v}}^c(t) = \vec{\mathbf{v}}^c_{\theta\mathrm{E}}$ (see Figure 4). Since the model of the motion is a first-order one, which ignores inertial forces, it allows the robot velocity to change instantaneously. This idealization is another source of error leading to uncertainty in control.

Friction in the contact space simply results in increasing the range of commanded velocities that cause the robot to stick to a contact edge. With no control error, a generalized damping motion with commanded velocity $\vec{\mathbf{v}}^c_\theta$ sticks to an edge E if $-\vec{\mathbf{v}}^c_\theta$ lies in the friction cone, i.e. if the magnitude of the angle between $-\vec{\mathbf{v}}^c_\theta$ and the outgoing normal $\vec{\nu}(E)$ is less or equal to ϕ. Otherwise, it slides along E. Figure 5 illustrates this behavior.

When the robot's configuration reaches a contact vertex X, the motion sticks at X if $-\vec{\mathbf{v}}_\theta^c$ lies in the friction cone at X. If $-\vec{\mathbf{v}}_\theta^c$ does not lie in the friction cone at X, the robot consistently moves away from X (in free or contact space) as if had just reached X by sliding along one of the two adjacent edges.

More formally, let $\vec{\mathbf{f}}_{reac}(\mathbf{q}, \vec{\mathbf{v}}_\theta^c)$ be the function that evaluates to the reaction force exerted by the contact space on the robot at configuration $\mathbf{q}$ when it is commanded along $\vec{\mathbf{v}}_\theta^c$. $\vec{\mathbf{f}}_{reac}(\mathbf{q}, \vec{\mathbf{v}}_\theta^c)$ is defined as follows:

- if $\mathbf{q} \in \mathcal{C}_{free}$, then $\vec{\mathbf{f}}_{reac}(\mathbf{q}, \vec{\mathbf{v}}_\theta^c) = 0$;
- if $\mathbf{q} \in \mathcal{C}_{contact}$, $\mathbf{q}$ is in the interior of an edge E, and $\vec{\mathbf{v}}_\theta^c$ projects positively on the outgoing normal of E, then $\vec{\mathbf{f}}_{reac}(\mathbf{q}, \vec{\mathbf{v}}_\theta^c) = 0$;
- if $\mathbf{q} \in \mathcal{C}_{contact}$, $\mathbf{q}$ is in the interior of an edge E, and $-\vec{\mathbf{v}}_\theta^c$ lies in the friction cone at the contact point, then $\vec{\mathbf{f}}_{reac}(\mathbf{q}, \vec{\mathbf{v}}_\theta^c) = -B\vec{\mathbf{v}}_\theta^c$;
- if $\mathbf{q} \in \mathcal{C}_{contact}$, $\mathbf{q}$ is in the interior of an edge E, and $-\vec{\mathbf{v}}_\theta^c$ projects positively on the outgoing normal of E, but lies outside the friction cone at the contact configuration, then $\vec{\mathbf{f}}_{reac}(\mathbf{q}, \vec{\mathbf{v}}_\theta^c) = \vec{\mathbf{f}}$, with $\vec{\mathbf{f}}$ being the solution of the following system of equations:

$$|angle(\vec{\nu}(E), \vec{\mathbf{f}})| = \phi$$
$$\vec{\nu}(E) \cdot \vec{\mathbf{f}} = \vec{\nu}(E) \cdot (-B\vec{\mathbf{v}}_\theta^c)$$
$$sign(angle(\vec{\nu}(E), \vec{\mathbf{f}})) = sign(angle(\vec{\nu}(E), -B\vec{\mathbf{v}}_\theta^c))$$

- if $\mathbf{q} \in \mathcal{C}_{contact}$, $\mathbf{q}$ is at a contact vertex, and $-\vec{\mathbf{v}}_\theta^c$ lies in the friction cone at this vertex, then $\vec{\mathbf{f}}_{reac}(\mathbf{q}, \vec{\mathbf{v}}_\theta^c) = -B\vec{\mathbf{v}}_\theta^c$;
- if $\mathbf{q} \in \mathcal{C}_{contact}$, $\mathbf{q}$ is at a contact vertex X, and $-\vec{\mathbf{v}}_\theta^c$ does not lie in the friction cone at X, then let $\vec{\mathbf{f}}_1$ and $\vec{\mathbf{f}}_2$ be the reaction forces that would be generated by the two edges abutting X; $\vec{\mathbf{f}}_{reac}(\mathbf{q}, \vec{\mathbf{v}}_\theta^c) = \max(\vec{\mathbf{f}}_1, \vec{\mathbf{f}}_2)$.

(When $\mathbf{q}$ is at a contact vertex, the actual reaction force may not be determined. Then the above definition of $\vec{\mathbf{f}}_{reac}$ is somewhat arbitrary, but it is consistent with our previous definitions.)

At every instant t during a motion commanded along $\vec{\mathbf{v}}_\theta^c$ we have that $\vec{\mathbf{f}}(t) = \vec{\mathbf{f}}_{reac}(\tau_\theta^c(t), \vec{\mathbf{v}}_\theta^c)$. The commanded trajectory τ_θ^c is thus defined by the following differential equation:

$$\frac{d\tau_\theta^c(t)}{dt} = \frac{1}{B}\vec{\mathbf{f}}_{reac}(\tau_\theta^c(t), \vec{\mathbf{v}}_\theta^c) + \vec{\mathbf{v}}_\theta^c$$

with the initial condition $\tau_\theta^c(0) = \mathbf{q}_s$.

The generalized damping model of motion remains applicable in higher-dimensional configuration spaces. However, when the robot is allowed to rotate, this model induces a significantly more complicated behavior.

1.3.3 Uncertainty in Control

Throughout the rest of the chapter, we assume generalized damping control. We denote a motion command by $\mathbf{M} = (\vec{\mathbf{v}}_\theta^c, \mathtt{TC})$, where $\vec{\mathbf{v}}_\theta^c$ is the commanded velocity.

Uncertainty in control is modeled as follows. Let $\vec{\mathbf{v}}_\theta^c$ be the commanded velocity. At any instant t during the motion, the **effective commanded velocity** is a vector $\vec{\mathbf{v}}_\theta(t) \in \mathcal{U}_v(\vec{\mathbf{v}}_\theta^c)$, such that the magnitude of the angle between $\vec{\mathbf{v}}_\theta(t)$ and $\vec{\mathbf{v}}_\theta^c$ is less than a fixed angle $\eta_v < \pi/2$. In other words, $\vec{\mathbf{v}}_\theta(t)$ lies in an open cone of angle $2\eta_v$ whose axis points along $\vec{\mathbf{v}}_\theta^c$. This cone is called the **control uncertainty cone**. In addition, the uncertainty on the velocity modulus is modeled by $\|\vec{\mathbf{v}}_\theta(t)\| \in \Delta_v(\vec{\mathbf{v}}_\theta^c)$, where $\Delta_v(\vec{\mathbf{v}}_\theta^c)$ is an interval[7] (v_1, v_2), with $0 < v_1 < \|\vec{\mathbf{v}}_\theta^c\| = 1 < v_2$. The orientation and the modulus of the effective commanded velocity can vary arbitrarily, but continuously, within these uncertainty bounds.

The motion of the robot is specified by the generalized damping equation (1) with $\vec{\mathbf{v}}_\theta(t)$ substituted for $\vec{\mathbf{v}}_\theta^c$, i.e. $\vec{\mathbf{f}}(t) = B(\vec{\mathbf{v}}(t) - \vec{\mathbf{v}}_\theta(t))$, where $\vec{\mathbf{v}}(t)$ is the actual robot velocity. For a given starting configuration $\mathbf{q}_s$, any continuous map $\vec{\mathbf{v}}_\theta : t \in [0, +\infty) \mapsto \vec{\mathbf{v}}_\theta(t) \in \mathcal{U}_v(\vec{\mathbf{v}}_\theta^c)$ determines a unique actual trajectory $\tau : t \in [0, +\infty) \mapsto \tau(t) \in \mathcal{C}_{valid}$, such that:

$$\frac{d\tau(t)}{dt} = \frac{1}{B}\vec{\mathbf{f}}_{reac}(\tau(t), \vec{\mathbf{v}}_\theta(t)) + \vec{\mathbf{v}}_\theta(t)$$

and $\tau(0) = \mathbf{q}_s$. The only configurations in τ where the actual velocity $\vec{\mathbf{v}}(t) = d\vec{\tau}(t)/dt$ varies discontinuously are those where the trajectory

[7]This interval should not include 0, since according to our model of uncertainty it would allow the robot to stay motionless. However, the lower bound v_1 can be chosen arbitrarily small.

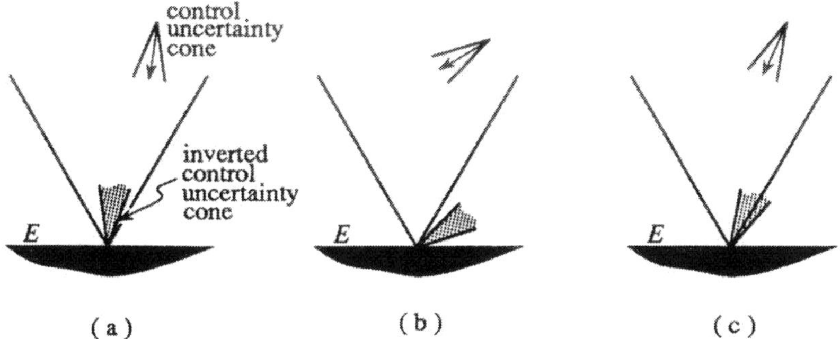

Figure 6. Assume that all the effective commanded velocities lying in the control uncertainty cone project positively on the ingoing normal of a contact edge E. Three cases are possible. If the inverted control uncertainty cone (shown grey) is completely contained in the friction cone (Figure a), the motion is guaranteed to stick along the edge. If the two cones have a null intersection (Figure b), the motion is guaranteed to slide along the edge. If the two cones intersect as in Figure c, the motion may slide or stick.

switches from free to contact space, and conversely. Since there is a finite number of contact edges, τ is piecewise of class C^1.

According to the above model of uncertainty, if $\mathcal{A}$ is in free space, it moves along a trajectory whose tangent at any configuration is contained in the control uncertainty cone anchored at this configuration. If $\mathcal{A}$'s configuration is in a contact edge E, it may move away, slide, or stick, depending on the orientation of the effective commanded velocity $\vec{\mathbf{v}}_\theta$ relative to the outgoing normal of E. In particular, if all vectors lying in the control uncertainty cone project positively on $\vec{\nu}(E)$, then $\mathcal{A}$ is guaranteed to move away from the edge. Instead, if the inverted control uncertainty cone is entirely contained in the friction cone, $\mathcal{A}$ sticks to the edge; if it is completely outside the friction cone and all the vectors in the control uncertainty cone project positively on the ingoing normal of the edge, $\mathcal{A}$ is guaranteed to slide in the direction of the projection of $\vec{\mathbf{v}}_\theta^c$ into E. Figure 6 illustrates the sticking and sliding conditions with uncertainty in control.

We denote the set of all the possible reaction forces that can be generated at configuration $\mathbf{q}$ when the commanded velocity is $\vec{\mathbf{v}}_\theta^c$ by $F_{reac}(\mathbf{q}, \theta)$. We

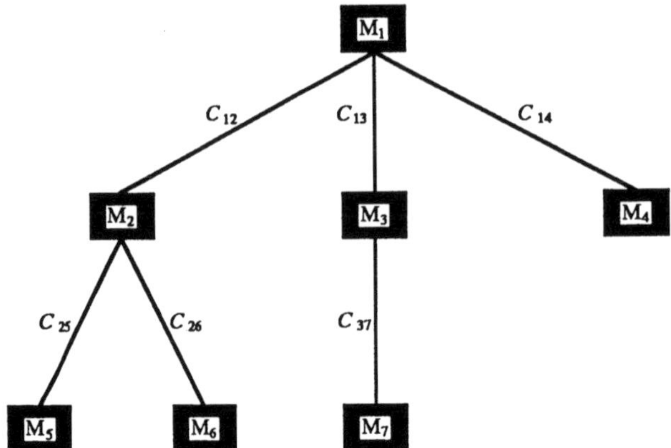

Figure 7. A conditional strategy is described by a tree. The nodes of the tree are labeled by motion commands and its arcs are labeled by conditions. The execution of the strategy consists of sequentially executing the motion commands labeling the nodes of a path (connecting the root to a leaf of the tree) whose arcs are labeled by conditions evaluating to `true`. One possible execution of the strategy depicted in the figure is the following. The motion command $\mathbf{M}_1$ is executed first. Then, if the condition C_{13} evaluates to `true` and C_{12} and C_{14} evaluate to `false`, the motion command $\mathbf{M}_3$ is executed. Since it has a single successor, C_{37} should evaluate to `true` (otherwise the strategy returns failure) and $\mathbf{M}_7$ is executed next.

have that:

$$F_{reac}(\mathbf{q}, \theta) = \bigcup_{\vec{\mathbf{v}}_\theta \in \mathcal{U}_v(\vec{\mathbf{v}}_\theta^c)} \vec{\mathbf{f}}_{reac}(\mathbf{q}, \vec{\mathbf{v}}_\theta).$$

A straightforward algorithm computes $F_{reac}(\mathbf{q}, \theta)$ in $O(1)$ time.

1.4 Motion Strategies

We describe a **motion strategy** by a tree (Figure 7). The nodes of the tree are motion commands and its arcs are conditions. The execution of the strategy consists of sequentially executing the motion commands labeling the nodes contained in a path connecting the root to a leaf of the tree. After a motion command has been executed, the conditions associated with its outgoing arcs are evaluated and the next command executed is such that the condition of its ingoing arc evaluates to `true`.

If several conditions evaluate to **true**, one of the corresponding commands (any one) is executed next. If no condition evaluates to **true**, the execution of the strategy terminates with failure. The execution of the strategy terminates with success after the motion command labeling a leaf of the tree has been executed.

A motion strategy described by a tree containing a single path is called a **sequential strategy**. A motion strategy described by a tree with multiple paths is called a **conditional strategy**. The **number of steps** of a strategy is the maximal number of motion commands that the strategy may have to execute before it can return success.

We assume that the geometry of $\mathcal{C}_{valid}$, the friction angle ϕ, and the uncertainty regions $\mathcal{U}_q$, $\mathcal{U}_f$ and $\mathcal{U}_v$ are given as described above. Given two regions $\mathcal{I}$ and $\mathcal{G}$ of $\mathcal{C}_{valid}$, called the **initial region** and the **goal region**, respectively, the *motion planning problem with uncertainty in control and sensing* is to generate a motion strategy Σ whose execution is guaranteed to return success with $\mathcal{A}$'s configuration actually in $\mathcal{G}$, whenever $\mathcal{A}$'s initial configuration is in $\mathcal{I}$ and the input $\mathcal{C}_{valid}$, $\phi, \mathcal{U}_q, \mathcal{U}_f$, and $\mathcal{U}_v$ are a correct description of the reality. This strategy is said to be **strongly guaranteed**. A complete planner should construct such a strategy whenever one exists and indicate that no such strategy exists otherwise.

Canny and Reif [Canny and Reif, 1987] have shown that the three-dimensional version of the above problem ($\mathcal{C} = \mathbf{R}^3$) is non-deterministic exponential time hard (NEXPTIME-hard) when the number of faces in $\mathcal{C}_{valid}$ is not fixed, which strongly suggests that finding plans can take double exponential time in the worst case[8]. They have also shown that the number of steps in a strategy may grow exponentially with the complexity of $\mathcal{C}_{valid}$. Previously, Natarajan [Natarajan, 1986] had shown that the three-dimensional problem is also PSPACE-hard. On the other hand, no lower-bound time complexity result has been established so far for the problem in two dimensions. There are, however, several upper-bound results obtained with algorithms that we will describe later in this chapter.

[8] Canny and Reif have also shown that in three dimensions, determining whether a given motion command with a given starting configuration is guaranteed to terminate in a goal is NP-hard.

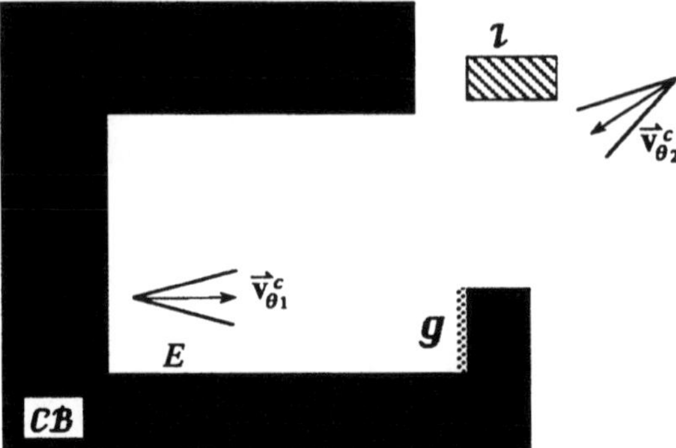

Figure 8. This figure illustrates a sequential two-step motion strategy. The initial region $\mathcal{I}$ is a rectangle in free space and the goal region $\mathcal{G}$ is a contact edge. The first motion is commanded along $\mathbf{v}^c_{\theta_2}$; the second along $\vec{\mathbf{v}}^c_{\theta_1}$.

Until Section 6, we will only consider sequential strategies. Figure 8 illustrates such a strategy. The initial region $\mathcal{I}$ is a rectangle in free space and the goal region $\mathcal{G}$ is an edge in $\mathcal{C}_{contact}$. The strategy consists of two motion commands with commanded velocities $\vec{\mathbf{v}}^c_{\theta_2}$ and $\vec{\mathbf{v}}^c_{\theta_1}$. The command along $\vec{\mathbf{v}}^c_{\theta_2}$ is executed first. (The reverse order on the subscripts results from the backchaining algorithm described in the next section.) With an appropriate termination condition, the motion commanded along $\vec{\mathbf{v}}^c_{\theta_2}$ and executed from within $\mathcal{I}$ is guaranteed to terminate in the horizontal edge denoted by E in the figure; with an appropriate termination condition, the motion commanded along $\vec{\mathbf{v}}^c_{\theta_1}$ and executed from within E is guaranteed to terminate in $\mathcal{G}$. (Note that the conditions labeling the arcs of a strongly guaranteed sequential strategy are irrelevant. They necessarily evaluate to `true`, since otherwise the strategy would not be strongly guaranteed. Hence, we just ignore them.)

Remark: There are problems which admit no strongly guaranteed strategy. This does not mean, however, that the goal $\mathcal{G}$ cannot be achieved from $\mathcal{I}$. Other strategies may achieve $\mathcal{G}$ from $\mathcal{I}$ without guaranteeing success. We might then be interested in generating a **weakly guaranteed** strategy, i.e. a strategy that is guaranteed to either succeed or fail recognizably. This means that the strategy is guaranteed (1) to terminate, (2) to return success only if the goal has actually been achieved, and (3) to return failure only if the goal has not been achieved yet and the remaining non-executed steps in the strategy (if any) have no chance to fortuitously achieve the goal recognizably. Even in cases where strongly guaranteed strategies exist, it may sometimes be preferable to generate weakly guaranteed ones, since the latter are usually simpler (they do not systematically consider the worst cases). The concept of weakly guaranteed strategy has been investigated in [Donald, 1987a and 1988b] under the name of Error Detection and Recovery (EDR) strategy. ∎

2 Preimage Backchaining

In this section we introduce a general approach to motion planning with uncertainty, called *preimage backchaining*. This approach will be investigated in detail in the rest of the chapter. It was originally proposed in [Lozano-Pérez, Mason and Taylor, 1984].

Let $\mathcal{T}$ be a subset of $\mathcal{C}_{valid}$ and $\mathbf{M} = (\vec{\mathbf{v}}_\theta^c, \texttt{TC})$ be a motion command. A **preimage** of $\mathcal{T}$ for $\mathbf{M}$ is defined as any subset $\mathcal{P}$ of $\mathcal{C}_{valid}$ such that: if $\mathcal{A}$'s configuration is in $\mathcal{P}$ when the execution of $\mathbf{M}$ starts, then it is guaranteed that $\mathcal{A}$ will reach $\mathcal{T}$ and that it will be in $\mathcal{T}$ when TC terminates the motion.

Example: Consider the point-in-hole problem illustrated in Figure 9.a. The region $\mathcal{T}$ is the bottom edge of the depression (the "hole") of the C-obstacle region. Let the commanded velocity be $\vec{\mathbf{v}}_\theta^c$ as shown in the figure. We make the following assumptions: contact space is frictionless; the modulus of the sensed force is accurate; the error in the orientation of the sensed force is bounded by $\varepsilon_f < \pi/4$; and the depth of the hole is greater than $2\rho_q$. The region shown with a striped contour in Figure 9.b is a preimage $\mathcal{P}$ of $\mathcal{T}$ for the generalized damping motion command $\mathbf{M} = (\vec{\mathbf{v}}_\theta^c, \texttt{TC})$, with TC terminating the motion when the position sensor measures a configuration consistent with $\mathcal{T}$ and the force sensor measures a non-zero force consistent with the outgoing normal of $\mathcal{T}$. More formally, TC is defined as:

$$\texttt{TC} \equiv [\mathbf{q}^*(t) \in \mathcal{T} \oplus \mathcal{U}_q(\mathbf{0})] \bigwedge [\|\vec{\mathbf{f}}^*(t)\| > 0 \wedge |angle(\vec{\mathbf{f}}^*(t), \vec{\nu}(\mathcal{T}))| < \varepsilon_f]$$

where:

- $\mathbf{q}^*(t)$ and $\vec{\mathbf{f}}^*(t)$ denote the sensed configuration and force at time t, respectively,

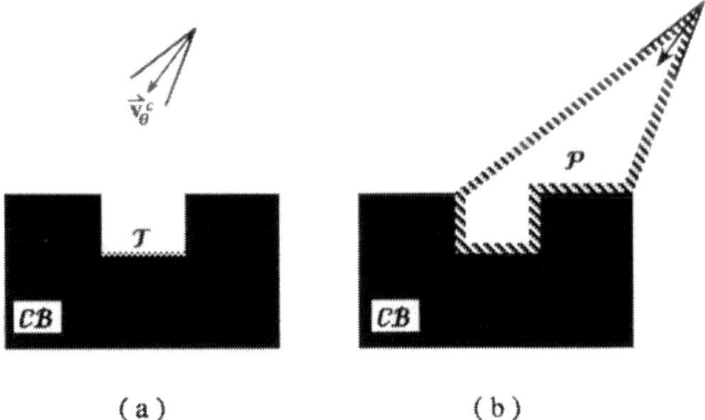

Figure 9. The point-in-hole problem consists of moving the robot's configuration into the edge $\mathcal{T}$ at the bottom of the hole (Figure a). The region shown in striped contour in Figure b depicts a preimage of $\mathcal{T}$ for the generalized damping motion command $(\vec{\mathbf{v}}_\theta^c, \mathtt{TC})$. The commanded velocity $\vec{\mathbf{v}}_\theta^c$ and the control uncertainty cone are shown in the figure. The termination condition TC is defined in the text.

- $\mathcal{T} \oplus \mathcal{U}_q(\mathbf{0})$ is the edge $\mathcal{T}$ "grown" by ρ_q. (The symbol $\oplus$ denotes the Minkowski operator for affine set addition.)

The execution of **M** generates a trajectory that (in free space) is locally contained in the control uncertainty cone and (in contact space) slides over edges that are not orthogonal to a direction contained in the control uncertainty cone. Thus, it is guaranteed to reach $\mathcal{T}$ whenever the starting configuration of $\mathcal{A}$ is in $\mathcal{P}$. Since $\varepsilon_f < \pi/4$, the second condition of TC, $\|\vec{\mathbf{f}}^*(t)\| > 0 \wedge |angle(\vec{\mathbf{f}}^*(t), \vec{\nu}(\mathcal{T}))| < \varepsilon_f$, guarantees that the motion will terminate in one of the three horizontal contact edges. Since the depth of the hole is greater than $2\rho_q$, the first condition, $\mathbf{q}^*(t) \in \mathcal{T} \oplus \mathcal{U}_q(\mathbf{0})$, allows TC to distinguish between the bottom edge $\mathcal{T}$ and the two horizontal side edges bordering the entrance of the hole. Hence, TC will stop the motion in $\mathcal{T}$, and $\mathcal{P}$ is a preimage of $\mathcal{T}$ for **M**. ∎

Let PREIMAGE be an algorithm that, given a commanded velocity $\vec{\mathbf{v}}_\theta^c$ and a region $\mathcal{T} \subset \mathcal{C}_{valid}$, returns a region $\mathcal{P}$ and a termination condition TC such that $\mathcal{P}$ is a preimage of $\mathcal{T}$ for the motion command $(\vec{\mathbf{v}}_\theta^c, \mathtt{TC})$. Given the initial and goal regions, $\mathcal{I}$ and $\mathcal{G}$, **preimage backchaining** essentially proceeds according to the following algorithm:

```
procedure PB;
    begin
        P_0 ← G; i ← 0;
        while I ⊈ P_i do
            begin
                i ← i + 1;
                select θ_i in S^1;
                P_i ← PREIMAGE(v^c_θi, P_{i-1});
            end;
    end;
```

Since it selects a commanded velocity at each iteration, the backchaining process in fact consists of searching an infinite tree of preimages. If the search terminates successfully, its outcome is a sequence of preimages $\mathcal{P}_1, \mathcal{P}_2, ..., \mathcal{P}_r$ such that:

- $\mathcal{P}_i$, $\forall i \in [1, r]$, is the preimage of $\mathcal{P}_{i-1}$ (with $\mathcal{P}_0 = \mathcal{G}$) computed by PREIMAGE for the selected commanded velocity $\vec{\mathbf{v}}^c_{\theta_i}$ and a termination condition $\mathtt{TC}_i$ returned by PREIMAGE;

- $\mathcal{I} \subseteq \mathcal{P}_r$.

The reverse sequence of the motion commands, $(\mathbf{M}_r, \mathbf{M}_{r-1}, ..., \mathbf{M}_1)$, with $\mathbf{M}_i = (\vec{\mathbf{v}}^c_{\theta_i}, \mathtt{TC}_i)$, is the generated (sequential, strongly guaranteed) r-step motion strategy.

Example: Figure 10 illustrates an idealized preimage backchaining process with the example previously given in Figure 8. First, a commanded velocity $\vec{\mathbf{v}}^c_{\theta_1}$ is selected and the preimage $\mathcal{P}_1$ of $\mathcal{G}$ is computed (Figure 10.a). This preimage does not contain $\mathcal{I}$ and is taken as an intermediate goal. Second, a commanded velocity $\vec{\mathbf{v}}^c_{\theta_2}$ is selected and the preimage $\mathcal{P}_2$ of $\mathcal{P}_1$ is computed (Figure 10.b). $\mathcal{P}_1$ is so thin at one end that the termination condition of this motion command can reliably recognize $\mathcal{P}_1$'s achievement only by detecting contact (using force sensing) in the edge denoted by E, which is part of $\mathcal{P}_1$. Since $\mathcal{P}_2$ contains $\mathcal{I}$, the planning problem is solved with a sequence of two motions. Interestingly, the preimage backchaining process has resulted in identifying and using E as an intermediate "landmark" to help the reliable progression of the robot toward $\mathcal{G}$. Of course, in practice, the backchaining process would probably not generate such a strategy directly; rather, it would try several commanded velocities, some unsuccessfully. ∎

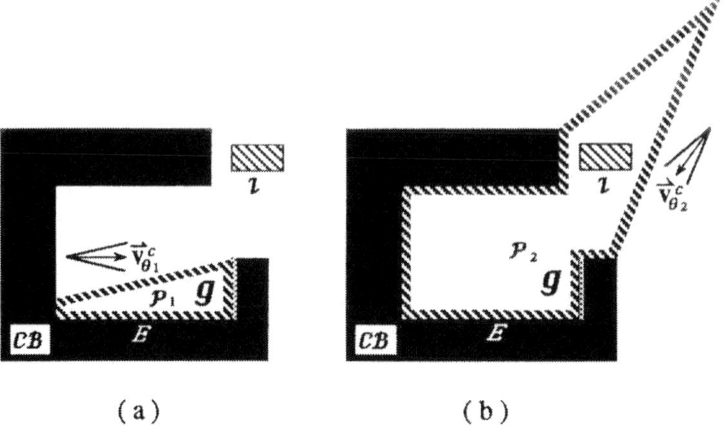

Figure 10. This figure illustrates an ideal preimage backchaining process with the example of Figure 8. First, the commanded velocity $\vec{\mathbf{v}}^c_{\theta_1}$ is selected and the preimage $\mathcal{P}_1$ of $\mathcal{G}$ is computed. This preimage does not contain $\mathcal{I}$. It is taken as an intermediate goal. The commanded velocity $\vec{\mathbf{v}}^c_{\theta_2}$ is selected and the preimage $\mathcal{P}_2$ of $\mathcal{P}_1$ is computed. $\mathcal{P}_2$ contains $\mathcal{I}$ and the planning problem is solved.

The preimage backchaining algorithm PB given above is obviously oversimplified. In order to construct an effective planner, the following three issues must be investigated:

- *Computation of maximal preimages:* For a given commanded velocity and a given region $\mathcal{T}$, the ideal algorithm PREIMAGE would compute the maximal preimage of $\mathcal{T}$ over all possible termination conditions and return a termination condition able to recognize the achievement of the goal. Indeed, a large preimage has more chance to include the initial region $\mathcal{I}$ than a smaller preimage; in addition, if it is considered recursively as an intermediate goal, a large preimage has more chance to admit large preimages than a smaller one. Thus, considering maximal preimages may reduce the size of the search space; it may also produce shorter strategies. On the other hand, using non-maximal preimages may not guarantee that a motion strategy will be ultimately generated, even if one exists. But do maximal preimages exist?

- *Choice of the commanded velocities:* The problem of generating a sequential strategy can be transformed into the combinatorial problem of searching a tree by selecting commanded velocities from a finite set of values in S^1. The root of this tree is the goal region $\mathcal{G}$, and every other node is a preimage region. Each arc is a motion command, connecting a region to its preimage computed by PREIMAGE for this command. However, a predefined discretization of S^1 may not guarantee the planner to generate a sequential strategy whenever one exists. Is there a way to discretize S^1 in a such a way that we are sure not to miss any important velocity orientation?

- *Generation of conditional strategies:* Searching a tree of preimages, as suggested above only permits the planner to generate sequential strategies. There exist problems whose only solutions are conditional strategies (see Section 6). A conditional branching occurs when a motion is guaranteed to achieve one of several intermediate goals, without knowing which one in advance, in such a way that it is guaranteed at planning time that it will be possible to recognize at execution time which one has been achieved. For these problems, another search space must be considered, one which allows the planner to compute the preimages of sets of goals, rather than of individual goals. How should this space be defined and searched?

These issues are addressed in the next sections (Sections 3 through 8).

Note that the preimage backchaining approach is applicable even when there is no uncertainty. It also applies with control models other than generalized damping. Only the computation of the preimages (which we have not described so far) and the parameters to discretize in order to build the search space would be changed. In fact, preimage backchaining is based on a very general planning idea. A preimage can be regarded as a precondition (ideally, the weakest one) whose satisfaction before executing an action guarantees that some postcondition will be satisfied after the action is executed. The same idea has been used previously as a tool to analyze algorithms [Hoare, 1969] and as the basis of the "goal regression" planning method in Artificial Intelligence [Waldinger, 1977] [Nilsson, 1980].

In the following, the word "goal" designates either the goal region $\mathcal{G}$ of the motion planning problem, or any preimage taken as an intermediate goal. A goal is generically denoted by $\mathcal{T}$.

3 Formal Definition of a Preimage

The notion of a preimage of a goal $\mathcal{T}$ for a motion command $\mathbf{M} = (\vec{\mathbf{v}}_\theta^c, \mathtt{TC})$ combines two basic concepts, known as **goal reachability** and **goal recognizability** [Erdmann, 1984].

Goal reachability relates to the fact that any trajectory obtained by executing a motion commanded along $\vec{\mathbf{v}}_\theta^c$ from a preimage of $\mathcal{T}$ should reach $\mathcal{T}$. Given a starting configuration, $\vec{\mathbf{v}}_\theta^c$ only determines a **commanded trajectory** τ_θ^c. Due to uncertainty in control, any execution produces an **actual trajectory** τ that is slightly different. The planner must be certain that all the possible actual trajectories consistent with both $\vec{\mathbf{v}}_\theta^c$ and $\mathcal{U}_v$ will reach $\mathcal{T}$ at some instant.

Goal reachability, however, is not enough, since the termination condition could just let the robot traverse the goal without stopping it or, instead, it could halt the motion prematurely before the goal has been reached. The planner must also be certain that the termination condition TC will stop $\mathcal{A}$ in $\mathcal{T}$. This is the goal recognizability issue. TC is an observer of the actual trajectory τ being executed. Since sensing is imperfect, it perceives the actual trajectory as an **observed trajectory** τ^* which is most likely to be neither the commanded one, nor the actual one. Thus, the problem for the planner is to be sure that, for every possible observed trajectory τ^*, (1) TC becomes `true` at some instant t_f, and (2) when TC first becomes `true`, all the actual trajectories τ consistent with τ^* (i.e. the trajectories which may be observed as τ^* given the uncertainty in sensing) have reached the goal, i.e. $\tau(t_f) \in \mathcal{T}$.

This discussion leads us to define the notion of trajectory directory:

DEFINITION 1: *The* **directory of actual trajectories** *for a region $\mathcal{P}$ and a commanded velocity $\vec{\mathbf{v}}_\theta^c$ is the set $\mathcal{D}_\theta(\mathcal{P})$ of all the trajectories starting inside $\mathcal{P}$ and consistent with $\vec{\mathbf{v}}_\theta^c$, the control uncertainty $\mathcal{U}_c$, and the C-obstacle region. More formally, $\mathcal{D}_\theta(\mathcal{P})$ is the set of all the trajectories $\tau : t \in [0, +\infty) \mapsto \tau(t) \in \mathcal{C}_{valid}$ piecewise of class C^1, such that $\tau(0) \in \mathcal{P}$ and (see Subsection 1.3.2):*

$$\frac{d\vec{\tau}(t)}{dt} = \frac{1}{B}\vec{\mathbf{f}}_{reac}(\tau(t), \vec{\mathbf{v}}_\theta(t)) + \vec{\mathbf{v}}_\theta(t)$$

with $\vec{\mathbf{v}}_\theta(t) \in \mathcal{U}_v(\vec{\mathbf{v}}_\theta^c)$ being of class C^0.

We have that:

- $\mathcal{D}_\theta(\mathcal{P}) = \bigcup_{\mathbf{q}_s \in \mathcal{P}} \mathcal{D}_\theta(\{\mathbf{q}_s\})$,

- $\mathcal{D}_\theta(\mathcal{P}_1 \cup \mathcal{P}_2) = \mathcal{D}_\theta(\mathcal{P}_1) \cup \mathcal{D}_\theta(\mathcal{P}_2)$.

With each trajectory τ in $\mathcal{D}_\theta(\mathcal{P})$ we associate two functions, $\vec{\mathbf{v}}_\tau(t)$ and $\vec{\mathbf{f}}_\tau(t)$. They respectively evaluate at each instant t to the effective commanded velocity that produces τ and to the reaction force exerted on the robot. We have that: $\vec{\mathbf{f}}_\tau(t) = \vec{\mathbf{f}}_{reac}(\tau(t), \vec{\mathbf{v}}_\tau(t))$.

We represent an observed trajectory τ^* by a pair $(\mathbf{q}_{\tau^*}(t), \vec{\mathbf{f}}_{\tau^*}(t))$, where $\mathbf{q}_{\tau^*}$ and $\vec{\mathbf{f}}_{\tau^*}$ are functions mapping the elapsed time t since the beginning of the motion to the sensed configuration and the sensed force at this instant.

The consistency between an actual and an observed trajectory is defined as follows:

DEFINITION 2: *An actual trajectory τ and an observed trajectory $\tau^* = (\mathbf{q}_{\tau^*}, \vec{\mathbf{f}}_{\tau^*})$ are* **consistent** *if and only if:*

$$\forall t \in [0, +\infty) : (\tau(t), \vec{\mathbf{f}}_\tau(t)) \in \mathcal{U}_q(\mathbf{q}_{\tau^*}(t)) \times \mathcal{U}_f(\vec{\mathbf{f}}_{\tau^*}(t)).$$

$\mathcal{K}^*_{traj}(\tau) \stackrel{\text{def}}{=} \{\tau^* \ / \ \tau^* \text{ and } \tau \text{ are consistent}\}$ *is the set of all the possible observations of τ.*

The termination condition $\mathtt{TC} = \mathtt{tp}(t, \mathbf{q}^*[0,t], \vec{\mathbf{f}}^*[0,t])$ applies to the observed trajectory. The directory $\mathcal{D}_\theta(\mathcal{P})$ contains all the possible actual trajectories. Each trajectory $\tau \in \mathcal{D}_\theta(\mathcal{P})$ may be observed by TC as any trajectory in $\mathcal{K}^*_{traj}(\tau)$. We define $\mathtt{Achieve}(\mathcal{T}, \mathrm{M}, \mathcal{P})$ as the condition expressing that the execution of M is guaranteed to terminate in $\mathcal{T}$ whenever the starting configuration is in $\mathcal{P}$:

DEFINITION 3: *Let* $\mathrm{M} = (\vec{\mathbf{v}}^c_\theta, \mathtt{tp}(t, \mathbf{q}^*[0,t], \vec{\mathbf{f}}^*[0,t]))$.

$$\begin{aligned} \mathtt{Achieve}(\mathcal{T}, \mathrm{M}, \mathcal{P}) \stackrel{\text{def}}{=} \ & \forall \tau \in \mathcal{D}_\theta(\mathcal{P}), \forall \tau^* \in \mathcal{K}^*_{traj}(\tau) : \\ & .\ \exists t \text{ verifying } \mathtt{tp}(t, \mathbf{q}_{\tau^*}[0,t], \vec{\mathbf{f}}_{\tau^*}[0,t]); \\ & .\ \tau(t_f) \in \mathcal{T}, \text{ where :} \\ & \qquad t_f = \inf\{t \ / \ \mathtt{tp}(t, \mathbf{q}_{\tau^*}[0,t], \vec{\mathbf{f}}_{\tau^*}[0,t])\}. \end{aligned}$$

This yields the following definition of a preimage:

DEFINITION 4: *A* **preimage** *of a goal $\mathcal{T}$ for the motion command* $\mathbf{M}$ *is any region $\mathcal{P}$ verifying* $\texttt{Achieve}(\mathcal{T}, \mathbf{M}, \mathcal{P})$.

The following properties are direct consequences of this definition:

- If $\mathcal{P}$ is a preimage of $\mathcal{T}$ for $\mathbf{M}$, then any subset of $\mathcal{P}$ is also a preimage of $\mathcal{T}$ for $\mathbf{M}$.

- If $\mathcal{P}_1$ and $\mathcal{P}_2$ are both preimages of $\mathcal{T}$ for $\mathbf{M}$, then $\mathcal{P}_1 \cup \mathcal{P}_2$ is also a preimage of $\mathcal{T}$ for $\mathbf{M}$.

These two properties directly yield the notion of maximal preimage of a goal for a given motion command:[9]

DEFINITION 5: *The region*

$$\mathcal{P}_{\mathbf{M}}^{max}(\mathcal{T}) = \{\mathbf{q}_s \ / \ \texttt{Achieve}(\mathcal{T}, \mathbf{M}, \{\mathbf{q}_s\})\}$$

is the **maximal preimage** *of $\mathcal{T}$ for* $\mathbf{M}$.

In general:

$$\mathcal{P}_{\mathbf{M}}^{max}(\mathcal{T}_1) \cup \mathcal{P}_{M}^{max}(\mathcal{T}_2) \subseteq \mathcal{P}_{\mathbf{M}}^{max}(\mathcal{T}_1 \cup \mathcal{T}_2).$$

Indeed, there may be configurations from which a motion command is guaranteed to reach and terminate in one of the two regions $\mathcal{T}_1$ and $\mathcal{T}_2$, without knowing which one in advance.

Given a commanded velocity $\vec{\mathbf{v}}_\theta^c$, the size of the maximal preimage of a goal $\mathcal{T}$ depends on the ability of the termination condition to recognize the achievement of $\mathcal{T}$. We will analyze this ability for several termination conditions in Section 7. Before that, we describe two practical methods which compute preimages for relatively simple termination conditions and we explore the construction of a motion planner using these two methods.

[9] This notion of maximal preimage is defined for a *fixed* motion command, hence a fixed termination condition. We will see that the maximal preimage for a goal $\mathcal{T}$ may not exist if we allow the termination condition to *vary*.

4 Preimage Computation

4.1 Separation of Reachability and Recognizability

As suggested in Section 2, for a given commanded velocity $\vec{\mathbf{v}}_\theta^c$ and a given goal region $\mathcal{T}$, the ideal PREIMAGE algorithm would compute the maximal preimage of $\mathcal{T}$ over all the possible termination conditions and return the corresponding termination condition. However, we will see later that: (1) if we allow the termination condition to vary, there does not always exist a maximal preimage; (2) if there exists one, it may not be unique; and (3) if there exists a unique maximal preimage, it may depend in a subtle fashion on the knowledge used by the termination condition. Most of the difficulties related to the computation of "large" preimages are due to the interdependence between goal reachability and goal recognizability, i.e. to the fact that deciding at each instant whether the robot is in the goal, or not, may depend on past history.

Erdmann [Erdmann, 1984] suggested approaching the preimage computation problem by restricting the recognition power of the termination condition so that the reachability and recognizability issues can be treated separately. The basic idea is to construct a termination condition which (1) may evaluate to `true` only if the robot is in the goal $\mathcal{T}$ and (2) is guaranteed to evaluate to `true` when a certain subset $\mathcal{S} \subseteq \mathcal{T}$ is attained. Then it remains to compute a region $\mathcal{L}$ from where the robot is guaranteed to reach $\mathcal{S}$ when commanded along some given velocity $\vec{\mathbf{v}}_\theta^c$. $\mathcal{L}$ is called a *backprojection* of $\mathcal{S}$ for $\vec{\mathbf{v}}_\theta^c$ (see Subsections 4.2 and 4.3). It is a preimage of $\mathcal{T}$.

In Subsections 4.4 and 4.5 we describe two methods based on this general idea. Both methods are easily implementable and can serve as building blocks for simple motion planners. However, since they restrict the recognition power of the termination conditions, such planners are not complete.

4.2 Notion of Backprojection

A backprojection is defined as follows [Erdmann, 1984]:

DEFINITION 6: *Let:*

$$\mathtt{Reach}(\mathcal{S}, \theta, \mathcal{L}) \stackrel{\text{def}}{=} \forall \tau \in \mathcal{D}_\theta(\mathcal{L}), \exists t \geq 0 : \tau(t) \in \mathcal{S}.$$

A **backprojection** *of $\mathcal{S}$ for $\vec{\mathbf{v}}^c_\theta$ is any region $\mathcal{L}$ verifying* Reach($\mathcal{S},\theta,\mathcal{L}$).

If two regions $\mathcal{L}_1$ and $\mathcal{L}_2$ are backprojections of $\mathcal{S}$ for $\vec{\mathbf{v}}^c_\theta$, then their union $\mathcal{L}_1 \cup \mathcal{L}_2$ is also a backprojection of $\mathcal{S}$ for $\vec{\mathbf{v}}^c_\theta$. Hence:

DEFINITION 7: *The region* $\{\mathbf{q}_s \in \mathcal{C}_{valid}$ / Reach($\mathcal{S},\theta,\{\mathbf{q}_s\}$)$\}$ *is the* **maximal backprojection** *of $\mathcal{S}$ for $\vec{\mathbf{v}}^c_\theta$. It is denoted by $\mathcal{L}_\theta(\mathcal{S})$.*

From now on, we will only consider backprojections that are maximal. Hence, we will usually omit this qualifier.

A backprojection of a goal differs from a preimage by the fact that it ignores the need to recognize the achievement of the goal. Clearly, any preimage of a goal $\mathcal{T}$ is included in the backprojection of $\mathcal{T}$ for the same commanded velocity. Hence, the backprojection of $\mathcal{T}$ for any given commanded velocity is an upper bound of the maximal preimage, if any, of the same goal for the same commanded velocity, over all possible termination conditions.

As with maximal preimages, in general:

$$\mathcal{L}_\theta(\mathcal{S}_1) \cup \mathcal{L}_\theta(\mathcal{S}_2) \subseteq \mathcal{L}_\theta(\mathcal{S}_1 \cup \mathcal{S}_2).$$

Indeed, there may be configurations from which a motion commanded along $\vec{\mathbf{v}}^c_\theta$ is guaranteed to reach one of the two regions $\mathcal{S}_1$ and $\mathcal{S}_2$, without knowing which one in advance.

Figure 11 shows the backprojections (striped contours) for a single contact edge $\mathcal{S}$ for two commanded velocities $\vec{\mathbf{v}}^c_{\theta_1}$ and $\vec{\mathbf{v}}^c_{\theta_2}$.

4.3 Computation of a Backprojection

Given a commanded velocity $\vec{\mathbf{v}}^c_\theta$, the following algorithm computes the maximal backprojection of a goal region[10] $\mathcal{S}$ consisting of a single contact edge[11] [Erdmann, 1984]:

---

[10] Throughout this subsection, we call the region $\mathcal{S}$ a goal, although it is more likely to be a subset of a goal.

[11] If $\mathcal{S}$ is a portion of a contact edge, the algorithm applies by treating the two endpoints of $\mathcal{S}$ as additional vertices of $\mathcal{C}_{contact}$.

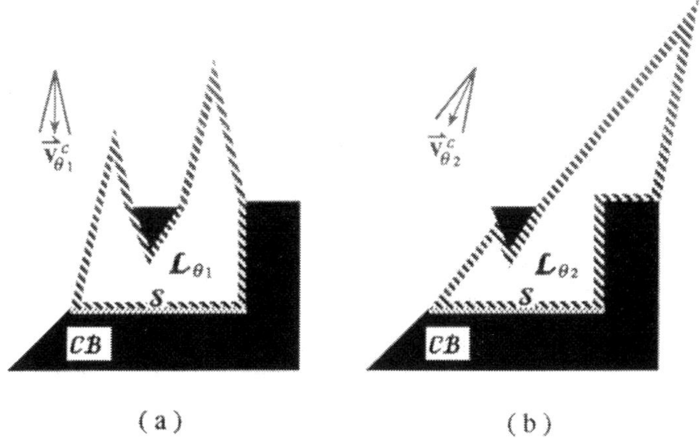

Figure 11. The regions outlined in striped contour in both figures are the maximal backprojections $\mathcal{L}_{\theta_1}(\mathcal{S})$ and $\mathcal{L}_{\theta_2}(\mathcal{S})$ of the edge $\mathcal{S}$ for the commanded velocities $\vec{\mathbf{v}}^c_{\theta_1}$ and $\vec{\mathbf{v}}^c_{\theta_2}$, respectively. Any motion commanded along $\vec{\mathbf{v}}^c_{\theta_1}$ (resp. $\vec{\mathbf{v}}^c_{\theta_2}$) and starting from within $\mathcal{L}_{\theta_1}(\mathcal{S})$ (resp. $\mathcal{L}_{\theta_2}(\mathcal{S})$) is guaranteed to reach $\mathcal{S}$.

1. Consider every vertex in $\mathcal{C}_{contact}$. Mark every non goal vertex at which sticking is possible (i.e. such that the friction cone intersects the inverted control uncertainty cone). Mark every goal vertex if sticking is possible in the non goal edge abutting this vertex. Mark every goal vertex if sliding away from the vertex in the non goal edge is possible.

2. At every marked vertex erect two rays parallel to the sides of the inverted control uncertainty cone. Compute the intersection of these rays among themselves and with $\mathcal{C}_{contact}$. Ignore each ray beyond the first intersection. The remaining segment in each ray is called a **free edge.** The marked vertex at which the ray was erected is called the **anchor vertex** of the free edge.

3. Trace out the backprojection region. To that purpose, trace the edge $\mathcal{S}$ in the direction that leaves the C-obstacle region on the right-hand side, until an endpoint of $\mathcal{S}$ is attained; then, let all the contact and free edges incident at this point be sorted clockwise in a circular list; trace the first edge following $\mathcal{S}$ in this list; proceed in the same way until $\mathcal{S}$ is encountered again.

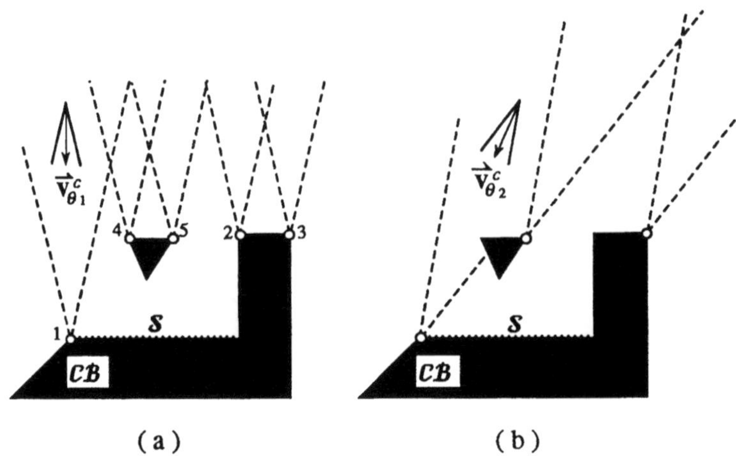

Figure 12. This figure illustrates the computation of the two backprojections shown in Figure 11, by Erdmann's algorithm. Figure a shows both the vertices which are marked at step 1 for the commanded velocity $\vec{\mathbf{v}}^c_{\theta_1}$ (small circles), and the rays that are erected at step 2 from these points. Vertex 1 is marked because it is possible to slide away from $\mathcal{S}$. Vertices 2 through 5 are marked because they are non goal vertices where sticking is possible. Figure b shows the marked vertices and the erected rays for the commanded velocity $\vec{\mathbf{v}}^c_{\theta_2}$.

This algorithm requires the backprojection to be bounded, which can be achieved by imposing that $\mathcal{C}_{valid}$ or (if $\eta_v \neq 0$) the C-obstacle region be bounded. In addition, the algorithm does not cover some particular cases, which we will discuss below.

The construction of the backprojections shown in Figure 11 by this algorithm is illustrated in Figure 12. The vertices marked at step 1 are depicted as small circles. Erecting the inverted control uncertainty cone at the marked vertices produces constraints — the free edges — which prevent the backprojection from containing configurations from where a motion may reach a vertex or an edge in which it can stick or slide away from the goal with no possibility of returning to it. Notice that the algorithm only constructs the boundary of the backprojection of $\mathcal{S}$. The region lying within this boundary and the portions of contact edges contained in this boundary are part of the backprojection. Assuming that the control uncertainty cone is open, each free edge in the boundary is

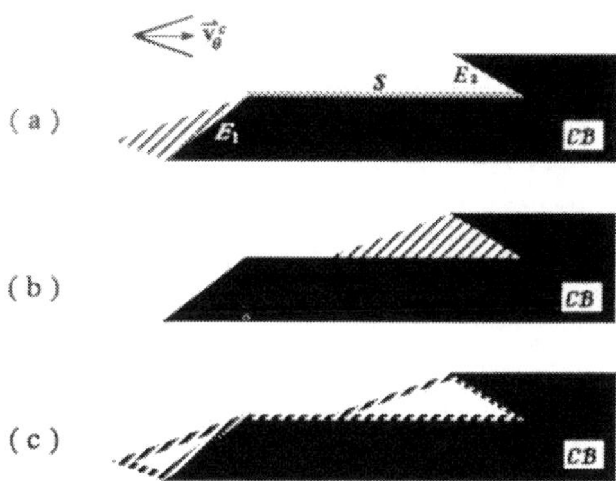

Figure 13. This figure illustrates special cases which may occur when some of the directions contained in the control uncertainty cone project negatively on the ingoing normal of the goal edge $\mathcal{S}$.

also part of the backprojection, with the exception of its anchor vertex if the latter is a non goal vertex.

The above algorithm treats only the case where all the directions contained in the control uncertainty cone project positively on the ingoing normal of $\mathcal{S}$. In the case where some (or all) directions project negatively, we could make the algorithm simply return $\mathcal{S}$ as the backprojection. This is correct, since $\mathcal{S}$ is always a (possibly non-maximal) backprojection of itself, but it is too conservative, as we illustrate in Figure 13.

The striped triangle in Figure 13.a contains starting configurations from which the motion along $\vec{\mathbf{v}}_\theta^c$ is guaranteed to reach the left vertex of $\mathcal{S}$ by sliding along the edge E_1 abutting this vertex (assuming frictionless contact space for simplicity) and therefore is part of $\mathcal{L}_\theta(\mathcal{S})$. The striped triangle in Figure 13.b is also part of $\mathcal{L}_\theta(\mathcal{S})$, since all the motions starting in it ultimately reach the right vertex of $\mathcal{S}$ eventually by sliding along E_2. Figure 13.c shows $\mathcal{L}_\theta(\mathcal{S})$. It contains the two previous triangles and the edge $\mathcal{S}$. (Notice that each of the two triangles is a backprojection of one of the two endpoints of $\mathcal{S}$. Removing these two vertices from the goal, i.e. considering the open edge, rather than the closed one, as the goal, would result in $\mathcal{S}$ being its own backprojection.) Erdmann [Erdmann, 1986] described a modification of his original algorithm for handling the cases illustrated in Figure 13 in a general way.

Erdmann's algorithm has been generalized by Canny and Donald [Donald, 1987a] to an algorithm that generates the maximal backprojection of any region $\mathcal{S}$ described as the union of contact edges and polygonal regions in $\mathcal{C}_{valid}$. The principle of the algorithm is the following. It sweeps a line L across the plane perpendicularly to the commanded velocity $\vec{\mathbf{v}}_\theta^c$. The sweep starts at a position of L where it is tangent to $\mathcal{S}$, with $\mathcal{S}$ lying entirely on the side of L pointed to by the vector $-\vec{\mathbf{v}}_\theta^c$. The sweep proceeds in the direction of the vector $-\vec{\mathbf{v}}_\theta^c$. During the sweep, the algorithm maintains the status of L, i.e. the description of its intersection with $\mathcal{C}_{contact}$, $\mathcal{S}$ and free edges erected from contact vertices and goal vertices. This status changes qualitatively at events occurring when L passes through a vertex of $\mathcal{C}_{contact}$, a vertex of $\mathcal{S}$, the intersection of two free edges, the intersection of a free edge and $\mathcal{C}_{contact}$, or the intersection of a free edge and $\mathcal{S}$. At each of these events, the algorithm updates the status of the line and the list of future events. During the sweep, the algorithm traces out the backprojection. Sweep stops when the last event is encountered.

The key idea underlying this algorithm is that the backprojection's boundary consists of straight segments and that at every event there exists a local criterion (which does not depend on what will happen later during the sweep) for deciding how to trace the backprojection's boundary. This criterion leads to erect rays (yielding free edges) from both contact and goal vertices, in order to prevent the backprojection from containing configurations from where a motion may reach a vertex or an edge in which it may stick or slide away from the goal with no possibility of returning to it. The algorithm does not require the backprojection to be bounded.

Figures 14 and 15 illustrate and comment on the operations of this sweep-line method on two examples.

Let n be the number of vertices in $\mathcal{C}_{contact}$. Assume the combined complexity of $\mathcal{C}_{contact}$ and $\mathcal{S}$, i.e. the number of vertices in $\mathcal{C}_{contact}$, plus the number of vertices of $\mathcal{S}$, plus the number of intersections between $\mathcal{S}$ and $\mathcal{C}_{contact}$, to be $O(n)$. The sweep-line algorithm erects $O(n)$ rays. The contour of $\mathcal{L}_\theta(\mathcal{S})$ consists of goal edges (edges bounding $\mathcal{S}$), free edges and contact edges. Each erected ray can contribute at most one edge to the backprojection. Each goal edge contributes at most one backprojection edge. Each non goal edge in contact space can contribute one or

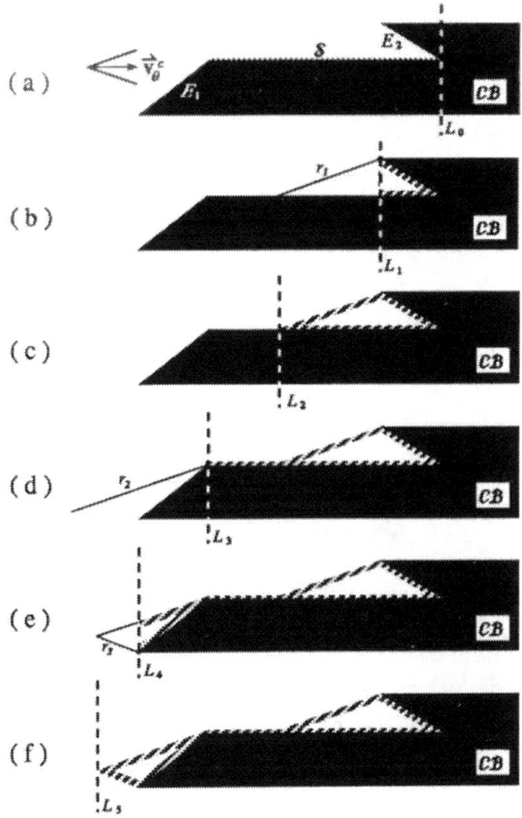

Figure 14. This figure illustrates the computation of a backprojection using the sweep-line algorithm. The goal region is the edge $\mathcal{S}$. Figure a shows the position of the line L when the sweep starts. Figures b through f show its position at the successive critical events. In each figure the striped contour depicts the portion of backprojection constructed so far. The first event (Figure b) occurs when L crosses the non goal vertex bounding the edge E_2. The algorithm erects a ray denoted by r_1 from this vertex. No configurations on the left-hand side of this ray (with the direction of $\vec{\mathbf{v}}_\theta^c$ as the reference) belong to the backprojection of $\mathcal{S}$. The intersection of r_1 with $\mathcal{S}$ produces the next event (Figure c). The following two events are due to the vertices bounding the sliding edge E_1 (Figures d and e). The rays r_2 and r_3 are erected at these vertices. No configurations on the left-hand side of r_2, nor on the right-hand side of r_3, belong to the backprojection. The intersection of the two rays forms the closing event (Figure f).

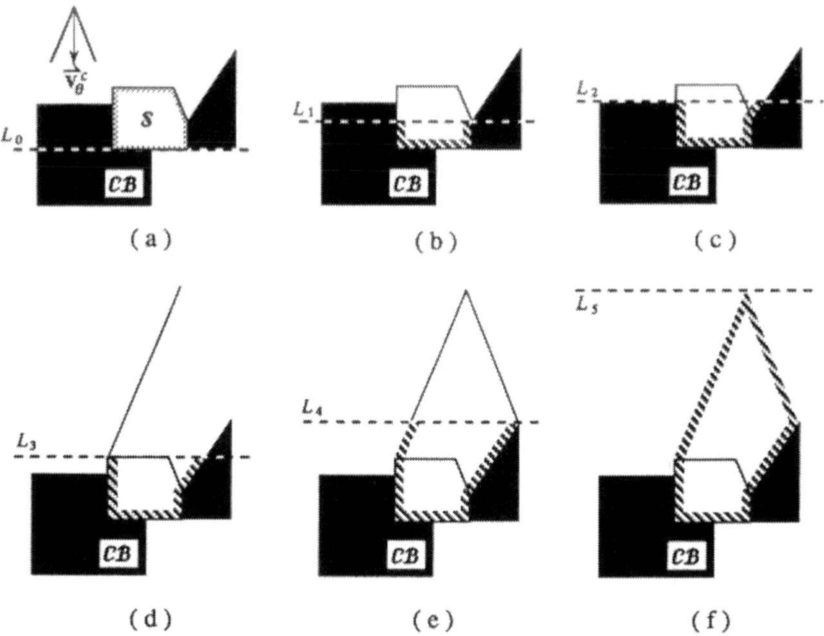

Figure 15. This figure is another illustration of the sweep-line algorithm in the case where $\mathcal{S}$ is a polygonal region in $\mathcal{C}_{valid}$. In Figure d, notice the erection of a ray from a goal vertex located in free space.

several backprojection edges, but whenever it contributes k backprojection edges, it also "consumes" $O(k)$ free edges. Hence, the boundary of the backprojection consists of $O(n)$ edges. Since each ray is interrupted at its first intersection with a goal edge, a contact edge, or another ray, the backprojection's boundary is computed in $O(n \log n)$ time.

4.4 Backprojection from Sticking Edges

Backprojection from sticking edges is a simple method for computing preimages. Given a commanded velocity $\vec{\mathbf{v}}_\theta^c$ and a goal $\mathcal{T}$, it consists of:

1. Determining the subset $\mathcal{T}^s$ of $\mathcal{T} \cap \mathcal{C}_{contact}$ in which motions commanded along $\vec{\mathbf{v}}_\theta^c$ are guaranteed to stick.

2. Computing the backprojection $\mathcal{L}_\theta(\mathcal{T}^s)$.

The computation of $\mathcal{T}^s$ is simply done by comparing the relative orientation of each contact edge in $\mathcal{T}$ with $\vec{\mathbf{v}}_\theta^c$. $\mathcal{T}^s$ is made of every edge E in $\mathcal{T} \cap \mathcal{C}_{contact}$ such that:[12]

$$|angle(-\vec{\mathbf{v}}_\theta^c, \vec{\nu}(E))| \leq \phi - \eta_v.$$

(Notice that guaranteed sticking in an edge is possible only if $\phi > \eta_v$, i.e. if the friction cone is larger than the control uncertainty cone.)

Assume that $\mathcal{T}^s$ is not empty (otherwise, the constructed preimage is the empty set). We still have to construct the termination condition that recognizes sticking. Indeed, while sticking "physically" terminates a motion, the robot must recognize sticking in order to "logically" terminate the motion and execute the next motion command in the motion strategy or return success. One way to construct the termination condition is to project $\mathcal{L}_\theta(\mathcal{T}^s)$ on a line parallel to $\vec{\mathbf{v}}_\theta^c$. Let D be the length of the projection. Since the friction cone is larger than the control uncertainty cone, the actual velocity $\vec{\mathbf{v}}$ of the robot has a positive component along $\vec{\mathbf{v}}_\theta^c$ everywhere in $\mathcal{L}_\theta(\mathcal{T}^s)$, i.e. $\vec{\mathbf{v}} \cdot \vec{\mathbf{v}}_\theta^c > 0$. By examining all the contact edges in the boundary of $\mathcal{L}_\theta(\mathcal{T}^s)$, one can compute the minimal value v_{min} of $\vec{\mathbf{v}} \cdot \vec{\mathbf{v}}_\theta^c$ in $\mathcal{L}_\theta(\mathcal{T}^s)$. A possible termination condition is:

$$\mathtt{TC} \ \equiv \ t > \frac{D}{v_{min}}.$$

The region $\mathcal{L}_\theta(\mathcal{T}^s)$ is a preimage of $\mathcal{T}$ for this termination condition, which will be referred to in the rest of the chapter as the *sticking termination condition.*

Let $\mathcal{T}$ be input as a list of $O(n)$ contact edges. The computation of $\mathcal{T}^s$ and $\mathcal{L}_\theta(\mathcal{T}^s)$ takes $O(n)$ time and $O(n \log n)$ time, respectively. The construction of the expression TC takes $O(n)$ time.

Figure 16 illustrates a strategy generated by a simple planner which computes preimages using the backprojection from sticking edges method [Latombe, Lazanas and Shekhar, 1989]. The planner constructs and

[12] We do not include isolated sticking vertices in $\mathcal{T}^s$, but we could easily do it. In practice, only the isolated concave sticking vertices are interesting, since the sticking behavior at the other isolated sticking vertices is not stable under small displacements. Sticking at a concave vertex may in fact be easier than sticking along a contact edge; hence, it may yield large preimages.

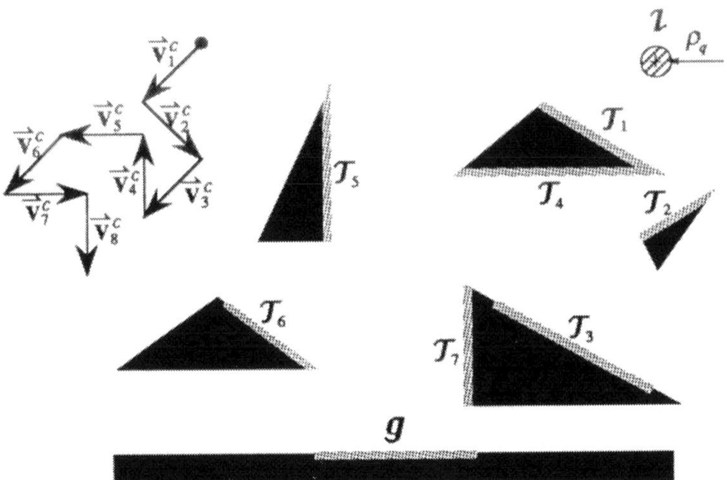

Figure 16. This figure illustrates a strategy produced by a simple planner using the backprojection-from-sticking-edges method. The initial region $\mathcal{I}$ is the disc at the top-right and the goal region is $\mathcal{G}$. The strategy consists of 8 motion commands whose commanded velocities, $\vec{v}_1^c$ through $\vec{v}_8^c$, are shown on the left-hand side. These velocities were selected from the set $\{k\pi/4\}_{k=0,\dots,7}$. Each region shown as a bold grey line segment and denoted by $\mathcal{T}_i$ $(i = 1, \dots, 7)$ represents the sticking subset (for $\vec{v}_i^c$) of the preimage of $\mathcal{T}_{i+1}$ $(\mathcal{T}_8 = \mathcal{G})$.

searches a graph of preimages by considering commanded velocities of orientations $\{k\pi/4\}_{k=0,\dots,7}$. The goal region $\mathcal{G}$ is a subset of a contact edge. The initial region $\mathcal{I}$ is the disc of radius ρ_q shown at the top right of the figure[13]. The generated strategy consists of 8 motion commands. The commanded velocities $\vec{v}_1^c$ through $\vec{v}_8^c$ of these commands are depicted on the left-hand side of the figure. $\mathcal{T}_i$ $(i \in [1, 7])$ denotes the sticking subset (for $\vec{v}_i^c$) of the preimage of $\mathcal{T}_{i+1}$ $(\mathcal{T}_8 = \mathcal{G})$. Hence, a motion along $\vec{v}_{i+1}^c$ starting from within $\mathcal{T}_i$ terminates by sticking in $\mathcal{T}_{i+1}$. The example of Figure 16 is a typical example where the backprojection-from-sticking-edges method works well. The C-obstacles are such that the robot can "bounce" from one sticking edge to another. In contrast, the method works poorly or fails when the C-obstacles are far apart and when the goal lies mostly (or completely) in free space.

[13]This initial region corresponds to the case where the robot senses its initial configuration before planning.

4.5 Backprojection from Goal Kernel

Backprojection from goal kernel is another method for computing preimages. Given a commanded velocity $\vec{\mathbf{v}}_\theta^c$ and a goal $\mathcal{T}$, it consists of:

1. Identifying a subset $\chi_\theta(\mathcal{T})$ of $\mathcal{T}$ whose achievement by a motion commanded along $\vec{\mathbf{v}}_\theta^c$ is recognizable by a termination condition whose only arguments are $\mathbf{q}^*(t)$ and $\vec{\mathbf{f}}^*(t)$, independent of the starting region.

2. Computing the backprojection $\mathcal{L}_\theta(\chi_\theta(\mathcal{T}))$.

The subset $\chi_\theta(\mathcal{T})$ is called the *θ-kernel* of $\mathcal{T}$.

We say that a configuration $\mathbf{q} \in \mathcal{C}_{valid}$ is **θ-consistent** with $\mathbf{q}^*$ and $\vec{\mathbf{f}}^*$ if and only if, when the robot commanded along $\vec{\mathbf{v}}_\theta^c$ is at configuration $\mathbf{q}$, $\mathbf{q}^*$ and $\vec{\mathbf{f}}^*$ are possible sensed configuration and force given the geometry of $\mathcal{C}_{valid}$, the friction angle ϕ and the uncertainty in sensing and control. The direction of the commanded velocity plays an important role in θ-consistency because the force that is sensed in contact space depends on it. (The robot may be in contact space without sensing any reaction force, if it moves tangentially to a C-obstacle.) We define $\mathcal{K}_\theta(\mathbf{q}^*, \vec{\mathbf{f}}^*)$ as the set of configurations $\mathbf{q}$ that are θ-consistent with $\mathbf{q}^*$ and $\vec{\mathbf{f}}^*$.

Assume for simplification that the uncertainty in the moduli of the commanded velocity and the sensed force is zero[14], i.e. $\Delta_v(\vec{\mathbf{v}}_\theta^c) = \{1\}$ and $\Delta_f(\vec{\mathbf{f}}^*) = \{\|\vec{\mathbf{f}}^*\|\}$. Then:

- If $\|\vec{\mathbf{f}}^*\| = 0$, $\mathcal{K}_\theta(\mathbf{q}^*, \vec{\mathbf{f}}^*)$ consists of all the valid configurations $\mathbf{q}$ which are distant from $\mathbf{q}^*$ by less than ρ_q and lie either in free space, or in contact space at a point where the actual reaction force may be zero:

$$\mathcal{K}_\theta(\mathbf{q}^*, \vec{0}) = \{\mathbf{q} \in \mathcal{C}_{valid} \; / \; \|\mathbf{q} - \mathbf{q}^*\| < \rho_q \; , \; \vec{0} \in F_{reac}(\mathbf{q}, \theta)\}.$$

  ($F_{reac}(\mathbf{q}, \theta)$ is the set of possible reaction forces at $\mathbf{q}$ when the commanded velocity is $\vec{\mathbf{v}}_\theta^c$. See Subsection 1.3.3.)

[14] Removing these assumptions is straightforward, but complicates slightly the construction of $\mathcal{K}_\theta(\mathbf{q}^*, \vec{\mathbf{f}}^*)$.

- If $||\vec{\mathbf{f}}^*|| \neq 0$, $\mathcal{K}_\theta(\mathbf{q}^*, \vec{\mathbf{f}}^*)$ consists of all the contact configurations $\mathbf{q}$ which are distant from $\mathbf{q}^*$ by less than ρ_q and which may generate a non-zero reaction force $\vec{\mathbf{f}}$ whose angle with $\vec{\mathbf{f}}^*$ is less than ε_f:

$$\mathcal{K}_\theta(\mathbf{q}^*, \vec{\mathbf{f}}^*) = \{\mathbf{q} \in \mathcal{C}_{contact} \;/\; ||\mathbf{q} - \mathbf{q}^*|| < \rho_q \;,\; \exists \vec{\mathbf{f}} \neq \vec{0} \in F_{reac}(\mathbf{q}, \theta) : \; |angle(\vec{\mathbf{f}}, \vec{\mathbf{f}}^*)| < \varepsilon_f\}.$$

The termination condition defined as:

$$\mathcal{K}_\theta(\mathbf{q}^*(t), \vec{\mathbf{f}}^*(t)) \subseteq \mathcal{T}$$

is such that when it evaluates to `true` it is guaranteed that the robot is in the goal. Its arguments are $\mathbf{q}^*(t)$ and $\vec{\mathbf{f}}^*(t)$, i.e. the sensed configuration and force at the time t when it is evaluated. Its predicate's knowledge is the geometry of $\mathcal{C}_{valid}$, the commanded velocity, the friction angle ϕ, and the uncertainty in control and in sensing (i.e. $\mathcal{U}_v$, $\mathcal{U}_q$ and $\mathcal{U}_f$). In the rest of the chapter, this condition will be referred to as the *instantaneous sensing termination condition.*

It remains now to define a subset of $\mathcal{T}$ such that if it is attained by a motion, it is guaranteed that the above termination condition evaluates to `true`. This subset is precisely the kernel of the goal.

Let $\mathbf{q}_1$ and $\mathbf{q}_2$ be two configurations in $\mathcal{C}_{valid}$. $\mathbf{q}_1$ and $\mathbf{q}_2$ are said to be **θ-distinguishable** if and only if:

$$\{(\mathbf{q}^*, \vec{\mathbf{f}}^*) \;/\; \mathbf{q}_1, \mathbf{q}_2 \in \mathcal{K}_\theta(\mathbf{q}^*, \vec{\mathbf{f}}^*)\} = \emptyset.$$

In other words, if two configurations $\mathbf{q}_1$ and $\mathbf{q}_2$ are θ-distinguishable, it is guaranteed that during a motion commanded along $\vec{\mathbf{v}}_\theta^c$ there is no instant when the sensed configuration and force are θ-consistent with both $\mathbf{q}_1$ and $\mathbf{q}_2$. Two configurations $\mathbf{q}_1$ and $\mathbf{q}_2$ which are not θ-distinguishable are **θ-confusable.**

The **θ-kernel** of $\mathcal{T}$ is:

$$\chi_\theta(\mathcal{T}) = \{\mathbf{q} \in \mathcal{T} \;/\; \forall \mathbf{q}' \in \mathcal{C}_{valid} \backslash \mathcal{T} : \mathbf{q} \text{ and } \mathbf{q}' \text{ are } \theta\text{-distinguishable}\}.$$

Again notice the importance of θ. Without taking the commanded velocity into account, we would be forced to distinguish between configurations using position sensing only, since no configuration in contact space

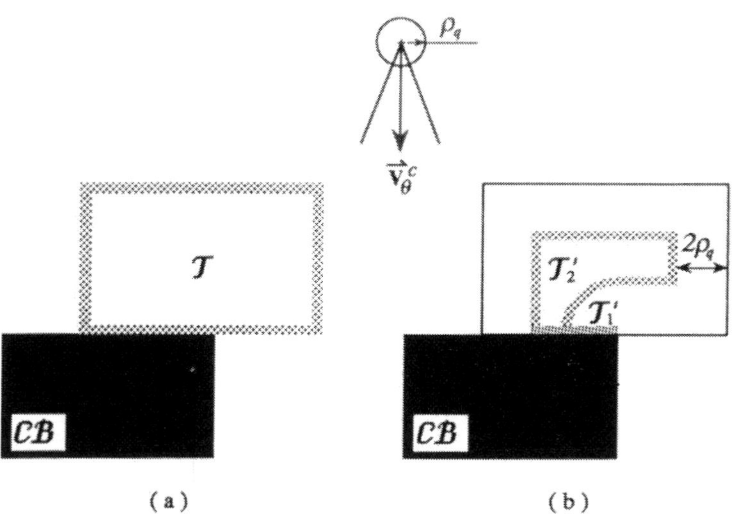

Figure 17. The goal $\mathcal{T}$ is a rectangle in $\mathcal{C}_{valid}$. The commanded velocity $\vec{\mathbf{v}}^c_\theta$ points downward. The θ-kernel of $\mathcal{T}$ is $\mathcal{T}_1' \cup \mathcal{T}_2'$. $\mathcal{T}_1'$ is a subset of a contact edge. $\mathcal{T}_2'$ is a generalized polygon.

is guaranteed to generate a non-zero reaction force under all commanded velocities. This would simply lead to isotropically shrinking the goal $\mathcal{T}$ in order to get its kernel.

The computation of $\chi_\theta(\mathcal{T})$ is not difficult (see [Latombe, Lazanas and Shekhar, 1989]). In general, the θ-kernel of a goal consisting of edges in $\mathcal{C}_{contact}$ and polygons in $\mathcal{C}_{valid}$ is a region made of both edges in $\mathcal{C}_{contact}$ and generalized polygons in $\mathcal{C}_{valid}$. Such a kernel is illustrated in Figure 17. In this example, the workspace contains a single rectangular C-obstacle. Figure 17.a displays the goal region $\mathcal{T}$, a rectangle in $\mathcal{C}_{valid}$. The θ-kernel, with $\vec{\mathbf{v}}^c_\theta$ pointing downward, is the region $\mathcal{T}_1' \cup \mathcal{T}_2'$ (Figure 17.b). $\mathcal{T}_1'$ is a subset of a contact edge. $\mathcal{T}_2'$ is a generalized polygon. Indeed, a configuration in $\mathcal{T}_1'$ is θ-confusable only with contact configurations contained in the goal. A configuration in $\mathcal{T}_2' \backslash \mathcal{T}_1'$ is θ-confusable only with configurations inside the goal or inside the C-obstacle (but the latter are not achievable). Any configuration outside $\mathcal{T}_1' \cup \mathcal{T}_2'$ is θ-confusable with a configuration in $\mathcal{C}_{valid}$ not located in the goal.

We can compute the backprojection of a region $\chi_\theta(\mathcal{T})$ containing generalized polygons using the Canny-Donald's sweep-line algorithm sketched

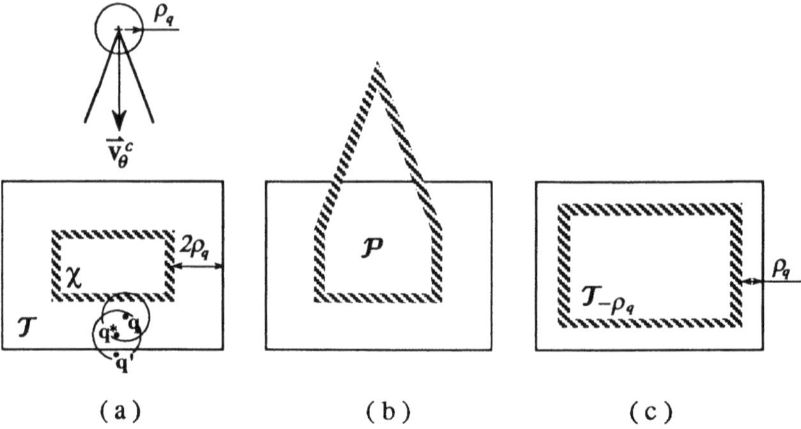

Figure 18. The goal $\mathcal{T}$ is a rectangle in a configuration space empty of C-obstacles. The commanded velocity $\vec{v}_\theta^c$ points along the ingoing normal of the top edge of the rectangle. Figure a shows the kernel χ of $\mathcal{T}$ (striped contour). It is independent of the direction θ since force sensing is useless in the absence of C-obstacles. Figure b shows the preimage $\mathcal{P}$ computed by the backprojection from kernel method. Figure c illustrates the termination condition. The rectangle shown with striped contour depicts the set of sensed configurations at which the termination condition evaluates to `true`.

in Subsection 4.3. The positions of the sweep-line L where it is tangent to the circular edges of $\mathcal{T}$ must be included as additional events. The boundary of the constructed backprojection consists of straight and circular edges. Preimage backchaining may hence lead to computing the kernel of a goal consisting of contact edges and generalized polygons in $\mathcal{C}_{valid}$. One can verify that this kernel also consists of contact edges and generalized polygons in $\mathcal{C}_{valid}$.

The region $\mathcal{L}_\theta(\chi_\theta(\mathcal{T}))$ is the computed preimage of $\mathcal{T}$. Indeed, a motion commanded along $\vec{v}_\theta^c$ from this region is guaranteed to reach $\chi_\theta(\mathcal{T})$ (if it is not terminated before). By definition of $\chi_\theta(\mathcal{T})$, it is then guaranteed that the termination condition $\mathcal{K}_\theta(\mathbf{q}^*(t), \vec{\mathbf{f}}^*(t)) \subseteq \mathcal{T}$ evaluates to `true` at some instant during the motion. The condition certainly evaluates to `true` when $\chi_\theta(\mathcal{T})$ is attained, but it may become `true` before. When the condition becomes `true`, even if the kernel has not been attained yet, the definition of $\mathcal{K}_\theta$ guarantees that the robot is in the goal.

Example: Figure 18 shows the preimage computed by this method in the simple case where the goal $\mathcal{T}$ is a rectangle and $\mathcal{C}_{valid} = \mathcal{C}$ (i.e. there is no C-obstacle). The commanded velocity points along the ingoing normal of an edge of the goal. The kernel, which is independent of the commanded velocity (since force sensing is useless), is the rectangle denoted by χ and obtained by "shrinking" $\mathcal{T}$ by $2\rho_q$ as shown in Figure 18.a. (We assume that each of $\mathcal{T}$'s edges is longer than $4\rho_q$.) In order to understand this transformation, consider a configuration $\mathbf{q}$ (see Figure 18.a). This configuration may be sensed as any configuration in the disc of radius ρ_q centered at $\mathbf{q}$. Any sensed configuration $\mathbf{q}^*$ in this disc may be interpreted as a configuration $\mathbf{q}'$ in the disc of radius ρ_q centered at $\mathbf{q}^*$. In order to be sure that this second disc is completely contained in $\mathcal{T}$, the goal has to be shrunk by $2\rho_q$. Figure 18.b shows the computed preimage $\mathcal{P}$ of $\mathcal{T}$. Figure 18.c depicts the set of sensed configurations (the rectangle denoted by $\mathcal{T}_{-\rho_q}$ and obtained by shrinking $\mathcal{T}$ by ρ_q) at which the termination condition $\mathcal{K}(\mathbf{q}^*(t)) \subseteq \mathcal{T}$ evaluates to `true`. (We write $\mathcal{K}_\theta(\mathbf{q}^*(t), \vec{\mathbf{f}}^*(t)) = \mathcal{K}(\mathbf{q}^*(t))$ since, in this particular example, neither the commanded velocity nor force sensing plays any role in the notion of consistency.) ■

Backprojecting from the goal's θ-kernel usually generates preimages which are significantly larger than those produced by backprojecting from the goal's sticking edges. In particular, it can produce preimages of goals made of regions lying in the free space and/or non-sticking contact edges. There are situations, however, where backprojecting from sticking edges produces larger preimages. Figure 19 shows one of them. The goal $\mathcal{T}$ is a sticking edge for the commanded velocity $\vec{\mathbf{v}}^c_\theta$ and none of the two edges adjacent to $\mathcal{T}$ are distinguishable from $\mathcal{T}$ using force sensing. Hence, the θ-kernel is obtained by shrinking $\mathcal{T}$ by $2\rho_q$ at both its endpoints. The backprojection of the shrunk edge (the θ-kernel) is smaller than the backprojection of the full edge (the sticking edge).

One can combine the two methods by: (1) computing the union S of the sticking edges and the θ-kernel for the selected $\vec{\mathbf{v}}^c_\theta$; and (2) computing the backprojection of S for $\vec{\mathbf{v}}^c_\theta$. This backprojection is the computed preimage $\mathcal{P}$. The termination condition is the disjunction of the two termination conditions given above, i.e. :

$$\left(t > \frac{D}{v_{min}}\right) \bigvee \left(\mathcal{K}_\theta(\mathbf{q}^*(t), \vec{\mathbf{f}}^*(t)) \subseteq \mathcal{T}\right).$$

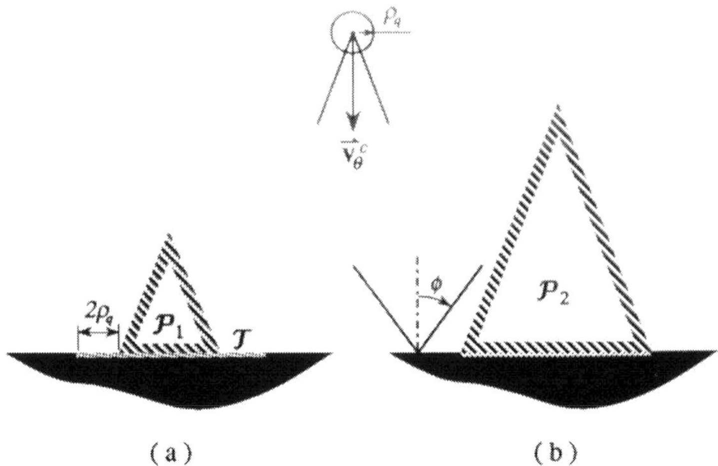

Figure 19. Backprojecting from the goal's θ-kernel often generates preimages which are significantly larger than those produced by backprojecting from the goal's sticking edges. The above figure shows that this is not always true. Figure a displays the preimage $\mathcal{P}_1$ of $\mathcal{T}$ computed by the backprojection-from-kernel method for the commanded velocity $\vec{v}_\theta^c$. Figure b displays the preimage $\mathcal{P}_2$ computed by the backprojection-from-sticking-edges method for the same commanded velocity.

Note that the preimage computed by combining the two methods may be larger than the union of the preimages computed by each of the two methods.

5 Selection of the Commanded Velocity

At this point, we can compute preimages for rather simple termination conditions. In order to plan a motion strategy, we still have to select commanded velocities. As suggested in Section 2, we can construct a tree of preimages by selecting the commanded velocities' orientations from a finite set of values in S^1. Each arc in the tree connects a region to its preimage for a motion commanded along one of these velocities. Figure 16 illustrates a strategy generated by searching such a tree. This naive approach is not very satisfactory, however; indeed, if a sequential motion strategy consisting of motion commands using the sticking or the instantaneous sensing termination condition exists, it may not always be found by the approach.

In this section we present another approach to the selection of the commanded velocities. The approach rests on the observation that when the orientation of the commanded velocity varies over S^1, the topology of a preimage computed as described in the previous section changes at a finite number of critical orientations. We first show how this idea can be used to construct a one-step strategy. Then, we consider the problem of generating a multi-step strategy. We only treat the case of the sticking termination condition. The results can be extended to the instantaneous sensing termination condition by noticing that when θ varies over S^1, the θ-kernel of a goal $\mathcal{T}$ may change only at a finite number of orientations.

5.1 One-Step Strategy

Consider a problem with $\mathcal{I}$ for initial region and $\mathcal{G}$ for goal region. Let $\mathcal{I}$ be a polygon in $\mathcal{C}_{valid}$ and $\mathcal{G}$ a collection of contact edges. Throughout the section, we assume that both $\mathcal{I}$ and $\mathcal{G}$ have constant complexity. We wish to construct a single motion command guaranteed to achieve $\mathcal{G}$ (by sticking in it) from $\mathcal{I}$, if one such command exists. In principle, this requires us to construct the backprojection of the sticking edges in $\mathcal{G}$ for every commanded velocity and to test whether it fully contains $\mathcal{I}$. We show below that this construction is only required at a finite number of critical orientations, where the backprojection topology changes.

5.1.1 Nondirectional Backprojection

DEFINITION 8: *The* **nondirectional backprojection** *of a region $\mathcal{S} \subset \mathcal{C}_{valid}$ is the set:*

$$\mathcal{L}(\mathcal{S}) = \bigcup_{\theta \in S^1} (\mathcal{L}_\theta(\mathcal{S}) \times \{\theta\}) \subset \mathcal{C}_{valid} \times S^1.$$

Let $\mathcal{S}$ consist of a finite collection of contact edges. Consider a backprojection $\mathcal{L}_\theta = \mathcal{L}_\theta(\mathcal{S})$ at some noncritical value of θ. (The notion of critical value of θ will be defined below.) The boundary of $\mathcal{L}_\theta$ consists of contact edges and free edges (see Subsection 4.3). Each free edge E is anchored at a contact vertex and is parallel to one of the two sides of the negated control uncertainty cone. If the commanded velocity's orientation varies slightly to $\theta + \delta\theta$, so that the interval $[\theta, \theta + \delta\theta]$ contains no critical orientation, the contact edges remain fixed, while the free edges rotate by $\delta\theta$, each around its anchor vertex. Hence, between θ and $\theta + \delta\theta$, the nondirectional backprojection is bounded by algebraic surfaces of two types, *contact surfaces* and *free surfaces*. The contact surfaces are liftings of the corresponding contact edges in $\mathcal{C}_{valid} \times S^1$. Each free surface is a ruled surface generated by a free edge rotating about its anchor vertex.

As we let θ vary over a larger interval, the topology of the backprojection may change at a finite number of orientations θ_i, where one of the following events occurs [Donald, 1988a]:

- A free edge of $\mathcal{L}_{\theta_i}$ ends at a contact vertex. When this event happens, a segment connecting two contact vertices (a link of the visibility graph of $\mathcal{C}_{valid}$) is parallel to a side of the control uncertainty cone.

- The sliding/sticking behavior on a contact edge of $\mathcal{L}_{\theta_i}$ or a contact edge abutting a vertex of $\mathcal{L}_{\theta_i}$ changes at θ_i.

- A vertex of $\mathcal{L}_{\theta_i}$ where two free edges meet lies in a contact edge.

This leads us to define the following **critical orientations**:

- A **vgraph-critical** orientation is an orientation where one side of the control uncertainty cone is parallel to a link of the visibility graph of $\mathcal{C}_{valid}$.

- A **sliding-critical** orientation is an orientation where the sliding/sticking behavior of a contact edge changes.

- A **vertex-critical** orientation is an orientation where the intersection of two rays erected at two contact vertices, each parallel to one side of the control uncertainty cone, lies in a contact edge.

The topology of $\mathcal{L}_\theta$ may change only at these orientations.

There are $O(n^2)$ edges in the visibility graph of $\mathcal{C}_{valid}$ (see Chapter 4). Hence, there are $O(n^2)$ vgraph-critical orientations. The sliding/sticking behavior on a contact edge changes at most four times. Hence, there are $O(n)$ sliding-critical orientations. At every contact vertex, let us erect two rays, each parallel to one side of the control uncertainty cone. These $O(n)$ rays intersect at $O(n^2)$ points. When θ varies over S^1, the various rays rotate together. Therefore, their intersection points describe $O(n^2)$ circles which intersect C_{valid} at $O(n^3)$ points. Hence, there are $O(n^3)$ vertex-critical orientations. In total, there are thus $O(n^3)$ critical orientations.

Let $\theta_{c_1}, \theta_{c_2}, \ldots$ be the critical orientations and θ_{nc_i} be any noncritical orientation in the interval $(\theta_{c_i}, \theta_{c_{i+1}})$. We can represent a nondirectional backprojection as a finite ordered set $\{\mathcal{L}_{\theta_{c_1}}, \mathcal{L}_{\theta_{nc_1}}, \mathcal{L}_{\theta_{c_2}}, \ldots\}$ of backprojections at alternately critical and noncritical orientations. At each noncritical orientation θ_{nc_i}, every edge bounding $\mathcal{L}_{\theta_{nc_i}}$ is labeled according to its type (contact or free edge). Since each backprojection has $O(n)$ bounding edges, the space complexity of this representation is $O(n^4)$. We can compute all the critical orientations in $O(n^3)$ time. For each critical and each noncritical orientation, it takes $O(n \log n)$ time to compute the backprojection. Thus, the representation can be computed in $O(n^4 \log n)$ time.

The above representation and its computation can be significantly improved. First, as noticed in [Donald, 1988a], only $O(n^2)$ vertex-critical orientations need to be considered, yielding a representation of size $O(n^3)$ for the nondirectional backprojection. In addition, we need not compute and represent a full backprojection at each new critical orientation. This has to be done only for one of them. At every other orientation, it suffices to represent the changes of the backprojection with respect to the backprojection at the previous orientation. This yields a representation of the nondirectional backprojection of size $O(n^2)$ which can be incrementally computed in $O(n^2 \log n)$ time [Briggs, 1989].

5.1.2 Application

Assume for a moment that the set $\mathcal{G}^s$ of sticking edges in the goal region $\mathcal{G}$ is constant, i.e. independent of the commanded velocity. This is equivalent to assuming that there exists a termination condition TC that evaluates to `true` if and only if the robot is in $\mathcal{G}^s$.

Under this assumption the nondirectional backprojection $\mathcal{L}(\mathcal{G}^s)$ can be regarded as a *nondirectional preimage* of $\mathcal{G}$ for the condition TC. This means that any slice $\mathcal{L}_\theta(\mathcal{G}^s)$ is a preimage of $\mathcal{G}$ for the motion command $(\vec{v}^c_\theta, \mathrm{TC})$. The problem of generating a one-step motion strategy can be solved by checking whether there exists an orientation θ such that $\mathcal{I} \subseteq \mathcal{L}_\theta(\mathcal{G}^s)$ and if there is, returning this orientation. We say that a commanded velocity orientation θ_{g_i} is **initial-region-critical** when a ray parallel to a side of the negated control uncertainty cone erected at a contact vertex contains a vertex of $\mathcal{I}$. Since there are $O(n)$ such rays and $\mathcal{I}$ has constant complexity, there are $O(n)$ initial-region-critical orientations. For each such orientation θ_{g_i}, we can check $\mathcal{L}_{\theta_{g_i}}$ for containment of $\mathcal{I}$ in $O(n)$ time. If there exists an orientation θ_{g_i} such that $\mathcal{I} \subseteq \mathcal{L}_{\theta_{g_i}}$, then a motion commanded along $\vec{\mathbf{v}}^c_{\theta_{g_i}}$ achieves $\mathcal{G}$ from $\mathcal{I}$; otherwise, there is no guaranteed one-step strategy.

Therefore, planning a one-step motion strategy takes $O(n^4 \log n)$ time. This time can be improved to $O(n^2 \log n)$ using Briggs' algorithm for computing nondirectional backprojections.

In reality, the set $\mathcal{G}^s$ of sticking edges in $\mathcal{G}$ is not constant and depends on the orientation θ. However, each edge in $\mathcal{G}$ is guaranteed to be a sticking edge over a single closed connected interval in S^1. By computing these sticking intervals for all the edges in $\mathcal{G}$ and intersecting them, we partition S^1 into a finite set of disjoint (open/closed) intervals $\{(\theta^{min}_{s_i}, \theta^{max}_{s_i})\}_{i=1,2,\ldots}$, such that in each interval $(\theta^{min}_{s_i}, \theta^{max}_{s_i})$ the set $\mathcal{G}^s_i$ of sticking edges in $\mathcal{G}$ remains constant. The nondirectional preimage of $\mathcal{G}$ for the sticking termination condition is:

$$\bigcup_{i=1,2,\ldots} \left(\bigcup_{\theta \in (\theta^{min}_{s_i}, \theta^{max}_{s_i})} (\mathcal{L}_\theta(\mathcal{G}^s_i) \times \{\theta\}) \right).$$

For a goal $\mathcal{G}$ of constant size, there are $O(1)$ intervals $(\theta^{min}_{s_i}, \theta^{max}_{s_i})$, and a representation of size $O(n^4)$ (resp. $O(n^2)$ with the improved algorithm) of the above nondirectional preimage can be computed in $O(n^4 \log n)$ (resp. $O(n^2 \log n)$) time. From this representation, we can compute the initial-region-critical orientations as above and generate a one-step strategy in $O(n^2)$ time.

A slightly more efficient algorithm based on an analysis of the dependence of the problem on the number of connected components in the C-obstacle region has been proposed in [Friedman, Hershberger and Snoeyink, 1990]. This algorithm assumes that $\mathcal{C}_{valid}$ is externally bounded by a polygon. It generates a one-step strategy, if one exists, in $O(kn \log n)$ total time, where k is the number of connected components of the C-obstacle region.

It first constructs a data structure of size $O(kn)$ in $O(kn \log n)$ time. From this structure, which is comparable to a nondirectional preimage, it then computes a one-step strategy in anywhere between $O(k \log n)$ and $O(kn)$ time depending on how precisely the initial region $\mathcal{I}$ is described.

5.2 Multi-Step Strategy

In the previous subsection we computed the nondirectional preimage of a goal $\mathcal{G}$ in $\mathcal{C}_{valid} \times S^1$. One way of generalizing this approach to the multi-step case is by introducing the notion of a "multi-step nondirectional preimage" of $\mathcal{G}$ in $\mathcal{C}_{valid} \times S^1 \times ... \times S^1$. Instead of critical values of θ in S^1, we would get critical surfaces in $S^1 \times ... \times S^1$. Although dealing with such surfaces is not impossible a priori, it is certainly not simple.

Another approach to the generation of multi-step strategies has been proposed in [Donald, 1988a]. It makes use of the concept of forward projection introduced below and reduces the problem of planning an r-step motion strategy to a decision problem in the first-order theory of the reals. The problem is described as a Tarski sentence with r existentially quantified variables $\theta_1, \theta_2, ..., \theta_r$ representing the orientations of the commanded velocities in the strategy to be generated. This sentence expresses the facts that the motion along $\vec{\mathbf{v}}^c_{\theta_i}$ starts from where the motion along $\vec{\mathbf{v}}^c_{\theta_{i-1}}$ terminated, or from $\mathcal{I}$ if $i = 1$, and that it terminates by sticking, in $\mathcal{G}$ if $i = r$. Hence, the sentence is built by forward chaining from the initial region.

The *forward projection* of a subset $\mathcal{R}$ of valid space, for the commanded velocity $\vec{\mathbf{v}}^c_\theta$, is the set of all the configurations that may be reached by a motion commanded along $\vec{\mathbf{v}}^c_\theta$ and starting anywhere in $\mathcal{R}$, assuming no termination condition.

DEFINITION 9: *Let $\mathcal{R}$ be a subset of $\mathcal{C}_{valid}$. The* **forward projection** *of $\mathcal{R}$ for the commanded velocity $\vec{\mathbf{v}}^c_\theta$ is the region:*

$$\mathcal{F}_\theta(\mathcal{R}) = \{\tau(t) \;/\; t \in [0, +\infty) \;,\; \tau \in \mathcal{D}_\theta(\mathcal{R})\}$$

where $\mathcal{D}_\theta(\mathcal{R})$ is the directory of actual trajectories for $\mathcal{R}$ and $\vec{\mathbf{v}}^c_\theta$.

The forward projection of a polygonal region $\mathcal{R}$ in $\mathcal{C}_{valid}$ can be computed using a variant of the Canny-Donald sweep-line algorithm presented in

Subsection 4.3 [Donald, 1988a]. The sweep-line L is moved across the plane perpendicularly to the commanded velocity $\vec{\mathbf{v}}^c_\theta$. The sweep starts at a position of L where it is tangent to $\mathcal{R}$, with $\mathcal{R}$ entirely lying on the side of L pointed to by $\vec{\mathbf{v}}^c_\theta$. The sweep proceeds in the direction of $\vec{\mathbf{v}}^c_\theta$. Since the sweep proceeds monotonically in the θ direction, it requires the actual velocity to never project negatively onto the θ direction. This is achieved by imposing that the friction cone be larger than the control uncertainty cone, i.e. $\phi > \eta_v$. (This condition was previously assumed in order to make guaranteed sticking possible.)

Let $\texttt{Stick}_\theta(\mathbf{q})$ be the condition which evaluates to `true` when sticking is possible at $\mathbf{q}$ under a commanded velocity $\vec{\mathbf{v}}^c_\theta$. The set:

$$\mathcal{S}_\theta(\mathcal{R}) = \{\mathbf{q} \in \mathcal{F}_\theta(\mathcal{R}) \ / \ \texttt{Stick}_\theta(\mathbf{q})\}$$

is the set of all configurations in the forward projection of $\mathcal{R}$ where sticking is possible.

Let `Forward` and `StickForward` be two predicates defined as:

$$\begin{aligned} \forall \mathbf{q}_1, \mathbf{q}_2 \in \mathcal{C}_{valid}: \quad & \texttt{Forward}(\theta, \mathbf{q}_1, \mathbf{q}_2) && \equiv \quad [\mathbf{q}_2 \in \mathcal{F}_\theta(\{\mathbf{q}_1\})] \\ & \texttt{StickForward}(\theta, \mathbf{q}_1, \mathbf{q}_2) && \equiv \quad [\mathbf{q}_2 \in \mathcal{S}_\theta(\{\mathbf{q}_1\})]. \end{aligned}$$

The set $\{(\theta, \mathbf{q}_1, \mathbf{q}_2) \ / \ \texttt{Forward}(\theta, \mathbf{q}_1, \mathbf{q}_2)\}$ is semi-algebraic and it can be described by a polynomial-sized (in the number n of vertices of $\mathcal{C}_{contact}$) quantifier-free polynomial expression of the variables $\theta, x_1, y_1, x_2, y_2$, with $\mathbf{q}_1 = (x_1, y_1)$ and $\mathbf{q}_2 = (x_2, y_2)$. Moreover, this expression can be computed in time polynomial in n [Donald, 1988a]. This result extends to the set $\{(\theta, \mathbf{q}_1, \mathbf{q}_2) \ / \ \texttt{StickForward}(\theta, \mathbf{q}_1, \mathbf{q}_2)\}$, since:

$$\texttt{StickForward}(\theta, \mathbf{q}_1, \mathbf{q}_2) \equiv \texttt{Forward}(\theta, \mathbf{q}_1, \mathbf{q}_2) \wedge \texttt{Stick}_\theta(\mathbf{q}_2).$$

For a given integer r, we define the predicate r*`StickForward` as:

$$\begin{aligned} & r{*}\texttt{StickForward}(\theta_1, ..., \theta_r, \mathbf{q}_0, \mathbf{q}_1, ..., \mathbf{q}_r) \equiv \\ & \qquad \texttt{StickForward}(\theta_1, \mathbf{q}_0, \mathbf{q}_1) \wedge ... \wedge \texttt{StickForward}(\theta_r, \mathbf{q}_{r-1}, \mathbf{q}_r). \end{aligned}$$

The set:

$$\{(\theta_1, ..., \theta_r, \mathbf{q}_0, ..., \mathbf{q}_r) \ / \ r{*}\texttt{StickForward}(\theta_1, ..., \theta_r, \mathbf{q}_0, \mathbf{q}_1, ..., \mathbf{q}_r)\}$$

is a semi-algebraic set that can be described by a polynomial-sized (quantifier-free) polynomial expression.

We can now formulate the problem of planning an r-step motion strategy as the problem of deciding the satisfiability of the following Tarski sentence:

$$(\exists \theta_1, \ldots, \theta_r \in S^1)(\forall \mathbf{q}_0, \ldots, \mathbf{q}_r \in \mathcal{C}_{valid})$$
$$((\mathbf{q}_0 \in \mathcal{I}) \wedge r{*}\mathtt{StickForward}(\theta_1, \ldots, \theta_r, \mathbf{q}_0, \mathbf{q}_1, \ldots, \mathbf{q}_r) \Rightarrow \mathbf{q}_r \in \mathcal{G})$$

and constructing a set of values for θ_1 through θ_r that satisfies this sentence. We could solve this problem using the Collins decomposition, yielding a planning procedure requiring time double-exponential in r. However, another decision procedure given by Grigoryev takes double-exponential time only in the number of quantifier alternations (2 in the above sentence) [Grigoryev, 1988]. Hence, planning an r-step motion strategy can be done in $n^{r^{O(1)}}$ time [Donald, 1988a].

Using a special sort of termination condition, Friedman, Hershberger and Snoeyink proposed a polynomial algorithm for multi-step strategy planning [Friedman, Hershberger and Snoeyink, 1990].

6 Conditional Strategies

So far, we have merely considered the generation of sequential strategies. However, some planning problems only admit conditional strategies for solutions. In this section we study how the search space of the backchaining algorithm should be modified in order to generate such strategies.

Consider the "point-on-hill" example shown in Figure 20.a [Mason, 1984]. The configuration space contains a single, non-compact C-obstacle bounded by three edges, the top edge $\mathcal{G}$, the left edge E_1 and the right edge E_2 (both E_1 and E_2 are semi-infinite lines). The goal region is the top edge $\mathcal{G}$. The initial region $\mathcal{I}$ is all $\mathcal{C}_{valid}$. We assume perfect control, no position sensing, perfect force sensing, and frictionless edges. The problem admits no sequential strategy for solution. However, a motion with a vertical commanded velocity pointing downward and $\|\vec{\mathbf{f}}^*(t)\| > 0$ as the termination condition, is guaranteed to terminate in one of the three edges $\mathcal{G}$, E_1 or E_2. Although it is not possible to know in advance which one of these edges will be achieved, it is known at planning time that this will be decidable at execution time (thanks to force sensing). In addition, it is easy to plan a motion that starts in E_1 (resp. E_2) and terminates in $\mathcal{G}$. Hence, the point-on-hill problem admits a conditional

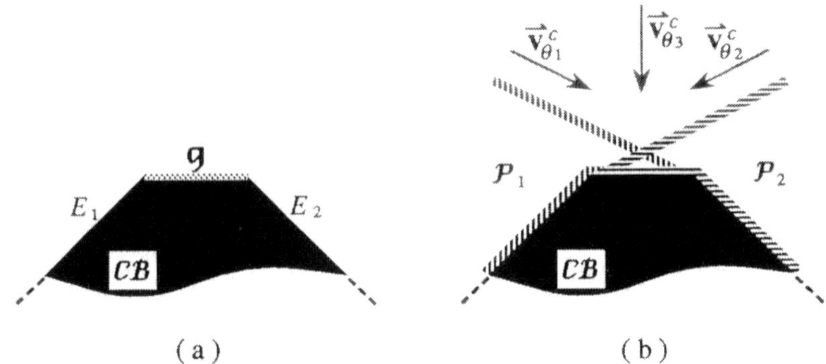

Figure 20. This figure illustrates the "point-on-hill" example. There is a single non-compact C-obstacle bounded by three edges, $\mathcal{G}$, E_1 and E_2 (Figure a). The initial region is $\mathcal{I} = \mathcal{C}_{valid}$ and the goal region is $\mathcal{G}$. We assume perfect control, no position sensing, perfect force sensing, and frictionless edges. The problem admits the following conditional strategy for solution (Figure b). First, move downward along $\vec{\mathbf{v}}^c_{\theta_3}$ until a non-zero reaction force is sensed. Then, based on the orientation of this force, decide which one of the edges $\mathcal{G}$, E_1 or E_2 has been achieved. If E_1 (resp. E_2) has been achieved, move along $\vec{\mathbf{v}}^c_{\theta_1}$ (resp. $\vec{\mathbf{v}}^c_{\theta_2}$) until the orientation of the sensed force is that of the outgoing normal of $\mathcal{G}$.

strategy for solution.

Planning a conditional strategy requires the backchaining process to compute preimages of goal sets of the form $\mathcal{ST} = \{\mathcal{T}_1, ..., \mathcal{T}_s\}$, where $\mathcal{T}_i$ ($i \in [1, s]$) is either the original goal or a previously computed preimage. Let $\mathbf{M} = (\vec{\mathbf{v}}^c_\theta, \mathtt{TC})$ be a motion command and $\mathtt{RC}_i$, $i = 1, ..., s$, be conditions called **recognition conditions**. A **preimage** of $\mathcal{ST} = \{\mathcal{T}_1, ..., \mathcal{T}_s\}$ for $\mathbf{M}$ and the $\mathtt{RC}_i$'s is any subset $\mathcal{P}$ of $\mathcal{C}_{valid}$ such that executing $\mathbf{M}$ from any configuration $\mathbf{q}_s \in \mathcal{P}$ is guaranteed to reach $\bigcup_{i=1}^{s} \mathcal{T}_i$ and terminate in it in such a way that, when the motion terminates, at least one recognition condition evaluates to `true` and, if the condition $\mathtt{RC}_i$ (for any $i \in [1, s]$) evaluates to `true`, $\mathcal{T}_i$ has been achieved.

In the example of Figure 20, the backchaining process would ideally proceed as follows:

- First, it would successively generate two preimages of $\mathcal{G}$, $\mathcal{P}_1$ and $\mathcal{P}_2$, for two motions commanded along $\vec{\mathbf{v}}^c_{\theta_1}$ and $\vec{\mathbf{v}}^c_{\theta_2}$ (see Figure 20.b). Both motions would have $angle(\vec{\mathbf{f}}^*(t), \vec{\nu}(\mathcal{G})) = 0$ as their termination conditions. Notice that $E_1 \subset \mathcal{P}_1$ and $E_2 \subset \mathcal{P}_2$.

- Second, it would generate $\mathcal{P} = \mathcal{C}_{valid}$ as the preimage of $\{\mathcal{G}, \mathcal{P}_1, \mathcal{P}_2\}$ for a motion commanded along $\vec{\mathbf{v}}^c_{\theta_3}$ (see Figure 20.b), with $\|\vec{\mathbf{f}}^*(t)\| > 0$ as the termination condition. The three conditions $angle(\vec{\mathbf{f}}^*(t), \vec{n}) = 0$, with $\vec{n} = \vec{\nu}(\mathcal{G})$, $\vec{\nu}(E_1)$ and $\vec{\nu}(E_2)$, would be the recognition conditions.

Thus, the generated strategy would be:

move along $\vec{\mathbf{v}}^c_{\theta_3}$ until $\|\vec{\mathbf{f}}^*(t)\| > 0$;
if $angle(\vec{\mathbf{f}}^*(t), \vec{\nu}(\mathcal{G})) = 0$
    then return;
    else if $angle(\vec{\mathbf{f}}^*(t), \vec{\nu}(E_1)) = 0$
        then move along $\vec{\mathbf{v}}^c_{\theta_1}$ until $angle(\vec{\mathbf{f}}^*(t), \vec{\nu}(\mathcal{G})) = 0$;
        else if $angle(\vec{\mathbf{f}}^*(t), \vec{\nu}(E_2)) = 0$
            then move along $\vec{\mathbf{v}}^c_{\theta_2}$ until $angle(\vec{\mathbf{f}}^*(t), \vec{\nu}(\mathcal{G})) = 0$;

The recognition conditions are the conditions labeling the arcs of the tree representing the strategy (see Figure 7).

More generally, in order to generate a conditional plan, the backchaining procedure operates as follows:

1. It creates a set of goals $\mathcal{SG}$ and initializes it to $\{\mathcal{G}\}$.

2. It selects a subset $\mathcal{ST}$ of $\mathcal{SG}$ and computes a preimage $\mathcal{P}$ of $\mathcal{ST}$.

3. If $\mathcal{I} \subseteq \mathcal{P}$, it exits with success. Otherwise, it inserts $\mathcal{P}$ in $\mathcal{SG}$ as a new goal and goes back to step 2.

The preimage of a set of goals $\mathcal{ST}$ can be computed by adapting the methods presented in Section 4. For example, the backprojection-from-goal-kernel method applies as follows. Let $\mathcal{ST} = \{\mathcal{T}_1, ..., \mathcal{T}_s\}$, with $s > 1$, and let $\vec{\mathbf{v}}^c_\theta$ be the selected commanded velocity. The θ-kernel $\chi_\theta(\mathcal{T}_i)$ of every goal $\mathcal{T}_i$, $i = 1, ..., s$, is first computed. A preimage of $\mathcal{ST}$ is then constructed as the backprojection of the union $\bigcup_{i=1}^{s} \chi_\theta(\mathcal{T}_i)$. The recognition condition $\mathtt{RC}_i$ is:

$$\mathcal{K}_\theta(\mathbf{q}^*(t), \vec{\mathbf{f}}^*(t)) \subseteq \mathcal{T}_i$$

for $i = 1$ to s. The termination condition is $\text{TC} \equiv \bigvee_{i=1}^{s} \text{RC}_i$.

A planner can be constructed by restricting the possible orientations of the commanded velocity to a finite set of predetermined orientations. However, the number of goal sets $\mathcal{ST}$ to be considered grows exponentially with the number of preimages generated since the beginning of the backchaining process. Additional techniques for selecting velocities and guiding the choice of goal sets would be needed.

Another empirical method for generating conditional strategies has been proposed[15] by Buckley [Buckley, 1987 and 1989b]. The method consists of partitioning $\mathcal{C}_{valid}$ into non-overlapping regions called *atoms*. A *state* of the robot is a set of atoms (possibly a singleton) so that no two atoms in this set are guaranteed to be distinguishable using sensing. A possible choice for the atoms is the free space and all the contact edges. For every pairs of atoms (A_1, A_2), the planner computes the range of commanded velocities that *may* achieve A_2 from A_1 with the sticking or the instantaneous sensing termination condition (to that purpose, one may use an adaptation of the algorithm given in Subsection 5.1.1 to compute the nondirectional forward projection of a region). Next, it intersects these ranges and constructs the range of commanded velocities that are guaranteed to achieve a state S_2 from a state S_1, for every pair of states S_1 and S_2 (see [Buckley, 1987] for details). The result is an AND/OR graph which remains to be searched for a solution AND subgraph describing a guaranteed strategy. The method assumes that the goal is an atom or a collection of atoms.

7 Power of a Termination Condition

So far, we have considered two relatively simple termination conditions — the sticking and instantaneous sensing termination conditions. There exist more sophisticated conditions which are more powerful at recognizing goal achievement. Such conditions may result in larger preimages, and hence simpler strategies. Some planning problems may also admit solutions only if such conditions are used.

The recognition power of a termination condition depends on the knowledge it uses. This knowledge may include both "dynamic" knowledge —

[15] See [Desai, 1988] and [Desai and Volz, 1989] for a related method.

i.e. the arguments (elapsed time and sensing) that the termination predicate can access during execution — and "static" knowledge — i.e. the information that is passed by the planner in the termination predicate. In the following subsections, we describe several termination conditions which illustrate these dependencies.

7.1 Termination Conditions with Initial State

The planner may include the knowledge of the preimage in the termination predicate. Then, at execution time, the termination condition can rule out interpretations of the sensory inputs that suggest configurations of $\mathcal{A}$ which are outside the range of configurations reachable from the preimage, i.e. the forward projection of the preimage (see Subsection 5.2). We formalize this idea and then show its application to two examples.

We say that a configuration $\mathbf{q} \in \mathcal{C}_{valid}$ is **θ-$\mathcal{P}$-consistent** with a sensed configuration $\mathbf{q}^*$ and a sensed force $\vec{\mathbf{f}}^*$ if and only if:

$$\mathbf{q} \in \mathcal{K}_\theta(\mathbf{q}^*, \vec{\mathbf{f}}^*) \cap \mathcal{F}_\theta(\mathcal{P})$$

with $\mathcal{K}_\theta(\mathbf{q}^*, \vec{\mathbf{f}}^*)$ and $\mathcal{F}_\theta(\mathcal{P})$ respectively defined as in Subsections 4.5 and 5.2. This definition captures the fact that if a motion with commanded velocity $\vec{\mathbf{v}}_\theta^c$ is known to begin in a region $\mathcal{P}$, then at any time t during the motion, the current configuration $\mathbf{q}$ is in both $\mathcal{K}_\theta(\mathbf{q}^*(t), \vec{\mathbf{f}}^*(t))$ and $\mathcal{F}_\theta(\mathcal{P})$. We define:

$$\mathcal{K}_{\theta,\mathcal{P}}(\mathbf{q}^*, \vec{\mathbf{f}}^*) \stackrel{\text{def}}{=} \mathcal{K}_\theta(\mathbf{q}^*, \vec{\mathbf{f}}^*) \cap \mathcal{F}_\theta(\mathcal{P}).$$

Two configurations $\mathbf{q}_1$ and $\mathbf{q}_2$ in $\mathcal{C}_{valid}$ are said to be **θ-$\mathcal{P}$-distinguishable** if and only if:

$$\{(\mathbf{q}^*, \mathbf{f}^*) \ / \ \mathbf{q}_1, \mathbf{q}_2 \in \mathcal{K}_{\theta,\mathcal{P}}(\mathbf{q}^*, \mathbf{f}^*)\} = \emptyset.$$

The **θ-$\mathcal{P}$-kernel** of $\mathcal{T}$ is defined as:

$$\chi_{\theta,\mathcal{P}}(\mathcal{T}) \stackrel{\text{def}}{=} \{\mathbf{q} \in \mathcal{T} \ / \ \forall \mathbf{q}' \in \mathcal{C}_{valid} \backslash \mathcal{T} : \mathbf{q} \text{ and } \mathbf{q}' \text{ are } \theta\text{-}\mathcal{P}\text{-distinguishable}\}.$$

When a preimage $\mathcal{P}$ of a goal $\mathcal{T}$ is constructed for a motion command $\mathbf{M}$, it is known, by the definition of the preimage backchaining process, that

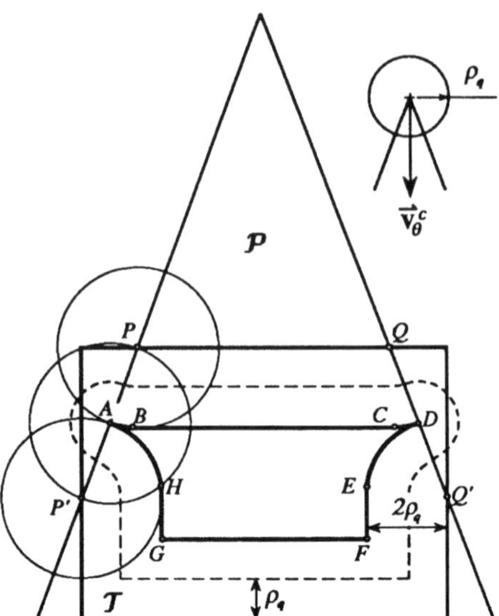

Figure 21. When a termination condition with initial state is used, a preimage of the rectangle $\mathcal{T}$ can be constructed as the backprojection of the region bounded by the generalized polygon whose vertices are A, B, C, D, E, F, G, and H. This preimage is larger than the preimage shown in Figure 18.b.

the execution of **M** will begin at a starting configuration in $\mathcal{P}$. (Indeed, either $\mathcal{P}$ includes the initial region $\mathcal{I}$, or the backchaining process will recursively find a way to achieve $\mathcal{P}$.) Hence, the region $\mathcal{P}$ defined as the backprojection of the θ-$\mathcal{P}$-kernel of $\mathcal{T}$, i.e. :

$$\mathcal{P} = \mathcal{L}_\theta(\chi_{\theta,\mathcal{P}}(\mathcal{T}))$$

is a preimage of $\mathcal{T}$ for the motion command $(\vec{\mathbf{v}}^c_\theta, \texttt{TC})$, where:

$$\texttt{TC} \equiv \mathcal{K}_{\theta,\mathcal{P}}(\mathbf{q}^*(t), \vec{\mathbf{f}}^*(t)) \subseteq \mathcal{T}.$$

This condition is called a termination condition with **initial state.**[16]

[16]This terminology differs from that used in various publications, including [Erdmann, 1984], where a state designates a pair of sensed configuration and force, and the initial state of a motion designates the pair of sensed configuration and force at time 0.

Example 1: Consider the example of Figure 18 again. The goal $\mathcal{T}$ is a rectangle in an empty configuration space. The region whose contour is labeled $ABCDEFGH$ in Figure 21 is a generalized polygon constructed as follows. The straight edge BC is at distance $2\rho_q$ from the top horizontal edge of $\mathcal{T}$. The circular edges AB and CD are circular arcs of radius $2\rho_q$ centered at P and Q, respectively. P (resp. Q) is selected in the top horizontal edge of $\mathcal{T}$ such that the intersection of $\mathcal{T}$ and a line passing through P (resp. Q) and parallel to the left (resp. right) side of the control uncertainty cone is a segment PP' (resp. QQ') of length $4\rho_q$. The circular edges AH and DE are circular arcs of radius $2\rho_q$ with centers at P' and Q', respectively. The straight edge GF is at distance $2\rho_q$ from the bottom edge of $\mathcal{T}$. The straight edges HG and EF are at distance $2\rho_q$ from the left and right edges of $\mathcal{T}$, respectively.

One can verify that the region thus outlined is the kernel $\chi_{\theta,\mathcal{P}}(\mathcal{T})$ with $\mathcal{P} = \mathcal{L}_\theta(\chi_{\theta,\mathcal{P}}(\mathcal{T}))$. In particular, assume that at some instant during the motion the actual configuration is the extreme point marked A in the figure. All the possible sensed configurations at this instant lie in the disc of radius ρ_q centered at A. If the forward projection is not taken into account, the set of all the interpretations of all these possible sensed configurations is the disc of radius $2\rho_q$ centered at A. The intersection of this disc with the forward projection $\mathcal{F}_\theta(\mathcal{P})$ is completely contained in $\mathcal{T}$. Hence, the point A and any configuration outside $\mathcal{T}$ are θ-$\mathcal{P}$-distinguishable, so that A belongs to $\chi_{\theta,\mathcal{P}}(\mathcal{T})$. The same sort of verification can be made with the other vertices B through H, the straight and circular edges connecting these points, and the interior of the outlined area. The resulting preimage is larger that that shown in Figure 18.b. The region outlined in a dotted line depicts the set of sensed configurations $\mathbf{q}^*(t)$ for which the termination condition $\mathcal{K}_{\theta,\mathcal{P}}(\mathbf{q}^*(t)) \subseteq \mathcal{T}$ evaluates to **true** ($\vec{\mathbf{f}}^*(t)$ is useless in this example). ■

Example 2: Another example is given in Figure 22. The goal region $\mathcal{T}$ is a subset of a contact edge (see Figure 22.a). The preimages $\mathcal{P}_1$ and $\mathcal{P}_2$ which are obtained by backprojecting from the θ-kernel and the θ-$\mathcal{P}_2$-kernel are shown in Figures 22.b and 22.c, respectively. The preimage of Figure 22.c is maximal over all possible termination conditions since it is identical to the maximal backprojection of $\mathcal{T}$. ■

Notice that there exists one condition of the form:

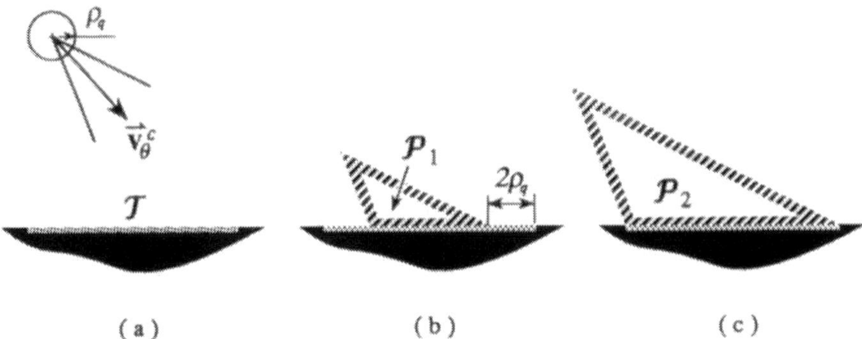

Figure 22. This figure is another illustration of the use of a termination condition with initial state. The goal $\mathcal{T}$ and the commanded velocity are shown in Figure a. The preimage $\mathcal{P}_1$ obtained by backprojecting from the θ-kernel of $\mathcal{T}$ is shown in Figure b. The preimage $\mathcal{P}_2$ obtained by backprojecting from the θ-$\mathcal{P}_2$-kernel of $\mathcal{T}$ is shown in Figure c.

$$\mathcal{K}_{\theta,\mathcal{P}}(\mathbf{q}^*(t), \vec{\mathbf{f}}^*(t)) \subseteq \mathcal{T}$$

for every subset $\mathcal{P}$ of $\mathcal{C}_{valid}$ (in the following we will say that the above expression is a termination condition *schema* parameterized by $\mathcal{P}$). It makes sense to use such a condition as the termination condition of a motion command **M** only if $\mathcal{P}$ is the preimage in which **M**'s execution starts. The apparent difficulty of using a termination condition with initial state is that the termination condition and therefore the preimage (see Definition 3) are defined in terms of the preimage itself. We will see in Subsection 8.2 that the problem of deciding whether a region $\mathcal{P}$ is a preimage, or not, for a motion command using a termination condition with initial state can be reduced to a decision problem in the first-order theory of the reals. This will be sufficient to define an effective planner capable of using termination conditions with initial state.

7.2 Termination Conditions with Final State

The power of a termination condition may be increased further by embodying the knowledge of itself in the termination predicate. Indeed, the termination condition can then rule out interpretations of the sensory inputs that suggest configurations of $\mathcal{A}$ beyond the limit where the motion would have been stopped by the termination condition [Latombe, Lazanas and Shekhar, 1989].

Any termination condition TC divides a forward projection $\mathcal{F}_\theta(\mathcal{P})$ into three regions, which we denote by F_1, F_2 and F_3:

- F_1 consists of all the configurations $\mathbf{q} \in \mathcal{F}_\theta(\mathcal{P})$ such that, for every possible sensed configuration $\mathbf{q}^*$ and sensed force $\vec{\mathbf{f}}^*$ θ-consistent with $\mathbf{q}$, TC evaluates to `false`.
- F_2 consists of all the configurations $\mathbf{q} \in \mathcal{F}_\theta(\mathcal{P})$ such that, for every possible sensed configuration $\mathbf{q}^*$ and sensed force $\vec{\mathbf{f}}^*$ θ-consistent with $\mathbf{q}$, TC evaluates to `true`.
- $F_3 = \mathcal{F}_\theta(\mathcal{P}) \backslash (F_1 \cup F_2)$, i.e. any configuration in F_3 may non-deterministically produce sensed configurations and forces for which TC evaluates to either `true` or `false`.

At every instant during a motion starting in $\mathcal{P}$ and commanded along $\vec{\mathbf{v}}_\theta^c$, if the current configuration is in F_1, the motion keeps going; if it is in F_2, the motion stops; if it is in F_3, the motion may either continue or stop. In general, there exist subsets of F_1, F_2 and F_3 that are inaccessible from $\mathcal{P}$ because reaching them would require to previously traverse F_2, where the motion would have been terminated. This is precisely why making the termination condition know itself may increase its recognition power.

Given a termination condition TC, a motion starting in $\mathcal{P}$ and commanded along $\vec{\mathbf{v}}_\theta^c$ can reach a configuration $\mathbf{q}_a$ if and only if there exists a configuration $\mathbf{q}_s \in \mathcal{P}$ and a trajectory $\tau \in \mathcal{D}_\theta(\mathcal{P})$ such that:

- $\tau(0) = \mathbf{q}_s$,
- $\exists t_a \in [0, +\infty) : \tau(t_a) = \mathbf{q}_a$,
- $\forall t \in [0, t_a)$, $\tau(t) \notin F_2$.

Hence, τ connects $\mathbf{q}_s$ to $\mathbf{q}_a$ without traversing F_2, except at $\mathbf{q}_a$ itself (if $\mathbf{q}_a \in F_2$). Let us denote by $\mathcal{F}_{\theta,\mathrm{TC}}(\mathcal{P}) \subseteq \mathcal{F}_\theta(\mathcal{P})$ the set of all the configurations $\mathbf{q}_a$ which can be reached by executing the motion command $\mathbf{M} = (\vec{\mathbf{v}}_\theta^c, \mathrm{TC})$ from within $\mathcal{P}$.

We say that a configuration $\mathbf{q} \in \mathcal{C}_{valid}$ is θ-$\mathcal{P}$-TC-**consistent** with a sensed configuration $\mathbf{q}^*$ and a sensed force $\vec{\mathbf{f}}^*$ if and only if:

$$\mathbf{q} \in \mathcal{K}_\theta(\mathbf{q}^*, \vec{\mathbf{f}}^*) \cap \mathcal{F}_{\theta,\mathrm{TC}}(\mathcal{P}).$$

We define:

$$\mathcal{K}_{\theta,\mathcal{P},\mathrm{TC}}(\mathbf{q}^*, \vec{\mathbf{f}}^*) \stackrel{\text{def}}{=} \mathcal{K}_\theta(\mathbf{q}^*, \vec{\mathbf{f}}^*) \cap \mathcal{F}_{\theta,\mathrm{TC}}(\mathcal{P}).$$

Two configurations $\mathbf{q}_1$ and $\mathbf{q}_2$ in $\mathcal{C}_{valid}$ are said to be θ-$\mathcal{P}$-TC-**distinguishable** if and only if:

$$\{(\mathbf{q}^*, \vec{\mathbf{f}}^*) \;/\; \mathbf{q}_1, \mathbf{q}_2 \in \mathcal{K}_{\theta,\mathcal{P},\mathrm{TC}}(\mathbf{q}^*, \vec{\mathbf{f}}^*)\} = \emptyset.$$

The θ-$\mathcal{P}$-TC-**kernel** of a goal $\mathcal{T}$ is defined as:

$$\chi_{\theta,\mathcal{P},\mathrm{TC}}(\mathcal{T}) = \{\mathbf{q} \in \mathcal{T} \;/\; \forall \mathbf{q}' \in \mathcal{C}_{valid} \backslash \mathcal{T} : \mathbf{q} \text{ and } \mathbf{q}' \text{ are } \theta\text{-}\mathcal{P}\text{-TC-distinguishable}\}.$$

A preimage $\mathcal{P}$ of a goal $\mathcal{T}$ is the backprojection of the θ-$\mathcal{P}$-TC-kernel of $\mathcal{T}$, with:

$$\mathcal{K}_{\theta,\mathcal{P},\mathrm{TC}}(\mathbf{q}^*(t), \vec{\mathbf{f}}^*(t)) \subseteq \mathcal{T}$$

as the termination condition of the motion commanded along $\vec{\mathbf{v}}_\theta^c$. This condition is called a termination condition with initial and **final** states.

Example: We show below that using a termination condition with initial and final states makes it possible to construct a preimage larger than that of Figure 21.

Consider Figure 23. Let R (resp. S) be the point in the top horizontal edge of $\mathcal{T}$ such that the intersection of $\mathcal{T}$ and a line passing through R (resp. S) and parallel to the left (resp. right) side of the control uncertainty cone is a segment RR' (resp. SS') of length $2\rho_q$. The line $R'S'$ that forms the upper portion of the striped contour consists of two circular arcs of radius $2\rho_q$, with respective centers R and S, and a straight segment at distance $2\rho_q$ from the top edge of the goal $\mathcal{T}$. The rest of the striped contour is the lower part of $\mathcal{T}$'s boundary. The region thus outlined is the kernel $\chi_{\theta,\mathcal{P},\mathrm{TC}}(\mathcal{T})$ with:

$$\mathcal{P} = \mathcal{L}_\theta(\chi_{\theta,\mathcal{P},\mathrm{TC}}(\mathcal{T}))$$

and:

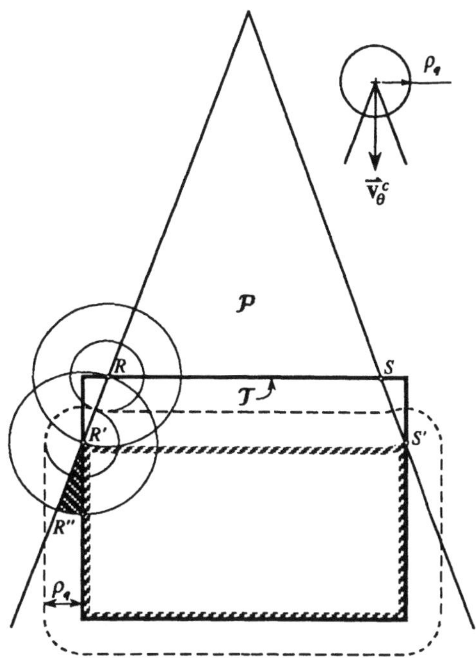

Figure 23. The preimage of the rectangular goal $\mathcal{T}$ for the commanded velocity $\vec{\mathbf{v}}_\theta^c$ and the termination condition with initial and final state is the region $\mathcal{P}$ which consists of both the region with a striped outline and the triangle on top of it. The region with a striped outline is the θ-$\mathcal{P}$-TC-kernel of $\mathcal{T}$. The region outlined in a dotted line depicts the set of sensed configurations for which the termination condition $\mathcal{K}_{\theta,\mathcal{P},\text{TC}}(\mathbf{q}^*(t)) \subseteq \mathcal{T}$ evaluates to **true** ($\vec{\mathbf{f}}^*(t)$ is useless).

$$\text{TC} \equiv [\mathcal{K}_{\theta,\mathcal{P},\text{TC}}(\mathbf{q}^*(t)) \subseteq \mathcal{T}]$$

(force sensing is useless). The region outlined in a dotted line depicts the set of sensed configurations $\mathbf{q}^*(t)$ for which the above termination condition evaluates to `true`.

One can verify that the region $\mathcal{P}$ is a preimage of $\mathcal{T}$. Indeed, all the motions starting from $\mathcal{P}$ are guaranteed to reach the goal (if they are not stopped before) and none of them can terminate before it has attained the goal. In addition, none of these motions can leave the goal without terminating.

In order to see how the termination condition uses the knowledge of itself, assume that at some instant during the motion the actual configuration is the point denoted by R'. All the possible sensed configurations at this instant lie in the disc of radius ρ_q centered at R'. If neither the forward projection nor the termination condition are taken into account, the set of all the interpretations of all these sensed configurations is the disc of radius $2\rho_q$ centered at R'. The intersection of this disc with the forward projection $\mathcal{F}_\theta(\mathcal{P})$ contains a sector that is not contained in $\mathcal{T}$. This sector (the striped area in Figure 23) can only be attained from $\mathcal{P}$ by crossing the segment marked $R'R''$. Since for any configuration in this segment, the termination condition evaluates to `true`, the sector cannot be attained. ■

A termination condition with final state is specified as a recursive function of itself. We do not know as yet whether a motion planner using this termination condition can be effectively built.

7.3 Termination Conditions with Time

The modulus of the robot's velocity can only vary in $\Delta_v = (v_1, v_2)$. Hence, at any given instant during the execution of a motion commanded along $\vec{\mathbf{v}}^c_\theta$ and starting from within a region $\mathcal{P}$, the robot is confined to a subset of the forward projection $\mathcal{F}_\theta(\mathcal{P})$. We define the **forward projection of** $\mathcal{P}$ **at time** t for the commanded velocity $\vec{\mathbf{v}}^c_\theta$ as the region:

$$\mathcal{F}_\theta(\mathcal{P}, t) = \{\tau(t) \mid \tau \in \mathcal{D}_\theta(\mathcal{P})\}.$$

We say that two configurations $\mathbf{q}_1$ and $\mathbf{q}_2$ are **θ-$\mathcal{P}$-distinguishable at time** t if and only if:

$$\{(\mathbf{q}^*, \vec{\mathbf{f}}^*) \mid \mathbf{q}_1, \mathbf{q}_2 \in \mathcal{K}_\theta(\mathbf{q}^*, \vec{\mathbf{f}}^*) \cap \mathcal{F}_\theta(\mathcal{P}, t)\} = \emptyset.$$

The **θ-$\mathcal{P}$-kernel** of $\mathcal{T}$ **at time** t is the set of all the configurations in $\mathcal{T}$ which are θ-$\mathcal{P}$-distinguishable at time t from all the configurations in $\mathcal{C}_{valid} \backslash \mathcal{T}$. A preimage can be computed by backprojecting from this kernel. The corresponding termination condition is:

$$\mathcal{K}_\theta(\mathbf{q}^*(t), \vec{\mathbf{f}}^*(t)) \cap \mathcal{F}_\theta(\mathcal{P}, t) \subseteq \mathcal{T}.$$

Example: Consider the point-in-hole example shown in Figure 24. The goal $\mathcal{T}$ is the horizontal contact edge at the bottom of the hole and the

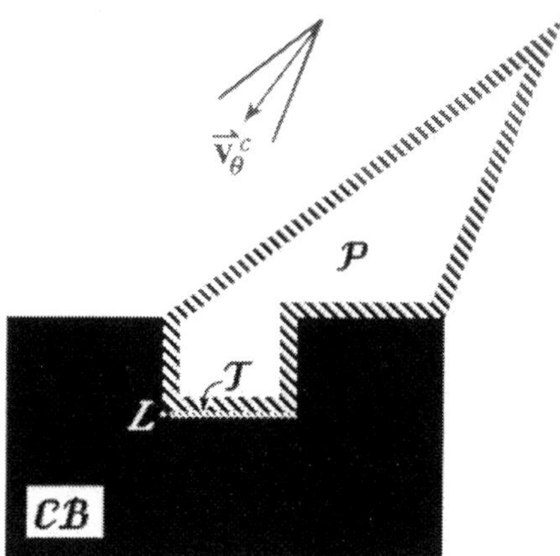

Figure 24. The region $\mathcal{P}$ (striped contour) is the backprojection $\mathcal{L}_\theta(\mathcal{T})$. Any motion starting in $\mathcal{P}$ is guaranteed to reach $\mathcal{T}$ and, after some bounded time T, to stick at the left vertex L of $\mathcal{T}$. Hence, $\mathcal{P}$ is a preimage of $\mathcal{T}$ for the commanded velocity $\vec{\mathbf{v}}^c_\theta$ and the termination condition $t \geq T$.

commanded velocity $\vec{\mathbf{v}}^c_\theta$ points downward toward the left as shown in the figure. Assume no position sensing and no force sensing. (This simply means that $\mathcal{K}_\theta(\mathbf{q}^*, \vec{\mathbf{f}}^*)$ is identical to $\mathcal{C}_{valid}$ in the expressions given above. Hence, it can be removed from them.) Consider the backprojection $\mathcal{P} = \mathcal{L}_\theta(\mathcal{T})$ (shown with striped contour in the figure). Any motion starting in $\mathcal{P}$ is guaranteed to reach $\mathcal{T}$ and, after some bounded time, to stick at the left vertex L of $\mathcal{T}$. Indeed, there exists T such that for all $t \geq T$, $\mathcal{F}_\theta(\mathcal{P}, t) = \{L\}$. Hence, $\mathcal{P}$ is a preimage of $\mathcal{T}$ and a "compiled" form of the termination condition $\mathcal{F}_\theta(\mathcal{P}, t) \subseteq \mathcal{T}$ is $t \geq T$. ■

The sticking termination condition given in Subsection 4.4 is an instance of a termination condition with time.

7.4 Termination Conditions with Sensing History

A termination condition may have position and force sensing history for argument, rather than just instantaneous sensing (see Subsection 1.3.1). Then it can continuously intersect the set $\mathcal{K}_\theta(\mathbf{q}^*(t), \vec{\mathbf{f}}^*(t))$ and the current forward projection. This intersection is a new region whose forward projection is taken as the current forward projection at the next instant. At every instant, if the computed intersection is contained in the goal, the termination condition evaluates to `true`.

A procedural (and perhaps more explicit) way of expressing this termination condition consists of continuously pruning the trajectory directory by removing every trajectory τ that is not consistent with current sensing [Mason, 1984]. Then, the termination condition is the following piece of code, which is (idealistically) evaluated at every instant t (D is initialized to $\mathcal{D}_\theta(\mathcal{P})$ at the beginning of the motion, where $\mathcal{P}$ is the preimage from which the motion starts):

> for every τ in D do
>   if $(\tau(t), \vec{\mathbf{f}}_\tau(t)) \not\in \mathcal{U}_q(\mathbf{q}^*(t)) \times \mathcal{U}_f(\vec{\mathbf{f}}^*(t))$ then $D \leftarrow D \backslash \{\tau\}$;
> $\mathcal{Q} \leftarrow \{\tau(t) \;/\; \tau \in D\}$;
> if $\mathcal{Q} \subseteq \mathcal{T}$ then return `true`; else return `false`;

(This is not an executable algorithm since the set D is usually infinite.)

Example: Consider the motion planning problem depicted in Figure 25. The initial region $\mathcal{I}$ consists of two configurations $\mathbf{q}_{s,1}$ and $\mathbf{q}_{s,2}$. The goal region $\mathcal{G}$ consists of two configurations $\mathbf{q}_{g,1}$ and $\mathbf{q}_{g,2}$. The commanded velocity $\vec{\mathbf{v}}^c_\theta$ is pointing vertically downward. We assume perfect control. Both the distances between $\mathbf{q}_{s,1}$ and $\mathbf{q}_{s,2}$, and between $\mathbf{q}_{g,1}$ and $\mathbf{q}_{g,2}$, are smaller than $2\rho_q$, so that neither the two initial configurations, nor the two goal configurations are guaranteed to be distinguishable at execution time. Since control is perfect, the trajectory directory $\mathcal{D}_\theta(\{\mathbf{q}_{s,1}, \mathbf{q}_{s,2}\})$ consists of two trajectories. The two trajectories are guaranteed to be distinguishable during some interval of time before any of the two goal configurations is attained. Hence, the set $\{\mathbf{q}_{s,1}, \mathbf{q}_{s,2}\}$ is a preimage of $\{\mathbf{q}_{g,1}, \mathbf{q}_{g,2}\}$ for $\mathbf{M} = (\vec{\mathbf{v}}^c_\theta, \mathtt{TC})$, where TC is a termination condition with sensing history. This example does not, however, require continuous sensing history. It could also be solved by generating a conditional strategy in which no motion command uses a termination condition with sensing history. The pertinent "sensing history" is then incorporated into the conditional branching of the strategy. ∎

The termination condition procedurally defined above is quite general — it uses sensing history, initial state, and time — but does not use final state. It is not known as yet whether an effective motion planner using this condition can be constructed.

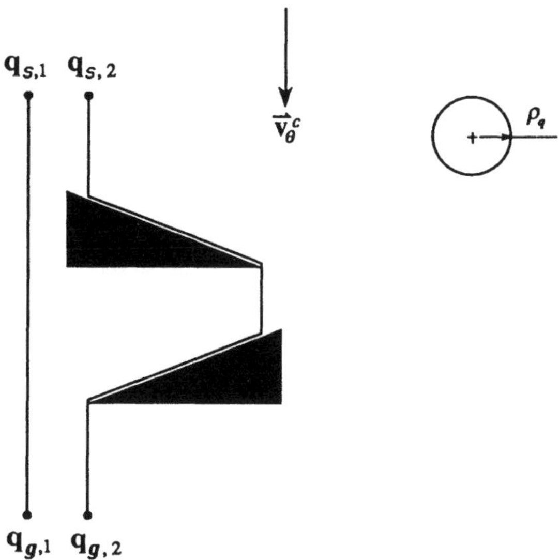

Figure 25. The initial region $\mathcal{I}$ consists of two configurations $\mathbf{q}_{s,1}$ and $\mathbf{q}_{s,2}$. The goal region $\mathcal{G}$ also consists of two configurations $\mathbf{q}_{g,1}$ and $\mathbf{q}_{g,2}$. Both the distances between $\mathbf{q}_{s,1}$ and $\mathbf{q}_{s,2}$, and between $\mathbf{q}_{g,1}$ and $\mathbf{q}_{g,2}$, are smaller than $2\rho_q$, so that neither the two initial configurations, nor the two goal ones are guaranteed to be distinguishable at execution time. We assume perfect control. The commanded velocity is pointing vertically downward. The trajectory directory $\mathcal{D}_\theta(\{\mathbf{q}_{s,1}, \mathbf{q}_{s,2}\})$ consists of two trajectories. The two trajectories are guaranteed to be distinguishable during some interval of time. Hence, the set $\{\mathbf{q}_{s,1}, \mathbf{q}_{s,2}\}$ is a preimage of $\{\mathbf{q}_{g,1}, \mathbf{q}_{g,2}\}$ for $\mathbf{M} = (\vec{\mathbf{v}}^c_\theta, \mathtt{TC})$, with TC being a termination condition with sensing history.

7.5 On Maximal Preimages

The above list of termination conditions illustrates the dependence of the recognition power of a termination condition on the knowledge it uses. Some of the given examples show that in some cases the maximal preimage of a goal over all possible termination conditions may exist. Consider for instance Figure 24. The constructed preimage is identical to the maximal backprojection. Since any preimage is necessarily contained in the maximal backprojection, if a preimage is identical to the maximal backprojection, it is the maximal preimage over all possible termination conditions.

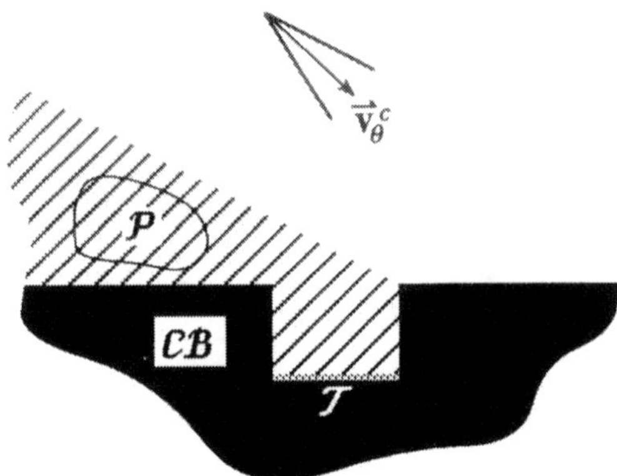

Figure 26. This figure shows a variant of the point-in-hole problem where the C-obstacle is not compact. We assume no sensing. Any bounded subset $\mathcal{P}$ of the striped region is a preimage of $\mathcal{T}$ for a termination condition with time, since any motion starting in $\mathcal{P}$ is guaranteed to stick in the right corner of the hole after some finite amount of time. The striped region, however, is infinite and is not a preimage. Hence, in this example, the goal $\mathcal{T}$ admits no maximal preimage over all possible termination conditions.

Unfortunately, there are also examples showing that the maximal preimage of a goal over all possible termination conditions may not exist, or that, instead, it may exist in quantity. These examples derive from the fact that if $\mathcal{P}_1$ and $\mathcal{P}_2$ are two preimages of the same goal for the same commanded velocity $\vec{v}_\theta^c$, but for two different termination conditions, their union is not a preimage in general. Indeed, the termination condition corresponding to $\mathcal{P}_1$ (resp. $\mathcal{P}_2$) may embody the knowledge of $\mathcal{P}_1$ (resp. $\mathcal{P}_2$), but no termination predicate embodying the knowledge of $\mathcal{P}_1 \cup \mathcal{P}_2$ may be guaranteed to recognize goal achievement by a motion starting from within $\mathcal{P}_1 \cup \mathcal{P}_2$.

Example 1: Consider the point-in-hole problem with a non-compact C-obstacle, as depicted in Figure 26. We assume no sensing. The commanded velocity points downward toward the right as shown in the figure. Every bounded subset $\mathcal{P}$ of the non-compact striped region is a preimage of the edge $\mathcal{T}$, with a termination condition of the form $t > T_{\mathcal{P}}$. Indeed, by waiting a sufficient amount of time, it is guaranteed that the motion will stick into the bottom-right corner of the hole. However, the

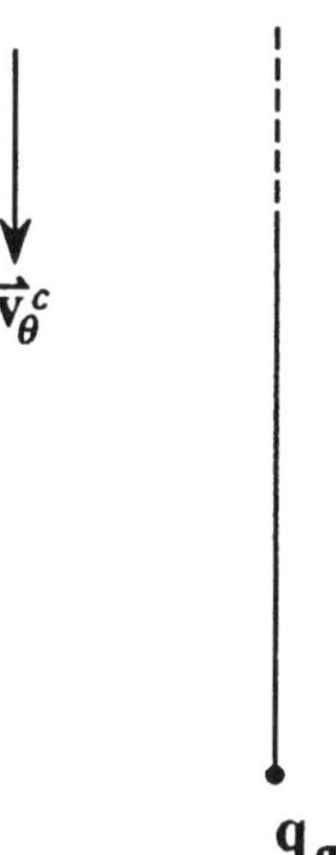

Figure 27. The goal is a single configuration $\mathbf{q}_g$ in an empty configuration space. We assume perfect control and no sensing. The commanded velocity is vertical and points downward. Each point in the vertical half-line above $\mathbf{q}_g$ is a maximal preimage (for some termination condition with time). Hence, in this example, there exists an infinity of maximal preimages of the goal.

union of all such preimages is the striped region itself, which is not a preimage, since no finite duration T is long enough to guarantee that the goal has been achieved by a motion starting anywhere in the striped region. Hence, there is no maximal preimage. (In this example, the absence of maximal preimage comes from the non-compactness of $\mathcal{C}_{contact}$. Another exhibited cause of non-existence of maximal preimage is non-compactness of the goal [Erdmann, 1984].) ■

Example 2: Consider the point-in-point problem depicted in Figure 27. The goal is a single configuration $\mathbf{q}_g$ in an empty configuration space. The commanded velocity is vertical and points downward. Assume perfect control and no position sensing. Each point in the vertical half-line above $\mathbf{q}_g$ is a maximal preimage. There is a continuous infinity of them. ■

If there existed a maximal preimage of a goal over all possible termination conditions, this preimage (assuming it were also computable) could be used as a description of all the other preimages, since any subset of a preimage is also a preimage. Therefore, the possible absence or multiplicity of maximal preimages (over all termination conditions) is unfortunate because it eliminates one natural representation of all possible preimages. This does not mean, however, that the set of all the preimages of a goal cannot be finitely represented. For instance, in the first example above, the striped region admits a finite description which is also a description of all the preimages of $\mathcal{T}$, since every bounded subset of this region is a preimage. In the second example, the half-line provides a finite description of all the preimages of $\mathbf{q}_g$, since every individual point in this half-line is a preimage.

In Subsection 8.2 we will use a Tarski sentence to finitely describe all the pertinent preimages of a goal and define a very general effective motion planner, despite the possible absence or multiplicity of maximal preimages.

8 Grand Scheme

In Section 2 we gave a simple introduction to preimage backchaining. Then we analyzed important aspects of the approach — preimage computation, choice of commanded velocities, generation of conditional strategies, and power of termination conditions. Now we revisit the approach and we describe the grand preimage backchaining scheme as was originally envisioned in [Lozano-Pérez, Mason and Taylor, 1984]. We first give an abstract planning procedure which formalizes this grand scheme, but ignores computability issues. Next, we present an effective planning method which under some restrictions can generate the same plans as the abstract procedure.

8.1 Abstract Planning Procedure

Let us consider a set of goals $\{\mathcal{T}_\alpha\}$ (goal set) which may possibly be infinite and not even countable. A region $\mathcal{P}$ is a **preimage** of $\{\mathcal{T}_\alpha\}$ for a motion command $\mathbf{M} = (\vec{\mathbf{v}}^c_\theta, \mathtt{TC})$ if and only if any motion starting in $\mathcal{P}$ is guaranteed to terminate inside a member of the goal set $\{\mathcal{T}_\alpha\}$ in such a way that the termination condition TC returns the goal $\mathcal{T}_\alpha$ which has been achieved. (Hence, TC incorporates the recognition conditions introduced in Section 6.)

Let $\mathtt{TC}^\infty_{\theta,\mathcal{P},\{\mathcal{T}_\alpha\}}$ be the "most powerful termination condition schema".

This schema, which is parameterized by θ, $\mathcal{P}$ and $\{\mathcal{T}_\alpha\}$, is such that if $\mathcal{P}$ is a preimage of $\{\mathcal{T}_\alpha\}$ for a motion command $(\vec{\mathbf{v}}_\theta^c, \mathrm{TC})$, then it is also a preimage of $\{\mathcal{T}_\alpha\}$ for the motion command $(\vec{\mathbf{v}}_\theta^c, \mathrm{TC}^\infty_{\theta,\mathcal{P},\{\mathcal{T}_\alpha\}})$. Conversely, if $\mathcal{P}$ is not a preimage of $\{\mathcal{T}_\alpha\}$ for the motion command $(\vec{\mathbf{v}}_\theta^c, \mathrm{TC}^\infty_{\theta,\mathcal{P},\{\mathcal{T}_\alpha\}})$, then it is not a preimage of $\{\mathcal{T}_\alpha\}$ for any other motion command $(\vec{\mathbf{v}}_\theta^c, \mathrm{TC})$ with the same value of θ.

Now, let us denote by $\{\mathcal{P}_{\beta_\theta}\}$ the set of all the preimages $\mathcal{P}$ of $\{\mathcal{T}_\alpha\}$ for a motion commanded along $\vec{\mathbf{v}}_\theta^c$ with $\mathrm{TC}^\infty_{\theta,\mathcal{P},\{\mathcal{T}_\alpha\}}$ for termination condition[17]. If $\mathcal{P} \in \{\mathcal{P}_{\beta_\theta}\}$, then every subset of $\mathcal{P}$ is also a preimage, hence is an element of the set $\{\mathcal{P}_{\beta_\theta}\}$.

In its most general form, the preimage backchaining approach is described by the following abstract procedure:

```
procedure PB({T_α});
   begin
      if I ∉ {T_α} then
         begin
            compute {P_βθ} for all θ ∈ S^1;
            PB(∪_{θ∈S^1} {P_βθ});
         end;
   end;
```

The initial call to this procedure is PB($\{\mathcal{G}_\alpha\}$), with $\{\mathcal{G}_\alpha\}$ being the set whose elements are $\mathcal{G}$ and all its subsets. In addition to the above operations, the procedure PB associates the direction θ with every computed preimage $\mathcal{P}_{\beta_\theta}$. (In order to avoid restricting this procedure to planar translational motions of the robot, one can regard θ as an encoding of the commanded velocity in an arbitrary velocity space S^1.)

Assume that the procedure PB terminates after r ($r \geq 0$) recursive calls. Then it has generated a sequence of $r+1$ goal sets, which we denote by $\bigcup_{\theta \in S^1}\{\mathcal{P}^{(k)}_{\beta_\theta}\}$, $k = 0, 1, \ldots, r$ (k designates the recursion at which the goal set was generated), with $\bigcup_{\theta \in S^1}\{\mathcal{P}^{(0)}_{\beta_\theta}\} = \{\mathcal{G}_\alpha\}$. This sequence determines a conditional strategy to be executed as follows:

1. If $r = 0$, the robot is already in the goal.

[17]This termination condition is different for every preimage $\mathcal{P}$ in the set $\{\mathcal{P}_{\beta_\theta}\}$.

2. Otherwise, let $\mathcal{I}_r = \mathcal{I}$; for $k = r, r-1, ..., 1$ do:

- set θ_k to any of the commanded velocity orientations associated with $\mathcal{I}_k \in \bigcup_{\theta \in S^1} \{\mathcal{P}^{(k)}_{\beta_\theta}\}$;
- execute the motion command $(\vec{\mathbf{v}}^c_{\theta_k}, \mathrm{TC}^\infty_{\theta_k, \mathcal{I}_k, \bigcup_{\theta \in S^1} \{\mathcal{P}^{(k-1)}_{\beta_\theta}\}})$;
- when the motion terminates, set $\mathcal{I}_{k-1}$ to the goal of $\bigcup_{\theta \in S^1} \{\mathcal{P}^{(k-1)}_{\beta_\theta}\}$ whose achievement has been recognized by the termination condition.

The procedure PB is obviously not an executable algorithm, since the goal sets which are considered and constructed at each recursion are typically noncountable. In addition, the strategies that it generates make use of the termination condition $\mathrm{TC}^\infty_{\theta, \mathcal{P}, \{\mathcal{T}_\alpha\}}$, which may not be computable. The fact that the procedure PB described above is not directly implementable does not mean, however, that one cannot define an effective algorithm capable of generating the same plans as PB.

8.2 Planning Method

In this subsection we describe an effective planning method that can generate "almost" the same motion strategies as the abstract procedure PB. This method, which is due to Canny [Canny, 1989], consists of formulating the planning problem as a decision problem in the first-order theory of the reals. It imposes two restrictions: first, it requires the forward projection of any configuration at any time to have an algebraic description; second, it uses a termination predicate that is less powerful than $\mathrm{TC}^\infty_{\theta, \mathcal{P}, \{\mathcal{T}_\alpha\}}$. Under these two restrictions, the method is complete and generates an r-step strategy in $O(2^{2^r})$ time whenever one such strategy exists (otherwise it terminates in the same time and returns failure). The double-exponential dependence comes from the Collins decomposition that is applied to eliminate the quantifiers from the Tarski sentence describing the r-step strategy. It is not known as yet whether there exists an effective planning method having the full power of the abstract procedure PB.

8.2.1 Feasible Trajectories

Let us consider a motion commanded along the velocity $\vec{\mathbf{v}}^c_\theta$. We assume that rather than varying continuously, the effective commanded velocity during the motion takes p successive values $\vec{\mathbf{v}}_\theta^{(1)}, ..., \vec{\mathbf{v}}_\theta^{(p)}$ in $\mathcal{U}_v(\vec{\mathbf{v}}_\theta^c)$, where p is a fixed prespecified integer. Let $t^{(k)}$, $k = 1, ..., p-1$, denote the times at which the effective control velocity switches from $\vec{\mathbf{v}}_\theta^{(k)}$ to $\vec{\mathbf{v}}_\theta^{(k+1)}$. This assumption is aimed at meeting the first restriction mentioned above (algebraicity of the forward projections). It is certainly not the only one possible, but it is a simple one.

Under this assumption, let $\tau : t \in [0, +\infty) \mapsto \tau(t) \in \mathcal{C}_{valid}$ be an actual trajectory that can be generated by a motion commanded along $\vec{\mathbf{v}}_\theta^c$. This trajectory consists of straight segments. The actual velocity is constant along each of these segments. The robot shifts from one segment to the next at each of the instants $t^{(1)}, ..., t^{(p-1)}$, and also when it switches from free to contact space, and conversely. The switch points from free to contact space and from contact to free space are algebraically defined by the parameters $\vec{\mathbf{v}}_\theta^{(1)}, ..., \vec{\mathbf{v}}_\theta^{(p)}, t^{(1)}, ..., t^{(p-1)}$, and by the algebraic description of $\mathcal{C}_{valid}$. There are at most pn such points, where n is the number of contact edges.

Let $\mathbf{s}_\tau(t)$ denote the pair $(\tau(t), \vec{\mathbf{f}}_\tau(t))$, where $\vec{\mathbf{f}}_\tau(t)$ is the reaction force exerted on the robot at time t when it moves along the trajectory τ (see Section 3). This force is algebraically determined by the configuration and the effective commanded velocity at time t (see Subsection 1.3.2).

Let $\overline{\mathcal{D}}_\theta(\{\mathbf{s}_0\})$ be the directory[18] of all feasible actual trajectories τ such that $\mathbf{s}_\tau(0) = \mathbf{s}_0$, which can be generated with the commanded velocity $\vec{\mathbf{v}}_\theta^c$. Each element of $\overline{\mathcal{D}}_\theta(\{\mathbf{s}_0\})$ is a list $(\vec{\mathbf{v}}_\theta^{(1)}, ..., \vec{\mathbf{v}}_\theta^{(p)}, t^{(1)}, ..., t^{(p-1)})$ in:

$$\{(\vec{\mathbf{v}}_\theta^{(1)}, ..., \vec{\mathbf{v}}_\theta^{(p)}, t^{(1)}, ..., t^{(p-1)}) \; / \\ \vec{\mathbf{v}}_\theta^{(1)}, ..., \vec{\mathbf{v}}_\theta^{(p)} \in \mathcal{U}_v(\vec{\mathbf{v}}_\theta^c) \; ; \; t^{(1)}, ..., t^{(p-1)} \in [0, +\infty) \; ; \; t^{(1)} \leq t^{(2)} \leq ... \leq t^{(p-1)}\}$$

which uniquely determines an actual trajectory τ such that $(\tau(0), \vec{\mathbf{f}}_\tau(0)) = \mathbf{s}_0$. $\tau(t)$ is defined as an algebraic expression involving both these $2p-1$ parameters and the variable t.

We now define the forward projection of $\mathbf{s}_0$ at time t as:

$$\overline{\mathcal{F}}_\theta(\{\mathbf{s}_0\}, t) = \bigcup_{\tau \in \overline{\mathcal{D}}_\theta(\{\mathbf{s}_0\})} \mathbf{s}_\tau(t).$$

[18] We denote this directory by $\overline{\mathcal{D}}_\theta$ rather than $\mathcal{D}_\theta$ (see Section 3) in order to indicate that the argument of $\overline{\mathcal{D}}_\theta$ is a set of configuration-force pairs, while the argument of $\mathcal{D}_\theta$ is a set of configurations.

For any region $\mathcal{X} \subseteq \mathcal{C}_{valid}$, both $\tilde{\mathcal{X}}_\theta$ and $\widetilde{(\mathcal{X})}_\theta$ denote the set defined as:

$$\tilde{\mathcal{X}}_\theta = \widetilde{(\mathcal{X})}_\theta = \{(\mathbf{q}, \vec{\mathbf{f}}) \;/\; \mathbf{q} \in \mathcal{X}; \vec{\mathbf{f}} \in F_{reac}(\mathbf{q}, \theta)\}$$

where $F_{reac}(\mathbf{q}, \theta)$ is the set of all the possible reaction forces exerted on the robot at configuration $\mathbf{q}$ when it is commanded along $\vec{\mathbf{v}}_\theta^c$ (see Subsection 1.3.3).

Given a region $\mathcal{R}$ in $\mathcal{C}_{valid}$, we define:

$$\overline{\mathcal{D}}_\theta(\tilde{\mathcal{R}}_\theta) = \bigcup_{\mathbf{s}_0 \in \tilde{\mathcal{R}}_\theta} \overline{\mathcal{D}}_\theta(\{\mathbf{s}_0\})$$

and:

$$\overline{\mathcal{F}}_\theta(\tilde{\mathcal{R}}_\theta, t) = \bigcup_{\mathbf{s}_0 \in \tilde{\mathcal{R}}_\theta} \overline{\mathcal{F}}_\theta(\{\mathbf{s}_0\}, t) = \bigcup_{\tau \in \overline{\mathcal{D}}_\theta(\tilde{\mathcal{R}}_\theta)} \mathbf{s}_\tau(t).$$

If $\mathcal{R}$ is a semi-algebraic set, so are $\overline{\mathcal{D}}_\theta(\tilde{\mathcal{R}}_\theta)$ and $\overline{\mathcal{F}}_\theta(\tilde{\mathcal{R}}_\theta, t)$.

The only requirements for the method described below are that: (1) every feasible trajectory τ be such that, for every t, $\tau(t)$ is defined by an algebraic expression of t and a finite number of parameters, and (2) $\mathcal{D}_\theta(\{\mathbf{s}_0\})$, for every $\mathbf{s}_0$, be a semi-algebraic subset of this parameter space. Restricting the effective commanded velocity to a finite number of successive values is one specific way to achieve these requirements.

8.2.2 Termination Condition

Given a set $\mathcal{Y}$, each element of which is a pair configuration-force, we denote by $\mathcal{Q}[\mathcal{Y}]$ the set defined as $\mathcal{Q}[\mathcal{Y}] = \{\mathbf{q} \;/\; \exists \vec{\mathbf{f}} : (\mathbf{q}, \vec{\mathbf{f}}) \in \mathcal{Y}\}$. In particular, for any $\mathcal{R} \subseteq \mathcal{C}_{valid}$, we have that: $\mathcal{Q}[\tilde{\mathcal{R}}_\theta] = \mathcal{R}$.

Let $\mathcal{T}$ be a goal described as a semi-algebraic set. Let $\vec{\mathbf{v}}_\theta^c$ be the commanded velocity of a motion intended to achieve this goal from a region $\mathcal{P}$ also described as a semi-algebraic set. This command uses:

$$\mathcal{Q}[\overline{\mathcal{F}}_\theta(\tilde{\mathcal{P}}_\theta \cap \mathcal{U}_s(\mathbf{s}^*(0)), t) \cap \mathcal{U}_s(\mathbf{s}^*(t))] \subseteq \mathcal{T}$$

as the termination condition, where:

- $\mathbf{s}^*(t) = (\mathbf{q}^*(t), \vec{\mathbf{v}}^c(t))$ represents the pair of sensed configuration and force at time t,

- $\mathcal{U}_s(\mathbf{s}^*) = \mathcal{U}_q(\mathbf{q}^*) \times \mathcal{U}_f(\vec{\mathbf{f}}^*)$, with $\mathbf{s}^* = (\mathbf{q}^*, \vec{\mathbf{f}}^*)$.

This termination condition is less powerful than $\text{TC}^{\infty}_{\theta,\mathcal{P},\{\mathcal{T}_\alpha\}}$ since it uses neither continuous sensing history, nor final state. It nevertheless uses initial state, time, and sensing at times 0 and t, which makes it quite powerful.

From now on, we denote the set $\overline{\mathcal{F}}_\theta(\widetilde{\mathcal{P}}_\theta \cap \mathcal{U}_s(\mathbf{s}^*(0)), t)$ by $\overline{\mathcal{F}}_{\mathcal{P},\mathbf{s}^*(0),\theta}(t)$.

The region $\mathcal{P}$ is a preimage of $\mathcal{T}$ for the motion command specified above if and only if the following *preimage condition* holds:

$$\begin{gathered}\forall \mathbf{s}_1 \in \widetilde{\mathcal{P}}_\theta \ , \ \forall \mathbf{s}_1^* \in \mathcal{U}_s(\mathbf{s}_1) \ , \ \forall \tau \ , \ \exists t \geq 0 \ , \ \forall \mathbf{s}_1'^* : \\ (\tau \in \overline{\mathcal{D}}_\theta(\widetilde{\mathcal{P}}_\theta \cap \mathcal{U}_s(\mathbf{s}_1^*)) \ \wedge \mathbf{s}_1'^* \in \mathcal{U}_s(\mathbf{s}_\tau(t))) \Rightarrow \mathcal{Q}[\overline{\mathcal{F}}_{\mathcal{P},\mathbf{s}_1^*,\theta}(t) \cap \mathcal{U}_s(\mathbf{s}_1'^*)] \subseteq \mathcal{T}.\end{gathered} \tag{2}$$

This is a Tarski sentence, and therefore we can decide whether $\mathcal{P}$ is a preimage of $\mathcal{T}$.

This result can be applied to the generation of a one-step motion strategy. Given an initial region $\mathcal{I}$ and a goal region $\mathcal{G}$, both described as semi-algebraic sets, there exists a single motion of commanded velocity $\vec{\mathbf{v}}_\theta^c$ that achieves $\mathcal{G}$ from $\mathcal{I}$ if and only if the sentence:

$$\begin{gathered}\forall \mathbf{s}_1 \in \widetilde{\mathcal{I}}_\theta \ , \ \forall \mathbf{s}_1^* \in \mathcal{U}_s(\mathbf{s}_1) \ , \ \forall \tau \ , \ \exists t \geq 0 \ , \ \forall \mathbf{s}_1'^* : \\ (\tau \in \overline{\mathcal{D}}_\theta(\widetilde{\mathcal{I}}_\theta \cap \mathcal{U}_s(\mathbf{s}_1^*)) \ \wedge \mathbf{s}_1'^* \in \mathcal{U}_s(\mathbf{s}_\tau(t))) \Rightarrow \mathcal{Q}[\overline{\mathcal{F}}_{\mathcal{I},\mathbf{s}_1^*,\theta}(t) \cap \mathcal{U}_s(\mathbf{s}_1'^*)] \subseteq \mathcal{G}\end{gathered}$$

is satisfiable. We can eliminate all the quantifiers using the Collins decomposition. It remains a polynomial expression in θ whose solutions are all the guaranteed one-step strategies. The decomposition also gives a sample value of θ.

But, in order to generate multi-step strategies, we still lack a first-order sentence that describes all the possible preimages of $\mathcal{G}$ so that they can be passed recursively as intermediate goals to the planner, as suggested by the procedure PB presented in the previous subsection. A major idea introduced by Canny [Canny, 1989] is that it is not necessary to consider all these preimages. Indeed, only those which can be attained recognizably from the initial region $\mathcal{I}$ by some motion strategy have to be considered. These preimages are called **recognizable preimages**. As shown below, they can be determined by forward chaining from $\mathcal{I}$. It turns out that under the assumptions made above the set of regions that can be recognizably achieved from $\mathcal{I}$ can be algebraically described.

8.2.3 Two-Step Strategy

Consider the problem of planning a two-step strategy to achieve a semi-algebraic goal region $\mathcal{G}$ from a semi-algebraic initial region $\mathcal{I}$. Let $\vec{\mathbf{v}}^c_{\theta_1}$ and $\vec{\mathbf{v}}^c_{\theta_2}$ denote the commanded velocities of the first and the second step, respectively.

Assume for an instant that we have constructed the set $\{\mathcal{P}_{\beta_{\theta_2}}\}$ of all the preimages of $\mathcal{G}$ for all the directions $\theta_2 \in S^1$. We derive from the preimage condition (2) that in order to achieve some preimage $\mathcal{P} \in \{\mathcal{P}_{\beta_{\theta_2}}\}$ from $\mathcal{I}$ with a motion commanded along θ_1, we must have that:

$$\begin{gathered}\forall \mathbf{s}_1 \in \widetilde{\mathcal{I}}_{\theta_1}\ ,\ \forall \mathbf{s}_1^* \in \mathcal{U}_s(\mathbf{s}_1)\ ,\ \forall \tau_1\ ,\ \exists t_1 \geq 0\ ,\ \forall \mathbf{s}_1'^*\ ,\ \exists \mathcal{P} \in \{\mathcal{P}_{\beta_{\theta_2}}\} : \\ (\tau_1 \in \overline{\mathcal{D}}_{\theta_1}(\widetilde{\mathcal{I}}_{\theta_1} \cap \mathcal{U}_s(\mathbf{s}_1^*))\ \wedge \mathbf{s}_1'^* \in \mathcal{U}_s(\mathbf{s}_{\tau_1}(t_1))) \Rightarrow \mathcal{Q}[\overline{\mathcal{F}}_{\mathcal{I},\mathbf{s}_1^*,\theta_1}(t_1) \cap \mathcal{U}_s(\mathbf{s}_1'^*)] \subseteq \mathcal{P}.\end{gathered} \tag{3}$$

This expression is not a first-order sentence since the variable $\mathcal{P}$ does not denote a real, but rather a preimage of $\mathcal{G}$. The satisfiability of such a second-order sentence is known to be undecidable.

Now, given the commanded velocity $\vec{\mathbf{v}}^c_{\theta_1}$, the smallest regions whose achievement from $\mathcal{I}$ can be recognized by the termination condition at time t_1 are regions of the form $\mathcal{Q}[\overline{\mathcal{F}}_{\mathcal{I},\mathbf{s}^*(0),\theta_1}(t_1) \cap \mathcal{U}_s(\mathbf{s}^*(t_1))]$. There is no gain in considering a non-recognizable preimage of $\mathcal{G}$. Hence, we can rewrite the sentence (3) into the following first-order sentence:

$$\begin{gathered}\forall \mathbf{s}_1 \in \widetilde{\mathcal{I}}_{\theta_1}\ ,\ \forall \mathbf{s}_1^* \in \mathcal{U}_s(\mathbf{s}_1)\ ,\ \forall \tau_1\ ,\ \exists t_1 \geq 0\ ,\ \forall \mathbf{s}_1'^* : \\ (\tau_1 \in \overline{\mathcal{D}}_{\theta_1}(\widetilde{\mathcal{I}}_{\theta_1} \cap \mathcal{U}_s(\mathbf{s}_1^*))\ \wedge \mathbf{s}_1'^* \in \mathcal{U}_s(\mathbf{s}_{\tau_1}(t_1))) \Rightarrow P_1(\mathbf{s}_1^*, \theta_1, t_1, \mathbf{s}_1'^*)\end{gathered}$$

where $P_1(\mathbf{s}_1^*, \theta_1, t_1, \mathbf{s}_1'^*)$ is an expression that is satisfied when the region $\mathcal{Q}[\overline{\mathcal{F}}_{\mathcal{I},\mathbf{s}_1^*,\theta_1}(t_1) \cap \mathcal{U}_s(\mathbf{s}_1'^*)]$ is a preimage of $\mathcal{G}$, hence a recognizable preimage. This expression can be constructed from the preimage condition (2):

$$\begin{aligned}&P_1(\mathbf{s}_1^*, \theta_1, t_1, \mathbf{s}_1'^*) \equiv \\ &\exists \theta_2 \in S^1\ ,\ \forall \mathbf{s}_2 \in \widetilde{(K_1)}_{\theta_2}\ ,\ \forall \mathbf{s}_2^* \in \mathcal{U}_s(\mathbf{s}_2)\ ,\ \forall \tau_2\ ,\ \exists t_2 \geq 0\ ,\ \forall \mathbf{s}_2'^* : \\ &\qquad (\tau_2 \in \overline{\mathcal{D}}_{\theta_2}(\widetilde{(K_1)}_{\theta_2} \cap \mathcal{U}_s(\mathbf{s}_2^*))\ \wedge\ \mathbf{s}_2'^* \in \mathcal{U}_s(\mathbf{s}_{\tau_2}(t_2))) \\ &\qquad\quad \Rightarrow \mathcal{Q}[\overline{\mathcal{F}}_{K_1,\mathbf{s}_2^*,\theta_2}(t_2) \cap \mathcal{U}_s(\mathbf{s}_2'^*)] \subseteq \mathcal{G}\end{aligned}$$

where $H_1 = \overline{\mathcal{F}}_{\mathcal{I},\mathbf{s}_1^*,\theta_1}(t_1) \cap \mathcal{U}_s(\mathbf{s}_1'^*)$ and $K_1 = \mathcal{Q}[H_1]$.

This leads us to describe the existence of a two-step motion strategy by the following Tarski sentence:

$$
\begin{array}{l}
\exists\theta_1 \in S^1\ ,\ \forall \mathbf{s}_1 \in \widetilde{\mathcal{I}}_{\theta_1}\ ,\ \forall \mathbf{s}_1^* \in \mathcal{U}_s(\mathbf{s}_1)\ ,\ \forall \tau_1\ ,\ \exists t_1 \geq 0\ ,\ \forall \mathbf{s}_1'^*\ : \\
\quad (\tau_1 \in \overline{\mathcal{D}}_{\theta_1}(\widetilde{\mathcal{I}}_{\theta_1} \cap \mathcal{U}_s(\mathbf{s}_1^*))\ \wedge\ \mathbf{s}_1'^* \in \mathcal{U}_s(\mathbf{s}_{\tau_1}(t_1))) \\
\qquad \Rightarrow \exists\theta_2 \in S^1\ ,\ \forall \mathbf{s}_2 \in \widetilde{(K_1)}_{\theta_2}\ ,\ \forall \mathbf{s}_2^* \in \mathcal{U}_s(\mathbf{s}_2)\ ,\ \forall \tau_2\ ,\ \exists t_2 \geq 0\ ,\ \forall \mathbf{s}_2'^*\ : \\
\qquad\quad (\tau_2 \in \overline{\mathcal{D}}_{\theta_2}(\widetilde{(K_1)}_{\theta_2} \cap \mathcal{U}_s(\mathbf{s}_2^*))\ \wedge\ \mathbf{s}_2'^* \in \mathcal{U}_s(\mathbf{s}_{\tau_2}(t_2))) \\
\qquad\qquad \Rightarrow \mathcal{Q}[\overline{\mathcal{F}}_{K_1,\mathbf{s}_2^*,\theta_2}(t_2) \cap \mathcal{U}_s(\mathbf{s}_2'^*)] \subseteq \mathcal{G}.
\end{array}
$$

We remove the first quantifier $\exists\theta_1 \in S^1$ and perform quantifier elimination on this sentence, yielding all the possible orientations of the commanded velocity for the first step of the strategy. If this set is not empty, i.e. if the above sentence is satisfiable, one value θ_1 is arbitrarily selected and the motion command $(\vec{\mathbf{v}}_{\theta_1}^c, P_1(\mathbf{s}_1^*, \theta_1, t, \mathbf{s}^*(t)))$ is executed, with $\mathbf{s}_1^*$ being the pair of sensed configuration and force at time $t = 0$ (of the first motion). Let t_1 be the instant when the termination condition $P_1(\mathbf{s}_1^*, \theta_1, t, \mathbf{s}^*(t))$ evaluates to `true`. At this instant, the motion has achieved a recognized preimage of $\mathcal{G}$. We remove the first quantifier $\exists\theta_2 \in S^1$ from the expression defining $P_1(\mathbf{s}_1^*, \theta_1, t_1, \mathbf{s}_1'^*)$ and we replace the variables $\mathbf{s}_1^*, \theta_1, t_1, \mathbf{s}_1'^*$ by their known values. It is guaranteed that the obtained expression is satisfiable. Its remaining quantifiers are eliminated yielding all the possible orientations of the second commanded velocity in the strategy. One value θ_2 is arbitrarily selected and the motion command $(\vec{\mathbf{v}}_{\theta_2}^c, \mathcal{Q}[\overline{\mathcal{F}}_{K_1,\mathbf{s}_2^*,\theta_2}(t) \cap \mathcal{U}_s(\mathbf{s}^*(t))] \subseteq \mathcal{G})$ is executed, with $\mathbf{s}_2^*$ being the pair of sensed configuration and force at time $t = 0$ (of the second motion). It is guaranteed to terminate in the goal $\mathcal{G}$.

Therefore, the above planning method forward chains from $\mathcal{I}$ in order to construct all the possible recognizable sets achievable from $\mathcal{I}$ and backward chains from $\mathcal{G}$ in order to determine which ones among these sets are preimages. The method gives a Tarski sentence that implicitly defines all the possible two-step strategies. Solving this sentence provides the commanded velocity of the first step and a sentence that defines the second commanded velocity. This sentence, which depends on parameters that are all known when the first motion terminates, allows the robot to select the second commanded velocity at execution time.

8.2.4 Multi-Step Strategy

Consider the problem of planning an r-step motion strategy to achieve $\mathcal{G}$ from $\mathcal{I}$. The above method extends in a straightforward fashion.

Let $K_k(\mathbf{s}_1^*, \theta_1, t_1, \mathbf{s}_1'^*, \mathbf{s}_2^*, \theta_2, ..., \theta_k, t_k, \mathbf{s}_k'^*)$, $k = 0, 1, ..., r$, denote a recognizable set after the execution of k motion commands. Let it be abbreviated to K_k. It is recursively defined as:

$$K_0 = \mathcal{I} \quad \text{and} \quad K_k = \mathcal{Q}[H_k], \quad \text{for } k = 1, ..., r,$$

with:

$$H_k = \overline{\mathcal{F}}_{K_{k-1}, \mathbf{s}_k^*, \theta_k}(t_k) \cap \mathcal{U}_s(\mathbf{s}_k'^*).$$

Now, let $P_k(\mathbf{s}_1^*, \theta_1, t_1, \mathbf{s}_1'^*, \mathbf{s}_2^*, \theta_2, ..., \theta_k, t_k, \mathbf{s}_k'^*)$ — we abbreviate it to P_k — be the condition which is satisfied when K_k is a preimage of $\mathcal{G}$ for a sequence of $r - k$ motion commands along the directions $\theta_{k+1}, ..., \theta_r$. P_k, $k = 0, ..., r - 1$, is recursively defined as follows:

$$\begin{aligned}
&P_k \equiv \\
&\exists\theta_{k+1} \in S^1,\ \forall \mathbf{s}_{k+1} \in (\widetilde{K_k})_{\theta_{k+1}},\ \forall \mathbf{s}_{k+1}^* \in \mathcal{U}_s(\mathbf{s}_{k+1}),\ \forall \tau_{k+1},\ \exists t_{k+1} \geq 0\ ,\ \forall \mathbf{s}_{k+1}'^* : \\
&\qquad (\tau_{k+1} \in \overline{\mathcal{D}}_{\theta_{k+1}}((\widetilde{K_k})_{\theta_{k+1}} \cap \mathcal{U}_s(\mathbf{s}_{k+1}^*)) \wedge \mathbf{s}_{k+1}'^* \in \mathcal{U}_s(\mathbf{s}_{\tau_{k+1}}(t_{k+1}))) \Rightarrow P_{k+1}
\end{aligned}$$

with:

$$P_r \equiv [K_r \subseteq \mathcal{G}].$$

The sentence $\forall \mathbf{s}_0^* : P_0(\mathbf{s}_0^*)$ describes the existence of an r-step motion strategy. It is satisfiable if and only if there is one such strategy. If this is the case, a guaranteed strategy is executed by considering the sentences P_0 through P_{r-1} in sequence. P_k gives the guaranteed commanded velocity and the termination condition P_{k+1} for the $(k+1)^{\text{th}}$ motion command.

The sentence $\forall \mathbf{s}_0^* : P_0(\mathbf{s}_0^*)$ involves r existential quantifiers alternating with universal quantifiers. Hence, it requires double-exponential time in r to determine its satisfiability (using, say, Collins' method).

It is interesting to note that the existential quantifiers apply to the robot's motions, while the universal quantifiers apply to the actions of nature (measured through the sensors). This alternation, which is classical in game theory, is investigated in [Taylor, Mason and Goldberg, 1987].

This alternation can be eliminated by removing sensing, for instance by using the sticking termination condition. Using this termination condition also results in using the timeless forward projection (defined as in Subsection 5.2) instead of the forward projection with time. Thus, we no longer need the somewhat arbitrary assumption that the effective commanded velocity takes at most p successive values during any actual motion. The method reduces to that presented in Subsection 5.2, which only requires time single-exponential in r.

The method can also be adapted to the following termination condition:

$$\mathcal{F}_\theta(\mathcal{P} \cap \mathcal{K}_\theta(\mathbf{q}^*(0), \vec{\mathbf{f}}^*(0))) \cap \mathcal{K}_\theta(\mathbf{q}^*(t), \vec{\mathbf{f}}^*(t)) \subseteq \mathcal{T}$$

where $\mathcal{F}_\theta(\mathcal{P})$ is the timeless forward projection of $\mathcal{P}$ for the commanded velocity $\vec{\mathbf{v}}_\theta^c$ and $\mathcal{K}_\theta(\mathbf{q}^*, \vec{\mathbf{f}}^*)$ is the set of configurations $\mathbf{q}$ θ-consistent with the sensed values $\mathbf{q}^*$ and $\vec{\mathbf{f}}^*$ (see Subsection 4.5). Like the sticking termination condition, this condition (which is slightly more powerful than the instantaneous sensing termination condition used in Subsection 4.5) does not require the effective commanded velocity to take p successive values. But it does involve sensing and yields a double-exponential time planner.

9 On Uncertainty in Model

Throughout this chapter, we have assumed that the geometry of $\mathcal{C}_{valid}$ was exactly known. In this last section, we briefly explore the effect of uncertainty in the geometry of $\mathcal{C}_{valid}$.

Consider the point-in-hole example shown in Figure 28. Assuming that all the angles of the C-obstacle are right angles and that the global orientation of the C-obstacle is perfectly known, the dimensions shown in Figure 28.a suffice to define the geometry of $\mathcal{C}_{valid}$. Consider the case where all these dimensions, except d (width of the hole), are exactly known. Assume further that $d \in (d_{min}, d_{max})$.

One way to represent this uncertainty is to extend the configuration space into a **generalized configuration space** by adding one dimen-

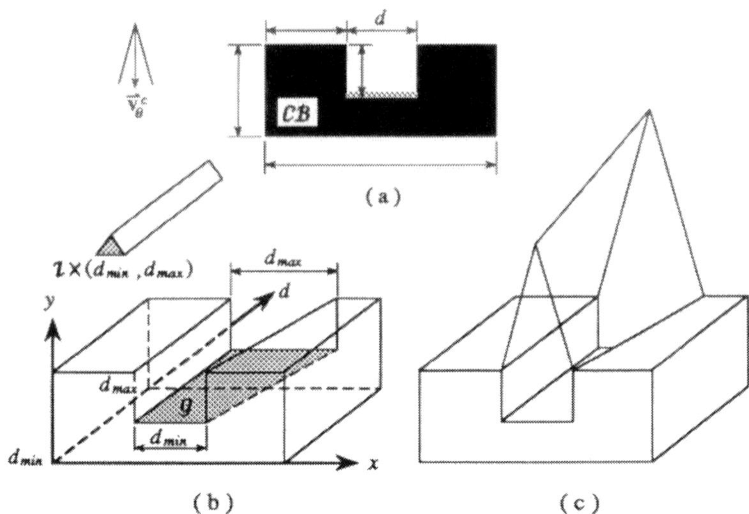

Figure 28. $\mathcal{C}_{valid}$ contains a single C-obstacle $\mathcal{CB}$ (Figure a). We assume that the geometry of $\mathcal{C}_{valid}$ is exactly known, except for the hole's width d, which is only known to be in the interval (d_{min}, d_{max}). This uncertainty is represented by adding a new dimension to the configuration space. We obtain the generalized configuration space shown in Figure b. The C-obstacle is now a three-dimensional region and $\mathcal{CB}$'s hole maps to a groove. The goal region $\mathcal{G}$ is the bottom face of the groove. A generalized preimage of this region for the commanded velocity $\vec{\mathbf{v}}_\theta^c$ is shown in Figure c.

sion corresponding to the d parameter (see Figure 28.b) [Donald, 1987a]. Each cross-section of the generalized configuration space perpendicular to the d-axis is the configuration space for a particular value of d. In the generalized configuration space, the hole becomes a groove and the goal $\mathcal{G}$ is the bottom face of this groove. The initial region is the cylinder $\mathcal{I} \times (d_{min}, d_{max})$. A strongly guaranteed strategy should reach and terminate in $\mathcal{G}$. (Notice that, when this strategy is executed, the robot moves within a single cross-section at constant d. We do not know which one in advance. Even during execution, the robot will usually not collect sufficient information to determine in which cross-section it is moving.)

The concept of generalized configuration space extends to the case where there are several imprecisely known parameters in the description of $\mathcal{C}_{valid}$, by simply adding a new dimension for each such parameter. We can define a preimage of a goal in the generalized configuration space as we did in a regular configuration space and the principle of the preimage backchaining approach still applies.

Assume as in the above example that there is a single parameter d, whose value is not known exactly. A preimage in the generalized configuration space — let us call it a *generalized preimage* — is a union of preimages computed at various values of d for the same commanded velocity (see Figure 28.c). Let us assume that we compute these preimages by backprojecting from the sticking edges in the goal. When the parameter d varies over its uncertainty interval, the topology of the preimage (computed at a specific value of d) changes at a finite number of critical values of d, and hence we can compute and represent the corresponding generalized preimage in much the same way as we did for nondirectional backprojections (Subsection 5.1.1). In the example of Figure 28, the only critical value of d is $d = 0$ (if $0 \in (d_{min}, d_{max})$), where the topology of $\mathcal{C}_{valid}$ changes. As another example, consider the case where the geometry of the C-obstacle region is exactly known, but its orientation is completely unknown (for instance, this could be the problem of meshing two planar gears [Donald, 1987a]). In this case, for any goal and any commanded velocity, the generalized preimage computed by backprojecting from the goal sticking edges is exactly the nondirectional preimage computed in Subsection 5.1.1 with the orientation of the C-obstacle region substituted for the orientation of the commanded velocity.

The planning method described in Subsection 8.2 is independent of the dimension of the planning space. Therefore, it also applies to motion planning with uncertainty in the input model, as long as this uncertainty involves a finite number of parameters [Canny, 1989].

Exercises

1: Justify the construction of the friction cone at a contact vertex, which is illustrated in Figure 3.

2: Assume perfect generalized damping control. $\mathcal{C}_{valid}$ is defined as in Section 1. Let $\mathbf{q}_s$ be the starting configuration of two motions commanded along two different velocities $\vec{\mathbf{v}}^c_{\theta_1}$ and $\vec{\mathbf{v}}^c_{\theta_2}$. Let τ_1 and τ_2 be the respective trajectories followed by these motions. Show that τ_1 and τ_2

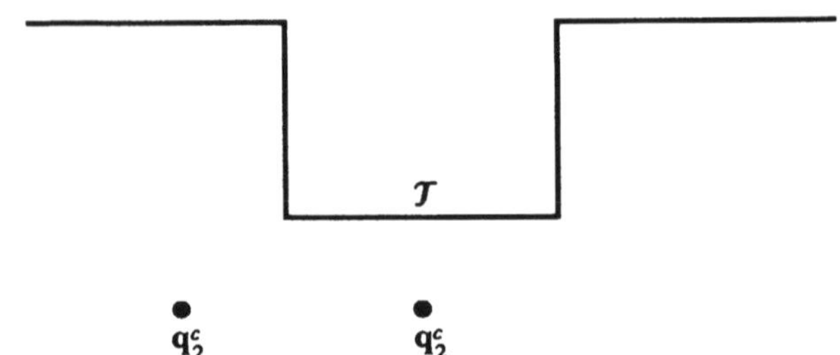

Figure 29. The configurations $\mathbf{q}_1^c$ and $\mathbf{q}_2^c$ represents two desired configurations given to a robot with generalized spring control (Exercise 4).

have no loops, and that they can meet (and possibly separate again), but never cross each other [Friedman, Hershberger and Snoeyink, 1990].

3: Consider the point-in-hole example of Figure 9. Assume non-zero friction in contact space. What is the union of all the preimages of the goal $\mathcal{T}$ when the orientation of the commanded velocity varies over S^1?

4: Generalized damping control is only one possible model of force-compliant motion control. Another interesting model is *generalized spring* control [Salisbury, 1980]. In this model, the control statement is parameterized by a desired configuration $\mathbf{q}^c$ and the robot's commanded trajectory τ^c is defined by the following equation:

$$\vec{\mathbf{f}}(t) = B\vec{\mathbf{v}}^c(t) - K(\mathbf{q}^c - \tau^c(t))$$

where B and K are 2×2 matrices and $\vec{\mathbf{v}}^c(t) = d\vec{\tau}^c(t)/dt$ is the velocity of the robot at time t along the commanded trajectory.

Assume here that B and K are constant diagonal matrices, each with equal positive diagonal coefficients, and that $B = K$. In free space, $\vec{\mathbf{f}}(t) = 0$, so that $\vec{\mathbf{v}}^c(t) = \mathbf{q}^c - \tau^c(t)$; hence, the robot moves along a straight trajectory pointing toward $\mathbf{q}^c$. In contact space, if the vector $-(\mathbf{q}^c - \tau^c(t))$ points into the friction cone at $\tau^c(t)$, then the robot sticks, otherwise it slides.

Consider the point-in-hole example shown in Figure 29. Assume perfect generalized spring control and non-zero friction. Construct the preimage of the bottom edge $\mathcal{T}$ for each of the two desired configurations $\mathbf{q}_1^c$ and

$\mathbf{q}_2^c$ shown in the figure. (Assume perfect position sensing so that in both cases the termination condition is $\mathbf{q}^*(t) \in \mathcal{T}$.)

5: Give an example of a planning problem which cannot be solved using the preimage backchaining approach, although the goal is achievable by executing a finite (but unbounded) number of motion commands [Mason, 1984].

6: A *weak preimage* of a goal for a motion command $\mathbf{M} = (\vec{\mathbf{v}}_\theta^c, \mathtt{TC})$ is any region $\mathcal{P} \subseteq \mathcal{C}_{valid}$ such that a motion commanded along $\vec{\mathbf{v}}_\theta^c$ and starting from within $\mathcal{P}$ may fortuitously attain the goal and terminate in the goal. Consider the point-in-hole of Figure 9. Construct the maximal weak preimage of the goal $\mathcal{T}$. Consider any initial region $\mathcal{I}$ contained in this weak preimage. Construct a one-step weakly guaranteed strategy (see the remark at the end of Subsection 1.4) that either achieves $\mathcal{T}$ from $\mathcal{I}$ or fails recognizably.

7: Show the following two properties:

- If $\mathcal{P}$ is a preimage of $\mathcal{T}$ for $\mathbf{M}$, then any subset of $\mathcal{P}$ is also a preimage of $\mathcal{T}$ for $\mathbf{M}$.
- If $\mathcal{P}_1$ and $\mathcal{P}_2$ are both preimages of $\mathcal{T}$ for $\mathbf{M}$, then $\mathcal{P}_1 \cup \mathcal{P}_2$ is also a preimage of $\mathcal{T}$ for $\mathbf{M}$.

8: Implement a simple preimage backchaining motion planner using the backprojection from sticking edges for computing preimages and Erdmann's algorithm for computing backprojections. Select the commanded velocities from a predefined finite set.

9: Consider the three goals $\mathcal{T}_1$, $\mathcal{T}_2$ and $\mathcal{T}_3$ shown in Figure 30. The commanded velocity $\vec{\mathbf{v}}_\theta^c$ is pointing vertically downward. The position uncertainty disc is as shown in the figure. Assume perfect force sensing. Construct the θ-kernel of each of the three goals.

10: Consider the goal region $\mathcal{T}$ in Figure 17.a. Construct all the θ-kernels of $\mathcal{T}$ when θ varies over S^1.

11: Propose an algorithm for computing the θ-kernel of a goal described as a polygon in $\mathcal{C}_{valid}$.

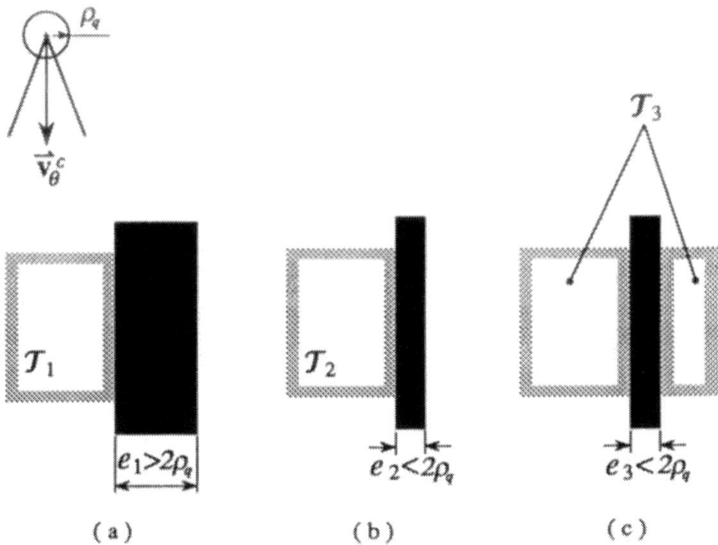

Figure 30. Each of the goal regions, $\mathcal{T}_1$, $\mathcal{T}_2$ and $\mathcal{T}_3$, consists of one or two polygons in valid space. The radius of the position uncertainty disc is ρ_q. The commanded velocity is pointing vertically downward (Exercise 9).

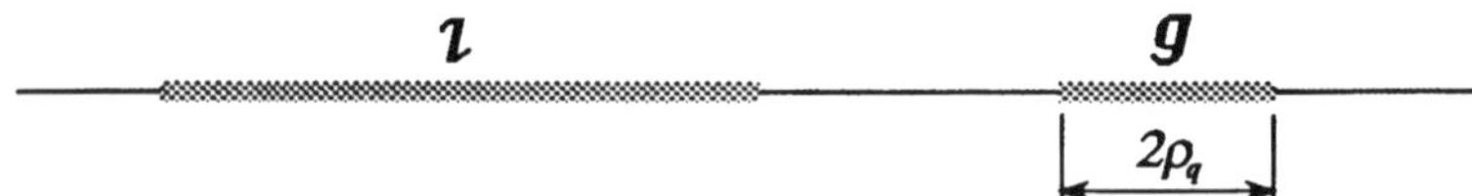

Figure 31. The robot moves in a horizontal line. The initial region is $\mathcal{I}$ and the goal region is $\mathcal{G}$. The goal region is a closed segment whose length is exactly $2\rho_q$, which is the same as for the open interval representing the uncertainty in position sensing. The goal can be achieved recognizably using a termination condition with final state (Exercise 13).

12: Consider Figure 23. Show the regions F_1, F_2 and F_3 defined as in Subsection 7.2.

13: Assume that the robot's configuration space is a horizontal line (i.e. $\mathcal{C} = \mathbf{R}$). There is no C-obstacle (i.e. $\mathcal{C}_{valid} = \mathbf{R}$). The uncertainty in position sensing is an open segment of length $2\rho_q$ centered at the

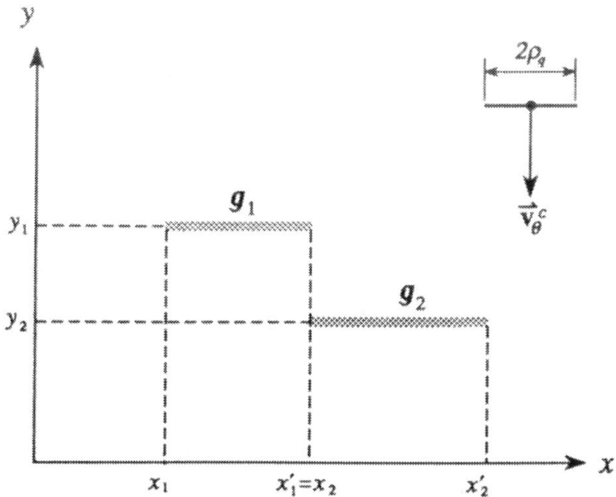

Figure 32. The goal consists of two horizontal segments in a configuration space with no C-obstacle (Exercise 14).

measured configuration. There are two possible velocities: "rightward" and "leftward". The goal region $\mathcal{G}$ is a closed segment of length $2\rho_q$ as shown in Figure 31. The initial region $\mathcal{I}$ is a segment of finite length on the left of $\mathcal{G}$. The selected commanded velocity is "rightward". Construct a computable form of the termination condition with final state that is guaranteed to terminate the motion in $\mathcal{G}$. What are the preimages of $\mathcal{G}$ for this termination condition?

14: Let $\mathcal{C} = \mathcal{C}_{valid} = \mathbf{R}^2$ (there is no obstacle). Assume perfect robot control and perfect position sensing in the vertical direction. Hence, $\mathcal{U}_q(\mathbf{q}^*)$ is an open horizontal segment of length $2\rho_q$ centered at the measured configuration $\mathbf{q}^*$. The goal region is $\mathcal{G} = \mathcal{G}_1 \cup \mathcal{G}_2$ (see Figure 32), where:

- $\mathcal{G}_1$ is a horizontal line segment connecting (x_1, y_1) to (x'_1, y_1),
- $\mathcal{G}_2$ is a horizontal line segment connecting (x_2, y_2) to (x'_2, y_2), with $x_2 = x'_1$.

The commanded velocity is pointing vertically downward.

Construct a preimage of $\mathcal{G}$ that is maximal over all possible termination conditions and specify the corresponding termination condition. Is this maximal preimage unique? Characterize all the preimages that are maximal over all possible termination conditions [Erdmann, 1984].

15: Consider the Tarski sentence of Subsection 8.2.3 which expresses the existence of a two-step strategy to achieve a goal $\mathcal{G}$. Rewrite this sentence by using the termination condition:

$$\mathcal{F}_\theta(\mathcal{P} \cap \mathcal{K}_\theta(\mathbf{q}^*(0), \vec{\mathbf{f}}^*(0))) \cap \mathcal{K}_\theta(\mathbf{q}^*(t), \vec{\mathbf{f}}^*(t)) \subseteq \mathcal{T}$$

where $\mathcal{F}_\theta(\mathcal{P})$ is the timeless forward projection defined as in Subsection 5.2 and $\mathcal{K}_\theta(\mathbf{q}^*, \vec{\mathbf{f}}^*)$ is the set of configurations $\mathbf{q}$ θ-consistent with the sensed values $\mathbf{q}^*$ and $\vec{\mathbf{f}}^*$ (see Subsection 4.5). Show that the new sentence remains a Tarski sentence even if we drop the condition that the effective commanded velocity takes a fixed number of successive values during any actual motion.

16: Consider an omnidirectional mobile robot equipped with odometric and proximity range sensors. Discuss the application of the preimage backchaining approach to the generation of motion strategies for this robot.

17: An approach to motion planning with uncertainty, known as *skeleton refinement*, has been presented in [Taylor, 1976], [Lozano-Pérez, 1976], [Brooks, 1982] and [Pertin-Troccaz and Puget, 1987]. Compare this approach to preimage backchaining.

Chapter 11

Movable Objects

So far we have considered workspaces populated by one or several robots and by obstacles which are either stationary and not movable, or moving along predetermined trajectories. In this chapter we introduce a third kind of object: the *movable object.* Such an object is like a stationary obstacle, except that its location can be changed by a robot. The most usual way for a robot to change the location of a movable object, in fact the only way that we will study in some depth in this chapter, is by grasping it and moving with it.

The presence of movable objects has a considerable impact on robot motion planning. For example, even when there exists no feasible path for a robot between two configurations in a given arrangement of the workspace, it may be possible for the robot to open one by transferring some movable objects to other locations and attain the goal configuration despite initial obstructions. The goal assigned to the robot may also be some desired arrangement of the movable objects, for example the arrangement of a set of parts in an assembly product.

As suggested above, movable objects yield a sophisticated form of motion planning which we call *manipulation planning.* A manipulation plan is a sequence of motions, each intended to achieve an intermediate goal which is usually not specified in the input problem. For example, the purpose of a path may be to grasp an object. The grasp configuration of the object, i.e. the geometric relationship between the robot and the object, is an intermediate goal that the planner must determine automatically.

In Section 1 we illustrate manipulation planning with a simple example. In Section 2 we propose a formal definition of the planning problem and in Section 3 we describe a theoretical solution due to Laumond and Alami [Laumond and Alami, 1988]. The existence of this solution serves mainly as a proof that manipulation planning is a decidable problem.

We then describe more specific methods addressing simpler forms of the manipulation planning problem (Section 4). In particular, we develop an exact cell decomposition method in the case where the robot is a disc moving in a two-dimensional workspace containing polygonal obstacles and a single movable object modeled as a disc. The method is an adaptation introduced in [Laumond and Alami, 1989] of a method originally proposed in [Schwartz and Sharir, 1983c] for coordinating the motion of two circular robots.

In the first four sections of the chapter, we simplify the way robots can grasp movable objects in order to focus on motion planning. However, grasping is a major subproblem in dealing with movable objects and it deserves specific attention. In Section 5 we identify the important issues raised by this subproblem and we survey major results related to them.

Because it requires the generation of multiple paths and the selection of intermediate goals, planning with movable obstacles is subject to combinatorial explosion. This suggests the design of a hierarchical planner that attempts to solve problems at successive levels of abstraction by iteratively considering more and more constraints. In Section 6 we illustrate hierarchical planning in the context of an important manipulation planning problem: assembly planning.

Most of this chapter deals with very recent research. A consequence is that several topics are not studied in much depth; for some of them we limit our presentation to pointers to the relevant literature. This chapter is nevertheless important because it contributes in closing the gap between basic motion planning and higher-level task planning, leading the way toward integrated planning systems for autonomous robots.

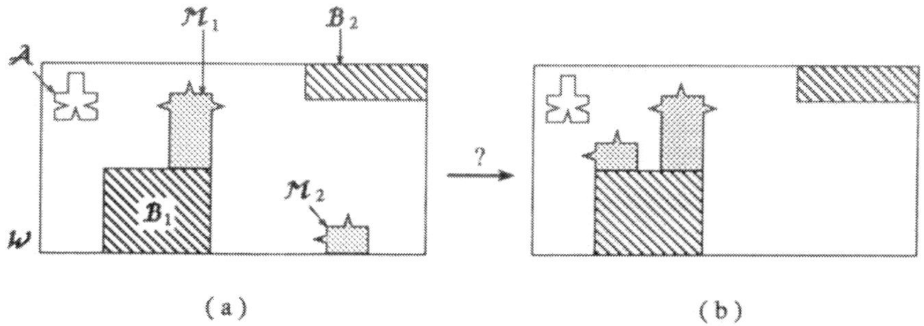

Figure 1. The robot $\mathcal{A}$ is a rigid object that can translate freely (at fixed orientation) in the rectangular workspace $\mathcal{W}$. $\mathcal{B}_1$ and $\mathcal{B}_2$ are fixed, non-movable obstacles. $\mathcal{M}_1$ and $\mathcal{M}_2$ are movable objects. Figure a depicts the initial arrangement of the objects. Figure b depicts the goal arrangement.

1 Introductory Example

In this section we briefly illustrate motion planning with movable objects. We make use of the example[1] depicted in Figure 1 (the problem) and Figure 2 (a solution).

The workspace $\mathcal{W}$ is a rectangle containing a robot $\mathcal{A}$, two fixed obstacles $\mathcal{B}_1$ and $\mathcal{B}_2$, and two movable objects $\mathcal{M}_1$ and $\mathcal{M}_2$. A movable object can be moved by the robot provided that it has been grasped; otherwise it remains stationary. The robot can grasp a movable object whenever a triangular cavity of $\mathcal{A}$ fits a triangular protrusion of the movable object.

The initial arrangement of the objects is shown in Figure 1.a and the goal arrangement in Figure 1.b. The only difference between the two arrangements is $\mathcal{M}_2$'s position.

In order to solve this problem, the planner must generate a sequence of paths of $\mathcal{A}$. Along some of these paths, called the *transit paths*, the robot moves alone (typically to a position where it grasps a movable object). Along the other paths, called the *transfer paths*, the robot carries a movable object to a new position where it ungrasps it. Figure 2 illustrates a solution of the problem which consists of a sequence of 13 paths. (If only the goal position of $\mathcal{M}_2$ is important, the paths 11, 12 and 13 are not necessary.)

[1]This example is drawn from [Alami, Siméon and Laumond, 1989].

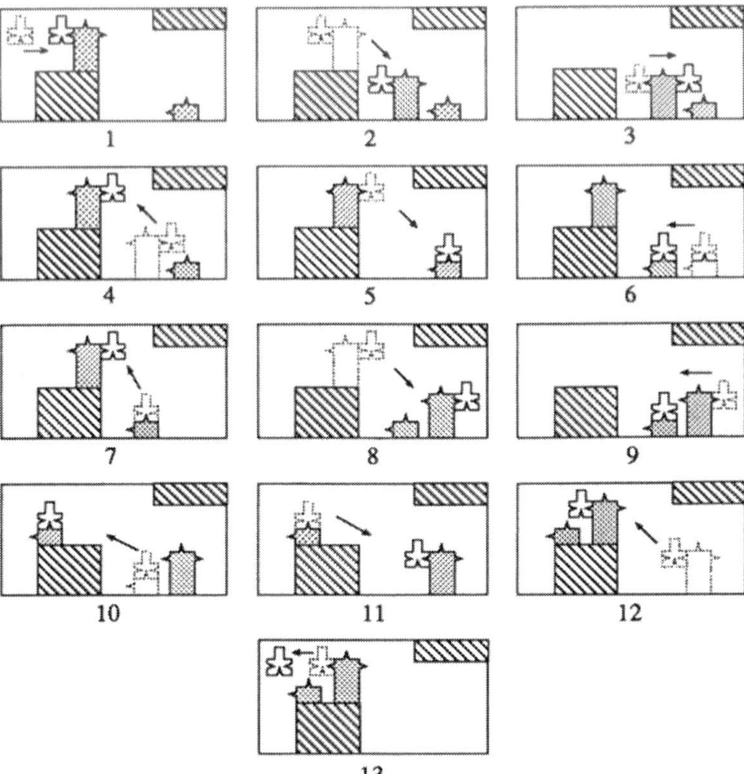

Figure 2. This figure illustrates a sequence of 13 paths that allows the robot to transform the initial arrangement of Figure 1.a into the goal arrangement of Figure 1.b. There are 7 transit paths along which the robot moves alone (paths 1, 3, 5, 7, 9, 11, and 13) and 6 transfer paths along which it transports either $\mathcal{M}_1$ (paths 2, 4, 8, and 12) or $\mathcal{M}_2$ (paths 6 and 10). Along each transit path the robot touches no obstacle or movable object, except at the extremities of the path where it may touch a single movable object in order to grasp or ungrasp it. Along each transfer path the robot touches no obstacle or movable object (other than the transported one); the transported object touches no obstacle or movable object, except at the extremities of the path.

The generation of the solution shown in Figure 2 requires that the planner overcome a number of difficulties. To achieve the goal arrangement, $\mathcal{A}$ must reach $\mathcal{M}_2$, grasp it and carry it to its goal position. But, in the initial arrangement, $\mathcal{M}_1$ prevents $\mathcal{A}$ from reaching $\mathcal{B}$. Thus $\mathcal{A}$ must first grasp $\mathcal{M}_1$ and move it elsewhere. However, there is no position where $\mathcal{A}$ can move $\mathcal{M}_1$ so that it can subsequently move $\mathcal{M}_2$ to its goal position (we assume that the robot can only translate). Therefore, $\mathcal{A}$ has to perform a series of motions during which it successively moves $\mathcal{M}_1$ and $\mathcal{M}_2$. In addition, every time $\mathcal{A}$ decides to move an object, it must compute a grasping position on the object. The example of Figure 1 has been simplified by restricting the set of possible grasps for each movable object to a finite one, but in general there are infinitely many ways for a robot to grasp an object. Furthermore, $\mathcal{A}$ should ungrasp a movable object only at a position where it will remain stable (we assume that each object in $\mathcal{W}$ is subject to a gravitational force oriented downward).

This example shows that the solution of a motion planning problem with movable objects typically consists of an alternate sequence of transit and transfer paths. The goal of each path (except the last one) is not given in the input definition of the problem and thus has to be devised by the planner. Obviously, the sequence of paths can also be seen as a single path τ in the *composite configuration space* of the robot $\mathcal{A}$ and the various movable objects. However, in this space, τ is locally constrained to lie in some adequate slice having the same dimension as $\mathcal{A}$'s configuration space. For instance, a subset of τ that corresponds to a transit path must lie in a slice where all the movable objects are stationary. A subset of τ that corresponds to a transfer path must lie in a slice where all the movable objects, except the transported one, are stationary and the position of the transported movable object is fixed relative to the robot. The path τ may change slices at any stable arrangement of the objects where two slices intersect.

2 Problem Formulation

We now give a precise definition of the motion planning problem with movable objects. This definition rests on some simplifications that we will discuss in Subsection 2.3.

2.1 Formal Definition

Configurations. Consider the case of a single robot $\mathcal{A}$ moving in a workspace $\mathcal{W} = \mathbf{R}^N$, $N = 2$ or 3, among stationary, non-movable obstacles $\mathcal{B}_1, ..., \mathcal{B}_q$ and **movable objects** $\mathcal{M}_1, ..., \mathcal{M}_r$.

Assume that $\mathcal{A}$ is a free-flying rigid object. As in the previous chapters, a Cartesian frame $\mathcal{F}_{\mathcal{A}}$ (resp. $\mathcal{F}_{\mathcal{W}}$) is attached to $\mathcal{A}$ (resp. $\mathcal{W}$). $\mathcal{A}$'s configuration space is $\mathcal{C}$.

Assume also that all the movable objects $\mathcal{M}_j$ are rigid. We attach a frame $\mathcal{F}_{\mathcal{M}_j}$ to every $\mathcal{M}_j$, $j = 1, \ldots, r$. The origin $O_{\mathcal{M}_j}$ of $\mathcal{F}_{\mathcal{M}_j}$ is the **reference point** of $\mathcal{M}_j$. We define the configuration $\mathbf{q}_{\mathcal{M}_j}$ of $\mathcal{M}_j$ as a specification of the position and orientation of $\mathcal{F}_{\mathcal{M}_j}$ with respect to $\mathcal{F}_{\mathcal{W}}$. The **configuration space** of $\mathcal{M}_j$ is denoted by $\mathcal{C}_{\mathcal{M}_j}$.

We define the **configuration of $\mathcal{A}$ relative to** $\mathcal{M}_j$ as the configuration of $\mathcal{A}$ if $\mathcal{F}_{\mathcal{W}}$ coincided with the current $\mathcal{F}_{\mathcal{M}_j}$, i.e. as a specification of the position and orientation of $\mathcal{F}_{\mathcal{A}}$ with respect to $\mathcal{F}_{\mathcal{M}_j}$. For a given configuration $\mathbf{q}$ of $\mathcal{A}$, the configuration of $\mathcal{A}$ relative to $\mathcal{M}_j$ depends on the current configuration $\mathbf{q}_{\mathcal{M}_j}$ of $\mathcal{M}_j$. We denote it by $\mathbf{q}/\mathbf{q}_{\mathcal{M}_j}$.

Grasp Sets. Unlike a robot, a movable object $\mathcal{M}_j$ cannot move by itself. It can only be displaced by the robot $\mathcal{A}$. We assume that $\mathcal{A}$ can displace $\mathcal{M}_j$ only if it has grasped it rigidly. In order to grasp $\mathcal{M}_j$, $\mathcal{A}$ must first move to a grasping configuration, or **grasp**, of $\mathcal{M}_j$ and then activate the GRASP operation[2]. A grasp is any configuration $\mathbf{q}$ of $\mathcal{A}$ such that $\mathbf{q}/\mathbf{q}_{\mathcal{M}_j} \in \mathcal{G}_j$, where $\mathcal{G}_j$ is a predefined set called the **grasp set** of $\mathcal{M}_j$. We also write $\mathbf{q} \in \mathcal{G}_j(\mathbf{q}_{\mathcal{M}_j})$. $\mathcal{G}_j$ may be given to the planner or it may be computed from the input models of $\mathcal{A}$ and $\mathcal{M}_j$ (see Section 5). Once $\mathcal{M}_j$ has been grasped, if the robot $\mathcal{A}$ moves, the object $\mathcal{M}_j$ moves with it as if $\mathcal{A}$ and $\mathcal{M}_j$ were forming a single rigid object, i.e. in such a way that the relative configuration $\mathbf{q}/\mathbf{q}_{\mathcal{M}_j}$ remains constant during the motion. In order to ungrasp $\mathcal{M}_j$, $\mathcal{A}$ simply activates the UNGRASP operation.

We assume that $\mathcal{A}$ can only grasp one movable object at a time.

Object Stability. At every instant, the set of movable objects which are not currently grasped by $\mathcal{A}$ must be at stable configurations under the gravitational force. The stability of an object $\mathcal{M}_j$ at a configuration $\mathbf{q}_{\mathcal{M}_j}$

[2]In general achieving a grasp configuration is not sufficient to grasp an object. It is also necessary to affix the object to the robot, for example, by exerting a gripping force or activating a suction device. We assume here that the GRASP operation is a parameterless switch-type operation requiring no additional motion. The reverse of the GRASP operation is the UNGRASP operation.

typically depends on the distribution of the obstacles $\mathcal{B}_i$ in $\mathcal{W}$ and on the configurations $\mathbf{q}_{\mathcal{M}_k}$ of the other movable objects (e.g. $\mathcal{M}_j$ could be stable by "leaning against" an obstacle or a movable object). Let STABLE (resp. STABLE$_j$) denote the subset of $\prod_{k=1}^{r} \mathcal{C}_{\mathcal{M}_k}$ (resp. $\prod_{k\neq j} \mathcal{C}_{\mathcal{M}_k}$) at which all the movable objects $\mathcal{M}_k$, $k \in [1,r]$ (resp. $k \in [1,r]$, $k \neq j$), are stable with no movable object overlapping another one or an obstacle. Hence, in order to ungrasp an object $\mathcal{M}_j$, $\mathcal{A}$ must be at a configuration $\mathbf{q} \in \mathcal{G}(\mathbf{q}_{\mathcal{M}_j})$ such that $(\mathbf{q}_{\mathcal{M}_1}, ..., \mathbf{q}_{\mathcal{M}_j}, ..., \mathbf{q}_{\mathcal{M}_r}) \in$ STABLE, with $\mathbf{q}_{\mathcal{M}_k}$, $k = 1$ to r, being the current configurations of the movable objects. Similarly, in order to grasp $\mathcal{M}_j$, $\mathcal{A}$ must be at a configuration $\mathbf{q} \in \mathcal{G}(\mathbf{q}_{\mathcal{M}_j})$ such that $(\mathbf{q}_{\mathcal{M}_1}, ..., \mathbf{q}_{\mathcal{M}_{j-1}}, \mathbf{q}_{\mathcal{M}_{j+1}}, ..., \mathbf{q}_{\mathcal{M}_r}) \in$ STABLE$_j$.

Remark: The computation of STABLE and STABLE$_j$ will not be investigated below. The computational analysis of the stability of an arrangement of objects under the gravitational forces (and possibly other external forces) is a difficult problem that has been addressed in a few publications only, including [Blum, Griffith and Neumann, 1970] [Fahlman, 1974] [Boneschanscher et al., 1988] and [Palmer, 1989]. In particular, it has been shown in [Palmer, 1989] that the problem of determining whether an arrangement of polygons in a vertical plane subject to gravity is at a static equilibrium point is NP-hard. An implemented algorithm is described in [Boneschanscher et al., 1988] which tests the stability of an arrangement of polyhedra on a table with a simplified model of friction in the six-dimensional space of translations and rotations. The algorithm takes $O(rn^3)$ time in the average case, where r is the number of polyhedra and n is the number of vertices. ∎

When $\mathcal{W}$ is the horizontal plane, gravity is irrelevant. Then STABLE is simply the subset of $\prod_{k=1}^{r} \mathcal{C}_{\mathcal{M}_k}$ where no movable object overlaps another one or an obstacle. Similarly, STABLE$_j$ is the subset of $\prod_{k\neq j} \mathcal{C}_{\mathcal{M}_k}$ where no movable object in $\{\mathcal{M}_k\}_{k\neq j}$ intersects another one or an obstacle.

Paths. Let $\mathcal{C}^\star$ denote the set product $\mathcal{C} \times \mathcal{C}_{\mathcal{M}_1} \times ... \times \mathcal{C}_{\mathcal{M}_r}$, i.e. the **composite configuration space** of $\mathcal{A}, \mathcal{M}_1, ..., \mathcal{M}_r$. An element of $\mathcal{C}^\star$ is a configuration $\mathbf{q}^\star = (\mathbf{q}, \mathbf{q}_{\mathcal{M}_1}, ..., \mathbf{q}_{\mathcal{M}_r})$ that determines the positions and orientations of the frames $\mathcal{F}_{\mathcal{A}}, \mathcal{F}_{\mathcal{M}_1}, ..., \mathcal{F}_{\mathcal{M}_r}$ relative to $\mathcal{F}_{\mathcal{W}}$. Let $\pi_{\mathcal{A}}$ and $\pi_{\mathcal{M}_j}$ be the projection operators defined as follows:

$$\begin{array}{lll} \pi_{\mathcal{A}}: & \mathbf{q}^\star = (\mathbf{q}, \mathbf{q}_{\mathcal{M}_1}, ..., \mathbf{q}_{\mathcal{M}_r}) \in \mathcal{C}^\star & \mapsto \quad \mathbf{q} \in \mathcal{C}, \\ \pi_{\mathcal{M}_j}: & \mathbf{q}^\star = (\mathbf{q}, \mathbf{q}_{\mathcal{M}_1}, ..., \mathbf{q}_{\mathcal{M}_r}) \in \mathcal{C}^\star & \mapsto \quad \mathbf{q}_{\mathcal{M}_j} \in \mathcal{C}_{\mathcal{M}_j}. \end{array}$$

The subset of $\mathcal{C}^*$ where no object in $\{\mathcal{A}, \mathcal{M}_1, ..., \mathcal{M}_r\}$ intersects with any other object in $\{\mathcal{A}, \mathcal{B}_1, ..., \mathcal{B}_q, \mathcal{M}_1, ..., \mathcal{M}_r\}$ is the (open) **free** subset of $\mathcal{C}^*$. It is denoted by $\mathcal{C}^*_{free}$.

We allow any movable object to touch any obstacle and any other movable object. We also allow the robot to touch a movable object in order to grasp it. We impose the constraint that any contact between any two objects can be broken by an arbitrarily small displacement of one of the objects relative to the other[3]. This corresponds to imposing that at any time the configuration $\mathbf{q}^*$ of $\{\mathcal{A}, \mathcal{M}_1, ..., \mathcal{M}_r\}$ be in $cl(\mathcal{C}^*_{free})$.

We assume that the robot $\mathcal{A}$ can execute two types of paths that we call *transit paths* and *transfer paths*, respectively. During a transit path, $\mathcal{A}$ moves alone; it touches no object, except possibly at the endpoints of the path where it may touch a single movable object. During a transfer path, $\mathcal{A}$ and a grasped movable object $\mathcal{M}_j$ move together as one rigid object; the composite object $\mathcal{A} \cup \mathcal{M}_j$ touches no other object, except possibly at the endpoints of the path where $\mathcal{M}_j$ may touch obstacles and/or other movable objects. A *manipulation path* is an alternate sequence of transit paths and transfer paths (for various movable objects).

Formal Definitions. The notions introduced in the previous paragraph yield the following definitions:

DEFINITION 1: *An* $\mathcal{A}$**-free configuration** *is any configuration* $\mathbf{q}^* \in cl(\mathcal{C}^*_{free})$ *verifying:*

1. $\mathcal{A}(\pi_{\mathcal{A}}(\mathbf{q}^*))$ *intersects no obstacle and no movable object,*

2. $(\pi_{\mathcal{M}_1}(\mathbf{q}^*), ..., \pi_{\mathcal{M}_r}(\mathbf{q}^*)) \in$ STABLE.

DEFINITION 2: *An* $\mathcal{A} \cup \mathcal{M}_j$**-free configuration** *is any configuration* $\mathbf{q}^* \in cl(\mathcal{C}^*_{free})$ *verifying:*

1. $\mathcal{A}(\pi_{\mathcal{A}}(\mathbf{q}^*))$ *intersects no obstacle and no movable object, except* $\mathcal{M}_j$*,*

2. $\mathcal{M}_j(\pi_{\mathcal{M}_j}(\mathbf{q}^*))$ *intersects no obstacle and no movable object,*

[3]In particular, this constraint excludes any configuration where a movable object is "inserted" into another one.

3. $\pi_{\mathcal{A}}(\mathbf{q}^\star)/\pi_{\mathcal{M}_j}(\mathbf{q}^\star) \in \mathcal{G}_j$,

4. $(\pi_{\mathcal{M}_1}(\mathbf{q}^\star), ..., \pi_{\mathcal{M}_{j-1}}(\mathbf{q}^\star), \pi_{\mathcal{M}_{j+1}}(\mathbf{q}^\star), ..., \pi_{\mathcal{M}_r}(\mathbf{q}^\star)) \in \text{STABLE}_j$.

DEFINITION 3: *A* **transit path** *is a continuous map* $\tau : [0,1] \rightarrow cl(C^\star_{free})$ *such that:*

1. *when s varies in* $[0,1]$, $(\pi_{\mathcal{M}_1}(\tau(s)), ..., \pi_{\mathcal{M}_r}(\tau(s)))$ *stays constant,*

2. $\forall s \in (0,1) : \tau(s)$ *is an* $\mathcal{A}$*-free configuration,*

3. *when* $s \in \{0,1\}$, $\mathcal{A}(\pi_{\mathcal{A}}(\tau(s)))$ *touches at most one movable object.*

DEFINITION 4: *A* **transfer path** *of* $\mathcal{M}_j$ *is a continuous map* $\tau : [0,1] \rightarrow cl(C^\star_{free})$ *such that:*

1. *when s varies in* $[0,1]$, $(\pi_{\mathcal{M}_1}(\tau(s)), ..., \pi_{\mathcal{M}_{j-1}}(\tau(s)), \pi_{\mathcal{M}_{j+1}}(\tau(s)), ..., \pi_{\mathcal{M}_r}(\tau(s)))$ *and* $\pi_{\mathcal{A}}(\tau(s))/\pi_{\mathcal{M}_j}(\tau(s))$ *stay constant,*

2. $\forall s \in (0,1) : \tau(s)$ *is an* $\mathcal{A} \cup \mathcal{M}_j$*-free configuration,*

3. *when* $s \in \{0,1\}$, $\mathcal{A}(\pi_{\mathcal{A}}(\tau(s)))$ *touches no object other than* $\mathcal{M}_j$.

Every transit and transfer path lies in some submanifold of $cl(C^\star_{free})$ whose dimension is equal to the dimension m of $\mathcal{C}$.

DEFINITION 5: *A* **manipulation path** *is a finite alternate sequence* $(\tau_1, ..., \tau_{2p+1})$, *with* $p \in \{0,1,2,...\}$, *of transit paths* $\tau_1, \tau_3, ...\tau_{2p+1}$ *and transfer paths* $\tau_2, \tau_4, ..., \tau_{2p}$ *such that for every* $l \in [1, 2p]$, $\tau_l(1) = \tau_{l+1}(0)$. *At the beginning (resp. at the end) of every transfer path,* $\mathcal{A}$ *activates the GRASP (resp. UNGRASP) operation.*

Let τ_l and τ_{l+1} be two successive paths in the sequence defining a manipulation path. Let $\mathbf{q}^\star_l = \tau_l(1) = \tau_{l+1}(0)$. The above definitions entail that $\mathbf{q}^\star_l$ is a configuration in $cl(C^\star_{free})$ verifying:

1. $\mathcal{A}(\pi_{\mathcal{A}}(\mathbf{q}^\star_l))$ intersects no obstacle and no movable object except one, say $\mathcal{M}_j$,

2. $\pi_{\mathcal{A}}(\mathbf{q}^\star_l)/\pi_{\mathcal{M}_j}(\mathbf{q}^\star_l) \in \mathcal{G}_j$,

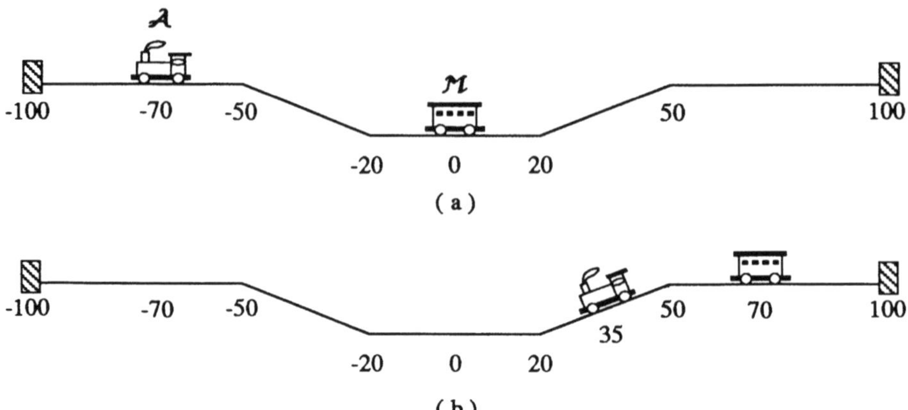

Figure 3. The robot $\mathcal{A}$ is a locomotive on a single railway connecting two symmetrical hills. There is a single movable object $\mathcal{M}$, a wagon on the same railway. Figure a shows the initial configuration of the objects and Figure b shows the goal configuration.

3. $(\pi_{\mathcal{M}_1}(\mathbf{q}_l^\star), ..., \pi_{\mathcal{M}_r}(\mathbf{q}_l^\star)) \in$ STABLE,

4. $(\pi_{\mathcal{M}_1}(\mathbf{q}_l^\star), ..., \pi_{\mathcal{M}_{j-1}}(\mathbf{q}_l^\star), \pi_{\mathcal{M}_{j+1}}(\mathbf{q}_l^\star), ..., \pi_{\mathcal{M}_r}(\mathbf{q}_l^\star)) \in \mathrm{STABLE}_j$.

A manipulation path may contain several transfer paths of the same object $\mathcal{M}_j$.

Planning with Movable Objects. A motion planning problem with movable objects, or manipulation planning problem, is specified by an initial $\mathcal{A}$-free configuration $\mathbf{q}_{init}^\star$ and a goal $\mathcal{A}$-free configuration $\mathbf{q}_{goal}^\star$. A solution to this problem is any manipulation path $(\tau_1, ..., \tau_{2p+1})$ with $\tau_1(0) = \mathbf{q}_{init}^\star$ and $\tau_{2p+1}(1) = \mathbf{q}_{goal}^\star$.

Often, the goal configuration is not uniquely defined. Instead, a subset $\mathcal{Q}_{goal}^\star$ of $\mathcal{A}$-free configurations is specified as the goal region. Then $\tau_{2p+1}(1)$ is only required to be a configuration in $\mathcal{Q}_{goal}^\star$.

2.2 Simple Example

We illustrate the above formulation with a very simple example drawn from [Alami, Siméon and Laumond, 1989]. This example is depicted in Figure 3.

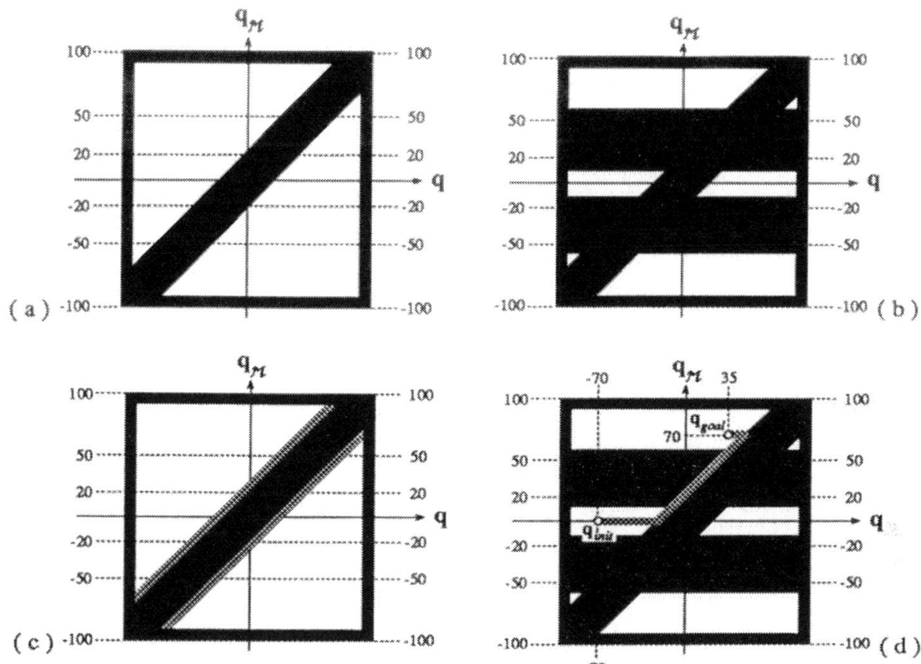

Figure 4. This figure depicts various spaces related to the problem illustrated in Figure 3. Figure a shows the free subset $\mathcal{C}^*_{free}$ of the composite configuration space $\mathcal{C}^* = \mathcal{C} \times \mathcal{C}_{\mathcal{M}}$; it consists of two open triangular regions (white regions in the figure). Figure b shows the subset of $cl(\mathcal{C}^*_{free})$ where the wagon $\mathcal{M}$ is stable; it consists of six closed regions (white regions). Figure c shows the subset of $cl(\mathcal{C}^*_{free})$ where the locomotive $\mathcal{A}$ can grasp (i.e. hook up to) $\mathcal{M}$; it consists of two line segments (thick grey lines). Figure d shows a manipulation path (thick grey lines) connecting the initial to the goal configuration.

The robot $\mathcal{A}$ is a locomotive moving along a single railway connecting two symmetrical hills. Hence, the configuration space $\mathcal{C}$ of $\mathcal{A}$ is an interval in $\mathbf{R}$. There is a single movable object $\mathcal{M}$, a wagon rolling along the same track as $\mathcal{A}$. Hence, the configuration space $\mathcal{C}_{\mathcal{M}}$ of $\mathcal{M}$ is also an interval in $\mathbf{R}$. There is no obstacle.

The reference points $O_{\mathcal{A}}$ and $O_{\mathcal{M}}$ are chosen as the center points of $\mathcal{A}$ and $\mathcal{M}$, respectively. We represent the configuration $\mathbf{q}$ (resp. $\mathbf{q}_{\mathcal{M}}$) of $\mathcal{A}$ (resp. $\mathcal{M}$) by the curvilinear abscissa of $O_{\mathcal{A}}$ (resp. $O_{\mathcal{M}}$) along the track, with the point of the track half-way between the two hills as the abscissae origin. Assuming that $\mathcal{A}$ and $\mathcal{M}$ have the same length, the composite configuration space $\mathcal{C}^{\star} = \mathcal{C} \times \mathcal{C}_{\mathcal{M}}$ is a square in $\mathbf{R}^2$ centered at $(0,0)$. The free subset $\mathcal{C}^{\star}_{free}$ of $\mathcal{C}^{\star}$ consists of two open triangular regions shown in Figure 4.a (white regions).

When it is not grasped, the wagon $\mathcal{M}$ is stable only if its wheels lie completely in any of the three horizontal portions of the railway. Figure 4.b shows the subset of $cl(\mathcal{C}^{\star}_{free})$ where $\mathcal{M}$ is at a stable configuration. This subset consists of six closed regions contained in three horizontal strips.

In order to grasp $\mathcal{M}$, $\mathcal{A}$ has to move to a configuration where it touches $\mathcal{M}$, either on its right-hand side, or its left-hand side. Hence, the grasp set $\mathcal{G}$ of $\mathcal{M}$ consists of two elements defined by:

$$\mathbf{q}/\mathbf{q}_{\mathcal{M}} \in \mathcal{G} \Leftrightarrow \|\mathbf{q} - \mathbf{q}_{\mathcal{M}}\| = L$$

where L denotes the common length of $\mathcal{A}$ and $\mathcal{M}$. Figure 4.c shows the subset of $cl(\mathcal{C}^{\star}_{free})$ where $\mathcal{A}$ can grasp $\mathcal{M}$. It consists of two segments shown as thick grey lines.

Let the initial and goal configurations be those depicted in Figure 3. A manipulation path between these two configurations is shown in Figure 4.d. It consists of a transit path at the end of which $\mathcal{A}$ grasps $\mathcal{M}$, followed by a transfer path during which $\mathcal{A}$ moves $\mathcal{M}$ to its final configuration (at this configuration $\mathcal{A}$ ungrasps $\mathcal{M}$), followed by a transit path at the end of which $\mathcal{A}$ attains its own goal configuration.

2.3 Extensions

The formulation of the manipulation planning problem given in Subsection 2.1 rests on several simplifications and can be extended in many ways. For example, we may replace the rigid object $\mathcal{A}$ by an articulated robot and/or we may introduce several robots in $\mathcal{W}$. We may also combine stationary and moving obstacles. In principle, such extensions could be handled using developments presented in Chapter 8 with perhaps some additional difficulties (for example, some robots could maintain an unstable assembly of objects temporarily stable, while other robots are bringing new objects that will make the completed assembly stable).

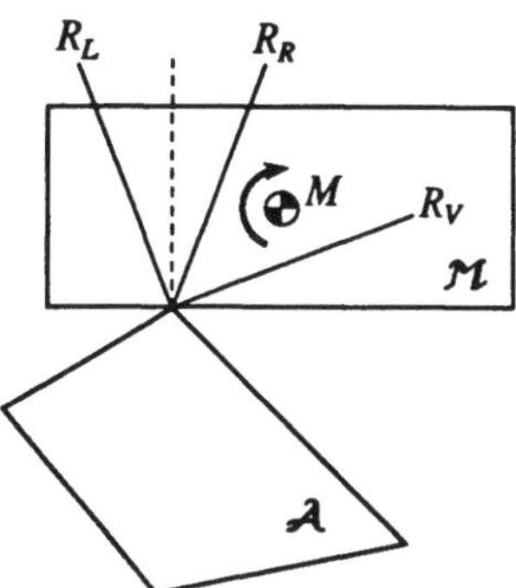

Figure 5. The robot $\mathcal{A}$ pushes the movable object $\mathcal{M}$. R_L and R_R are the edges of the friction cone at the contact point. R_V is the ray directed along the linear velocity of $\mathcal{A}$. The positions of these three rays relative to the center of mass M of $\mathcal{M}$ vote on the direction of rotation of $\mathcal{M}$. In the figure's case, $\mathcal{M}$ would rotate clockwise (with $\mathcal{A}$ sliding in contact with $\mathcal{M}$) [Mason, 1986].

Another extension consists of allowing arrangements of movable objects such that the contacts between these objects cannot be broken by arbitrarily small displacements. This would allow movable objects to be inserted into one another, which is a typical situation in assembly products. But this extension raises the difficulty that the robot's paths have to be planned in spaces of different dimensions and it is not easy to determine when it is appropriate to quit one space for another. This issue has been separately investigated to some extent in [Valade, 1985] [Koutsou, 1986] [Laugier and Théveneau, 1986] and [Desai, 1988]. The proposed approach is to represent the robot's contact space as a collection of connected subsets (submanifolds) of various dimensions together with the adjacency relation between these subsets. Planning consists of searching the graph associated with this relation, and constructing paths in the traversed subsets. Graph searching typically works backward and generates a sequence of adjacent subsets of increasing dimensions (since every increase in dimension corresponds to breaking a contact). However, this approach ignores uncertainty, and its combination with preimage backchaining (see Chapter 10) would be more realistic.

An interesting extension is to allow the robot $\mathcal{A}$ to move a movable object $\mathcal{M}$ without having to grasp it. For example, $\mathcal{M}$ could lie on a horizontal plane P and $\mathcal{A}$ could push $\mathcal{M}$ through a contact point. The instantaneous motion of $\mathcal{M}$ then results from the combined effect of the force applied by $\mathcal{A}$, the frictional forces between $\mathcal{A}$ and $\mathcal{M}$ and between $\mathcal{M}$ and the plane P, and the inertial forces. The mechanics of pushing has been investigated in [Mason, 1982 and 1986]. Under the assumptions that there is a single contact point between $\mathcal{A}$ and $\mathcal{M}$ (a vertex of $\mathcal{A}$ against a face of $\mathcal{M}$) and that the frictional forces dominate the inertial forces, Mason showed the following result. Draw three rays in P from the projection of the contact point into P: the two edges of the friction cone and a ray parallel to the linear velocity of $\mathcal{A}$. The positions of these rays relative to the projection of the center of mass of $\mathcal{M}$ into P vote on the direction of rotation of $\mathcal{M}$, with ties resulting in pure translation. For instance, if two rays pass to the left of the projection of the center of mass, then $\mathcal{M}$ rotates clockwise (Figure 5). Pushing has also been investigated in [Peshkin, 1986] and [Brost, 1988].

Another way for $\mathcal{A}$ to move an object $\mathcal{M}$ that rests on a plane P is to tilt the plane so that gravity makes $\mathcal{M}$ slide. Erdmann and Mason developed a planner that generates sequences of tray-tilting operations for orienting an Allen wrench in a tray [Erdmann and Mason, 1986].

Both pushing and tilting operations cause motions of $\mathcal{M}$ which are difficult to model with precision because various parameters (especially friction and the distribution of the support forces over the plane P) are imperfectly known, if at all. This imprecision suggests using a planning approach such as preimage backchaining (see [Erdmann and Mason, 1986]). A different approach is proposed in [Taylor, Mason and Goldberg, 1987]. Each operation is modeled with multiple possible outputs (e.g. tilting the tray yields different possible orientations of the Allen wrench). Planning consists of searching an AND/OR graph [Nilsson, 1980] whose nodes represent states of $\mathcal{M}$ (e.g. an orientation of the Allen wrench against an edge of the tray), with each AND-arc representing the various possible outputs of an operation and each OR-arc representing the various operations applicable to a state.

A movable object can also be moved within a robot's gripper, if this gripper is a dexterous hand made of multiple articulated fingers (Figure 6). Motions of the movable object can be achieved through various coordinated motions of the fingers, including rolling motions, sliding motions and finger relocation [Fearing, 1986]. A comprehensive formulation of the motion planning problem for an object held by a dexterous hand has

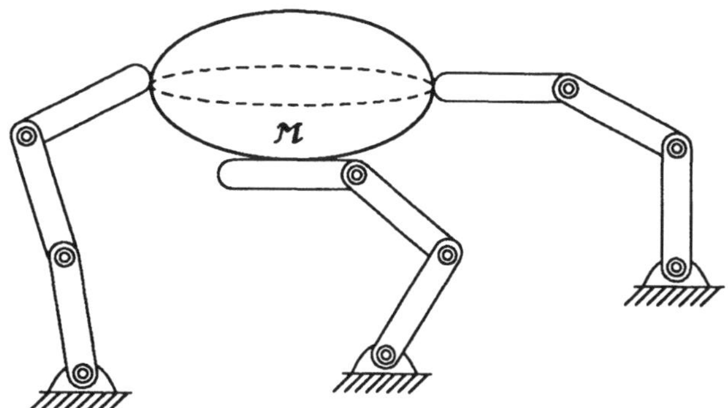

Figure 6. A dexterous hand consists of several multi-joint fingers attached to a "palm" (not shown in the figure). An object $\mathcal{M}$ can be moved within the hand through various motions of the fingers, e.g. rolling motions, sliding motions and finger relocation.

been given in [Li, Canny and Sastry, 1989]

3 General Approach

We present below a theoretical solution of the manipulation planning problem as stated in Subsection 2.1. This solution was originally proposed in [Laumond and Alami, 1988].

Let STABLE* be the subset of $cl(\mathcal{C}^\star_{free})$ where all the movable objects are in stable configurations and the robot $\mathcal{A}$ either intersects no obstacle or movable object, or touches a single movable object at a grasp configuration. STABLE* contains in particular all the $\mathcal{A}$-free configurations.

Let GRASP$^\star_j$, $j \in [1, r]$, be the subset of $cl(\mathcal{C}^\star_{free})$ where $\mathcal{A}$ grasps $\mathcal{M}_j$ while intersecting no other movable object and no obstacle, and the movable objects other than $\mathcal{M}_j$ are in stable configurations. GRASP$^\star_j$ contains in particular all the $\mathcal{A} \cup \mathcal{M}_j$-free configurations.

Let GRASP$^\star = \bigcup_{j=1}^{r}$ GRASP$^\star_j$.

Any manipulation path τ is a path in STABLE$^\star \cup$ GRASP*. However, any path in STABLE$^\star \cup$ GRASP* between the initial configuration $\mathbf{q}^\star_{init}$ and the goal configuration $\mathbf{q}^\star_{goal}$ is not necessarily a manipulation path. Indeed, a manipulation path consists of successive transit and transfer paths which, in particular, must satisfy the following constraints:

- A transit path lies in STABLE* and leaves the configurations of the movable objects unchanged.

- A transfer path lies in $\text{GRASP}_j^\star$, for some $j \in [1, r]$, and leaves both the configuration of $\mathcal{A}$ relative to $\mathcal{M}_j$ and the configurations of the movable objects $\mathcal{M}_k$, $k \neq j$, unchanged.

Each transit or transfer path starts and ends in STABLE* ∩ GRASP*.

Case of a Single Movable Object. Let us first assume that there is a single movable object, i.e. $r = 1$. Then, it can be shown that if there exists a path τ in STABLE* ∩ GRASP* between two configurations, this path can be transformed into a manipulation path between the same two configurations; the manipulation path can be as close as one wants to τ. The time required for computing the manipulation path is finite, though it depends on the geometry of the workspace and the objects.

This property, called the *reduction property* [Laumond and Alami, 1988], is critical to the rest of the method. Indeed, it allows us to reduce the original manipulation planning problem to the following three subproblems:

1. Analyze the connectivity of STABLE* ∩ GRASP* and determine its connected components.

2. Analyze the connectivity of the connected components of STABLE* ∩ GRASP* by transit paths in STABLE*.

3. Analyze the connectivity of the connected components of STABLE* ∩ GRASP* by transfer paths in GRASP*.

Assume that $\mathcal{C}_{free}^\star$, STABLE* and GRASP* are described as semi-algebraic sets. Let $\mathcal{P}$ be the collection of all the polynomials defining these sets. The connected components of STABLE* ∩ GRASP* can be computed from any $\mathcal{P}$-invariant Collins decomposition of $\mathcal{C}^\star$.

On the other hand, the connectivity of two components of STABLE* ∩ GRASP* by a transit or a transfer path can be determined from appropriate $\mathcal{P}$-invariant Collins decompositions of $\mathcal{C}^\star$. These decompositions simply require that the list of the configuration parameters be ordered in such a way that the parameters which remain constant along the path to be constructed appear first, so that the path can be searched in a cylinder where the other parameters can vary freely. The parameters which remain constant are the parameters of $\mathbf{q}_{\mathcal{M}_1}$ during a transit path and the parameters of $\mathbf{q}/\mathbf{q}_{\mathcal{M}_1}$ during a transfer path.

A **manipulation graph** *MG* is then constructed as follows:

- *MG*'s nodes are the connected components of STABLE* ∩ GRASP*, the initial configuration $\mathbf{q}^*_{init}$ and the goal configuration[4] $\mathbf{q}^*_{goal}$.
- Two nodes N_1 and N_2 of *MG* are connected by a link if and only if there exists a transit path or a transfer path which connects a configuration in the component of STABLE* ∩ GRASP* corresponding to N_1 to a configuration in the component corresponding to N_2.

One can show that there exists a manipulation path between two $\mathcal{A}$-free configurations $\mathbf{q}^*_{init}$ and $\mathbf{q}^*_{goal}$ of STABLE* if and only if there exists a path between the two corresponding nodes in *MG*.

Case of Several Movable Objects. The construction of *MG* can be extended to the case where there are several movable objects. Then we consider every set STABLE* ∩ GRASP^*_j, $j \in [1, r]$, separately. If there exists a path τ in a slice of STABLE* ∩ GRASP^*_j where every object $\mathcal{M}_k$, $k \neq j$, remains stationary, then τ can be transformed into a manipulation path connecting the same two configurations (indeed, the movable objects $\mathcal{M}_k$ simply act as stationary obstacles). Thus we can construct a manipulation graph MG_j for $\mathcal{M}_j$.

Assume that C^*_{free}, STABLE* and GRASP^*_j, for $j = 1$ to r, are described as semi-algebraic sets. Let $\mathcal{P}$ be the collection of all the polynomials defining these sets. We compute the $\mathcal{P}$-invariant Collins decomposition of C^*, with the configuration parameters of $\mathbf{q}_{\mathcal{M}_k}$, $k \neq j$, occurring before those of $\mathbf{q}_{\mathcal{M}_j}$ and $\mathbf{q}$. This decomposition yields a collection of cells in $\mathcal{C}_{\mathcal{M}_1} \times \ldots \times \mathcal{C}_{\mathcal{M}_{j-1}} \times \mathcal{C}_{\mathcal{M}_{j+1}} \times \ldots \times \mathcal{C}_{\mathcal{M}_r}$ such that over each cell κ in this collection the manipulation graph of $\mathcal{M}_j$ — let us call it MG^κ_j — remains unchanged. For every such cell κ, we construct MG^κ_j.

[4] Or the connected components of the goal region $\mathcal{Q}^*_{goal}$.

The entire manipulation graph MG is obtained by taking the union of all the MG_j^κ, for all $j \in [1, r]$. In addition, we connect every node N_1 of every $MG_{j_1}^\kappa$ to every node N_2 of every $MG_{j_2}^\kappa$, $j_1 \neq j_2$, if there exists a transit path between a configuration in the subset of STABLE$^\star \cap$GRASP$_{j_1}^\star$ corresponding to N_1 and a configuration in the subset of STABLE$^\star \cap$GRASP$_{j_2}^\star$ corresponding to N_2. The existence of such a path is determined from the $\mathcal{Q}$-invariant Collins decomposition of $\mathcal{C}^\star$, where $\mathcal{Q}$ is the collection of all the polynomials defining $\mathcal{C}_{free}^\star$ and all the subsets corresponding to the nodes of the graphs MG_j^κ. This decomposition requires the configuration parameters to be ordered so that the parameters of $\mathbf{q}_{\mathcal{M}_1}, \ldots, \mathbf{q}_{\mathcal{M}_r}$ appear before those of $\mathbf{q}$.

Again, one can show that there exists a manipulation path between any two $\mathcal{A}$-free configurations $\mathbf{q}_{init}^\star$ and $\mathbf{q}_{goal}^\star$ of STABLE* if and only if there exists a path between the two corresponding nodes of MG.

Therefore, assuming that all the objects in $\mathcal{W}$ are given as semi-algebraic sets and that both STABLE* and every GRASP$_j^\star$ ($j \in [1, r]$) are semi-algebraic sets, the problem of deciding whether the manipulation planning problem stated in Subsection 2.1 has a solution, or not, can be solved in polynomial time if the number r of movable objects is fixed. The time required to solve the planning problem itself is more difficult to analyze. This results from the fact that the search of MG may produce a path including subpaths lying in STABLE$^\star \cap$ GRASP*. The time required to transform this path into a manipulation path, though finite, depends on the shape of the workspace and, hence, is not easily bounded.

The extension of the above method to the case where $\mathcal{A}$ is a manipulator arm is described in [Laumond and Alami, 1988].

4 Specific Algorithms

In this section we present specific algorithms addressing simpler forms of the problem stated in Subsection 2.1. In Subsection 4.1 we consider the case where each movable object has a finite number of grasps and a finite number of stable configurations. In Subsections 4.2 and 4.3 we consider the case of a single movable object in a polygonal workspace. In Subsection 4.2 both the robot and the movable object are discs. In Subsection 4.3 both the robot and the movable object are convex polygons.

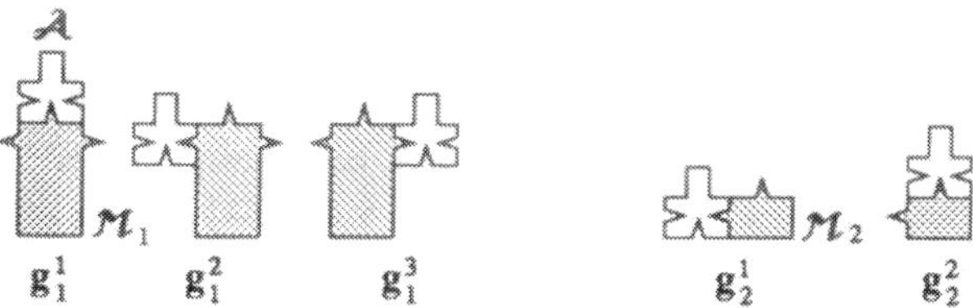

Figure 7. The robot $\mathcal{A}$ can grasp a movable object ($\mathcal{M}_1$ or $\mathcal{M}_2$) when a small triangular cavity of $\mathcal{A}$ matches a triangular protrusion of the movable object. This yields two finite grasp sets, $\mathcal{G}_1$ and $\mathcal{G}_2$. $\mathcal{G}_1 = \{\mathbf{g}_1^1, \mathbf{g}_1^2, \mathbf{g}_1^3\}$ and $\mathcal{G}_2 = \{\mathbf{g}_2^1, \mathbf{g}_2^2\}$.

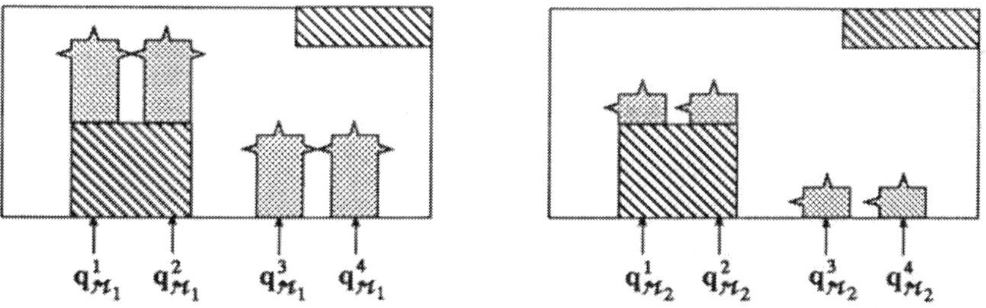

Figure 8. Each movable object $\mathcal{M}_j$ has four stable configurations $\mathbf{q}_{\mathcal{M}_j}^1$ through $\mathbf{q}_{\mathcal{M}_j}^4$. Hence, the set STABLE consists of all the pairs $(\mathbf{q}_{\mathcal{M}_1}^a, \mathbf{q}_{\mathcal{M}_2}^b)$, with $a \in [1,4]$, $b \in [1,4]$, and $a \neq b$.

4.1 Finite Sets of Grasps and Stable Configurations

Consider again the example of Section 1. In this example we restrict the possible grasps of the movable objects to finite sets, $\mathcal{G}_1$ and $\mathcal{G}_2$, as shown in Figure 7. We have:

$$\mathcal{G}_1 = \{\mathbf{g}_1^1, \mathbf{g}_1^2, \mathbf{g}_1^3\} \quad \text{and} \quad \mathcal{G}_2 = \{\mathbf{g}_2^1, \mathbf{g}_2^2\}.$$

We also restrict the set of stable configurations of the movable objects to finite sets illustrated in Figure 8. We have:

- $\text{STABLE}_1 = \{\mathbf{q}_{\mathcal{M}_1}^1, \ldots, \mathbf{q}_{\mathcal{M}_1}^4\}$,
- $\text{STABLE}_2 = \{\mathbf{q}_{\mathcal{M}_2}^1, \ldots, \mathbf{q}_{\mathcal{M}_2}^4\}$,
- $\text{STABLE} = (\text{STABLE}_1 \times \text{STABLE}_2) \setminus \{(\mathbf{q}_{\mathcal{M}_1}^j, \mathbf{q}_{\mathcal{M}_2}^j)\}_{j=1,2,3,4}$.

With these restrictions, the set STABLE$^\star \cap$ GRASP* consists of a finite number of configurations in $\mathcal{C}^\star$. These configurations, together with the initial and the goal configurations $\mathbf{q}^\star_{init}$ and $\mathbf{q}^\star_{goal}$ are the nodes of the manipulation graph *MG* defined in Section 3. Two nodes of *MG* corresponding to two configurations $\mathbf{q}^\star_1$ and $\mathbf{q}^\star_2$ are connected by a link if and only if either one of the following two conditions is verified:

- For $k = 1$ and 2, $\pi_{\mathcal{M}_k}(\mathbf{q}^\star_1) = \pi_{\mathcal{M}_k}(\mathbf{q}^\star_2)$ and there exists a transit path from $\mathbf{q}^\star_1$ to $\mathbf{q}^\star_2$.

- For $j = 1$ or 2, $\pi_{\mathcal{A}}(\mathbf{q}^\star_1)/\pi_{\mathcal{M}_j}(\mathbf{q}^\star_1) \in \mathcal{G}_j$ and $\pi_{\mathcal{A}}(\mathbf{q}^\star_1)/\pi_{\mathcal{M}_j}(\mathbf{q}^\star_1) = \pi_{\mathcal{A}}(\mathbf{q}^\star_2)/\pi_{\mathcal{M}_j}(\mathbf{q}^\star_2)$, and there exists a transfer path of $\mathcal{M}_j$ from $\mathbf{q}^\star_1$ to $\mathbf{q}^\star_2$.

Figure 9 shows a subset of the manipulation graph. The links in dashed lines correspond to transit paths and those in plain lines correspond to transfer paths. Checking the existence of a link between two nodes of *MG* can be done using a standard path planning method in a two-dimensional configuration space. During a transit path, $\mathcal{M}_1$ and $\mathcal{M}_2$ are treated as stationary obstacles. During a transfer path, say of $\mathcal{M}_1$, $\mathcal{A} \cup \mathcal{M}_1$ is treated as the rigid moving object and $\mathcal{M}_2$ is treated as a stationary obstacle. Each path is therefore computed in an appropriate two-dimensional slice of $\mathcal{C}^\star$. Three such slices are associated with the corresponding links of *MG* in Figure 9. Notice that several transit or transfer paths may be constructed in the same slice.

The graph *MG* can be constructed incrementally while it is searched. A simple cost function combining the length of the path of $\mathcal{A}$ generated so far and the distance between the current configuration of every object and its goal configuration can be used to guide the search.

4.2 Discs in Polygonal Workspace

In this subsection we consider the case of two discs (the robot and a movable object) in a polygonal environment. We describe a planning method originally proposed in [Laumond and Alami, 1989]. It is based on an exact cell decomposition method given in [Schwartz and Sharir, 1983c] for planning the coordinated paths of two independent circular robots. A manipulation graph is constructed from this decomposition.

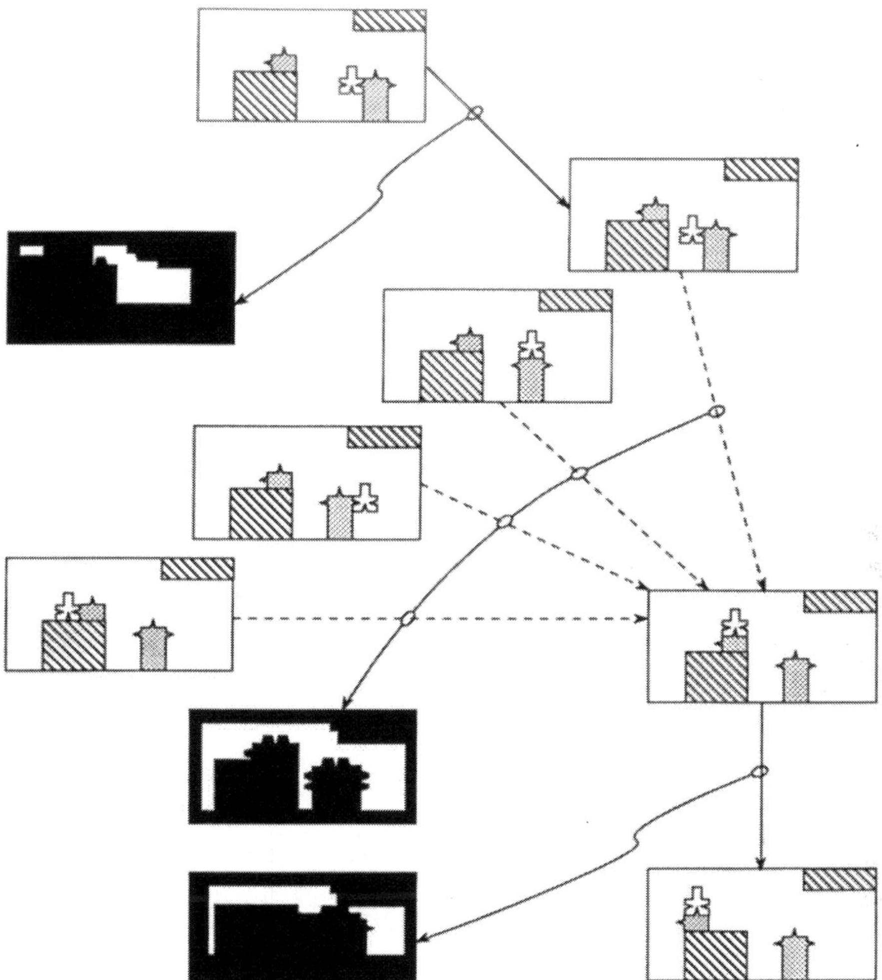

Figure 9. This figure displays a small subset of the manipulation graph MG for the problem of Section 1. Links in dashed lines correspond to transit paths. Links in plain lines correspond to transfer paths. Each path is computed in a two-dimensional slice of $\mathcal{C}^{\star}$. The corresponding slice of $\mathcal{C}^{\star}$ is associated with every link in the displayed subset of MG.

Problem. We consider the case of a circular robot $\mathcal{A}$ (a disc) translating in $\mathcal{W} = \mathbf{R}^2$. The obstacles form a regular polygonal region $\mathcal{B}$. There is a single movable object, a disc $\mathcal{M}$. The radii of $\mathcal{A}$ and $\mathcal{M}$ are $r_{\mathcal{A}}$ and $r_{\mathcal{M}}$, respectively.

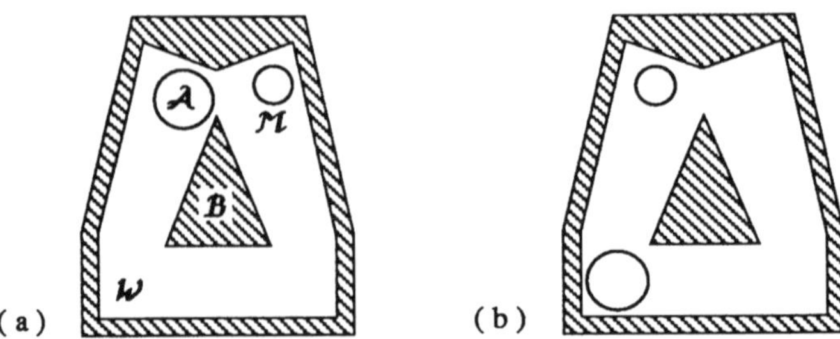

Figure 10. The robot $\mathcal{A}$ is a disc in a polygonal workspace $\mathcal{W}$. There is a single movable object, the disc $\mathcal{M}$. The initial configuration $\mathbf{q}^{\star}_{init}$ is shown in Figure a. The goal configuration $\mathbf{q}^{\star}_{goal}$ is shown in Figure b.

$\mathcal{A}$ can grasp $\mathcal{M}$ by touching it and activating the GRASP operation. Hence, the number of grasps is infinite. When $\mathcal{A}$ has grasped $\mathcal{M}$, the two objects move as a single one, $\mathcal{A} \cup \mathcal{M}$, with a fixed orientation in $\mathcal{W}$. We assume that there is no gravitational force (i.e. $\mathcal{W}$ is horizontal), so that every configuration of $\mathcal{M}$ is stable. We require that $\mathcal{A}$ never intersect the obstacle region $\mathcal{B}$. $\mathcal{M}$ may touch $\mathcal{B}$ in such a way that the contact can be broken by an arbitrarily small displacement of $\mathcal{M}$.

Figure 10 illustrates a particular instance of a planning problem under the above conditions. The initial and goal configurations are depicted in Figures 10.a and 10.b, respectively. A solution to this problem is shown in Figure 11. Figures 11.a, 11.c, and 11.e show transit motions. Figures 11.b and 11.d correspond to sequences of alternate small transit and transfer paths.

Notations. We represent a configuration $\mathbf{q}$ of $\mathcal{A}$ by the coordinates (x, y) of the center of $\mathcal{A}$ in $\mathcal{F}_{\mathcal{W}}$, a configuration $\mathbf{q}_{\mathcal{M}}$ of $\mathcal{M}$ by the coordinates $(x_{\mathcal{M}}, y_{\mathcal{M}})$ of the center of $\mathcal{M}$ in $\mathcal{F}_{\mathcal{W}}$, and a configuration $\mathbf{q}^{\star} = (\mathbf{q}, \mathbf{q}_{\mathcal{M}})$ by $(x, y, x_{\mathcal{M}}, y_{\mathcal{M}})$. Hence, $\mathcal{C} = \mathbf{R}^2$, $\mathcal{C}_{\mathcal{M}} = \mathbf{R}^2$ and $\mathcal{C}^{\star} = \mathbf{R}^4$. The initial configuration is $\mathbf{q}^{\star}_{init} = (\mathbf{q}_{init}, \mathbf{q}_{\mathcal{M},init})$ and the goal one is $\mathbf{q}^{\star}_{goal} = (\mathbf{q}_{goal}, \mathbf{q}_{\mathcal{M},goal})$.

Let $\mathcal{C}_{free}$ denote the open subset of $\mathcal{C}$ where $\mathcal{A}$ does not intersect $\mathcal{B}$. $\mathcal{C}_{free}$ is obtained by growing $\mathcal{B}$ isotropically by $\mathcal{A}$'s radius $r_{\mathcal{A}}$ (see Figure 6 in Chapter 2).

There is no gravitational force applying to $\mathcal{M}$ so that STABLE is the

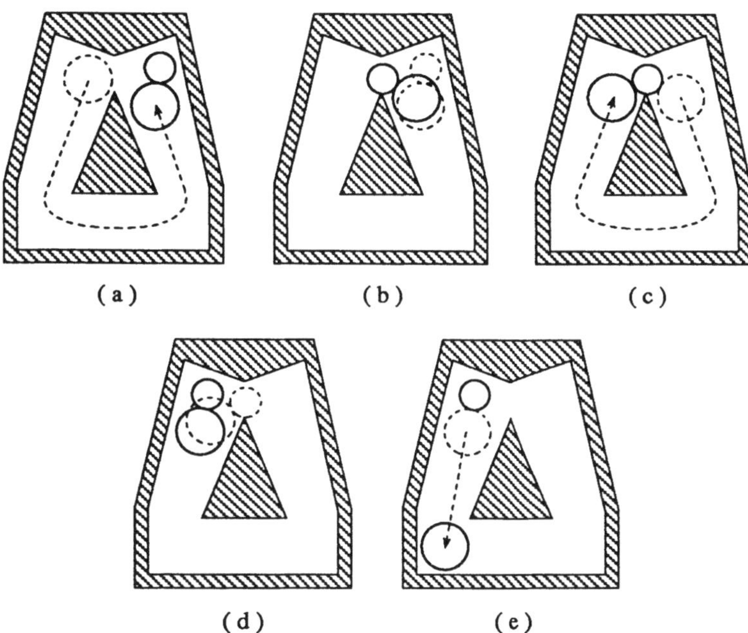

Figure 11. This series of diagrams illustrates a solution for the problem of Figure 10. $\mathcal{A}$ moves around the obstacle in the middle of the workspace in order to grasp $\mathcal{M}$ (Figure a). It moves $\mathcal{M}$ through the narrow passage (which it cannot traverse itself) and ungrasps it (Figure b). It returns to the other side of the narrow passage, regrasps $\mathcal{M}$ (Figure c), moves it to its goal configuration (Figure d) and ungrasps it. $\mathcal{A}$ finally moves to its goal configuration (Figure e).

closed subset[5] of $\mathcal{C}_{\mathcal{M}}$ where $int(\mathcal{M})$ does not intersect $int(\mathcal{B})$. STABLE is obtained by growing $\mathcal{B}$ isotropically by $\mathcal{M}$'s radius $r_{\mathcal{M}}$. If the obtained set is not regular, we "regularize" it by taking the closure of its interior.

For any $\mathbf{q}_{\mathcal{M}} \in$ STABLE, let $\mathcal{C}_{free}(\mathbf{q}_{\mathcal{M}})$ designate the subset of $\mathcal{C}_{free}$ where $\mathcal{A}$ does not intersect $\mathcal{M}(\mathbf{q}_{\mathcal{M}})$, i.e. :

$$\mathcal{C}_{free}(\mathbf{q}_{\mathcal{M}}) = \{\mathbf{q} \in \mathcal{C}_{free} \ / \ \mathcal{A}(\mathbf{q}) \cap \mathcal{M}(\mathbf{q}_{\mathcal{M}}) = \emptyset\}.$$

$\mathcal{C}_{free}(\mathbf{q}_{\mathcal{M}})$ is the complement in $\mathcal{C}_{free}$ of the closed disc of radius $r_{\mathcal{A}}+r_{\mathcal{M}}$ centered at $\mathbf{q}_{\mathcal{M}}$. The circle bounding this disc is denoted by $\rho(\mathbf{q}_{\mathcal{M}})$.

[5]Following the terminology of Subsection 2.1, we keep calling this subset STABLE, though this name is no longer very relevant here.

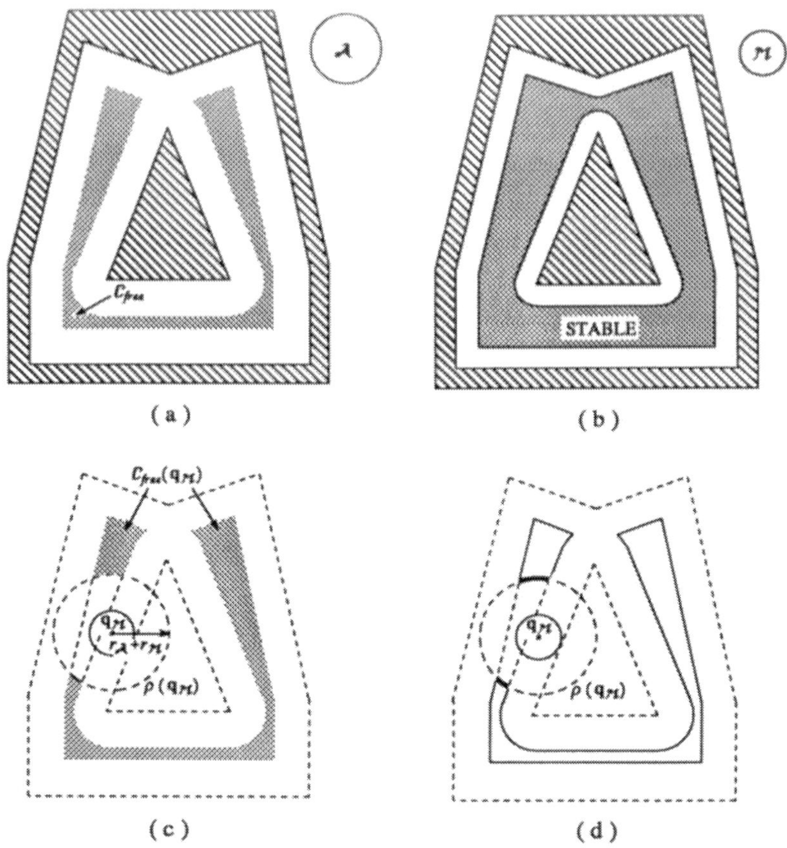

Figure 12. Figures a, b, c and d show the sets $\mathcal{C}_{free}$, STABLE, $\mathcal{C}_{free}(\mathbf{q}_{\mathcal{M}})$ and $\mathcal{C}_{grasp}(\mathbf{q}_{\mathcal{M}})$, respectively, for the example of Figure 10.

Let $\mathcal{C}_{grasp}(\mathbf{q}_{\mathcal{M}})$ designate the subset of $\mathcal{C}_{free}$ where $\mathcal{A}$ and $\mathcal{M}(\mathbf{q}_{\mathcal{M}})$ are in contact, i.e. $\mathcal{C}_{grasp}(\mathbf{q}_{\mathcal{M}}) = \mathcal{C}_{free} \cap \rho(\mathbf{q}_{\mathcal{M}})$. $\mathcal{C}_{grasp}(\mathbf{q}_{\mathcal{M}})$ is the (possibly empty) union of a finite number of disjoint open circular intervals.

Figure 12 displays the sets $\mathcal{C}_{free}$, STABLE, $\mathcal{C}_{free}(\mathbf{q}_{\mathcal{M}})$ and $\mathcal{C}_{grasp}(\mathbf{q}_{\mathcal{M}})$, the last two being constructed for a particular $\mathbf{q}_{\mathcal{M}} \in$ STABLE, for the example of Figure 10. $\mathcal{C}_{free}$ consists of a single (simply-)connected component. STABLE consists of a single (multiply-)connected component. Both $\mathcal{C}_{free}(\mathbf{q}_{\mathcal{M}})$ and $\mathcal{C}_{grasp}(\mathbf{q}_{\mathcal{M}})$ have two connected components.

We have:

- STABLE* = $\{(\mathbf{q}, \mathbf{q}_{\mathcal{M}}) \mathbin{/} \mathbf{q}_{\mathcal{M}} \in$ STABLE and $\mathbf{q} \in \mathcal{C}_{free}(\mathbf{q}_{\mathcal{M}}) \cup \mathcal{C}_{grasp}(\mathbf{q}_{\mathcal{M}})\}$,
- GRASP* = $\{(\mathbf{q}, \mathbf{q}_{\mathcal{M}}) \mathbin{/} \mathbf{q}_{\mathcal{M}} \in$ STABLE and $\mathbf{q} \in \mathcal{C}_{grasp}(\mathbf{q}_{\mathcal{M}})\}$.

Hence, GRASP* $\subset$ STABLE* and STABLE* $\cup$ GRASP* = STABLE*.

GRASP* lies in a three-dimensional submanifold of $\mathbf{R}^4$ defined by:

$$(x - x_{\mathcal{M}})^2 + (y - y_{\mathcal{M}})^2 = (r_{\mathcal{A}} + r_{\mathcal{M}})^2.$$

The general approach of Section 3 leads us to construct the manipulation graph by (1) determining the connected components of GRASP*, and (2) analyzing their connectivity by transit paths[6]. To that purpose, we make use of a more specific decomposition of STABLE* into cells than the Collins decomposition. This decomposition, which is presented in the following paragraphs, consists of first decomposing the two-dimensional set STABLE $\subset \mathcal{C}_{\mathcal{M}}$ into two-dimensional "noncritical" regions, and then lifting these regions into four-dimensional cells of STABLE*. The decomposition of STABLE $\subset \mathcal{C}_{\mathcal{M}}$ is based on a symbolic labeling of the edges bounding $\mathcal{C}_{free}(\mathbf{q}_{\mathcal{M}})$.

Labeling. We label every edge and every convex vertex of $\mathcal{B}$ by a distinct symbol (say, a letter). The boundary of $\mathcal{C}_{free}$ consists of line segments and circular arcs. Each such element Z is given the label associated with the unique open edge or vertex touched by $\mathcal{A}$ when its center lies in Z (see Figure 13.a). The elements of the boundary of $\mathcal{C}_{free}(\mathbf{q}_{\mathcal{M}})$ are labeled in the same way, with the elements brought by the circle $\rho(\mathbf{q}_{\mathcal{M}})$ labeled by ρ (see Figure 13.b).

Each simply-connected component of $\mathcal{C}_{free}(\mathbf{q}_{\mathcal{M}})$ is labeled by the counterclockwise circular list of the labels attached to the elements of its boundary. For instance, in Figure 13.b there are two such components designated by K and K'; they are given the labels $(a\rho inb)$ and $(ahgfednmlkji\rho)$, respectively. A multiply-connected component K of $\mathcal{C}_{free}(\mathbf{q}_{\mathcal{M}})$ would be labeled by a set of circular lists, with each list corresponding to a connected component of K's boundary.

Each connected component of $\mathcal{C}_{grasp}(\mathbf{q}_{\mathcal{M}})$ is a circular arc in $\rho(\mathbf{q}_{\mathcal{M}})$. It is also a subset of the boundary of a unique component K of $\mathcal{C}_{free}(\mathbf{q}_{\mathcal{M}})$. It

[6]The transfer paths are necessarily contained in GRASP* $\subset$ STABLE*. We do not have to consider them in order to construct the links of the manipulation graph.

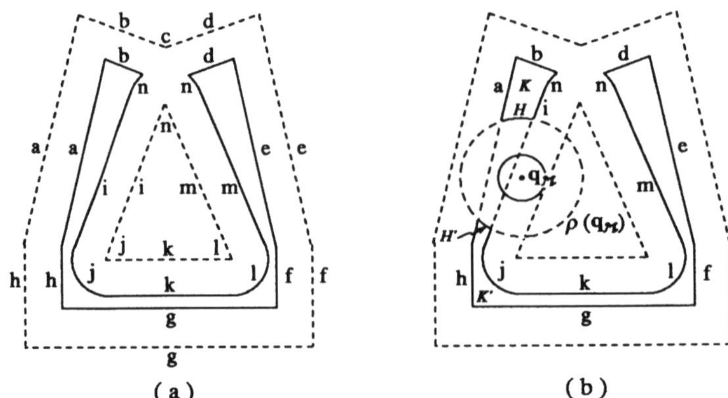

Figure 13. This figure illustrates the labeling of the elements (line segments and circular arcs) forming the boundaries of $\mathcal{C}_{free}$ and $\mathcal{C}_{free}(\mathbf{q}_{\mathcal{M}})$. $\mathcal{C}_{free}(\mathbf{q}_{\mathcal{M}})$ consists of two maximally connected regions K and K' that are labeled by $(a\rho inb)$ and $(ahgfednmlkji\rho)$, respectively. $\mathcal{C}_{grasp}(\mathbf{q}_{\mathcal{M}})$ consists of two circular arcs H and H' that are labeled by $(a\rho i)$ and $(i\rho a)$, respectively.

is labeled by a triplet that is a sublist of the label of K centered at ρ. For example, in Figure 13.b there are two such components designated by H and H' which are included in the boundaries of K and K', respectively. H is labeled by $(a\rho i)$ and H' by $(i\rho a)$.

We denote the set of labels of all the components of $\mathcal{C}_{free}(\mathbf{q}_{\mathcal{M}})$ (resp. $\mathcal{C}_{grasp}(\mathbf{q}_{\mathcal{M}})$) by $\lambda(\mathbf{q}_{\mathcal{M}})$ (resp. $\mu(\mathbf{q}_{\mathcal{M}})$). For every $L \in \lambda(\mathbf{q}_{\mathcal{M}})$, let $\varphi(\mathbf{q}_{\mathcal{M}}, L)$ be the unique component of $\mathcal{C}_{free}(\mathbf{q}_{\mathcal{M}})$ that is labeled by L. Similarly, for every $M \in \mu(\mathbf{q}_{\mathcal{M}})$, let $\psi(\mathbf{q}_{\mathcal{M}}, M)$ be the unique component of $\mathcal{C}_{grasp}(\mathbf{q}_{\mathcal{M}})$ that is labeled by M. By construction, for each label $M \in \mu(\mathbf{q}_{\mathcal{M}})$, there exists a unique label $L \in \lambda(\mathbf{q}_{\mathcal{M}})$ such that M is a sublist of L. $\psi(\mathbf{q}_{\mathcal{M}}, M)$ is a subset of the boundary of $\varphi(\mathbf{q}_{\mathcal{M}}, L)$.

Noncritical Regions of STABLE. A configuration $\mathbf{q}_{\mathcal{M}} \in$ STABLE is called **critical** if $\rho(\mathbf{q}_{\mathcal{M}})$ passes through a vertex of the boundary of $\mathcal{C}_{free}$ or is tangent to this boundary.

The set of critical configurations in STABLE consists of the following curve segments (Figure 14):

- Straight segments lying at distance $2r_{\mathcal{A}} + r_{\mathcal{M}}$ from the edges of $\mathcal{B}$ (β_1 and β_3 in Figure 14.a);

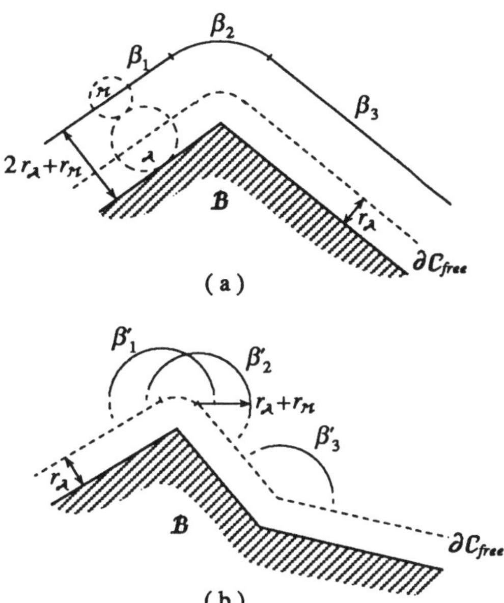

Figure 14. This figure shows the various kinds of critical curves in STABLE. ($\partial\mathcal{C}_{free}$ denotes the boundary of $\mathcal{C}_{free}$).

- Circular arcs of radius $2r_{\mathcal{A}} + r_{\mathcal{M}}$ centered at convex vertices of $\mathcal{B}$ (β_2 in Figure 14.a);
- Circular arcs of radius $r_{\mathcal{A}} + r_{\mathcal{M}}$ centered at the vertices of $\mathcal{C}_{free}$, i.e. at the various extremities of the line segments and circular arcs forming the boundary of $\mathcal{C}_{free}$ (β'_1, β'_2 and β'_3 in Figure 14.b).

These curve segments are called the **critical curves**. Their removal decomposes STABLE into a finite number of disjoint connected open regions called the **noncritical regions**. These regions verify the following proposition [Schwartz and Sharir, 1983c] [Laumond and Alami, 1988]:

PROPOSITION 1: *Let R be any noncritical region of* STABLE. *For all $\mathbf{q}_{\mathcal{M}} \in R$ the sets $\lambda(\mathbf{q}_{\mathcal{M}})$ (resp. $\mu(\mathbf{q}_{\mathcal{M}})$) are identical; for each $L \in \lambda(\mathbf{q}_{\mathcal{M}})$ (resp. $M \in \mu(\mathbf{q}_{\mathcal{M}})$), the function $\varphi(\mathbf{q}_{\mathcal{M}}, L)$ (resp. $\psi(\mathbf{q}_{\mathcal{M}}, M)$) is continuous (in the Hausdorff topology[7] of sets) when $\mathbf{q}_{\mathcal{M}}$ varies over*

[7]This topology is the metric topology induced by the Hausdorff distance between two sets (see Section 7 of Chapter 2).

R.

Therefore, for each noncritical region R of STABLE, we can write $\lambda(R) = \lambda(\mathbf{q}_{\mathcal{M}})$ and $\mu(R) = \mu(\mathbf{q}_{\mathcal{M}})$, for any $\mathbf{q}_{\mathcal{M}} \in R$.

The set of critical curves constructed as above may include "redundant" curve segments (see Subsection 2.1 of Chapter 5) whose crossing produces no change in $\lambda(\mathbf{q}_{\mathcal{M}})$ and $\mu(\mathbf{q}_{\mathcal{M}})$. These redundant segments may easily be removed once the functions λ and μ have been computed.

Cells. We now lift the noncritical region into cells of $cl(\mathcal{C}^{\star}_{free})$.

DEFINITION 6: *For every noncritical region R of* STABLE, *for every label $L \in \lambda(R)$, the region:*

$$fcell(R, L) = \bigcup_{\mathbf{q}_{\mathcal{M}} \in R} \varphi(\mathbf{q}_{\mathcal{M}}, L) \times \{\mathbf{q}_{\mathcal{M}}\}$$

is called an **f-cell** *(for "free cell"). It is an open connected subset of $\mathcal{C}^{\star}_{free}$, hence a four-dimensional set.*

Let F be the union of all the f-cells. We have that $cl(F) \cap \mathcal{C}^{\star}_{free} = \mathcal{C}^{\star}_{free}$.

Let FG be the non-directed graph defined as follows:

- The set of nodes of FG is the set of all the f-cells defined as above.
- Two nodes $fcell(R, L)$ and $fcell(R', L')$ are adjacent in FG if and only if R and R' are adjacent (i.e. their boundaries share an open portion of a critical curve) and L and L' contain a common sublist S of at least two labels if $\rho \notin S$ and three labels if $\rho \in S$.

Let $\mathbf{q}^{\star}_1 = (\mathbf{q}_1, \mathbf{q}_{\mathcal{M},1})$ and $\mathbf{q}^{\star}_2 = (\mathbf{q}_2, \mathbf{q}_{\mathcal{M},2})$ be two configurations in $\mathcal{C}^{\star}_{free}$. Let R_1 and R_2 be the two noncritical regions of STABLE which contain $\mathbf{q}_{\mathcal{M},1}$ and $\mathbf{q}_{\mathcal{M},2}$, respectively. Let L_1 and L_2 be the labels of $\lambda(R_1)$ and $\lambda(R_2)$ such that $\mathbf{q}_1 \in \varphi(\mathbf{q}_{\mathcal{M},1}, L_1)$ and $\mathbf{q}_2 \in \varphi(\mathbf{q}_{\mathcal{M},2}, L_2)$. There exists a path[8] in $\mathcal{C}^{\star}_{free}$ between $\mathbf{q}^{\star}_1$ and $\mathbf{q}^{\star}_2$ if and only if $fcell(R_1, L_1)$ and f-$cell(R_2, L_2)$ are connected by a path in the graph FG. In fact, searching

[8] This path may not be a transit, a transfer or a manipulation path.

FG would allow us to plan a free path for the two discs $\mathcal{A}$ and $\mathcal{M}$ if both of them were robots.

We now construct a representation of the connectivity of GRASP* by "projecting" the f-cells of $C^{\star}_{free}$ onto GRASP*.

DEFINITION 7: *For every noncritical region R of* STABLE, *for every label $M \in \mu(R)$, the region:*

$$ccell(R, M) = \bigcup_{\mathbf{q}_{\mathcal{M}} \in R} \psi(\mathbf{q}_{\mathcal{M}}, M) \times \{\mathbf{q}_{\mathcal{M}}\}$$

is called a **c-cell** *(for "contact cell"). It is a connected three-dimensional subset of $C^{\star}_{free}$ contained in* GRASP*.

Recall that for each $M \in \mu(R)$ there exists a unique $L \in \lambda(R)$ such that $\psi(\mathbf{q}_{\mathcal{M}}, M)$ is a subset of the boundary of $\varphi(\mathbf{q}_{\mathcal{M}}, L)$. Hence, for this same L, $ccell(R, M)$ is contained in the boundary of $fcell(R, L)$.

Let CG be the non-directed graph defined as follows:

- The set of nodes of CG is the set of all the c-cells defined as above.
- Two nodes $ccell(R, M)$ and $ccell(R', M')$ are adjacent in CG if and only if the f-cells of which they form part of the boundaries are adjacent in FG and M and M' differ by at most one label.

Let $\mathbf{q}_1^{\star} = (\mathbf{q}_1, \mathbf{q}_{\mathcal{M},1})$ and $\mathbf{q}_2^{\star} = (\mathbf{q}_2, \mathbf{q}_{\mathcal{M},2})$ be two configurations in GRASP*. Let R_1 and R_2 be the two noncritical regions of STABLE which contain $\mathbf{q}_{\mathcal{M},1}$ and $\mathbf{q}_{\mathcal{M},2}$, respectively. Let M_1 and M_2 be the labels of $\mu(R_1)$ and $\mu(R_2)$ such that $\mathbf{q}_1 \in \psi(\mathbf{q}_{\mathcal{M},1}, M_1)$ and $\mathbf{q}_2 \in \psi(\mathbf{q}_{\mathcal{M},2}, M_2)$. There exists a path in GRASP* between $\mathbf{q}_1^{\star}$ and $\mathbf{q}_2^{\star}$ if and only if $ccell(R_1, M_1)$ and $ccell(R_2, M_2)$ are connected by a path in the graph CG.

The connectivity of the connected components of GRASP* by transit paths is easily determined using the following remark. Let $\mathbf{q}_{\mathcal{M}}$ be a noncritical configuration of STABLE. There exists a transit path between any two configurations $\mathbf{q}_1^{\star} = (\mathbf{q}_1, \mathbf{q}_{\mathcal{M}})$ and $\mathbf{q}_2^{\star} = (\mathbf{q}_2, \mathbf{q}_{\mathcal{M}})$ of GRASP* (i.e. $\mathbf{q}_1$ and $\mathbf{q}_2$ are in $C_{grasp}(\mathbf{q}_{\mathcal{M}})$) if and only if there exists a label $L \in \lambda(\mathbf{q}_{\mathcal{M}})$ that contains two sublists M_1 and M_2 such that $\mathbf{q}_1 \in \psi(\mathbf{q}_{\mathcal{M}}, M_1)$ and $\mathbf{q}_2 \in \psi(\mathbf{q}_{\mathcal{M}}, M_2)$.

Manipulation Graph. The manipulation graph MG is built as follows:

- The nodes of MG are those of CG, plus $\mathbf{q}^{\star}_{init} = (\mathbf{q}_{init}, \mathbf{q}_{\mathcal{M},init})$ and $\mathbf{q}^{\star}_{goal} = (\mathbf{q}_{goal}, \mathbf{q}_{\mathcal{M},goal})$.

- Two nodes $ccell(R, M)$ and $ccell(R', M')$ of MG are adjacent if and only if they are adjacent in CG, or if $R = R'$ and there exists $L \in \lambda(R)$ such that M and M' are both sublists of L.

- A node $\mathbf{q}^{\star} = (\mathbf{q}, \mathbf{q}_{\mathcal{M}})$, with $\mathbf{q}^{\star} = \mathbf{q}^{\star}_{init}$ or $\mathbf{q}^{\star}_{goal}$, and a node $ccell(R, M)$ are adjacent in MG if and only if $\mathbf{q}_{\mathcal{M}} \in R$, there exists $L \in \lambda(R)$ such that $\mathbf{q} \in \varphi(\mathbf{q}_{\mathcal{M}}, L)$, and M is a sublist of L.

There exists a manipulation path between $\mathbf{q}^{\star}_{init}$ and $\mathbf{q}^{\star}_{goal}$ if and only if $\mathbf{q}^{\star}_{init}$ and $\mathbf{q}^{\star}_{goal}$ are connected by a path in MG. The size of MG is $O(n^4)$, where n is the number of vertices of the obstacle region $\mathcal{B}$, and the time required to construct and search MG is also $O(n^4)$ [Schwartz and Sharir, 1983c] [Laumond and Alami, 1988]. As mentioned in Section 3, this is the time complexity of deciding whether the planning problem has a solution. The (finite) cost of transforming the path in MG into a manipulation path depends on the shape of $\mathcal{B}$.

4.3 Polygons in Polygonal Workspace

Wilfong [Wilfong, 1988a] addressed the problem of planning the path of a convex polygonal robot $\mathcal{A}$ (which can only translate) in a two-dimensional polygonal workspace in the presence of convex polygonal movable objects $\mathcal{M}_j$. In this problem, $\mathcal{A}$ can grasp a movable object $\mathcal{M}_j$ if an edge of $\mathcal{A}$ coincides exactly with an edge of $\mathcal{M}_j$. (Hence, the grasp set $\mathcal{G}_j$ of each object $\mathcal{M}_j$ is necessarily finite.) If $\mathcal{M}_j$ has been grasped by $\mathcal{A}$, then $\mathcal{A}$ and $\mathcal{M}_j$ move together as one rigid object. In addition, $\mathcal{A}$ can move movable objects by "pushing" them. $\mathcal{M}_k$ is pushed by $\mathcal{A}$ if:

- An edge E of $\mathcal{A}$ coincides with an edge of $\mathcal{M}_k$ and the motion of $\mathcal{A}$ is in the direction of the outgoing normal of E.

- $\mathcal{M}_j$ is being grasped or pushed by $\mathcal{A}$, and an edge E_j of $\mathcal{M}_j$ coincides with an edge of $\mathcal{M}_k$ and the motion of $\mathcal{A}$ is in the direction of the outgoing normal of E_j.

When an object is pushed it keeps a constant position relative to $\mathcal{A}$. In addition, every position of $\mathcal{M}_j$ is stable (there is no gravitational force).

Wilfong showed that the above problem is PSPACE-hard if the goal positions of the movable objects are specified and NP-hard otherwise. He also considered the simpler case where there is only one movable object. In this case, using an exact cell decomposition method, he proposed an algorithm that runs in $O(n^2)$ time after $O(n^3 \log^2 n)$ preprocessing time when the goal configuration $\mathbf{q}^{\star}_{goal}$ is uniquely specified (n is the total number of vertices of the obstacles, the robot and the movable object).

5 Grasp Planning

A typical way for a robot to move a movable object is by grasping it. In the previous sections we simplified grasp planning considerably, either by predefining a finite set of possible grasps for each movable object, or by posing simple grasp conditions. The problem is usually much more complicated. In general, the robot is equipped with a multi-fingered gripper and it must select the position of the fingers on the object so that various sorts of physical and geometrical constraints be satisfied. There typically are infinitely many candidate grasps to consider.

In this section we analyze the grasp planning problem for a multi-fingered gripper. This problem raises a variety of issues which extend far beyond the scope of this book. For this reason, we restrict our presentation to a survey of important issues and results. (See [Pertin-Troccaz, 1989] for a broad survey of publications related to grasping.)

5.1 Grasp Constraints

Let us consider a robot gripper with multiple fingers. It is an articulated object made of several rigid components connected by joints and forming a tree-like kinematic structure. The root of this tree is the gripper's component — let us call it the palm — to which the fingers are connected. A very classical gripper (see Figure 15.a) is one with two parallel rigid fingers that translate simultaneously in opposite direction, with respect to the palm. A more complicated gripper (see Figure 15.b) is made of several independent articulated fingers.

We designate the gripper by $\mathcal{H}$ (for "hand"). A configuration $\mathbf{q}_{\mathcal{H}}$ of $\mathcal{H}$ is

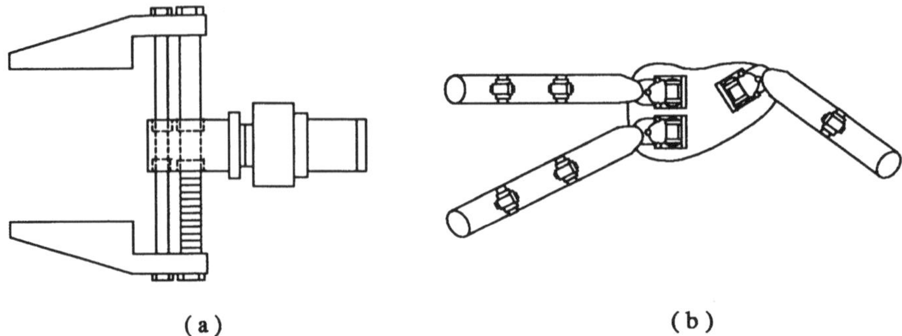

(a) (b)

Figure 15. Figure a shows a gripper with two parallel fingers that translate simultaneously in opposite directions. Figure b shows a gripper with multiple independent fingers.

a specification of the position and orientation of each of its components in the workspace $\mathcal{W}$. Let $\mathcal{M}$ designate the object to grasp. Grasp planning consists of finding a configuration $\mathbf{g} = \mathbf{q}_{\mathcal{H}}/\mathbf{q}_{\mathcal{M}}$ of the gripper relative to $\mathcal{M}$ that satisfies various conditions which we present below. Such a configuration is called a **grasp**.

Force Closure. $\mathcal{M}$ must remain in equilibrium under a variety of external forces (e.g. the gravitational force). Equilibrium requires that the gripping forces and torques exerted by the fingers cancel the external forces and torques. When the external forces are not well-known in advance, it is suitable that the grasp be "force-closure". Assuming no limit on the strength of the fingers, this means that the fingers can exert altogether arbitrary force and torque on $\mathcal{M}$, or, equivalently, any force and torque applied on the object can be resisted by the fingers.

Stability. $\mathcal{M}$ must remain stable in the gripper, that is it should not twist or slip if an external force/torque is applied. In theory, force closure guarantees that $\mathcal{H}$ can resist any motion of $\mathcal{M}$, but it assumes that the fingers are perfectly rigid, which is not the case in practice and not even always desirable (in particular, perfectly rigid fingers could be damaging to the object). A grasp should be such that a slight displacement of the object with respect to $\mathcal{H}$ cause $\mathcal{H}$'s fingers to apply forces that tend to restore the relative configuration $\mathbf{q}_{\mathcal{H}}/\mathbf{q}_{\mathcal{M}}$ to its desired value $\mathbf{g}$.

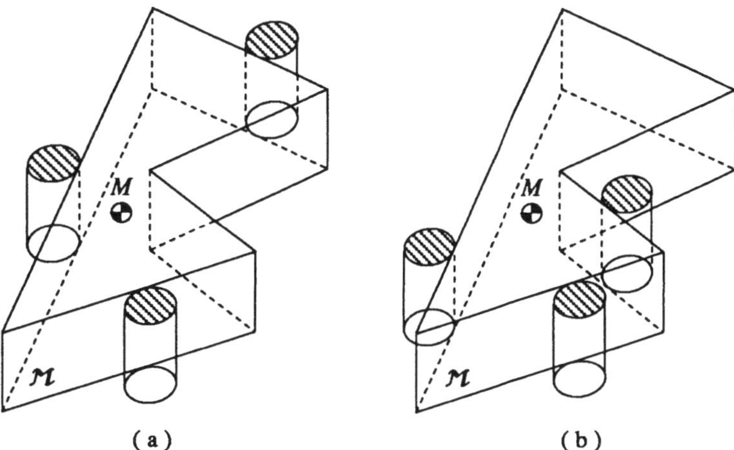

Figure 16. A prism $\mathcal{M}$ is grasped by a three-fingered gripper. M is the center of mass of the object. Assume some friction between the cylindrical fingers and the object. If the gravitational force is the dominant external force applied to the object, the grasp shown in Figure a requires the fingers to exert smaller forces on the object than the grasp of Figure b.

Tolerance to Positioning Errors. A gripper is usually unable to achieve a planned grasp $\mathbf{g}$ exactly. Therefore, the force closure and stability conditions should hold in some neighborhood of $\mathbf{g}$ such that when it aims at achieving $\mathbf{g}$, the gripper is guaranteed to attain some grasp in this neighborhood. Hence, even if the positions of the fingers on the surface bounding $\mathcal{M}$ are not exactly the planned ones, the achieved grasp is still force-closure and stable. Such an error-tolerant grasp can also accommodate some slipping of the fingers on the object.

Minimization of Gripping Forces. The force closure condition says nothing about the actual forces that the fingers would have to exert on $\mathcal{M}$ in order to oppose an external force applied to $\mathcal{M}$. In practice, the moduli of the forces and torques that can be exerted by a finger are bounded. In addition, we would like to minimize these forces in order to avoid unnecessary stress and deformation of both $\mathcal{M}$ and the fingers. This goal is particularly relevant if we have a good prior idea of the dominant external force. For instance, in Figure 16, the first grasp is "better" than the second if the dominant external force is the gravitational force.

Geometric Feasibility of the Grasp. Finally, a grasp must be geometrically feasible. The object $\mathcal{M}$ acts as a potential obstacle which may prevent the various components of the gripper from attaining their grasping locations. In addition, a grasp should allow a transfer motion of the robot between the initial and the final configuration of $\mathcal{M}$ along this path. In order for such a path to exist, the entire robot together with its gripper and the carried object should not collide with any other object in the workspace. If the robot is a manipulator arm, its joints should also remain within the limits set by the mechanical stops.

5.2 Synthesis of Force-Closure Grasps

Most of the basic results related to planning force-closure grasps have been obtained with a simplified model of the gripper. In this model the gripper is seen as a set of p points f_i, $i \in [1, p]$, called the **fingertips**. The configuration of the gripper is defined as the positions of these p points.

Several types of primitive fingertips can be considered [Nguyen, 1988]:

- A **frictionless fingertip** f_i can only exert a force (of any modulus) on a face F (an edge in a two-dimensional workspace) of $\mathcal{M}$ that is normal to F at the point of contact of f_i.

- A **hardcontact fingertip** f_i can exert any force on a face F (an edge in a two-dimensional workspace) of $\mathcal{M}$ that points into the friction cone normal to F at the point of contact of f_i.

- A **softcontact fingertip** f_i can exert any force on a face F of $\mathcal{M}$ that points into the friction cone normal to F at the point of contact of f_i and any torque about the axis of the friction cone. (This type of fingertip is meaningful only in a three-dimensional workspace.)

More complicated fingertips can be mechanically described as combinations of several contacts of the above types. For instance, an edge in frictionless contact with a face is equivalent to two distinct frictionless fingertips.

An important question relative to force closure is to determine the *smallest number of fingertips* that are needed to achieve a force-closure grasp of any object. This question has been addressed in several publications, including[9] [Lakshminarayana, 1978] [Ohwovoriole, 1980] [Salisbury and Roth, 1983] [Mishra, Schwartz and Sharir, 1987] [Markenscoff, Ni and Papadimitriou, 1990]. The following general results have been obtained (see [Markenscoff, Ni and Papadimitriou, 1990]):

- *Force closure with frictionless fingertips requires 4 fingertips (this is the lowest number possible) in a two-dimensional workspace. A general upper bound of 12 fingertips has been shown in three dimensions, but a bound of 7 can be achieved under quite general conditions which are satisfied, for example, by all polyhedra. These bounds hold only for objects that have no rotational symmetry. Objects with rotational symmetry (e.g. discs in two dimensions, spheres and cylinders in three dimensions) admit no force-closure grasps with frictionless fingertips.*

- *In order to achieve a force-closure grasp with hardcontact fingers, 3 fingertips are necessary and sufficient in two dimensions for any object; 4 fingertips are necessary and sufficient in three dimensions.*

The proofs of these results which are presented in [Markenscoff, Ni and Papadimitriou, 1990] are constructive and yield general algorithms for computing force-closure grasps.

More specific results have been obtained by Nguyen for two-fingered grippers [Nguyen, 1988]:

- *In two dimensions a grasp of $\mathcal{M}$ with two hardcontact fingertips f_1 and f_2 is force-closure if and only if the line segment $f_1 f_2$ points within the two friction cones at the contact points of f_1 and f_2 with $\mathcal{M}$ (Figure 17).*

- *The same result applies in three dimensions for a grasp with two softcontact fingertips.*

These results yield simple algorithms for computing force-closure grasps of polygons and polyhedra with two (hardcontact or softcontact) fingertips. Consider, for example, two hardcontact fingertips f_1 and f_2 and

[9] Initial results on this question were obtained by Reuleaux and Somoff at the end of the 19th century.

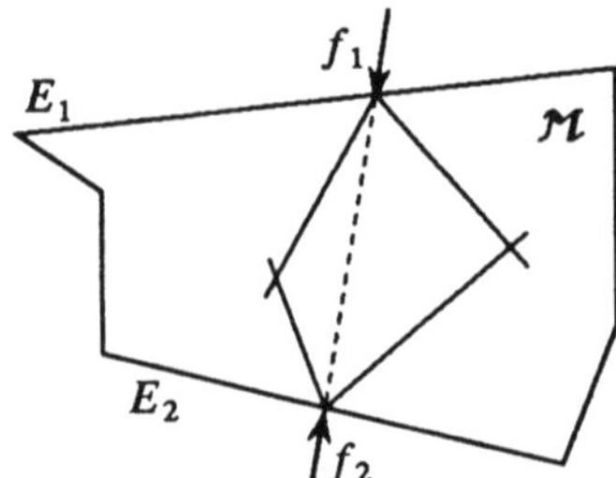

Figure 17. The object $\mathcal{M}$ is grasped with two hardcontact fingertips f_1 and f_2. The line segment $f_1 f_2$ is contained in the quadrilateral obtained by intersecting the two negated friction cones at f_1 and f_2. Such a grasp is force-closure [Nguyen, 1988].

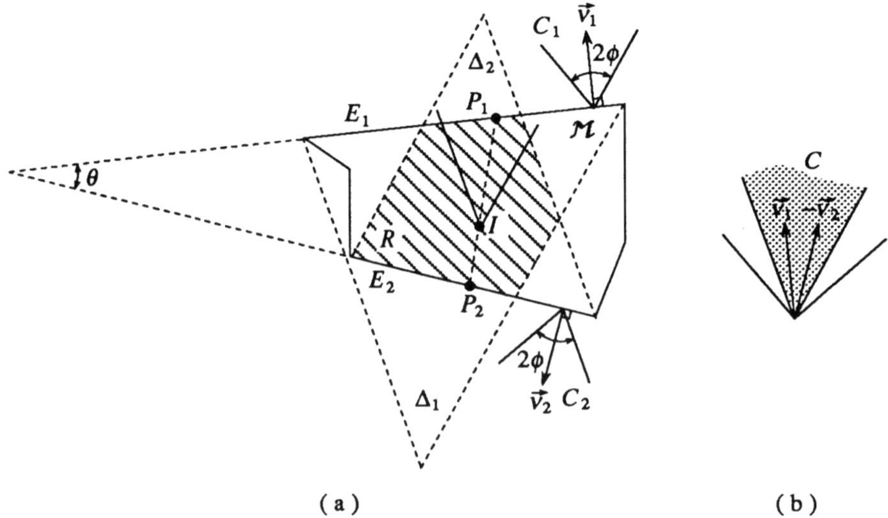

Figure 18. C_1 and C_2 are the friction cones associated with the hardcontact fingers f_1 and f_2 when they are in the edges E_1 and E_2 of $\mathcal{M}$ (Figure a). $C = C_1 \cap -C_2$ (Figure b). The striped region R in Figure a is obtained by intersecting the triangles Δ_1 and Δ_2 and $\mathcal{M}$. I is any point in R. From I we construct the points P_1 and P_2, which determine a force-closure grasp.

two edges E_1 and E_2 in a polygon $\mathcal{M}$ (see Figure 18.a). Assume that the angle θ made by the two edges is smaller than the angle 2ϕ of the friction cones associated with the hardcontact fingers (otherwise there does not exist a force-closure grasp with f_1 and f_2 in E_1 and E_2). Let C_1 (resp. C_2) be the cone of angle 2ϕ whose axis points along the outgoing normal $\vec{\nu}_1$ (resp. $\vec{\nu}_2$) of E_1 (resp E_2). Let $C = C_1 \cap -C_2$ (see Figure 18.b). We construct the triangle Δ_1 (resp. Δ_2) having one edge coinciding with E_1 (resp. E_2) and the other two edges parallel to the edges of C (Figure 18.a). Let R be the intersection of Δ_1, Δ_2 and $\mathcal{M}$, if any (if the intersection is empty, there exists no force-closure grasp with f_1 and f_2 in E_1 and E_2). Pick any point I in R, erect the cone C at I, pick any point P_1 in the intersection of this cone with E_1, draw the line passing through P_1 and I, and let P_2 be the intersection of this line with the edge E_2 (Figure 18.a). The grasp that consists of placing f_1 at P_1 and f_2 at P_2 is force-closure.

Most implemented grasp planners [Laugier and Pertin, 1983] [Wolter, Voltz and Woo, 1984] [Tournassoud, Lozano-Pérez and Mazer, 1987] are limited to grippers with two parallel parallelepipedic fingers covered with rubber (creating friction) and polyhedral objects[10]. These planners produce candidate grasps where the fingers are in contact with (1) two parallel faces of the object whose projections normal to the surfaces overlap (so that the faces are "in front of" each other), or (2) a face and a parallel edge whose projections overlap, or (3) a face and a vertex whose projection is interior to the face. They also insure a sufficient area (resp. length) of contact between a face (resp. an edge) and a finger in order to guarantee a "good quality" contact (i.e. a contact that can be modeled by a softcontact fingertip). This simple ad hoc construction of candidate grasps is supported by Nguyen's results given above.

5.3 Synthesis of Stable Grasps

Stability is usually investigated by modeling each finger as a set of independent linear and angular springs, each one being characterized by a stiffness relation of the form $\vec{f} = K(x - x_0)$ [Nguyen, 1989]. More precisely, the fingertip f_i of each finger is connected through a set of springs to a base point b_i attached to the gripper's palm. For a frictionless fingertip f_i the set of springs simply consists of one linear spring normal to the face (edge in two dimensions) of contact at f_i. For a hardcontact fingertip it contains three linear springs (two in two dimensions) capable

[10]The planner described in [Laugier and Pertin, 1983] also allows objects to be bounded by cylindrical and spherical faces.

of generating forces parallel to the tangent plane (line) of the object at the contact point. For a softcontact fingertip it consists of three linear springs and an angular spring whose axis is the normal to the face of contact passing through f_i.

Let a grasp of $\mathcal{M}$ be defined by the positions of the fingertips f_i and the base points b_i in $\mathcal{F}_{\mathcal{M}}$. Assume that it is an equilibrium grasp, i.e. the forces and torques applied by the springs through the fingertips cancel the external forces exerted on $\mathcal{M}$ (if any). Let the object $\mathcal{M}$ move slightly with respect to the base points b_i. If the fingertips are frictionless they will move accordingly to maintain the springs perpendicular to the surfaces of contact. If the fingertips are hardcontact or softcontact ones, we assume that the forces exerted by the finger springs point in the interiors of the friction cones at the contact points. Hence, as long as $\mathcal{M}$'s displacement is small enough, the fingertips do not slide in $\mathcal{M}$'s boundary. Thus, in both cases (frictionless fingertips and hard/softcontact fingertips), the system is locally conservative. The equilibrium grasp is stable if it is a strict local minimum of the potential energy stored in the fingers' springs [Hanafusa and Asada, 1977] [Baker, Fortune and Grosse, 1985] [Nguyen, 1989]. Then, any sufficiently small displacement of $\mathcal{M}$ with respect to the base points b_i will cause the springs to generate forces and torques aimed at restoring the object to its equilibrium configuration.

Nguyen showed the following result which relates force closure to stability [Nguyen, 1989]: *If a grasp defined by the positions of p fingertips in the boundary of an object $\mathcal{M}$ is force-closure, then the set of p virtual springs for which the grasp is stable always exists and can be computed in time linear in p.* The proof of this result supports a simple heuristic for facilitating grasping with two fingers: cover the fingers with soft rubber. This heuristic is implicitly used by most implemented grasp planners.

It should be noted that stable grasps do not necessarily correspond to force-closure grasps. For example, let $\mathcal{M}$ be a polygon such that the maximal inscribed circle C touches the polygon at three points, with less than π between any two successive points (see Figure 19). Consider a grasp with three frictionless fingertips f_1, f_2 and f_3 located at these three points. The line of action of the force that can be applied by each finger contains the center of C. Thus, provided that the vector sum of the three forces applied by the fingertips is zero, the grasp is in equilibrium in the absence of external forces. This *triangular grasp*

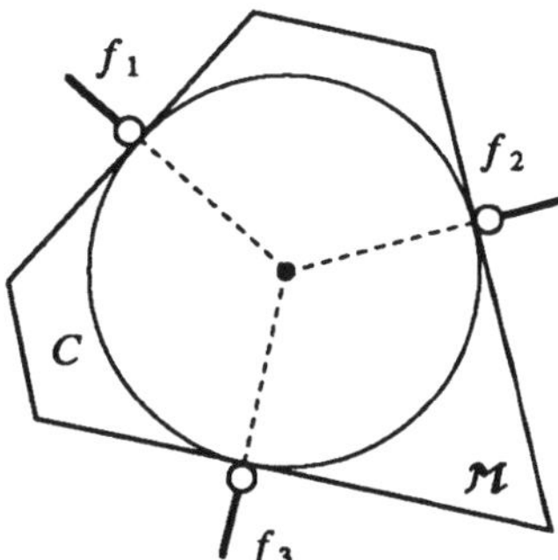

Figure 19. The maximal circle C inscribed in the polygon $\mathcal{M}$ touches $\mathcal{M}$ at three points such that no two of them are antipodal points of C. In the absence of external forces, a grasp with three frictionless fingers f_1, f_2 and f_3 located at these points (triangular grasp) is an equilibrium grasp provided that the three fingers generate a net force of zero. This grasp is also stable. However, it is not force-closure.

is also stable [Baker, Fortune and Grosse, 1985]. At the equilibrium position, the three fingertips cannot exert any torque around the center of C. Hence, the grasp is not force-closure. Such a grasp is potentially less resistant to external forces.

5.4 Tolerance to Positioning Errors

One way to construct a p-fingered grasp that is tolerant to some positioning errors is to construct p regions e_i of non-zero length (two-dimensional workspace) or area (three-dimensional workspace) in $\mathcal{M}$'s boundary such that if every fingertip f_i is placed anywhere in e_i the grasp is force-closure (thus it can also be made stable by adjusting the stiffness of the virtual springs).

Algorithms for constructing such regions e_i have been proposed in [Nguyen, 1988] for grasps in two dimensions with two hardcontact fingertips and four frictionless fingertips, and in three dimensions with three softcontact fingertips and seven frictionless fingertips. For example, consider again the case of a grasp with two hardcontact fingertips f_1 and f_2 on two edges E_1 and E_2 of a polygon $\mathcal{M}$ (Figure 18). In Subsection 5.2 we described a simple technique for computing a force-closure grasp. This technique requires the construction of two triangles Δ_1 and Δ_2 in-

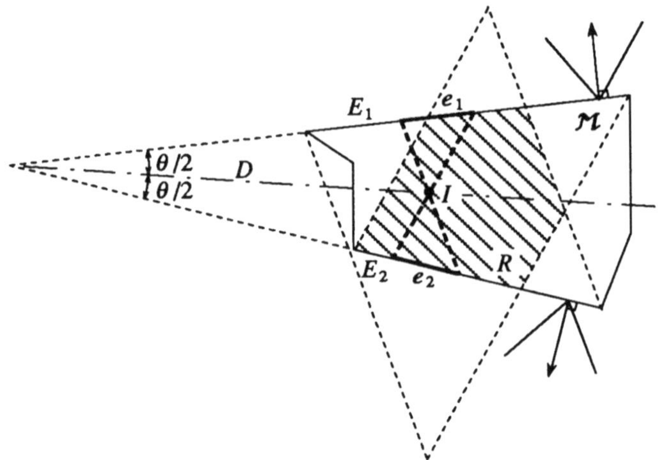

Figure 20. This figure is in the continuation of Figure 18. The point I is now taken anywhere in the intersection of R and the bisector D of the angle made by the edges E_1 and E_2. The two-sided cone $C \cup -C$ is erected at I. It intersects E_1 and E_2 at two segments e_1 and e_2, respectively. Any independent placement of f_1 in e_1 and f_2 in e_2 is a force-closure grasp. The choice of I in $R \cap D$ maximizes the length of e_1 and e_2.

tersecting at a region R. Let D be the bisector of the angle formed by E_1 and E_2 (if the two edges are parallel, D is the line at mid-distance between them). If D intersects R, pick any point I in the intersection and erect the two-sided cone $C \cup -C$ at I (Figure 20). The intersection of this two-sided cone with E_1 and E_2 generates two edge segments $e_1 \subseteq E_1$ and $e_2 \subseteq E_2$. The independent placements of f_1 in e_1 and f_2 in e_2 form a force-closure grasp. By taking I in the intersection of D and R, we maximize the length of e_1 and e_2. If D does not intersect R, we should choose a point I in R that minimizes the distance between D and R.

5.5 Minimization of Gripping Forces

Most implemented grasp planners address this problem by means of a simple heuristic. Recall from above that these planners [Laugier and Pertin, 1983] [Wolter, Voltz and Woo, 1984] [Tournassoud, Lozano-Pérez and Mazer, 1987] are limited to grippers with two parallel fingers and polyhedral objects. Let F_1 and F_2 denote the contact areas between the fingers and the object when a grasp has been achieved. The heuristic used by most of these planners consists of minimizing the distance between the projection of the center of mass of $\mathcal{M}$ on each finger and the contact area F_i, $i = 1, 2$. Candidate grasps are typically sorted using this heuristic.

A more theoretical attack of the question is reported in [Markenscoff and Papadimitriou, 1989]. The authors define "form semiclosure" to be a grasp with frictionless fingers that can balance any force through the center of mass of the object. They propose mathematical optimization techniques to construct the form semiclosure of a polygon with four fingers[11] that minimizes the worst-case forces needed to balance any unit force acting on the center of mass of the object.

Ji [Ji, 1987] also addressed the problem of minimizing gripping forces in a multi-fingered gripper.

5.6 Checking Geometric Feasibility

At this point we can assume that we have constructed a candidate grasp set of the object $\mathcal{M}$. We may possibly have several candidate subsets sorted according to the "quality" of the grasps they contain measuring how well they satisfy the conditions studied in the previous subsections.

Given a grasp (sub)set $\mathcal{G}$, it first seems natural to consider the search of a geometrically feasible grasp in $\mathcal{G}$ as a subproblem of finding a transfer path τ of $\mathcal{M}$ between two configurations $\mathbf{q}_{\mathcal{M},1}$ and $\mathbf{q}_{\mathcal{M},2}$ of $\mathcal{M}$. Assuming that the gripper $\mathcal{H}$ is part of a robot $\mathcal{A}$, the initial configuration of τ is $\mathbf{q}_1^\star = (\mathbf{q}_1, \mathbf{q}_{\mathcal{M},1})$ and its goal configuration is $\mathbf{q}_2^\star = (\mathbf{q}_2, \mathbf{q}_{\mathcal{M},2})$. The robot's configurations $\mathbf{q}_1$ and $\mathbf{q}_2$ are not uniquely defined (only $\mathbf{q}_{\mathcal{M},1}$ and $\mathbf{q}_{\mathcal{M},2}$ are), but both $\mathbf{q}_1/\mathbf{q}_{\mathcal{M},1}$ and $\mathbf{q}_2/\mathbf{q}_{\mathcal{M},2}$ are constrained to be in $\mathcal{G}$. Thus, τ should connect a configuration $\mathbf{q}_1^\star$ in a region of $\mathcal{C}^\star$ determined by $\mathcal{G}$ and $\mathbf{q}_{\mathcal{M},1}$ to a configuration $\mathbf{q}_2^\star$ in a region determined by $\mathcal{G}$ and $\mathbf{q}_{\mathcal{M},2}$. In theory, we know how to plan a path when the initial and goal configurations are incompletely specified. In practice, however, the complexity of finding a geometrically feasible grasp in a set $\mathcal{G}$ led several authors to use the following two-step approach (in particular, see

[11] Four fingers are necessary and sufficient for constructing a form-semiclosure grasp of any polygon. For some polygons, however, three fingers are enough, but then the form-semiclosure grasp is unique.

[Laugier and Pertin, 1983]):

1. The gripper $\mathcal{H}$ is first considered without the robot carrying it. A grasp is selected in $\mathcal{G}$ that satisfies the accessibility constraints imposed on $\mathcal{H}$ by both $\mathcal{M}$ and the objects located in $\mathcal{M}$'s environment at its initial configuration $\mathbf{q}_{\mathcal{M},1}$ and goal configuration $\mathbf{q}_{\mathcal{M},2}$.

2. A transfer path is computed for the entire robot with the gripper holding $\mathcal{M}$ at the selected grasp. If the computation is successful, i.e. if it returns a path, the selected grasp is accepted; otherwise, step 1 is executed again in order to generate an alternative grasp.

These two steps are motivated by the fact that the most restrictive geometric constraints often result from the potential interactions between the gripper and the objects at the endpoints of the transfer path. Except in very cluttered environments, the constraints on the rest of the transfer path and/or the rest of the robot are usually less obstructive. If the mechanical stops of the robot limit the range of possible orientations of the gripper significantly, it is suitable to also incorporate these constraints in the first step [Tournassoud, Lozano-Pérez and Mazer, 1987]. If the two steps fail to find a feasible grasp in $\mathcal{G}$, another grasp set, if any, should be considered.

In the rest of this subsection we illustrate the first step of this approach with an example. We consider the case where the gripper $\mathcal{H}$ consists of a parallelepipedic palm and two parallel parallelepipedic fingers (Figure 21). The two fingers can only translate in opposite directions at the same velocity. Their internal faces (i.e. the gripping faces) are denoted by J_1 and J_2. They exactly face each other. A Cartesian frame $\mathcal{F}_{\mathcal{H}}$ is attached to $\mathcal{H}$. We select its origin $O_{\mathcal{H}}$ at mid-distance between J_1 and J_2 (this choice is independent of the distance between the two fingers) and the x-axis parallel to J_1 and J_2 and perpendicular to the palm. $\mathcal{M}$ is a polyhedron shown in Figure 22. Four faces of $\mathcal{M}$ designated by F_1, F_2, F_3 and F_4 are of importance in the example.

The parallel faces F_1 and F_2 of $\mathcal{M}$ determine a candidate grasp subset denoted by $\mathcal{G}(F_1, F_2)$. A grasp $\mathbf{g}$ in $\mathcal{G}(F_1, F_2)$ is any configuration of the gripper such that J_1 and F_1, and J_2 and F_2 are coplanar[12], and

[12]We assume that the distance between F_1 and F_2 is achievable by the gripper, otherwise $\mathcal{G}(F_1, F_2)$ is empty.

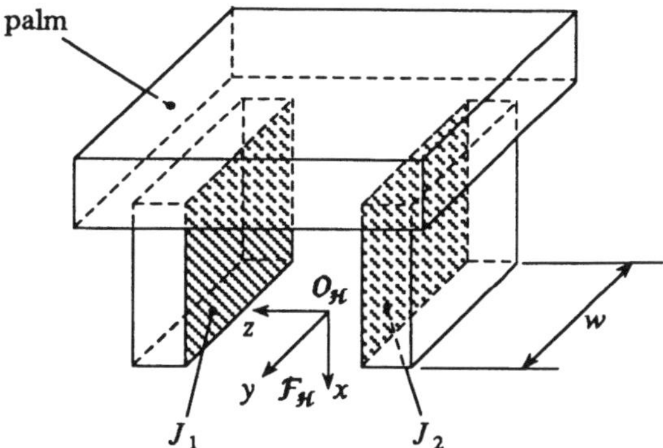

Figure 21. This gripper consists of two parallel parallelepipedic fingers that translate in opposite directions at the same velocity. The gripping faces of the fingers are denoted by J_1 and J_2. A Cartesian frame $\mathcal{F}_{\mathcal{H}}$ is attached to the gripper with the origin $O_{\mathcal{H}}$ located at equal distance between J_1 and J_2. The x-axis points in the same direction as the fingers.

the intersection of the perpendicular projections of J_1, F_1 and F_2 into a plane parallel to all these faces has a non-zero area (actually, imposing a minimal area would be more realistic). The plane GP equidistant from F_1 and F_2 is called the **grasp plane** of $\mathcal{G}(F_1, F_2)$. A coordinate system is placed in this plane.

A grasp **g** in $\mathcal{G}(F_1, F_2)$ is completely determined by a triplet (x_g, y_g, θ_g), where:

- (x_g, y_g) are the coordinates of $O_{\mathcal{H}}$ in GP (by construction, at any grasp in $\mathcal{G}(F_1, F_2)$, $O_{\mathcal{H}}$ lies in GP).

- θ_g is the angle between some fixed direction in GP and the x-axis of $\mathcal{F}_{\mathcal{H}}$.

Assume that the gripper will reach (and depart from) a grasp $\mathbf{g} = (x_g, y_g, \theta_g)$ by a pure translation so that $O_{\mathcal{H}}$ traces a straight line segment of orientation θ_g in GP. In most cases, this assumption is reasonable, although one can easily construct examples where it prevents the planner from reaching (or depart from) a grasp, though a feasible grasp exists.

Let S_1 and S_2 be the two infinite slices through the workspace which would contain the two parallelepipedic fingers of $\mathcal{H}$ if a grasp in $\mathcal{G}(F_1, F_2)$

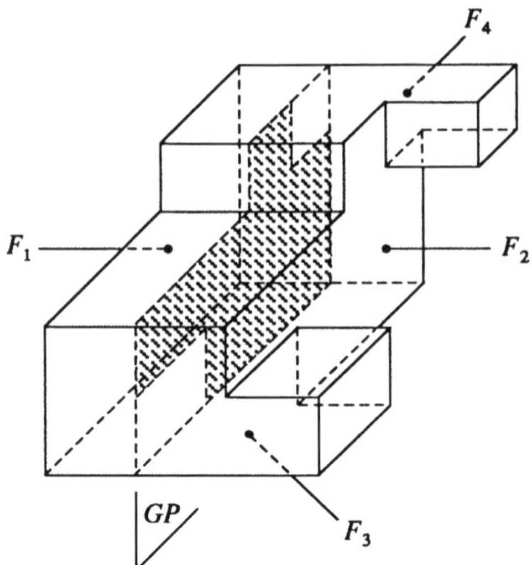

Figure 22. The object $\mathcal{M}$ is a polyhedron. The grasping set considered in the text is the set $\mathcal{G}(F_1, F_2)$ determined by the faces F_1 and F_2. $\mathcal{M}$ rests on a horizontal table both at its initial configuration and at its goal configuration. The support face at the initial (resp. goal) configuration is F_3 (resp. F_4). The grasp plane *GP* is the plane equidistant from F_1 and F_2. The striped area is the intersection of the projections of F_1 and F_2 into *GP*.

was achieved[13] (see Figure 23.a). The intersection of these slices with $\mathcal{M}$ and the objects (obstacles and movable objects) in $\mathcal{M}$'s environment at its initial and goal configurations is projected into the grasp plane *GP*. The obtained projection is called the "obstacle to the fingers in the grasp plane" and is denoted by $\mathcal{B}_{\mathcal{J}}$ (figure 23.b). Figure 23 shows the construction of $\mathcal{B}_{\mathcal{J}}$ when $\mathcal{M}$ rests on a table (modeled as a half-space obstacle) both at its initial configuration $\mathbf{q}_{\mathcal{M},1}$ and at its goal configuration $\mathbf{q}_{\mathcal{M},2}$, with no other object in its environment. At $\mathbf{q}_{\mathcal{M},1}$, $\mathcal{M}$ rests on the face F_3 (the bottom face in Figure 22); at $\mathbf{q}_{\mathcal{M},2}$, it rests on the face F_4 (the back face in Figure 22).

Assume that a candidate direction θ_g has been selected. Let $\mathcal{J}_{\theta_g}(x, y)$ denote the projection of J_1 (equivalently, J_2) in the grasp plane when the

[13] Actually, the thickness of the slices should be slightly greater than the thickness of the fingers to take into account the fact that the gripper will reach the grasp with an opening greater than the distance between F_1 and F_2.

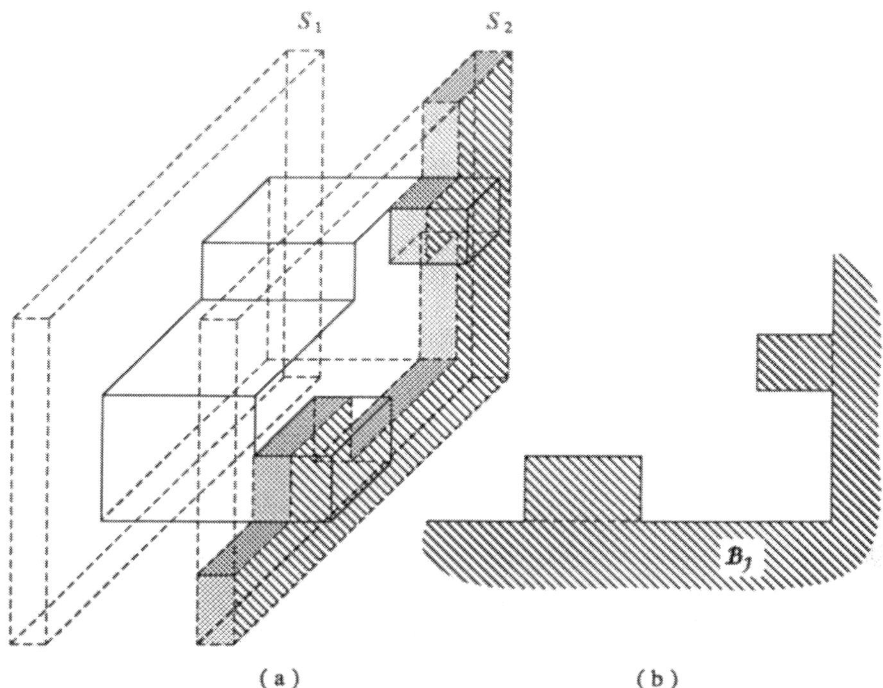

Figure 23. S_1 and S_2 are the infinite slices that would contain the fingers of the gripper if a grasp of $\mathcal{G}(F_1, F_2)$ was achieved (Figure a). Their intersection with $\mathcal{M}$ and the objects in $\mathcal{M}$'s environment at its initial and goal configurations project on the grasp plane according to a region $\mathcal{B}_\mathcal{J}$ representing the obstacle to the fingers in the grasp plane (Figure b). In our example, the only ojects in $\mathcal{M}$'s environment at its initial and goal configurations are the table on which it rests; the support faces are F_3 and F_4, respectively (see Figure 22).

gripper's orientation is θ_g and $O_\mathcal{H}$'s coordinates are (x, y). $\mathcal{J}_{\theta_g}(x, y)$ is a rectangle which can only translate in *GP*. (In fact, with our assumptions, it can only move along a line of orientation θ_g, but we do not know this line yet.) Let us choose $O_\mathcal{H}$ as the reference point of $\mathcal{J}_{\theta_g}$. At a grasp **g**, the point $O_\mathcal{H}$ (equivalently, the configuration of $\mathcal{J}_{\theta_g}$) should lie in a region $GZ(\theta_g)$ called the **grasp zone**. This region is defined as follows: if the orientation of the gripper is θ_g and $O_\mathcal{H}$ lies in $GZ(\theta_g)$, then the projections of J_1, F_1 and F_2 on *GP* have non-zero intersection. $GZ(\theta_g)$ is easily computed by growing the intersection of the projections of F_1 and F_2 inversely to the shape of $\mathcal{J}_{\theta_g}$, i.e. :

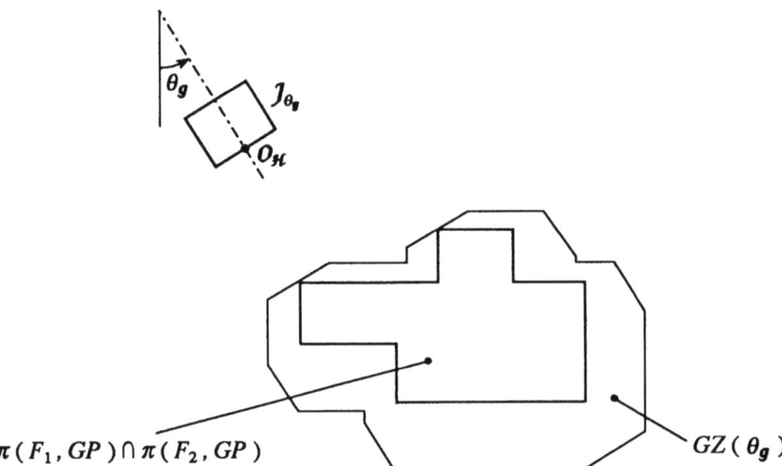

Figure 24. The grasp zone $GZ(\theta_g)$ is defined as follows: if the orientation of the gripper is θ_g and $O_{\mathcal{H}}$ lies in $GZ(\theta_g)$, then the projections of J_1, F_1 and F_2 into the grasp plane have a non-zero intersection. $GZ(\theta_g)$ is constructed by growing the intersection of the projections of F_1 and F_2 .

$$GZ(\theta_g) = [\pi(F_1, GP) \cap \pi(F_2, GP)] \ominus \mathcal{J}_{\theta_g}(0,0)$$

with $\pi(F_i, GP)$ denoting the projection of F_i into GP and $\ominus$ being the Minkowski difference operator. The intersection $\pi(F_1, GP) \cap \pi(F_2, GP)$ is the striped area shown in Figure 22. The construction of $GZ(\theta_g)$ is illustrated in Figure 24.

The region $\mathcal{B}_{\mathcal{J}}$ is an obstacle to the motion of $\mathcal{J}_{\theta_g}$ in GP. We can compute the corresponding C-obstacle as (Figure 25):

$$\mathcal{CB}_{\mathcal{J}}(\theta_g) = \mathcal{B}_{\mathcal{J}} \ominus \mathcal{J}_{\theta_g}(0,0).$$

Since we restrict the possible paths of $O_{\mathcal{H}}$ to straight lines of orientation θ_g, this C-obstacle should be extended by the "shadow" that it projects in the direction θ_g. (In Figure 25 this shadow is already contained in $\mathcal{CB}_{\mathcal{J}}(\theta_g)$.) The extended C-obstacle $\mathcal{CB}^s_{\mathcal{J}}(\theta_g)$ is formally defined as:

$$\mathcal{CB}^s_{\mathcal{J}}(\theta_g) = \{(x', y') \,/\, (x', y') = (x, y) + \lambda(\cos\theta_g, \sin\theta_g), \text{ with } (x, y) \in \mathcal{CB}_{\mathcal{J}}(\theta_g) \text{ and } \lambda \geq 0\}.$$

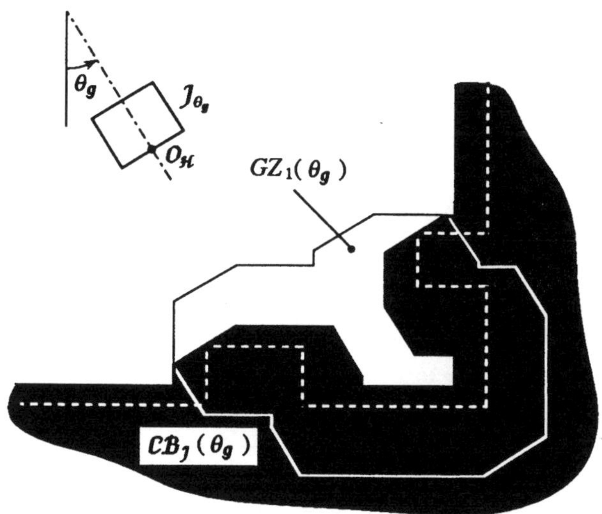

Figure 25. The region $\mathcal{B}_{\mathcal{J}}$ constructed in Figure 23 is the obstacle to the motion of $\mathcal{J}_{\theta_g}$ in *GP*. The corresponding C-obstacle $C\mathcal{B}_{\mathcal{J}}(\theta_g)$ restricts the possible location of $O_{\mathcal{H}}$ to a subset $GZ_1(\theta_g)$ of the grasp zone.

The region $GZ_1(\theta_g) = GZ(\theta_g) \backslash C\mathcal{B}^s_{\mathcal{J}}(\theta_g)$ is the subset of the grasp zone that remains accessible after the obstacles to the fingers have been considered.

We still have to consider the obstacles to the palm of $\mathcal{H}$. The computation is similar to the above one. Let S be the infinite slice through the workspace which would contain the palm of $\mathcal{H}$ if a grasp of $\mathcal{G}(F_1, F_2)$ was achieved. We intersect S with $\mathcal{M}$ and the objects in $\mathcal{M}$'s environment at its initial and goal configurations. The projection of this intersection into *GP* forms the obstacle $\mathcal{B}_{\mathcal{P}}$ (the "obstacle to the palm in the grasp plane"). The projection of the palm into *GP* is a translating rectangle $\mathcal{P}_{\theta_g}(x, y)$ with $O_{\mathcal{H}}$ as the reference point. The extended C-obstacle $C\mathcal{B}^s_{\mathcal{P}}(\theta_g)$ is computed and removed from $GZ_1(\theta_g)$. We thus obtain a refined subset $GZ_2(\theta_g)$ of the grasp zone.

Any grasp $\mathbf{g} = (x_g, y_g, \theta_g)$, with $(x_g, y_g) \in GZ_2(\theta_g)$, if $GZ_2(\theta_g) \neq \emptyset$, can then be returned as a selected grasp to be checked for the existence of a transfer path (step 2 of the approach). A reasonable heuristic is to return the centroid of $GZ_2(\theta_g)$.

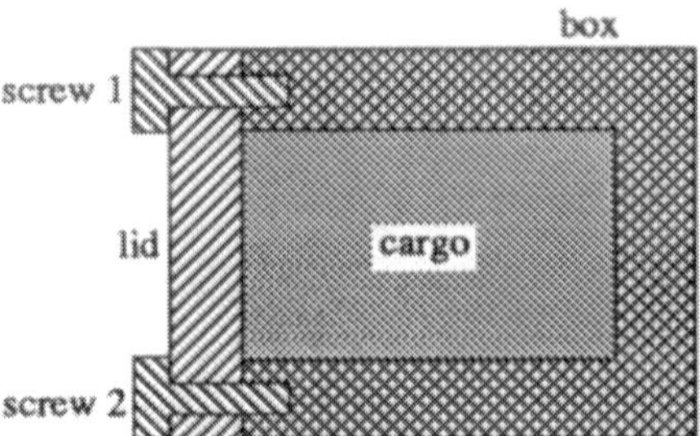

Figure 26. The composite object shown in this figure is a crate made of a box, a cargo, a lid and two screws. Throughout Section 6 this example is referred to as the "crate example".

The computation of $GZ_1(\theta_g)$ and $GZ_2(\theta_g)$ assumes the prior selection of a direction θ_g. One may discretize the range S^1 values of θ_g and try each value successively. Another approach consists of constructing the three-dimensional region $\bigcup_{\theta_g \in S^1} GZ_2(\theta_g) \times \{\theta_g\} \subset \mathbf{R}^2 \times S^1$. This can be done by noticing that the topology of $GZ_2(\theta_g)$ changes only at a finite number of critical values of θ_g. The equations of the edges and vertices of $GZ_2(\theta_g)$ can be expressed as continuous functions of θ_g between any two successive critical values. Optimization techniques can then be used to select a grasp in the three-dimensional region thus constructed.

6 Assembly Planning

The complexity of motion planning with movable objects and the fact that a manipulation path consists of a sequence of transit and transfer paths achieving intermediate goals suggests that we attempt to solve the problem at successive levels of abstraction. In fact, this is already the idea underlying the grasp planning approach presented at the end of the previous section. This idea has also been explored to some extent in the context of assembly planning.

Let us consider a set of parts $\mathcal{M}_1$ through $\mathcal{M}_r$ (movable objects) and a connected composite object made of these parts (see Figure 26). Assembly planning consists of planning a sequence of motions for a robot in order to move the parts from an initial configuration where they are mutually separated to a final configuration where they are assembled together. Assembly planning requires the determination of a sequence

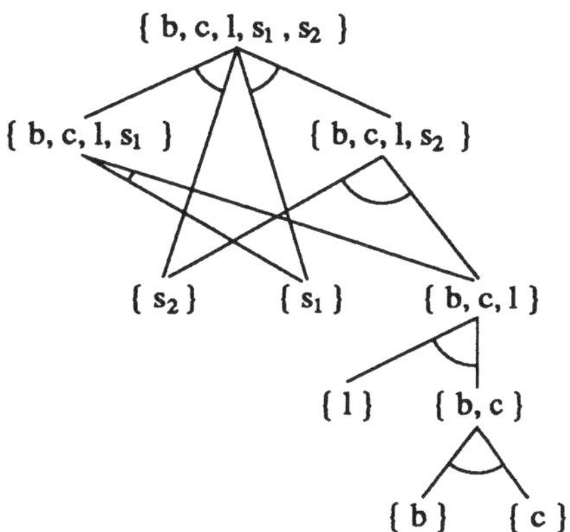

Figure 27. AND/OR graphs are a relatively compact representation of the possible assembly sequences of a composite object. This figure shows a subgraph of the AND/OR graph for the crate example. It only represents the two assembly sequences that are geometrically feasible. Each AND-arc is displayed with a brace connecting its links.

of transit and transfer motions together with the grasps of the parts.

Several publications have addressed the assembly planning problem, e.g. [Bourjault, 1984] [De Fazio and Whitney, 1987] [Wolter, 1988] [Homem de Mello, 1989] [Hoffman, 1989] [Wilson and Rit, 1990] [Baldwin, 1990] [Homem de Mello and Sanderson, 1990] [Lee and Shin, 1990]. Natarajan [Natarajan, 1988] studied the computational complexity of the problem and proposed new measures to capture the inherent complexity of assembling a composite object. One such measure is the number of "hands" that is required to hold the parts. Natarajan showed that a composite object made of k polygons in the plane may require up to k hands. This means that $k-1$ parts may have to move simultaneously (the support of the remaining fixed part is considered as one hand). Some studies in Computational Geometry are also directly relevant to the assembly planning problem (e.g. [Guibas and Yao, 1983] [Toussaint and El Gindy, 1984]).

A classical abstraction of the assembly planning problem consists of regarding the various parts $\mathcal{M}_j$ as independent moving objects (i.e. as robots), thus ignoring the actual robots that transport them, and planning an **assembly sequence** specifying the order in which the parts are to be moved to their goal configurations. One (relatively) compact representation of the possible assembly sequences is the AND/OR graph defined as follows. Each node in the graph corresponds to a subassembly that might be reached in the process of assembling the composite object. The entire composite object is associated with the root of the graph and each distinct part $\mathcal{M}_j$, $j \in [1, r]$, is associated with a leaf of the graph (there are exactly r leaves). An AND-arc represents the operation of putting child subassemblies together to make a parent. Multiple AND-arcs abutting the same parent node give different ways of creating the same subassembly. Each maximal AND-subtree of the graph represents one possible assembly sequence (usually it is a partial ordering) for the entire composite object. Figure 27 shows a subgraph of the AND/OR graph for the crate example of Figure 26 [Wilson and Rit, 1990]. This subgraph contains two AND-subtrees representing two assembly sequences (the only ones which are geometrically feasible). These two sequences differ only by the order in which the two screws are assembled.

The construction of the full AND/OR graph supposes that the parts are assembled directly to their final configurations and that they stay there for all subsequent operations. Thus, the geometry of a subassembly is uniquely defined by the set of its components. It is also often required that each subassembly associated with a node of the AND/OR graph be connected and that each parent subassembly be obtained by mating exactly two subassemblies[14]. Despite these assumptions (which may lead to disregarding all the feasible assembly sequences in some cases), the size of the AND/OR graph remains exponential in r (the number of parts).

The AND/OR graph can be pruned by applying simple feasibility tests. For example, let us specify a **terminal path** by a unit vector $(\vec{t}, \vec{\omega})$ in $\mathbf{R}^6$ (for a three-dimensional workspace). When moving along this path, an object translates along $\vec{t}$ while rotating around an axis parallel to $\vec{\omega}$ and passing through some fixed point (for instance, the origin of the frame $\mathcal{F}_\mathcal{W}$ attached to the workspace), with $\|\vec{\omega}\|/\|\vec{t}\|$ being the amount of rotation per unit of translation. If $\|\vec{\omega}\| = 0$, the path is a pure translation; if $\|\vec{t}\| = 0$, it is a pure rotation. Any motion of a rigid object can

[14]This requirement corresponds to assuming that two "hands" (e.g. a gripper and a table) are sufficient for assembling the parts.

be instantaneously seen as a motion along a terminal path for some uniquely defined $(\vec{t}, \vec{\omega})$ [Bottema and Roth, 1979]. Two subassemblies S_1 and S_2 can be assembled into a parent subassembly only if there exists a terminal path of non-zero length of S_1 relative to S_2 such that the interiors of S_1 and S_2 along that path do not overlap [Wilson and Rit, 1990]. In the crate example of Figure 26, there is no terminal path for the subassemby $\{s_1, s_2, b\}$ consisting of the two screws and the box relative to the subassembly $\{c, l\}$ consisting of the cargo and the lid. Hence, $\{s_1, s_2, b\}$ cannot be assembled with $\{c, l\}$.

The above necessary condition yields a simple test for pruning the AND/OR graph. Assume for simplification that the parts are polygons in a two-dimensional workspace and that they can only translate. With this assumption $\vec{t} \in \mathbf{R}^2$ and $\|\vec{\omega}\| = 0$. Consider two subassemblies S_1 and S_2. Each contact between an edge of S_1 and an edge of S_2 restricts the set of possible terminal path directions $\vec{t}$ to a half-plane. We can consider all the contacts and intersect the corresponding half-planes. If the resulting intersection is empty, there is no terminal path. In the crate example, there are three contacts between the cargo and the box yielding three half-planes that intersect along a unique direction. In contrast, the contacts between the lid, on the one hand, and the screws and the box, on the other hand, yield four half-planes with zero intersection; the corresponding AND-arc can be removed from the AND/OR graph.

In fact, the above test can be modified in order to be used in a more efficient way to directly construct a reduced AND/OR graph, rather than pruning the larger graph [Wilson, 1990]. This is done by noticing that the set of terminal path orientations of a subassembly S_1 relative to a subassembly S_2 (the five-dimensional unit sphere S^5 for a three-dimensional workspace; the unit circle S^1 in a two-dimensional workspace where parts can only translate) can be partitioned into a collection of cones such that the set of parts in S_2 that prevent S_1 from moving is constant within each cone. If there is no feasible terminal path of S_1 relative to S_2, then opening a path makes it necessary to choose a cone, remove from S_2 the parts that prevent S_1 from moving along a direction in that cone, and include these parts in S_1. The construction of the reduced AND/OR graph consists of treating each part in the composite object as a seed subassembly and "growing" every seed assembly recursively by checking whether a terminal path exists and including additional parts if this is not the case.

The reduced AND/OR graph can be pruned further by applying additional tests. For example, one may require that each subassembly associated with a node be stable under the gravitational force. In order for a subassembly S obtained by bringing two subassemblies S_1 and S_2 together to be stable, the set of vectors $\vec{t}$ of the feasible terminal paths must include no direction whose angle with the gravity direction is less than the friction angle ϕ between the parts[15] [Wilson and Rit, 1990]. Notice that fixtures could be used to prevent an unstable subassembly S from falling apart. However, the automatic design of fixtures in the context of assembly planning is still an open research topic (see [Ingrand and Latombe, 1984] [Bausch and Youcef-Toumi, 1990]).

Experiments with implemented systems have demonstrated that the application of such simple tests results in an AND/OR graph containing a relatively small number of AND-trees. The feasibility of each of these trees may then be checked more completely by using motion planning methods.

Exercises

1: Compare motion planning with movable objects to motion planning with multiple robots.

2: Design an algorithm for determining all the stable orientations of a three-dimensional polyhedron (with uniform mass density) on a horizontal plane.

3: Consider the case of a single movable object in Section 3. Show that any path τ in $STABLE^* \cap GRASP^*$ can be transformed into a manipulation path τ' connecting the same two configurations as τ and as close as one wishes to τ.

4: Consider the case of a finite number of grasps and a finite number of stable configurations (as in Subsection 4.1). Evaluate the number of nodes in MG, the number of links corresponding to transit paths,

[15]This condition is necessary if the relation between S_1 and S_2 is clamping-free.

the number of links corresponding to transfer paths, and the number of distinct two-dimensional slices of $\mathcal{C}^\star$ in which these paths lie.

5: Formulate the Hanoi Tower problem as a manipulation problem. Show how the method described in Subsection 4.1 applies to this problem.

6: Compare manipulation planning as described in Sections 2, 3, and 4 to the type of planning performed by Strips [Nilsson, 1980].

7: Consider the case of a polygonal workspace with a polygonal robot $\mathcal{A}$ and polygonal movable objects. Assume that $\mathcal{A}$ can only translate and that each movable objects has a finite number of grasps and a finite number of stable positions, as in Subsection 4.1. Implement a planner capable of generating manipulation paths under these conditions.

8: Consider the planning example of Figure 10. Draw some critical curves in STABLE. Consider any noncritical region R bounded by these curves. Verify that $\lambda(\mathbf{q}_{\mathcal{M}})$ and $\mu(\mathbf{q}_{\mathcal{M}})$ remain constant over R.

9: Show that the triangular grasp depicted in Figure 19 is stable.

10: The triangular grasp of Figure 19 is stable, but not force-closure. Discuss the drawbacks of stable, but non-force-closure grasps.

11: At the end of Subsection 5.2 we mentioned grasp planners for two-fingered grippers that generate candidate grasp sets for a polyhedral object $\mathcal{M}$ from pairs of features of $\mathcal{M}$: (1) two parallel faces, (2) a face and a parallel edge, and (3) a face and a vertex. Describe other combinations of features that could be used to generate candidate grasps. Discuss how they can effectively be used in a grasp planner.

12: Give an algorithm for computing the extended C-obstacle $\mathcal{CB}^{s}_{\mathcal{J}}(\theta_g)$ defined in Subsection 5.6.

13: Design and implement a simple grasp planner for a gripper with two parallel gripper and polyhedral objects. Discuss the limitations of your planner.

14: Discuss various forms of regrasping, e.g. : by putting down an object on a table and regrasping it at a different grasp with the same gripper, by passing an object from a gripper to another, or by changing grasp within the same multi-fingered gripper. In particular, describe circumstances in which regrasping is necessary or useful. Propose ideas for planning a regrasping sequence. (In particular, see [Tournassoud, Lozano-Pérez and Mazer, 1987].)

15: Discuss in more detail than we did the method sketched at the end of Subsection 5.6 to compute the region $\bigcup_{\theta_g \in S^1} GZ_2(\theta_g) \times \{\theta_g\}$.

16: Develop in detail the method sketched in Section 6 for partitioning a composite object into two separable subassemblies by recursively growing seed subassemblies. (Assume a two-dimensional workspace with parts that can only translate. Discuss how the principle of the method extends to a three-dimensional workspace with parts that can both translate and rotate.)

Prospects

Over the past few years, Robot Motion Planning has grown into a major research field. This growth is attested to by the increasing number of publications in a large variety of journals and conferences. A unique feature of this field has been its ability to attract the interest of a diversity of researchers with backgrounds in Computer Science, Mathematics and Mechanical Engineering. Thanks to the combined work of these researchers, an impressive set of theoretical and practical results has been obtained.

Although these results have still made few inroads in industrial applications, this should change soon. Several graphic simulators already include collision checking capabilities and it will not take long before they also offer path planning tools. The availability of relatively fast path planning algorithms is recent and there is always some delay between the design of new algorithms and their incorporation into commercial products. While the need for motion planning techniques grows (e.g. in manufacturing, construction, service robots, space and undersea work, mechanical design, and process planning), more students are being trained in them at various universities. These students will very soon be in positions where they can contribute to the development of technologies, systems and products embedding these techniques.

Despite substantial progress, many problems remain for future research. These include motion planning with uncertainty (results reported in

Chapter 10 are obviously too limited), motion planning with flexible objects, minimum-time motion planning with dynamic constraints (see [Canny et al., 1988] [Donald and Xavier, 1989] [Jacobs et al., 1989]), assembly planning, dexterous manipulation planning, etc. Various combinations of problems which have so far been considered separately (e.g. planning with kinematic constraints *and* uncertainty *and* moving obstacles) will have to be explored.

Very basic questions related to the correctness and the stability of exact planning methods when implemented on finite-precision computers will have to be addressed in depth before these methods can be used in practice [Hoffmann, 1988]. Given the high worst-case complexity of general motion planning problems, future research should aim at identifying more tractable, but still useful, problems, or equivalently, at producing incomplete methods with well-defined and reasonable failure conditions. Algorithms with lower average complexity should also be a target. Substantial experiments will increasingly be needed to measure the effective efficiency and reliability of these methods.

The interaction of advanced motion planning algorithms with modules doing task planning, motion control and perception will have to be investigated further. New issues are likely to emerge from this investigation. Eventually, real-time motion planners will have to be developed and directly embedded in robot controllers.

Massively-parallel connectionist architectures may provide new opportunities to be explored (see [Mel, 1990]). These architectures may well push the field from using algebraic representations (as is mostly the case now) to using "distributed" representations such as grids (bitmaps). They may also motivate the investigation of new paradigms for motion planning, e.g. paradigms based on learning (see [Dufay and Latombe, 1984]) and randomization (see Section 5 of Chapter 7, and [Erdmann, 1989 and 1990]).

A recent trend has been to adapt motion planning techniques for automating the design of objects. For instance, motion planning with uncertainty can be used to design part feeders (see [Erdmann and Mason, 1986] and [Natarajan, 1989]). In [Zhu and Latombe, 1990] standard path planning techniques are used to design three-dimensional pipe layouts. Motion planning techniques for nonholonomic vehicles may be integrated in systems for designing commercial facility layouts. Grasp planning techniques may help in designing fixtures for machine-tools, and assembly planning techniques in analyzing and improving the manufacturability of assembly products. One can also envision the application of motion planning algorithms in biochemistry to synthesize organic molecules.

In fact, this trend and other ongoing research in various fields, including Computational Geometry, Computer-Aided Design, Computer Graphics and Computer Vision, should converge toward the development of a comprehensive computational theory of *spatial reasoning.* We see this development as a major undertaking for the next two decades or more. Ultimately, it will have a radical impact on the way computers are used in Engineering as well as in other domains.

Appendix A

Basic Mathematics

This appendix defines some mathematical terms that are used in the book. The terms are given in alphabetic order. They correspond to well-known definitions and results, and are included for the reader's convenience only.

Affine Space: Let E be a vector space. The **affine space** associated with E is a set A such that:

- Every pair $(a, b) \in A \times A$ determines a unique vector $\vec{ab} \in E$.

- Every pair $(a, \vec{x}) \in A \times E$ determines a unique element $b \in A$ such that $\vec{ab} = \vec{x}$.

- $\forall a, b, c \in A : \vec{ac} = \vec{ab} + \vec{bc}$.

The elements of A are called **points**. In general, one of them — call it O — is arbitrarily taken as the **origin**. This point determines the bijection $a \in A \mapsto \vec{x} = \vec{Oa} \in E$. It allows us to define the operations $+$ and $-$ in A:

- For any $a, b \in A$, $c = a + b$ is the element of A corresponding to $\vec{Oc} = \vec{Oa} + \vec{Ob}$. The operation is commutative.

- For any $a \in A$, $c = -a$ is the element of A corresponding to $\vec{Oc} = -\vec{Oa}$.

Bijective Map (or Bijection): Let E and F be two sets. A map $f : E \to F$ is **bijective** if and only if it is both surjective and injective.

Boundary of a Set: Let U be any subset of a topological space E. A point $x \in E$ is called a **boundary point** of U if and only if neither U, nor $E \backslash U$ is a neighborhood of x.

The set of boundary points of U is called the **boundary** of U. It is denoted by ∂U.

Class C^k: Let $f : E \subseteq \mathbf{R} \to F \subseteq \mathbf{R}$ be a function.

f is said to be of **class** C^k if all its derivatives of order less than or equal to k exist and are continuous functions on E.

f is said to be **piecewise of class** C^k if all its derivatives of order less than or equal to k exist and are continuous functions on E, except possibly at a finite number of points in any bounded subset of E.

Closure of a Set: Let U be a subset of a topological space E. The set of points of E which are not exterior points of U is called the **closure** of U. It is denoted by $cl(U)$.

Compact Set: Let E be a topological space and A a subset of E. A is **compact** if every open cover of A includes a finite subcover. In other words, if $\mathcal{U} = \{U_\alpha\}_{\alpha \in J}$ is an arbitrary open cover of A, hence all U_α are open subsets of E and $\bigcup_{\alpha \in J} U_\alpha = E$, then there is a finite number of elements of J, say $\alpha_1, ..., \alpha_r$, such that $U_{\alpha_1} \cup ... \cup U_{\alpha_r} = E$.

A subset of $\mathbf{R}^n$ is compact if and only if it is both closed and bounded.

Connectedness: A topological space is **connected** if it is not the union of two non-empty, open, disjoint subspaces.

Continuous Map: Let E and F be two topological spaces. A map

$f : E \to F$ is **continuous** if and only if the inverse image of any open set (in the topology of F) is an open set (in the topology of E).

Convex Set: A set $S \subset \mathbf{R}^n$ is **convex** if for any two points $x_1, x_2 \in S$ the segment $\{x \;/\; x = \lambda x_1 + (1-\lambda)x_2 \;;\; 0 \leq \lambda \leq 1\}$ is entirely contained in S.

Diffeomorphism: Let $f : X \subset \mathbf{R}^n \to Y \subset \mathbf{R}^m$ be a map between two subsets of two Euclidean spaces. It is a **diffeomorphism** if f is bijective and smooth, and the inverse map $f^{-1} : Y \to X$ is also smooth. X and Y are said to be **diffeomorphic** if such a map exists.

Distance: Let E be a set. A **distance** or **metric** on E is a function $d : E \times E \to \mathbf{R}$ verifying the following axioms:

- $\forall x, y \in E : d(x, y) \geq 0$ (positiveness).
- $d(x, y) = 0 \Rightarrow x = y$ (non-degeneracy).
- $\forall x, y \in E : d(x, y) = d(y, x)$ (symmetry).
- $\forall x, y, z \in E : d(x, y) + d(y, z) \geq d(x, z)$ (triangle inequality).

The pair (E, d) is called a **metric space**. When there is no ambiguity about the distance, E alone is called the metric space.

Let $\mathbf{x} = (x_1, x_2, ..., x_n)$ and $\mathbf{y} = (y_1, y_2, ..., y_n)$ be two points of $\mathbf{R}^n$. The map:

$$\|\mathbf{x} - \mathbf{y}\| = [(y_1 - x_1)^2 + (y_2 - x_2)^2 + ... + (y_n - x_n)^2]^{1/2}$$

defines a distance in $\mathbf{R}^n$ called the **Euclidean distance**. The set $\mathbf{R}^n$ equipped with this distance is a Euclidean space.

Equivalence (Relation of): A **relation of equivalence** on a set E is any subset $\mathcal{R} \subseteq E \times E$ such that:

- $\forall x \in E : (x, x) \in \mathcal{R}$ (reflexivity).
- $\forall x, y \in E : (x, y) \in \mathcal{R} \Rightarrow (y, x) \in \mathcal{R}$ (symmetry).
- $\forall x, y, z \in E : (x, y) \in \mathcal{R}$ and $(y, z) \in \mathcal{R} \Rightarrow (x, z) \in \mathcal{R}$ (transitivity).

Exterior of a Set: Let U be any subset of a topological space E. A point $x \in E$ is called an **exterior point** of U if and only if $E \backslash U$ is a neighborhood of x.

The set of the exterior points of U is called the **exterior** of U.

Group: A set G together with a binary operation $\diamond : G \times G \rightarrow G$ has the structure of a **group** if and only if:

- $\diamond$ is associative, i.e. $x \diamond (y \diamond z) = (x \diamond y) \diamond z$ for all x, y and z in G.
- There exists $e \in G$ such that $x \diamond e = x$, for all x in G. The element e is called the *identity*.
- For each element x in G there exists an element x' in G such that $x \diamond x' = x' \diamond x = e$. The elements x and x' are said to be the *inverses* of each other.

In addition, G is said to be a *commutative* or *abelian* group if and only if $x \diamond y = y \diamond x$ for all x and y in G.

Homeomorphism: Let $f : X \subset \mathbf{R}^n \rightarrow Y \subset \mathbf{R}^m$ be a map between two subsets of two topological spaces. It is a **homeomorphism** if f is bijective and continuous, and the inverse map $f^{-1} : Y \rightarrow X$ is also continuous (in the subspace topologies of X and Y). X and Y are said to be **homeomorphic** if such a map exists.

Implicit Function Theorem: Let $f : \mathbf{R}^n \times \mathbf{R}^m \rightarrow \mathbf{R}^m$ be a smooth map in an open subset X of $\mathbf{R}^n \times \mathbf{R}^m$. Suppose $(a, b) \in X$, with $a \in \mathbf{R}^n$, $b \in \mathbf{R}^m$, and $f(a, b) = 0$. Let M be the $m \times m$ matrix:

$$\left(D_{n+j} f^i(a, b)\right)_{1 \leq i, j \leq m}$$

where $D_{n+j} f^i(a, b)$ denotes the first derivative at (a, b) of the ith component of f with respect to the $(n + j)^{\text{th}}$ variable of f.

If *det* $M \neq 0$, there is an open set $A \subset \mathbf{R}^n$ containing a and an open set $B \subset \mathbf{R}^m$ containing b, with the following property: For each $x \in A$ there is a unique $g(x) \in B$ such that $f(x, g(x)) = 0$. The function g is smooth.

Injective Map: Let E and F be two sets. A map $f : E \to F$ is **injective** if and only if:

$$\forall x, x' \in E : \ f(x) = f(x') \ \Rightarrow \ x = x'$$

or, equivalently:

$$\forall x, x' \in E : \ x \neq x' \ \Rightarrow \ f(x) \neq f(x').$$

Inner Product: Let E be a vector space over $\mathbf{R}$. An **inner product** $< , >$ on E is a map of $E \times E$ into $\mathbf{R}$:

$$(\vec{x}, \vec{y}) \in E \times E \ \mapsto < \vec{x}, \vec{y} > \in \mathbf{R}$$

such that:

- $\forall \vec{x} \in E :< \vec{x}, \vec{x} > \geq 0$.
- $\forall \vec{x} \in E : < \vec{x}, \vec{x} > = 0 \ \Leftrightarrow \ \vec{x} = \vec{0}$.
- $\forall \vec{x}, \vec{y} \in E :< \vec{x}, \vec{y} > = < \vec{y}, \vec{x} >$.
- $\forall \vec{x_1}, \vec{x_2}, \vec{y} \in E, \forall \lambda_1, \lambda_2 \in \mathbf{R}$:

$$< \lambda_1 \vec{x_1} + \lambda_2 \vec{x_2}, \vec{y} > = \ \lambda_1 < \vec{x_1}, \vec{y} > + \lambda_2 < \vec{x_2}, \vec{y} > .$$

Interior of a Set: Let U be any subset of a topological space E. A point $x \in E$ is called an **interior point** of U if and only if U is a neighborhood of x.

The set of the interior points of U is called the **interior** of U. It is denoted by $int(U)$.

Isomorphism: Let E be a set equipped with a binary operation $\diamond : E \times E \to E$. Let F be a set equipped with a binary operation $\triangle : F \times F \to F$. Let $f : E \to F$ be a bijective map between the two sets.

The map f is said to be an **isomorphism** if and only if:

$$\forall x, y \in E : f(x \diamond y) = f(x) \triangle f(y).$$

Since f is bijective, if it is an isomorphism, we also have:

$$\forall x, y \in F : f^{-1}(x \triangle y) = f^{-1}(x) \diamond f^{-1}(y).$$

If there exists an isomorphism f between E and F, E and F are said to be **isomorphic**.

Manifold: A topological space M is a **manifold** (resp. a **smooth manifold**) if every point $x \in M$ has an open neighborhood homeomorphic (resp. diffeomorphic) to an open ball of $\mathbf{R}^m$, for some m independent of x. The number m is the **dimension** of the manifold.

Manifold with Boundary: A subset M of a topological space X is an m-dimensional **manifold with boundary** if every point $x \in M$ has a neighborhood V (in the topology of X) such that the set $V \cap M$ is homeomorphic to either an open ball of $\mathbf{R}^m$ or a closed half-space of $\mathbf{R}^m$.

Measure Zero (Set of): A subset E of $\mathbf{R}^n$ has **measure zero** if it can be covered by a countable number of rectangloids (Cartesian products of n intervals in $\mathbf{R}$) with arbitrarily small total volume.

Metric: See **Distance**.

Metric Topology: Let (E, d) be a metric space. The topology on E induced by the distance d, called the **metric topology**, is the set $\mathcal{O}$ of all subsets $V \subset E$ such that:

$$\forall x \in V, \exists \varepsilon > 0 : B_\varepsilon(x) \subset V$$

where $B_\varepsilon(x) = \{y \in E \;/\; d(x, y) < \varepsilon\}$ is the open ball of radius ε centered at x.

Two distances d_1 and d_2 in E *induce the same topology* if and only if for any pair x and x' of elements of E:

- $\forall \varepsilon > 0$, $\exists \eta > 0$ such that: $d_1(x, x') < \eta \Rightarrow d_2(x, x') < \varepsilon$, and

- $\forall \eta > 0$, $\exists \varepsilon > 0$ such that: $d_2(x, x') < \varepsilon \Rightarrow d_1(x, x') < \eta$.

Minkowski Operators: Let X be an affine space whose origin is O, and A and B be any two subsets of X. The Minkowski operators $\oplus$

(addition) and $\ominus$ (subtraction) are defined by:

$$\begin{aligned} A \oplus B &= \{x \mid x = a + b, a \in A, b \in B\}, \\ \ominus B &= \{-b \mid b \in B\}, \\ A \ominus B &= A \oplus (\ominus B). \end{aligned}$$

Neighborhood: Let E be a topological space and x an element of E. A subset $U \subset E$ is called a **neighborhood** of x if and only if there is an open set V (in the topology of E) such that $x \in V \subset U$.

Norm: Let E be a vector space over $\mathbf{R}$. A **norm** on E is a map:

$$\vec{x} \in E \mapsto \|\vec{x}\| \in \mathbf{R}$$

such that:

- $\forall \vec{x} \in E : \|\vec{x}\| \geq 0.$
- $\|\vec{x}\| = 0 \Leftrightarrow \vec{x} = \vec{0}.$
- $\forall \vec{x} \in E, \forall \lambda \in \mathbf{R} : \|\lambda \vec{x}\| = |\lambda| \; \|\vec{x}\|.$
- $\forall \vec{x}, \vec{y} \in E : \|\vec{x} + \vec{y}\| \leq \|\vec{x}\| + \|\vec{y}\|.$

If an inner product $< , >$ has been specified on E, then the map:

$$\vec{x} \in E \mapsto \sqrt{< \vec{x}, \vec{x} >} \in \mathbf{R}$$

defines a norm on E. By definition, a **Euclidean vector space** is a vector space in which one has defined an inner product and the norm is the norm associated with this inner product.

Outer Product: Let $\vec{x}$ and $\vec{y}$ be two vectors of the Euclidean space $\mathbf{R}^2$ whose components in an orthonormal basis are x_1 and x_2, and y_1 and y_2, respectively. The **outer product** of $\vec{x}$ and $\vec{y}$, denoted by $\vec{x} \wedge \vec{y}$, is the scalar z defined by:

$$z = x_1 y_2 - x_2 y_1.$$

Let $\vec{x}$ and $\vec{y}$ be two vectors of the Euclidean space $\mathbf{R}^3$, whose components in an orthonormal basis β are x_1, x_2 and x_3, and y_1, y_2 and y_3, respectively. The **outer product** of $\vec{x}$ and $\vec{y}$, denoted by $\vec{x} \wedge \vec{y}$, is the vector $\vec{z} \in \mathbf{R}^3$, whose components z_1, z_2 and z_3 in β are defined by:

$$z_1 = x_2y_3 - x_3y_2,$$
$$z_2 = x_3y_1 - x_1y_3,$$
$$z_3 = x_1y_2 - x_2y_1.$$

Path-Connectedness: A topological space E is **path-connected** if any two elements of E are connected by a path in E.

In a manifold, connectedness and path-connectedness are two equivalent notions.

Quotient Topology: Let X be a topological space and $\sim$ an equivalence relation on X. The quotient space $X/\sim$ can be given the structure of a topological space as follows: A subset U of $X/\sim$ is open if and only if the subset of X containing the elements of the equivalent classes in U is an open subset of X. This topology is called the **quotient topology** induced by the topology of X and the relation $\sim$.

Regular Set: Let Y be a subset of a topological space X. Y is said to be a **regular set** if and only if it is equal to the closure of its interior, i.e. $Y = cl(int(Y))$.

Retraction: Let X be a topological space and A be a subset of it. A **retraction** from X onto A is a surjective continuous map $X \rightarrow A$ whose restriction to A is the identity.

Subspace Topology: Let $(E, \mathcal{O})$ be a topological space and F a subset of E. The set $\mathcal{O}' = \{X \cap F \ / \ X \in \mathcal{O}\}$ is a topology on F and is called the **subspace topology**. The topological space $(F, \mathcal{O}')$ is called a **subspace** of $(E, \mathcal{O})$.

Surjective Map: Let E and F be two sets. A map $f : E \rightarrow F$ is **surjective** if and only if:

$$\forall y \in F, \exists x \in E : y = f(x).$$

f is also said to be a map of E **onto** F.

Topology: A **topology** on a set E is a set $\mathcal{O}$ of subsets of E such that:

- Any union of subsets in $\mathcal{O}$ belongs to $\mathcal{O}$.

- The intersection of any two subsets in $\mathcal{O}$ belongs to $\mathcal{O}$.

- $\emptyset$ and E belong to $\mathcal{O}$.

The pair $(E, \mathcal{O})$ is called a **topological space**. When there is no ambiguity about the topology, E alone is called the topological space.

The elements of $\mathcal{O}$ are called **open sets**. Every set $E \backslash U$, where $U \in \mathcal{O}$, is called a **closed set**.

Vector Space: A set E is a **vector space** over $\mathbf{R}$ if it has:

- A binary operation $E \times E \to E$, called $+$, with which it is an abelian group, i.e. :
  . $\forall x, y \in E : x + y = y + x$,
  . $\forall x, y, z \in E : x + (y + z) = (x + y) + z$,
  . $\exists e \in E, \forall x \in E : x + e = x$,
  . $\forall x \in E, \exists x' \in E : x + x' = e$.

- An operation $E \times \mathbf{R} \to E$, called $\times$, such that:
  . $\forall x, y \in E, \forall \lambda \in \mathbf{R} : \lambda \times (x + y) = \lambda \times x + \lambda \times y$,
  . $\forall x \in E, \forall \lambda, \mu \in \mathbf{R} : (\lambda + \mu) \times x = \lambda \times x + \mu \times x$,
  . $\forall x \in E, \forall \lambda, \mu \in \mathbf{R} : (\lambda\mu) \times x = \lambda \times (\mu \times x)$,
  . $\forall x \in E : 1 \times x = x$.

Often, $\lambda \times x$ is written λx, the elements of E are denoted by $\vec{x}$, and the identity element e in E is written $\vec{0}$.

Weierstrass' Approximation Theorem: Any continuous function $f : \Delta \subset \mathbf{R} \to \mathbf{R}$, where Δ is a compact interval of $\mathbf{R}$, is the limit of a sequence of polynomials that uniformly converges in Δ.

Appendix B

Computational Complexity

This appendix briefly reviews important concepts in Computational Complexity. A more comprehensive presentation of these concepts can be found in various books such as [Garey and Johnson, 1979] [Aho, Hopcroft and Ullman, 1983] and [Sedgewick, 1988].

The most important (and most commonly used) measure of the performance of an algorithm A is its **running time**, i.e. the time required by its execution. This time should be evaluated as a function of the input of A.

In general, we want to estimate the running time, not as a function of the exact input, which would be too specific, but as a function of one or a few parameters measuring the **size** of the input. In motion planning, typical parameters for measuring the size of a problem are the dimension of the configuration space, the number of polynomial constraints defining the shape of the objects, and the maximal degree of the polynomials used to express these constraints.

Let us assume that the size of the input to A is measured by a single parameter denoted by n. Let $T(n)$ denote the running time of A as a function of n. $T(n)$ is still not a well-defined function, since several inputs of the same size may result in different running times. For instance, one could construct a sorting algorithm that first checks whether the input list is already sorted, and proceeds further only if the list is not sorted. The running time of this algorithm obviously depends on whether the input list is sorted or not. Usually, $T(n)$ measures the maximum running time required by A for any input of size n, i.e. the running time in the **worst case**.

We must be aware that worst-case analysis is sometimes misleading. When the worst case is unlikely to happen in practice, it may be more appropriate to determine the average running time of A. But average-case analysis is often considerably more complicated than worst-case analysis, and it requires a realistic probabilistic distribution of the possible inputs to be selected. One may also be interested in best-case evaluation.

Let us assume that $T(n)$ stands for the running time of A in the worst case. The exact expression of $T(n)$ depends on many implementation factors and usually presents little interest. Instead, the important, implementation-independent notion is the rate of growth of $T(n)$ when n becomes large (it is also called the asymptotic behavior of $T(n)$). Let $O(f(n))$ denote the set of real functions $g(n)$ such that there exist positive constants α and n_0 verifying $g(n) \leq \alpha\ f(n)$, for all $n \geq n_0$. We say that the running time of A is $O(f(n))$, or, equivalently, that the **time complexity** of A is $O(f(n))$, if and only if $T(n) \in O(f(n))$. Since $O(f(n))$ only determines an upper bound on the actual growth of $T(n)$, its evaluation should not be too conservative in order to be reasonably informative.[1]

An algorithm of complexity $O(f_1(n))$ is said to be more complex than (resp. as complex as) an algorithm of complexity $O(f_2(n))$ if $O(f_2(n)) \subset O(f_1(n))$ (resp. $O(f_2(n)) = O(f_1(n))$). However, the important distinction is between algorithms that require polynomial time and those that require exponential time. A **polynomial-time** algorithm is defined to be one whose time complexity is $O(n^q)$, for some $q \geq 0$. An **exponential-time** algorithm is any algorithm whose complexity cannot be so bounded[2]. The rate of growth of an exponential-time algorithm

[1]It is sometimes useful to also consider the lower bound on the growth rate of $T(n)$. It is denoted by $\Omega(n)$.

[2]Note that an algorithm of time complexity $O(n^{\log n})$ is often said to be time

is such that, in the worst cases, this algorithm can at best be applied to problems of small size. Inversely, a polynomial-time algorithm has a much smaller rate of growth. Thus, a problem is often not considered solved as long as a polynomial-time algorithm is not known for it.

The reader is cautioned that a polynomial-time algorithm may not be a practical solution of a problem, since the multiplicative constant and/or the polynomial degree might be quite large. On the other hand, there exist exponential-time algorithms which are practical (e.g. the simplex algorithm). In addition, two algorithms with the same complexity $O(f(n))$ may behave very differently for small values of n. An algorithm of time complexity $O(f_1(n))$ may also run faster than an algorithm of time complexity $O(f_2(n)) \subset O(f_1(n))$, for sufficiently small values of n. Sometimes, these small values of n correspond to most of the practical cases.

The above notions can easily be generalized to inputs whose sizes are characterized by several parameters. They can also be extended to performance measures other than time. For example, saying that the space complexity of an algorithm is $O(f(n))$ means that the amount of memory space required to run the algorithm is a function $S(n)$ belonging to $O(f(n))$.

Now, rather than focusing on a particular algorithm, we may consider a problem $\mathcal{P}$ and attempt to characterize its intrinsic computational complexity (time, space), i.e. the smallest complexity of known and unknown algorithms that solve $\mathcal{P}$. Indeed, if we can show that solving a problem inherently requires a number of steps exponential in its size, it is futile to spend much effort finding a polynomial-time algorithm for the problem. It is probably more rewarding to try to understand the problem better and to make it simpler, or more specific. In fact, the important issue is to determine whether a problem is **tractable**, i.e. it admits a polynomial-time solution, or **intractable**, i.e. it admits only exponential-time solutions. Unfortunately, determining the complexity of a problem is a far more delicate issue than evaluating the complexity of a particular algorithm.

The usual approach to the analysis of the complexity of a problem is to establish the membership of the problem in a class of problems by showing its equivalence to a problem already in that class. P, NP, and

exponential, although $n^{\log n}$ is not usually considered an exponential function of n.

PSPACE are among the most frequently considered classes of problems:

- Class P corresponds to the problems that can be solved by an algorithm of polynomial time complexity on a deterministic Turing machine.
- Class NP corresponds to the problems that can be solved by an algorithm of polynomial time complexity on a nondeterministic Turing machine, i.e. a machine that has the ability to perform an unlimited number of independent computational sequences in parallel.
- Class PSPACE corresponds to the problems that can be solved by an algorithm of polynomial space complexity on a nondeterministic (hence, on a deterministic) machine.

It has been proven that P $\subseteq$ NP $\subseteq$ PSPACE. Although there is strong evidence that this inclusion relation is proper, this has not been proven yet.

Therefore, if a problem is known to be in NP or in PSPACE, it is reasonable to believe that the time required to solve this problem is inherently exponential in the size of the problem. If a problem is known to be in P, it is recognized to be tractable.

Let C be a class of problem (e.g. P, NP or PSPACE). A problem $\mathcal{P}$ is said to be **hard** for C, or C-**hard**, if every problem in C is reducible to $\mathcal{P}$ by a polynomial time transformation, but $\mathcal{P}$ is not necessarily in C. This intuitively means that $\mathcal{P}$ is at least as hard as any problem in C. Indeed, the existence of an algorithm for $\mathcal{P}$ implies the existence of an algorithm of the same class of complexity for any problem in C. A problem $\mathcal{P}$ that is both in C and hard for C is said to be **complete** for C, or C-**complete**.

Locating precisely a problem in the hierarchy of classes given above provides evidence of a **lower bound** on the computational complexity of a problem. Indeed, every algorithm for this problem can be expected to have a complexity higher than or equal to this bound. Inversely, exhibiting an algorithm that solves a problem provides an **upper bound** on the inherent complexity of the problem.

Appendix C

Graph Searching

Motion planning methods transform a "continuous" problem (e.g. finding a path in a manifold) into the "discrete" problem of searching a graph (e.g. the visibility graph, the Voronoi diagram, the connectivity graph) between an initial node and a goal node. Various methods have been developed for searching graphs. They are described in detail in various textbooks such as [Nilsson, 1980] [Aho, Hopcroft and Ullman, 1983] [Sedgewick, 1988] and [Brassard and Bratley, 1988]. We briefly present a few of them in this appendix.

Let $G = (X, A)$ be a finite directed graph[1] consisting of a set of nodes X and a set of arcs A. G may or may not be connected. Let n be the number of nodes and r be the number of arcs in G. In the worst case, $r \in O(n^2)$. We assume below that $r \geq n$, which is the usual case.

There are many ways to represent G. We assume below that the *adjacency list representation* is used. The adjacency list for a node N is a list, in some order, of all the nodes N' adjacent to N (a node N' is *adjacent* to N if there is an arc of G connecting N to N'). This representation of

[1] An undirected graph can be transformed into a directed one by replacing every link by two opposite arcs.

G requires $O(n + r)$ space.

1 Basic Methods

One simple method for searching G is the **depth-first search** method. Assume that all the nodes of G are initially marked *unvisited*. Depth-first search starts with the initial node N_{init} and operates recursively, until the goal node N_{goal} is attained or all the nodes that can be reached from N_{init} have been visited. Each recursion consists of marking the current node N (initially N_{init}) *visited* and searching G from an *unvisited* node adjacent to N. During the search, the algorithm constructs a tree T as follows. Whenever a node N is considered and leads to an *unvisited* node N', N' is added to T together with an arc connecting N' to N. T is the *spanning tree* of the subset of G so far visited. If N_{goal} is ultimately attained, a path is constructed by tracing the arcs in T from N_{goal} to N_{init}. The method takes $O(r)$ time. Indeed, since each node is marked *visited* the first time it is visited, each arc is traversed only once.

Another simple method for searching G is the **breadth-first search** method. This method uses a queue of nodes denoted by OPEN that is initialized to the queue containing N_{init} only. All the nodes of G are initially marked *unvisited*. Breadth-first search works iteratively. While OPEN is not empty, it considers the first node N in OPEN and removes it from OPEN. If any of the nodes adjacent to N is N_{goal}, it terminates. Otherwise, it marks *visited* every *unvisited* node N' adjacent to N and places it at the end of OPEN. During the search, a spanning tree T of the subset of G so far visited is constructed as in the depth-first search method. If N_{goal} is ultimately attained, a path is constructed by tracing the arcs in T from N_{goal} to N_{init}. The method takes $O(r)$ time. Indeed, every node visited is placed in OPEN once, so that the while-loop is executed once for each node. Each arc is examined once.

2 A^* Algorithm

In many cases, arcs in G are labeled by costs (numbers in $\mathbf{R}^+$) and the cost of a path is defined as the sum of the costs attached to its arcs. In these cases, we are interested in generating a minimum-cost path between the initial node and the goal node. A classical algorithm is the

A^* algorithm [Hart, Nilsson and Raphael, 1968] [Nilsson, 1980]. Under a simple condition specified below, A^* is guaranteed to return a path of minimum cost whenever a path exists, and to return failure otherwise.

A^* explores G iteratively by following paths originating at N_{init}. At the beginning of every iteration, there are some nodes that the algorithm has already visited, and there may be others that are still unvisited. For each visited node N, the previous iterations have produced one or several paths connecting N_{init} to N, but the algorithm only memorizes a representation of a path of minimum cost (among those so far constructed). At any instant, the set of all such paths forms a spanning tree T of the subset of G so far explored. T is represented by associating to each visited node N (except N_{init}) a pointer to its parent node in the current T.

A^* assigns a cost function $f(N)$ to every node N in the current T. This function is an estimate of the cost of the minimum-cost path in G connecting N_{init} to N_{goal} and constrained to go through N. It is computed as follows:

$$f(N) = g(N) + h(N)$$

where:

- $g(N)$ is the cost of the path between N_{init} and N in the current T,
- $h(N)$ is a *heuristic* estimate of the cost $h^*(N)$ of the minimum-cost path between N and N_{goal} in G.

The A^* algorithm is given below. The inputs consist of G, N_{init}, N_{goal}, k, and h, where $k : X \times X \rightarrow \mathbf{R}^+$ is the partial function specifying the cost of each arc in G. All the nodes of G are initially marked *unvisited*. The algorithm makes use of a list denoted by OPEN that contains nodes of G sorted by the values of the function f. The list OPEN supports the following operations:

- FIRST(OPEN): remove the node of OPEN with the smallest value of f and return it,
- INSERT(N,OPEN): insert node N in OPEN,
- DELETE(N,OPEN): remove node N from OPEN,

- **MEMBER**(N,OPEN): evaluate to **true** if N is in OPEN and to **false** otherwise,

- **EMPTY**(OPEN): evaluate to **true** if OPEN is empty and to **false** otherwise.

At each iteration, A^* explores the nodes adjacent to the node returned by FIRST(OPEN). Initially, both the tree T and the list OPEN are empty.

```
procedure A*(G,N_init,N_goal,k,h);
  begin
    install N_init into T;
    INSERT(N_init,OPEN); mark N_init visited;
    while ¬EMPTY(OPEN) do
      begin
        N ← FIRST(OPEN);
        if N = N_goal then exit while-loop;
        for every node N' adjacent to N in G do
          if N' is not visited then
            begin
              add N' to T with a pointer toward N;
              INSERT(N',OPEN); mark N' visited;
            end;
          else if g(N') > g(N) + k(N, N') then
            begin
              modify T by redirecting the pointer of N' toward N;
              if MEMBER(N',OPEN) then DELETE(N',OPEN);
              INSERT(N',OPEN);
            end;
      end;
    if ¬EMPTY(OPEN) then
      return the constructed path by tracing the pointers in T from N_goal
      back to N_init;
    else return failure;
  end;
```

The examination of the nodes adjacent to a node N in the main loop of the algorithm is called the *expansion* of N.

The inner loop in the algorithm may require some explanation. When the expansion of N produces a *visited* node N', it may happen that the new way of attaining N' provides a path that is less costly than any of the previously generated paths from N_{init} to N'. Then, the algorithm updates T by redirecting the pointer issued from N'. If N' is in OPEN, its position in the list must be updated according to the new value of $f(N')$. If N' is not in OPEN, it may happen that the better path discovered between N_{init} and N' also provides better paths for attaining previously visited successors of N' which are currently not successors of N' in T; rather than doing the modification immediately, the algorithm reinserts N' in OPEN, so that its successors will eventually be reconsidered later. (Hence, at any instant, the tree T may not strictly contain the minimum-cost paths discovered so far.)

A variant of the above algorithm consists of inserting new nodes in OPEN without sorting them by the values of f. Then, at every iteration, OPEN must be entirely scanned in order to identify a node N with the smallest value of the function f.

The heuristic function h is said to be **admissible** if and only if it satisfies:

$$\forall N \in G: \quad 0 \leq h(N) \leq h^*(N).$$

Under the condition that h is admissible, A^* is guaranteed to generate a minimum-cost path between the initial and goal nodes whenever these two nodes are connected by a path in G, and to return failure otherwise.

A trivial admissible heuristic function is $h_1(N) = 0$ for all N. A^* together with this "non-informed" heuristic function is known as the Dijkstra's algorithm. It explores G in a kind of breadth-first fashion. Another admissible heuristic function $h_2(N)$, applicable when the nodes of G are points in $\mathbf{R}^n$ and $k(N_1, N_2)$ is the length of a path joining N_1 and N_2, is the Euclidean distance from N to N_{goal}. Intuitively, h_2 seems better than h_1. This intuition is formalized as follows. If h_1 and h_2 are two admissible heuristic functions such that:

$$\forall N \in G: \quad h_2(N) \geq h_1(N)$$

h_2 is said to be **more informed** than h_1. It can be proven that if h_2 is more informed than h_1, then every node of G expanded by A^* using h_2 is also expanded by A^* using h_1 [Nilsson, 1980].

The heuristic function h is said to be **locally consistent** (or to satisfy the **monotone restriction**) if and only if for every pair of nodes N and N', such that N' is adjacent to N, it verifies:

$$0 \leq h(N) \leq h(N') + k(N, N').$$

It is easy to verify that a locally consistent heuristic function is also admissible.

If h is locally consistent, then whenever A^* selects a node N for expansion, it has already found an optimal path from N_{init} to N [Nilsson, 1980]. This means that once a node is removed from OPEN, it will never be reinserted into OPEN. Both the admissible heuristic functions h_1 and h_2 defined above are locally consistent.

The length of OPEN is $O(n)$. Assuming that a locally consistent heuristic function is used, the time complexity of A^* is $O(n^2)$, if OPEN is not sorted by the values of f. Indeed, there are $O(n)$ iterations. At each iteration, the identification of the node with the smallest f requires $O(n)$ time and the treatment of its children also requires $O(n)$ time. If OPEN is sorted, A^* requires $O(r \log n)$ time (where r is the number of arc of G). Indeed, each arc of G may be traversed once and every node may be inserted in OPEN once. Every insertion costs $O(\log n)$.

The output-sensitive time complexity of A^*, i.e. its time complexity as a function of the length of the minimum-cost path, is explored in [Pohl, 1977], [Gaschnig, 1979] and [Pearl, 1983].

Appendix D

Sweep-Line Algorithm

In several places of this book we refer to algorithmic techniques known as "sweep-line" techniques in order to solve a variety of geometrical problems. Although the details of the techniques vary from one problem to another, their basic principle remains the same. In this appendix we expose this principle on an intersection computation problem. Our presentation is drawn from [Preparata and Shamos, 1985] and corresponds to an algorithm initially proposed in [Bentley and Ottmann, 1979].

Line sweeping is of interest for two reasons. On the one hand, it often allows us to construct algorithms that surpass more straightforward algorithms in time efficiency. On the other hand, it provides a convenient framework for organizing possibly complicated processing.

Consider the following problem:

> Given n line segments, each described by the coordinates (x, y) of its endpoints, compute all their intersection points and sort these points by their abscissae.

A straightforward algorithm solves this problem by considering each pair of line segments in turn. For each pair, it determines whether the two

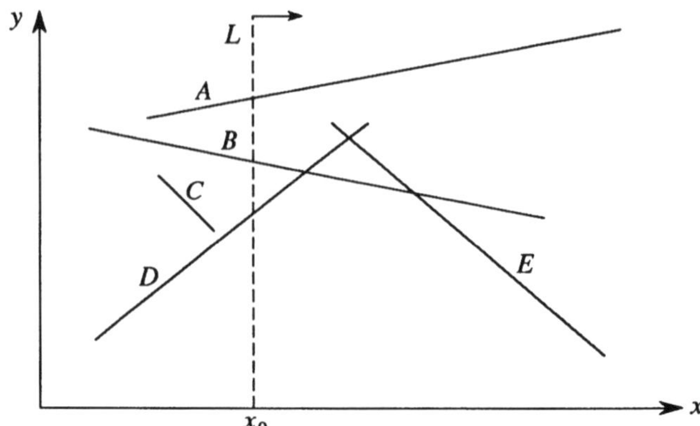

Figure 1. This figure shows an arrangement of 5 line segments (A, B, C, D, E) in the plane. The sweep-line technique consists of moving a vertical line L across the plane from left to right. At any given abscissa, the status of the line is the sorted list of its intersections with the segments. For instance, at x_0, it is $(D\ B\ A)$. The status changes at a finite number of events. Each event occurs either when L encounters an endpoint of a line segment or the intersection point of two segments.

---

segments intersect and, if this is the case, it calculates the coordinates of the intersection point. Let c be the number of intersection points. In the worst case $c \in O(n^2)$, but often c is much smaller. The c intersection points are computed in time $\Theta(n^2)$, i.e. in time exactly proportional to n^2. Sorting these points then takes time $O(c \log c)$. Hence, the overall algorithm requires $O(n^2 + c \log c)$ time. The sweep-line algorithm described below provides a more efficient solution when c is significantly smaller than n^2, which is usually the case.

Line sweeping consists of moving a vertical line L (i.e. a line parallel to the y-axis) from left to right across the planar arrangement of the n line segments. (The line L could be moved from right to left as well; it could also be horizontal, i.e. parallel to the x-axis, and be swept upward or downward.) At any given abscissa x_0, the line L intersects some subset of the line segments, called the **active** segments at x_0. The **status** $\mathcal{S}$ of L is defined as the list of active segments at the current abscissa. $\mathcal{S}$ is sorted by the ordinates of the intersections of L with the active segments. Consider for instance the line segments shown in Figure 1. At abscissa x_0, we have $\mathcal{S} = (D\ B\ A)$. If two segments X_1 and X_2 are adjacent in

$\mathcal{S}$, with X_1 occurring before X_2, X_1 is said to be *below* X_2, and X_2 is said to be *above* X_1.

During the sweep, $\mathcal{S}$ changes only at a finite number of "critical" abscissae x_c, each being one of the following types:

1. x_c is the abscissa of the left endpoint of a line segment X. X is added to $\mathcal{S}$.

2. x_c is the abscissa of the right endpoint of a line segment X. X is removed from $\mathcal{S}$.

3. x_c is the abscissa of the intersection point of two line segments X_1 and X_2. X_1 and X_2 exchange places in $\mathcal{S}$.

In the following we assume that no line segment is vertical and that no two critical abscissae are the same. Simple adaptations of the algorithm can handle such coincidences.

The sweep-line algorithm makes use of two data structures:

- One is the current status $\mathcal{S}$ of the sweep-line L. It supports the following operations:
  . INSERT($X, \mathcal{S}$): insert X in $\mathcal{S}$,
  . DELETE($X, \mathcal{S}$): remove X from $\mathcal{S}$,
  . ABOVE($X, \mathcal{S}$): return the segment above X in $\mathcal{S}$, if any,
  . BELOW($X, \mathcal{S}$): return the segment below X in $\mathcal{S}$, if any.

- The other data structure, denoted by $\mathcal{E}$, is the queue of scheduled future events. It supports the following operations:
  . FIRST($\mathcal{E}$): remove the smallest abscissa from $\mathcal{E}$ and return it,
  . INSERT($x, \mathcal{E}$): insert abscissa x in $\mathcal{E}$,
  . **MEMBER**($x, \mathcal{E}$): determine if abscissa x is a member of $\mathcal{E}$.

Each of these two data structures is implemented as a balanced tree [Aho, Hopcroft and Ullman, 1983] [Sedgewick, 1988], so that each of the above operations can be done in time at most logarithmic in the number of elements in the data structure.

Initially, both $\mathcal{S}$ and $\mathcal{E}$ are empty. The algorithm first places the abscissae of the $2n$ endpoints of the line segments in $\mathcal{E}$. Then, it operates iteratively. At each iteration, it removes the smallest abscissa from $\mathcal{E}$ (i.e. the next event) and updates $\mathcal{S}$ and $\mathcal{E}$ according to the type of this event, i.e. left endpoint, right endpoint, intersection point. Appropriate information is attached to each abscissa so that the type of the event and the relevant information can be retrieved. Updating $\mathcal{S}$ consists of adding a line segment to $\mathcal{S}$, removing one, or interchanging the order of two line segments. Every pair of line segments that are made adjacent in $\mathcal{S}$ by this modification are then checked for intersection. If they intersect and if this intersection point is new, its abscissa is inserted in $\mathcal{E}$. Every time an abscissa is removed from $\mathcal{E}$, if it is the abscissa of the intersection point of two line segments, then it is inserted in a queue $\mathcal{L}$. When the algorithm terminates, $\mathcal{L}$ is the sorted list of the abscissae of the intersection points.

Notice that the algorithm may "discover" the same intersection point several times. For example, in Figure 1, the intersection between the segments D and B is discovered a first time when the line L encounters the left endpoint of B. It discovers it a second time when the line encounters the right endpoint of C. The algorithm avoids inserting the same abscissa several times in $\mathcal{E}$ by checking that the intersection point is a new one.

A more formal expression of the sweep-line algorithm is the following:

```
procedure LINESWEEP;
begin
    place the abscissae of the 2n endpoints of the line segments into ℰ;
    𝒮 ← ∅; ℒ ← ∅; 𝒬 ← ∅;
    [𝒬 is a queue used internally by the procedure]
    while ℰ ≠ ∅ do
        begin
            x ← FIRST(ℰ);
            p ← endpoint or intersection point of which x is the abscissa;
            if p is a left endpoint then
                begin
                    X ← segment of which p is endpoint;
                    INSERT(X, 𝒮);
                    X₁ ← ABOVE(X, 𝒮);
                    X₂ ← BELOW(X, 𝒮);
                    if X₁ intersects X then INSERT((X₁, X), 𝒬);
```

```
                if X2 intersects X then INSERT((X, X2), Q);
            end;
        else
            if p is a right endpoint then
                begin
                    X ← segment of which p is endpoint;
                    X1 ← ABOVE(X, S);
                    X2 ← BELOW(X, S);
                    if X1 intersects X2 then INSERT((X1, X2), Q);
                    DELETE(X, S);
                end;
            else [p is an intersection point]
                begin
                    L ← INSERT(x, L);
                    (X1, X2) ← pair of segments intersecting at p;
                    [with X1 = ABOVE(X2) on the left of p]
                    X3 ← ABOVE(X1, S);
                    X4 ← BELOW(X2, S);
                    if X3 intersects X2 then INSERT((X3, X2), Q);
                    if X1 intersects X4 then INSERT((X1, X4), Q);
                    interchange X1 and X2 in S;
                end;
        while Q ≠ ∅ do
            begin
                (X, X') ← FIRST(Q);
                x ← abscissa of the intersection point of X and X';
                if ¬ MEMBER(x, E) then INSERT(x, E);
            end;
    end;
end;
```

Placing the $2n$ abscissae in $\mathcal{E}$ takes time $O(n \log n)$. The first three blocks in the main while-loop of the procedure are mutually exclusive and take time $O(n)$ each. Let c be the number of intersections of the n line segments. The main while-loop is executed exactly $2n + c$ times. The total number of intersection pairs produced by the first three blocks in the main while-loop over the $2n + c$ iterations is $O(n + c)$. Thus, the inner while-loop is executed $O(n + c)$ times. Each execution of this loop takes time $O(\log(n + c)) = O(\log n)$, since $c \in O(n^2)$. Therefore, the overall sweep-line procedure takes time $O((n + c) \log n)$ with $c \in O(n^2)$. If c is smaller than $O(n^2)$ this complexity is better than that of the straightforward algorithm given above. In some specific cases, the difference may be much greater. For instance, if we were only interested in computing the left-most intersection along each segment and interrupting the segment beyond this intersection (this is basically what we do when we compute the backprojection of a goal region in Chapter 10), then the number of intersections to be computed would be $c \in O(n)$ and the time complexity of the sweep-line algorithm would be $O(n \log n)$, while the straightforward algorithm would require time $\Theta(n^2)$.

In the book, we use several variants of the above algorithm. For instance, in order to compute the intersection of generalized polygons (Chapter 4), we sweep a line across a planar arrangement of straight and circular curve segments. For constructing a visibility graph (Chapter 4), the sweep-line is a half-line (ray) rotating about its endpoint. In order to construct a backprojection (Chapter 10), we sweep a line and construct the boundary of a region; the action to be taken at each event is more involved than in the above algorithm. In all these applications, the computation is **monotonic**, i.e. at any moment the sweep-line divides the space into two half-spaces; the solution to the problem already established in one of the two half-spaces will not be affected by the computation to be carried out later in the other half-space. Furthermore, there exists a finite set of discrete events that divide the plane into parallel strips (or angular sectors, if the sweep-line is a rotating line), such that between any two consecutive events the relevant description of the intersection of the sweep-line with the geometrical arrangement being swept, i.e. the status of the sweep-line, remains unchanged. Finally, events can be classified into a finite number of types (in the above example, these are left endpoint, right endpoint, and intersection of two segments), each requiring the same processing algorithm to be applied.

Sweep techniques can be adapted to spaces of higher dimensions. Generally speaking, suppose that we have an arrangement of geometric objects in $\mathbf{R}^m$ ($m \geq 2$). Space sweep consists of moving a hyperplane of dimension $m-1$ along the mth axis, thus reducing the original m-dimensional static problem into an $(m-1)$-dimensional dynamic problem [Guibas and Stolfi, 1989] (see the Silhouette Method, Section 4 of Chapter 4).

References

Aho, A.V., Hopcroft, J.E. and Ullman, J.D. (1983) *Data Structures and Algorithms*, Addison-Wesley, Reading, MA.

Ahuja, N., Chien, R.T., Yen, R. and Bridwell, N. (1980) "Interference Detection and Collision Avoidance Among Three Dimensional Objects," *Proceedings of the First AAAI Conference*, Stanford, 44-48.

Akman,V. (1987) *Unobstructed Shortest Paths in Polyhedral Environments*, Lecture Notes in Computer Science, 251, Springer-Verlag, Berlin.

Akritas, A.G. (1980) "The Fastest Exact Algorithms for the Isolation of the Real Roots of a Polynomial Equation," *Computing*, 24, 299-313.

Alami, R., Siméon, T. and Laumond, J.P. (1989) "A geometrical Approach to Planning Manipulation Tasks. The Case of Discrete Placements and Grasps," *Preprints of the Fifth International Symposium of Robotics Research*, Tokyo, 113-119.

Alexander, H.L. (1987) *Experiments in Control of Satellite Manipulators*, Ph.D. Dissertation, Department of Electrical Engineering, Stanford University.

Alexander, J.C. (1984) "On the Motion of a Trailer-Truck," Problem 84-12, *SIAM Review*, 26, 579; Solutions, *SIAM Review*, 27 (1985), 578-579.

Alexander, J.C. and Maddocks, J.H. (1988) "On the Maneuvering of

Vehicles," *SIAM Journal of Applied Mathematics*, 48(1), 38-51.

Ambler, A.P. and Popplestone, R.J. (1975) "Inferring the Positions of Bodies From Specified Spatial Relationships," *Artificial Intelligence*, 6(2), 157-174.

Anderson, R.F. and Orey, S. (1976) "Small Random Perturbations of Dynamical Systems with Reflecting Boundary," *Nagoya Mathematics Journal*, 60, 189-216.

Arnold, V.I. (1978) *Mathematical Methods of Classical Mechanics*, Springer-Verlag, New York.

Arnon, D.S. (1979) *A Cellular Decomposition Algorithm for Semi-Algebraic Sets*, Technical Report No. 353, Department of Computer Science, University of Wisconsin.

Arnon, D.S. (1988) "Geometric Reasoning With Logic and Algebra," *Artificial Intelligence*, 37(1-3), 37-60.

Asano, T., Asano, T., Guibas, L., Hershberger, J. and Imai, H. (1986) "Visibility of Disjoint Polygons," *Algorithmica*, 1(1), 49-63.

Avnaim, F. and Boissonnat, J.D. (1988) *Polygon Placement Under Translation and Rotation*, Technical Report No. 889, INRIA, Sophia-Antipolis, France.

Avnaim, F., Boissonnat, J.D. and Faverjon, B. (1988) *A Practical Exact Motion Planning Algorithm for Polygonal Objects Amidst Polygonal Obstacles*, Technical Report No. 890, INRIA, Sophia-Antipolis, France.

Ayala, D., Brunet, P., Juan, R. and Navazo, I. (1985) "Object Representation by Means of Nonminimal Division Quadtrees and Octrees," *ACM Transactions on Graphics*, 4(1), 41-59.

Bajaj, C. and Dey, T.K. (1988) *Convex Decompositions of Simple Polyhedra*, Technical Report No. CSD-TR-833, Computer Sciences Department, Purdue University.

Baker, B.S., Fortune, S. and Grosse, E. (1985) "Stable Prehension with a Multi-Fingered Hand," *Proceedings of the IEEE International Conference on Robotics and Automation*, St. Louis, MO, 570-575.

Baldwin, D.F. (1990) *Algorithmic Methods and Software Tools for the*

Generation of Mechanical Assembly Sequences, Master's Thesis, Department of Mechanical Engineering, MIT, MA.

Bañon, J. (1990) "Implementation and Extension of the Ladder Algorithm," *Proceedings of the IEEE International Conference on Robotics and Automation*, Cincinnati, OH, 1548-1553.

Barraquand, J., Langlois, B. and Latombe, J.C. (1989a) "Robot Motion Planning with Many Degrees of Freedom and Dynamic Constraints," *Preprints of the Fifth International Symposium of Robotics Research*, Tokyo, 74-83.

Barraquand, J., Langlois, B. and Latombe, J.C. (1989b) *Numerical Potential Field Techniques for Robot Path Planning*, Report No. STAN-CS-89-1285, Department of Computer Science, Stanford University.

Barraquand, J. and Latombe, J.C. (1989a) *Robot Motion Planning: A Distributed Representation Approach*, Report No. STAN-CS-89-1257, Department of Computer Science, Stanford University. (To appear in *International Journal of Robotics Research.*)

Barraquand, J. and Latombe, J.C. (1989b) "On Non-Holonomic Mobile Robots and Optimal Maneuvering," *Revue d'Intelligence Artificielle*, 3(2), Hermes, Paris, 77-103.

Barraquand, J. and Latombe, J.C. (1990a) "A Monte-Carlo Algorithm for Path Planning with Many Degrees of Freedom," *Proceedings of the IEEE International Conference of Robotics and Automation*, Cincinnati, OH, 1712-1717.

Barraquand, J. and Latombe, J.C. (1990b) *Controllability of Mobile Robots with Kinematic Constraints*, Technical Report No. STAN-CS-90-1317, Department of Computer Science, Stanford University.

Bausch, J.J. and Youcef-Toumi, K. (1990) "Kinematic Methods for Automated Fixture Reconfiguration Planning," *Proceedings of the IEEE International Conference of Robotics and Automation*, Cincinnati, OH, 1396-1401.

Bentley, J.L. and Ottman, T.A. (1979) "Algorithms for Reporting and Counting Geometric Intersections," *IEEE Transaction on Computers*, 28, 643-647.

Binford, T.O. (1971) "Visual Perception by Computers," *Proceedings of the IEEE Conference on Systems Science and Cybernetics*, Miami.

Bishop, R.L. and Goldberg, S.I. (1980) *Tensor Analysis on Manifolds*, Dover Publications, New York.

Blum, M., Griffith, A. and Neumann, B. (1970) *A Stability Test for Configurations of Blocks*, AI Memo 188, Artificial Intelligence Laboratory, MIT.

Boaz, M. and Roach, J. (1985) "An Octree Representation for Three-Dimensional Motion and Collision Detection," *Proceedings of the SIAM Conference on Geometrical Modeling and Robotics*, Albany, NY.

Boneschanscher, N., Van der Drift, H., Buckley, S.J., and Taylor, R.H. (1988) "Subassembly Stability," *Proceedings of the Seventh National Conference on Artificial Intelligence*, AAAI 88, 780-785.

Bottema, O. and Roth, B. (1979) *Theoretical Kinematics*, North-Holland, New York.

Bourjault, A. (1984) *Contribution à une Approche Méthodologique de l'Assemblage Automatisé: Elaboration Automatique des Séquences Opératoires*, Thesis Dissertation, Université de Franche-Comté, Besançon, France.

Brassard, G. and Bratley, P. (1988) *Algorithmics – Theory and Practice*, Prentice Hall, Englewood Cliffs, NJ.

Briggs, A.J. (1989) "An Efficient Algorithm for One-Step Planar Compliant Motion Planning with Uncertainty," *Proceedings of the Fifth ACM Symposium on Computational Geometry*, 187-196.

Brooks, R.A. (1982) "Symbolic Error Analysis and Robot Planning," *International Journal of Robotics Research*, 1(4), 29-68.

Brooks, R.A. and Lozano-Pérez, T. (1982) *A Subdivision Algorithm in Configuration Space for Findpath with Rotation*, AI Memo 684, Artificial Intelligence Laboratory, MIT; also published in *IEEE Transactions on Systems, Man and Cybernetics*, SMC-15(2), 1985, 224-233.

Brooks, R.A. (1983a) "Solving the Find-Path Problem by Good Representation of Free Space," *IEEE Transactions on Systems, Man and*

Cybernetics, SMC-13(3), 190-197.

Brooks, R.A. (1983b) "Planning Collision-Free Motions for Pick-and-Place Operations," *International Journal of Robotics Research*, 2(4), 19-44.

Brooks, R.A. and Lozano-Pérez, T. (1983) "A Subdivision Algorithm in Configuration Space for Findpath with Rotation," *Proceedings of the 8th International Conference on Artificial Intelligence*, Karlsruhe, FRG, 799-806.

Brost, R.C. (1988) "Automatic Grasp Planning in the Presence of Uncertainty," *International Journal of Robotics Research*, 7(1), 3-17.

Brost, R.C. (1989) "Computing Metric and Topological Properties of Configuration-Space Obstacles," *Proceedings of the IEEE International Conference on Robotics and Automation*, Scottsdale, AZ, 170-176.

Buckley, C.E. (1985) *The Application of Continuum Methods to Path Planning*, Ph.D. Dissertation, Department of Mechanical Engineering, Stanford University.

Buckley, S.J. (1987) *Planning and Teaching Compliant Motion Strategies*, Ph.D. Dissertation, Department of Electrical Engineering and Computer Science, MIT.

Buckley, S.J. (1989a) "Fast Motion Planning for Multiple Moving Robots," *Proceedings of the IEEE International Conference on Robotics and Automation*, Scottsdale, AZ, 322-326.

Buckley, S.J. (1989b) "Planning Compliant Motion Strategies," *International Journal of Robotics Research*, 8(5), 28-44.

Burdick, J.W. (1988) *Kinematic Analysis and Design of Redundant Robot Manipulators*, Ph.D. Dissertation, Report No. STAN-CS-88-1207, Department of Computer Science, Stanford University.

Cai, C. (1988) *Instantaneous Robot Motion with Contact Between Surfaces*, Report No. STAN-CS-88-1191, Department of Computer Science, Stanford University.

Canny, J.F. and Reif, J. (1987) "New Lower Bound Techniques for Robot Motion Planning Problems," *Proceedings of the 28th IEEE Symposium on Foundations of Computer Science*, Los Angeles, 49-60.

Canny, J.F. (1988) *The Complexity of Robot Motion Planning*, MIT Press, Cambridge, MA.

Canny, J.F. and Donald, B.R. (1988) "Simplified Voronoi Diagrams," *Discrete and Computational Geometry*, Springer-Verlag, Vol. 3, 219-236.

Canny, J.F., Donald, B.R., Reif, J.H. and Xavier, P. (1988) "On the Complexity of Kinodynamic Planning," *Proceedings of the 29th IEEE Symposium on Foundations of Computer Science*, White Plains, NY, 306-316.

Canny, J.F. (1989) "On Computability of Fine Motion Plans," *Proceedings of the IEEE International Conference on Robotics and Automation*, Scottsdale, AZ, 177-182.

Castells, C. and Melville, R. (1985) *An Unusual Algorithm for the Minimum Spanning Circle Problem*, Technical Report, Electrical Engineering and Computer Science, John Hopkins University, Baltimore, MD.

Cerny, V. (1985) "Thermodynamical Approach to the Traveling Salesman Problem: An Efficient Simulation Algorithm," *Journal of Optimization Theory and Applications*, 45(1), 41-51.

Chatila, R. (1982) "Path Planning and Environment Learning in a Mobile Robot System," *Proceedings of the European Conference on Artificial Intelligence*, Orsay, France.

Chazelle, B. and Dobkin, D.P. (1979) "Decomposing a Polygon into Convex Parts," *Proceedings of the 11th Annual ACM Symposium on Theory of Computing*, 38-48.

Chazelle, B. (1984) "Convex Partitions of Polyhedra: A Lower Bound and Worst-Case Optimal Algorithm," *SIAM Journal on Computing*, 13(3), 488-507.

Chazelle, B. and Dobkin, D.P. (1985) "Optimal Convex Decomposition," in *Computational Geometry*, edited by Toussaint, G.T., North-Holland, New York, 63-133.

Chazelle, B. (1987) "Approximation and Decomposition of Shapes," in [Schwartz and Yap, 1987], 145-185.

Cheung, E. and Lumelsky, V. (1990) "Motion Planning for a Whole-Sensitive Robot Arm Manipulator," *Proceedings of the IEEE International Conference on Robotics and Automation*, Cincinnati, OH, 344-349.

Chien, R.T., Zhang, L. and Zhang, B. (1984) "Planning Collision-Free Paths for Robotic Arm Among Obstacles," *IEEE Transactions on Pattern Analysis and Machine Intelligence*, PAMI-6(1), 91-96.

Choi, W., Zhu, D.J. and Latombe, J.C. (1989) "Contingency-Tolerant Motion Planning and Control," *Proceedings of the IEEE/RSJ International Workshop on Intelligent Robots and Systems*, IROS'89, Tsukuba, Japan, 78-85.

Collins, G.E. (1975) "Quantifier Elimination for Real Closed Fields by Cylindrical Algebraic Decomposition," *Lecture Notes in Computer Science*, 33, Springer-Verlag, New York, 135-183.

Craig, J.J. (1986) *Introduction to Robotics. Mechanics and Control*, Addison-Wesley, Reading, MA.

Cutkosky, M. (1985) *Grasping and Fine Manipulation*, Kluwer Academic Publishers, Boston.

De Fazio, T.L. and Whitney, D.E. (1987) "Simplified Generation of All Mechanical Assembly Sequences," *IEEE Transactions of Robotics and Automation*, RA-3(6), 640-658; errata in RA-6(6), 705-708.

Desai, R.S. (1988) *On Fine Motion in Mechanical Assembly in Presence of Uncertainty*, Ph.D. Dissertation, Department of Mechanical Engineering, University of Michigan.

Desai, R.S. and Volz, R.A. (1989) "Identification and Verification of Termination Conditions in Fine Motion in Presence of Sensor Errors and Geometric Uncertainties," *Proceedings of the IEEE International Conference on Robotics and Automation*, Scottsdale, AZ, 800-807.

Dobkin, D.P. and Kirkpatrick, D.G. (1985) "A Linear Algorithm for Determining the Separation of Convex Polyhedra," *Journal of Algorithms*, 6, 381-392.

Donald, B.R. (1984) *Motion Planning with Six Degrees of Freedom*, Report No. AI-TR-791, Artificial Intelligence Laboratory, MIT.

Donald, B.R. (1987a) *Error Detection and Recovery for Robot Motion Planning with Uncertainty*, Ph.D. Dissertation, Department of Electrical Engineering and Computer Science, MIT.

Donald, B.R. (1987b) "A Search Algorithm for Motion Planning with Six Degrees of Freedom," *Artificial Intelligence*, 31(3), 295-353.

Donald, B.R. (1988a) "The Complexity of Planar Compliant Motion Planning Under Uncertainty," *Proceedings of the Fourth ACM Symposium on Computational Geometry*, 309-318.

Donald, B.R. (1988b) "A Geometric Approach to Error Detection and Recovery for Robot Motion Planning with Uncertainty," *Artificial Intelligence*, 223-271.

Donald, B.R. and Xavier, P. (1989) "A Provably Good Approximation Algorithm for Optimal-Time Trajectory Planning," *Proceedings of the IEEE International Conference on Robotics and Automation*, Scottsdale, AZ, 958-963.

Dooley, J.R. and McCarthy, J.M. (1990) "Parameterized Descriptions of the Joint Space Obstacles for a 5R Closed Chain Robot," *Proceedings of the IEEE International Conference on Robotics and Automation*, Cincinnati, OH, 1536-1541.

Dorst, L., Mandhyan, I. and Trovato, K. (1989) "The Geometrical Structure of Path Planning Problems," *Preprints of the Second International Conference on Intelligent Autonomous Systems*, Amsterdam, The Netherlands.

Dufay, B. and Latombe, J.C. (1984) "An Approach to Automatic Robot Programming Based on Inductive Learning," *International Journal of Robotics Research*, 3(4), 3-20.

Edelsbrunner, H. (1987) *Algorithms in Combinatorial Geometry*, Monographs on Theoretical Computer Science, Springer-Verlag, Berlin.

Erdmann, M. (1984) *On Motion Planning With Uncertainty*, Technical Report 810, AI Laboratory, MIT.

Erdmann, M. (1986) "Using Backprojections for Fine Motion Planning with Uncertainty," *International Journal of Robotics Research*, 5(1), 19-45.

Erdmann, M. and Lozano-Pérez, T. (1986) *On Multiple Moving Objects*, AI Memo No. 883, Artificial Intelligence Laboratory, MIT.

Erdmann, M. and Mason, M.T. (1986) "An Exploration of Sensorless Manipulation," *Proceedings of the IEEE International Conference of Robotics and Automation*, San Francisco, 1569-1574.

Erdmann, M. (1989) *On Probabilistic Strategies for Robot Tasks*, Ph.D. Dissertation, Department of Electrical Engineering and Computer Science, MIT.

Erdmann, M. (1990) "Randomization in Robot Tasks," *Proceedings of the IEEE International Conference on Robotics and Automation*, Cincinnati, OH, 1744-1749.

Fahlman, S.E. (1974) "A Planning System for Robot Construction Tasks," *Artificial Intelligence*, 5(1), 1-49.

Faverjon, B. (1984) "Obstacle Avoidance Using an Octree in the Configuration Space of a Manipulator," *Proceedings of the IEEE International Conference on Robotics and Automation*, Atlanta, 504-512.

Faverjon, B. (1986) "Object Level Programming of Industrial Robots," *Proceedings of the IEEE International Conference on Robotics and Automation*, San Francisco, 1406-1412.

Faverjon, B. and Tournassoud, P. (1987) "A Local Based Approach for Path Planning of Manipulators with a High Number of Degrees of Freedom," *Proceedings of the IEEE International Conference on Robotics and Automation*, Raleigh, NC, 1152-1159.

Faverjon, B. and Tournassoud, P. (1989) "A Practical Approach to Motion Planning for Manipulators with Many Degrees of Freedom," *Preprints of the Fifth International Symposium of Robotics Research*, Tokyo, 65-73.

Fearing, R.S. (1986) "Simplified Grasping and Manipulation with Dextrous Robot Hands," *IEEE Transactions of Robotics and Automation*, RA-2(4), 188-195.

Fortune, S. (1986) "A Sweepline Algorithm For Voronoi Diagrams," *Proceedings of the Second ACM Symposium on Computational Geometry*, 313-322.

Fortune, S. and Wilfong, G. (1988) "Planning Constrained Motion," *Proceedings of the Fourth ACM Symposium on Computation Geometry*, 445-459.

Friedman, J., Hershberger, J. and Snoeyink, J. (1990) "Input-Sensitive Compliant Motion in the Plane" *Proceedings of the Second Scandinavian Workshop on Algorithmic Theory*, Bergen, Norway.

Fujimura, K. and Samet, H. (1989) "A Hierarchical Strategy for Path Planning Among Moving Obstacles," *IEEE Transactions on Robotics and Automation*, 5(1), 61-69.

Garey, M.R., Johnson, D.S., Preparata, F.P. and Tarjan, R.E. (1978) "Triangulating a Simple Polygon," *Information Processing Letter*, 7(4), 175-180.

Garey, M.R. and Johnson, D.S. (1979) *Computers and Intractability. A Guide to the Theory of NP-Completeness*, W.H. Freeman and Company, New York.

Gaschnig, J. (1979) *Performance Measurement and Analysis of Certain Search Algorithms*, Report No. CMU-CS-79-124, Department of Computer Science, Carnegie-Mellon University.

Ge, Q.J. and McCarthy, J.M. (1990) "An Algebraic Formulation of Configuration Space Obstacles for Spatial Robots," *Proceedings of the IEEE International Conference on Robotics and Automation*, Cincinnati, OH, 1542-1547.

Genesereth, M.R. and Nilsson, N.J. (1987) *Logical Foundations of Artificial Intelligence*, Morgan Kaufmann, Los Altos, CA.

Georgeff, M.P. (1987) "Planning," *Annual Review of Computer Science*, Annual Reviews Inc., 2, 359-400.

Germain, F. (1984) *Planification de Trajectoires sans Collision*, DEA Report, LIFIA Laboratory, National Polytechnic Institute, Grenoble (in French).

Ghosh, S.K. and Mount, D.M. (1987) "An Output Sensitive Algorithm for Computing Visibility Graphs," *Proceedings of the 28th IEEE Symposium on Foundations of Computer Science*, Los Angeles, 11-19.

Gilbert, E.G. and Johnson, D.W. (1985) "Distance Functions and their Application to Robot Path Planning in the Presence of Obstacles," *IEEE Transactions of Robotics and Automation*, RA-1(1), 21-30.

Gilbert, E.G., Johnson, D.W. and Keerthi, S.S. (1988) "A Fast Procedure for Computing the Distance Between Complex Objects in Three-Dimensional Space," *IEEE Transactions of Robotics and Automation*, RA-4(2), 193-203.

Gouzènes, L. (1984a) *De la Perception à l'Action en Robotique d'Assemblage: Contribution à la Modélisation et à l'Algorithmique de Base*, Thesis Dissertation, University Paul Sabatier, Toulouse (in French).

Gouzènes, L. (1984b) "Strategies for Solving Collision-Free Trajectories Problems for Mobile and Manipulator Robots," *International Journal of Robotics Research*, 3(4), 51-65.

Gowda, I.G., Kirkpatrick, D.G., Lee, D.T. and Naamad, A. (1983) "Dynamic Voronoi Diagram," *IEEE Transactions of Information Theory*, IT-29, 724-731.

Greenwood, D.T. (1965) *Principles of Dynamics*, Prentice-Hall, Englewood, NJ.

Grigoryev, D. (1988) "Complexity of Deciding Tarski Algebra," *Journal of Symbolic Computation.*

Grünbaum, B. (1967) *Convex Polytopes*, Wiley-Intersciences, New York.

Guibas, L., Ramshaw, L. and Stolfi, J. (1983) "A Kinematic Framework for Computational Geometry," *Proceedings of the IEEE Symposium on Foundations of Computer Science*, 100-111.

Guibas, L.J. and Yao, F.F. (1983) "On Translating a Set of Rectangles," *Advances in Computing Research*, 1, JAI Press, Greenwich, CT, 235-260.

Guibas, L. and Seidel, R. (1986) "Computing Convolution by Reciprocal Search," *Proceedings of the ACM Symposium on Computational Geometry*, Yorktown Heights, NY, 90-99.

Guibas, L.J., Sharir, M. and Sifrony, S. (1988) *On the General Motion Planning Problem with Two Degrees of Freedom*, Technical Report No.

359, Department of Computer Science, New York University.

Guibas, L.J. and Stolfi, J. (1989) *Ruler, Compass, and Computer: The Design and Analysis of Geometric Algorithms*, Technical Report No. 37, Systems Research Center, DEC, Palo Alto, CA.

Guillemin, V. and Pollack, A. (1974) *Differential Topology*, Prentice-Hall, Englewood Cliffs, NJ.

Hamilton, W.R. (1969) *Elements of Quaternion*, 3rd edition, Chelsea Publishing Co., New York.

Hanafusa, H. and Asada, H. (1977) "Stable Prehension by a Robot Hand with Elastic Fingers," *Proceedings of the Seventh International Symposium on Industrial Robots*, 361-368.

Hart, P.E., Nilsson, N.J. and Raphael, B. (1968) "A Formal Basis for the Heuristic Determination of Minimum Cost Paths," *IEEE Transactions on Systems, Science and Cybernetics*, SSC-4(2), 100-107.

Hayward, V. (1986) "Fast Collision Detection Schema by Recursive Decomposition of a Manipulator Workspace," *Proceedings of the IEEE International Conference on Robotics and Automation*, San Francisco, 1044-1049.

Herman, M. (1986) "Fast, Three-Dimensional, Collision-Free Motion Planning," *Proceedings of the IEEE International Conference on Robotics and Automation*, San Francisco, 1056-1063.

Hermann, R. and Krener, A.J. (1977) "Nonlinear Controllability and Observability," *IEEE Transactions on Automatic Control*, AC-22(5), 728-740.

Hoare, C.A.R. (1969) "An Axiomatic Basis for Computer Programming," *Communications of the ACM*, 12, 576-580.

Hoffman, R. (1989) "Automated Assembly in a CSG Domain," *Proceedings of the IEEE International Conference on Robotics and Automation*, Scottsdale, AZ, 210-214.

Hoffmann, C.M. (1988) *The Problem of Accuracy and Robustness in Geometric Computation*, Technical Report CSD-TR-771, Department of Computer Sciences, Purdue University.

Hogan, N. (1985) "Impedance Control: An Approach to Manipulation," *ASME Journal of Dynamic Systems, Measurement, and Control*, 107, 1-7.

Homem de Mello, L.S. (1989) *Task Sequence Planning for Robotic Assembly*, Ph.D. Dissertation, Electrical and Computer Engineering Department, Carnegie-Mellon University.

Homem de Mello, L.S. and Sanderson, A.C. (1990) "AND/OR Graph Representation of Assembly Plans," *IEEE Transactions on Robotics and Automation*, RA-6(2), 188-199.

Hopcroft, J.E., Joseph, D.A. and Whitesides, S.H. (1984) "Movement Problems for 2-Dimensional Linkages," *SIAM Journal on Computing*, 13(3), 610-629; also in [Schwartz, Sharir and Hopcroft, 1987], 282-303.

Hopcroft, J.E., Schwartz, J.T. and Sharir, M. (1984) "On the Complexity of Motion Planning for Multiple Independent Objects: PSPACE-Hardness of the 'Warehouseman's Problem'," *International Journal of Robotics Research*, 3(4), 76-88.

Hopcroft, J.E., Joseph, D.A. and Whitesides, S.H. (1985) "On the Movement of Robot Arms in 2-Dimensional Bounded Regions," *SIAM Journal on Computing*, 14(2), 315-333; also in [Schwartz, Sharir and Hopcroft, 1987], 304-329.

Hopcroft, J.E. and Wilfong, G. (1986a) "Motion of Objects in Contact," *International Journal of Robotics Research*, 4(4), 32-46.

Hopcroft, J.E. and Wilfong, G.T. (1986b) "Reducing Multiple Object Motion Planning to Graph Searching," *SIAM Journal on Computing*, 15(3), 768-785.

Hwang, Y.K. and Ahuja, N. (1988) "Path Planning Using a Potential Field Representation," *Proceedings of the IEEE International Conference on Robotics and Automation*, Philadelphia, PA, 648-649.

Hwang, Y.K. (1990) "Boundary Equations of Configuration Obstacles for Manipulators," *Proceedings of the IEEE International Conference on Robotics and Automation*, Cincinnati, OH, 298-303.

Ingrand, F. and Latombe, J.C. (1984) "Functional Reasoning for Automatic Fixture Design," *Thirteenth CAM-I Annual Technical Conference*,

Clearwater Beach, FL.

Inoue, H. (1974) *Force Feedback in Precise Assembly Tasks*, Report AIM-308, Artificial Intelligence Laboratory, MIT.

Isodori, A. (1989) *Nonlinear Control Systems*, Second Edition, Springer-Verlag.

Jackins, C.L. and Tanimoto, S.L. (1980) "Octrees and their Use in Representing Three-Dimensional Objects," *Computer Graphics and Information Processing*, 14(3).

Jacobs, P. and Canny, J. (1989) "Planning Smooth Paths for Mobile Robots," *Proceedings of the International IEEE Conference on Robotics and Automation*, Scottsdale, AZ, 2-7.

Jacobs, P., Heinzinger, G., Canny, J. and Paden, B. (1989) *Planning Guaranteed Near-Time-Optimal Trajectories for a Manipulator in a Cluttered Workspace*, Report ESRC 89-20, Engineering Systems Research Center, University of California, Berkeley.

Jänich, K. (1984) *Topology*, Springer-Verlag, New York.

Ji, Z. (1987) *Dexterous Hands: Optimizing Grasp by Design and Planning*, Ph.D. Dissertation, Department of Mechanical Engineering, Stanford University.

Joseph, D.A. and Plantiga, W.H. (1985) "On the Complexity of Reachability and Motion Planning Questions," *Proceedings of the First ACM Symposium on Computational Geometry*, Baltimore, 62-66.

Kambhampati, S. and Davis, L.S. (1986) "Multiresolution Path Planning for Mobile Robots," *IEEE Transactions of Robotics and Automation*, RA-2 (3), 135-145.

Kant, K. and Zucker, S.W. (1986a) "Toward Efficient Trajectory Planning: Path Velocity Decomposition," *International Journal of Robotics Research*, 5, 72-89.

Kant, K. and Zucker, S.W. (1986b) *Planning Smooth Collision-Free Trajectories: Path, Velocity and Splines in Free Space*, Technical Report, Computer Vision and Robotics Laboratory, McGill University, Montréal.

Kantabutra, V. and Kosaraju, S.R. (1986) "New Algorithms for Multilink Robot Arms," *Journal of Computer and System Sciences*, 32(1), 136-153.

Keil, J.M. and Sack, J.R. (1985) "Minimum Decompositions of Polygonal Objects," in *Computational Geometry*, edited by Toussaint, G., North-Holland, New York, 197-216.

Khatib, O. (1980) *Commande Dynamique dans l'Espace Opérationnel des Robots Manipulateurs en Présence d'Obstacles*, Thesis Dissertation, Ecole Nationale Supérieure de l'Aéronautique et de l'Espace, Toulouse (in French).

Khatib, O. (1986) "Real-Time Obstacle Avoidance for Manipulators and Mobile Robots," *International Journal of Robotics Research*, 5(1), 90-98.

Khosla, P. and Volpe, R. (1988) "Superquadric Artificial Potentials for Obstacle Avoidance and Approach," *Proceedings of the IEEE International Conference of Robotics and Automation*, Philadelphia, PA, 1178-1784.

Kirkpatrick, D.G. (1979) "Efficient Computation of Continuous Skeletons," *Proceedings of the 20th Symposium on Foundations of Computer Science*, 18-27.

Kirkpatrick, S., Gelatt, C.D. and Vecchi, M.P. (1983) "Optimization by Simulated Annealing," *Science*, 220, 671-680.

Koditschek, D.E. (1987) "Exact Robot Navigation by Means of Potential Functions: Some Topological Considerations," *Proceedings of the IEEE International Conference on Robotics and Automation*, Raleigh, NC, 1-6.

Koditschek, D.E. (1989) "Robot Planning and Control Via Potential Functions," in *The Robotics Review 1*, edited by Khatib, O., Craig, J.J. and Lozano-Pérez, T., MIT Press.

Koutsou, A. (1986) *Planning Motion in Contact to Achieve Parts Mating*. Ph.D. Dissertation, Department of Computer Science, University of Edinburgh, UK.

Krogh, B.H. (1984) *A Generalized Potential Field Approach to Obstacle Avoidance Control*, SME Paper MS84-484.

Lakshminarayana, K. (1978) *The Mechanics of Form Closure*, ASME

Paper No. 78-DET-32.

Latombe, J.C. (1984) "Automatic Synthesis of Robot Programs from CAD Specifications," in *Robotics and Artificial Intelligence*, edited by Brady, M., Gerhardt, L.A. and Davidson, H.F., NATO ASI Series, Springer-Verlag, Berlin, 199-217.

Latombe, J.C., Laugier, C., Lefebvre, J.M. and Mazer, E. (1984) "The LM Robot Programming System," in *Robotics Research 2*, edited by Hanafusa, H. and Inoue, H., MIT Press.

Latombe, J.C., Lazanas, A. and Shekhar, S. (1989) *Robot Motion Planning with Uncertainty in Control and Sensing*, Report No. STAN-CS-89-1292, Department of Computer Science, Stanford University. (To appear in *Artificial Intelligence*.)

Laugier, C. and Pertin, J. (1983) "Automatic Grasping: A Case Study in Accessibility Analysis," in *Advanced Software in Robotics*, edited by Danthine, A. and Géradin, M., North-Holland, New York, 201-214.

Laugier, C. and Germain, F. (1985) "An Adaptative Collision-Free Trajectory Planner," *Proceedings of the International Conference on Advanced Robotics*, Tokyo, Japan.

Laugier, C. and Théveneau, P. (1986) "Planning Sensor-Based Motions for Part-Mating Using Geometric Reasoning Techniques," *Proceedings of European Conference on Artificial Intelligence*, Brighton, UK.

Laumond, J.P. (1986) "Feasible Trajectories for Mobile Robots with Kinematic and Environment Constraints," *Preprints of the International Conference on Intelligent Autonomous Systems*, Elsevier Science Publishers B.V., Amsterdam, The Netherlands, 346-354.

Laumond, J.P. (1987a) "Obstacle Growing in a Non-Polygonal World," *Information Processing Letters*, 25(1), 41-50.

Laumond, J.P. (1987b) "Finding Collision-Free Smooth Trajectories for a Non-Holonomic Mobile Robot," *Proceedings of the 10th International Joint Conference on Artificial Intelligence*, Milan, Italy, 1120-1123.

Laumond, J.P. and Alami, R. (1988) *A Geometrical Approach to Planning Manipulation Tasks: The Case of a Circular Robot and a Movable Circular Object Amidst Polygonal Obstacles*, Technical Report No.

88314, LAAS, Toulouse.

Laumond, J.P. and Alami, R. (1989) *A Geometrical Approach to Planning Manipulation Tasks in Robotics*, Technical Report No. 89261, LAAS, Toulouse.

Laumond, J.P. and Siméon, T. (1989) *Motion Planning for a Two Degrees of Freedom Mobile Robot with Towing*, Technical Report, LAAS/CNRS No. 89-148, LAAS, Toulouse.

Lee, D.T. (1978) *Proximity and Reachability in the Plane*, Ph.D. Dissertation, Coordinated Science Laboratory, University of Illinois, Urbana, IL.

Lee, D.T. and Drysdale, R.L. (1981) "Generalization of Voronoi Diagrams in the Plane," *SIAM Journal on Computing*, 10, 73-87.

Lee, S. and Shin, Y.G. (1990) "Assembly Planning Based on Subassembly Extraction," *Proceedings of the IEEE International Conference on Robotics and Automation*, Cincinnati, OH, 1606-1611.

Lengyel, J., Reichert, M., Donald, B.R. and Greenberg, D.P. (1990) "Real-Time Robot Motion Planning Using Rasterizing Computer Graphics Hardware," *Proceedings of SIGGRAPH'90*, Dallas.

Leven, D. and Sharir, M. (1985a) "An Efficient and Simple Motion Planning Algorithm for a Ladder Moving in Two-Dimensional Space Amidst Polygonal Barriers," *Proceedings of the First ACM Symposium on Computational Geometry*, 211-227.

Leven, D. and Sharir, M. (1985b) *Planning a Purely Translational Motion for a Convex Object in Two-Dimensional Space Using Generalized Voronoi Diagrams*, Technical Report No. 34/85, The Eskenasy Institute of Computer Science, Tel Aviv University.

Leven, D. and Sharir, M. (1987) "Intersection and Proximity Problems and Voronoi Diagrams," in [Schwartz and Yap, 1987], 187-228.

Li, Z. and Canny, J.F. (1989) *Robot Motion Planning With Non-Holonomic Constraints*, Memo UCB/ERL M89/13, Electronics Research Laboratory, University of California, Berkeley.

Li, Z., Canny, J.F. and Sastry, S.S. (1989) "On Motion Planning for

Dexterous Manipulation, Part I: The Problem Formulation," *Proceedings of the IEEE International Conference of Robotics and Automation*, Scottsdale, AZ, 775-780.

Liebermann, L.I. and Wesley, M.A. (1977) "AUTOPASS: An Automatic Programming System for Computer Controlled Mechanical Assembly," *IBM Journal of Research and Development*, 21(4), 321-333.

Lingas, A. (1982) "The Power of Non-Rectilinear Holes," *Proceedings of the 9th Colloquium on Automata, Languages and Programming*, Aarhus, LNCS Springer-Verlag, 369-383.

Lozano-Pérez, T. (1976) *The Design of a Mechanical Assembly System*, Technical Report AI-TR 397, Artificial Intelligence Laboratory, MIT.

Lozano-Pérez, T. and Wesley, M.A. (1979) "An Algorithm for Planning Collision-Free Paths Among Polyhedral Obstacles," *Communications of the ACM*, 22(10), 560-570.

Lozano-Pérez, T. (1981) "Automatic Planning of Manipulator Transfer Movements," *IEEE Transactions on Systems, Man and Cybernetics*, SMC-11(10), 681-698.

Lozano-Pérez, T. (1983) "Spatial Planning: A Configuration Space Approach," *IEEE Transactions on Computers*, C-32(2), 108-120.

Lozano-Pérez, T., Mason, M.T. and Taylor, R.H. (1984) "Automatic Synthesis of Fine-Motion Strategies for Robots," *International Journal of Robotics Research*, 3(1), 3-24.

Lozano-Pérez, T. (1987) "A Simple Motion-Planning Algorithm for General Robot Manipulators," *IEEE Transactions of Robotics and Automation*, RA-3(3), 224-238.

Lozano-Pérez, T. et al. (1987) "Handey: A Robot System that Recognizes, Plans, and Manipulates," *Proceedings of the IEEE International Conference on Robotics and Automation*, Raleigh, North Carolina, 843-849.

Lumelsky, V. (1987a) "Algorithmic Issues of Sensor-Based Robot Motion Planning," *Proceedings of the 26th Conference on Decision and Control*, Los Angeles, CA, 1796-1801.

Lumelsky, V. (1987b) "Effect of Kinematics on Motion Planning for Planar Robot Arms Moving Amidst Unknown Obstacles," *IEEE Transactions of Robotics and Automation*, RA-3(3), 207-223.

Maddila, S. and Yap, C.K. (1986) "Moving a Polygon Around the Corner in a Corridor," *Proceedings of the ACM Symposium on Computational Geometry*, Yorktown Heights, NY, 187-192.

Markenscoff, X. and Papadimitriou C.H. (1989) "Optimum Grip of a Polygon," *International Journal of Robotics Research*, 8(2), 17-29.

Markenscoff, X., Ni, L. and Papadimitriou C.H. (1990) "The Geometry of Grasping," *International Journal of Robotics Research*, 9(1), 61-74.

Mason, M.T. (1981) "Compliance and Force Control for Computer Controlled Manipulators," *IEEE Transactions on Systems, Man, and Cybernetics*, SMC-11, 6, 418-432.

Mason, M.T. (1982) *Manipulator Grasping and Pushing Operations*, Report No. TR-690, Artificial Intelligence Laboratory, MIT.

Mason, M.T. (1984) "Automatic Planning of Fine Motions: Correctness and Completeness," *Proceedings of the IEEE International Conference on Robotics and Automation*, Atlanta, GA, 492-503.

Mason, M.T. (1986) "Mechanics and Planning of Manipulator Pushing Operations," *International Journal of Robotics Research*, 5(3), 53-71.

Massey, W.S. (1967) *Algebraic Topology: An Introduction*, Springer-Verlag, New York.

Mazer, E. (1983) "LM-Geo: Geometric Programming of Assembly Robots," in *Advanced Software in Robotics*, edited by Danthine, A. and Géradin, M., North-Holland, New York, 99-110.

Meagher, D. (1982) "Geometric Modeling Using Octree Encoding," *Computer Graphics and Image Processing*, 19, 129-147.

Mel, B.W. (1990) *Connectionist Robot Motion Planning*, Academic Press.

Mishra, B., Schwartz, J.T. and Sharir, M. (1987) "On the Existence and Synthesis of Multifinger Positive Grips," *Algorithmica*, 2, 541-558.

Mitchell, J.S.B. (1986) *Planning Shortest Paths*, Ph.D. Dissertation, Department of Operations Research, Stanford University.

Mitchell, J.S.B., Mount, D.M. and Papadimitriou, C.H. (1987) "The Discrete Geodesic Problem," *SIAM Journal of Computing*, 16(4), 647-668.

Natarajan, B.K. (1986) *The Complexity of Fine Motion Planning*, Technical Report No. TR-86-734, Department of Computer Science, Cornell University, Ithaca, NY.

Natarajan, B.K. (1988) "On Planning Assemblies," *Proceedings of the Fourth ACM Symposium on Computational Geometry*, 299-308.

Natarajan, B.K. (1989) "Some Paradigms for the Automated Design of Parts Feeders," *International Journal of Robotics Research*, 8(6), 98-109.

Nguyen, V.D. (1988) "Constructing Force-Closure Grasps," *International Journal of Robotics Research*, 7(3), 3-16.

Nguyen, V.D. (1989) "Constructing Stable Grasps," *International Journal of Robotics Research*, 8(1), 26-37.

Nilsson, N.J. (1969) "A Mobile Automaton: An Application of Artificial Intelligence Techniques," *Proceedings of the 1st International Joint Conference on Artificial Intelligence*, Washington D.C., 509-520.

Nilsson, N.J. (1980) *Principles of Artificial Intelligence*, Morgan Kaufmann, Los Altos, CA.

O'Donnell, P.A. and Lozano-Pérez, T. (1989) "Deadlock-Free and Collision-Free Coordination of Two Robot Manipulators," *Proceedings of the IEEE International Conference on Robotics and Automation*, 484-489.

Ó'Dúnlaing, C. and Yap, C.K. (1982) "A Retraction Method for Planning the Motion of a Disc," *Journal of Algorithms*, 6, 104-111; also in [Schwartz, Sharir and Hopcroft, 1987], 187-192.

Ó'Dúnlaing, C., Sharir, M. and Yap, C.K. (1983) "Retraction: A New Approach to Motion Planning," *Proceedings of the 15th ACM Symposium on the Theory of Computing*, Boston, 207-220.

Ó'Dúnlaing, C., Sharir, M. and Yap, C.K. (1984a) "Generalized Voronoi

Diagrams for Moving a Ladder: I. Topological Analysis," Technical Report No. 32, Robotics Laboratory, Courant Institute, New-York University.

Ó'Dúnlaing, C., Sharir, M. and Yap, C.K. (1984b) "Generalized Voronoi Diagrams for Moving a Ladder: II. Efficient Construction of the Diagram," Technical Report No. 32, Robotics Laboratory, Courant Institute, New-York University.

Ó'Dúnlaing, C. (1987) "Motion Planning With Inertial Constraints," *Algorithmica*, 2(4), 431-475.

Ohwovoriole, M.S. (1980) *An Extension of Screw Theory and its Application to the Automation of Industrial Assemblies*, Report No. STAN-CS-80-809, Department of Computer Science, Stanford University.

Palmer, R.S. (1989) *Computational Complexity of Motion and Stability of Polygons*, Ph.D. Dissertation, Technical Report TR 89-1004, Department of Computer Science, Cornell University, NY.

Papadimitriou, C.H. (1985) "An Algorithm for Shortest-Path Motion in Three Dimensions," *Information Processing Letters*, 20, 259-263.

Papoulis, A. (1965) Probability, Random Variables, and Stochastic Processes, McGraw-Hill.

Pasquier, M. (1989) *Planification de Trajectoires pour un Robot Manipulateur*, Thesis Dissertation, Institut National Polytechnique, Grenoble (in French).

Paul, R.P. (1981) *Robot Manipulators: Mathematics, Programming, and Control*, MIT Press, Cambridge, MA.

Pearl, J. (1983) "Knowledge Versus Search: A Quantitative Analysis using A^*," *Artificial Intelligence*, 20, 1-13.

Pertin-Troccaz, J. and Puget, P. (1987) "Dealing with Uncertainty in Robot Planning Using Program Proving Techniques," *Proceedings of the Fourth International Symposium on Robotics Research*, Santa-Cruz, CA, 455-466.

Pertin-Troccaz, J. (1989) "Grasping: A State of the Art," in *The Robotics Review 1*, edited by Khatib, O., Craig, J.J. and Lozano-

Pérez, T., MIT Press.

Peshkin, M.A. (1986) *Planning Robotic Manipulation Strategies for Sliding Objects*, Ph.D Dissertation, Department of Physics, Carnegie-Mellon University.

Pohl, I. (1977) "Practical and Theoretical Considerations in Heuristic Search Algorithms," in *Machine Intelligence 8: Machine Representations of Knowledge*, edited by E.Elcock and D.Michie, Ellis Horwood, Chichester, UK, 55-72.

Ponce, J. and Chelberg, D. (1987) "Finding the Limbs and Cusps of Generalized Cylinders," *International Journal of Computer Vision*, 1, 195-210.

Popplestone, R.J., Ambler, A.P. and Bellos, I.M. (1980) "An Interpreter for a Language for Describing Assemblies," *Artificial Intelligence*, 14(1), 79-107.

Preparata, F.P. and Hong, S.J. (1977) "Convex Hulls of Finite Sets of Points in Two and Three Dimensions," *Communications of the ACM*, 20(2), 87-93.

Preparata, F.P. and Shamos, M.I. (1985) *Computational Geometry: An Introduction*, Springer-Verlag, New York.

Raibert, M.H. and Craig, J.J. (1981) "Hybrid Position/Force Control of Manipulators," *ASME Journal of Dynamic Systems, Measurement and Control*, 102.

Reif, J.H. (1979) "Complexity of the Mover's Problem and Generalizations," *Proceedings of the 20th IEEE Symposium on Foundations of Computer Science*, 421-427.

Reif, J.H. and Sharir, M. (1985) "Motion Planning in the Presence of Moving Obstacles," *Proceedings of the 25th IEEE Symposium on Foundations of Computer Science*, 144-154.

Reif, J.H. and Storer, J. (1985) *Shortest Paths in Euclidean Space with Polyhedral Obstacles*, Technical report CS-85-121, Computer Science Department, Brandeis University.

Reif, J.H. (1987) "Complexity of the Generalized Mover's Problem," in

[Schwartz, Sharir and Hopcroft, 1987], 267-281.

Requicha, A. (1977) *Mathematical Models of Rigid Solid Objects*, Technical Memo No. 28, Production Automation Project, University of Rochester, NY.

Rimon, E. and Koditschek, D.E. (1988) "Exact Robot Navigation Using Cost Functions: The Case of Distinct Spherical Boundaries in E^n," *Proceedings of the IEEE International Conference on Robotics and Automation*, Philadelphia, 1791-1796.

Rimon, E. and Koditschek, D.E. (1989) "The Construction of Analytic Diffeomorphisms for Exact Robot Navigation on Star Worlds," *Proceedings of the IEEE International Conference on Robotics and Automation*, Scottsdale, AZ, 21-26.

Rimon, E. and Koditschek, D.E. (1990) "Exact Robot Navigation in Geometrically Complicated but Topologically Simple Spaces," *Proceedings of the IEEE International Conference on Robotics and Automation*, Cincinnati, OH, 1937-1942.

Rohnert, H. (1986) "Shortest Paths in the Plane with Convex Polygonal Obstacles," *Information Processing Letters*, 23, 71-76.

Salisbury, J.K. (1980) "Active Stiffness Control of a Manipulator in Cartesian Coordinates," *Proceedings of the 19th IEEE Conference on Decision and Control.*

Salisbury, J.K. and Roth, B. (1983) "Kinematic and Force Analysis of Articulated Mechanical Hands," *ASME Journal of Mechanisms, Transmissions, and Automation in Design*, 105, 35-41.

Samet, H. (1980) "Region Representation: Quadtrees from Boundary Codes," *Communications of the ACM*, 23(3), 163-170.

Schachter, B. (1978) "Decomposition of Polygons into Convex Sets," *IEEE Transactions on Computers*, C-27, 1078-1082.

Schwartz, J.T. and Sharir, M. (1983a) "On the Piano Movers' Problem: I. The Case if a Two-Dimensional Rigid Polygonal Body Moving Amidst Polygonal Barriers," *Communications on Pure and Applied Mathematics*, 36, 345-398; also in [Schwartz, Sharir and Hopcroft, 1987], 1-50.

Schwartz, J.T. and Sharir, M. (1983b) "On the 'Piano Movers' Problem: II. General Techniques for Computing Topological Properties of Real Algebraic Manifolds," *Advances in Applied Mathematics*, Academic Press 4, 298-351; also in [Schwartz, Sharir and Hopcroft, 1987], 51-96.

Schwartz, J.T. and Sharir, M. (1983c) "On the Piano Movers' Problem: III. Coordinating the Motion of Several Independent Bodies: The Special Case of Circular Bodies Moving Amidst Polygonal Barriers," *International Journal of Robotics Research*, 2(3), 46-75; also in [Schwartz, Sharir and Hopcroft, 1987], 97-140.

Schwartz, J.T. and Sharir, M. (1983d) "On the Piano Movers' Problem: V. The Case of a Rod Moving in Three-Dimensional Space Amidst Polyhedral Obstacles," *Communications on Pure and Applied Mathematics*, 37, 815-848; also in [Schwartz, Sharir and Hopcroft, 1987], 154-186.

Schwartz, J.T., Sharir, M. and Hopcroft, J. (1987) *Planning, Geometry, and Complexity of Robot Motion*, Ablex, Norwood, NJ.

Schwartz, J.T. and Yap, C.K. (1987) *Algorithmic and Geometric Aspects of Robotics*, Lawrence Erlbaum Associates, Hillsdale, NJ.

Sechen, C. (1988) *VLSI Placement and Global Routing Using Simulated Annealing*, Kluwer Academic Publishers.

Sedgewick, P. (1988) *Algorithms*, 2nd edition, Addison Wesley.

Serra, J. (1982) *Image Analysis and Mathematical Morphology*, Academic Press.

Sharir, M. and Ariel-Sheffi, E. (1983) "On the Piano Movers' Problem: IV. Various Decomposable Two-Dimensional Planning Problems," *Communications on Pure and Applied Mathematics*, 37, 479-493; also in [Schwartz, Sharir and Hopcroft, 1987], 140-153.

Sharir, M. (1987) "Efficient Algorithms for Planning Purely Translational Collision-Free Motion in Two and Three Dimensions," *Proceedings of the IEEE International Conference on Robotics and Automation*, Raleigh, NC, 1326-1331.

Sharir, M. and Sifrony, S. (1988) "Coordinated Motion Planning for Two Independent Robots," *Proceedings of the Fourth ACM Symposium on Computational Geometry.*

Shekhar, S. and Khatib, O. (1987) "Force Strategies in Real-Time Fine Motion Assembly," *Proceedings of the ASME Winter Annual Meeting*, Boston.

Shiller, Z. and Dubowsky, S. (1989) "Robot Path Planning with Obstacles, Actuators, Gripper, and Payload Constraints," *International Journal of Robotics Research*, 8(6), 3-18.

Sifrony, S. and Sharir, M. (1986) "A New Efficient Motion Planning Algorithm for a Rod in Two-Dimensional Polygonal Space," *Proceedings of the Second ACM Symposium on Computational Geometry*, Waterloo, Canada, 178-186; also published in *Algorithmica*, 2, 367-402.

Siméon, T. (1989) *Génération Automatique de Trajectoires Sans Collision et Planification de Tâches de Manipulation en Robotique*, Thesis Dissertation, University Paul Sabatier, Toulouse (in French).

Spivak, M. (1965) *Calculus on Manifolds*, Benjamin/Cummings, Menlo Park, CA.

Spivak, M. (1979) *A Comprehensive Introduction to Differential Geometry*, Publish or Perish, Wilmington, DE.

Suh, S.H. and Shin, K.G. (1988) "A Variational Dynamic Programming Approach to Robot-Path Planning With a Distance-Safety Criterion," *IEEE Transactions of Robotics and Automation*, 4(3), 334-349.

Sun, K. and Lumelsky, V. (1989) "Motion Planning with Uncertainty for a 3D Cartesian Robot Arm," *Preprints of the Fifth International Symposium on Robotics Research*, Tokyo, 57-64.

Sussman, H. and Jurdjevic, V. (1972) "Controllability of Nonlinear Systems," *Journal of Differential Equations*, 12, 95-116.

Takahashi, O. and Schilling, R.J. (1989) "Motion Planning in a Plane Using Generalized Voronoi Diagrams," *IEEE Transactions on Robotics and Automation*, RA-5(2), 143-150.

Tarski, A. (1951) *A Decision Method for Elementary Algebra and Geometry*, University of California Press, Berkeley, CA.

Taylor, R.H. (1976) *Synthesis of Manipulator Control Programs from Task-Level Specifications*, Ph.D. Dissertation, Department of Computer

Science, Stanford University.

Taylor, R.H., Mason, M.T. and Goldberg, K.Y. (1987) "Sensor-Based Manipulation Planning as a Game with Nature," *Proceedings of the Fourth International Symposium on Robotics Research*, Santa-Cruz, CA, 421-429.

Tilove, R.B. (1990) "Local Obstacle Avoidance for Mobile Robots Based on the Method of Artificial Potentials," *Proceedings of the IEEE International Conference on Robotics and Automation*, Cincinnati, OH, 566-571.

Tournassoud, P. (1986) "A Strategy for Obstacle Avoidance and its Application to Multi-Robot Systems," *Proceedings of the IEEE International Conference on Robotics and Automation*, San Francisco, 1224-1229.

Tournassoud, P., Lozano-Pérez, T. and Mazer, E. (1987) "Regrasping," *Proceedings of the IEEE International Conference on Robotics and Automation*, Raleigh, NC, 1924-1928.

Tournassoud, P. and Jehl, O. (1988) "Motion Planning for a Mobile Robot with a Kinematic Constraint," *Proceedings of the IEEE International Conference on Robotics and Automation*, Philadelphia, 1785-1790.

Toussaint, G.T. and El Gindy, H.A. (1984) "Separation of Two Monotone Polygons in Linear Time," *Robotica*, 2.

Udupa, S. (1977) *Collision Detection and Avoidance in Computer Controlled Manipulators*, Ph.D. Dissertation, Department of Electrical Engineering, California Institute of Technology.

Valade, J.M. (1985) "Geometric Reasoning and Automatic Synthesis of Assembly Trajectory," *Proceedings of the International Conference on Advanced Robotics*, ICAR'85, 43-50.

Volpe, R. and Khosla, P. (1987) "Artificial Potentials with Elliptical Isopotential Contours for Obstacle Avoidance," *Proceedings of the 26 IEEE Conference on Decision and Control*, Los Angeles, 180-185.

Waldinger, R.J. (1977) "Achieving Several Goals Simultaneously," in *Machine Intelligence 8: Machine Representations of Knowledge*, edited by Elcock, E. and Michie, D., Ellis Horwood, Chichester, UK, 94-136.

Welzl, E. (1985) "Constructing the Visibility Graph for n Line Segments in $O(n^2)$ Time," *Information Processing Letters*, 20, 167-171.

Whitesides, S.H. (1985) "Computational Geometry and Motion Planning," in *Computational Geometry*, edited by Toussaint, G.T., North-Holland, Amsterdam, 377-427.

Whitney, D. (1977) "Force Feedback Control of Manipulator Fine Motions," *ASME Journal of Dynamic Systems, Measurement, and Control*, 91-97.

Whitney, D.E. (1985) "Historical Perspectives and State of the Art in Robot Force Control," *Proceedings of IEEE International Conference on Robotics and Automation*, St. Louis, MO, 262-268.

Wilfong, G. (1988a) "Motion Planning in the Presence of Movable Obstacles," *Proceedings of the Fourth ACM Symposium on Computational Geometry*, 279-288.

Wilfong, G. (1988b) "Motion Planning for an Autonomous Vehicle," *Proceedings of the IEEE International Conference on Robotics and Automation*, 529-533.

Wilson, R.H. and Rit, J.F. (1990) "Maintaining Geometric Dependencies in an Assembly Planner," *Proceedings of the IEEE International Conference on Robotics and Automation*, Cincinnati, OH, 890-895.

Wilson, R.H (1990) "Efficiently Partitioning an Assembly," *Proceedings of the IASTED International Symposium on Robotics and Manufacturing*, Santa Barbara, CA.

Wingham, M. (1977) *Planning How to Grasp Objects in a Cluttered Environment*, Master Thesis, University of Edinburgh.

Wolter, J.D., Volz, R.A. and Woo, A.C. (1984) *Automatic Generation of Gripping Positions*, Technical Report, Center for Robotics and Integrated Manufacturing, University of Michigan, Ann Arbor.

Wolter, J.D. (1988) *On the Automatic Generation of Plans for Mechanical Assembly*, Ph.D. Dissertation (Computer, Information and Control Engineering), University of Michigan, Ann Arbor.

Yannakakis, M.Z., Papadimitriou, C.H. and Kung, H.T. (1979) "Locking

Policies: Safety and Freedom for Deadlock," *Proceedings of the 20th Annual Symposium on Foundations of Computer Science*, 286-297.

Yap, C.K. (1985) *An $O(n \log n)$ Algorithm for the Voronoi Diagram of a Set of Simple Curve Segments*, Technical Report No. 43, Robotics Laboratory, Courant Institute, New-York University.

Yap, C.K. (1987a) "Algorithmic Motion Planning," in [Schwartz and Yap, 1987], 95-143.

Yap, C.K. (1987b) "How to Move a Chair Through a Door," *IEEE Transactions of Robotics and Automation*, RA-3(3), 173-181.

Zhu, D.J. and Latombe, J.C. (1989) *Constraint Reformulation and Graph Searching techniques in Hierarchical Path Planning*, Report No. STAN-CS-89-1279, Department of Computer Science, Stanford University. (To appear in *IEEE Transactions on Robotics and Automation.*)

Zhu, D.J. and Latombe, J.C. (1990) *Mechanization of Spatial Reasoning for Automatic Pipe Layout Design*, Technical Report, Department of Computer Science, Stanford University.

Index

D

E

F

Made in the USA
Lexington, KY
20 August 2019